Isabelle Guyon, Steve Gunn, Masoud Nikravesh, Lotfi A. Zadeh (Eds.)

Feature Extraction

Studies in Fuzziness and Soft Computing, Volume 207

Editor-in-chief
Prof. Janusz Kacprzyk
Systems Research Institute
Polish Academy of Sciences
ul. Newelska 6
01-447 Warsaw
Poland
E-mail: kacprzyk@ibspan.waw.pl

Isabelle Guyon
Steve Gunn
Masoud Nikravesh
Lotfi A. Zadeh
(Eds.)

Feature Extraction

Foundations and Applications

 Springer

Isabelle Guyon
Clopinet
955 Creston Road
94708 Berkeley, USA
E-mail: isabelle@clopinet.com

Steve Gunn
School of Electronics
and Computer Sciences
University of Southampton
SO17 1BJ Southampton
Highfield, United Kingdom
E-mail: srg@ecs.soton.ac.uk

Masoud Nikravesh
Department of Electrical
Engineering & Computer
Science – EECS
University of California
94720 Berkeley, USA
E-mail: nikravesh@cs.berkeley.edu

Lotfi A. Zadeh
Division of Computer Science
Lab. Electronics Research
University of California
Soda Hall 387
94720-1776 Berkeley, CA, USA
E-mail: zadeh@cs.berkeley.edu

Library of Congress Control Number: 2006928001

ISSN print edition: 1434-9922
ISSN electronic edition: 1860-0808
ISBN-10 3-540-35487-5 Springer Berlin Heidelberg New York
ISBN-13 978-3-540-35487-1 Springer Berlin Heidelberg New York

Springer is a part of Springer Science+Business Media
springer.com
© Springer-Verlag Berlin Heidelberg 2006
Printed in The Netherlands

Typesetting: by the authors and techbooks using a Springer LATEX macro package
Cover design: Erich Kirchner, Heidelberg

Printed on acid-free paper   SPIN: 10966471   89/techbooks   5 4 3 2 1 0

To our friends and foes

Foreword

Everyone loves a good competition. As I write this, two billion fans are eagerly anticipating the 2006 World Cup. Meanwhile, a fan base that is somewhat smaller (but presumably includes you, dear reader) is equally eager to read all about the results of the NIPS 2003 Feature Selection Challenge, contained herein. Fans of Radford Neal and Jianguo Zhang (or of Bayesian neural networks and Dirichlet diffusion trees) are gloating "I told you so" and looking for proof that their win was not a fluke. But the matter is by no means settled, and fans of SVMs are shouting "wait 'til next year!" You know this book is a bit more edgy than your standard academic treatise as soon as you see the dedication: "To our friends and foes."

Competition breeds improvement. Fifty years ago, the champion in 100m butterfly swimming was 22 percent slower than today's champion; the women's marathon champion from just 30 years ago was 26 percent slower. Who knows how much better our machine learning algorithms would be today if Turing in 1950 had proposed an effective competition rather than his elusive Test?

But what makes an effective competition? The field of Speech Recognition has had NIST-run competitions since 1988; error rates have been reduced by a factor of three or more, but the field has not yet had the impact expected of it. Information Retrieval has had its TREC competition since 1992; progress has been steady and refugees from the competition have played important roles in the hundred-billion-dollar search industry. Robotics has had the DARPA Grand Challenge for only two years, but in that time we have seen the results go from complete failure to resounding success (although it may have helped that the second year's course was somewhat easier than the first's).

I think there are four criteria that define effective technical competitions:

1. The task must be approachable. Non-experts should be able to enter, to see some results, and learn from their better-performing peers.
2. The scoring must be incremental. A pass-fail competition where everyone always fails (such as the Turing Test) makes for a boring game and discourages further competition. On this score the Loebner Prize, despite its

faults, is a better competition than the original Turing Test. In one sense, everyone failed the DARPA Grand Challenge in the first year (because no entrant finished the race), but in another sense there were incremental scores: the distance each robot travelled, and the average speed achieved.

3. The results should be open. Participants and spectators alike should be able to learn the best practices of all participants. This means that each participant should describe their approaches in a written document, and that the data, auxiliary programs, and results should be publicly available.

4. The task should be relevant to real-world tasks. One of the problems with early competitions in speech recognition was that the emphasis on reducing word error rates did not necessarily lead to a strong speech dialog system—you could get almost all the words right and still have a bad dialog, and conversely you could miss many of the words and still recover. More recent competitions have done a better job of concentrating on tasks that really matter.

The Feature Selection Challenge meets the first three criteria easily. Seventy five teams entered, so they must have found it approachable. The scoring did a good job of separating the top performers while keeping everyone on the scale. And the results are all available online, in this book, and in the accompanying CD. All the data and Matlab code is provided, so the Challenge is easily reproducible. The level of explication provided by the entrants in the chapters of this book is higher than in other similar competitions. The fourth criterion, real-world relevance, is perhaps the hardest to achieve. Only time will tell whether the Feature Selection Challenge meets this one. In the mean time, this book sets a high standard as the public record of an interesting and effective competition.

Palo Alto, California   *Peter Norvig*
January 2006

Preface

Feature extraction addresses the problem of finding the most compact and informative set of features, to improve the efficiency or data storage and processing. Defining feature vectors remains the most common and convenient means of data representation for classification and regression problems. Data can then be stored in simple tables (lines representing "entries", "data points, "samples", or "patterns", and columns representing "features"). Each feature results from a quantitative or qualitative measurement, it is an "attribute" or a "variable". Modern feature extraction methodology is driven by the size of the data tables, which is ever increasing as data storage becomes more and more efficient.

After many years of parallel efforts, researchers in Soft-Computing, Statistics, Machine Learning, and Knowledge Discovery, who are interested in predictive modeling are uniting their effort to advance the problem of feature extraction. The recent advances made in both sensor technologies and machine learning techniques make it possible to design recognition systems, which are capable of performing tasks that could not be performed in the past. Feature extraction lies at the center of these advances with applications in the pharmaco-medical industry, oil industry, industrial inspection and diagnosis systems, speech recognition, biotechnology, Internet, targeted marketing and many of other emerging applications.

The present book is organized around the results of a benchmark that took place in 2003. Dozens of research groups competed on five large feature selection problems from various application domains: medical diagnosis, text processing, drug discovery, and handwriting recognition. The results of this effort pave the way to a new generation of methods capable of analyzing data tables with million of lines and/or columns.

Part II of the book summarizes the results of the competition and gathers the papers describing the methods used by the top ranking participants. Following the competition, a NIPS workshop took place in December 2003 to discuss the outcomes of the competition and new avenues in feature extraction. The contributions providing new perspectives are found in Part III of the book. Part I provides all the necessary foundations to understand the recent advances made in Parts II and III. The book is complemented by appendices and by a web site. The appendices include fact sheets summarizing the methods used in the competition, tables of results of the competition, and a summary of basic concepts of statistics.

This book is directed to students, researchers, and engineers. It presents recent advances in the field and complements an earlier book (Liu and Motoda, 1998), which provides a thorough bibliography and presents methods of historical interest, but explores only small datasets and ends before the new era of kernel methods. Readers interested in the historical aspects of the problem are also directed to (Devijver and Kittler, 1982). A completely novice reader will find all the necessary elements to understand the material of the book presented in the tutorial chapters of Part I. The book can be used as teaching material for a graduate class in statistics and machine learning, Part I supporting the lectures, Part II and III providing readings, and the CD providing data for computer projects.

Zürich, Switzerland   *Isabelle Guyon*
Southampton, UK   *Steve Gunn*
Berkeley, California   *Masoud Nikravesh and Lofti A. Zadeh*
November 2005

References

P.A. Devijver and J. Kittler. *Pattern Recognition: A Statistical Approach*. Prentice-Hall, 1982.

H. Liu and H. Motoda. *Feature Extraction, Construction and Selection: A Data Mining Perspective*. Kluwer Academic, 1998.

Contents

Part III New Perspectives in Feature Extraction

Appendix A Elementary Statistics

Appendix B Feature Selection Challenge Datasets

Appendix C Feature Selection Challenge Fact Sheets

List of Abbreviations and Symbols

$\mathcal{D}$	a dataset
$\boldsymbol{X}$	a sample of input patterns
F	feature space
Y	a sample of output labels
$\ln$	logarithm to base e
$\log_2$	logarithm to base 2
$\boldsymbol{x}^T \boldsymbol{x}'$	inner product between vectors $\boldsymbol{x}$ and $\boldsymbol{x}'$
$\|.\|$	Euclidean norm
n	number of input variables
N	number of features
m	number of training examples
$\boldsymbol{x}_k$	input vector, $k = 1 \ldots m$
$\boldsymbol{\phi}_k$	feature vector, $k = 1 \ldots m$
$x_{k,i}$	input vector elements, $i = 1 \ldots n$
$\phi_{k,i}$	feature vector elements, $i = 1 \ldots n$
y_i	target values, or (in pattern recognition) classes
$\boldsymbol{w}$	input weight vector or feature weight vector
w_i	weight vector elements, $i = 1 \ldots n$ or $i = 1 \ldots N$
b	constant offset (or threshold)
h	VC dimension
$\mathcal{F}$	a concept space
$f(.)$	a concept or target function
$\mathcal{G}$	a predictor space
$g(.)$	a predictor function function (real valued of with values in $\{-1, 1\}$ for classification)
$s(.)$	a non linear squashing function (e.g. sigmoid)
$\rho_f(\boldsymbol{x}, y)$	margin function equal to $yf(x)$
$l(\boldsymbol{x}; y; f(\boldsymbol{x}))$	loss function
$R(g)$	risk of g, i.e. expected fraction of errors
$R_{emp}(g)$	empirical risk of g, i.e. fraction of training errors
$R(f)$	risk of f

$R_{emp}(f)$	empirical risk of f
$k(\boldsymbol{x}, \boldsymbol{x}')$	Mercer kernel function (real valued)
$\boldsymbol{A}$	a matrix (use capital letters for matrices)
$\boldsymbol{K}$	matrix of kernel function values
α_k	Lagrange multiplier or pattern weights, $k = 1 \ldots m$
$\boldsymbol{\alpha}$	vector of all Lagrange multipliers
ξ_i	slack variables
$\boldsymbol{\xi}$	vector of all slack variables
C	regularization constant for SV Machines
$\mathbf{1}$	vector of ones $[11 \ldots 1]^T$

An Introduction to Feature Extraction

Isabelle Guyon[1] and André Elisseeff[2]

[1] ClopiNet, 955 Creston Rd., Berkeley, CA 94708, USA. isabelle@clopinet.com
[2] IBM Research GmbH, Zürich Research Laboratory, Säumerstrasse 4, CH-8803 Rüschlikon, Switzerland. ael@zurich.ibm.com

This chapter introduces the reader to the various aspects of feature extraction covered in this book. Section 1 reviews definitions and notations and proposes a unified view of the feature extraction problem. Section 2 is an overview of the methods and results presented in the book, emphasizing novel contributions. Section 3 provides the reader with an entry point in the field of feature extraction by showing small revealing examples and describing simple but effective algorithms. Finally, Section 4 introduces a more theoretical formalism and points to directions of research and open problems.

1 Feature Extraction Basics

In this section, we present key notions that will be necessary to understand the first part of the book and we synthesize different notions that will be seen separately later on.

1.1 Predictive Modeling

This book is concerned with problems of predictive modeling or supervised machine learning. The latter refers to a branch of computer Science interested in reproducing human learning capabilities with computer programs. The term machine learning was first coined by Samuel in the 50's and was meant to encompass many intelligent activities that could be transferred from human to machine. The term "machine" should be understood in an abstract way: not as a physically instantiated machine but as an automated system that may, for instance, be implemented in software. Since the 50's machine learning research has mostly focused on finding relationships in data and analyzing the processes for extracting such relations, rather than building truly "intelligent systems".

I. Guyon and A. Elisseeff: *An Introduction to Feature Extraction*, StudFuzz **207**, 1–25 (2006)
www.springerlink.com

Machine learning problems occur when a task is defined by a series of cases or examples rather than by predefined rules. Such problems are found in a wide variety of application domains, ranging from engineering applications in robotics and pattern recognition (speech, handwriting, face recognition), to Internet applications (text categorization) and medical applications (diagnosis, prognosis, drug discovery). Given a number of "training" examples (also called data points, samples, patterns or observations) associated with desired outcomes, the machine learning process consists of finding the relationship between the patterns and the outcomes using solely the training examples. This shares a lot with human learning where students are given examples of what is correct and what is not and have to infer which rule underlies the decision. To make it concrete, consider the following example: the data points or examples are clinical observations of patient and the outcome is the health status: healthy or suffering from cancer.[3] The goal is to predict the unknown outcome for new "test" examples, e.g. the health status of new patients. The performance on test data is called "generalization". To perform this task, one must build a predictive model or *predictor*, which is typically a function with adjustable parameters called a "learning machine". The training examples are used to select an optimum set of parameters.

We will see along the chapters of this book that enhancing learning machine generalization often motivates feature selection. For that reason, classical learning machines (e.g. Fisher's linear discriminant and nearest neighbors) and state-of-the-art learning machines (e.g. neural networks, tree classifiers, Support Vector Machines (SVM)) are reviewed in Chapter 1. More advanced techniques like ensemble methods are reviewed in Chapter 5. Less conventional neuro-fuzzy approaches are introduced in Chapter 8. Chapter 2 provides guidance on how to assess the performance of learning machines.

But, before any modeling takes place, a data representation must be chosen. This is the object of the following section.

1.2 Feature Construction

In this book, data are represented by a fixed number of features which can be binary, categorical or continuous. Feature is synonymous of input variable or attribute.[4] Finding a good data representation is very domain specific and related to available measurements. In our medical diagnosis example, the features may be symptoms, that is, a set of variables categorizing the health status of a patient (e.g. fever, glucose level, etc.).

---

[3]The outcome, also called target value, may be binary for a 2-class classification problem, categorical for a multi-class problem, ordinal or continuous for regression.

[4]It is sometimes necessary to make the distinction between "raw" input variables and "features" that are variables constructed for the original input variables. We will make it clear when this distinction is necessary.

Human expertise, which is often required to convert "raw" data into a set of useful features, can be complemented by *automatic feature construction* methods. In some approaches, feature construction is integrated in the modeling process. For examples the "hidden units" of artificial neural networks compute internal representations analogous to constructed features. In other approaches, feature construction is a preprocessing. To describe preprocessing steps, let us introduce some notations. Let x be a pattern vector of dimension n, $x = [x_1, x_2, ...x_n]$. The components x_i of this vector are the original features. We call x' a vector of transformed features of dimension n'. Preprocessing transformations may include:

- *Standardization:* Features can have different scales although they refer to comparable objects. Consider for instance, a pattern $x = [x_1, x_2]$ where x_1 is a width measured in meters and x_2 is a height measured in centimeters. Both can be compared, added or subtracted but it would be unreasonable to do it before appropriate normalization. The following classical centering and scaling of the data is often used: $x'_i = (x_i - \mu_i)/\sigma_i$, where μ_i and σ_i are the mean and the standard deviation of feature x_i over training examples.
- *Normalization:* Consider for example the case where x is an image and the x_i's are the number of pixels with color i, it makes sense to normalize x by dividing it by the total number of counts in order to encode the distribution and remove the dependence on the size of the image. This translates into the formula: $x' = x/\|x\|$.
- *Signal enhancement.* The signal-to-noise ratio may be improved by applying signal or image-processing filters. These operations include baseline or background removal, de-noising, smoothing, or sharpening. The Fourier transform and wavelet transforms are popular methods. We refer to introductory books in digital signal processing (Lyons, 2004), wavelets (Walker, 1999), image processing (R. C. Gonzalez, 1992), and morphological image analysis (Soille, 2004).
- *Extraction of local features:* For sequential, spatial or other structured data, specific techniques like convolutional methods using hand-crafted kernels or syntactic and structural methods are used. These techniques encode problem specific knowledge into the features. They are beyond the scope of this book but it is worth mentioning that they can bring significant improvement.
- *Linear and non-linear space embedding methods:* When the dimensionality of the data is very high, some techniques might be used to project or embed the data into a lower dimensional space while retaining as much information as possible. Classical examples are Principal Component Analysis (PCA) and Multidimensional Scaling (MDS) (Kruskal and Wish, 1978). The coordinates of the data points in the lower dimension space might be used as features or simply as a means of data visualization.

- *Non-linear expansions:* Although dimensionality reduction is often summoned when speaking about complex data, it is sometimes better to increase the dimensionality. This happens when the problem is very complex and first order interactions are not enough to derive good results. This consists for instance in computing products of the original features x_i to create monomials $x_{k_1} x_{k_2} ... x_{k_p}$.
- *Feature discretization.* Some algorithms do no handle well continuous data. It makes sense then to discretize continuous values into a finite discrete set. This step not only facilitates the use of certain algorithms, it may simplify the data description and improve data understanding (Liu and Motoda, 1998).

Some methods do not alter the space dimensionality (e.g. signal enhancement, normalization, standardization), while others enlarge it (non-linear expansions, feature discretization), reduce it (space embedding methods) or can act in either direction (extraction of local features).

Feature construction is one of the key steps in the data analysis process, largely conditioning the success of any subsequent statistics or machine learning endeavor. In particular, one should beware of not losing information at the feature construction stage. It may be a good idea to add the raw features to the preprocessed data or at least to compare the performances obtained with either representation. We argue that it is always better to err on the side of being too inclusive rather than risking to discard useful information. The medical diagnosis example that we have used before illustrates this point. Many factors might influence the health status of a patient. To the usual clinical variables (temperature, blood pressure, glucose level, weight, height, etc.), one might want to add diet information (low fat, low carbonate, etc.), family history, or even weather conditions. Adding all those features seems reasonable but it comes at a price: it increases the dimensionality of the patterns and thereby immerses the relevant information into a sea of possibly irrelevant, noisy or redundant features. How do we know when a feature is relevant or informative? This is what "feature selection" is about and is the focus of much of this book.

1.3 Feature Selection

We are decomposing the problem of feature extraction in two steps: feature construction, briefly reviewed in the previous section, and feature selection, to which we are now directing our attention. Although feature selection is primarily performed to select relevant and informative features, it can have other motivations, including:

1. *general data reduction*, to limit storage requirements and increase algorithm speed;
2. *feature set reduction*, to save resources in the next round of data collection or during utilization;

3. *performance improvement*, to gain in predictive accuracy;
4. *data understanding*, to gain knowledge about the process that generated the data or simply visualize the data

Several chapters in Part I are devoted to feature selection techniques. Chapter 3 reviews *filter* methods. Filters are often identified to feature ranking methods. Such methods provide a complete order of the features using a relevance index. Methods for computing ranking indices include correlation coefficients, which assess the degree of dependence of individual variables with the outcome (or target). A variety of other statistics are used, including classical test statistics (T-test, F-test, Chi-squared, etc.) More generally, methods that select features without optimizing the performance of a predictor are referred to as "filters". Chapter 6 presents information theoretic filters.

Chapter 4 and Chapter 5 are devoted to *wrappers* and *embedded* methods. Such methods involve the predictor as part of the selection process. Wrappers utilize a learning machine as a "black box" to score subsets of features according to their predictive power. Embedded methods perform feature selection in the process of training and are usually specific to given learning machines. Wrappers and embedded methods may yield very different feature subsets under small perturbations of the dataset. To minimize this effect, Chapter 7 explains how to improve feature set stability by using ensemble methods.

A critical aspect of feature selection is to properly assess the quality of the features selected. Methods from classical statistics and machine learning are reviewed in Chapter 2. In particular, this chapter reviews hypothesis testing, cross-validation, and some aspects of experimental design (how many training examples are needed to solve the feature selection problem.)

A last, it should be noted that it is possible to perform feature construction and feature selection simultaneously, as part of a global optimization problem. Chapter 6 introduces the reader to methods along this line.

1.4 Methodology

The chapters of Part I group topics in a thematic way rather than in a methodological way. In this section, we present a unified view of feature selection that transcends the old cleavage filter/wrapper and is inspired by the views of (Liu and Motoda, 1998).

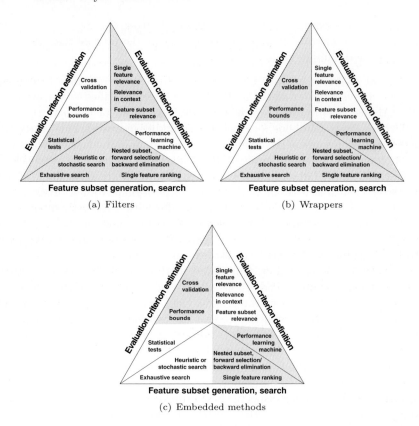

(a) Filters

(b) Wrappers

(c) Embedded methods

Fig. 1. *The three principal approaches of feature selection.* The shades show the components used by the three approaches: filters, wrappers and embedded methods.

There are four aspects of feature extraction:

- feature construction;
- feature subset generation (or search strategy);
- evaluation criterion definition (e.g. relevance index or predictive power);
- evaluation criterion estimation (or assessment method).

The last three aspects are relevant to feature selection and are schematically summarized in Figure 1.

Filters and *wrappers* differ mostly by the *evaluation criterion*. It is usually understood that filters use criteria not involving any learning machine, e.g. a relevance index based on correlation coefficients or test statistics, whereas wrappers use the performance of a learning machine trained using a given feature subset.

Both filter and wrapper methods can make use of *search strategies* to explore the space of all possible feature combinations that is usually too large to be explored exhaustively (see Chapter 4.) Yet filters are sometimes assimilated to feature ranking methods for which feature subset generation is trivial since only single features are evaluated (see Chapter 3). Hybrid methods exist, in which a filter is used to generate a ranked list of features. On the basis of the order thus defined, nested subsets of features are generated and computed by a learning machine, i.e. following a wrapper approach. Another class of *embedded methods* (Chapter 5) incorporate feature subset generation and evaluation in the training algorithm.

The last item on the list, *criterion estimation*, is covered in Chapter 2. The difficulty to overcome is that a defined criterion (a relevance index or the performance of a learning machine) must be estimated from a *limited amount of training data*. Two strategies are possible: "in-sample" or "out-of-sample". The first one (in-sample) is the "classical statistics" approach. It refers to using all the training data to compute an empirical estimate. That estimate is then tested with a statistical test to assess its significance, or a performance bound is used to give a guaranteed estimate. The second one (out-of-sample) is the "machine learning" approach. It refers to splitting the training data into a training set used to estimate the parameters of a predictive model (learning machine) and a validation set used to estimate the learning machine predictive performance. Averaging the results of multiple splitting (or "cross-validation") is commonly used to decrease the variance of the estimator.

2 What is New in Feature Extraction?

As of 1997, when a special issue on relevance including several papers on variable and feature selection was published (Blum and Langley, 1997, Kohavi and John, 1997), few domains explored used more than 40 features. The situation has changed considerably in the past few years. We organized in 2001 a first NIPS workshop, the proceedings of which include papers exploring domains with hundreds to tens of thousands of variables or features (Guyon and Elisseeff, 2003). Following this workshop, we organized a feature selection competition, the results of which were presented at a NIPS workshop in 2003. The present book is the outcome of the latter.

Part II of the book describes the methods used by the best ranking participants. Chapter II summarizes the results of the competition. Five datasets were used that were chosen to span a variety of domains (biomarker discovery, drug discovery, handwriting recognition, and text classification) and difficulties (the input variables are continuous or binary, sparse or dense; one dataset has unbalanced classes.) One dataset was artificially constructed to illustrate a particular difficulty: selecting a feature set when no feature is informative individually. We chose datasets that had sufficiently many examples to create a large enough test set and obtain statistically significant results (Guyon, 2003). We introduced a number of random features called *probes* to make the task more difficult and identify the algorithms capable of filtering them out.

The challenge winning methods are described in Chapter 10. The authors use a combination of Bayesian neural networks (Neal, 1996) and Dirichlet diffusion trees (Neal, 2001). Two aspects of their approach were the same for all data sets: (1) reducing the number of features used for classification to no more than a few hundred, either by selecting a subset of features using simple univariate significance tests, or by Principal Component Analysis; (2) applying a classification method based on Bayesian learning, using an Automatic Relevance Determination (ARD) prior that allows the model to determine which of the features are most relevant (MacKay, 1994, Neal, 1996). Bayesian neural network learning with computation by Markov chain Monte Carlo (MCMC) is a well developed technology (Neal, 1996). Dirichlet diffusion trees are a new Bayesian approach to density modeling and hierarchical clustering.

A wide variety of other methods presented in Part II performed nearly as well. For feature selection, filter methods proved quite effective. Four of the top entrants explore successfully the use of Random Forests (RF)[5] as a filter (Chapter 11, Chapter 15, and Chapter 12). Simple correlation coefficients also performed quite well (Chapter 13, Chapter 14, Chapter 20, and Chapter 23), as well as information theoretic ranking criteria (Chapter 22 and Chapter 24). Some of the recently introduced embedded methods using a Support Vector Machine (SVM) or a related kernel method were applied with success (Chapter 12, Chapter 13, Chapter 16, Chapter 18, Chapter 19, and Chapter 21). Among the most innovative methods, Chapter 17 and Chapter 29 present a margin-based feature selection method inspired by the Relief algorithm (Kira and Rendell, 1992).

As far as classifier choices are concerned, the second best entrants (Chapter 11) use the simple regularized least square kernel method as classifier. Many of the other top entrants use regularized kernel methods with various loss functions, including kernel partial least squares (KPLS) (Chapter 21), vanilla Support Vector machines (SVM) (Chapter 12, Chapter 20, Chapter 22, Chapter 23 and Chapter 24), transductive SVM (Chapter 13), Bayesian SVM (Chapter 18), Potential SVM (Chapter 19), and 1-norm SVM (Chapter 16). Two other entrants used neural networks like the winners (Chapter 14 and Chapter 26). Other methods includes Random Forests (RF) (Chapter 15), Naïve Bayes (Chapter 24 and Chapter 25) and simple nearest neighbors (Chapter 17).

Part III of the book devotes several chapters to novel approaches to feature construction. Chapter 27 provides a unifying framework to many methods of linear and non-linear space embedding methods. Chapter 28 proposes a method for constructing orthogonal features for an arbitrary loss. Chapter 31 gives an example of syntactic feature construction: protein sequence motifs.

---

[5]Random Forests are ensembles of tree classifiers.

3 Getting Started

Amidst the forest of methods, the reader who is getting started in the field may be lost. In this section, we introduce basic concepts and briefly describe simple but effective methods. We illustrate with small two-dimensional classification problems (Figure 2) some special cases.

One approach to feature selection is to rank features according to their individual relevance (Section 3.1.) Such feature ranking methods are considered fast and effective, particularly when the number of features is large and the number of available training examples comparatively small (e.g. 10,000 features and 100 examples.) In those cases, methods that attempt to search extensively the space of feature subsets for an optimally predictive can be much slower and prone to "overfitting" (perfect predictions may be achieved on training data, but the predictive power on test data will probably be low.)

However, as we shall see in some other examples (Section 3.2 and 3.3), there are limitations to individual feature ranking, because of the underlying feature independence assumptions made by "univariate" methods:

- features that are not individually relevant may become relevant in the context of others;
- features that are individually relevant may not all be useful because of possible redundancies.

So-called "multivariate" methods take into account feature dependencies. Multivariate methods potentially achieve better results because they do not make simplifying assumptions of variable/feature independence.

3.1 Individual Relevance Ranking

Figure 2-a shows a situation in which one feature (x_1) is relevant individually and the other (x_2) does not help providing a better class separation. For such situations individual feature ranking works well: the feature that provides a good class separation by itself will rank high and will therefore be chosen.

The **Pearson correlation coefficient** is a classical relevance index used for individual feature ranking. We denote by x_j the m dimensional vector containing all the values of the jth feature for all the training examples, and by y the m dimensional vector containing all the target values. The Pearson correlation coefficient is defined as:

$$C(j) = \frac{|\sum_{i=1}^{m}(x_{i,j} - \bar{x}_j)(y_i - \bar{y})|}{\sqrt{\sum_{i=1}^{m}(x_{i,j} - \bar{x}_j)^2 \sum_{i=1}^{m}(y_i - \bar{y})^2}} , \tag{1}$$

where the bar notation stands for an average over the index i. This coefficient is also the absolute value of the cosine between vectors x_i and y, after they have been centered (their mean subtracted). The Pearson correlation coefficient may be used for regression and binary classification problems. For

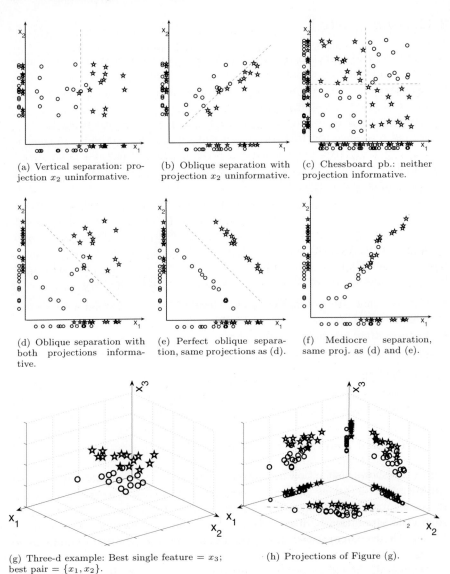

(a) Vertical separation: projection x_2 uninformative.

(b) Oblique separation with projection x_2 uninformative.

(c) Chessboard pb.: neither projection informative.

(d) Oblique separation with both projections informative.

(e) Perfect oblique separation, same projections as (d).

(f) Mediocre separation, same proj. as (d) and (e).

(g) Three-d example: Best single feature = x_3; best pair = $\{x_1, x_2\}$.

(h) Projections of Figure (g).

Fig. 2. *Small classification examples.* One class is represented by circles and the other by stars. The horizontal axis represents one feature and the vertical axis the other. In the last example we have a third feature. We represent each class by circles or stars. We show the projections of the classes on the axes as superimposed circles and stars.

multi-class problems, one can use instead the closely related Fisher coefficient. The Pearson correlation coefficient is also closely related to the T-test statistic, and the Naïve Bayes ranking index. See Chapter 3 for details and for other examples of ranking criteria.

Rotations in feature space often simplify feature selection. Figure 2-a is obtained from Figure 2-d by a 45 degree rotation. One notices that to achieve the same separation, two features are needed in Figure 2-d, while only one is needed in Figure 2-a. Rotation is a simple linear transformation. Several preprocessing methods such as principal component analysis (PCA) perform such linear transformations, which permit reducing the space dimensionality and exhibit better features.

The notion of **relevance is related to the objective** being pursued. A feature that is irrelevant for classification may be relevant for predicting the class conditional probabilities. Such is the case of feature x_2 in Figure 2-a. The examples of the two classes are drawn from overlapping Gaussian distributions whose class centers are aligned with axis x_1. Thus, $P(y|\mathbf{x})$ is not independent of x_2, but the error rate of the optimum Bayes classifier is the same whether feature x_2 is kept or discarded. This points to the fact that density estimation is a harder problem than classification and usually requires more features.

3.2 Relevant Features that are Individually Irrelevant

In what follows, we justify the use of multivariate methods, which make use of the predictive power of features considered jointly rather than independently.

A helpful feature may be irrelevant by itself. One justification of multivariate methods is that features that are individually irrelevant may become relevant when used in combination. Figure 2-b gives an example of a linear separation in which an individually irrelevant feature helps getting a better separation when used with another feature.[6] This case occurs in real world examples: feature x_1 might represent a measurement in an image that is randomly offset by a local background change; feature x_2 might be measuring such local offset, which by itself is not informative. Hence, feature x_2 might be completely uncorrelated to the target and yet improve the separability of feature x_1, if subtracted from it.

Two individually irrelevant features may become relevant when used in combination. The case of Figure 2-c, known as the "chessboard problem", illustrates this situation.[7] In the feature selection challenge (see Part II), we proposed a problem that generalizes this case in a higher dimension space: The MADELON dataset is built from clusters placed on the vertices of a hypercube in five dimensions and labeled at random.

The **Relief method** is a classical example of multivariate filter. Most multivariate methods rank subsets of features rather than individual features.

---

[6] It is worth noting that the x_2 projection is the same in Figures 2-a and 2-b.

[7] This is a small 2x2 chessboard. This problem is analogous to the famous XOR problem, itself a particular case of the parity problem.

12   Isabelle Guyon and André Elisseeff

Still, there exist multivariate relevance criteria to rank individual features according to their relevance *in the context of others*. To illustrate this concept, we give as example a ranking index for classification problems derived from the Relief algorithm (Kira and Rendell, 1992):

$$C(j) = \frac{\sum_{i=1}^{m} \sum_{k=1}^{K} |x_{i,j} - x_{M_k(i),j}|}{\sum_{i=1}^{m} \sum_{k=1}^{K} |x_{i,j} - x_{H_k(i),j}|} \ , \tag{2}$$

Notations will be explained shortly. The Relief algorithm uses an approach based on the K-nearest-neighbor algorithm. To evaluate the index, we first identify in the original feature space, for each example x_i, the K closest examples of the same class $\{x_{H_k(i)}\}, k = 1...K$ (nearest hits) and the K closest examples of a different class $\{x_{M_k(i)}\}$ (nearest misses.)[8] Then, in projection on feature j, the sum of the distances between the examples and their nearest misses is compared to the sum of distances to their nearest hits. In Equation 2, we use the ratio of these two quantities to create an index independent of feature scale variations. The Relief method works for multi-class problems.

3.3 Redundant Features

Another justification of multivariate methods is that they take into account feature redundancy and yield more compact subsets of features. Detecting redundancy cannot be done by analyzing only the feature projections, as univariate methods do. This point is illustrated in the following examples.

Noise reduction can be achieved with features having identical projected distributions. In Figure 2-d, the two features look similar if we compare their projected distributions. Yet they are not completely redundant: the two-dimensional distribution shows a better class separation than the one achievable with either feature. In this example the data points of the two classes are generated from Gaussian distributions with equal variance σ^2. In projection on either feature, the distance d between the two classes is identical. The signal to noise ratio of each individual feature is therefore d/σ. In projection on the first diagonal, the distance between the two classes is $d\sqrt{2}$, hence the signal-to-noise ratio is improved by $\sqrt{2}$. Adding n features having such class conditional independence would result in an improvement of the signal-to-noise ratio by $\sqrt{n}$.

Correlation does NOT imply redundancy. Figures 2-e and Figure 2-f show even more striking examples in which the feature projections are the same as in Figure 2-d. It is usually thought that feature correlation (or anticorrelation) means feature redundancy. In Figure 2-f, the features are correlated and indeed redundant: the class separation is not significantly improved by using two features rather than one. But in Figure 2-e, despite that two

---

[8] All features are used to compute the closest examples.

features have similar projections and are anti-correlated, they are not redundant at all: a perfect separation is achieved using the two features while each individual feature provides a poor separation.

3.4 Forward and Backward Procedures

Having recognized the necessity of selecting features in the context of other features and eliminating redundancy, we are left with a wide variety of algorithms to choose from. Among wrapper and embedded methods (Chapter 4 and Chapter 5), greedy methods (forward selection or backward elimination) are the most popular. In a forward selection method one starts with an empty set and progressively add features yielding to the improvement of a performance index. In a backward elimination procedure one starts with all the features and progressively eliminate the least useful ones. Both procedures are reasonably fast and robust against overfitting. Both procedures provide nested feature subsets. However, as we shall see, they may lead to different subsets and, depending on the application and the objectives, one approach may be preferred over the other one. We illustrate each type of procedure with examples of algorithms.

Forward or backward? In Figures 2-g and h, we show an example in three dimensions illustrating differences of the forward and backward selection processes. In this example, a forward selection method would choose first x_3 and then one of the two other features, yielding to one of the orderings x_3, x_1, x_2 or x_3, x_2, x_1. A backward selection method would eliminate x_3 first and then one of the two other features, yielding to one of the orderings x_1, x_2, x_3 or x_2, x_1, x_3. Indeed, on Figure 2-h, we see that the front projection in features x_1 and x_2 gives a figure similar to Figure 2-e. The last feature x_3 separates well by itself, better than x_1 or x_2 taken individually. But, combined with either x_1 or x_2, it does not provide as good a separation as the pair $\{x_1, x_2\}$. Hence, the forward selection ordering yields a better choice if we end up selecting a single feature (the top ranking x_3), but the backward selection method will give better results if we end up selecting two features (the top ranking x_1 and x_2). Backward elimination procedures may yield better performances but at the expense of possibly larger feature sets. However if the feature set is reduced too much, the performance may degrade abruptly. In our previous example, choosing the top ranking feature by backward selection would be much worse than choosing x_3 as given by the forward approach.

Forward selection algorithm examples are now provided. The *Gram-Schmidt orthogonalization* procedure is a simple example of *forward selection* method (see Chapter 2 for details and references.) The first selected feature has largest cosine with the target. For centered features, this is equivalent to selecting first the feature most correlated to the target (Equation 1.) The subsequent features are selected iteratively as follows:

- the remaining features and the target are projected on the null space of the features already selected;

- the feature having largest cosine with the target in that projection is added to the selected features.

The procedure selects the features that incrementally decrease most the least-square error of a linear predictor. One can stop the procedure using a statistical test or by cross-validation (Chapter 2.) This procedure has the advantage to be described in few lines of codes and it performs well in practice. We give a Matlab implementation of the algorithm in Appendix A. It is worth noting the similarity with the Partial Least Square (PLS) method (see e.g. (Hastie et al., 2000)): both methods involve iteratively the computation of the correlation of the (projected) input features with the target, followed by a new projection on the null space of the features selected; the difference is that, in Gram-Schmidt, original input features are selected while in PLS the features selected are constructed as a weighted sum of the original features, the weights being given by the correlation to the target.

Another more advanced example of forward selection method is "Random Forests" or RF. Ensembles of decision trees (like Random Forests (Breiman, 2001)) select features in the process of building a classification or regression tree. A free RF software package is available from `http://www.stat.berkeley.edu/users/breiman/RandomForests/` and a Matlab interface from `http://sunsite.univie.ac.at/statlib/matlab/RandomForest.zip`.

Backward elimination algorithm examples are now provided. The *recursive feature elimination Support Vector Machine (RFE-SVM)* is a simple example of *backward elimination* method (see Chapter 5 for details and references.) For the linear SVM having decision function $f(\boldsymbol{x}) = \boldsymbol{w} \cdot \boldsymbol{x} + b$, the method boils down to simply iteratively removing the feature x_i with the smallest weight in absolute value $|w_i|$ and retraining the model.[9] At the expense of some sub-optimality, the method can be sped up by removing several features at a time at each iteration. The method can also be extended to the non-linear SVM (Chapter 5.) SVMs are described in Chapter 1 and numerous free software packages are available (see `http://www.kernel-machines.org/`) which makes this approach rather simple in terms of implementation.

RFE is a weight pruning method according to the smallest change in objective function. It follows the same paradigm than the Optimal Brain Damage procedure (OBD), which is used to prune weights in neural networks and can be used for feature selection. OBD also bears resemblance with the Automatic Relevance Determination (ARD) Bayesian method used by the winners of the competition (see Chapter 7 and Chapter 10 for details.)

3.5 Recapitulation

Table 1 summarizes the methods mentioned in this section. We recommend to try the methods in order of increasing statistical complexity:

---

[9]RFE usually works best on standardized features, see Section 1.2.

Table 1. *Frequently used feature selection methods.* We use the abbreviations: RLSQ=regularized least square; RFE=recursive feature elimination; SVM=support vector machine; OBD=optimum brain damage; ARD=automatic relevance determination; RF=random forest. We call m the number of training examples and n the number of features. The computational complexity main vary a lot depending on the implementation and should be taken with caution.

Feature selection	Matching classifier	Computational complexity	Comments
Pearson (Eq. 1)	Naïve bayes	nm	Feature ranking filter. Linear univariate. Makes independence assumptions between features. Low computational and statistical complexity.
Relief (Eq. 2)	Nearest neighbors	nm^2	Feature ranking filter. Non-linear multivariate. Statistical complexity monitored by the number of neighbors.
Gram-Schmidt (Sec. 3.4)	linear RLSQ	fnm	Forward selection, stopped at f features. Linear multivariate. The statistical complexity of RLSQ monitored by the regularization parameter or "ridge".
RFE-SVM (Sec. 3.4)	SVM	$\max(n,m)m^2$	Backward elimination. Multivariate, linear or non-linear. Statistical complexity monitored by kernel choice and "soft-margin" constraints.
OBD/ARD	Neural Nets	$\min(n,m)nmh$	Backward elimination. Non-linear multivariate. Statistical complexity monitored by the number h of hidden units and the regularization parameter or "weight decay".
RF	RF	$t\sqrt{n}\,m\log m$	Ensemble of t tree classifiers, each preforming forward selection. Non-linear multivariate.

1. **Univariate methods** making independence assumptions between variables. Feature selection: Ranking with the Pearson correlation coefficient. Classifier: Naïve Bayes.
2. **Linear multivariate methods.** Feature selection: Gram-Schmidt forward selection or RFE with linear SVM. Predictors: Linear SVM or linear regularized least-square model (RLSQ.)

3. **Non-linear multivariate methods.** Feature selection: Relief, RFE, OBD or ARD combined with non-linear models. Predictors: Nearest neighbors, non-linear SVM or RLSQ, neural network, RF.

Computational complexity is also sometimes of consideration. We have added to Table 1 some orders of magnitude of the computational complexity of the feature selection process. This does not include the assessment part determining the optimum number of features to be selected. Justifications of our estimates are provided in Appendix B.

4 Advanced Topics and Open Problems

This book presents the status of a rapidly evolving field. The applications are driving this effort: bioinformatics, cheminformatics, text processing, speech processing, and machine vision provide machine learning problems in very high dimensional spaces, but often comparably few examples (hundreds). It may be surprising that there is still a profusion of feature selection methods and that no consensus seems to be emerging. The first reason is that there are several statements of the feature selection problem. Other reasons include that some methods are specialized to particular cases (e.g. binary inputs or outputs), some methods are computationally inefficient so they can be used only for small numbers of features, some methods are prone to "overfitting" so they can be used only for large numbers of training examples.

The fact that simple methods often work well is encouraging for practitioners. However, this should not hide the complexity of the problems and the challenges ahead of us to improve on the present techniques and consolidate the theory. Inventing a new algorithm is a good way to be acquainted with the problems. But there exist already so many algorithms that it is difficult to improve significantly over the state of the art without proceeding in a principled way. This section proposes some formal mathematical statements of the problems on which new theories can be built.

Let us first introduce some notations. A pattern is a feature vector $x = [x_1, x_2, ...x_n]$, which is an instance of a random vector $X = [X_1, X_2, ...X_n]$. For each assignment of values, we have a probability $P(X = x)$. We assume that the values are discrete for notational simplicity. The target is a random variable Y taking values y. The dependency between X and Y is governed by the distribution $P(X = x, Y = y) = P(Y = y|X = x)P(X = x)$. When we write $P(X, Y) = P(Y|X)P(X)$, we mean that the equality hold true for all the values taken by the random variables. Let V be some subset of X. Let X^{-i} be the subset of X excluding x_i and V^{-i} be some subset of X^{-i}.

4.1 Relevant Features

We start with the notion of relevant feature. We first define irrelevance as a consequence of random variable independence and then define relevance by contrast. First we assume the knowledge of the data distributions, which in reality are unknown. We then discuss what can be done in the finite sample case.

Definition 1 (Surely irrelevant feature). *A feature X_i is surely irrelevant iff for all subset of features $\boldsymbol{V}^{-i}$ including $\boldsymbol{X}^{-i}$,*

$$P(X_i, Y | \boldsymbol{V}^{-i}) = P(X_i | \boldsymbol{V}^{-i}) P(Y | \boldsymbol{V}^{-i}).$$

Since we care little about cases that occur with zero or small probability it seems natural to measure irrelevance in probability e.g. with the Kullback-Leibler divergence between $P(X_i, Y | \boldsymbol{V}^{-i})$ and $P(X_i | \boldsymbol{V}^{-i}) P(Y | \boldsymbol{V}^{-i})$:

$$MI(X_i, Y | \boldsymbol{V}^{-i}) = \sum_{\{X_i, Y\}} P(X_i, Y | \boldsymbol{V}^{-i}) \log \frac{P(X_i, Y | \boldsymbol{V}^{-i})}{P(X_i | \boldsymbol{V}^{-i}) P(Y | \boldsymbol{V}^{-i})}$$

The sum runs over all possible values of the random variables X_i and Y. We note that the expression obtained is the conditional mutual information. It is therefore a function of $n - 1$ variables.[10] In order to derive a score that summarizes how relevant feature X_i is, we average over all the values of $\boldsymbol{V}^{-i}$:

$$EMI(X_i, Y) = \sum_{\boldsymbol{V}^{-i}} P(\boldsymbol{V}^{-i}) MI(X_i, Y | \boldsymbol{V}^{-i})$$

We define then:

Definition 2 (Approximately irrelevant feature). *A feature X_i is approximately irrelevant, with level of approximation $\epsilon > 0$ or ϵ-relevant, iff, for all subset of features $\boldsymbol{V}^{-i}$ including $\boldsymbol{X}^{-i}$,*

$$EMI(X_i, Y) \leq \epsilon.$$

When $\epsilon = 0$, the feature will be called **almost surely irrelevant**.

With that statement, conditional mutual information comes as a natural relevance ranking index and we may define relevance by contrast to irrelevance. The practical use of our definitions to perform feature selection is computationally expensive since it requires considering all subset of features $\boldsymbol{V}^{-i}$ and summing over all the values of $\boldsymbol{V}^{-i}$. However if we assume that features X_i and X_j for all $i \neq j$, are independent, the average conditional mutual information is the same as the mutual information between X_i and Y:

$$EMI(X_i, Y) = MI(X_i, Y)$$

This motivates the following definition:

---

[10] Recall that n is the total number of features.

Definition 3 (Individually irrelevant feature). *A feature X_i is individually irrelevant iff for some relevance threshold epsilon ≥ 0*

$$MI(X_i, Y) \leq \epsilon.$$

The derivation of this definition justifies of the use of mutual information as a feature ranking index (see Chapter 6.)

The **finite sample case** is now discussed. In practical cases, we do not have access to the probability distributions $P(X)$ and $P(Y|X)$, but we have training examples drawn from these distributions. We define a new notion of probable approximate irrelevance. At the same time, we replace in our definition the criteria $EMI(X_i, Y)$ or $MI(X_i, Y)$ by a generic non-negative index $C(i)$ whose expected value is zero for irrelevant features. We write our index as $C(i, m)$ to emphasize that it is an empirical index computed from m training examples.

Definition 4 (Probably approximately irrelevant feature). *A feature i is probably approximately irrelevant with respect to an index C estimated with m examples, with level of approximation $\epsilon \geq 0$ and risk $\delta \geq 0$ iff*

$$P(C(i, m) > \epsilon(\delta, m)) \leq \delta.$$

Clearly, for relevant features, we do not know the probability distribution of $C(i, m)$ across different drawings of the training set of size m, so it does not seem that we have progressed very much. However, we may be able to make some assumptions about the distribution of C for irrelevant features. Following the paradigm of hypothesis testing, we call the distribution of C for irrelevant features the "null" distribution. For a given candidate feature i, the null hypothesis is that this feature is irrelevant. We will reject this null hypothesis if $C(i, m)$ departs significantly from zero. Using the "null" distribution and a chosen risk δ, we can compute the significance threshold $\epsilon(\delta, m)$. This method of assessing the statistical significance of feature relevance is further developed in Chapter 2.

Discussion. Many definitions of relevance have been provided in the literature. Kohavi and John (Kohavi and John, 1997) make a distinction between strongly and weakly relevant features. We recall below those definitions:

A feature X_i is strongly relevant iff there exists some values x_i, y and v_i with $P(X_i = x_i, \mathbf{X}^{-i} = \mathbf{v}_i) > 0$ such that: $P(Y = y|X_i = x_i, \mathbf{X}^{-i} = \mathbf{v}_i) \neq P(Y = y|\mathbf{X}^{-i} = \mathbf{v}_i)$. A feature X_i is weakly relevant iff it is not strongly relevant and if there exists a subset of features $\mathbf{V}^{-i}$ for which there exists for values x_i, y and v_i with $P(X_i = x_i, \mathbf{V}_i = \mathbf{v}_i) > 0$ such that: $P(Y = y|X_i = x_i, \mathbf{V}_i = \mathbf{v}_i) \neq P(Y = y|\mathbf{V}_i = \mathbf{v}_i)$.

Our asymptotic definitions of relevance are similarly based on conditioning. Kohavi and John's introduction of strong and weak relevance seems to have been guided by the need to account for redundancy: the strongly relevant feature that is needed on its own and cannot be removed, and the weakly relevant feature that is redundant with other relevant features and can therefore be omitted if similar features are retained. Our approach separates the notion of redundancy from that of relevance: A feature is relevant if it contains some information about the target. Since our definition of relevance is less specific, we introduce in Section 4.2 the notion of *sufficient feature subset*, a concept to extract a minimum subset of relevant feature and therefore to rule out redundancy when required.

4.2 Sufficient Feature Subset

In the previous section, we have provided formal definitions for the notion of feature relevance. As outlined in section 3.3, relevant features may be redundant. Hence, a ranking of features in order of relevance does not allow us to extract a minimum subset of features that are sufficient to make optimal predictions. In this section, we propose some formal definitions of feature subset sufficiency. We introduce the additional notation $\bar{V}$ for the subset that complements a set of feature V in X: $X = [V, \bar{V}]$.

Definition 5 (Surely sufficient feature subset). *A subset V of features is surely sufficient iff, for all assignments of values to its complementary subset $\bar{V}$,*

$$P(Y|V) = P(Y|X).$$

As in the case of the definition of feature relevance, since we care little about cases that occur with zero or small probability, it seems natural to measure sufficiency in probability. We define a new quantity:

$$DMI(V) = \sum_{\{v,\bar{v},y\}} P(X = [v,\bar{v}], Y = y) \log \frac{P(Y = y|X = [v,\bar{v}])}{P(Y = y|V = v)}.$$

This quantity, introduced in (Koller and Sahami, 1996), is the expected value over $P(X)$ of the Kullback-Leibler divergence between $P(Y|X)$ and $P(Y|V)$. It can be verified that:

$$DMI(V) = MI(X,Y) - MI(V,Y).$$

Definition 6 (Approximately sufficient feature subset). *A subset V of features is approximately sufficient, with level of approximation $\epsilon \geq 0$, or ϵ-sufficient, iff,*

$$DMI(V) \leq \epsilon.$$

If $\epsilon = 0$ the subset V will be called **almost surely sufficient**.

Definition 7 (Minimal approximately sufficient feature subset). *A subset **V** of features is minimal approximately sufficient , with level of approximation $\epsilon \geq 0$ iff it is ϵ-sufficient and there does not exist other ϵ-sufficient subsets of smaller size.*

From our definition, it follows that a minimal approximately sufficient feature subset is a solution (probably not unique) to the optimization problem:

$$\min_{\boldsymbol{V}} \|\boldsymbol{V}\|_0 \ \text{ such that } \ DMI(\boldsymbol{V}) \leq \epsilon,$$

where $\|\boldsymbol{V}\|_0$ denotes the number of features selected. Such optimization problem can be transform via the use of a Lagrange multiplier $\lambda > 0$ into:

$$\min_{\boldsymbol{V}} \|\boldsymbol{V}\|_0 + \lambda \ DMI(\boldsymbol{V}).$$

Noting that $MI(\boldsymbol{X}, Y)$ is constant, this is equivalent to:

$$\min_{\boldsymbol{V}} \|\boldsymbol{V}\|_0 - \lambda \ MI(\boldsymbol{V}, Y).$$

We recover the feature selection problem stated in Chapter 6: find the smallest possible feature subset that maximizes the mutual information between the feature subset and the target.

We remark that the quantity $\|\boldsymbol{V}\|_0$ is discrete and therefore difficult to optimize. It has been suggested (Tishby et al., 1999) to replace it by $MI(\boldsymbol{X}, \boldsymbol{V})$. As noted in section 3.1, the prediction of posterior probabilities is a harder problem than classification or regression. Hence, we might want to replace the problem of maximizing mutual information by that of minimizing a given risk functional, e.g. the classification error rate. The formulation of the "zero-norm" feature selection method follows this line of thoughts (see Chapter 5.)

4.3 Variance of Feature Subset Selection

If the data have redundant features, different subsets of features can be equally efficient. For some applications, one might want to purposely generate alternative subsets that can be presented to a subsequent stage of processing. Still one might find this variance undesirable because (i) variance is often the symptom of a "bad" model that does not generalize well; (ii) results are not reproducible; and (iii) one subset fails to capture the "whole picture".

One method to "stabilize" variable selection developed in Chapter 7 is to use ensemble methods. The feature selection process may be repeated e.g. with sub-samples of the training data. The union of the subsets of features selected may be taken as the final "stable" subset. An index of relevance of individual features can be created considering how frequently they appear in the selected subsets.

This approach has shown great promises but the following limitation is worth mentioning: when one feature that is highly relevant by itself is complemented by many alternative features having weak individual relevance, the highly relevant feature will easily emerge from the procedure while the weak features will be difficult to differentiate from irrelevant features. This may be detrimental to performances.

4.4 Suggested Problems

Before closing this chapter, we would like to describe some research directions that we believe deserve attention.

More theoretically grounded algorithms. A lot of popular algorithms are not principled and it is difficult to understand what problem they seek to solve and how optimally they solve it. It is important to start with a clean mathematical statement of the problem addressed (see Sections 4.1 and 4.2 for preliminary guidelines.) It should be made clear how optimally the chosen approach addresses the problem stated. Finally, the eventual approximations made by the algorithm to solve the optimization problem stated should be explained. An interesting topic of research would be to "retrofit" successful heuristic algorithms in a theoretical framework.

Better estimation of the computational burden. Computational considerations are fairly well understood. But, even though the ever increasing speed of computers lessens the importance of algorithmic efficiency, it remains essential to estimate the computational burden of algorithms for feature selection problems. The computational time is essentially driven by the search strategy and the evaluation criterion. Several feature selection methods require examining a very large number of subsets of features, and possibly, all subsets of features, i.e. 2^n subsets. Greedy methods are usually more parsimonious and visit only of the order of n or n^2 subsets. The evaluation criterion may also be expensive as it may involve training a classifier or comparing every pairs of examples or features. Additionally, the evaluation criterion may involve one or several nested cross-validation loops. Finally, ensemble methods offer performance increases at the expense of additional computations.

Better performance assessment of feature selection. The other important question to be addressed is of statistical nature: some methods require more training examples than others to select relevant features and/or obtain good predictive performances. The danger of "overfitting" is to find features that "explain well" the training data, but have no real relevance or no predictive power. Making theoretical predictions on the number of examples needed to "solve" the feature selection problem is essential both to select an appropriate feature selection method and to plan for future data acquisition. Initial results to tackle this problem are found e.g. in (Almuallim and Dietterich, 1991) and (Ng, 1998).

The sagacious reader will have noticed that we did not treat the finite sample case in Section 4.2 for "sufficient feature subsets". There is still a lack of adequate formalism. We argue that in the finite sample case, feature subsets that are NOT sufficient may yield better performance than sufficient subsets (even if they are minimal and contain no irrelevant feature) because further reducing the space dimensionality may help reducing the risk of overfitting. In line with the "wrapper" methodology (Kohavi and John, 1997), it might be necessary to introduce a notion of "efficient feature subset": a subset providing best expected value of the risk when the learning machine is trained with a finite number m of examples. One central issue is to devise performance bounds characterizing efficient feature subsets.

Other challenges. Although we have made an effort in this introduction and in the book to cover a large number of topics related to feature extraction, we have not exhausted all of them. We briefly list some other topics of interest.

- *Unsupervised variable selection.* Several authors have attempted to perform feature selection for clustering applications (see, e.g., Xing and Karp, 2001, Ben-Hur and Guyon, 2003, and references therein). For supervised learning tasks, one may want to pre-filter a set of most significant variables with respect to a criterion which does not make use of y to lessen the problem of overfitting.
- *Selection of examples.* The dual problems of feature selection/construction are those of example selection/construction. Mislabeled examples may induce the choice of wrong variables, so it may be preferable to perform jointly the selection of variables and examples.
- *Reverse engineering the system.* Our introduction focuses on the problem of constructing and selecting features useful to build a good predictor. Unraveling the causal dependencies between variables and reverse engineering the system that produced the data is a far more challenging task (see, e.g., Pearl, 2000) that goes beyond the scope of this book.

5 Conclusion

We have presented in this introductions many aspects of the problem of feature extraction. This book covers a wide variety of topics and provides access to stimulating problems, particularly via the feature selection challenge, which is the object of Part II of the book. Simple but effective solutions have been presented as a starting point. The reader is now invited to study the other chapters to discover more advanced solutions. We have indicated a number of open problems to challenge the reader to contribute to this rapidly evolving field.

Acknowlegments

We are grateful to Eugene Tuv for providing us information on the computational complexity of RF.

References

H. Almuallim and T. G. Dietterich. Learning with many irrelevant features. In *Proceedings of the Ninth National Conference on Artificial Intelligence (AAAI-91)*, volume 2, pages 547–552, Anaheim, California, 1991. AAAI Press.

A. Ben-Hur and I. Guyon. Detecting stable clusters using principal component analysis. In M.J. Brownstein and A. Kohodursky, editors, *Methods In Molecular Biology*, pages 159–182. Humana Press, 2003.

A. Blum and P. Langley. Selection of relevant features and examples in machine learning. *Artificial Intelligence*, 97(1-2):245–271, December 1997.

Leo Breiman. Random forests. *Machine Learning*, 45(1):5–32, 2001.

I. Guyon. Design of experiments of the NIPS 2003 variable selection benchmark. *http://www.nipsfsc.ecs.soton.ac.uk/papers/NIPS2003-Datasets.pdf*, 2003.

I. Guyon and A. Elisseeff. An introduction to variable and feature selection. *JMLR*, 3:1157–1182, March 2003.

T. Hastie, R. Tibshirani, and J. Friedman. *The Elements of Statistical Learning, Data Mining, Inference and Prediction*. Springer Verlag, 2000.

K. Kira and L. Rendell. A practical approach to feature selection. In D. Sleeman and P. Edwards, editors, *International Conference on Machine Learning*, pages 249–256, Aberdeen, July 1992. Morgan Kaufmann.

R. Kohavi and G. John. Wrappers for feature selection. *Artificial Intelligence*, 97 (1-2):273–324, December 1997.

D. Koller and M. Sahami. Toward optimal feature selection. In *13th International Conference on Machine Learning*, pages 284–292, July 1996.

J. Kruskal and M. Wish. *Multidimensional Scaling*. Sage Publications, 1978.

H. Liu and H. Motoda. *Feature Extraction, Construction and Selection: A Data Mining Perspective*. Kluwer Academic, 1998.

R. G. Lyons. *Understanding Digital Signal Processing*. Prentice Hall, 2004.

D. J. C. MacKay. Bayesian non-linear modeling for the energy prediction competition. *ASHRAE Transactions*, 100:1053–1062, 1994.

R. M. Neal. Defining priors for distributions using dirichlet diffusion trees. Technical Report 0104, Dept. of Statistics,University of Toronto, 2001.

R. M. Neal. *Bayesian Learning for Neural Networks*. Number 118 in Lecture Notes in Statistics. Springer-Verlag, New York, 1996.

A. Y. Ng. On feature selection: learning with exponentially many irrelevant features as training examples. In *15th International Conference on Machine Learning*, pages 404–412. Morgan Kaufmann, San Francisco, CA, 1998.

J. Pearl. *Causality*. Cambridge University Press, 2000.

R. E. Woods R. C. Gonzalez. *Digital Image Processing*. Prentice Hall, 1992.

P. Soille. *Morphological Image Analysis*. Springer-Verlag, 2004.

N. Tishby, F. C. Pereira, and W. Bialek. The information bottleneck method. In *Proc. of the 37th Annual Allerton Conference on Communication, Control and Computing*, pages 368–377, 1999.

J. S. Walker. *A primer on wavelets and their scientific applications*. Chapman and Hall/CRC, 1999.

E.P. Xing and R.M. Karp. Cliff: Clustering of high-dimensional microarray data via iterative feature filtering using normalized cuts. In *9th International Conference on Intelligence Systems for Molecular Biology*, 2001.

A Forward Selection with Gram-Schmidt Orthogonalization

```
function idx = gram_schmidt(X, Y, featnum)
%idx = gram_schmidt(X, Y, featnum)
% Feature selection by Gram Schmidt orthogonalization.
% X          -- Data matrix (m, n), m patterns, n features.
% Y          -- Target vector (m,1).
% featnum    -- Number of features selected.
% idx        -- Ordered indices of the features (best first.)

[m, N]=size(X);
if nargin<3 | isempty(featnum), featnum=min(m,N); end
idx=zeros(1,featnum); w=zeros(1,featnum);
rss=zeros(1,featnum);        % Residual sum of squares
colid=1:N;                   % Original feature numbering
n=N;
% Main loop over features
for k=1:featnum
    fprintf('\nTraining on feature set size: %d\n', N-n+1);
    % Normalize
    XN=sqrt(sum(X.^2));      % Norms of the feature vectors
    XN(XN==0)=eps;
    X_norma = X./repmat(XN, m,1); % Normalized feature matrix
    % Project onto Y
    y_proj = sum(repmat(Y, 1, n).*X_norma);
    ay_proj=abs(y_proj);
    % Find the direction of maximum projection
    [maxval, maxidx] = max(ay_proj); % Dir. of max. proj.
    idx(k)=colid(maxidx);            % Index of that feature
    % Update the model
    w(k)=y_proj(maxidx)/XN(maxidx);  % Weight of that feature
    Y_proj = w(k)*X(:,maxidx); % Proj. Y on dir. X(:,maxidx)
    Y_residual = Y - Y_proj;         % New residual
    rss(k) = sum(Y_residual.^2);     % Residual error model
    % Compute the residual X vectors
    X_proj = sum(repmat(X_norma(:,maxidx),1,n).*X);
    X_residual = X-repmat(X_proj, m, 1).* ...
                            repmat(X_norma(:,maxidx), 1, n);
    % Change the matrix to iterate
    Y=Y_residual;
    X=X_residual(:, [1:maxidx-1,maxidx+1:n]);
    colid=colid([1:maxidx-1,maxidx+1:n]);
    n=n-1;
    fprintf('Training mse: %5.2f\n', rss(k)/m);
    fprintf('Features selected:\nidx=[');
    fprintf('%d ',idx(1:k));
    fprintf(']\n');
end
```

B Justification of the Computational Complexity Estimates

Pearson. For the Pearson method, for each feature, the number of operations needed is proportional to the number m or examples and the coefficient needs to be evaluated for all n features. Sorting will cost only of the order of $n \log(n)$ operations in non-pathological cases, so the computations will be dominated by the computation of the coefficient, i.e. of the order of mn operations. Determining the optimum number of features may require running a cross-validation experiment, which we do not take into account.

Relief. For Relief, we first need to determine for each example its "nearest hit" and "nearest miss", which require of the order of m^2 comparisons using a distance measure requiring of the order of n operations. The computation of the ranking index itself is less expensive (a number of calculations proportional to nmK). Hence, we quote a computational complexity of the order of nm^2.

Gram-Schmidt. For linear least square classifiers, the computations are dominated by solving a system of m equations of n variables, which costs of the order of $\min(n, m)nm$. But for the Gram-Schmidt procedure, we can stop at an upper bound of the desired number of features f, which is never to exceed $\min(n, m)$, so, we get a computational complexity of fnm.

Recursive Feature Elimination-SVM. For the SVM, some implementations are faster than others, but a complexity of the order of $\max(n, m)m^2$ can be assumed, since it costs of the order of nm^2 to compute the kernel matrix and of the order of m^3 to invert it. For RFE, the SVM will be retrained several times with a decreasing number of features. The number of iterations is n if we remove the features one by one and $\log_2(n)$ if we reduce the number of features by a factor of 2 at every iteration. Assuming that we take this last strategy, by summing the series, we see that we only approximately double the amount of computations. Hence, the overall computational complexity of RFE-SVM is of the order of $\max(n, m)m^2$.

Neural net with OBD. For a neural network, assuming that the number of hidden units h is very small compared to n, a forward propagation and a backward propagation cost of the order of mn times h. We assume that of the order of $\min(n, m)$ iterations are needed for convergence. The pruning process (Optimum Brain Damage or OBD) requires the same additional computational burden as for RFE. Hence, the overall computational complexity of OBD is of the order of $\min(n, m)nmh$.

Random Forest (RF). For RF, the initial sort for each variable for a single tree takes of the order of $m \log m$ computations. Additionally, for a "balanced" (middle splits) tree of depth $\log m$, it takes approximately $m \log m$ split evaluations (if splits are near the end of the predictor range, the complexity could increase to m^2.) The default number of variables considered at every split by RF is $\sqrt{n}$. Hence, for t trees the overall computational complexity is of the order of $t\sqrt{n}m \log m$.

Part I

Feature Extraction Fundamentals

Chapter 1

Learning Machines

Norbert Jankowski and Krzysztof Grabczewski

Department of Informatics, Nicolaus Copernicus University, Toruń, Poland
norbert,kgrabcze@phys.uni.torun.pl

1.1 Introduction

Learning from data may be a very complex task. To satisfactorily solve a variety of problems, many different types of algorithms may need to be combined. Feature extraction algorithms are valuable tools, which prepare data for other learning methods. To estimate their usefulness one must examine the whole complex processes they are parts of.

The goal of the chapter is to present a short survey of different approaches to learning from data with a special emphasis on solving classification (and approximation) problems, where feature extraction plays a particularly important role. We address this review to readers who know the basics of the field and would like to get quickly acquainted with techniques they are less familiar with. For novice readers we recommend textbooks by Duda et al. (2001), Mitchell (1997), Bishop (1995), Haykin (1994), Cherkassky and Mulier (1998), Schalkoff (1992), de Sá (2001), Hastie et al. (2001), Ripley (1996), Schölkopf and Smola (2001), Friedman et al. (2001).

Our tutorial starts with the mathematical statement of the learning problem. Then, it presents two general induction principles: risk minimization and Bayesian learning, that are widely applied in this review and the next chapters. Classification algorithms are then discussed in more details, including: Naïve Bayes, Linear Discriminant Analysis, kernel methods, Neural Networks, similarity based approaches and Decision Trees.

1.2 The Learning Problem

The term *learning machines* encompasses many kinds of computational intelligence systems capable of gathering knowledge by means of data analysis. The algorithms are sometimes divided into different groups (not necessarily disjoint) such as *machine learning*, *soft computing* or more uniform types: *neural networks*, *decision trees*, *evolutionary algorithms* etc. Learning from

N. Jankowski and K. Grabczewski: *Learning Machines*, StudFuzz **207**, 29–64 (2006)
www.springerlink.com

data may be treated as searching for the most adequate model (hypothesis) describing a given data. A *learning machine* is an algorithm which determines a *learning model*, which can be seen as a function:

$$f : \mathcal{X} \to \mathcal{Y}. \tag{1.1}$$

The function transforms objects from the data domain $\mathcal{X}$ to the set $\mathcal{Y}$ of possible target values. The data domain and the set of target values are determined by the definition of the problem for which the f is constructed.

The learning model f usually depends on some *adaptive parameters* sometimes also called *free parameters*. In this context, learning can be seen as a process in which a learning algorithm searches for parameters of the model f, which solve a given task.

The learning algorithm learns from a *sequence* $\mathcal{D}$ of data, defined in the space $\mathcal{X}$ or in $\mathcal{X} \times \mathcal{Y}$:

$$\mathcal{D} = \{\mathbf{x}_1, \mathbf{x}_2, \dots, \mathbf{x}_m\} = X \tag{1.2}$$
$$\mathcal{D} = \{\langle \mathbf{x}_1, y_1 \rangle, \langle \mathbf{x}_2, y_2 \rangle, \dots, \langle \mathbf{x}_m, y_m \rangle\} = \langle X, Y \rangle \tag{1.3}$$

Usually X has a form of a sequence of multidimensional vectors. An alternative statement of the problem defines X as a sequence of object names and provides a matrix of values describing similarity between the objects.

The definition (1.2) states an *unsupervised learning* problem (learning without teacher), where learning algorithms may be base only on values $\mathbf{x}_i$ (called *inputs*) from the data domain. Unsupervised learning is used for example in clustering, self-organization, auto-association and some visualization algorithms.

In the case of definition (1.3) learning algorithms use pairs $\langle \mathbf{x}_i, y_i \rangle$ where y_i is the desired *output* value for $\mathbf{x}_i$. Such learning is called *supervised* (with teacher). When $\mathcal{Y}$ is a set of several symbols (the number $|\mathcal{Y}|$ of elements of $\mathcal{Y}$ is usually significantly smaller than the number of vectors in the training data set), the learning problem is called a *classification* task and $\mathcal{Y}$ is called the set of *class labels*. If $|\mathcal{Y}| = 2$, then we deal with *binary classification* and if $|\mathcal{Y}| > 2$ — with multi–class problems. For convenience it is often assumed that $\mathcal{Y} = \{-1, +1\}$, $\mathcal{Y} = \{0, 1\}$ or $\mathcal{Y} = \{1, \dots, c\}$. Other examples of supervised learning are *approximation* (or *regression*) and *time series prediction*. In such cases $\mathcal{Y} = \mathbb{R}$ or $\mathcal{Y} = \mathbb{R} \times \dots \times \mathbb{R}$.

Risk minimization

Many learning algorithms perform a minimization of *risk!expected* (or more precisely: a measure approximating the expected risk), defined by:

$$R[f] = \int_{\mathcal{X} \times \mathcal{Y}} l(f(\mathbf{x}), y) \, dP(\mathbf{x}, y), \tag{1.4}$$

where $l(\cdot)$ is a *loss function*, and P is the data distribution. Note that in the case of unsupervised learning the above equation does not depend on y and $\mathcal{Y}$.

The loss function may be defined in one of several ways. In the case of classification it may be

$$l_c(f(\mathbf{x}), y) = \begin{cases} 0 & f(\mathbf{x}) = y, \\ 1 & f(\mathbf{x}) \neq y. \end{cases} \tag{1.5}$$

For binary classification with $\mathcal{Y} = \{-1, +1\}$ the *soft margin* loss proposed by Bennett and Mangasarian (1992) may be used:

$$l_b(f(\mathbf{x}), y) = max\{0, 1 - yf(\mathbf{x})\}. \tag{1.6}$$

The most popular loss function designed for regression (however it is also commonly used for classification tasks) is the *squared loss*:

$$l_s(f(\mathbf{x}), y) = (f(\mathbf{x}) - y)^2. \tag{1.7}$$

Another popular loss function dedicated to regression is called *ϵ-insensitive loss*:

$$l_\epsilon(f(\mathbf{x}), y) = max\{0, |f(\mathbf{x}) - y| - \epsilon\}. \tag{1.8}$$

It may be seen as an extension of the soft margin loss.

In the case of $\mathcal{Y} = \{0, 1\}$ and $f(\mathbf{x}) \in (0, 1)$, a possible choice is the *cross-entropy* (Kullback and Leibler, 1951) loss function

$$l_{ce}(f(\mathbf{x}), y) = -y \log f(\mathbf{x}) - (1 - y) \log(1 - f(\mathbf{x})). \tag{1.9}$$

In practice, the distribution $P(\mathbf{x}, y)$, crucial for the integration of (1.4), is usually unknown. Thus, the expected risk is replaced by *risk!empirical*

$$R_{emp}[f] = \frac{1}{m} \sum_{i=1}^{m} l(f(\mathbf{x}_i), y_i), \tag{1.10}$$

which is the average of errors over the set of data pairs $\langle \mathbf{x}_i, y_i \rangle$. The empirical risk used with the squared loss function is the well known *mean squared error* (MSE) function:

$$R_{MSE}[f] = \frac{1}{m} \sum_{i=1}^{m} l_s(f(\mathbf{x}_i), y_i) = \frac{1}{m} \sum_{i=1}^{m} (f(\mathbf{x}_i) - y_i)^2. \tag{1.11}$$

The *sum squared error* (SSE) is equal to $m \cdot R_{MSE}[f]$. Hence, minimization of MSE is equivalent to minimization of SSE (up to the constant m). In practice, yet another formula is used ((1.11) with m replaced by 2 in the denominator), because of a convenient form of its derivative.

Minimization of the empirical risk measured on a training data sample does not guarantee good *generalization* which is the ability to accurately estimate the target values for *unseen data*[1]. Even if the empirical risk $R_{emp}[f]$ is small, f may provide poor generalization i.e. for a representative sample $\mathcal{D}' = \{\langle \mathbf{x}'_1, y'_1\rangle, \langle \mathbf{x}'_2, y'_2\rangle, \ldots, \langle \mathbf{x}'_{m'}, y'_{m'}\rangle\}$ of test patterns the empirical risk

$$R_{test}[f] = \frac{1}{m'} \sum_{i=1}^{m'} l(f(\mathbf{x}'_i), y'_i) \qquad (1.12)$$

may be much higher. The $R_{test}[f]$ is commonly called *test error*.

Poor generalization of a model accompanied by high accuracy on the training data is called *overfitting* and is often caused by too large complexity of the model (too many adaptive parameters). One of possible ways to protect the model against overfitting is to add a *regularization term*. Regularization was originally proposed for ill-posed problems by Tikhonov (1963), Tikhonov and Arsenin (1977):

$$R_{reg}[f] = R_{emp}[f] + \lambda\Omega(f). \qquad (1.13)$$

$\Omega(f)$ is a *regularizer* designated to control the complexity of f. More details on regularization are presented in section 1.3.7.

Bayesian learning

The Bayes theorem defines the relationship between the *posterior probability*[2] $P(f|\mathcal{D})$ and *prior probability* $P(f)$ of given hypothesis f:

$$P(f|\mathcal{D}) = \frac{P(\mathcal{D}|f)P(f)}{P(\mathcal{D})}. \qquad (1.14)$$

It may be used in learning algorithms in a number of ways.

Some Bayesian learning approaches estimate parameters of posterior probabilities $P(z|\mathcal{D})$ using the *marginalization*[3] scheme:

$$P(z|\mathcal{D}) = \int_{\mathcal{H}} P(z, h|\mathcal{D}) \, \mathrm{d}h = \int_{\mathcal{H}} P(z|h)P(h|\mathcal{D}) \, \mathrm{d}h. \qquad (1.15)$$

The $P(h|\mathcal{D})$ plays the role of model weighting factors. From the Bayes rule we have:

$$P(h|\mathcal{D}) = \frac{P(\mathcal{D}|h)P(h)}{\int P(\mathcal{D}|h)P(h) \, \mathrm{d}h}. \qquad (1.16)$$

---

[1] *Unseen* means not used in the learning process.

[2] In the case of discrete distribution P denotes the probability, otherwise it is the density function.

[3] Marginalization is an integration over all possible values of unknown parameters of given density distribution function.

Computing the integral (1.15) is rather hard and is usually solved by some approximations (Duda et al., 2001, Bishop, 1995, Neal, 1996).

Another group of algorithms aims in finding the hypothesis which maximizes the *a posteriori* probability (MAP):

$$f_{MAP} = \operatorname*{argmax}_{f \in H} \ P(f|\mathcal{D}) = \operatorname*{argmax}_{f \in H} \ P(\mathcal{D}|f)P(f). \tag{1.17}$$

The main difference between the previous approach (1.15) and MAP is that instead of considering the entire distribution $P(h|\mathcal{D})$ of functions a single solution is used.

Assuming equal *a priori* probabilities for all the hypotheses $f \in H$ leads to the definition of the *maximum likelihood* (ML) hypothesis:

$$f_{ML} = \operatorname*{argmax}_{f \in H} \ P(\mathcal{D}|f). \tag{1.18}$$

With additional assumptions that the training examples are identically and independently distributed and correspond to a target function g with some normally distributed noise $\epsilon \sim N(0, \sigma)$ (i.e. $\mathbf{y} = g(\mathbf{x}) + \epsilon$), the maximization of $P(\mathcal{D}|f)$ is equivalent to maximization of $\prod_{i=1}^{m} P(y_i|x_i, f)$ and to minimization of the negation of its logarithm which is exactly the minimization of MSE.

The following equivalent formulation of (1.17):

$$f_{MAP} = \operatorname*{argmin}_{f \in H} \ [-\log_2 P(\mathcal{D}|f) - \log_2 P(f)], \tag{1.19}$$

lends itself to using the information theory language. In this framework, the optimal model is the one which minimizes the sum of the description length of the hypothesis f (assuming the optimal hypotheses coding) and the description length of the data $\mathcal{D}$ under the assumption that f holds (also assuming the optimal coding[4]). If the symbol of $L_C(X)$ is used to denote the length of the description of X using coding C, then for a hypotheses coding C_H and data coding C_f (assuming the hypothesis f) the *Minimum Description Length* (MDL) principle (Rissanen, 1978) can be formulated as:

$$f_{MDL} = \operatorname*{argmin}_{f \in H} \ [L_{C_f}(\mathcal{D}) + L_{C_H}(f)]. \tag{1.20}$$

It confirms the expectations that shorter descriptions (simpler models) should be preferred over sophisticated ones. The function being minimized in (1.19) and (1.20) can be seen as a risk function $(L_{C_f}(\mathcal{D}) = -\log_2 P(\mathcal{D}|f))$ with a regularizer $(L_{C_H}(f) = -\log_2 P(f))$.

The MAP, ML and MDL approaches deal with the problem of selection of the optimal hypothesis describing a given data set. Another problem is how to

---

[4]Shannon and Weaver (1949) proved that the optimal code assigns $\log_2(P(i))$ bits to encode message i.

assign proper class labels to given data objects. The optimal choice is defined by the *Bayes Optimal Classifier*:

$$BOC(\mathbf{x}) = \underset{y\in\mathcal{Y}}{\operatorname{argmax}}\ P(y|\mathbf{x}). \tag{1.21}$$

Given a set of possible hypotheses H and a training data set $\mathcal{D}$ the most probable class of a new data vector $\mathbf{x}$ can be determined in a manner similar to (1.15) as:

$$BOC(\mathbf{x}|\mathcal{D}, H) = \underset{y\in\mathcal{Y}}{\operatorname{argmax}} \sum_{f\in H} P(y|f,\mathbf{x})P(f|\mathcal{D}). \tag{1.22}$$

Anyways, the *BOC* formula is useless in most real applications. To calculate the probabilities for each of the classes one needs to know the probabilistic structure of the problem and to examine all the candidate hypotheses, which is usually impossible (even in quite simple cases).

The error rate of BOC is called *Bayes error* and it is the smallest possible error.

More detailed theoretical description of the problem can be found for example in (Mitchell, 1997).

Deeper analysis of the MSE risk function (most commonly used in regression and classification tasks) reveals that for given data point $\langle x, y\rangle$ the expected value of the risk over all the training data sets of a given size, can be decomposed into *bias* and *variance* terms (Bishop, 1995, Duda et al., 2001):

$$E_\mathcal{D}(f(x)-y)^2 = \underbrace{(E_\mathcal{D}(f(x)-y))^2}_{\text{bias}^2} + \underbrace{E_\mathcal{D}(f(x)-E_\mathcal{D}f(x))^2}_{\text{variance}}. \tag{1.23}$$

The bias component represents the discrepancy between the target value and the average model response (over different training data sets) while the other component corresponds to the variance of the models trained on different samples. The decomposition reveals the so-called *bias–variance trade-off*: flexible (complex) models are usually associated with a low bias, but at the price of a large variance. Conversely simple models (like linear models) may result in a lower variance but may introduce unreasonable bias. There is no simple solution of this dilemma and good model must balance between bias and variance to keep the generalization as high as possible.

It is important to remember that regardless of the type of the error function being used, obtaining high level of generalization requires building as simple models as possible. Model complexity should be increased only when simpler models do not offer satisfactory results. This rule is consistent with the medieval rule called *Ockham's razor* and other ideas such as MDL principle or regularization.

1.3 Learning Algorithms

There is no commonly used taxonomy of the algorithms that learn from data. Different point of views stress different aspects of learning and group the methods in different ways. Learning algorithms have different theoretical background and model the knowledge with miscellaneous data structures. Their model building strategies may be founded on the achievements of statistics or different kinds of optimization methods. The optimal (or suboptimal) models can be determined with *search strategies* (from simple, blind search methods, through heuristically augmented search and evolutionary computation, to global optimization techniques such as simulated annealing), *gradient descent* methods, *mathematical programming* (linear and quadratic programming find especially numerous applications) or other tools including *fuzzy logic* (see chapter Chapter 8) and *rough sets theory* (Pawlak, 1982).

To increase the adaptation capability of computational intelligence systems, *ensembles* of many models are used (see Chapter 7 and Kuncheva (2004)). Such ensembles may be *homogeneous* (multiple models of the same kind) or *heterogeneous* (taking advantage of different methodologies in a single learning process) Jankowski and Grabczewski.

Classifier decision borders

Classification models divide the feature space into disjoint regions assigned to class labels. Different classifiers provide different kinds of borders between the regions (*decision borders*).

Figure 1.1 illustrates two examples of two-dimensional classification problems. Four different solutions for each of the data sets are depicted.

For both tasks, the top-left plot presents an example of linear decision borders. Many learning algorithms yield models which discriminate with linear functions only, e.g. linear discrimination methods (see section 1.3.2) and simple neural networks (see section 1.3.4).

The top-right plots show decision borders perpendicular to the axes of the feature space – most common solutions of decision tree algorithms (see section 1.3.6).

In the bottom-left plots the classes are embraced with quadratic curves. Such shapes can be obtained for example with simple neural networks with radial basis functions (see section 1.3.4) and other nonlinear methods.

The bottom-right plots present decision borders of maximum classification accuracy. Many learning algorithms are capable of finding models with so complicated borders, but usually it does not provide a desirable solution because of overfitting the training data – the accuracy reaches maximum for the sample used for training, but not for the whole of the data.

Some adaptive models can separate more than two classes in a single learning process. Other algorithms must be run several times and their results must be appropriately combined to construct a final classifier. The combination is

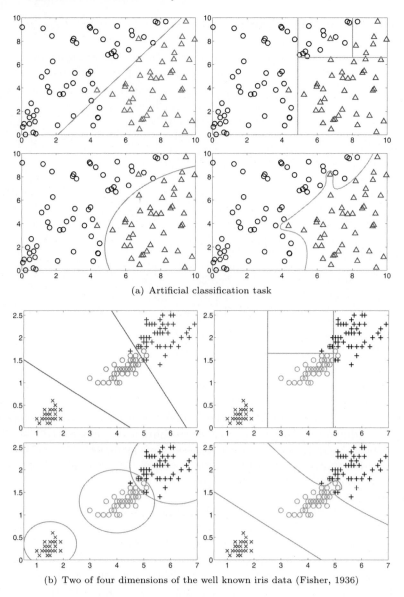

(a) Artificial classification task

(b) Two of four dimensions of the well known iris data (Fisher, 1936)

Fig. 1.1. Example solutions to two classification tasks: top-left – with linear decision borders, top-right – specific for decision trees, bottom-left – with centered decision areas and bottom-right – of maximum accuracy.

not always straightforward because of possible conflicts between the combined models.

Generalization and model selection

To determine the generalization ability of a model one would need to measure the average risk for the set of all possible data objects. In real life applications it is not feasible, so we estimate the risk using a test set (equation (1.12)). When doing this we should be aware of the danger of testing models on a single test set (for example resulting from a rigid partition of the set of all available data to the training and test parts). Model selection based on testing trained models on a single test set does not get rid of the danger of overfitting. A more accurate estimation of the empirical risk can be obtained with *K-fold cross-validation* (CV). In this technique we split the set of available data into n parts and perform n training and test processes (each time the test set is one of the parts and the training set consists of the rest of the data). The average test risk can be a good estimate of real generalization ability of the tested algorithm, especially when the whole cross-validation is performed several times (each time with different data split) and n is appropriately chosen[5]. To get a good estimate of generalization ability of a learning machine, it is important to analyze not only the average test error, but also its variance, which can be seen as a measure of *stability* (for more see Chapter 2).

The *Ockham's razor* and other issues discussed in section 1.2 suggest that accurate models which are simple should provide stability and good generalization. So, if we have several models of similar accuracy, we should prefer stable models and the simplest ones (linear over nonlinear, those with the smallest number of parameters etc.). Even if more complex models are more accurate it is often worth to resigning from high learning accuracy and selecting less accurate but more stable or simpler models. The bottom-right images of figure 1.1 present highly accurate models for the training data, but providing poor generalization.

Another technique to obtain models of good generalization is regularization (see section 1.3.7). Preserving large classification margins may also be seen as a kind of regularization and has been successfully used in SVM methodology (see section 1.3.3). Models with optimized margins may reach high level of generalization despite the complexity of their decision borders.

In the case of high-dimensional feature spaces the information about the function being modelled is often contained within a small-dimensional subspace and the rest of the features play a role of a noise, making learning difficult. Hence, the methods of feature selection can be very helpful in the pursuit of high accuracy and good generalization.

---

[5]Setting n to the number of data objects yields a special case of cross-validation called *leave-one-out* which, although sometimes used, is not a good estimate (see e.g. (Kohavi, 1995)).

1.3.1 Naïve Bayes Classifiers

Feasible Bayes classifiers, in contrary to the abstract Bayes Optimal Classifier, must pay the price of losing the optimality guarantee. One of the well known simplifications of the method is *naïve Bayes*. Its reduced complexity is the result of the assumption that the random variables corresponding to particular features of the data space are independent. It makes the definition of the maximum *a posteriori* class as simple as:

$$NBC(\mathbf{x}) = \operatorname*{argmax}_{y \in \mathcal{Y}} P(y|\mathbf{x}) = \operatorname*{argmax}_{y \in \mathcal{Y}} \frac{P(\mathbf{x}|y)P(y)}{P(\mathbf{x})}$$

$$= \operatorname*{argmax}_{y \in \mathcal{Y}} P(y) \prod_i P(x_i|y). \qquad (1.24)$$

The formula can be easily used if we know the probabilities of observing each class and of observing a particular feature value among the populations of vectors representing the classes. In real world applications the probabilities can only be estimated on the basis of a given training data set.

In the case of discrete features the probabilities $P(x_i|y)$ can be evaluated as the relevant frequencies of observing x_i values among the vectors of class y. This is the most common application of Naïve Bayes Classifiers. The frequencies are good estimators when the training data set is large enough to reflect real distributions. Otherwise some corrections to the frequencies calculations (e.g. *Laplace correction* or *m-estimate*) are strongly recommended (Provost and Domingos, 2000, Kohavi et al., 1997, Cestnik, 1990, Zadrozny and Elkan, 2001). The corrections are also fruitfully used in decision tree algorithms.

The features with continuous values can be first discretized to use the same method of probability estimation. An alternative is to assume some precise distribution of the *a priori* class probabilities. Most commonly the normal distribution is assumed:

$$P(x_i|y) \propto N(\mu_i^y, \sigma_i^y), \qquad (1.25)$$

where N is the normal density function while μ_i^y and σ_i^y are respectively: the mean value and the standard deviation of the i-th feature values observed among the training data vectors belonging to class y.

In practice the assumption of independence of different features and of the normality of the distributions may be far from true – in such cases also the Naïve Bayes Classifier may be far from optimal.

Some other approaches to the problem of estimation of the conditional probabilities have also been proposed. One of the examples (John and Langley, 1995) gets rid of the assumption of normal distributions by using a kernel density estimation technique.

1.3.2 Linear Discriminant Methods

Linear discriminant methods determine linear functions, which divide the domain space into two regions (see the top-left plot in figure 1(a)). Learning processes adjust the parameters of linear models to obtain an optimal correspondence between the half-spaces of the feature space and the data categories (classes). *Linear discriminant* functions are defined by linear combinations of the argument vector components:

$$f(\mathbf{x}) = \mathbf{w}^T \mathbf{x} + b, \tag{1.26}$$

where $\mathbf{w}$ is a *weight vector* and $-b$ defines the *threshold*.

If a vector $\mathbf{x}$ satisfies $f(\mathbf{x}) > 0$, then the model assigns the label of the positive category to it, otherwise the label of the negative class is assigned. The instances for which $f(\mathbf{x}) = 0$ define the hyperplane, which splits the whole space into the two regions.

For a given training data set such separating hyperplane may not exist. In such a case we say that the problem is *linearly inseparable*.

In a multicategory case it is not possible to separate different classes with a single linear discriminant function. One of the solutions to this is to form a linear discriminant function for each class to separate the samples of the i-th class from the rest of the training data. Another solution is to compose one linear model for each pair of classes. Both ways may result in such a partition of the space that for some regions there is no simple way to determine the winner class, because the half-spaces corresponding to different classes overlap. A good combination of linear discriminants is the *linear machine* which, given a linear discriminant function for each class $y \in \mathcal{Y}$:

$$f_y(\mathbf{x}) = \mathbf{w}_y^T \mathbf{x} + b_y, \tag{1.27}$$

provides a reasonable scheme of label assignment eligible for each point of the space (except for the decision borders):

$$k = \arg\max_{y \in \mathcal{Y}} f_y(\mathbf{x}). \tag{1.28}$$

The linear discriminant function can be determined in different ways. The idea of Fisher's linear discriminant (Fisher, 1936) lies in maximization of

$$m_{-1} - m_{+1} = \mathbf{w}^T (\mathbf{m}_{-1} - \mathbf{m}_{+1}), \tag{1.29}$$

where $\mathbf{m}_{\pm 1}$ are the means calculated for the two classes:

$$\mathbf{m}_{\pm 1} = \frac{1}{|\{j : y_j = \pm 1\}|} \sum_{\{j : y_j = \pm 1\}} \mathbf{x}_j. \tag{1.30}$$

Maximization of (1.29) can be seen as the maximization of the distance between the projected averages ($m_{\pm 1}$). Such criterion could strongly depend on the directions of the largest spread of the data. This is why the final Fisher criterion is

$$J(\mathbf{w}) = \frac{(m_{-1} - m_{+1})^2}{s_{-1}^2 + s_{+1}^2}, \tag{1.31}$$

where $s_{\pm 1}^2$ is the within-class variance. It results that the $\mathbf{w}$ should be

$$\mathbf{w} \propto \mathbf{S_w}^{-1}(\mathbf{m}_{-1} - \mathbf{m}_{+1}) \qquad (1.32)$$

where

$$\mathbf{S_w} = \sum_{\{j:y_j=-1\}} (\mathbf{x}_j - \mathbf{m}_{-1})(\mathbf{x}_j - \mathbf{m}_{-1})^T + \sum_{\{j:y_j=+1\}} (\mathbf{x}_j - \mathbf{m}_{+1})(\mathbf{x}_j - \mathbf{m}_{+1})^T$$

$$(1.33)$$

Equation (1.32) defines the direction of the discriminant:

$$f(\mathbf{x}) = \mathbf{w}^T(\mathbf{x} - \bar{\mathbf{x}}) \qquad (1.34)$$

where $\bar{\mathbf{x}}$ is the average of all the training samples.

Most linear discriminant learning methods are based on the gradient descent procedure. Its general scheme is presented in the following algorithm. To simplify the notation we define $\mathbf{w} = [b, w_1, \ldots, w_n]^T$ and $\bar{\mathbf{x}} = [1 \ \mathbf{x}]^T$.

Gradient descent procedure

```
k:=0; init w₀; do
    w_{k+1} := w_k − η(k)∇J(w_k);
    k:=k+1;
while not stop−criterion(k, θ, ∇J(w_k),
η(k)) return w;
```

A typical definition of the `stop-criterion` is $|\eta(k)\nabla J(\mathbf{w}_k) < \theta|$, where θ is a user-defined parameter. The $\eta(k)$ controls the speed of learning. Sometimes $\eta(k)$ is a constant scalar below 1 and sometimes a decreasing function, such as $\eta(k) = \eta(0)/k$. In the second order forms $\eta(k)$ can be equal to $\frac{||\nabla J(\mathbf{w}_k)||^2}{\nabla J(\mathbf{w}_k)^T \mathbf{H} \nabla J(\mathbf{w}_k)}$ ($\mathbf{H}$ is the Hessian of $J(\mathbf{w}_k)$). If $\eta(k)$ is defined as $\mathbf{H}^{-1}$, then the *Newton descent algorithm* is obtained.

Another source of flexibility of the gradient descent algorithm is the definition of $J(\mathbf{w}_k)$. Moreover, the algorithm may work in *batch* or *on-line* mode. In the batch mode the $\nabla J(\mathbf{w}_k)$ is calculated using the whole training data set, and in the online mode $\nabla J(\mathbf{w}_k)$ is defined for a single (current) vector. The *Least Mean Square* (LMS) algorithm defines the $J(\mathbf{w}_k)$ as:

$$J(\mathbf{w}_k) = \sum_i (\mathbf{w}_k^T \bar{\mathbf{x}}_i - y_i)^2, \qquad (1.35)$$

and modifies the weights (in the online version) as follows:

$$\mathbf{w}_{k+1} = \mathbf{w}_k + \eta(k)(y_p - \mathbf{w}_k^T \bar{\mathbf{x}}_p)\bar{\mathbf{x}}_p. \qquad (1.36)$$

The p index is chosen randomly or is equal to $(k \mod m) + 1$.

Another definition of $J(\mathbf{w}_k)$ is the perceptron criterion:

$$J(\mathbf{w}_k) = \sum_{p \in \mathcal{P}_k} -y_p \mathbf{w}^T \bar{\mathbf{x}}_p \tag{1.37}$$

where $\mathcal{P}_k$ is the set of indexes of misclassified vectors. Then the weight changes are:

$$\mathbf{w}_{k+1} = \mathbf{w}_k + \eta(k) \sum_{p \in \mathcal{P}_k} y_p \bar{\mathbf{x}}_p. \tag{1.38}$$

Relaxation procedures use squared versions of the perceptron criterion:

$$J(\mathbf{w}_k) = \sum_{p \in \mathcal{P}_k} (\mathbf{w}_k^T \bar{\mathbf{x}}_p)^2 \quad \text{or} \quad J(\mathbf{w}_k) = \sum_{p \in \mathcal{P}'_k} \frac{(\mathbf{w}_k^T \bar{\mathbf{x}}_p - \beta)^2}{||\bar{\mathbf{x}}||^2} \tag{1.39}$$

where $\mathcal{P}'_k = \{i : \mathbf{w}_k^T \bar{\mathbf{x}}_i \leq \beta\}$ and β is a constant defining a margin. A similar idea of margins plays a prominent role in Support Vector Machine algorithms which, used with linear kernels, are an efficient linear discrimination methodology capable of optimizing the margins (see below for more).

An exhaustive description of several linear discriminant methods can be found in (Duda et al., 2001, Guyon and Stork, 2000).

1.3.3 Support Vector Machines

Support Vector Machines (SVMs) were introduced by Boser et al. (1992). Initially SVMs were constructed to solve binary classification and regression problems. Today, there are several more areas where the SVM *framework* has been successfully applied (Schölkopf and Smola, 2001), for example: novelty detection (data set consistency) (Schölkopf et al., 2001, Schölkopf and Smola, 2001), clustering (Ben-Hur et al., 2001), feature selection (Fung and Mangasarian, 2004, Guyon et al., 2002, Weston et al., 2001, Schölkopf and Smola, 2001), feature extraction (kernel PCA) (Schölkopf and Smola, 2001), kernel Fisher discriminant (Schölkopf and Smola, 2001).

In application to classification problems SVMs can produce models with different kinds of decision borders – it depends on the parameters used (especially on the kernel type). The borders can be linear (like in top-left plots of figure 1.1) or highly nonlinear (may resemble the bottom-right images of figure 1.1). Here the complexity of the borders not necessarily announces poor generalization, because the margin optimization (described below) takes care for proper placement of the border.

SVMs minimize the empirical risk function (1.10) with soft margin loss function (1.6) for classification problems or with ϵ-insensitive loss function (1.8) for regression problems. For connections of risk function and regularization with SVMs compare section 1.3.7 and especially table 1.1.

Optimal hyperplane

The construction of the *optimal hyperplane* is the fundamental idea of SVM. The optimal hyperplane separates different classes with maximal margin (the distance between the hyperplane and the closest training data point). Such goal can be defined as maximization of the minimum distance between vectors and the hyperplane:

$$\max_{\mathbf{w},b} \quad \min\{||\mathbf{x} - \mathbf{x}_i|| \ : \ \mathbf{w}^T\mathbf{x} + b = 0, \quad i = 1,\dots,m\}. \tag{1.40}$$

The $\mathbf{w}$ and b can be rescaled in such a way that the point closest to the hyperplane $\mathbf{w}^T\mathbf{x} + b = 0$, lies on a hyperplane $\mathbf{w}^T\mathbf{x} + b = \pm 1$. Hence, for every $\mathbf{x}_i$ we get: $y_i[\mathbf{w}^T\mathbf{x}_i + b] \geq 1$, so the width of the margin is equal to $2/||\mathbf{w}||$. The goal (1.40) can be restated as the optimization problem of *objective function* $\tau(\mathbf{w})$:

$$\min_{\mathbf{w},b} \ \tau(\mathbf{w}) = \frac{1}{2}||\mathbf{w}||^2 \tag{1.41}$$

with the following constraints:

$$y_i[\mathbf{w}^T\mathbf{x}_i + b] \geq 1 \quad i = 1,\dots,m. \tag{1.42}$$

To solve it a Lagrangian is constructed:

$$L(\mathbf{w},b,\boldsymbol{\alpha}) = \frac{1}{2}||\mathbf{w}||^2 - \sum_{i=1}^{m}\alpha_i(y_i[\mathbf{x}_i^T\mathbf{w} + b] - 1), \tag{1.43}$$

where $\alpha_i > 0$ are Lagrange multipliers. Its minimization leads to:

$$\sum_{i=1}^{m}\alpha_i y_i = 0, \qquad \mathbf{w} = \sum_{i=1}^{m}\alpha_i y_i \mathbf{x}_i. \tag{1.44}$$

According to the Karush-Kuhn-Thucker (KKT) conditions (Schölkopf and Smola, 2001):

$$\alpha_i(y_i[\mathbf{x}_i^T\mathbf{w} + b] - 1) = 0, \quad i = 1,\dots,m. \tag{1.45}$$

The non-zero α_i correspond to $y_i[\mathbf{x}_i^T\mathbf{w} + b] = 1$. It means that the vectors which lie on the margin play the crucial role in the solution of the optimization problem. Such vectors are called *support vectors*.

After some substitutions the optimization problem can be transformed to the *dual optimization problem*:

$$\max_{\boldsymbol{\alpha}} \quad W(\boldsymbol{\alpha}) = \sum_{i=1}^{m}\alpha_i - \frac{1}{2}\sum_{i,j=1}^{m}\alpha_i\alpha_j y_i y_j \mathbf{x}_i^T\mathbf{x}_j \tag{1.46}$$

with constraints:

$$\alpha_i \geq 0 \quad i = 1, \ldots, m, \qquad \sum_{i=1}^{m} \alpha_i y_i = 0. \tag{1.47}$$

Using the solution of this problem the decision function can be written as:

$$f(\mathbf{x}) = \text{sgn} \left(\sum_{i=1}^{m} \alpha_i y_i \mathbf{x}^T \mathbf{x}_i + b \right). \tag{1.48}$$

The kernel trick

The dot product $\mathbf{x}^T \mathbf{x}'$ in (1.46) and (1.48) can be replaced by a kernel function $k(\mathbf{x}, \mathbf{x}') = \phi(\mathbf{x})^T \phi(\mathbf{x}')$ (Boser et al., 1992). It extends the linear discriminant SVM to a nonlinear machine. The new decision function is:

$$f(\mathbf{x}) = \text{sgn} \left(\sum_{i=1}^{m} \alpha_i y_i k(\mathbf{x}, \mathbf{x}_i) + b \right). \tag{1.49}$$

The dot product is the simplest kernel and may be generalized to the polynomial kernel

$$k_p(\mathbf{x}, \mathbf{x}') = [\gamma(\mathbf{x}^T \mathbf{x}') + \theta]^q, \tag{1.50}$$

where q is an integer and $\theta = 0$ or $\theta = 1$. Probably the most powerful is the Gaussian kernel

$$k_G(\mathbf{x}, \mathbf{x}') = \exp[-\gamma ||\mathbf{x} - \mathbf{x}'||^2]. \tag{1.51}$$

The SVM decision function (1.49) with Gaussian kernel is equivalent to RBF networks (1.70). Another popular kernel is the hyperbolic tangent

$$k_t(\mathbf{x}, \mathbf{x}') = \tanh(\gamma[\mathbf{x}^T \mathbf{x}'] + \theta). \tag{1.52}$$

Soft margin hyperplane

The construction of optimal hyperplane is impossible if the data set (transformed by kernels if kernels are used) is not linearly separable. To solve this problem Cortes and Vapnik (1995) introduced the *soft margin* hyperplane technique using *slack variables* ξ_i ($\xi_i \geq 0$):

$$y_i[\mathbf{w}^T \mathbf{x}_i + b] \geq 1 - \xi_i \quad i = 1, \ldots, m. \tag{1.53}$$

This leads to a new optimization problem:

$$\min_{\mathbf{w}, b, \xi} \quad \frac{1}{2} ||\mathbf{w}||^2 + C \sum_{i=1}^{m} \xi_i \tag{1.54}$$

with constraints (1.53). It defines a *Support Vector Classifier* (SVC) with the C parameter (C-SVC) controlling the balance between training accuracy and the margin width (C must be greater than 0).

The dual optimization problem for C-SVC is defined as

$$\max_{\boldsymbol{\alpha}} \quad W(\boldsymbol{\alpha}) = \sum_{i=1}^{m} \alpha_i - \frac{1}{2}\sum_{i,j=1}^{m} \alpha_i \alpha_j y_i y_j \mathbf{x}_i^T \mathbf{x}_j \tag{1.55}$$

with constraints:

$$0 \le \alpha_i \le C \quad i=1,\ldots,m, \qquad \sum_{i=1}^{m}\alpha_i y_i = 0. \tag{1.56}$$

ν-SVC

Schölkopf and Smola proposed the ν-SVM (Schölkopf and Smola, 2001) justifying that the C parameter of C-SVC is not intuitive. They defined a new primary optimization problem as:

$$\min_{\mathbf{w},b,\boldsymbol{\xi},\rho} \quad \tau(\mathbf{w},\boldsymbol{\xi},\rho) = \frac{1}{2}||\mathbf{w}||^2 - \nu\rho + \frac{1}{m}\sum_{i=1}^{m}\xi_i \tag{1.57}$$

with constraints:

$$y_i[\mathbf{x}_i^T\mathbf{w}+b] \ge \rho - \xi_i, \qquad \xi_i \ge 0, \qquad \rho \ge 0 \tag{1.58}$$

If the ρ after the optimization procedure is greater than 0 then ν has an interesting interpretation: it is the upper bound of the fraction of vectors within the margin and the lower bound of the fraction of vectors which are support vectors.

The ν-SVM has also been stated for regression tasks (Schölkopf et al., 2000).

Regression with SVM (ε-SVR)

The starting point to define the SVM for regression (SVR) is the $\epsilon-insensitive$ *error function* which is defined by (1.8).

The goal of regression can be defined as minimization of

$$\frac{1}{2}||\mathbf{w}||^2 + C\sum_{i=1}^{m}|y_i - f(\mathbf{x}_i)|_\epsilon. \tag{1.59}$$

The primary optimization problem formulation is based on two types of slack variables $\boldsymbol{\xi}$ and $\boldsymbol{\xi}^*$: the former for $f(\mathbf{x}_i) - y_i > \epsilon$ and the latter for $y_i - f(\mathbf{x}_i) > \epsilon$. It can be stated as:

$$\min_{\mathbf{w},\boldsymbol{\xi},\boldsymbol{\xi}^*,b} \quad \tau(\mathbf{w},\boldsymbol{\xi},\boldsymbol{\xi}^*) = \frac{1}{2}||\mathbf{w}||^2 + C\sum_{i=1}^{m}(\xi_i + \xi_i^*) \tag{1.60}$$

with constraints:

$$f(\mathbf{x}_i) - y_i \le \epsilon + \xi_i, \qquad y_i - f(\mathbf{x}_i) \le \epsilon + \xi_i^*, \qquad \xi_i,\xi_i^* \ge 0, \quad i=1,\ldots,m. \tag{1.61}$$

Quadratic programming problems

Each of the above algorithms solves its dual optimization problem by means of quadratic programming (QP). The first implementations of QP were significantly less effective than the recent ones. The dual optimization problems defined for SVC (1.46), C-SVC (1.55), ν-SVM, ϵ-SVR can be generalized to:

$$\min_{\alpha} \quad \frac{1}{2}\alpha^T Q\alpha + \mathbf{p}^T\alpha, \tag{1.62}$$

where $\mathbf{p}$ depend directly on dual optimization problem (for example in the case of optimization problem defined by (1.46) $\mathbf{p}$ is a vector of 1's).

The crucial point of the most effective implementations is the *decomposition* of the QP problem (Osuna et al., 1997, Joachims, 1998, Platt, 1998).

$$\max_{\alpha_B} \quad W(\alpha_B) = (\mathbf{p} - Q_{BR}\alpha_R)^T \alpha_B - \frac{1}{2}\alpha_B^T Q_{BB}\alpha_B, \tag{1.63}$$

where $[Q_{ij} = y_i y_j k(\mathbf{x}_i, \mathbf{x}_j))]$. The idea is that the vector α is divided into: the working part α_B and the fixed one α_R. At particular stage only the working part is being optimized while the fixed part does not change. During the optimization procedure the subset B is changed from time to time. The most interesting examples of decomposition algorithms are SVM^{light} (Joachims, 1998) and SMO (Platt, 1998) with modifications described in (Shevade et al., 2000). These two algorithms differ in the working set selection technique and in the stop criterion.

1.3.4 Artificial Neural Networks

Artificial neural networks (ANN) represent a very broad class of different algorithms designed for classification, regression, (auto-)associations, signal processing, time series prediction, clustering etc. A number of good books on neural networks may be recommended (Bishop, 1995, Haykin, 1994, Kohonen, 1995, Ripley, 1996, Zurada, 1992). Here only a few most popular concepts related to artificial neural networks, used in classification and regression problems, are presented.

First neural networks were proposed by McCulloch and Pitts (1943). Neural networks are built from *neurons* which are grouped in layers. Neurons may be connected in a number of ways. For an example see figure 1.3. A single neuron can be seen as an operational element which realizes a *transfer function* based on incoming signals $\mathbf{x}$. Transfer function is a superposition of *output function* $o(\cdot)$ and *activation function* $I(\mathbf{x})$ – compare figure 1.2. If all the transfer functions realized by neurons are linear, then also the neural network realizes a linear transformation. There is a variety of transfer functions, and their type strongly determines the properties of the network they compose (see the review by Duch and Jankowski (1999)). Two best known activation

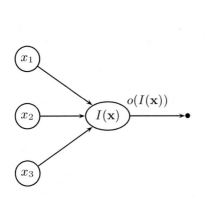

Fig. 1.2. A neuron. Transfer function $F(\mathbf{x}) = o(I(\mathbf{x}))$ is a superposition of output function $o(\cdot)$ and activation function $I(\cdot)$.

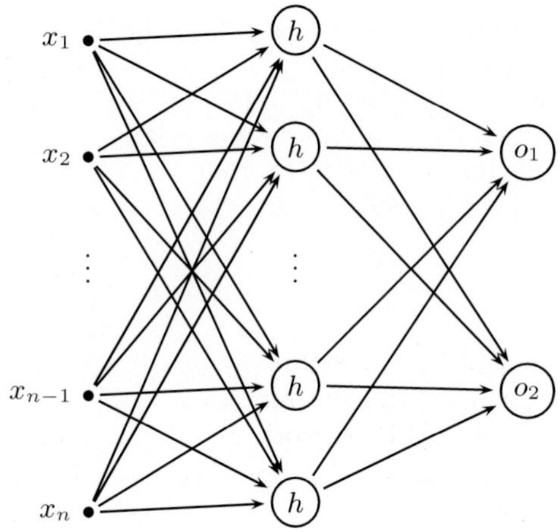

Fig. 1.3. An example of neural network. x_i denotes inputs, h – hidden neurons, o_j – outputs. This network is composed of input, one hidden and output layers.

functions are the *inner product* $(\mathbf{w}^t\mathbf{x})$ and the *Euclidean distance* $(||\mathbf{x} - \mathbf{t}||)$. The best known output functions are the *threshold function* (McCulloch and Pitts, 1943):

$$\Theta(I;\theta) = \begin{cases} -1 & I - \theta < 0, \\ 1 & I - \theta \geq 0, \end{cases} \tag{1.64}$$

the *logistic function*[6]: $\sigma(I) = 1/[1 + e^{-I}]$, and the *Gaussian function* (1.51).

Perceptrons (or *linear threshold unit*) are defined as the threshold output function with the inner product activation and were studied by Rosenblatt (1962):

$$F(\mathbf{x}; \mathbf{w}) = \Theta(\mathbf{x}^T\mathbf{w}; \theta). \tag{1.65}$$

The perceptron learning goes according to the following update rule:

$$\mathbf{w} = \mathbf{w} + \eta[y_i - F(\mathbf{x}_i; \mathbf{w})]\mathbf{x}_i. \tag{1.66}$$

The most common and successful neural network is the *multi-layer perceptron* (MLP). It is an extension of the concept of perceptron. MLP is a feedforward network with at least one hidden layer. The ith output of MLP (with l hidden layers) is defined by:

$$o_i(\mathbf{x}; \mathbf{w}) = \tilde{o}\left(\sum_j w_{ij}^{l+1}\phi_j^l\right), \tag{1.67}$$

---

[6]Logistic function is a special case of sigmoidal function.

where $\tilde{o}$ may be a linear or a nonlinear function, w_{ij}^k denotes the weight connecting jth neuron in kth layer with ith neuron in layer $k+1$ (we assume that the output is the layer number $l+1$) and ϕ_i^k is the total output of ith neuron in kth layer:

$$\phi_j^0 = x_j, \qquad \phi_i^k = \sigma\left(\sum_j w_{ij}^k \phi_j^{k-1})\right) \qquad k = 1, \ldots, l, \qquad (1.68)$$

where $\sigma(\cdot)$ is a sigma–shaped function, typically the logistic function or the hyperbolic tangent function.

The MLP network may be trained with the *back-propagation algorithm* (Rumelhart et al., 1986, Werbose, 1974). The weights (connections between neurons) are adapted according to the gradient-based *delta-rule*:

$$w_{ji} \leftarrow w_{ji} + \Delta w_{ji}, \qquad \Delta w_{ji} = -\eta \frac{\partial E}{\partial w_{ji}} \qquad (1.69)$$

where E is the mean squared error (1.11).

Neural networks with single hidden layer, using sigmoidal functions are universal approximators (Cybenko, 1989, Hornik et al., 1989), i.e. they can approximate any continuous function on a compact domain with arbitrary precision given sufficient number of neurons[7].

Decision borders of neural network classifiers using linear transfer functions only, are linear (like in top-left plots of figure 1.1). Nonlinear transfer functions introduces nonlinearity to decision borders which can get different shapes (for example similar to the bottom-left or bottom-right plot of figure 1.1).

The back-propagation algorithm has been accelerated and optimized in a number of ways. For example *Quickprop* (Fahlman, 1989), *conjugate gradient* (Fletcher and Reeves, 1964) or Newton method (see section 1.3.2) or Levenberg-Marquardt (Levenberg, 1944, Marquardt, 1963). Other types of algorithms are also in use. The cascade correlation (CC) is one of the best learning algorithms for the MLP networks. In CC the network is growing. For each new neuron the correlation between its output and the error of the network is maximized (Bishop, 1995).

Striving for as simple models as possible, we can eliminate the useless weights with miscellaneous *pruning techniques*, such as the *Optimal Brain Damage* (LeCun et al., 1990) and *Optimal Brain Surgeon* (Hassibi and Stork, 1993) or with *weight decay* regularization (Hinton, 1987).

Another popular group of neural networks is the family of *Radial Basis Function* (RBF) networks (Poggio and Girosi, 1990) applicable to both classification and regression problems. They are also universal approximators

---

[7]These mathematical results do not mean that sigmoidal functions always provide the optimal choice or that a good neural approximation is easy to find.

(Hartman et al., 1990, Park and Sandberg, 1991). These networks substantially differ from the MLPs: the neurons are based on radial functions instead of sigmoidal functions, and quite different learning algorithms are used. RBF networks can be defined by:

$$f_{RBF}(\mathbf{x}; \mathbf{w}) = \sum_i^K w_i G_i(\mathbf{x}) + b. \tag{1.70}$$

Here $G(\cdot)$ represents a radial function. The simplest type is the radial coordinate neuron ($||\mathbf{x} - \mathbf{t}||$, $\mathbf{t}$ is the center of the neuron), and the most often used one is the gaussian neuron which is nothing else but Gaussian kernel defined by (1.51). RBF network function (1.70) is, up to the sgn function, the same as the SVM model (1.49) with Gaussian kernel (1.51).

Typically the learning process of the RBF network is divided into two stages. The first one (usually unsupervised) determines the initial positions of radial neurons as centers of clusters obtained by a clustering algorithm (for example *k-means clustering* (Duda and Hart, 1973)) or as a random subset of the learning data. The second phase tunes the weights and (frequently) the centers and biases (γ in (1.51)) using a gradient descent algorithm (Poggio and Girosi, 1990, Bishop, 1995), *orthogonal least squares* (Chen et al., 1989) or EM algorithm (Bishop et al., 1996). Often, some regularization terms (compare section 1.3.7) are added to the MSE error function (Poggio and Girosi, 1990, Bishop, 1995) to avoid the overfitting of the network.

Some RBF networks are able to automatically adjust their architectures (the number of neurons and weights in hidden layer) during the learning process, for example RAN (Platt, 1991) or IncNet (Jankowski and Kadirkamanathan, 1997).

There are many different transfer functions (Duch and Jankowski, 1999) which can be used in MLP or RBF networks. Trying different functions (especially combined into a single heterogeneous network (Jankowski and Duch, 2001)) can significantly increase the generalization capability and at the same time reduce the network complexity.

1.3.5 Instance Based Learning

The *instance based* or *similarity based* methods are a large branch of the machine learning algorithms that were developed by the pattern recognition community. A primary example of such models is the *k nearest neighbors* (kNN) algorithm (Cover and Hart, 1967), which classifies data vectors on the basis of their *similarity* to some memorized *instances*. More precisely, for a given data vector it assigns the class label that appears most frequently among its k nearest neighbors. This may be seen as an approximation of the Bayes Optimal Classifier (1.21), where the probabilities are estimated on the basis of the analysis of the vicinity of the classified example.

An analysis of the unconditional probability of error of nearest neighbors models (Duda et al., 2001) brings its upper bound of twice the error rate of the Bayes Optimal Classifier. A tighter bound is given by $P_O(2 - \frac{c}{c-1}P_O)$, where P_O is the Bayes rate and c is the number of classes.

Methods restricting their analysis to a neighborhood of given object are called *local*. The locality causes that the shape of the decision border can be very complicated and has no simple and general characteristic. In the case of single neighbor classifier (1NN), the decision borders are the *Voronoi tesselation* of the feature space (see e.g. (Duda et al., 2001)).

There are many different possibilities of kNN algorithm implementation and application. For example, different distance measures can be used (possibly yielding quite different results and different decision borders), and the choice is not simple. There are some methods aiming at the optimal choice of the similarity measure (Wilson and Martinez, 1997). Heterogenous measures are recommended when the classified objects are described by both ordered and unordered features.

Also the choice of *neighbors* is not an unambiguous task. One can select k nearest neighbors or neighbors within a given radius. Neighbors influence on the decision may be weighted according to the distance ($1/(1 + dist)$ or $\max[1 - dist, 0]$), flexible k depending on the region may be used, best k may be estimated by means of cross-validation, etc. (for a systematic presentation of different aspects of such methods see (Duch, 2000)).

A group of algorithms is dedicated to prototype selection (instances among which search for neighbors is done). Comparisons and reviews of these methods can be found in (Wilson and Martinez, 2000, Grochowski and Jankowski, 2004). Algorithms that work well for prototype selection include DROP, Explore, LVQ, DEL, ICF and ENN.

1.3.6 Decision Trees

Decision trees are hierarchical models, very popular in classification tasks. Figure 1.4 presents two simple examples of such trees displayed in upside-down manner – the root is placed at the top and branches grow down. The tree nodes are described by logical conditions using single features. This is the most common technique in decision trees implementations and results in decision borders perpendicular to the axes of the feature space (see figure 1.1). Another shapes of borders are possible when, for instance, linear combinations of features or distances from some reference points are used in logical expressions assigned to the nodes, however perpendicular borders are often preferred, because of their comprehensibility.

In some fields experts' knowledge can be directly used to specify decision rules, but more often all that is given, is a set of classified data vectors, so the best way to find decision rules is to use computational intelligence tools.

Tree growing algorithms start with a given training data set as the root node and recursively split the nodes into several disjoint parts to separate

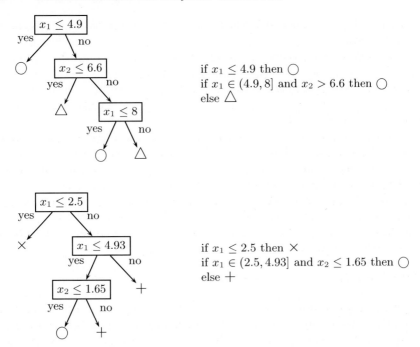

if $x_1 \leq 4.9$ then ◯
if $x_1 \in (4.9, 8]$ and $x_2 > 6.6$ then ◯
else △

if $x_1 \leq 2.5$ then ✕
if $x_1 \in (2.5, 4.93]$ and $x_2 \leq 1.65$ then ◯
else +

Fig. 1.4. Decision trees and rules corresponding to decision borders presented in the top-right plots of figures 1(a) and 1(b).

vectors assigned to different classes. Each split is described by a general logical rule, which in fact divides the whole feature space (not just the training set) into separate regions. Hence, decision trees can usually be presented in a form of logical rules (see figure 1.4). Such comprehensible models are advantageous in many fields (for instance in medical diagnosis) and provide information about particular feature relevance. The recursiveness of the processes makes feature selection a local task – useful features can be picked up even if they are valuable only in a subregion of the input space, while from the point of view of the whole space they seem to contain less information than other features.

Building optimal trees is a very complex problem, especially because the optimization of a particular split is not the same as the maximization of classification accuracy for the whole model (de Sá (2001) shows that it is true even when dealing with a family of quite simple trees).

A number of different decision tree construction methods has already been published. Some of them are: *Classification and Regression Trees* (CART) (Breiman et al., 1984), ID3 (Quinlan, 1986), C4.5 (Quinlan, 1993), *Separability of Split Value* (SSV) Trees (Grabczewski and Duch, 1999), *Fast Algorithm for Classification Trees* (FACT) (Loh and Vanichsetakul, 1988), *Quick, Unbiased,*

Efficient, Statistical Tree (QUEST) (Loh and Shih, 1997), Cal5 (Müller and Wysotzki, 1994).

The methods are based on different ideas and use different model structures. Some of them use dichotomic splits (CART, SSV), others allow for more complex branching. Some algorithms assume a particular data distribution and use parametric statistical tests (FACT, QUEST, Cal5), others make no such assumptions.

Nevertheless, all trees can be described with the same terminology including nodes, subnodes, subtrees, leaves, branches, branch length, tree depth, class labels assigned to nodes, data vectors falling into a node, and others. The definitions are quite natural, so we do not provide them here.

In general, a decision tree algorithm is defined by its three components: the method of splitting nodes into subnodes, the strategy of tree construction, which in most cases is just a search process[8], and the way of taking care of generalization (stopping criteria or pruning techniques).

Splitting criteria

Splitting nodes according to continuous and discrete inputs must be performed differently. This is why some methods (like ID3) can deal only with discrete features. Continuous attributes need to be converted to discrete before such algorithms can be applied. Good classification results can be obtained only in the company of good (external) discretization methods.

On the other side: methods like FACT and QUEST are designed to deal only with continuous inputs – here discrete features are converted to continuous by translating them into binary vectors of dimension equal to the number of possible discrete values and then, projected into a single dimension. The original Cal5 algorithm was also designed for continuous data, however it can be quite easily adapted to deal with discrete features.

The candidate splits of continuous features are usually binary (of the form $\{(-\infty, a], (a, \infty)\}$), however there exist some methods which split such input into several intervals (e.g. FACT and Cal5).

Most decision tree algorithms split the nodes with respect to the values of a single feature. Selecting the feature most eligible for the split is often closely bound up with the selection of the splitting points (CART, C4.5, SSV). An alternative strategy is applied in FACT, where the split feature is defined as the one that maximizes the value of F statistic known from the *analysis of variance* (ANOVA) method. The QUEST algorithm also calculates F statistic for continuous features, but for discrete ones it uses χ^2 statistic (to prevent

---

[8]Simple search techniques like *hill climbing* are most frequently used. The experience shows that increasing the computational efforts of the search method (e.g. using *beam search*) does not improve the generalization abilities of constructed trees (Quinlan and Cameron-Jones, 1995) – no reliable explanation of this fact is known, but it looks like more thorough search leads to solutions more specific to the training set.

from favoring the discrete attributes over continuous ones). Also Cal5 selects the split feature with a separate algorithms: either the amount of information about the classes is measured with entropy based formula (in this case the discretization must be performed first) or a coefficient is calculated for each feature to estimate their class separation abilities on the basis of mean squared distances within classes and between class centroids.

The most common split selection criterion is called *purity gain* or *impurity reduction*. For a split s and tree node N it is defined as:

$$\Delta I(s,N) = I(N) - \sum_i p_i I(N_i^s), \qquad (1.71)$$

where I is a node impurity[9] measure, N_i^s is the i-th subnode of N resulting from split s, and p_i is the estimated probability of falling into N_i^s provided that the data vector falls into N (usually the quotient of the numbers of training vectors in the relevant nodes). This criterion is used in CART with the impurity measure called *Gini index*:

$$I_G(N) = 1 - \sum_{y\in\mathcal{Y}} [P(y|N)]^2. \qquad (1.72)$$

Another impurity measure comes from the information theory and uses entropy:

$$I_E(N) = - \sum_{y\in\mathcal{Y}} P(y|N) \log_2 P(y|N). \qquad (1.73)$$

Applied with (1.71) it gives the *information gain* criterion. It is the idea of ID3, also available in CART. C4.5 uses a modification of this criterion (information gain divided by *split information*) known as the impurity *gain ratio*:

$$\Delta I'(s,N) = \frac{\Delta I_E}{SI(s,N)}, \qquad SI(s,N) = -\sum_i p_i \log_2 p_i. \qquad (1.74)$$

The idea of SSV criterion is to perform binary splits which separate as many pairs of training vectors belonging to different classes as possible, while separating the lowest possible number of pairs of vectors representing the same class:

$$SSV(s,N) = 2\cdot\sum_{y\in\mathcal{Y}} |N_1^s\cap N_y|\cdot|N_2^s\setminus N_y| - \sum_{y\in\mathcal{Y}} \min(|N_1^s\cap N_y|, |N_2^s\cap N_y|), \quad (1.75)$$

where N_y is the set of training vectors of class y falling into node N. Because of the second part of the formula, the SSV criterion does not fit to the impurity reduction scheme (1.71).

---

[9]A node is regarded as pure if all the samples belonging to it, represent the same class.

FACT implements a completely different idea of nodes splitting – a linear discrimination analysis splits each node to the number of parts equal to the number of classes represented within the node. QUEST uses quadratic discriminant analysis and performs binary splits – the *two-means* clustering of the class centers is used to group classes into two *superclasses*.

The authors of CART also group the classes in order to make dichotomic splits. They call the technique *twoing* and group vectors striving for superclasses of possibly similar counts.

Although SSV Trees are also binary, there is no need for twoing, because the SSV criterion definition guarantees good behavior also for multiclass data.

Cal5 also applies statistical methods for splitting continuous features into intervals. A single process is responsible for both discretization and deciding whether to further split the subnodes or not. The data vectors are sorted by the values of particular feature, and the intervals are created from $-\infty$ to ∞ testing statistical hypotheses to decide the positions of interval borders. After discretization the adjacent intervals can be merged: leaf-intervals are merged if they share the majority class, and non-leaves if they consist of vectors from the same set of classes (after discarding infrequent class labels by means of a statistical test).

Generalization

Building decision trees which maximally fit the training data usually ends up in overfitted, large trees with leaves classifying only a few cases, and thus does not provide general knowledge about the problem. To keep the trees simpler and more general the methods may use *stopping criteria*. A simple idea is to keep the numbers of vectors in the nodes above a specified threshold. Another way is to extend the separability criterion with a penalty term conforming to the Minimum Description Length principle. In both cases it is difficult to define criteria which yield results close to optimal. A more reasonable way is to use statistical tests to verify the significance of the improvement introduced by a split (such technique is a part of C4.5).

Stopping tree growth can save much time, but makes optimum solutions less probable. Usually, better generalization is obtained when building overfitted trees and *pruning* them. There are many different pruning methods. They may concern tree depth, number of vectors in a node etc. Their parameters can be chosen on the basis of a cross-validation test performed within the training data (like in CART, QUEST or SSV Tree).

Missing values

The training data set can be incomplete (some feature values for some data vectors may be inaccessible). Putting arbitrary values in the place of the missing ones should be taken into consideration only if no other methods can be used. Much caution about it is advised.

CART introduced the idea of *surrogate splits*, which is to collect some spare splits, maximally similar to the main one but using different features. The surrogate splits are used when the classified data vector does not provide the value necessary to check the main split condition.

When C4.5 is trained on incomplete data, the gains are scaled with a factor equal to the frequency of observing the feature values among the vectors falling into the node – each training sample has a weight associated with it, and the weights affect the p_i values of (1.74).

The SSV criterion simply does not include the vectors with missing values in calculations. Such vectors are regarded to fall into both subnodes. When an incomplete vector is to be classified by SSV Tree, all the branches of non-zero probability are checked and their leaves are treated as a single leaf to determine the dominating class.

Other decision tree ideas

There are many other decision tree algorithms, not mentioned here. Some of them are just slight modifications of the basic ones (e.g. NewID is an improved ID3), and some are quite original.

TDDT (*Top-Down Decision Trees*) is the algorithm available in the MLC++ library (Kohavi et al., 1996). It is similar to C4.5 and introduces some changes to protect against generating small nodes which are created by the information gain based strategies when discrete features have multiple values.

There are also some decision tree approaches, where node membership is decided by more than one feature. *Dipol criteria* (similar to SSV) have been used (Bobrowski and Krtowski, 2000) to construct decision trees where splitting conditions use linear combinations of features. *Linear Machine Decision Trees* (LMDT) (Utgoff and Brodley, 1991) use linear machines at tree nodes. They try some variable elimination, but in general such methods are not eligible for feature selection – instead they construct new features as combinations of the original ones.

Oblique Classifier (OC1) (Murthy et al., 1994) combines heuristic and non-deterministic methods to determine interesting linear combination of features. It searches for trees by *hill climbing*.

Option decision trees (Buntine, 1993) allow several alternative splits of the same node. The final classification is determined with relevant calculations and an analysis of probabilities.

On the basis of the SSV criterion heterogeneous decision trees have been proposed (Duch and Grabczewski, 2002). Their split conditions may concern distances from some reference vectors.

1.3.7 Regularization and Complexity Control

Regularization and model complexity control are very important, because they help us obtain most accurate and stable models for given data $\mathcal{D}$.

There are many possible definitions of the regularizer $\Omega(f)$ augmenting the empirical risk (1.13). One of them is the *weight decay* proposed by Hinton (1987), also called *ridge regression*, defined as square L_2-norm of parameters:

$$\Omega_{wd}[f] = ||\mathbf{w}||_2^2. \tag{1.76}$$

We assume that the model f is parameterized by $\mathbf{w}$ ($f = f(\mathbf{x}; \mathbf{w})$) or its superset. The regularizer used with the empirical risk yields:

$$R_{wd}[f] = R_{emp} + \lambda ||\mathbf{w}||_2^2, \tag{1.77}$$

but it may be used also with other risk functions. For example in the SV framework it is used in conjunction with loss function defined for classification (1.6), regression (1.8) or *empirical quantile function* (Schölkopf et al., 2001), to control the spread of the margin (see (1.41), (1.54) and (1.59)).

Regularization via the $\Omega_{wd}[f]$ term is equivalent to assuming a Gaussian prior distribution on the parameters of f in f_{MAP} (1.19). On the assumption of Gaussian priors for parameters w_i with zero mean ($P(w_i) \propto \exp[-w_i^2/\sigma_a^2]$) and independence assumption on parameters $\mathbf{w}$ we have

$$- \log_2 P(f) \quad \propto \quad - \ln \prod_i \exp[-w_i^2/\sigma_a^2] \quad \propto \quad ||\mathbf{w}||^2. \tag{1.78}$$

Ω_{wd} is also used with *penalized logistic regression* (PLR), where the loss is defined as $\log(1+exp[-yf(\mathbf{x})])$. On the other side, PLR may also be used with $||\mathbf{w}||_1$ regularizer, which in *regularized adaboost* is used with loss $\exp[-yf(\mathbf{x})]$ and in *lasso regression* with squared loss.

A variant of ridge regression is the *local ridge regression*:

$$\Omega_{lrr}[f] = \sum_i \lambda_i w_i^2, \tag{1.79}$$

intended for local smoothing of the model f (compare (Orr, 1996)). Commonly, the regularization parameters are determined with a cross-validation.

Weigend et al. (1990, 1991) proposed a *weight elimination* algorithm which can be seen as another regularization penalty:

$$\Omega_{we}(f) = \sum_i \frac{w_i^2/w_0^2}{1 + w_i^2/w_0^2}, \tag{1.80}$$

where w_0 is a constant. This regularizer is not so restrictive as Ω_{wd} and allows for some amount of parameters of large magnitude. For the weight elimination the λ from (1.13) may become a part of learning (Weigend et al., 1990, 1991).

A regularization of the form

$$\Omega_{mlp2ln}[f] = \sum_i w_i^2 (w_i - 1)^2 (w_i + 1)^2 \tag{1.81}$$

was used to build and learn a special type of MLP networks (MLP2LN). Such regularizer forces the weights to become close to 0, +1 or −1, which is very advantageous from the point of view of logical rule extraction (Duch et al., 1998).

A very interesting property of regularization was shown by Bishop (1995). He proved that learning with Tikhonov regularization (Tikhonov, 1963, Tikhonov and Arsenin, 1977) is equivalent to learning with noise.

Table 1.1 displays some well known algorithms in the context of loss functions and regularizers.

Algorithm	Loss	Regularizer
Weight decay/ridge regression	$(y - f(\mathbf{x}))^2$	$\|\|\mathbf{w}\|\|_2^2$
Original SV classifier	$\max\{0, 1 - yf(\mathbf{x})\}$	$\|\|\mathbf{w}\|\|_2^2$
SV for regression	$\max\{0, \|y - f(\mathbf{x})\| - \epsilon\}$	$\|\|\mathbf{w}\|\|_2^2$
Penalized logistic regression	$\log(1 + \exp[-yf(\mathbf{x})])$	$\|\|\mathbf{w}\|\|_2^2, \|\|\mathbf{w}\|\|_1$
Regularized adaboost	$\exp[-yf(\mathbf{x})]$	$\|\|\mathbf{w}\|\|_1$
Lasso regression	$(y - f(\mathbf{x}))^2$	$\|\|\mathbf{w}\|\|_1$
Local ridge regression	$(y - f(\mathbf{x}))^2$	$\sum_i \lambda_i w_i^2$
Weight elimination	$(y - f(\mathbf{x}))^2$	$\sum_i (w_i^2/w_0^2)/(1 + w_i^2/w_0^2)$
MLP2LN	$(y - f(\mathbf{x}))^2$	$\sum_i w_i^2 (w_i - 1)^2 (w_i + 1)^2$
Linear discrimination	miscellaneous	$\|\|\mathbf{w}\|\|_2^2$
Learning with noise	$(y - f(\mathbf{x}))^2$	noise

Table 1.1. Combinations of loss functions and regularizers.

Regularization may be used as embedded feature selection or a neuron pruning technique. Beside the regularization several other techniques were developed to control the complexity of learning machines. One of them uses the *cross-validation* technique (compare page 37) *for learning*: the submodels are learned and the influence of selected parameter(s) is measured and validated on the test part of the data. This type of complexity control is used for example in decision trees (see section 1.3.6) or (as mentioned above) for estimation of the adequate strength of regularization. A similar goal may be reached with Monte Carlo scheme used in place of the cross-validation randomization.

Controlling complexity of artificial neural networks may be done by adjusting the structure of neural network (the number of neurons and the weights of connections between neurons) to the complexity of considered problem. ANN's which can change their structure during learning are called *ontogenic*. For more information on ontogenic neural networks see (Fiesler, 1994, Platt,

1991, Jankowski and Kadirkamanathan, 1997, Adamczak et al., 1997, Le-
Cun et al., 1990, Hassibi and Stork, 1993, Finnoff et al., 1993, Orr, 1996,
Mézard and Nadal, 1989, Frean, 1990, Campbell and Perez, 1995, Fahlman
and Lebiere, 1990).

1.3.8 Complex Systems

Learning machines generate very different models, demonstrating large variability (Breiman, 1998). Different models combined into a single, larger model
facilitate data analysis from different points of view. A non-deterministic
learning process may produce different models even when trained several times
on the same data. A deterministic machine can also give different results when
trained on different data samples (e.g. generated in a cross-validation manner or with different bootstrap methods). Quite different models should be
expected when completely different learning techniques are applied. A number of different models can be used as a *committee* (or *ensemble*) – an averaged decision of several experts is likely to be more accurate and more
stable. Averaging of results can be done in several ways, compare the strategies of known ensembles such as *bagging* (Breiman, 1998), *adaboost* (Freund
and Schapire, 1996, 1997, Schapire et al., 1998), arcing (Breiman, 1996), re-
gionboost (Maclin, 1998), *stacking* (Wolpert, 1992), *mixture of local experts*
(Jacobs et al., 1991), *hierarchical mixture of experts* (Jordan and Jacobs, 1994)
and *heterogenous committees* (Jankowski and Grabczewski, Jankowski et al.,
2003). Cross-validation can be used to test the generalization abilities of models and to build committees at the same time. Sometimes (especially when the
validation results are unstable) it is more reasonable to combine the validated
models than to use their parameters to train a new model on the whole training data set. Chapter 7 presents more information on ensemble models.

Each expert has an area of *competence*, and the same applies to computational intelligence models. It is worth to analyze the competence of committee
members and to reflect it in committee decisions (Jankowski and Grabczewski,
Duch et al., 2002).

Some learning algorithms should be applied only to appropriately prepared
data (standardized, with preselected features or vectors, discretized etc.). In
such cases, the data preparation stage of learning should always be regarded
as a part of the system.

Recently a growing interest in *meta-learning* techniques may be observed,
aimed at finding the most successful learning algorithms and their parameters for given data. The methods include simple search procedures and more
advanced learning techniques for the meta level.

1.4 Some Remarks on Learning Algorithms

No algorithm is perfect or best suited for all the applications. Selection of the
most accurate algorithm for a given problem is a very difficult and complex

task. Exploring the space of possible models efficiently in the pursuit of optimal results requires a lot of knowledge about the advantages and dangers of applying different sorts of methods to particular domains.

First of all, the computational complexity of learning methods (with regard to the number of features, the number of vectors, data complexity and possibly other quantities) decides on their applicability to the task.

Learning processes should always be accompanied by validation tests. The subject is discussed in more detail in Chapter 2. Here, we will just point out one of the dangers of incorrect validation: when supervised feature selection (or other supervised data transformation) techniques are used as the first stage of classification (or regression) they must be treated as inseparable part of a complex model. Validation of classification models on pre-transformed data is usually much faster, but yields unreliable (and over-optimistic) results. Especially when the number of features describing data vectors is large (similar to the number of training vectors or bigger), it is easy to select a small number of features for which simple models demonstrate very good results of cross-validation tests, but their generalization abilities are illusory – this may be revealed when the feature selection is performed inside each fold of the cross-validation as a part of a complex system.

Another very important problem of learning methods is data incompleteness. Much care must be devoted to data analysis when missing data are replaced by some arbitrary values, some averages or even with techniques like *multiple imputation*, because the substitutions may strongly affect the results. The safest way to analyze incomplete data is to use methods which can appropriately exhibit the influence of the missing values on the model representation.

Acknowledgements

We are very grateful to Isabelle Guyon and Włodek Duch for their very precious suggestions and constructive comments.

References

R. Adamczak, W. Duch, and N. Jankowski. New developments in the feature space mapping model. In *Third Conference on Neural Networks and Their Applications*, pages 65–70, Kule, Poland, 1997. Polish Neural Networks Society.

A. Ben-Hur, D. Horn, H.T. Siegelman, and V. Vapnik. Suppor vector clustering. *Journal of Machine Learning Research*, 2:125–137, 2001.

K. P. Bennett and O. L. Mangasarian. Robust linear programming discrimination of two linearly inseparable sets. *Optimization Methods and Software*, 1:23–34, 1992.

C.M. Bishop. *Neural Networks for Pattern Recognition*. Oxford University Press, London, UK, 1995.

C.M. Bishop, M. Svensén, and C.K.I. Williams. EM optimization of latent-variable density models. In *Advances in Neural Information Processing Systems*, volume 8. MIT Press, Cambridge, MA, 1996.

L. Bobrowski and M. Krtowski. Induction of multivariate decision trees by using dipolar criteria. In D. A. Zighed, J. Komorowski, and J. M. ytkow, editors, *Principles of data mining and knowledge discovery: 5th European Conference: PKDD'2000*, pages 331–336, Berlin, 2000. Springer Verlag.

B.E. Boser, I. Guyon, and V. Vapnik. A training algorithm for optimal margin classifiers. In *Fifth Annual Workshop on Computational Learning Theory*, pages 144–152. ACM, 1992.

L. Breiman. Bias, variance, and arcing classifiers. Technical Report Technical Report 460, Statistics Department, University of California, Berkeley, CA 94720, April 1996.

L. Breiman. Bias-variance, regularization, instability and stabilization. In C. M. Bishop, editor, *Neural Networks and Machine Learning*, pages 27–56. Springer, 1998.

L. Breiman, J. H. Friedman, A. Olshen, and C. J. Stone. *Classification and regression trees*. Wadsworth, Belmont, CA, 1984.

W. Buntine. Learning classification trees. In D. J. Hand, editor, *Artificial Intelligence frontiers in statistics*, pages 182–201. Chapman & Hall,London, 1993. URL citeseer.nj.nec.com/buntine91learning.html.

C. Campbell and C.V. Perez. Target switching algorithm: a constructive learning procedure for feed-forward neural networks. *Neural Networks*, pages 1221–1240, 1995.

B. Cestnik. Estimating probabilities: A crucial task in machine learning. In *Proceedings of the Ninth European Conference on Artificial Intelligence*, pages 147–149, 1990.

S. Chen, S.A. Billings, and W. Luo. Orthogonal least squares methods and their application to non-linear system identification. *International Journal of Control*, 50:1873–1896, 1989.

V. Cherkassky and F. Mulier. *Learning from data*. Adaptive and learning systems for signal processing, communications and control. John Wiley & Sons, Inc., New York, 1998.

C. Cortes and V. Vapnik. Soft margin classifiers. *Machine Learning*, 20:273–297, 1995.

T.M. Cover and P.E. Hart. Nearest neighbor pattern classification. *IEEE Transactions on Information Theory*, 13(1):21–27, 1967.

G. Cybenko. Approximation by superpositions of a sigmoidal function. *Mathematics of Control, Signals, and Systems*, 2:303–314, 1989.

J. P. Marques de Sá. *Pattern Recognition. Concepts, Methods and Applications*. Springer Verlag, 2001.

W. Duch. Similarity based methods: a general framework for classification, approximation and association. *Control and Cybernetics*, 29:937–968, 2000.

W Duch and K. Grabczewski. Heterogeneous adaptive systems. In *Proceedings of the World Congress of Computational Intelligence*, Honolulu, May 2002.

W. Duch and N. Jankowski. Survey of neural transfer functions. *Neural Computing Surveys*, 2:163–212, 1999.

W. Duch, R. Adamczak, and K. Grabczewski. Extraction of logical rules from backpropagation networks. *Neural Processing Letters*, 7:1–9, 1998.

W. Duch, L. Itert, and K. Grudziński. Competent undemocratic committees. In L. Rutkowski and J. Kacprzyk, editors, *6th International Conference on Neural Networks and Soft Computing*, pages 412–417, Zakopane, Poland, 2002. Springer-Verlag.

R. O. Duda, P. E. Hart, and D. G. Stork. *Patter Classification*. John Wiley and Sons, New York, 2001.

R.O. Duda and P.E. Hart. *Pattern Classification and Scene Analysis*. Wiley, New York, 1973.

S.E. Fahlman. Fast-learning variations on back-propagation: An empirical study. In D. Touretzky, G. Hinton, and T. Sejnowski, editors, *Proceedings of the 1988 Connectionist Models Summer School*, pages 38–51, Pittsburg, 1989. Morgan Kaufmann, San Mateo.

S.E. Fahlman and C. Lebiere. The cascade-correlation learning architecture. In D.S. Touretzky, editor, *Advances in Neural Information Processing Systems 2*, pages 524–532, Denver, CO, 1990. Morgan Kaufmann, San Mateo.

E. Fiesler. Comparative bibliography of ontogenic neural networks. In *International Conference on Artificial Neural Networks*, pages 793–796, 1994.

W. Finnoff, F. Hergert, and H.G. Zimmermann. Improving model detection by nonconvergent methods. *Neural Networks*, 6(6):771–783, 1993.

R. A. Fisher. The use of multiple measurements in taxonomic problems. *Annals of Eugenics*, 7:179–188, 1936. Reprinted in *Contributions to Mathematical Statistics*, John Wiley & Sons, New York, 1950.

R. Fletcher and C.M. Reeves. Function minimization by conjugate gradients. *Computer journal*, 7:149–154, 1964.

M. Frean. *Small nets and short paths: optimizing neural computation*. PhD thesis, Center for cognitive science. University of Edinburgh, 1990.

Y. Freund and R.E. Schapire. Experiments with a new boosting algorithm. In *Machine Learning: Proceedings of Thirteenth International Conference*, pages 148–156, 1996.

Y. Freund and R.E. Schapire. A decision theoretic generalization of on-line learning and an application to boosting. *Journal of Computer and System Science*, 55 (1):119–139, 1997.

J. Friedman, T. Hastie, and R. Tibshirani. *The Elements of Statistical Learning: Data Mining, Inference, and Prediction*. Springer-Verlag, 2001.

G.M. Fung and O.L. Mangasarian. A feature selection newton method for support vector machine classification. *Comput. Optim. Appl.*, 28(2):185–202, 2004. ISSN 0926-6003. doi: http://dx.doi.org/10.1023/B:COAP.0000026884.66338.df.

K. Grabczewski and W. Duch. A general purpose separability criterion for classification systems. In *Proceedings of the 4th Conference on Neural Networks and Their Applications*, pages 203–208, Zakopane, Poland, June 1999.

M. Grochowski and N. Jankowski. Comparison of instances seletion algorithms II: Algorithms survey. In *Artificial Intelligence and Soft Computing*, pages 598–603, 2004.

I. Guyon and D.G. Stork. *Advances in large margin classifiers*, chapter Linear discriminant and support vector classiers, pages 147–169. MIT Press, 2000.

I. Guyon, J. Weston, and S. Barnhilland V. Vapnik. Gene selection for cancer classification using support vector machines. *Machine Learning*, 2002.

E.J. Hartman, J.D. Keeler, and J.M. Kowalski. Layered neural networks with gaussian hidden units as universal approximations. *Neural Computation*, 2:210–215, 1990.

B. Hassibi and D.G. Stork. Second order derivatives for network pruning: Optimal brain surgeon. In C.L. Giles, S.J. Hanson, and J.D. Cowan, editors, *Advances in Neural Information Processing Systems 5*, pages 164–171, San Mateo, CA, 1993. Morgan Kaufmann.

T. Hastie, R. Tibshirani, and J. Friedman. *The Elements of Statistical Learning: Data Mining, Inference, and Prediction.* Springer, 2001.

S. Haykin. *Neural Networks - A Comprehensive Foundation.* Maxwell MacMillian Int., New York, 1994.

G.E. Hinton. Learning translation invariant in massively parallel networks. In J.W. de Bakker, A.J. Nijman, and P.C. Treleaven, editors, *Proceedings of PARLE Conference on Parallel Architectures and Languages Europe*, pages 1–13, Berlin, 1987. Springer-Verlag.

K. Hornik, M. Stinchcombe, and H. White. Multilayer feedforward networks are universal approximators. *Neural Networks*, 2:359–366, 1989.

R.A. Jacobs, M.I. Jordan, S.J. Nowlan, and G.E. Hinton. Adaptive mixture of local experts. *Neural Computation*, 3:79–87, 1991.

N. Jankowski and W. Duch. Optimal transfer function neural networks. In *9th European Symposium on Artificial Neural Networks*, pages 101–106, Bruges, Belgium, 2001.

N. Jankowski and V. Kadirkamanathan. Statistical control of RBF-like networks for classification. In *7th International Conference on Artificial Neural Networks*, pages 385–390, Lausanne, Switzerland, October 1997. Springer-Verlag.

N. Jankowski, K. Grabczewski, and W. Duch. Ghostminer 3.0. FQS Poland, Fujitsu, Kraków, Poland, 2003.

Norbert Jankowski and Krzysztof Grabczewski. Heterogenous committees with competence analysis. In *Proceedings of the Fifth International conference on Hybrid Intelligent Systems.*

T. Joachims. *Advances in kernel methods — support vector learning*, chapter Making large-scale SVM learning practical. MIT Press, Cambridge, MA, 1998.

G. H. John and P. Langley. Estimating continuous distributions in bayesian classifiers. In *Proceedings of the 11th Conference on Uncertainty in Artificial Intelligence*, San Mateo, 1995. Morgan Kaufmann Publishers.

M.I. Jordan and R.A. Jacobs. Hierarchical mixtures of experts and the EM algorithm. *Neural Computation*, 6:181–214, 1994.

R. Kohavi. *Wrappers for performance enhancement and oblivious decision graphs.* PhD thesis, Stanford University, 1995.

R. Kohavi, D. Sommerfield, and J. Dougherty. Data mining using MLC++: A machine learning library in C++. In *Tools with Artificial Intelligence*, pages 234–245. IEEE Computer Society Press, 1996. http://www.sgi.com/tech/mlc.

R. Kohavi, B. Becker, and D. Sommerfield. Improving simple bayes. In *Proceedings of the European Conference on Machine Learning*, 1997. URL `citeseer.nj.nec.com/kohavi97improving.html`.

T. Kohonen. *Self-organizing maps.* Springer, Heidelberg Berlin, 1995.

S. Kullback and R.A. Leibler. On information and sufficiency. *Annals of Mathematical Statistics*, 22:76–86, 1951.

L.I. Kuncheva. *Combining Pattern Classifiers. Methods and Algorithms.* Wiley-Interscience, 2004.

Y. LeCun, J.S. Denker, and S.A. Solla. Optimal brain damage. In D.S. Touretzky, editor, *Advances in Neural Information Processing Systems 2*, pages 598–605, Denver, CO, 1990. Morgan Kaufmann, San Mateo.

K. Levenberg. A method for the solution of certain non-linear problems in least squares. *Quarterly Journal of Applied Mathematics*, II(2):164–168, 1944.

W.-Y. Loh and Y.-S. Shih. Split selection methods for classification trees. *Statistica Sinica*, 7:815–840, 1997.

W.-Y. Loh and N. Vanichsetakul. Tree-structured classification via generalized discriminant analysis (with discussion). *Journal of the American Statistical Association*, 83:715–728, 1988.

R. Maclin. Boosting classifiers regionally. In *Proceeding of AAAI*, 1998.

D.W. Marquardt. An algorithm for least-squares estimation of non-linear parameters. *Journal of the Society of Industrial and Applied Mathematics*, 11(2):431–441, 1963.

W. S. McCulloch and W. Pitts. A logical calculus of the ideas immanent in nervous activity. *Bulletin of Mathematical Biophysics*, 5:115–133, 1943.

M. Mézard and J.-P. Nadal. Learning in feedforward layered networks: The tiling algorithm. *Journal of Physics A*, 22:2191–2204, 1989.

T. Mitchell. *Machine learning*. McGraw Hill, 1997.

W. Müller and F. Wysotzki. Automatic construction of decision trees for classification. *Annals of Operations Research*, 52:231–247, 1994.

S. K. Murthy, S. Kasif, and S. Salzberg. A system for induction of oblique decision trees. *Journal of Artificial Intelligence Research*, 2:1–32, August 1994.

R.M. Neal. *Bayesian Learning for Neural Networks. Number 118 in Lecture Notes in Statistics.* Springer-Verlag, 1996.

M. Orr. Introduction to radial basis function networks. Technical report, Centre for Cognitive Science, University of Edinburgh, 1996.

E. Osuna, R. Freund, and F. Girosi. Training support vector machines: An application to face detection. In *CVPR'97*, pages 130–136, New York, NY, 1997. IEEE.

J. Park and I.W. Sandberg. Universal approximation using radial-basis-function networks. *Neural Computation*, 3(2):246–257, 1991.

Z. Pawlak. Rough sets. *International Journal of Computer and Information Sciences*, 11(5):341–356, 1982.

J. Platt. A resource-allocating network for function interpolation. *Neural Computation*, 3(2):213–225, 1991.

J. C. Platt. Fast training of support vector machines using sequential minimal optimization. In B. Schölkopf, C. J. C. Burges, and A. J. Smola, editors, *Advances in Kernel Methods - Support Vector Learning*. MIT Press, Cambridge, MA., 1998.

T. Poggio and F. Girosi. Networks for approximation and learning. *Proceedings of the IEEE*, 78(9):1481–1497, 1990.

F. Provost and P. Domingos. Well-trained PETs: Improving probability estimation trees. Technical Report IS-00-04, Stern School of Business, New York University, 2000.

J. Quinlan. Programs for machine learning, 1993.

J. Quinlan. Induction of decision trees. *Machine Learning*, 1:81–106, 1986.

J. R. Quinlan and R. M. Cameron-Jones. Oversearching and layered search in empirical learning. In *IJCAI*, pages 1019–1024, 1995. URL `citeseer.nj.nec.com/quinlan95oversearching.html`.

B. D. Ripley. *Pattern Recognition and Neural Networks*. Cambridge University Press, Cambridge, 1996.

J. Rissanen. Modeling by shortest data description. *Automatica*, 14:445–471, 1978.

F. Rosenblatt. *Principles of Neurodynamics*. Spartan, New York, 1962.

D. E. Rumelhart, G. E. Hinton, and R. J. Williams. Learning internal representations by error propagation. In J. L. McCleland D. E. Rumelhart, editor, *Parallel Distributed Processing: Explorations in Microstructure of Congnition*, volume 1: Foundations, pages 318–362. Cambridge, 1986.

R. Schalkoff. *Pattern Recognition: statistical, structural and neural approaches*. Wiley, 1992.

R.E. Schapire, Y. Freund, P. Bartlett, and W.S. Lee. Boosting the margin: A new explanation for the effectiveness of voting methods. *The Annals of Statistics*, 26 (5):1651–1686, 1998.

B. Schölkopf and A.J. Smola. *Learning with Kernels: Support Vector Machines, Regularization, Optimization, and Beyond*. MIT Press, 2001.

B. Schölkopf, A.J. Smola, R.C. Williamson, and P.L. Bartlett. New support vector algorithms. *Neural Computation*, 12:1207–1245, 2000.

B. Schölkopf, J.C. Platt, J. Shawe-Taylor, A.J. Smola, and R.C. Williamson. Estimating the support of a high-dimensional distribution. *Neural Computation*, 13 (7):1443–1471, 2001.

C.E. Shannon and W. Weaver. *The mathematical theory of communication*. University of Illinois Press, Urbana, 1949.

S.K. Shevade, S.S. Keerthi, C. Bhattacharyya, and K.R.K. Murthy. Improvements to the SMO algorithm for SVM regression. *IEEE Transactions on Neural Networks*, 11:1188–1194, Sept. 2000.

A.N. Tikhonov. On solving incorrectly posed problems and method of regularization. *Doklady Akademii Nauk USSR*, 151:501–504, 1963.

A.N. Tikhonov and V.Y. Arsenin. *Solutions of Ill-posed Problems*. W.H. Winston, Washington D.C., 1977.

P. E. Utgoff and C. E. Brodley. Linear machine decision trees. Technical Report UM-CS-1991-010, Department of Computer Science, University of Massachusetts, , 1991. URL `citeseer.nj.nec.com/utgoff91linear.html`.

A.S. Weigend, D.E. Rumelhart, and B.A. Huberman. Back–propagation, weight elimination and time series prediction. In *Proceedings of the 1990 Connectionist Models Summer School*, pages 65–80, Los Altos, Palo Alto, San Francisco, 1990. Morgan Kaufmann, San Mateo.

A.S. Weigend, D.E. Rumelhart, and B.A. Huberman. Generalization by weight elimination with application to forecasting. In *Advances in Neural Information Processing Systems 3*, pages 875–882, Los Altos, Palo Alto, San Francisco, 1991. Morgan Kaufmann, San Mateo.

P.J. Werbose. *Beyond regression: New tools for prediction and analysis in the bahavioral sciences*. PhD thesis, Harvard Univeristy, Cambridge, MA, 1974.

J. Weston, A. Elisseeff, and B. Schölkopf. Use of the ℓ_0-norm with linear models and kernel methods. Technical report, Biowulf Technologies, 2001.

D.R. Wilson and T.R. Martinez. Reduction techniques for instance-based learning algorithms. *Machine Learning*, 38:257–286, 2000.

D.R. Wilson and T.R. Martinez. Instance pruning techniques. In *14th International Conference on Machine Learning*, pages 403–411. Morgan Kaufmann, 1997. URL `citeseer.nj.nec.com/wilson97instance.html`.

D.H. Wolpert. Stacked generalization. *Neural Networks*, 5:241–249, 1992.

B. Zadrozny and C. Elkan. Obtaining calibrated probability estimates from decision trees and naive Bayesian classifiers. In *Proc. 18th International Conf. on Machine Learning*, pages 609–616. Morgan Kaufmann, San Francisco, CA, 2001. URL `citeseer.nj.nec.com/zadrozny01obtaining.html`.

J.M. Zurada. *Artificial neural systems*. West Publishing Company, 1992.

Chapter 2

Assessment Methods

Gérard Dreyfus[1] and Isabelle Guyon[2]

[1] École Supérieure de Physique et de Chimie Industrielles (ESPCI-Paristech), Laboratoire d'Électronique (CNRS UMR 7084), 10 rue Vauquelin, 75005 Paris - FRANCE Gerard.Dreyfus@espci.fr

[2] ClopiNet, 955 Creston Rd., Berkeley, CA 94708, USA. isabelle@clopinet.com

2.1 Introduction

This chapter aims at providing the reader with the tools required for a statistically significant assessment of feature relevance and of the outcome of feature selection. The methods presented in this chapter can be integrated in feature selection wrappers and can serve to select the number of features for filters or feature ranking methods. They can also serve for hyper-parameter selection or model selection. Finally, they can be helpful for assessing the confidence on predictions made by learning machines on fresh data. The concept of model complexity is ubiquitous in this chapter. Before they start reading the chapter, readers with little or old knowledge of basic statistics should first delve into Appendix A; for others, the latter may serve as a quick reference guide for useful definitions and properties. The first section of the present chapter is devoted to the basic statistical tools for feature selection; it puts the task of feature selection into the appropriate statistical perspective, and describes important tools such as hypothesis tests - which are of general use - and random probes, which are more specifically dedicated to feature selection. The use of hypothesis tests is exemplified, and caveats about the reliability of the results of multiple tests are given, leading to the Bonferroni correction and to the definition of the false discovery rate. The use of random probes is also exemplified, in conjunction with forward selection. The second section of the chapter is devoted to validation and cross-validation; those are general tools for assessing the ability of models to generalize; in the present chapter, we show how they can be used specifically in the context of feature selection; attention is drawn to the limitations of those methods.

G. Dreyfus and I. Guyon: *Assessment Methods*, StudFuzz **207**, 65–88 (2006)
www.springerlink.com

2.2 A Statistical View of Feature Selection: Hypothesis Tests and Random Probes

The present section first describes feature selection from a general statistical viewpoint. The rationale and principle of the random probe generation method will be discussed, together with the use of standard hypothesis tests.

2.2.1 A Statistical View of Feature Selection

The problem of feature selection can be viewed in the following, abstract way: assume that a relevance index $r(\mathbf{x}, \mathbf{y})$ has been defined, which provides a quantitative assessment of the relevance of a candidate variable x for modeling a quantity y; the latter may be either real (for regression problems) or categorical (for classification problems). A vector $\mathbf{x}$ of measured values of x and a vector $\mathbf{y}$ of measured values of y are available; their components are modeled as realizations of i.i.d. random variables X and Y. Hence $r(\mathbf{x}, \mathbf{y})$ can be viewed as a realization of a random variable $R(X, Y)$. Figure 2.1 shows the probability distributions of R for relevant features and for irrelevant features (both those distributions are unknown, of course.) In that framework, feature selection consists in setting a threshold r_0 and making the decision that all candidate features with relevance smaller than r_0 should be discarded. The probability of keeping a feature although it is irrelevant (false positive) and the probability of discarding a feature although it is relevant (false negative) are displayed on Figure 2.1.

Although it is conceptually simple, the above view of the problem is abstract since:

- the relevance index is to be defined, and
- both probability distribution functions are unknown.

The definition of relevance indices is extensively addressed in several chapters of this book (for a review, see Chapter 3.) A few illustrative examples are given in the present chapter, including univariate feature ranking criteria, and forward or backward selection criteria. As for the probability distributions, little is known of the probability distribution of relevant variables, but we shall see that by making some mild assumptions about the probability distributions of irrelevant variables, one can draw conclusions about variable relevance. In particular we introduce in the next section the random probe method, which provides means of estimating the fraction of falsely significant features.

2.2.2 Random Probes as Realizations of Irrelevant Variables

In (Stoppiglia, 1997) and (Oukhellou et al., 1998), it was suggested to generate "random probes", i.e. features that are not related to y, and compare the relevance of the probe features to that of the candidate features. Probe features can be generated in different ways, e.g.:

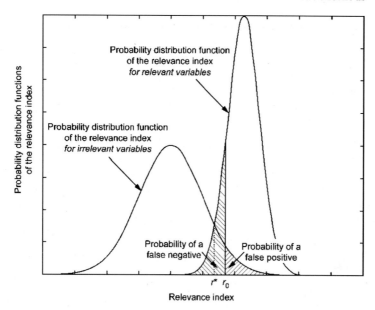

Fig. 2.1. Probability distribution functions of the relevance index for relevant and for irrelevant features, and probabilities of false negatives and false positives if it is decided that all features with relevance index smaller than r_0 are discarded and all features with relevance index larger than r_0 are kept. Both distributions are unknown in a real problem.

- generating random variables with a known probability distribution that is similar to the distribution of the candidate features (e.g. standard normal probes),
- randomly shuffling the components of the candidate feature vectors in the training data matrix (Bi et al., 2003).[3]

The probability distribution function of the relevance index of irrelevant features can either be estimated by generating a large number of probes, or be computed analytically (see examples in Section 2.2.3.) Therefore, the combination of

- the computation of a relevance index of the candidate features,
- the generation of random probes,

allows the model designer to effectively choose a threshold on the relevance index that guarantees an upper bound on the false positive rate (the rate of falsely significant features.) To that end, consider that the decision of keeping or discarding a feature is made on the basis of its relevance index being larger or smaller than a threshold r_0. The probability of a false positive (or false

---

[3]Permutation tests (Cox and Hinkley, 1974) use the complementary view of shuffling the target values rather than the features.

positive rate FPR) is the ratio of the number of false positives n_{fp} to the total number of irrelevant features n_{irr}. If the distribution of the probe features is similar to the distribution of irrelevant features, and if the number of generated probes n_p is large, the probability of a false positive is the ratio of the number of selected probes n_{sp} to the total number of probes:

$$FPR = \frac{n_{fp}}{n_{irr}} = \frac{n_{sp}}{n_p}. \tag{2.1}$$

The estimation of the probability of false negatives is less straightforward; an approach to that problem will be discussed in Section 2.2.6.

2.2.3 Hypothesis Testing

Variable selection and model selection involve making decisions (e.g. discarding a feature of a model.) Those decisions must be made on the basis of the limited amount of available data. Statistical tests are the appropriate technique in such situations because they allow making decisions in a principled way. Readers who are not familiar with the terminology and the mechanics of statistical tests are invited to read Appendix A before proceeding with the rest of this section.

In application to variable selection, we shall be interested in testing whether a variable is irrelevant or a subset of variables is irrelevant. Relevance indices will become test statistics. The false positive rate will become the p-value of the test.

To test the relevance of individual variables with so-called "univariate tests" (Section 2.2.4), we shall make mild assumptions about the distribution of irrelevant variables (e.g. that their relevance index is Gaussian distributed with zero mean.) The "probe" method will allow us to address cases in which analytic calculations of the probability distribution of the test statistic cannot be carried out. To test the relevance of a subset of variables (Section 2.2.5), we shall make the assumptions about the "null model", namely that a predictor built with the entire set of available variables can perfectly approximate the target function. We shall test whether a model built with a restricted number of variables can also perfectly approximate the target.

2.2.4 Univariate Tests of Variable Irrelevance

Many filter methods rank variables in order of relevance using an index R, which assesses the dependence on the target of each variable individually. The index R can be used as a test statistic (or conversely, some well known test statistics can be used as a basis for feature ranking.) The null hypothesis to be tested is H_0:"The variable is irrelevant". To formalize this hypothesis, something must be known about the distribution of R for irrelevant variables. At the very least, one must know the mean value r_0 of the index for irrelevant variables. We can then test H_0:"The expected value of R for a given feature is equal to r_0". Relevant variables may have a value of the expected value of R larger than r_0, in which case, the alternative hypothesis will be H_1:"The expected value of R for a given feature is larger then r_0" (i.e. the variable is relevant.) Knowing more about the distribution of R for irrelevant features will allow us to derive more powerful tests.

T-test criterion for binary classification problems

As a first example, assume that we have a classification problem with two classes A and B. Assume further that the distribution of examples of both classes is Gaussian. As a univariate ranking index, we will use a test statistic to compare the means μ_A and μ_B of the two Gaussian distributions in projection on a given variable. The *paired T-statistic* performs that task; a realization t of the statistic T is given by:

$$t = \frac{\mu_A - \mu_B}{s\sqrt{\frac{1}{m_A} + \frac{1}{m_B}}}, \quad \text{with} \quad s = \frac{(m_A - 1)s_A^2 + (m_B - 1)s_B^2}{m_A + m_B - 2},$$

where m_A and m_B are the number of examples of each distribution, and s_A and s_B the estimation of the standard deviations of the distributions obtained from the available examples. Under the null hypothesis that the means are equal, T has a Student distribution with $m_A + m_B - 2$ degrees of freedom.

Clearly, a variable for which the means of the two Gaussians are further apart is more informative. Since the polarity of the classes is interchangeable, it does not matter which mean is larger than the other. The absolute value of the T-statistic may be used directly as a ranking index: the larger its value, the more informative the feature. But the two-tailed p-value of t, which varies monotonically with $abs(t)$, may be also used as a ranking index. The smaller the p-value of the feature, the larger the discrepancy between the means, hence the more informative the feature. One can then set a threshold on the p-value above which features are considered statistically significant.

Note that when a large number of features are being tested simultaneously, corrections to the p-value must be applied (see Section 2.2.6.)

Wilcoxon test and AUC criterion for binary classification problems

If the assumption that the distribution of the features is Gaussian does not hold, then a non-parametric test, the Wilcoxon-Mann-Whitney rank sum test, can be conducted. The null hypothesis is that the two classes have the same probability distribution in projection on a candidate feature, but that distribution is not specified. The alternative hypothesis is that there is a location shift in the distributions of the two classes.

To compute the Wilcoxon statistic W_A, proceed as follows. With the values of the candidate feature for the two classes, a single set with $m_A + m_B$ elements is built; those values are ranked in ascending order: the smallest observation of the candidate feature is assigned rank 1, the next is assigned rank 2, etc. If two values have the same rank (the case of a "tie"), a fractional rank is assigned to both: if k examples have the same value of rank i, they are assigned rank $i + \frac{k-1}{2}$. Denote by W_A the sum of the ranks of the features that pertain to class A. If the null hypothesis that the cumulative distribution functions of the feature for class A and class B are identical is true, the expectation value of W_A is equal to $E_{W_A} = m_A(m_A + m_B + 1)/2$ and its variance to $var_{W_A} = m_A m_B (m_A + m_B + 1)/12$. Therefore, if w_A departs from E_{W_A} by more than a threshold $w_{\alpha/2}$, the null hypothesis must be rejected. The distribution of W_A is tabulated and is available in all statistics software packages. Furthermore, it has been shown that W_A is approximately normal if m_A is larger than 7; therefore, if that condition is obeyed, α can be chosen from the normal distribution. Similarly as for the T-test, the p-values of the test can be used for feature ranking.

The Wilcoxon statistic is closely related to the Mann-Whitney U_A-statistic. The latter is the sum, over the examples of both classes, of the number of times the score of an element of class A exceeds that of an element of class B. Mann and Whitney have shown that $W_A = U_A + m_A(m_A + 1)/2$. This implies that tests based on U_A are equivalent to tests based on W_A. It is worth noting that $U_A/(m_A m_B)$ is nothing but the AUC criterion: the area under the ROC curve, using the value of a single feature as class prediction score (Cortes and Mohri, 2004). The AUC criterion values are between 0 and 1. For irrelevant features, the expected value of the AUC is 0.5. Hence, $abs(AUC - 0.5)$ and the two-tailed p-values of the Mann-Whitney statistic can interchangeably be used as a ranking index. Setting a threshold on the p-values allows us to retain those features that are statistically significant.

Other univariate tests

Many other univariate tests are applicable to variable ranking (see Chapters 3 and 10. We briefly go over a few examples.

To test whether the values of a feature are likely to have been drawn from a "null" distribution representing the distribution of a random feature, one may use a permutation test: the null distribution is estimated by permuting the target values at random. For a binary classification problem, the (posterior) distribution of the variable is represented by all the values of a contingency table: a_1=number of +1 values for target +1, a_2=number of +1 values for target -1, a_3=number of -1 values for target +1, a_4=number of -1 values for target -1. The test consists in comparing a_1, a_2, a_3, and a_4 with their "expected" values e_1, e_2, e_3, and e_4 in the "null" distribution. The values e_1, e_2, e_3, and e_4 are estimated by computing the values of a_1, a_2, a_3, and a_4 for random permutations of the target values and averaging the results. It can be shown that $\sum_i (a_i - e_i)^2/e_i$ asymptotically obeys a χ^2 distribution, hence p-values can be computed from the tables of that distribution. In the case of categorical variables and continuous variables, histograms are being compared but the principle remains the same.

There is a test of significance for the Pearson correlation coefficient R, which has been applied to classification problems in the feature selection challenge (see Part II) and which is also applicable to regression problems. The p-value is computed by transforming the correlation to create a T statistic having $m - 2$ degrees of freedom, where m is the number of examples. The confidence bounds are based on an asymptotic normal distribution of $0.5 \ln((1 + R)/(1 - R))$, with an approximate variance equal to $1/(N - 3)$.

For problems in which the input is categorical and the output continuous or *vice versa* (*e.g.* for a multiclass classification problem), an ANOVA test may be used. The null hypothesis is the equality of the mean of the variable in the various categories. The test statistic is the ratio of the between-category variance and the pooled within-category variance, which is an F statistic (distributed according to the Fisher distribution.)

For problems in which both inputs and outputs are binary, an odds-ratio test may be performed. A realization of the test statistic is given by $\theta = (p_{11}p_{22}/p_{12}p_{21})$, where p_{ij} is the fraction of co-occurrences of variable and target values of categories "1" and "2". The independence between variable and target implies $\theta = 1$. If the null hypothesis of independence is true, the asymptotic distribution of $\ln \theta$ is the Normal law with zero mean and variance $1/p_{11} + 1/p_{22} + 1/p_{12} + 1/p_{21}$.

Non-monotonic transformations can be tested with a *runs test* (see Chapter 10.)

Many feature ranking criteria are not based on test statistics with a known tabulated distribution. For such cases, we must resort to using the false positive rate given in Equation 2.1 as a p-value on which we can set a threshold to perform a test of hypothesis.

2.2.5 Hypothesis Testing for Feature Subset Selection

In the previous section, we investigated univariate tests in application to feature ranking. In this section, we address problems of feature subset selection via methods of backward elimination or forward selection. We describe examples of linear regression problems.

Fisher's test for backward variable elimination

In its general form, a linear model can be written as:

$$g_n(\boldsymbol{x}, \boldsymbol{w}) = \sum_{i=1}^{n} w_i x_i,$$

where n is the number of variables (hence of parameters) of the model, x is the n-vector of the variables of the model.[4] We denote by g_n the m-vector of the model in observation space (of dimension m, where m is the number of observations in the training set.) We assume that Q candidate variables are available. The model with Q variables is termed the complete model.

It is further assumed that the quantity to be modeled is generated by $y = \sum_{i=1}^{Q} w_i x_i + \Omega$, where Ω is a Gaussian random variable with zero mean and variance σ^2. In other words, it is assumed that the complete model contains the regression. We denote by y the m-vector of the observations of the quantity to be modeled.

If one (or more) variable is irrelevant, the corresponding parameter of the model should be equal to zero. Therefore, the hypothesis that is tested is the fact that one or more parameters are equal to zero. Assume that it is desired to test the validity of the complete model against that of a sub-model with q parameters equal to zero. Following the definitions stated in Appendix A, the considered hypotheses are: H_0: the q parameters are equal to zero, H_1: the q parameters are not equal to zero.

Consider the random variable

$$R = \frac{m - Q - 1}{q} \frac{\|Y - G_{Q-q}\|^2 - \|Y - G_Q\|^2}{\|Y - G_Q\|^2}$$

where Y, G_Q and G_{Q-q} are Gaussian random vectors, of which y, g_Q and g_{Q-q} are known realizations. Then it can be shown (Seber, 1977) that, if the null hypothesis is true, R has a Fisher distribution with q and $(m - Q - 1)$ degrees of freedom. Note that $\|y - g_{Q-q}\|^2$ is the sum of the squared modeling errors made by the model with $Q - q$ parameters, while $\|y - g_Q\|^2$ is the sum of the squared modeling errors made by the model with Q parameters.

If the number of observations were infinite, R would be equal to zero since the training procedure would set the q parameters to zero, so that the models with Q and $Q - q$ parameters would be identical after training. Therefore, the null hypothesis will be rejected if the realization $r = \frac{m-Q-1}{q} \frac{\|y-g_{Q-q}\|^2-\|y-g_Q\|^2}{\|y-g_Q\|^2}$ is too large. A risk α is chosen, the corresponding value r_α is found, r (and its p-value $p(r)$) is computed, and the null hypothesis is rejected if $r > r_\alpha(p(r) < \alpha)$: in that case, it is concluded that the group of q features should not be discarded.

Thus, Fisher's test compares a sub-model to the complete model. Other tests, such as the Likelihood Ratio Test (Goodwin and Payne, 1977) and the Logarithm Determinant Ratio Test (Leontaritis and Billings, 1987) compare models that are not thus related. It is proved in (Söderström, 1977) that those tests are asymptotically equivalent to Fisher's test.

In principle, the complete model (with Q parameters) may be compared, using Fisher's test, to all 2^Q sub-models. In practice, this is usually not

---

[4]The method obviously also works for models linear in their parameters to select features derived from x rather than input variables.

computationally feasible and prone to overfitting (see Section 2.2.6.) It has been suggested in (Rivals and Personnaz, 2003) to use Fisher's test to progressively eliminate features in a backward elimination procedure.

A random probe test for orthogonal forward regression ranking

As an illustration, the present section describes the use of random probes in the case of forward selection. Forward selection of variables consists in adding variables one by one, assess their relevance by a suitable method and stop including variables when a predefined criterion is met. That is in contrast to backward elimination, whereby all candidate variables are considered at the beginning, and "useless" variables are eliminated.

The random probe method provides a conceptually simple illustration of the framework described by Figure 2.1. Two ingredients are necessary: a relevance index, and a "fake" set of features (the probes), generated as described in Section 2.2.2, which mimic irrelevant variables for the purpose of selection.

The method described below takes advantage of the fact that orthogonal forward regression (Chen et al., 1989) builds linear models by a sequence of simple computations that use the variables in order of decreasing relevance. For each new model, the relevance of the new variable is computed, and, based on the cumulative distribution function of the relevance of the probe features, the probability of that new variable being a false positive is either estimated or computed. The process is terminated when the probability of the new variable being a false positive becomes larger than a predefined threshold; the selected variables are subsequently used as features of a nonlinear model if necessary.

In the present illustration, the relevance index of a variable is its rank in a ranked list by order or increasing relevance: the higher the rank, the larger the relevance. Ranking of candidate variables for a linear model can be performed in a computationally efficient way by orthogonal forward regression: in observation space, each candidate feature is described by a vector whose components are the values of that input over the training set, and the quantity to be modeled is similarly described by an m-vector. Then the relevance of each candidate feature can be assessed as the angle φ_i between the vector representing that feature and the vector representing the output:

- if that angle is zero, i.e. if the output is proportional to input i, the latter fully explains the output;
- if that angle is $\pi/2$, i.e. if the output is fully uncorrelated to input i, the latter has no influence on the output.

In practice, the quantity $\cos^2 \varphi_i$ is computed as: $\cos^2 \varphi_i = \dfrac{\left(\boldsymbol{y}^T \boldsymbol{x}_i\right)^2}{\|\boldsymbol{y}\|^2 \|\boldsymbol{x}_i\|^2}$, where $\boldsymbol{y}$ denotes the vector describing the quantity to be modeled in observation space, and $\boldsymbol{x}_i$ the vector describing the i-th candidate feature. In order to rank the inputs in order of decreasing relevance, the following orthogonalization procedure can be used:

- select the candidate feature j that is most correlated to the quantity to be modeled: $\boldsymbol{x}_j = \arg\max_i \cos^2 \varphi_i$;
- project the output vector and all other candidate features onto the null space of the selected feature; compute the parameter pertaining to that feature;
- iterate in that subspace.

The orthogonalization step can advantageously be performed by the Gram-Schmidt or modified Gram-Schmidt procedure (Björck, 1967). The procedure terminates when all candidate features are ranked, or when a prescribed stopping condition is met. At that point, a model of the data is available.[5]

Random probes can be used to construct a stopping criterion in the following way. Different realizations of the "random probe" (considered as a random variable) are appended to the set of candidate features and ranked among them as previously described.[6] However, they should not be included in the selected feature set to build the model: they should be discarded once selected; a probe counter is simply incremented when a probe is encountered in the forward selection process. The rank of the random probe is a random variable and can serve as a relevance index. Since the random probe realizations are ranked among the candidate features, an estimate of the cumulative distribution function (*cdf*) of the rank can be computed as follows: for a total number of probes n_p, the probability that the probe rank is smaller than or equal to a given rank r_0 is estimated as the fraction n_{sp}/n_p of the number of realizations of the random probe n_{sp} that have a rank smaller than or equal to r_0. Alternatively, one can compute *analytically* the cumulative distribution function of the rank of the probe if the probe is drawn from a normal distribution. The details of that computation are provided in (Stoppiglia et al., 2003).

In either case, one must choose a risk α of retaining a candidate variable although it is less relevant than the probe, i.e. the probability α of the rank of a probe being smaller than the rank of a selected feature. Ranking is stopped when that threshold on the probe *cdf* is reached, i.e. when $n_{sp}/n_p \geq \alpha$.

Figure 2.2 displays the cumulative distribution function of the rank of the random probe, as estimated from 100 realizations of the random probe, and computed analytically, for a problem with 10 candidate features. The graph shows that, if a 10% risk is chosen ($\alpha = 0.1$), the first five features should be selected.

In that example, by construction, the number of irrelevant features is known to be $n_{irr} = 5$. Following Equation 2.1 the false positive rate is

---

[5]The procedure works also for models linear in their parameters (the original variables being replaced by derived features.) Thus, if the linear model is not satisfactory, a new model may be built, e.g. with a polynomial expansion (see the "Introduction" chapter.)

[6]For simplicity, we sometimes call "probes" the realizations of the "probe" random variable.

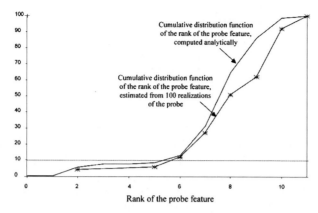

Fig. 2.2. Estimated and computed cumulative distribution function of the rank of a random probe for a problem with ten candidate features.

$FPR = n_{fp}/n_{irr} = n_{sp}/n_p$, so that the number of false positives can be estimated as 0.5 if a 10% risk is chosen.

It can be shown (Stoppiglia, 1997, Stoppiglia et al., 2003) that the random probe procedure in conjunction with ranking by OFR is a generalization of Fisher's test, with two differences: it does not rely on the assumption that the complete model contains the regression, and it does not make any assumption about the feature distribution. Academic and industrial examples of the use of the above method can be found e.g. in (Stoppiglia et al., 2003), together with examples of how to use hypothesis tests to assess the significance of the comparisons between performances of different models.

Note that, since the method uses forward selection, it can handle problems where the number of *candidate features* (non-probes) is larger than the number of examples. It fails if and only if the number of *features considered relevant* by the model designer is larger than the number of relevant variables.

2.2.6 Multiple Tests and False Discovery Rate

The danger of multiple testing

In the framework of feature selection, we have described statistical tests in which the null hypothesis is $\{H_0$: a candidate feature (or a group of features) is irrelevant$\}$ against the alternative hypothesis $\{H_1$: a candidate feature (or a group of features) is relevant$\}$. Our selection criterion is based on the test statistic, used as a ranking index. A Type-I error results in keeping a feature (or group of features) although it is not significant: the p-value is therefore the probability of a "false positive" feature, that is what we also called the false positive rate $FPR = n_{fp}/n_{irr}$. p-values can be either computed from the *cdf* of the ranking index for irrelevant features (if that one is known) or estimated as the fraction of selected random probes n_{sp}/n_p as the ranking index is varied. If a threshold α of 0.05 on the FPR is chosen, with risk less than 5% of being wrong the feature (or a group of features) being tested should be kept.

Assume now that n_{sc} candidate features have been selected among n_c candidate features, using a threshold on our ranking index. If that threshold corresponds to a p-value of 0.05, should we conclude that less than 5% of the features that should be discarded will actually be kept?

In fact we should not: the fraction of irrelevant features selected is probably much larger than that because we have implicitly performed *multiple tests*; we have been testing n_c features simultaneously, not just one feature. Generally, for feature ranking methods and for feature subset ranking methods (forward selection and backward elimination), n_c tests are being performed. For exhaustive search, all possible combinations of candidate features are considered, so 2^{n_c} statistical tests are effectively performed!

If a large number of statistical tests is necessary for making a decision, the following problem arises. Testing the null hypothesis with a threshold of $\alpha = 0.05$ results in 95% correct decisions. But, if two independent tests are performed, making a correct decision on both tests arises only with probability $0.95^2 \approx 0.9$; if three independent tests are performed, that probability drops to $0.95^3 \approx 0.86$. The probability of Type-I errors thus increases geometrically and becomes much larger than α. The simplest way to circumvent the problem is the *Bonferroni correction*, which consists in multiplying the p-value by the number n_t of tests, using the first order approximation $(1 - pval)^{n_t} \simeq 1 - n_t\, pval$. Alternatively, one can divide the threshold α used for each test by n_t to obtain a corrected threshold α/n_t for the multiple test. Unfortunately, that correction turns out to overestimate the p-value, so that performing the Bonferroni correction leads to be overly selective, whereas performing no correction at all leads to selecting too many features.

False discovery rate

Benjamini and Hochberg (1995) defined the False Discovery Rate (FDR) as the ratio of the number of falsely rejected hypotheses to the total number of rejected hypotheses. It can be shown that the FDR provides a method that is intermediate between the Bonferroni correction and no correction (Genovese and Wasserman, 2002). In the context of feature selection, it is the ratio of false positives n_{fp} to the total number of selected features n_{sc}:

$$FDR = \frac{n_{fp}}{n_{sc}}. \qquad (2.2)$$

The FDR is a very useful quantity: even if the probability of false positive $FPR = n_{fp}/n_{irr}$ (p-value of the test) has been bounded by a small value of α, the number n_{fp} of false positives may become quite large if the number n_{irr} of irrelevant features is large, so that the sole rate of false positives may lead to poor selection. Noting that the number of irrelevant features n_{irr}

is smaller or equal to the total number of candidate features n_c, $FPR \geq n_{fp}/n_c$. Combining with Equation 2.2 we can relate the FDR and FPR by the inequality:

$$FDR \leq FPR \frac{n_c}{n_{sc}}. \tag{2.3}$$

It may be useful to set an upper bound α to the FDR. To that effect, the procedure consists in ranking the n_t p-values resulting from n_t tests ($n_t = n_c$ in the case of feature ranking) in increasing order: $p^1 < p^2 < \cdots < p^i < \cdots < p^{n_t}$. The quantity $l_i = i\alpha/n_t$ is computed, and the threshold of acceptance of the null hypothesis is defined as $\delta = \max_i(p^i < l_i)$ (that is $\delta = n_{sc}\alpha/n_c$ in the case of feature ranking): in all tests generating p-values ranking lower than δ, the null hypothesis is rejected. The Bonferroni correction leads to rejecting the null hypothesis for tests generating p-values smaller than α/n_t, which is smaller than δ. Therefore, the Bonferroni correction rejects the null hypothesis less frequently, hence discards a larger number of features, than the FDR procedure. Conversely, if no correction is made, the threshold of acceptance is α, which is larger than δ: in that case, the null hypothesis is discarded more frequently, so that a larger number of features are accepted, thereby resulting in a larger number of false positives. From inequality (2.3) it can easily be verified that the FDR resulting from the procedure is smaller than α. Several extensions of that principle have been investigated. In the context of genome studies, Storey and Tibshirani (2003) define the q-value of a feature in analogy to its p-value, and provide a methodology for upper bounding the false discovery rate.

Estimation of the FDR with random probes

As explained in Section 2.2.2, the generation of random probes provides a way of estimating the probability of false positives (the FPR.) The false discovery rate can also be estimated by generating random probes. From inequality (2.3), and according to Equation 2.1, for a large number of random probes the following inequality holds:

$$FDR \leq \frac{n_{sp}}{n_p} \frac{n_c}{n_{sc}}. \tag{2.4}$$

We remind that in our notations n_{sp} is the number of selected probes, n_p the total number of probes, n_{sc} the number of selected candidates, and n_c the total number of candidates.

If the number of relevant features is a small fraction of the number of candidate features, as is often the case in genome studies for instance, then that upper bound is a good estimate of the false discovery rate. An alternative estimate of the false discovery rate can be derived if the number of false positives is on the same order as the number of false negatives: in that case, the number of irrelevant features can be approximated as $n_{irr} \approx n_c - n_{sc}$. Following again Equation 2.1, the number of false positive can be approximated as $n_{fp} \approx (n_c - n_{sc})n_{sp}/n_p$. It follows from Equation 2.2 that:

$$FDR \approx \frac{n_{sp}}{n_p} \frac{n_c - n_{sc}}{n_{sc}}. \qquad (2.5)$$

If all candidate features are selected, the number of false negatives is zero; under the above assumption, the number of false positives is also zero, so that the estimated FDR is zero.

2.3 A Machine Learning View of Feature Selection

In traditional approaches to modeling, the main issue is the estimation of the parameters of a model that is assumed to be "true", i.e. it is assumed that the family of functions within which the model is sought contains the target function itself. In practice, an analysis of the phenomenon of interest leads to a mathematical relation (a function, or a differential equation) that contains parameters; the latter have a physical meaning, and must be estimated from measurement data. The validity of the estimated parameter values must be assessed, e.g. confidence intervals for the parameters must be computed, but the validity of the mathematical function whose parameters are estimated is not questioned at the level of the statistical analysis itself.

By contrast, in machine learning approaches, the family of functions within which the model is sought has no specific *a priori* justification, so that the model is generally not true. Models may differ in various respects:

- sets of variables or features,
- model complexity: number of support vectors for SVM's, number of hidden neurons for neural nets, number of monomials for polynomial models, etc.,
- initial parameter values or stopping criteria.

The choice of the model itself is part of the statistical analysis: it is the purpose of the *model selection* step of the modeling process. As a consequence, the parameters of the model have no specific meaning, so that the accuracy of the estimation of the parameters is not an issue *per se*. The real issue is the generalization ability of the model, i.e. its ability to provide accurate estimates of the quantity of interest on fresh data (data that has not been used for estimating the parameters.)

In the present section, we first recall the general principles of sound model selection. Then various useful variants of cross-validation are described.

2.3.1 General Principles of Validation and Cross-Validation

In this section we introduce the vocabulary, notations and basic concepts of learning, generalization, overfitting, validation, and cross-validation.

In its simplest setting, model selection generally involves two data sets: a *training set* that is used for parameter estimation, and a *validation set* that is used for model selection (Stone, 1974). Model selection can usually not be performed by comparing performances on training data: adjusting a model very accurately to the training examples is always possible if the number of adjustable parameters is large enough and comparatively few training data points are available. If the learning problem is thus underdetermined and/or if the data are noisy, fitting the training data very accurately may result in a model with poor predictive performance on new data points (poor *generalization.*) That effect is known as *overfitting.* The validation set provides a means of comparing the generalization ability of various models and detecting overfitting: if the performance on the validation set is significantly poorer than the performance on the training set, the model exhibits overfitting. If many models are compared with the validation set, the performance of the best selected model on the validation set is an optimistic estimate of the prediction performance on new data; in a sense, learning is then performed on the validation set. A third set called *test set*, distinct from the training and validation set, is then needed to assess the final performance of the model.

In the following, we assume that the observed quantities can be modeled as realization of a random variable Y such that $Y = f(\boldsymbol{x}) + \Omega$, where Ω is the "noise", a random variable whose expected value is zero. A parameterized model $g(\boldsymbol{x}, \boldsymbol{\theta})$, with vector of parameters $\boldsymbol{\theta}$, is sought, such that $g(\boldsymbol{x}, \boldsymbol{\theta})$ is as close as possible to the unknown function $f(\boldsymbol{x})$. In that context overfitting means fitting both the deterministic part of the data - which is the function of interest - and the noise present in the data. Fitting the noise means that the model is very sensitive to the specific realization of the noise present in the data, hence will generalize poorly.

In the risk minimization framework (see Chapter 1), the generalization error is called the expected risk:

$$R[g] = \int l(y, g(\boldsymbol{x}, \boldsymbol{\theta})) \, dP(\boldsymbol{x}, y). \tag{2.6}$$

where $l(\cdot)$ is a *loss function*, and P is the data distribution. The goal is to minimize the expected risk.

In process or signal modeling, the squared loss $(y - g(\boldsymbol{x}, \boldsymbol{\theta}))^2$ is commonly used; for classification problems the 0/1 loss inflicts a penalty of 1 for classification errors and 0 otherwise. When the data distribution in unknown but m data samples are given, one can compute the empirical risk:

$$R_{emp}[g] = \frac{1}{m} \sum_{k=1}^{m} l(y_k, g(\boldsymbol{x}_k, \boldsymbol{\theta})). \tag{2.7}$$

The performance on the training set is thus the empirical risk measured on the m_T training examples; the performance on the validation set is the empirical risk measured on the m_V validation examples.

For the squared loss, the performance on the training set is usually called the training mean square error $T_{MSE} = (1/m_T) \sum_{k=1}^{m_T} r_k^2$, with $r_k = y^k - g(\boldsymbol{x}^k, \boldsymbol{\theta})$. Here, r_k is the modeling error on example k of the training set. Similarly, the performance on the validation set may be expressed by the validation mean square error $V_{MSE} = (1/m_V) \sum_{k=1}^{m_V} r_k^2$. Assuming m_V and $m_T \gg 1$, the ideal model, $g(\boldsymbol{x}, \boldsymbol{\theta}) = f(\boldsymbol{x})$, would yield values of the T_{MSE} and V_{MSE} that are equal to the estimated variance of the noise $(1/(m - 1)) \sum_{k=1}^{m} (y^k - f(\boldsymbol{x}^k))^2$. The "best" model will be the model that comes closest to that ideal situation, given the available data, the chosen features, and the model g.

For small datasets, a single split of that data into a training set and a validation set may provide a very inaccurate estimate of the generalization error. A preferred strategy consists in splitting the data into D subsets, training on $D - 1$ subsets, validating on the last subset, and repeating that D times with D different splits between training and validation data. That technique is known as "D-fold cross validation". Each example of the data set is present once, and only once, in the validation subset. The cross-validation performance is the average of the results on all the splits. For instance, the cross-validation mean square error is given by $CV_{MSE} = (1/m) \sum_{k=1}^{m} (r_k^V)^2$, where m is the total number of examples and r_k^V is the modeling error on example k when the latter is in the validation subset. Variants of that scheme will be considered in the next section.

Learning theory has addressed the problem of bounding the error made by approximating the expected risk with an empirical risk (see e.g. (Vapnik, 1998).) The bounds are of the type:

$$R[g] \le R_{emp}[g] + \epsilon \qquad (2.8)$$

The right hand side of the inequality is called "guaranteed risk". Such bounds may be used for various purposes:

- To avoid splitting the training data between a training set and a validation set: models are trained using all the available training data; the training examples are used to compute $R_{emp}[g]$; models are then selected by comparing the guaranteed risks. Alternatively, training and model selection may be performed in a single optimization process, which minimizes the guaranteed risk directly. Here the ratio of the complexity of the model and the number of training examples often appears in ϵ.
- To avoid reserving an additional test set to compute to final prediction performance: the validation error is used for $R_{emp}[g]$ and the bound provides a worst-case guaranteed generalization performance. Here, the ratio of the log of the number of models and the number of validation examples appears in ϵ.

2.3.2 Cross-Validation and Its Variants

The general D-fold cross-validation technique was described in the previous section. In the present section, we describe useful variants of that technique.

Leave-one-out

Leave-one-out is a special case of cross-validation, whereby the number of data splits is equal to the number of examples: in other words, each example is withdrawn in turn from the training set, and, for each left-out example

- training is performed on all examples except the left-out example,
- the modeling error on the left-out example is computed.

After m repetitions of that procedure (where m is the number of available examples), the "leave-one-out score" is computed as the average of the modeling error on the left-out examples. For instance, for the squared loss, we obtain

$$LOO_{MSE} = \frac{1}{m} \sum_{k=1}^{m} \left(r_k^{(-k)} \right)^2, \tag{2.9}$$

where $r_k^{(-k)}$ is the modeling error on example k when the latter is withdrawn from the training set. It can be shown (Vapnik, 1982) that the leave-one-out score is an unbiased estimator (as defined in Appendix A) of the generalization error, for a training set of $(m-1)$ examples. However, it has a large variance and is computationally expensive since the number of trainings is equal to the number of available examples. This last drawback is alleviated by the *virtual leave-one-out* technique, as discussed in the next section.

Virtual leave-one-out

The virtual leave-one-out technique relies on the assumption that, in general, the withdrawal of a single example from the training set will yield a model that is not substantially different from the model that is obtained by training on the whole set of available data. Therefore, the output of the model trained without example k can be written as a first-order Taylor expansion with respect to the parameters: $g\left(x, \theta^{(-k)}\right) \approx g\left(x, \theta\right) + Z\left(\theta^{(-k)} - \theta\right)$ where

- $g\left(x, \theta\right)$ is the m-vector of the predictions of model g when its input vector is x and its parameter vector, after training on the whole data set, is θ,
- $g\left(x, \theta^{(-k)}\right)$ is the m-vector of the predictions of the model trained when example k is withdrawn from the training set, with parameter vector $\theta^{(-k)}$,
- Z is the jacobian matrix of the model, i.e. the matrix whose columns are the values of the gradient of the model output with respect to each parameter of the model; hence, Z is an (m, q) matrix, where q is the number of parameters, and where x^j denotes the vector of variables pertaining to example j.

Then the approximate model is linear with respect to its parameters, so that results from linear modeling hold approximately, insofar as the Taylor expansion is accurate enough (Seber and Wild, 1989).

For the squared loss, the PRESS (Predicted Residual Sum of Squares) statistic can be used for assessing the leave-one-out error:

$$r_k^{(-k)} \simeq \frac{r_k}{1 - h_{kk}}$$

where $r_k^{(-k)}$ is the modeling error on example k when the latter is left out of the training set, r_k is the modeling error on example k when the latter belongs to the training set, and h_{kk} is the k-th diagonal element of matrix $\boldsymbol{H} = \boldsymbol{Z}(\boldsymbol{Z}^T \boldsymbol{Z})\boldsymbol{Z}^T$. The quantity h_{kk} is the leverage of example k. The above relation is exact for linear-in-their-parameters models (Allen, 1974) such as kernel machines, and approximate otherwise (Monari and Dreyfus, 2002, Oussar et al., 2004). Therefore, the leave-one-out score can be expressed as the *virtual leave-one-out* score

$$VLOO_{MSE} = \frac{1}{m} \sum_{k=1}^{m} \left(\frac{r_k}{1 - h_{kk}} \right)^2, \tag{2.10}$$

which is exact for linear-in-their-parameters machines, and approximate otherwise; it is obtained at a computational cost that is m times as small as the cost of computing LOO_{MSE} from Equation 2.9.

Ideally, the leave-one-out score should be of the order of the variance of the noise present in the measurements (assuming homoscedasticity, i.e. that the variance of the noise is identical for all measurements.) Therefore, for sound model selection, some estimate of the noise should be available, for instance by performing repeated measurements.

For classification, an approximate leave-one-out score for SVM classifiers has been derived (Opper and Winther, 2000), under the assumptions that, upon withdrawal of an example from the training set, (i) non-support vectors will remain non-support vectors, (ii) margin support vectors will remain margin support vectors, (iii) misclassified patterns will remain misclassified.

When several candidate models, with different features and/or different structures, have virtual leave-one-out scores that are on the same order of magnitude, the significance of the differences between the estimated performances of the models must be assessed by statistical tests (see for instance Anders and Korn (1999)).

2.3.3 Performance Error Bars and Tests of Significance

Given the widespread use of the generalization error estimates provided by cross-validation and its variants, it would be desirable to have an estimator of the variance of those estimates. Unfortunately, in general, no unbiased estimator of the variance of cross-validation exists (Bengio and Grandvalet, 2003).

Conversely, estimators can be found for simple validation methods where data points are used only once. For instance, for classification problems, if we call E the error rate computed with a single validation set (or test set) of m examples, the standard error of E is:

$$stderr(E) = \sqrt{\frac{E(1-E)}{m}}$$

This error bar can be used to determine the size of a test set needed to obtain good error rate estimates (see e.g. (Guyon et al., 1998).) If one desires that the error bar does not exceed $0.1E$, assuming that $E \ll 1$, the number of test examples should be larger than $100/E$. This rule-of-thumb was used to determine the size of the test sets of the challenge (see Part II.)

Determining an optimum size for the validation set is a more challenging task. One must then monitor the tradeoff between biasing the model unfavorably by withholding too many training examples and adding variance to the performance estimate by reserving too few validation examples. Furthermore, the complexity of the models and the number of models to be compared should be taken into account. To our knowledge, this is still an open problem.

Finally, it is often desirable to assess the significance of the difference between the performance of two models. For the challenge (Part II), the organizers have used the McNemar test, which is applicable to classification problems. To perform the test, one must keep track of the errors made by the two classifiers to be compared and compute for each pair of classifiers the number of errors that one makes and the other does not, n_1 and n_2. If the null hypothesis is true (both classifiers have identical error rate), $z = (n_2 - n_1)/\sqrt{n_1 + n_2}$ obeys approximately the standard Normal law. We compute p-values according to: $pval = 0.5 \ (1 - erf(z/\sqrt{2}))$. This allows us to make a decision: significantly better $(pval < 0.05)$, not significantly different $(0.05 \leq pval \leq 0.95)$, significantly worse $(pval > 0.95)$. For details, see (Guyon et al., 1998).

2.3.4 Bagging

When data are sparse, "bagging" (short for "bootstrap aggregating") can be used with advantage, for modeling and performance assessment. The idea underlying bagging (Breiman, 1996) is that averaging predictions over an ensemble of predictors will usually improve the generalization accuracy. That ensemble of predictors is trained from an ensemble of training sets that are obtained by bootstrap from the initial training data (Efron and Tibshirani, 1993): subsets are drawn randomly with replacement, so that an example may be altogether absent from some training sets, and present several times in others. Then different predictors are trained from those data sets; for process modeling, their predictions are averaged, while, for classification, majority voting is performed. Bagging and other ensemble methods are reviewed in more details in Chapter 7.

The generalization performance of bagged predictors can be assessed on *out-of-bag samples*: for each example i, the loss $l(y_i, g^{j(-i)}(\boldsymbol{x}_i))$ of each predictor $g^{j(-i)}$ whose training set does not contain example i is computed, and the generalization performance estimate is $(1/N_{j(-i)}) \sum_{j(-i)} l(y_i, g^{j(-i)}(\boldsymbol{x}_i))$, where $N_{j(-i)}$ is the number of training sets that do not contain example i (Wolpert and Macready, 1999).

In the same spirit, a given number M of D-fold cross-validation splits can be generated. In each cross-validation split, a given example appears $D-1$ times in the training sets and one time in the validation sets. Overall, each example appears M times in the validation sets and $M(D-1)$ times in the training sets; a total of MD models are trained. For each example, the average prediction (or classification) error made, on that example, by the D models that were not trained with that example, is computed. The procedure is repeated for each example, and the average error is computed.

2.3.5 Model Selection for Feature Selection

In this section, we describe some aspects of model selection that are more specific to feature selection. We remind the reader that techniques of feature selection involving the performance of a learning machines are known as "wrappers" (see the Introduction chapter and Chapter 4.)

Cross-validation for feature selection

Since cross-validation provides an estimate of the generalization ability of models, it is frequently used for feature selection. Actually, the prediction ability of models depends both on the features used and on the complexity of the model. Therefore, several cross-validation loops may be nested when feature selection and machine selection are performed simultaneously.

In Section 2.2.1, it was shown that the number of features can be selected from statistical principles such as upper bounding the number of false positives, or the false discovery rate, irrespective of the machine used. Then, cross-validation can be used *for model selection* (e.g. find the best kernel, the best number of hidden neurons, the best regularization hyperparameter, ...), with the features selected at the previous step.

A natural, but possibly dangerous, alternative would consist in ranking all features with the whole data set, by any appropriate ranking algorithm, and performing feature selection by cross-validation or leave-one-out:

- Rank candidate features using all training data and construct nested feature subsets.
- For each candidate feature subset,
  - perform D-fold cross-validation or leave-one-out.

- Select the feature subset that provides the best error estimate.

However, we must caution against this method, which is prone to overfitting because the validation data is used in the feature selection process (Ambroise and McLachlan, 2002).

It is more appropriate to *include the feature selection in the cross-validation loop*, determine the best number of features n_0, and then select n_0 features using the whole training set to produce the final model. In the end, that model must be tested on an independent test set. We summarize the procedure:

- Split data into training and validation sets.

- For each training subset:

  - rank all features and construct nested feature subsets.

  - For each feature subset:

    · train a model and compute its validation error.

- Average the validation errors of the subsets of same size.

- Find the number n_0 of features that provides the smallest average validation error.

- Rank the features on all training data and select the top n_0 features and train the final model.
- Test the final model with an independent test set.

For model selection, an inner cross-validation (or leave-one-out, or virtual leave-one-out) loop may be inserted in the "For each feature subset" loop; alternatively, model selection by cross-validation, leave-one-out or virtual leave-one-out, can be performed before and/or after feature selection

Guaranteed risk for feature selection

Performance bounds of the type of Equation 2.8 are useful in several respect for feature selection.

The chapter on embedded methods (Chapter 5) reviews several feature selection algorithms derived from performance bounds. Kernel methods such as support vector machines (SVMs) lend themselves particularly well to this methodology. In principle, reserving a validation set for feature selection may be avoided altogether, since the approach consists in minimizing a "guaranteed risk", which incorporates a penalization for the model complexity that should also penalize large feature set sizes. In practice, the number of features is sometimes determined by cross-validation.

Performance bounds also give us insight into the relative statistical complexity of feature selection strategies. In Section 2.2.6, we have already been warned that extensively searching the space of all possible feature subsets comes at a price: the statistical significance of the features selected will suffer. Similarly, when feature selection is performed in the context of machine learning using the performance of a model on a validation set, comparing many feature subsets comes at a price: the confidence interval on the generalization error will suffer. For instance, for classification problems, if N models (here corresponding to N feature subsets) are being compared with a validation set of size m_V, the error bar on the prediction increases monotonically with $\sqrt{\ln N / m_V}$ (see (Vapnik, 1982), Theorem 6.1.) Therefore, if all possible combinations of n_c candidate features are considered, $N = 2^{n_c}$; to keep the confidence interval from growing, the number of validation examples needed will have to scale with the number of candidate features. However, for feature ranking and nested feature subset strategies, only $N = n_c$ feature subsets are considered; a number of validation examples scaling with $\ln n_c$ will only be needed. Other algorithms enjoying a "logarithmic sample complexity" have been described (Langley, 1994, Ng, 1998).

Bagging for feature selection

Ensemble methods such as bagging can be used for feature ranking: features can be ranked for each model of the bag, and a "bagged feature ranking" is performed based on the frequency of occurrence of each feature at or above a given rank in the individual rankings. Several feature aggregation procedures are described in (Jong et al., 2004) and in Chapter 7.

2.4 Conclusion

The statistical assessment of the validity of selected features is an important part in the methodology of model design. In the present chapter, we have provided basic elements of such a methodology, relying on classical statistical tests, as well as on more recently developed methods. Some of the methods described here are of general use, but their application to feature selection has specific requirements that should be kept in mind. In view of the difficulty of the problem, there is room for much research effort in that area.

References

D.M. Allen. The relationship between variable selection and prediction. *Technometrics*, 16:125–127, 1974.

C. Ambroise and G. J. McLachlan. Selection bias in gene extraction on the basis of microarray gene-expression data. *PNAS*, (99):6562–6566, 2002.

U. Anders and O. Korn. Model selection in neural networks. *Neural Networks*, 12: 309–323, 1999.

J. Bengio and Y. Grandvalet. No unbiased estimator of the variance of K-fold cross-validation. *Journal of Machine Learning Research*, 5:1089–1105, 2003.

Y. Benjamini and Y. Hochberg. Controlling the false discovery rate: a practical and powerful approach to multiple testing. *J. Roy. Stat. Soc. B*, 85:289–300, 1995.

J. Bi, K.P. Bennett, M. Embrechts, C.M. Breneman, and M. Song. Dimensionality reduction via sparse support vector machines. *Journal of Machine Learning Research*, 3:1229–1243, 2003.

A. Björck. Solving linear least squares problems by gram-schmidt orthogonalization. *Nordisk Tidshrift for Informationsbehadlung*, 7:1–21, 1967.

L. Breiman. Bagging predictors. *Machine Learning*, 24:123–140, 1996.

S. Chen, S.A. Billings, and W. Luo. Orthogonal least squares methods and their application to non-linear system identification. *International Journal of Control*, 50:1873–1896, 1989.

C. Cortes and M. Mohri. Confidence intervals for area under ROC curve. In *Neural information Processing Systems 2004*, 2004.

D. R. Cox and D. V. Hinkley. *Theoretical Statistics*. Chapman and Hall/CRC, 1974.

B. Efron and R.J. Tibshirani. *Introduction to the bootstrap*. Chapman and Hall, New York, 1993.

C.R. Genovese and L. Wasserman. Operating characteristics and extensions of the false discovery rate procedure. *J. Roy. Stat. Soc. B*, 64:499–518, 2002.

G.C. Goodwin and R.L. Payne. *Dynamic system identification: experiment design and data analysis*. Academic Press, 1977.

I. Guyon, J. Makhoul, R. Schwartz, and V. Vapnik. What size test set gives good error rate estimates? *IEEE Transactions on Pattern Analysis and Machine Intelligence*, 20:52–64, 1998.

K. Jong, E. Marchiori, and M. Sebag. Ensemble learning with evolutionary computation: Application to feature ranking. In *8th International Conference on Parallel Problem Solving from Nature*, pages 1133–1142. Springer, 2004.

P. Langley. Selection of relevant features in machine learning, 1994.

I.J. Leontaritis and S.A. Billings. Model selection and validation methods for nonlinear systems. *International Journal of Control*, 45:311–341, 1987.

G. Monari and G. Dreyfus. Local overfitting control via leverages. *Neural Computation*, 14:1481–1506, 2002.

A. Y. Ng. On feature selection: learning with exponentially many irrelevant features as training examples. In *15th International Conference on Machine Learning*, pages 404–412. Morgan Kaufmann, San Francisco, CA, 1998.

M. Opper and O. Winther. *Advances in large margin classifiers*, chapter Gaussian processes and Support Vector Machines: mean field and leave-one-out, pages 311–326. MIT Press, 2000.

L. Oukhellou, P. Aknin, H. Stoppiglia, and G. Dreyfus. A new decision criterion for feature selection: Application to the classification of non destructive testing signatures. In *European SIgnal Processing COnference (EUSIPCO'98)*, Rhodes, 1998.

Y. Oussar, G. Monari, and G. Dreyfus. Reply to the comments on "local overfitting control via leverages" in "jacobian conditioning analysis for model validation". *Neural Computation*, 16:419–443, 2004.

I. Rivals and L. Personnaz. MLPs (mono-layer polynomials and multi-layer percep-trons) for non-linear modeling. *JMLR*, 2003.

G.A. Seber. *Linear regression analysis*. Wiley, New York, 1977.

G.A. Seber and C.J. Wild. *Nonlinear regression*. John Wiley and Sons, New York, 1989.

T. Söderström. On model structure testing in system identification. *International Journal of Control*, 26:1–18, 1977.

M. Stone. Cross-validatory choice and assessment of statistical predictions. *J. Roy. Stat. Soc. B*, 36:111–147, 1974.

H. Stoppiglia. *Méthodes Statistiques de Sélection de Modèles Neuronaux ; Applications Financières et Bancaires*. PhD thesis, l'Université Pierre et Marie Curie, Paris, 1997. (available electronically at http://www.neurones.espci.fr).

H. Stoppiglia, G. Dreyfus, R. Dubois, and Y. Oussar. Ranking a random feature for variable and feature selection. *Journal of Machine Learning Research*, pages 1399–1414, 2003.

J.D. Storey and R. Tibshirani. Statistical significance for genomewide studies. *Proc. Nat. Acad. Sci.*, 100:9440–9445, 2003.

V. Vapnik. *Statistical Learning Theory*. John Wiley & Sons, N.Y., 1998.

V.N. Vapnik. *Estimation of dependencies based on empirical data*. Springer, New-York, 1982.

D. Wolpert and W.G. Macready. An efficient method to estimate bagging's gener-alization error. *Machine Learning*, 35(1):41–55, 1999.

Chapter 3

Filter Methods

Włodzisław Duch

Department of Informatics, Nicolaus Copernicus University,
Grudziądzka 5, 87-100 Toruń, Poland, and
Department of Computer Science, School of Computer Engineering,
Nanyang Technological University, Singapore 639798
Google: Duch

3.1 Introduction to Filter Methods for Feature Selection

Feature ranking and feature selection algorithms may roughly be divided into three types. The first type encompasses algorithms that are built into adaptive systems for data analysis (predictors), for example feature selection that is a part of embedded methods (such as neural training algorithms). Algorithms of the second type are wrapped around predictors providing them subsets of features and receiving their feedback (usually accuracy). These wrapper approaches are aimed at improving results of the specific predictors they work with. The third type includes feature selection algorithms that are independent of any predictors, filtering out features that have little chance to be useful in analysis of data. These filter methods are based on performance evaluation metric calculated directly from the data, without direct feedback from predictors that will finally be used on data with reduced number of features. Such algorithms are usually computationally less expensive than those from the first or the second group. This chapter is devoted to filter methods.

The feature filter is a function returning a relevance index $J(S|\mathcal{D})$ that estimates, given the data $\mathcal{D}$, how relevant a given feature subset S is for the task Y (usually classification or approximation of the data). Since the data and the task are usually fixed and only the subsets S vary the relevance index may be written as $J(S)$. In text classification these indices are frequently called "feature selection metrics" (Forman, 2003), although they may not have formal properties required to call them a distance metric. Instead of a simple function (such as a correlation or information content) some algorithmic procedure may be used to estimate the relevance index (such as building of a decision tree or finding nearest neighbors of vectors). This means that also a wrapper or an embedded algorithm may be used to provide relevance estimation to a filter used with another predictor.

Relevance indices may be computed for individual features $X_i, i = 1 \ldots N$, providing indices that establish a ranking order $J(X_{i_1}) \leq J(X_{i_2}) \cdots \leq$

W. Duch: *Filter Methods*, StudFuzz **207**, 89–117 (2006)
www.springerlink.com

$J(X_{i_N})$. Those features which have the lowest ranks are filtered out. For independent features this may be sufficient, but if features are correlated many of important features may be redundant. Moreover, the best pair of features do not have to include a single best one (Toussaint, 1971, Cover, 1974). Ranking does not guarantee that the largest subset of important features will be found. Methods that search for the best subset of features may use filters, wrappers or embedded feature selection algorithms. Search methods are independent of the evaluation of feature subsets by filters, and are a topic of Chapter 5. The focus here is on filters for ranking, with only a few remarks on calculation of relevance indices for subsets of features presented in Sec. 3.8.

The value of the relevance index should be positively correlated with accuracy of any reasonable predictor trained for a given task Y on the data $\mathcal{D}$ using the feature subset $\mathcal{S}$. This may not always be true for all models, and on theoretical grounds it may be difficult to argue which filter methods are appropriate for a given data analysis model. There is little empirical experience in matching filters with classification or approximation models. Perhaps different types of filters could be matched with different types of predictors but so far no theoretical arguments or strong empirical evidence has been given to support such claim.

Although in the case of filter methods there is no direct dependence of the relevance index on the predictors obviously the thresholds for feature rejection may be set either for relevance indices, or by evaluation of the feature contributions by the final system. Features are ranked by the filter, but how many are finally taken may be determined using the predictor in a "wrapper setting". This "filtrapper" approach is computationally less expensive than the original wrapper approach because the evaluation of the predictor's performance (for example by a cross-validation test) is done only for a few preselected feature sets. There are also theoretical arguments showing that this technique is less prone to overfitting than pure wrapper methods (Ng, 1998). In some data mining applications (for example, analysis of large text corpora with noun phrases as features) even relatively inexpensive filter methods, with costs linear in the number of features, may be prohibitively slow.

Filters, as all other feature selection methods, may be divided into local and global types. Global evaluation of features takes into account all data in a context-free way. Context dependence may include different relevance for different tasks (classes), and different relevance in different areas of the feature space. Local classification methods, for example nearest neighbor methods based on similarity, may benefit more from local feature selection, or from filters that are constructed on demand using only data from the neighborhood of a given vector. Obviously taking too few data samples may lead to large errors in estimations of any feature relevance index and the optimal tradeoff between introduction of context and the reliability of feature evaluation may be difficult to achieve. In any case the use of filter methods for feature selection depends on the actual predictors used for data analysis.

In the next section general issues related to the filter methods are discussed. Section 3.3 is focused on the correlation based filtering, Sec. 3.4 on relevance indices based on distances between distributions and Sec. 3.5 on the information theory. In Section 3.6 the use of decision trees for ranking as well as feature selection is discussed. Reliability of calculation of different indices and bias in respect to the number of classes and feature values is very important and is treated in Section 3.7. This is followed by some remarks in Sec. 3.8 on filters for evaluation of feature redundancy. The last section contains some conclusions.

3.2 General Issues Related to Filters

What does it mean that the feature is relevant to the given task? *Artificial Intelligence* journal devoted in 1996 a special issue to the notion of relevance (Vol. 97, no. 1–2). The common-sense notion of relevance has been rigorously defined in an axiomatic way (see the review in (Bell and Wang, 2000)). Although such definitions may be useful for the design of filter algorithms a more practical approach is followed here. Kohavi and John (1996) give a simple and intuitive definition of relevance that is sufficient for the purpose of feature selection: a feature X is relevant in the process of distinguishing class $Y = y$ from others if and only if for some values $X = x$ for which $\mathcal{P}(X = x) > 0$ the conditional probability $\mathcal{P}(Y = y|X = x)$ is different than the unconditional probability $\mathcal{P}(Y = y)$. Moreover, a good feature should not be redundant, i.e. it should not be correlated with other features already selected. These ideas may be traced back to the test theory (Ghiselli, 1964) developed for psychological measurements.

The main problem is how to calculate the strength of correlations between features and classes (or more generally, between features and target, or output, values), and between features themselves. The Bayesian point of view is introduced below for the classification problems, and many other approaches to estimation of relevance indices are described in subsequent sections. Some of these approaches may be used directly for regression problems, others may require quantization of continuous outputs into a set of pseudo-classes.

Consider the simplest situation: a binary feature X with values $x = \{0, 1\}$ for a two class $y = \{+, -\}$ problem. For feature X the joint probability $\mathcal{P}(y, x)$ that carries full information about the relevance of this feature is a 2 by 2 matrix. Summing this matrix over classes ("marginalizing", as statisticians say) the values of $\mathcal{P}(x)$ probabilities are obtained, and summing over all feature values x gives *a priori* class probabilities $\mathcal{P}(y)$. Because class probabilities are fixed for a given dataset and they sum to $\mathcal{P}(y = +) + \mathcal{P}(y = -) = 1$ only two elements of the joint probability matrix are independent, for example $\mathcal{P}(y = -, x = 0)$ and $\mathcal{P}(y = +, x = 1)$. For convenience notation $\mathcal{P}(y_i, x_j) = \mathcal{P}(y = i, x = j)$ is used below.

The expected accuracy of the majority classifier (MC) $A_{\mathrm{MC}} = \max_y \mathcal{P}(y)$ is independent of the feature X because MC completely ignores information about feature values. The Bayesian Classifier (BC) makes optimal decisions based on the maximum *a posteriori* probability: if $x = x_0$ then for $\mathcal{P}(y_-, x_0) > \mathcal{P}(y_+, x_0)$ class y_- should always be selected, giving a larger fraction $\mathcal{P}(y_-, x_0)$ of correct predictions, and smaller fraction $\mathcal{P}(y_+, x_0)$ of errors. This is equivalent to the *Maximum-a-Posteriori* (MAP) rule: given $X = x$ select class that has greater posterior probability $\mathcal{P}(y|x) = \mathcal{P}(y, x)/\mathcal{P}(x)$. The Bayes error is given by the average accuracy of the MAP Bayesian Classifier (BC). For a single feature, the Bayes error is given by:

$$A_{\mathrm{BC}}(X) = \sum_{j=0,1} \max_i \mathcal{P}(y_i, x_j) = \sum_{j=0,1} \max_i \mathcal{P}(x_j|y_i)\mathcal{P}(y_i). \qquad (3.1)$$

Precise calculation of "real" joint probabilities $\mathcal{P}(y_i, x_j)$ or the conditional probabilities $\mathcal{P}(x_j|y_i)$ using observed frequencies require an infinite amount of the training data, therefore such Bayesian formulas are strictly true only in the asymptotic sense. The training set should be a large, random sample that represents the distribution of data in the whole feature space.

Because $A_{\mathrm{MC}}(X) \leq A_{\mathrm{BC}}(X) \leq 1$, a Bayesian relevance index scaled for convenience to the $[0, 1]$ interval may be taken as:

$$J_{\mathrm{BC}}(X) = (A_{\mathrm{BC}}(X) - A_{\mathrm{MC}}(X))/(1 - A_{\mathrm{MC}}(X)) \in [0, 1]. \qquad (3.2)$$

The $J_{\mathrm{BC}}(X)$ may also be called "a purity index", because it indicates how pure are the discretization bins for different feature values (intervals). This index is also called "the misclassifications impurity" index, and is sometimes used to evaluate nodes in decision trees (Duda et al., 2001).

Two features with the same relevance index $J_{\mathrm{BC}}(X) = J_{\mathrm{BC}}(X')$ may be ranked as equal, although their joint probability distributions $\mathcal{P}(y_i, x_j)$ may significantly differ. Suppose that $\mathcal{P}(y_-) > \mathcal{P}(y_+)$ for some feature X, therefore $A_{\mathrm{MC}}(X) = \mathcal{P}(y_-)$. For all distributions with $\mathcal{P}(y_-, x_0) > \mathcal{P}(y_+, x_0)$ and $\mathcal{P}(y_+, x_1) > \mathcal{P}(y_-, x_1)$ the accuracy of the Bayesian classifier is $A_{\mathrm{BC}}(X) = \mathcal{P}(y_-, x_0) + \mathcal{P}(y_+, x_1)$, and the error is $\mathcal{P}(y_+, x_0) + \mathcal{P}(y_-, x_1) = 1 - A_{\mathrm{BC}}(X)$. As long as these equalities and inequalities between joint probabilities hold (and $\mathcal{P}(y_i, x_j) \geq 0$) two of the probabilities may change, for example $\mathcal{P}(y_+, x_1)$ and $\mathcal{P}(y_+, x_0)$, without influencing $A_{\mathrm{BC}}(X)$ and $J_{\mathrm{BC}}(X)$ values. Thus the Bayesian relevance index is not sufficient to uniquely rank features even in the simplest, binary case. In fact most relevance indices cannot do that without additional conditions (see also Sec. 3.7).

This reasoning may be extended to multi-valued features (or continuous features after discretization (Liu et al., 2002)), and multi-class problems, leading to probability distributions that give identical J_{BC} values. The expected accuracy of a Bayesian classification rule is only one of several aspects that could be taken into account in assessment of such indices. In the statistical and pattern recognition literature various measures of inaccuracy (error rates, discriminability), imprecision (validity, reliability), inseparability and resemblance (resolution, refinement) are used (see (Hand, 1997, Duch et al., 2004) for extended discussion). Knowing the joint $\mathcal{P}(y, x)$ probabilities and using the MAP Bayesian Classifier rule confusion matrices $\mathcal{F}_{ij} = N(y_i, y_j)/m = M_{ij}/m$ may easily be constructed for each feature, representing the joint probability of predicting sample from class y_i when the true class was y_j:

$$\mathcal{F}(\text{true}, \text{predicted}) = \frac{1}{m} \begin{bmatrix} M_{++} & M_{+-} \\ M_{-+} & M_{--} \end{bmatrix} = \frac{1}{m} \begin{bmatrix} \text{TP} & \text{FN} \\ \text{FP} & \text{TN} \end{bmatrix} \tag{3.3}$$

where M_{++} is the number of hits or true positives (TP); M_{--} is the number of hits in the y_- class, or true negatives (TN); M_{-+} is the number of false alarms, or false positives (FP) (for example, healthy people predicted as sick), and M_{+-} is the number of misses, or false negatives (FN) (sick people predicted as healthy), and the number of samples m is the sum of all M_{ij}.

Confusion matrices have only two independent entries because each row has to sum to $\mathcal{F}_{+j} + \mathcal{F}_{-j} = \mathcal{P}(y_j)$, the *a priori* class probability (estimated as the fraction of all samples that belong to the class y_j). Class accuracies, or conditional probabilities that given a sample from class y it will be really classified as class y are usually taken as the two independent variables. In medical informatics $S_+ = \mathcal{F}_{++}/\mathcal{P}(y_+) = \mathcal{F}(y_+|y_+)$ is called sensitivity or true positive rate (in information retrieval the name recall or detection rate is used), and $S_- = \mathcal{F}_{--}/\mathcal{P}(y_-) = \mathcal{F}(y_-|y_-)$ is called specificity. These diagonal elements of the conditional confusion matrix $\mathcal{F}(y_i|y_i)$ reflect the type of errors that the predictor makes. For example, sensitivity shows how well sick people (class $y = +$) are correctly recognized by classification rule based on some feature (results of a medical test), and specificity shows how well healthy people (class $y = -$) are recognized as healthy by the same test. Generalization to the K-class case is obvious. Standard classifier accuracy is obtained as a trace of the $\mathcal{F}(y_i, y_j)$ matrix, or $\text{Acc} = \sum_i \mathcal{F}(y_i|y_i)\mathcal{P}(y_i)$. The arithmetic average of class accuracies $\mathcal{F}(y_i|y_i)$ is called a balanced accuracy

$$\text{Acc}_2 = \frac{1}{K} \sum_{i=1}^{K} \mathcal{F}(y_i|y_i). \tag{3.4}$$

The Balanced Error Rate BER=$1 - \text{Acc}_2$ is a particularly useful evaluation measure for unbalanced datasets. For feature ranking, using accuracy-based relevance indices, such as the $A_{\text{BC}}, J_{\text{BC}}$ indices, is equivalent to comparing $\mathcal{F}(y_+, y_+) - \mathcal{F}(y_+, y_-)$ (the probability of true positives minus false positives), while using balanced accuracy is equivalent to $\mathcal{F}(y_+|y_+) - \mathcal{F}(y_+|y_-)$ (true positives ratio minus false positives ratio), because terms that are constant for a given data will cancel during comparison. This difference may be rescaled, for example by using (Forman, 2003):

$$BNS = G^{-1}\left(\mathcal{F}(y_+|y_+)\right) - G^{-1}\left(\mathcal{F}(y_+|y_-)\right) \tag{3.5}$$

where $G^{-1}(\cdot)$ is the z-score, or the standard inverse cumulative probability function of a normal distribution. This index, called bi-normal separation index, worked particularly well in information retrieval (IR) (Forman, 2003). Another simple criterion used in this field is called the Odds Ratio:

$$\text{Odds} = \frac{\mathcal{F}(y_+|y_+)\mathcal{F}(y_-|y_-)}{\mathcal{F}(y_+|y_-)\mathcal{F}(y_-|y_+)} = \frac{\mathcal{F}(y_+|y_+)(1 - \mathcal{F}(y_-|y_+))}{(1 - \mathcal{F}(y_+|y_+))\mathcal{F}(y_-|y_+)} \quad (3.6)$$

where zero probabilities are replaced by small positive numbers.

Ranking of features may be based on some combination of sensitivity and specificity. The cost of not recognizing a sick person (low sensitivity) may be much higher than the cost of temporary hospitalization (low specificity). Costs of misclassification may also be introduced by giving a factor to specify that $\mathcal{F}_{+-}$ type of errors (false positive) are α times less important than $\mathcal{F}_{-+}$ type of errors (false negative). Thus instead of just summing the number of errors the total misclassification cost is $E(\alpha) = \alpha\mathcal{F}_{-+} + \mathcal{F}_{+-}$. For binary feature values the BC decision rule has no parameters, and costs $E(\alpha)$ are fixed for a given dataset. However, if the $\mathcal{P}(y,x)$ probabilities are calculated by discretization of some continuous variable z so that the binary value $x = \Theta(z - \theta)$ is calculated using a step function Θ, the values of sensitivity $\mathcal{F}(y_+|y_+;\theta)$ and specificity $\mathcal{F}(y_-|y_-;\theta)$ depend on the threshold θ, and the total misclassification cost $E(\alpha, \theta)$ can be optimized with respect to θ.

A popular way to optimize such thresholds (called also "operating points") of classifiers) is to use the receiver operator characteristic (ROC) curves (Hand, 1997, Swets, 1988). These curves show points $R(\theta) = (\mathcal{F}(y_+|y_-;\theta), \mathcal{F}(y_+|y_+;\theta))$ that represent a tradeoff between the false alarm rate $\mathcal{F}(y_+|y_-;\theta)$ and sensitivity $\mathcal{F}(y_+|y_+;\theta)$ (true positives rate). The Area Under the ROC curve (called AUC) is frequently used as a single parameter characterizing the quality of the classifier (Hanley and McNeil, 1982), and may be used as a relevance index for BC or other classification rules. For a single threshold (binary features) only one point $R = (\mathcal{F}(y_+|y_-), \mathcal{F}(y_+|y_+))$ is defined, and the ROC curve has a line segment connecting it with points $(0,0)$ and $(1,1)$. In this case AUC$= \frac{1}{2}(\mathcal{F}(y_+|y_+) + \mathcal{F}(y_-|y_-))$ is simply equal to the balanced accuracy Acc$_2$, ranking as identical all features that have the same difference between true positive and false positive ratios. In general this will not be the case and comparison of AUCs may give a unique ranking of features. In some applications (for example, in information retrieval) classifiers may have to work at different operating points, depending on the resources that may change with time. Optimization of ROC curves from the point of view of feature selection leads to filtering methods that may be appropriate for different operating conditions (Coetzee et al., 2001).

A number of relevance indices based on modified Bayesian rules may be constructed, facilitating feature selection not only from the accuracy, but also from the cost or confidence point of view. The confusion matrix $\mathcal{F}(y_1, y_2)$ for the two-class problems may be used to derive various combinations of accuracy and error terms, such as the harmonic mean of recall and precision called the

$F1$-measure,

$$J_F(X) = 2\mathcal{F}_{++}/(1 + \mathcal{F}_{++} - \mathcal{F}_{--}), \tag{3.7}$$

well-justified in information retrieval (van Rijsbergen, 1979). Selection of the AUC or balanced accuracy instead of the standard accuracy corresponds to a selection of the relative cost factor $\alpha = \mathcal{P}(y_-)/\mathcal{P}(y_+)$ (Duch et al., 2004). An index combining the accuracy and the error term $J(\gamma) = \mathcal{F}_{--} + \mathcal{F}_{++} - \gamma(\mathcal{F}_{-+} + \mathcal{F}_{+-}) = A - \gamma E$ does not favor one type of errors over another, but it may be used to optimize confidence and rejection rates of logical rules (Duch and Itert, 2002). For $\gamma = 0$ this leads to the A_{BC} Bayesian accuracy index, but for large γ a classification rule that maximizes $J(\gamma)$ may reduce errors increasing confidence in the rule at the expense of leaving some samples unclassified. Non-zero rejection rates are introduced if only significant differences between the $\mathcal{P}(y, x)$ values for different classes are kept, for example the feature is may be rejected if $|\mathcal{P}(y_+, x) - \mathcal{P}(y_-, x)| < \theta$ for all values of x.

From the Bayesian perspective one cannot improve the result of the maximum *a posteriori rule*, so why is the $J_{BC}(X)$ index rarely (if ever) used, and why are other relevance indices used instead? There are numerous theoretical results (Devroye et al., 1996, Antos et al., 1999) showing that for any method of probability density estimations from finite samples convergence may be very slow and no Bayes error estimate can be trusted. The reliability of $\mathcal{P}(y, x)$ estimates rapidly decreases with a growing number of distinct feature values (or continuous values), growing number of classes, and decreasing number of training samples per class or per feature value. Two features with the same $J_{BC}(X)$ index may have rather different distributions, but the one with lower entropy may be preferred. Therefore methods that compare distributions of feature and class values may have some advantages (Torkkola, 2003). An empirical study of simple relevance indices for text classification shows (Forman, 2003) that accuracy is rather a poor choice, with balanced accuracy (equivalent to comparison of AUCs for the two-class problems) giving much higher recall at similar precision. This is not surprising remembering that in the applications to text classification the number of classes is high and the data are usually very unbalanced ($\mathcal{P}(y_+)$ is very small).

Distribution similarity may be estimated using various distance measures, information theory, correlation (dependency) coefficients and consistency measures, discussed in the sections below. Some theoretical results relating various measures to the expected errors of the Bayesian Classifier have been derived (Vilmansen, 1973, Vajda, 1979) but theoretical approaches have met only with limited success and empirical comparisons are still missing. Features with continuous values should be discretized to estimate probabilities needed to compute the relevance indices (Liu and Setiono, 1997, Liu et al., 2002). Alternatively, the data may be fitted to a combination of some continuous one-dimensional kernel functions (Gaussian functions are frequently used), and integration may be used instead of summation.

The relevance indices $J(X)$ introduced above are global or context-free, evaluating the average usefulness of a single feature X. This may be sufficient in many applications, but for some data distributions and for complex domains features may be highly relevant in one area of the feature space and not relevant at all in some other area. Some feature selection algorithms (such as Relief described below) use local information to calculate global, averaged indices. Decision trees and other classification algorithms that use the "divide and conquer" approach hierarchically partitioning the whole feature space, need different subsets of features at different stages. Restricting calculations to the neighborhood $O(\mathbf{x})$ of some input vector $\mathbf{x}$, local or context-dependent, relevance indices $J(X, O(\mathbf{x}))$ are computed.

In multiclass problems or in regression problems features that are important for specific target values ("local" in the output space) should be recognized. For example, if the data is strongly unbalanced, features that are important for discrimination of the classes with small number of samples may be missed. In this case the simplest solution is to apply filters to multiple two-class problems. In case of regression problems filters may be applied to samples that give target values in a specific range.

3.3 Correlation-Based Filters

Correlation coefficients are perhaps the simplest approach to feature relevance measurements. In contrast with information theoretic and decision tree approaches they avoid problems with probability density estimation and discretization of continuous features and therefore are treated first.

In statistics "contingency tables" defined for pairs of nominal features X, Y are frequently analyzed to determine correlations between variables. They contain the numbers of times $M_{ij} = N(y_i, x_j)$ objects with feature values $Y = y_j, X = x_i$ appear in a database. In feature selection m training samples may be divided into subsets of M_{ij} samples that belong to class $y_i, i = 1 \ldots K$ and have a specific feature value x_j; summing over rows of the M_{ij} matrix marginal distribution $M_{i.}$ of samples over classes is obtained, and summing over columns distribution $M_{.j}$ of samples over distinct feature values x_j is obtained. The strength of association between variables X, Y is usually measured using χ^2 statistics:

$$\chi^2 = \sum_{ij} (M_{ij} - m_{ij})^2 / m_{ij}, \text{ where } m_{ij} = M_{i.} M_{.j} / m, \qquad (3.8)$$

Here m_{ij} represent the expected number of observations assuming X, Y independence. Terms with $m_{ij} = 0$ should obviously be avoided (using sufficient data to have non-zero counts for the number of samples in each class and each feature value), or replaced by a small number. If feature and target values were completely independent $m_{ij} = M_{ij}$ would be expected, thus large differences show strong dependence. To estimate the significance of the χ^2

test an incomplete gamma function $Q(\chi^2|\nu)$ is used (Press et al., 1988). The number of degrees of freedom ν is set to $K - 1$. This approach is justified from the statistical point of view only if the number of classes or the number of feature values are large. In contrast to the Bayesian indices the χ^2 results depend not only on the joint probabilities $\mathcal{P}(x_i, y_j) = N(x_i, y_j)/m$, but also on the number of samples m, implicitly including the intuition that estimation of probabilities from small samples is not accurate and thus the significance of small correlations is rather low. χ^2 statistics have been used in several discretization methods combined with feature selection (Liu and Setiono, 1997, Liu et al., 2002).

The linear correlation coefficient of Pearson is very popular in statistics (Press et al., 1988). For feature X with values x and classes Y with values y treated as random variables it is defined as:

$$\varrho(X, Y) = \frac{E(XY) - E(X)E(Y)}{\sqrt{\sigma^2(X)\sigma^2(Y)}} = \frac{\sum_i (x_i - \bar{x}_i)(y_i - \bar{y}_i)}{\sqrt{\sum_i (x_i - \bar{x}_i)^2 \sum_j (y_i - \bar{y}_i)^2}}. \tag{3.9}$$

$\varrho(X, Y)$ is equal to ± 1 if X and Y are linearly dependent and zero if they are completely uncorrelated. Some features may be correlated positively, and some negatively. Linear coefficient works well as long as the relation between feature values and target values is monotonic. Separation of the means of the class distributions leads to an even simpler criterion, called sometimes the "signal-to-noise ratio":

$$\mu(X, Y) = \frac{\mu(y_+) - \mu(y_-)}{(\sigma(y_+) + \sigma(y_-))}, \tag{3.10}$$

where $\mu(y_+)$ is the mean value for class y_+ vectors and $\sigma(y_+)$ is the variance for this class. For continuous targets a threshold $y < \theta$ divides vectors into y_+ and y_- groups. The square of this coefficient is similar to the ratio of between-class to within-class variances, known as the Fisher criterion (Duda et al., 2001). The two-sample T-test uses slightly different denominator (Snedecorand and Cochran, 1989):

$$T(X, Y) = \frac{\mu(y_+) - \mu(y_-)}{\sqrt{\sigma(y_+)^2/m_+ + \sigma(y_-)^2/m_-}}, \tag{3.11}$$

where $m_\pm$ is the number of samples in class $y_\pm$. For ranking absolute values $|\varrho(X, Y)|$, $|\mu(X, Y)|$ and $|T(X, Y)|$ are taken. Fukunaga (Fukunaga, 1990) contains an excellent analysis of such criteria.

How significant are differences in $\varrho(X, Y)$ and other index values? The simplest test estimating the probability that the two variables are correlated is:

$$\mathcal{P}(X \sim Y) = \mathrm{erf}\left(|\varrho(X, Y)|\sqrt{m/2}\right), \tag{3.12}$$

where erf is the error function. Thus for $m = 1000$ samples linear correlations coefficients as small as 0.02 lead to probabilities of correlation around 0.5.

This estimation may be improved if the joint probability of X, Y variables is binormal. The feature list ordered by decreasing values (descending order) of the $\mathcal{P}(X \sim Y)$ may serve as feature ranking. A similar approach is also taken with χ^2, but the problem in both cases is that for larger values of χ^2 or correlation coefficient, probability $\mathcal{P}(X \sim Y)$ is so close to 1 that ranking becomes impossible due to the finite numerical accuracy of computations. Therefore an initial threshold for $\mathcal{P}(X \sim Y)$ may be used in ranking only to determine how many features are worth keeping, although more reliable estimations may be done using cross-validation or wrapper approaches. An alternative is to use a permutation test, computationally expensive but improving accuracy for small number of samples (Cox and Hinkley, 1974) (see also Neal and Zhang Chapter 10).

If a group of k features has already been selected, correlation coefficients may be used to estimate correlation between this group and the class, including inter-correlations between the features. Relevance of a group of features grows with the correlation between features and classes, and decreases with growing inter-correlation. These ideas have been discussed in theory of psychological measurements (Ghiselli, 1964) and in the literature on decision making and aggregating opinions (Hogarth, 1977). Denoting the average correlation coefficient between these features and the output variables as $r_{ky} = \bar{\varrho}(\mathbf{X}_k, Y)$ and the average between different features as $r_{kk} = \bar{\varrho}(\mathbf{X}_k, \mathbf{X}_k)$ the group correlation coefficient measuring the relevance of the feature subset may be defined as:

$$J(\mathbf{X}_k, Y) = \frac{k r_{ky}}{\sqrt{k + (k-1)r_{kk}}}. \tag{3.13}$$

This formula is obtained from Pearson's correlation coefficient with all variables standardized. It has been used in the Correlation-based Feature Selection (CFS) algorithm (Hall, 1999) adding (forward selection) or deleting (backward selection) one feature at a time.

Non-parametric, or Spearman's rank correlation coefficients may be useful for ordinal data types. Other statistical tests of independence that could be used to define relevance indices, such as the Kolmogorov-Smirnov test based on cumulative distributions and G-statistics (Press et al., 1988).

A family of algorithms called Relief (Robnik-Sikonja and Kononenko, 2003) are based on the feature weighting, estimating how well the value of a given feature helps to distinguish between instances that are near to each other. For a randomly selected sample $\mathbf{x}$ two nearest neighbors, $\mathbf{x}_s$ from the same class, and $\mathbf{x}_d$ from a different class, are found. The feature weight, or the Relief relevance index $J_R(X)$ for the feature X, is increased by a small amount proportional to the difference $|X(\mathbf{x}) - X(\mathbf{x}_d)|$ because relevance should grow for features that separate vectors from different classes, and is decreased by a small amount proportional to $|X(\mathbf{x}) - X(\mathbf{x}_s)|$ because relevance should decrease for feature values that are different from features of nearby vectors from the same class. Thus $J_R(X) \leftarrow J_R(X) + \eta(|X(\mathbf{x}) - X(\mathbf{x}_d)| - |X(\mathbf{x}) - X(\mathbf{x}_s)|)$, where η is of the order of $1/m$. After a large number of iterations this index captures local correlations between feature values and their ability to help in discrimination of vectors from different classes. Variants include ratio of the average over all examples of the distance to the nearest miss and the average distance to the nearest hit, that self-normalizes the results (Guyon et al., 2003):

$$J_R(X) = \frac{E_x(|X(\mathbf{x}) - X(\mathbf{x}_d)|)}{E_x(|X(\mathbf{x}) - X(\mathbf{x}_s)|)}. \tag{3.14}$$

The ReliefF algorithm has been designed for multiclass problems and is based on the k nearest neighbors from the same class, and the same number of vectors from different classes. It is more robust in the presence of noise in the data, and includes an interesting approach to the estimation of the missing values. Relief algorithms represent quite original approach to feature selection, that is not based on evaluation of one-dimensional probability distributions (Robnik-Sikonja and Kononenko, 2003). Finding nearest neighbors assures that the feature weights are context sensitive, but are still global indices (see also (Hong, 1997) for another algorithm of the same type). Removing context sensitivity (which is equivalent to assuming feature independence) makes it possible to provide a rather complex formula for ReliefX:

$$J_{\mathrm{RX}}(Y, X) = \frac{GSx}{(1 - Sy)Sy}; \quad \text{where}$$

$$Sx = \sum_{i=1}^{K} \mathcal{P}(x_i)^2; \quad Sy = \sum_{j=1}^{M_Y} \mathcal{P}(y_j)^2 \tag{3.15}$$

$$G = \sum_j \mathcal{P}(y_j)(1 - \mathcal{P}(y_j)) - \sum_{i=1}^{K} \left(\frac{\mathcal{P}(x_i)^2}{Sx} \sum_j \mathcal{P}(y_j|x_i)(1 - \mathcal{P}(y_j|x_i)) \right).$$

The last term is a modified Gini index (Sec. 3.6). Hall (Hall, 1999) has used a symmetrized version of $J_{\mathrm{RX}}(Y, X)$ index (exchanging x and y and averaging) for evaluation of correlation between pairs of features. Relief has also been combined with a useful technique based on the successive Gram-Schmidt orthogonalization of features to the subset of features already created (Guyon et al., 2003). Connection to the Modified Value Difference Metric (MVDM) is mentioned in the next section.

3.4 Relevance Indices Based on Distances Between Distributions

There are many ways to measure dependence between the features and classes based on evaluating differences between probability distributions. A simple measure – a difference between the joint and the product distributions – has been proposed by Kolmogorov:

$$D_{\mathrm{K}}(Y, X) = \sum_i \sum_{j=1}^{K} |\mathcal{P}(y_j, x_i) - \mathcal{P}(x_i)\mathcal{P}(y_j)|. \qquad (3.16)$$

This is very similar to the χ^2 statistics except that the results do not depend on the number of samples. After replacing summation by integration this formula may be easily applied to continuous features, if probability densities are known or some kernel functions have been fitted to the data. It may reach zero for completely irrelevant features, and it is bounded from above:

$$0 \le D_{\mathrm{K}}(Y, X) \le 1 - \sum_i \mathcal{P}(x_i)^2, \qquad (3.17)$$

if the correlation between classes and feature values is perfect. Therefore this index is easily rescaled to the $[0, 1]$ interval. For two classes with the same *a priori* probabilities Kolmogorov measure reduces to:

$$D_{\mathrm{K}}(Y, X) = \frac{1}{2} \sum_i |\mathcal{P}(x_i|y = 0) - \mathcal{P}(x_i|y = 1)|. \qquad (3.18)$$

The expectation value of squared *a posteriori* probabilities is known as the average Euclidean norm of the conditional distribution, called also the Bayesian measure (Vajda, 1979):

$$J_{\mathrm{BM}}(Y, X) = \sum_i \mathcal{P}(x_i) \sum_{j=1}^{K} \mathcal{P}(y_j|x_i)^2, \qquad (3.19)$$

It measures concentration of the conditional probability distribution for different x_i values in the same way as the Gini index (Eq. 3.39) used in decision trees (Sec. 3.6).

The Kullback-Leibler divergence:

$$D_{\mathrm{KL}}((\mathcal{P}(X)||(\mathcal{P}(Y)) = \sum_i \mathcal{P}_Y(y_i) \log \frac{\mathcal{P}_Y(y_i)}{\mathcal{P}_X(x_i)} \ge 0, \qquad (3.20)$$

is used very frequently, although it is not a distance (it is not symmetric). The KL divergence may be applied to relevance estimation in the same way as the χ^2 statistics:

$$D_{\mathrm{KL}}(\mathcal{P}(X, Y)||\mathcal{P}(X)\mathcal{P}(Y)) = \sum_i \sum_{j=1}^{K} \mathcal{P}(y_j, x_i) \log \frac{\mathcal{P}(y_j, x_i)}{\mathcal{P}(x_i)\mathcal{P}(y_j)}. \qquad (3.21)$$

This quantity is also known as "mutual information" $MI(Y, X)$. The Kullback-Liebler measure is additive for statistically independent features. It is sensitive to the small differences in distribution tails, which may lead to problems, especially in multiclass applications where the relevance index is taken as the average value of KL divergences between all pairs of classes.

The Jeffreys-Matusita distance (JM-distance) provides a more robust criterion:

$$D_{\mathrm{JM}}(Y, X) = \sum_i \sum_{j=1}^{K} \left[\sqrt{\mathcal{P}(y_j, x_i)} - \sqrt{\mathcal{P}(x_i)\mathcal{P}(y_j)} \right]^2 . \qquad (3.22)$$

For Gaussian distributions D_{JM} is related to the Bhattacharya distance. Because $D_{JM} \leq 2(1 - \exp(-D_{KL}/8))$ an exponential transformation $J_{KL} = 1 - \exp(-D_{KL}/8)$ is sometimes defined, reaching zero for irrelevant features and growing to 1 for a very large divergences, or highly relevant features. There is some evidence that these distances are quite effective in remote sensing applications (Bruzzone et al., 1995).

The Vajda entropy is defined as (Vajda, 1979):

$$J_{\mathrm{V}}(Y, X) = \sum_i \mathcal{P}(x_i) \sum_{j=1}^{K} \mathcal{P}(y_j|x_i)(1 - \mathcal{P}(y_j|x_i)), \qquad (3.23)$$

and is simply equal to the $J_{\mathrm{V}}(Y, X) = 1 - J_{\mathrm{BM}}(Y, X)$. The error rate of the Bayesian Classifier is bounded by the Vajda entropy, $A_{\mathrm{BC}}(X) \leq J_{\mathrm{V}}(Y, X)$. Although many other ways to compare distributions may be devised they may serve as better relevance indicators only if tighter error bounds could be established.

In the memory-based reasoning the distance between two vectors X, X' with discrete elements (nominal or discretized), in a K class problem, is computed using conditional probabilities (Wilson and Martinez, 1997):

$$VDM(X, X'; Y)^2 = \sum_i \sum_{j=1}^{K} |\mathcal{P}(y_j|x_i) - \mathcal{P}(y_j|x_i')|^2 \qquad (3.24)$$

This formula may be used to evaluate feature similarity when redundant features are searched for.

3.5 Relevance Measures Based on Information Theory

Information theory indices are most frequently used for feature evaluation. Information (negative of entropy) contained in the class distribution is:

$$H(Y) = - \sum_{i=1}^{K} \mathcal{P}(y_i) \log_2 \mathcal{P}(y_i), \qquad (3.25)$$

where $\mathcal{P}(y_i) = m_i/m$ is the fraction of samples $\mathbf{x}$ from class $y_i, i = 1..K$. The same formula is used to calculate information contained in the discrete distribution of feature X values:

$$H(X) = - \sum_i \mathcal{P}(x_i) \log_2 \mathcal{P}(x_i). \tag{3.26}$$

Continuous features are discretized (binned) to compute information associated with a single feature or some kernel functions are fitted to approximate the density of X values and integration performed instead of summation. Information contained in the joint distribution of classes and features, summed over all classes, gives an estimation of the importance of the feature. Information contained in the joint distribution is:

$$H(Y, X) = - \sum_i \sum_{j=1}^{K} \mathcal{P}(y_j, x_i) \log_2 \mathcal{P}(y_j, x_i), \tag{3.27}$$

or for continuous features:

$$H(Y, X) = - \sum_{j=1}^{K} \int \mathcal{P}(y_j, x) \log_2 \mathcal{P}(y_j, x) dx, \tag{3.28}$$

where $\mathcal{P}(y_j, x_i), j = 1 \ldots K$ is the joint probability (density for continuous features) of finding the feature value $X = x_i$ for vectors $\mathbf{x}$ that belong to some class y_j and $\mathcal{P}(x_i)$ is the probability (density) of finding vectors with feature value $X = x_i$. Low values of $H(Y, X)$ indicate that vectors from a single class dominate in some intervals, making the feature more valuable for prediction.

Information is additive for the independent random variables. The difference $MI(Y, X) = H(Y) + H(X) - H(Y, X)$ may therefore be taken as "mutual information" or "information gain". Mutual information is equal to the expected value of the ratio of the joint to the product probability distribution, that is to the Kullback-Leibler divergence:

$$MI(Y, X) = - \sum_{i,j} \mathcal{P}(y_j, x_i) \log_2 \frac{\mathcal{P}(y_j, x_i)}{\mathcal{P}(y_j)\mathcal{P}(x_i)} = D_{KL}(\mathcal{P}(y_j, x_i)|\mathcal{P}(y_j)\mathcal{P}(x_i)).$$
$$\tag{3.29}$$

A feature is more important if the mutual information $MI(Y, X)$ between the target and the feature distributions is larger. Decision trees use closely related quantity called "information gain" $IG(Y, X)$. In the context of feature selection this gain is simply the difference $IG(Y, X) = H(Y) - H(Y|X)$ between information contained in the class distribution $H(Y)$, and information after the distribution of feature values is taken into account, that is the conditional information $H(Y|X)$. This is equal to $MI(Y, X)$ because $H(Y|X) = H(Y, X) - H(X)$. A standard formula for the information gain is easily obtained from the definition of conditional information:

$$IG(Y, X) = H(Y) - H(Y|X) = H(Y) + \sum_{ij} \mathcal{P}(y_j, x_i) \log_2 \mathcal{P}(y_j|x_i) \quad (3.30)$$

$$= H(Y) - \sum_{ij} \mathcal{P}(x_i) \left[-\mathcal{P}(y_j|x_i) \log_2 \mathcal{P}(y_j|x_i) \right],$$

where the last term is the total information in class distributions for subsets induced by the feature values x_i, weighted by the fractions $\mathcal{P}(x_i)$ of the number of samples that have the feature value $X = x_i$. Splits induced by tests in nodes of decision trees are usually not based directly on all attribute values and thus information gain in general is different from mutual information, but for the feature selection purposes these two quantities are identical.

It is not difficult to prove that the Bayes error A_{BC} is bounded from above by half of the value of the conditional information and from below by the Fano inequality,

$$\frac{H(Y|X) - 1}{\log_2 K} \le A_{\mathrm{BC}} \le \frac{1}{2} H(Y|X), \quad (3.31)$$

although the left side is usually negative and thus not useful. Minimizing $H(Y|X) = H(Y) - MI(Y, X)$, or maximizing mutual information, leads to an approximation of Bayes errors and optimal predictions. Error bounds are also known for the Renyi entropy that is somehow easier to estimate in on-line learning than the Shannon entropy (Erdogmus and Principe, 2004).

Various modifications of the information gain have been considered in the literature on decision trees (cf. (Quinlan, 1993)), aimed at avoiding bias towards the multivalued features. These modifications include:

$$IGR(Y, X) = MI(Y, X)/H(X), \quad (3.32)$$
$$D_H(Y, X) = 2H(Y, X) - H(Y) - H(X), \quad (3.33)$$
$$D_M(Y, X) = 1 - MI(Y, X)/H(Y, X), \quad (3.34)$$
$$J_{\mathrm{SU}}(Y, X) = 1 - \frac{D_H(Y, X)}{H(Y) + H(X)} = 2\frac{MI(Y, X)}{H(Y) + H(X)} \in [0, 1]. \quad (3.35)$$

where IGR is the information gain ratio, D_H is the entropy distance, D_M is the Mantaras distance (de Mantaras, 1991) and J_{SU} is the symmetrical uncertainty coefficient. The J_{SU} coefficient seems to be particularly useful due to its simplicity and low bias for multi-valued features (Hall, 1999).

The J-measure:

$$J_J(X) = \sum_i \mathcal{P}(x_i) \sum_j \mathcal{P}(y_j|x_i) \log \frac{\mathcal{P}(y_j|x_i)}{\mathcal{P}(y_j)}, \quad (3.36)$$

has been initially introduced to measure information content of logical rules (Smyth and Goodman, 1992), but it is applicable also to the feature selection (Kononenko, 1995).

Michie (1990) has defined an index called "average weight of evidence", based on plausibility, an alternative to entropy in information:

$$J_{\mathrm{WE}}(X) = \sum_{j=1}^{K} \sum_{i} \mathcal{P}(x_i) \left| \log \frac{\mathcal{P}(y_j|x_i)(1 - \mathcal{P}(y_j))}{(1 - \mathcal{P}(y_j|x_i))\mathcal{P}(y_j)} \right|. \qquad (3.37)$$

Minimum Description Length (MDL) is a general idea based on the Occkam's razor principle and Kolmogorov's algorithmic complexity (Li and Vitányi, 1993). The joint complexity of the theory inferred from the data and the length of the data encoded using this theory should be minimal. MDL has been applied to the construction of decision trees and the selection of features (Kononenko, 1995). As in the description of χ^2 test, m training samples are divided into subsets of M_{ij} samples that belong to class $y_j, j = 1 \ldots K$ and have a specific feature value $x_i, i = 1 \ldots M_x$. The number of bits needed for optimal encoding of the information about the class distribution for m training samples is estimated (this number is fixed for a given dataset), and the same estimation is repeated for each partitioning created by a feature value (or interval) x. Combinatorics applied to the information coding leads to the MDL formula expressed using binomial and multinomial coefficients $m!/m_1! \ldots m_K!$ in the following way (Kononenko, 1995, Hall, 1999):

$$MDL(Y, X) = \log_2 \frac{m!}{M_1! \ldots M_K!} + \log_2 \binom{m + K - 1}{K - 1} \qquad (3.38)$$

$$- \sum_{j=1}^{M_x} \log_2 \binom{M_{.j} + K - 1}{K - 1} - \sum_{j=1}^{M_x} \log_2 \frac{M_{.j}!}{M_{1j}! \ldots M_{Kj}!},$$

where $M_{i.}$ and $M_{.j}$ are marginal distributions calculated from the M_{ij} matrix. The final relevance index $J_{\mathrm{MDL}}(Y, X) \in [0, 1]$ is obtained by dividing this value by the first two terms representing the length of the class distribution description. A symmetrized version of MDL relevance index is used in (Hall, 1999), calculated by exchanging features and classes and averaging over the two values.

3.6 Decision Trees for Filtering

Decision trees select relevant features using top-down, hierarchical partitioning schemes. In the deeper branches of a tree only a small portion of all data is used and only local information is preserved. In feature selection global relevance is of greater importance. One way to achieve it is to create a single-level tree (for algorithms that allow for multiple splits), or a tree based on a single feature (for algorithms that use binary splits only) and evaluate their accuracy. An additional benefit of using decision trees for continuous features is that they provide optimized split points, dividing feature values into relatively pure bins. Calculation of probabilities $\mathcal{P}(x_j)$ and $\mathcal{P}(y_i|x_j)$ needed for the estimation of mutual information and other relevance indices becomes more accurate than with the naïve discretization based on the bins of equal width or bins with equal number of samples. Mutual information calculated after discretization based on a decision tree may be a few times larger than using naive discretization (Duch et al., 2003).

The 1R decision tree algorithm (Holte, 1993) is most appropriate for feature filtering because it creates only single level trees. Features are analyzed searching for a subset of values or a range of values for which vectors from a single class dominate. The algorithm has one parameter (called the "bucket size"), an acceptable level of impurity for each range of the feature values, allowing for reduction of the number of created intervals. Performance may be estimated using the $J_{\mathrm{BC}}(Y, X)$ index, and the optimal bucket size may be evaluated using cross-validation or bootstrap sampling that can help to avoid the bias for large number of intervals but will also increase computational costs.

The C4.5 tree (Quinlan, 1993) uses information gain to determine the splits and to select the most important features, therefore it always ranks as the most important features that are close to the root node. The CHAID decision tree algorithm (Kass, 1980) measures association between classes and feature values using χ^2 values, as in Eq. 3.8. Although the information gain and the χ^2 have already been mentioned as relevance indices the advantage of using decision trees is that automatic discretization of continuous features is performed.

The Gini impurity index used in the CART decision trees (Breiman et al., 1984) sums the squares of the class probability distribution for a tree node, $J_{\mathrm{Gini}}(Y) = 1 - \sum_i \mathcal{P}(y_i)^2$. Given a feature X a split into subsets with discrete feature values x_j (or values in some interval) may be generated and Gini indices in such subsets calculated. The gain is proportional to the average of the sum of squares of all conditional probabilities:

$$J_{\mathrm{Gini}}(Y, X) = \sum_j \mathcal{P}(x_j) \sum_i \mathcal{P}(y_i|x_j)^2 \in [0, 1], \qquad (3.39)$$

giving a measure of the probability concentration useful for feature ranking. This index is similar to the entropy of class distributions and identical with the Bayesian measure Eq. 3.19.

The Separability Split Value (SSV) criterion is used to determine splits in decision tree (Grąbczewski and Duch, 2000) and to discretize continuous features (Duch et al., 2004, 2001), creating a small number of intervals (or subsets) with high information content. It may also be used as feature relevance index. The best "split value" should separate the maximum number of pairs of vectors from different classes. Among all split values that satisfy this condition, the one that separates the smallest number of pairs of vectors belonging to the same class is selected. The *split value* for a continuous feature X is a real number s, while for a discrete feature it is a subset of all possible values of the feature. In all cases, the *left side* (*LS*) and the *right side* (*RS*) of a split value s is defined by a test $f(X, s)$ for a given dataset $\mathcal{D}$:

$$LS(s, f, \mathcal{D}) = \{\mathbf{x} \in \mathcal{D} : f(\mathbf{x}, s) = T\}\}$$
$$RS(s, f, \mathcal{D}) = \mathcal{D} - LS(s, f, \mathcal{D}), \tag{3.40}$$

where the typical test $f(\mathbf{x}, s)$ is true if the selected feature $x_i < s$ or (for discrete feature) $x_i \in \{s\}$. The *separability of a split value* s is defined for a given test f as:

$$\mathrm{SSV}(s, f) = 2 \sum_{i=1}^{K} |\mathrm{LS}(s, f, \mathcal{D}_i)| \cdot |\mathrm{RS}(s, f, \mathcal{D} - \mathcal{D}_i)| \tag{3.41}$$
$$- \sum_{i} \min \left(|\mathrm{LS}(s, f, \mathcal{D}_i)|, |\mathrm{RS}(s, f, \mathcal{D}_i)| \right),$$

where $\mathcal{D}_k$ is the subset of $\mathcal{D}$ vectors that belong to the class k. If several features separate the same number of pairs of training vectors the second term ranks higher the one that separates a lower number of pairs from the same class. This index has smilar properties to Gini and is easily calculated for both continuous and discrete features. For 10 or less feature values all subsets are checked to determine the simplest groupings, for a larger number of unique values the feature is treated as ordered and the best split intervals are searched for. In the feature selection applications of the SSV, splits are calculated and applied recursively to the data subsets $\mathcal{D}_k$, creating a single-feature tree. When pure nodes are obtained the algorithm stops and prunes the tree. The Bayesian Classifier rule is applied in each interval or for each subset created by this algorithm to calculate the $J_{\mathrm{SSV}}(Y, X)$ relevance index. More complex tree-based approaches to determine feature relevance use pruning techniques (Duch et al., 2004).

3.7 Reliability and Bias of Relevance Indices

How good are different relevance indices? Empirical comparisons of the influence of various indices are difficult because results depend on the data and the classifier. What works well for document categorization (Forman, 2003) (large number of classes, features and samples), may not be the best for bioinformatics data (small number of classes, large number of features and a few samples), or analysis of images. One way to characterize relevance indices is to see which features they rank as identical. If a monotonic function could transform one relevance index into another the two indices would always rank features in the same way. Indeed such relations may be established between some indices (see Sec. 3.4), allowing for clustering of indices into highly similar or even equivalent groups, but perhaps many more relations may be established.

The ranking order predicted by the mutual information and other information theoretic measures, and by the accuracy of the optimal Bayesian Classifier using information contained in a single feature, is not identical. It is easy to find examples of binary-valued features where BC and MI predictions are reversed. Consider three binary features with the following class distributions:

$$\mathcal{P}(Y,X) = \begin{pmatrix} 0.50\ 0.00 \\ 0.25\ 0.25 \end{pmatrix}, \mathcal{P}(Y,X') = \begin{pmatrix} 0.45\ 0.05 \\ 0.20\ 0.30 \end{pmatrix}, \mathcal{P}(Y,X'') = \begin{pmatrix} 0.41\ 0.09 \\ 0.10\ 0.40 \end{pmatrix}.$$

The J_{BC} relevance indices for the three distributions are 0.50, 0.50, 0.62, the MI values are 0.31, 0.21, 0.30, and the J_{Gini} indices are 0.97, 0.98, and 0.99. Therefore the ranking in descending order according of the Bayesian relevance is $X'', X = X'$, mutual information gives X, X'', X', and the Gini index predicts X, X', X''.

The differences between relevance indices are apparent if the contour plots showing lines of constant values of these three indices are created for probability distributions $\mathcal{P}(y,x) = \begin{pmatrix} a\ 0.5 - a \\ b\ 0.5 - b \end{pmatrix}$. These contour plots are shown in Fig. 3.1 in the (a, b) coordinates. The $J_{BC}(Y,X)$ index is linear, the $MI(Y,X)$ has logarithmic nonlinearity and the Gini index has stronger quadratic nonlinearity. For many distributions each index must give identical values. Unique ranking is obtained asking for "the second opinion", that is using pairs of indices if the first one gives identical values. In the example given above the Bayesian relevance index could not distinguish between X and X', but using mutual information for such cases will give a unique ranking X'', X, X'.

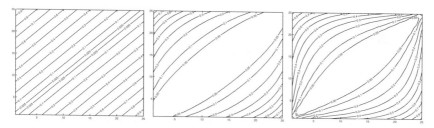

Fig. 3.1. Contours of constant values for BC relevance index (left), MI index (middle) and Gini index (right), in a, b coordinates.

Calculation of indices based on information theory for discrete features is straightforward, but for the continuous features the accuracy of entropy calculations based on simple discretization algorithms or histogram smoothing may be low. The literature on entropy estimation is quite extensive, especially in physics journals, where the concept of entropy has very wide applications (cf. (Holste et al., 1998)). The variance of the histogram-based mutual information estimators has been analyzed in (Moddemeijer, 1999). A simple and effective way to calculate mutual information is based on Parzen windows

(Kwak and Choi, 2002a). Calculation of mutual information between pairs of features and the class distribution is more difficult, but interesting approximations based on the conditional mutual information have been proposed recently to calculate it (Kwak and Choi, 2002b).

Filters based on ranking using many relevance indices may give similar results. The main differences between relevance indices of the same type is in their bias in relation to the number of distinct feature values, and in their variance in respect to the accuracy of their estimation for small number of samples. The issue of bias in estimating multi-valued features has initially been discussed in the decision tree literature (Quinlan, 1993). Gain-ratio and Mantaras distance have been introduced precisely to avoid favoring attributes with larger number of values (or intervals). Biases of 11 relevance indices, including information-based indices, Gini, J-measure, weight of evidence, MDL, and Relief, have been experimentally examined for informative and non-informative features (Kononenko, 1995). For the two-class problems biases for a large number of feature values are relatively small, but for many classes they become significant. For mutual information, Gini and J-measure approximately linear increase (as a function of the number of feature values) is observed, with steepness proportional to the number of classes. In this comparison indices based on the Relief (Sec. 3.3) and MDL (Sec. 3.5) came as the least biased. Symmetrical uncertainty coefficient J_{SU} has a similar low bias (Hall, 1999). Biases in evaluation of feature correlations have been examined by Hall (Hall, 1999).

Significant differences are observed in the accuracy and stability of calculation of different indices when discretization is performed. Fig. 3.2 shows convergence plots of 4 indices created for overlapping Gaussian distributions (variance=1, means shifted by 3 units), as a function of the number of bins of a constant width that partition the whole range of the feature values. Analytical values of probabilities in each bin were used to simulate infinite amount of data, renormalized to sum to 1. For small (4-16) number of bins errors as high as 8% are observed in the accuracy of J_{BC} Bayesian relevance index. Convergence of this index is quite slow and oscillatory. Mutual information (Eq. 3.21) converges faster, and the information gain ratio (Eq. 3.32) shows similar behavior as the Gini index (Eq. 3.39) and the symmetrical uncertainty coefficient J_{SU} (Eq. 3.35) that converge quickly, reaching correct values already for 8 bins (Fig. 3.2). Good convergence and low bias make this coefficient a very good candidate for the best relevance index.

3.8 Filters for Feature Selection

Relevance indices discussed in the previous sections treat each feature as independent (with the exception of Relief family of algorithms Sec. 3.3 and the group correlation coefficient Eq. 3.13), allowing for feature ranking. Those features that have relevance index below some threshold are filtered out as not

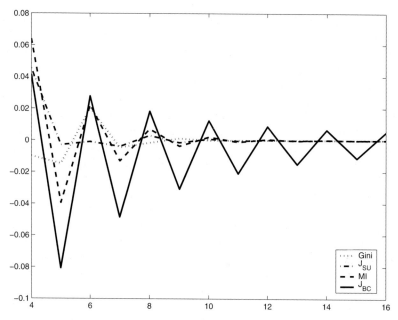

Fig. 3.2. Differences between the Gini, J_{SU}, MI, and J_{BC} indices and their exact value (vertical axis), as a function of the number of discretization bins (horizontal axis).

useful. Some feature selection algorithms may try to include interdependence between features. Given a subset of features $\mathbf{X}$ and a new candidate feature X with relevance index $J(X)$ an index $J(\{\mathbf{X}, X\})$ for the whole extended set is needed. In theory a rigorous Bayesian approach may be used to evaluate the gain in accuracy of the Bayesian classifier after adding a single feature. For k features the rule is:

$$A_{\mathrm{BC}}(\mathbf{X}) = \sum_{x_1, x_2, \ldots x_k} \max_i \mathcal{P}(y_i, x_1, x_2, \ldots x_k) \qquad (3.42)$$

where the sum is replaced by integral for continuous features.

This formula converges slowly even in one dimension (Fig. 3.2), so the main problem is how to reliably estimate the joint probabilities $\mathcal{P}(y_j, x_1, x_2 \ldots x_k)$. The density of training data $\propto \mathcal{P}(x)^k$ goes rapidly to zero with the growing dimensionality k of the feature space. Already for 10 binary features and less than 100 data samples less than 10% of 2^{10} bins are non-empty. Although various histogram smoothing algorithms may regularize probabilities, and hashing techniques may help avoiding high computational costs (Duch et al., 2003), a reliable estimation of $A_{\mathrm{BC}}(\mathbf{X})$ is possible only if the underlying distributions are fully known. This may be useful as a "golden standard" to calculate error bounds, as it is done for one-dimensional distributions, but it is not a practical method.

Calculating relevance indices for subsets selected from a large number of features it is not possible to include full interactions between all the features. Note however that most wrappers may evaluate full feature interactions, depending on the classification algorithm used. Approximations based on summing pair-wise interactions offer a computationally less expensive alternative. The CFS algorithm described in Sec. 3.3 is based on Eq. 3.13, calculating average correlation coefficients between features and classes and between different features. Instead of a ratio for some relevance indices that may measure correlation or dependency between features one may use a linear combination of the two terms: $J(Y, X; \mathcal{S}) = J(Y, X) - \beta \sum_{s \in \mathcal{S}} J(X, X_s)$, where the user-defined constant β is introduced to balance the importance of the relevance $J(Y, X)$ and the redundancy estimated by the sum of feature-feature relevancies. Such algorithm has been used with mutual information as relevance measure by Battiti (1994). In this way redundancy of features is (at least partially) taken into account and search for good subsets of features may proceed at the filter level. A variant of this method may use a maximum of the pair relevance $J(X, X_s)$ instead of the sum over all features $s \in \mathcal{S}$; in this case β is not needed and fewer features will be recognized as redundant.

The idea of inconsistency or conflict – a situation in which two or more vectors with the same subset of feature values are associated with different classes – leads to a search for subsets of features that are consistent (Dash and Liu, 2003, Almuallim and Dietterich, 1991). This is very similar to the indiscernability relations and the search for reducts in rough set theory (Swiniarski and Skowron, 2003). The inconsistency count is equal to the number of samples with identical features, minus the number of such samples from the class to which the largest number of samples belong (thus if there is only one class the index is zero). Summing over all inconsistency counts and dividing by the number of samples m the inconsistency rate for a given subset is obtained. This rate is an interesting measure of feature subset quality, for example it is monotonic (in contrast to most other relevance indices), decreasing with the increasing feature subsets. Features may be ranked according to their inconsistency rates, but the main application of this index is in feature selection.

3.9 Summary and Comparison

There are various restrictions on applications of the relevance indices discussed in the previous sections. For example, some correlation coefficients (such as the χ^2 or Pearson's linear correlation) require numerical features and cannot be applied to features with nominal values. Most indices require probabilities that are not so easy to estimate for continuous features, especially when the number of samples is small. This is usually achieved using discretization methods (Liu et al., 2002). Relevance indices based on decision trees may automatically provide such discretization, other methods have to rely on external algorithms.

In Table 3.1, information about the most popular filters is collected, including the formulas, the types of inputs X (binary, multivalued integer or symbolic, or continuous values), and outputs Y (binary for 2-class, multivalued integer for multiclass problems and continuous for regression).

The first method, Bayesian accuracy A_{BC}, is based on observed probabilities $\mathcal{P}(y_j, x_i)$ and provides a "golden standard" for other methods. Relations between the Bayesian accuracy and mutual information are known 3.31, and such relations may be inferred for other information-based indices, but in general theoretical results of this sort are difficult to find and many indices are simply based on heuristics. New methods are almost never tested against Bayesian accuracy for simple binary features and binary classes. Differences in ranking of features between major relevance indices presented in Sec. 3.7 are probably amplified in more general situations, but this issue has not been systematically investigated so far.

Other methods that belong to the first group of methods in Tab. 3.1 are somehow special. They are based on evaluation of confusion matrix elements and thus are only indirectly dependent on probabilities $\mathcal{P}(y_j, x_i)$. Confusion matrix may be obtained by any classifier, but using Bayesian approach for classification balanced accuracy, area-under-curve (AUC), F-measure, Bi-normal separation and odds ratio are still the best possible approaches, assuming specific costs of different type of errors.

Many variants of a simple statistical index based on separation of the class means exist. Although these indices are commonly applied to problems with binary targets extension to multiple target values is straightforward. In practice pair-wise evaluation (single target value against the rest) may work better, finding features that are important for discrimination of small classes. Feature values for statistical relevance indices must be numerical, but target values may be symbolic. Pearson's linear correlation coefficient can be applied only for numerical feature and target values, and its averaged (or maximum) version is used for evaluation of correlations with a subset of features. Decision-tree based indices are applicable also to symbolic values and may be computed quite rapidly. Some trees may capture the importance of a feature for a local subset of data handled by the tree nodes that lie several levels below the root. The Relief family of methods are especially attractive because they may be applied in all situations, have low bias, include interaction among features and may capture local dependencies that other methods miss.

Continuous target values are especially difficult to handle directly, but distance-based measures of similarity between distributions may handle them without problems. Kolmogorov distance and other relevance indices from this group may be expressed either by a sum of discrete probabilities or an integral over probability density functions. Bayesian measure, identical with the Gini index for discrete distributions, generalizes it to continuous features and continuous targets. The only exception in this group is the Value Difference Metric that has been specifically designed for symbolic data.

Table 3.1. Summary of the relevance measures suitable for filters. Features X and targets Y may be of the B = binary type (+), M = mutivalued, s (symbolic), or i (integer) only (symbolic implies integer), or C = continuous, real numbers (+). Methods that cannot work directly with continuous values need discretization.

Method		X			Y			Comments
Name	Formula	B	M	C	B	M	C	Comments
Bayesian accuracy	Eq. 3.1	+	s		+	s		Theoretically the golden standard, rescaled Bayesian relevance Eq. 3.2.
Balanced accuracy	Eq. 3.4	+	s		+	s		Average of sensitivity and specificity; used for unbalanced dataset, same as AUC for binary targets.
Bi-normal separation	Eq. 3.5	+	s		+	s		Used in information retrieval.
F-measure	Eq. 3.7	+	s		+	s		Harmonic of recall and precision, popular in information retrieval.
Odds ratio	Eq. 3.6	+	s		+	s		Popular in information retrieval.
Means separation	Eq. 3.10	+	i		+		+	Based on two class means, related to Fisher's criterion.
T-statistics	Eq. 3.11	+	i		+		+	Based also on the means separation.
Pearson correlation	Eq. 3.9	+	i	+		i	+	Linear correlation, significance test Eq. 3.12, or a permutation test.
Group correlation	Eq. 3.13	+	i	+		i	+	Pearson's coefficient for subset of features.
χ^2	Eq. 3.8	+	s		+		+	Results depend on the number of samples m.
Relief	Eq. 3.15	+	s	+	+	s	+	Family of methods, the formula is for a simplified version ReliefX, captures local correlations and feature interactions.
Separability Split Value	Eq. 3.41	+	s		+	s		Decision tree index.
Kolmogorov distance	Eq. 3.16	+	s		+		+	Difference between joint and product probabilities.
Bayesian measure	Eq. 3.16	+	s		+		+	Same as Vajda entropy Eq. 3.23 and Gini Eq. 3.39.
Kullback-Leibler divergence	Eq. 3.20	+	s		+		+	Equivalent to mutual information.
Jeffreys-Matusita distance	Eq. 3.22	+	s		+		+	Rarely used but worth trying.
Value Difference Metric	Eq. 3.22	+	s		+		+	Used for symbolic data in similarity-based methods, and symbolic feature-feature correlations.
Mutual Information	Eq. 3.29	+	s		+		+	Equivalent to information gain Eq. 3.30.
Information Gain Ratio	Eq. 3.32	+	s		+		+	Information gain divided by feature entropy, stable evaluation.
Symmetrical Uncertainty	Eq. 3.35	+	s		+		+	Low bias for multivalued features.
J-measure	Eq. 3.36	+	s		+		+	Measures information provided by a logical rule.
Weight of evidence	Eq. 3.37	+	s		+		+	So far rarely used.
MDL	Eq. 3.38	+	s		+	s		Low bias for multivalued features.

Indices based on information theory may also be used for continuous features and targets if probability densities are defined. Information gain ratio and symmetrical uncertainty coefficient are especially worth recommending, sharing low bias with the MDL approach (Sec. 3.5), and converging in a stable and quick way to their correct values.

3.10 Discussion and Conclusions

Filters provide the cheapest approach to the evaluation of feature relevance. For a very large number of features they are indispensable, and only after filtering out most of the features other, more expansive feature selection methods become feasible.

Many approaches to filters discussed in the preceding sections show that there is no simple answer to the question: which relevance index is the best to construct a good filter? If there is sufficient data and joint probabilities may be estimated in a reliable way there is no reason why Bayesian relevance J_{BC} should not be used. After all other relevance indices, and in particular indices based on the theory of information, are only approximations to the Bayesian relevance. Unfortunately this index seems to be the most difficult to estimate reliably (see Fig. 3.2), leaving room for other approaches. In some applications including costs of different types of misclassifications (Sec. 3.2) is a better choice of relevance index, leading to the balanced accuracy (Eq. 3.4), F-measure or optimization of ROC curves. Evaluation of all such quantities will suffer from the same problem as evaluation of the Bayesian relevance J_{BC}, and therefore other, approximate but more reliable methods should be studied.

Different approaches to relevance evaluation lead to a large number of indices for ranking and selection. Certainly more papers with new versions of relevance indices for information filters will be published, but would they be more useful? As noted in the book on CART (Breiman et al., 1984) the splitting criteria do not seem to have much influence on the quality of decision trees, so in the CART tree an old index known as Bayesian measure J_{BM} (Eq. 3.19) or Vajda Entropy (Eq. 3.23) has been employed, under the new name "Gini". Perhaps the actual choice of feature relevance indices also has little influence on performance of filters. For many applications a simple approach, for example using a correlation coefficient, may be sufficient.

Not all options have been explored so far and many open questions remain. Similarities, and perhaps equivalence up to monotonic transformation of relevance indices, should be established. The reliability of estimation of relevance indices – with the exception of entropy estimations – is not known. Biases towards multi-valued features of several indices have been identified but their influence on ranking is not yet clear. Little effort has been devoted so far towards cost-sensitive feature selection. In this respect the accuracy of

Bayesian classification rules and other performance metrics related to logical rules are worth investigating.

Not much attention has been paid towards specific class-oriented and local, context-dependent filters. Some problems (especially in bioinformatics) require the simultaneous identification of several features that may individually have quite poor relevance. The paradigmatic benchmark problems of this sort are the parity problems, starting from the XOR. Only context-dependent local feature selection methods (like Relief, or filter methods applied to vectors in a localized feature space region) seem to be able to deal with such cases. Although our knowledge of filter-based feature selection has significantly grown in recent years still much remains to be done in this field.

Acknowledgment

I am grateful to Jacek Biesiada for drawing my attention to many papers in this field, to Norbert Jankowski and Krzysztof Grąbczewski for critically reading this paper, and to Isabelle Guyon for many helpful remarks. This work has been supported by the grant from the Polish Committee of Scientific Research 2005-2007.

References

H. Almuallim and T.G. Dietterich. Learning with many irrelevant features. In *Proceedings of the 9th National Conference on Artificial Intelligence (AAAI-91)*, pages 547–552, 1991.

A. Antos, L. Devroye, and L. Gyorfi. An extensive empirical study of feature selection metrics for text classification. *IEEE Transactions on Pattern Analysis and Machine Intelligence*, 21(7):643–645, 1999.

R. Battiti. Using mutual information for selecting features in supervised neural net learning. *IEEE Trans. on Neural Networks*, 5:537–550, 1994.

D.A. Bell and H. Wang. A formalism for relevance and its application in feature subset selection. *Machine Learning*, 41:175–195, 2000.

L. Breiman, J.H. Friedman, R.A. Olshen, and C.J. Stone. *Classification and Regression Trees*. Wadsworth and Brooks, Monterey, CA, 1984.

L. Bruzzone, F. Roli, and S.B. Serpico. An extension of the jeffreys-matusita distance to multiclass cases for feature selection. *IEEE Transactions on Geoscience and Remote Sensing*, 33(6):1318–1321, 1995.

F.M. Coetzee, E. Glover, L. Lawrence, and C.L Giles. Feature selection in web applications by roc inflections and powerset pruning. In *Proceedings of 2001 Symp. on Applications and the Internet (SAINT 2001)*, pages 5–14, Los Alamitos, CA, 2001. IEEE Computer Society.

T.M. Cover. The best two independent measurements are not the two best. *IEEE Transactions on Systems, Man, and Cybernetics*, 4:116–117, 1974.

D.R. Cox and D.V. Hinkley. *Theoretical Statistics*. Chapman and Hall/CRC Press, Berlin, Heidelberg, New York, 1974.

M. Dash and H. Liu. Consistency-based search in feature selection. *Artificial Intelligence*, 151:155–176, 2003.

R.L. de Mantaras. A distance-based attribute selection measure for decision tree induction. *Machine Learning Journal*, 6:81–92, 1991.

L. Devroye, L. Gyrfi, and G. Lugosi. *A Probabilistic Theory of Pattern Recognition*. Springer, Berlin, Heidelberg, New York, 1996.

W. Duch, R. Adamczak, and K. Grąbczewski. A new methodology of extraction, optimization and application of crisp and fuzzy logical rules. *IEEE Transactions on Neural Networks*, 12:277–306, 2001.

W. Duch and L. Itert. A posteriori corrections to classification methods. In L. Rutkowski and J. Kacprzyk, editors, *Neural Networks and Soft Computing*, pages 406–411. Physica Verlag, Springer, Berlin, Heidelberg, New York, 2002.

W. Duch, R. Setiono, and J. Zurada. Computational intelligence methods for understanding of data. *Proceedings of the IEEE*, 92(5):771–805, 2004.

W. Duch, T. Winiarski, J. Biesiada, and A. Kachel. Feature ranking, selection and discretization. In *Proceedings of Int. Conf. on Artificial Neural Networks (ICANN)*, pages 251–254, Istanbul, 2003. Bogazici University Press.

R.O. Duda, P.E. Hart, and D.G. Stork. *Patter Classification*. John Wiley & Sons, New York, 2001.

D. Erdogmus and J.C. Principe. Lower and upper bounds for misclassification probability based on renyis information. *Journal of VLSI Signal Processing Systems*, 37(2-3):305–317, 2004. ISSN 1533-7928.

G. Forman. An extensive empirical study of feature selection metrics for text classification. *Journal of Machine Learning Research*, 3:1289–1305, 2003.

K. Fukunaga. *Introduction to Statistical Pattern Recognition. 2nd ed.* Academic Press, Boston, 1990.

E.E. Ghiselli. *Theory of psychological measurement*. McGrawHill, New York, 1964.

K. Grąbczewski and W. Duch. The separability of split value criterion. In *Proceedings of the 5th Conf. on Neural Networks and Soft Computing*, pages 201–208, Zakopane, Poland, 2000. Polish Neural Network Society.

I. Guyon, H.-M. Bitter, Z. Ahmed, M. Brown, and J. Heller. Multivariate non-linear feature selection with kernel multiplicative updates and gram-schmidt relief. In *BISC FLINT-CIBI 2003 workshop, Berkeley, Dec. 2003*, 2003.

M.A. Hall. *Correlation-based Feature Subset Selection for Machine Learning*. PhD thesis, Department of Computer Science, University of Waikato, Waikato, N.Z., 1999.

D.J. Hand. *Construction and assessment of classification rules*. J. Wiley and Sons, Chichester, 1997.

J.A. Hanley and B.J. McNeil. The meaning and use of the area under a receiver operating characteristic (roc) curve. *Radiology*, 143:29–36, 1982.

R.M. Hogarth. Methods for aggregating opinions. In H. Jungermann and G. de Zeeuw, editors, *Decision Making and Change in Human Affairs*. D. Reidel Publishing, Dordrecht, Holland, 1977.

D. Holste, I. Grosse, and H. Herzel. Bayes' estimators of generalized entropies. *J. Physics A: Math. General*, 31:2551–2566, 1998.

R.C. Holte. Very simple classification rules perform well on most commonly used datasets. *Machine Learning*, 11:63–91, 1993.

S.J. Hong. Use of contextual information for feature ranking and discretization. *IEEE Transactions on Knowledge and Data Engineering*, 9:718–730, 1997.

G.V. Kass. An exploratory technique for investigating large quantities of categorical data. *Applied Statistics*, 29:119–127, 1980.

R. Kohavi and G. John. Wrappers for feature subset selection. *Artificial Intelligence*, 97(1-2):273–324, 1996.

I. Kononenko. On biases in estimating the multivalued attributes. In *Proceedings of IJCAI-95, Montreal*, pages 1034–1040, San Mateo, CA, 1995. Morgan Kaufmann.

N. Kwak and C-H. Choi. Input feature selection by mutual information based on parzen window. *IEEE Transactions on Pattern Analysis and Machine Intelligence*, 24:1667–1671, 2002a.

N. Kwak and C-H. Choi. Input feature selection for classification problems. *IEEE Transactions on Neural Networks*, 13:143–159, 2002b.

M. Li and P. Vitányi. *An Introduction to Kolmogorov Complexity and Its Applications*. Text and Monographs in Computer Science. Springer, Berlin, Heidelberg, New York, 1993.

H. Liu, F. Hussain, C.L. Tan, and M. Dash. Discretization: An enabling technique. *Journal of Data Mining and Knowledge Discovery*, 6(4):393–423, 2002.

H. Liu and R. Setiono. Feature selection and discretization. *IEEE Transactions on Knowledge and Data Engineering*, 9:1–4, 1997.

D. Michie. Personal models of rationality. *J. Statistical Planning and Inference*, 21: 381–399, 1990.

R. Moddemeijer. A statistic to estimate the variance of the histogram based mutual information estimator based on dependent pairs of observations. *Signal Processing*, 75:51–63, 1999.

A.Y. Ng. On feature selection: learning with exponentially many irrelevant features as training examples. In *Proceedings of the 15th International Conference on Machine Learning*, pages 404–412, San Francisco, CA, 1998. Morgan Kaufmann.

W.H. Press, S.A. Teukolsky, W.T. Vetterling, and B.P. Flannery. *Numerical recipes in C. The art of scientific computing*. Cambridge University Press, Cambridge, UK, 1988.

J.R. Quinlan. *C4.5: Programs for Machine Learning*. Morgan Kaufman, San Mateo, CA, 1993.

M. Robnik-Sikonja and I. Kononenko. Theoretical and empirical analysis of relieff and rrelieff. *Machine Learning*, 53:23–69, 2003.

P. Smyth and R.M. Goodman. An information theoretic approach to rule induction from databases. *IEEE Transactions on Knowledge and Data Engineering*, 4: 301–316, 1992.

G.W. Snedecorand and W.G. Cochran. *Statistical Methods, 8th ed.* Iowa State University Press, Berlin, Heidelberg, New York, 1989.

J.A. Swets. Measuring the accuracy of diagnostic systems. *Proceedings of the IEEE*, 240(5):1285–1293, 1988.

R.W. Swiniarski and A. Skowron. Rough set methods in feature selection and recognition. *Pattern Recognition Letters*, 24:833–849, 2003.

K. Torkkola. Feature extraction by non parametric mutual information maximization. *Journal of Machine Learning Research*, 3:1415–1438, 2003. ISSN 1533-7928.

G.T. Toussaint. Note on optimal selection of independent binary-valued features for pattern recognition. *IEEE Transactions on Information Theory*, 17:618–618, 1971.

I. Vajda. *Theory of statistical inference and information*. Kluwer Academic Press, London, 1979.

C.J. van Rijsbergen. *Information Retrieval.* Butterworths, London, 1979.

T.R. Vilmansen. Feature evaluation with measures of probabilistic dependence. *IEEE Transactions on Computers,* 22:381–388, 1973.

D.R. Wilson and T.R. Martinez. Improved heterogeneous distance functions. *Journal of Artificial Intelligence Research,* 6:1–34, 1997.

Chapter 4

Search Strategies

Juha Reunanen

ABB, Web Imaging Systems
P.O. Box 94, 00381 Helsinki, Finland
Juha.Reunanen@iki.fi

4.1 Introduction

In order to make a search for good variable subsets, one has to know which subsets are good and which are not. In other words, an evaluation mechanism for an individual variable subset needs to be defined first.

The evaluation mechanism can be based on any of the more or less simple filter-type criteria discussed in Chapter 3. However, if the evaluation scheme utilizes the particular predictor architecture for which the variables are being selected and that will ultimately be used, then we have a wrapper approach. A common choice for performing the evaluation in a wrapper method is cross-validation (Chapter 2). While the emphasis is on wrappers, the algorithms presented in this chapter can be used also if the actual predictor is not involved in the evaluation.

Once an evaluation method for the variable subsets has been defined, one can start the search for the best subset. A *search strategy* defines the order in which the variable subsets are evaluated. In this chapter, some of the strategies that have been proposed are discussed. The intent is to describe a few well-known algorithms at such a level that implementation should be rather straightforward, and to mention just a few others to widen the scope a little bit. In other words, the purpose is not to give a comprehensive list of all the existing search strategies.

The chapter is organized as follows: The search strategies are covered in Sects. 4.2 to 4.5. Then, Sect. 4.6 discusses the different levels of testing required when using the strategies. Finally, Sect. 4.7 concludes the chapter by tackling the obvious question: which strategy is the best?

4.2 Optimal Results

Traditionally, researchers have been worried about being unable to find an optimal subset, one for which the error estimation procedure, using the data

J. Reunanen: *Search Strategies*, StudFuzz **207**, 119–136 (2006)
www.springerlink.com   © Springer-Verlag Berlin Heidelberg 2006

available, yields a value no larger than for any other subset. In the following, a strategy that is able to find such a subset is called an optimal strategy.

4.2.1 Exhaustive Search

A simple way to find the best subset is to evaluate every possible subset. This approach is called exhaustive search: with n candidate variables, there are $2^n - 1$ subsets to go through. Unfortunately, even a moderate number of variables easily makes it impossible to evaluate all the subsets in a practical amount of time.

If it is known beforehand that exactly d variables out of n should be chosen, then one only needs to go through the subsets of size d. This results in $\binom{n}{d}$ subsets, which is still too much for most values of d.

4.2.2 Branch and Bound

As an exhaustive search is normally not possible, it would be nice to search only a part of the subset space and still find the best subset. Unfortunately, this cannot be guaranteed in general (Cover and van Campenhout, 1977). However, if a certain subset size d is desired and the evaluation function is known to be monotonic,[1] then an optimal strategy potentially running a lot faster than exhaustive search does exist. The algorithm is called branch and bound, and it was proposed for variable selection by Narendra and Fukunaga (1977).

The strategy is based on the fact that once a subset S consisting of more than d variables has been evaluated, we know, thanks to the monotonicity property, that no subset of it can be better. Thus, unless S excels the currently best known subset S' of target size d, the subsets of S need not be evaluated at all, because there is no way the evaluation result for any of them could exceed the score of S'.

The algorithm still has an exponential worst case complexity, which may render the approach infeasible when a large number of candidate variables is available. A remedy compromising the optimality guaranty was suggested already by Narendra and Fukunaga. However, the number of subsets to evaluate can be decreased also without such a compromise (Yu and Yuan, 1993, Somol et al., 2004).

Branch and bound is not very useful in the wrapper model, because the evaluation function is typically not monotonic. While methods like the RBABM (Kudo and Sklansky, 2000) are able to relax the requirement slightly, the problem is that typical predictor architectures evaluated for example using cross-validation provide no guaranties at all regarding the monotonicity.

---

[1]The evaluation function is monotonic if the addition of a variable never makes a subset worse.

4.3 Sequential Selection

Because of the difficulties with optimal strategies, researchers have already for a long time tried to come up with search strategies that find reasonably good variable subsets without going through all of them. During the last years, the strategies have become increasingly complex. This and the next two sections describe some of these "suboptimal" approaches.

4.3.1 Sequential Pruning

Sequential pruning of variables, introduced by Marill and Green (1963), seems to have been the first search method suggested for variable subset selection. Here, the method will be referred to as sequential backward selection (SBS). SBS starts with the variable set that consists of all the candidate variables. During one step of the algorithm, each variable still left in the set is considered to be pruned. The results of the exclusion of each variable are compared to each other using the evaluation function $J(\cdot)$. The step is finished by actually pruning the variable whose removal yields the best results. Steps are taken and variables are pruned until a prespecified number of variables is left, or until the results get too poor.

SBS is summarized in Fig. 4.1. In the notation used, S and S' are arrays of Boolean values representing variable subsets. The binary value S_j indicates whether the jth variable is selected in S. For the sake of clarity, it is assumed here that the evaluation function $J(\cdot)$ carries the dataset and any evaluator parameters with it. The interpretation for the return values of $J(\cdot)$ is such that the smaller the value, the better the subset.

The algorithm shown in Fig. 4.1 goes on until there are no variables left in S. To stop the search earlier, the **while** statement with the guarding expression $k > 1$ could be modified. However, just breaking the execution at any point of time is the same thing and hence the algorithms and especially the control flow expressions will not be further complicated by additional parameters. This applies to most of the algorithms presented in this chapter.

4.3.2 Sequential Growing

Sequential growing of the variable set is similar to SBS but starts with an empty set and proceeds by adding variables (Whitney, 1971). In what follows, the method will be called sequential forward selection (SFS). During one step, each candidate variable that is not yet part of the current set is included into the current set, and the resulting set is evaluated. At the end of the step, the variable whose inclusion resulted in the best evaluation is inserted in the current set. The algorithm proceeds until a prespecified number of variables is selected, or until no improvement of the results is observed anymore. SFS is summarized in Fig. 4.2.

```
function SBS(n, J)              Returns a set of variable subsets of different sizes (B)
begin        n                 J(·) is the function used to evaluate different subsets
  S := (1, ..., 1);            Start with the full set of all n variables
  k := n;
  B := ∅;                      Initialize the set of best variable sets found
  while k > 1                  Repeat for as long as there are branches to compare
    R := ∅;                    Initialize the set of evaluations of different branches
    for each {j | S_j = 1}     Repeat for each possible branch
      S' := S;                 Copy the variable set
      S'_j := 0;               Prune the jth variable
      R(j) := J(S');           Evaluate the branch
    end;
    k := k - 1;
    j := argmin R(·);          Find the best branch
    S_j := 0;                  Take the best branch
    B(k) := S;                 Store the newly found subset
  end;
  return B;
end;
```

Fig. 4.1. Sequential backward selection algorithm

With typical evaluator functions, SFS executes faster than SBS. This is due to the fact that in the beginning of the search, when both algorithms have lots of possible branches to evaluate at each step (up to n), SFS evaluates very small variable sets, which is often faster than evaluating the almost full sets which SBS has to do. It is true that in the end of the search SFS has to evaluate almost full sets while SBS deals with small subsets, but in the end of the search there are so few options left to be considered that the subset sizes in the beginning of the search simply count more.

On the other hand, SFS evaluates the variables in the context of only those variables that are already included in the set. Thus, it may not be able to detect a variable that is beneficial not by itself but only with the information carried by some other variables. While examples of such data can easily be constructed (Guyon and Elisseeff, 2003), empirical evidence does not clearly suggest the superiority of one or the other (Aha and Bankert, 1996, Jain and Zongker, 1997, Kudo and Sklansky, 2000).

```
function SFS(n, J)              Returns a set of variable subsets of different sizes (B)
begin         n
  S := (0, . . . , 0);          Start with an empty set
  k := 0;
  B := ∅;                       Initialize the set of best variable sets found
  while k < n − 1               Repeat for as long as there are branches to compare
    R := ∅;                     Initialize the set of evaluations of different branches
    for each {j | S_j = 0}      Repeat for each possible branch
      S' := S;                  Copy the variable set
      S'_j := 1;                Add the jth variable
      R(j) := J(S');            Evaluate the branch
    end;
    k := k + 1;
    j := argmin R(·);           Find the best branch
    S_j := 1;                   Take the best branch
    B(k) := S;                  Store the newly found subset
  end;
  return B;
end;
```

Fig. 4.2. Sequential forward selection algorithm

4.4 Extensions to Sequential Selection

Over the years, researchers have proposed a plethora of modifications to the basic sequential search strategies. In this section, a few examples of them are discussed.

4.4.1 Generalized Sequential Selection

Simple SFS and SBS can be easily generalized (Kittler, 1978). In GSFS(g) (GSBS(g)), the inclusion (exclusion) of a set of g variables at a time is evaluated. When there are $n - k$ candidate variables left to be included (excluded), this results into $\binom{n-k}{g}$ evaluations in one step of the algorithm. This is typically much more than the plain $n - k$ evaluations in SFS, even if g is only 2. On the other hand, the algorithms do not take as many steps as SFS and SBS, because they select more variables at a time, thus getting to a prespecified number of variables using a smaller number of steps. However, the combinatorial explosion in the complexity of one step outweighs the reduction in the number of steps.

The increased computational cost of GSFS(g) may in some applications be justified by the fact that it is able to perform well in situations where there are up to g variables of which none is any good by itself, but all of them together provide enough information for good predictions. It is easy to give pathological examples where this is the case and the simple algorithms selecting one variable at a time fail (Jain and Zongker, 1997, Guyon and Elisseeff, 2003). However, this does not mean that real-world datasets would often exhibit such phenomena.

4.4.2 Backtracking During Search

The search methods described so far start with the initial variable set and then progress straightforwardly towards the final set. In particular, if a variable is added in SFS or GSFS (or removed in SBS or GSBS), it will never be removed (added) later. This results in what has often been called the *nesting effect* in variable selection literature: bad decisions made in the beginning of the search cannot be corrected later, as the successive variable sets are always nested so that one set is a proper subset of the other.

The nesting effect has been fought by allowing backtracking during the search. In "plus ℓ-take away r" selection (Stearns, 1976), referred here to as PTA(ℓ, r), each step of the algorithm is divided into two substeps. In the first substep, SFS is run to include ℓ new variables. The second substep consists of running SBS to exclude r variables from those that have already been selected. If $\ell > r$, the PTA(ℓ, r) algorithm starts with an empty variable set and the first substep. For $\ell < r$, the initial variable set should include all the candidate variables, and the algorithm is supposed to be started from the second substep. A forward version of PTA(ℓ, r) is given in Fig. 4.3. It is assumed that when a subset is currently stored in the set of the best subsets B, its evaluation score is readily available and needs not be recalculated. Further, it needs to be defined that if $B(k)$ is undefined, then $R(j) < J(B(k))$ evaluates to true.

A straightforward generalization of the PTA(ℓ, r) algorithm just described would be to run GSFS(ℓ) and GSBS(r) instead of SFS and SBS, and this is how Kudo and Sklansky (2000) describe the GPTA(ℓ, r) algorithm. However, Kittler (1978) actually took the generalization a little bit further by running GSFS(ℓ_i) for several integer values of ℓ_i between 0 and ℓ, as well as GSBS(r_i) for several integers r_i between 0 and r. Thus, the steps are split into smaller steps, which reduces computational complexity. If all ℓ_i and r_i are equal to one, then the algorithm is reduced to the nongeneralized PTA(ℓ, r) algorithm.

4.4.3 Beam Search

The strategies discussed so far are *greedy* in the sense that they are only interested in the best subset that can be found amongst the candidates currently being evaluated. However, sometimes there might be several promising branches, and it may feel like a bad idea to restrict oneself to choosing only

function PTA(n, J, ℓ, r) *Returns a set of variable subsets of different sizes (B)*

begin

$\quad S := (\overbrace{0, \ldots, 0}^{n});$ *Start with an empty set*

$\quad k := 0;$

$\quad B := \emptyset;$ *Initialize the set of best variable sets found*

$\quad$ **while** $k < n - \ell$ *Repeat for as long as there are branches to compare*

$\quad\quad t := k + \ell;$ *Add ℓ variables*

$\quad\quad$ **while** $k < t$

$\quad\quad\quad R := \emptyset;$ *Initialize the set of evaluations of different branches*

$\quad\quad\quad$ **for each** $\{j \mid S_j = 0\}$ *Repeat for each possible branch*

$\quad\quad\quad\quad S' := S;$

$\quad\quad\quad\quad S'_j := 1;$ *Add the jth variable*

$\quad\quad\quad\quad R(j) := J(S');$ *Evaluate the branch*

$\quad\quad\quad$ **end**;

$\quad\quad\quad k := k + 1;$

$\quad\quad\quad j := \text{argmin } R(\cdot);$ *Find the best branch*

$\quad\quad\quad S_j := 1;$ *Take the best branch*

$\quad\quad\quad$ **if** $R(j) < J(B(k))$ *Is the new subset the best one of its size so far?*

$\quad\quad\quad\quad B(k) := S;$ *If so, store it*

$\quad\quad\quad$ **end**;

$\quad\quad$ **end**;

$\quad\quad t := k - r;$ *Remove r variables*

$\quad\quad$ **while** $k > t$

$\quad\quad\quad R := \emptyset;$ *Initialize the set of evaluations of different branches*

$\quad\quad\quad$ **for each** $\{j \mid S_j = 1\}$ *Repeat for each possible branch*

$\quad\quad\quad\quad S' := S;$

$\quad\quad\quad\quad S'_j := 0;$ *Prune the jth variable*

$\quad\quad\quad\quad R(j) := J(S');$ *Evaluate the branch*

$\quad\quad\quad$ **end**;

$\quad\quad\quad k := k - 1;$

$\quad\quad\quad j := \text{argmin } R(\cdot);$ *Find the best branch*

$\quad\quad\quad S_j := 0;$ *Take the best branch*

$\quad\quad\quad$ **if** $R(j) < J(B(k))$ *Is the new subset the best one of its size so far?*

$\quad\quad\quad\quad B(k) := S;$ *If so, store it*

$\quad\quad\quad$ **end**;

$\quad\quad$ **end**;

$\quad$ **end**;

$\quad$ **return** $B;$

end;

Fig. 4.3. A forward plus ℓ–take away r search algorithm ($\ell > r$)

---

function FBS(n, J, q)	*Returns a set of variable subsets of different sizes (B)*
begin	
$Q(1) := (\overbrace{0, \ldots, 0}^{n})$;	*Start with just the empty set in the queue*
$B := \emptyset$;	*Initialize the set of best variable sets found*
while $Q \neq \emptyset$	*Repeat until the queue is empty*
$S := Q(1)$;	*Pick the best set in the queue*
remove$(Q, 1)$;	*Remove the set from the queue*
for each $\{j \mid S_j = 0\}$	*Repeat for each possible branch*
$S' := S$;	
$S'_j := 1$;	*Add the jth variable*
$R := J(S')$;	*Evaluate the new subset*
$k := \sum_j S'_j$;	*Get the number of variables selected*
if $R < J(B(k))$	*Is the new subset the best one of its size so far?*
$B(k) := S'$;	*If so, store it*
end;	
$\# := $ length(Q);	*Get the number of subsets in the queue*
if $\# < q$	*Is the queue full?*
insert(Q, S', R);	*No: add the new subset to the queue*
elseif $R < J(Q(\#))$	*Is S' better than the worst subset in the queue?*
remove$(Q, \#)$;	*Yes: Remove the worst subset (the last item),*
insert(Q, S', R);	*and add the new subset*
end;	
end;	
end;	
return B;	
end;	

Fig. 4.4. Forward beam search

one of them. It could be desirable to be able to later return to branches other than the one that happens to be chosen first.

In beam search (Siedlecki and Sklansky, 1988, Aha and Bankert, 1996), a list of the most interesting branches (i.e., subsets) that have not yet been explored is maintained. If the length of the list is restricted to one, then the algorithm is reduced to basic sequential selection. If, however, more subsets can be maintained on the list, then the algorithm can, after finishing the search in one branch, jump back and examine the other branches too.

A forward version of the beam search algorithm is illustrated in Fig. 4.4. Here, the queue function insert is supposed to maintain the order of the queue: the best subset stored is always the first one on the list. It is also assumed that the evaluation scores for the subsets stored in the queue, in addition to those found in the current set of best subsets, are readily available and thus require no recomputation.

4.4.4 Floating Search

The concept of floating search methods was introduced by Pudil et al. (1994). In sequential forward floating selection (SFFS) and sequential backward floating selection (SBFS), the backtracking like that in $\text{PTA}(\ell, r)$ is not limited to ℓ or r variables, but can go on for as long as better results than those found previously can be obtained.

SFFS (SBFS) consists of two different and alternating phases. The first phase is just one step of SFS (SBS). The second phase consists of performing SBS (SFS), whose steps are taken for as long as the obtained variable set is the best one of its size found so far. When this is no longer the case, the first phase takes place again.

In the original version of the algorithm, there was a minor bug, which was pointed out and corrected by Somol et al. (1999). After backtracking for a while, the algorithm goes to the first phase. Now it might be that performing the first phase gives a subset that is worse than a previously found subset of the same size. The flaw here is that the original algorithm follows this less promising search branch, although a better one has already been found. This can be remedied by abruptly changing the current variable set to be the best one of the same size that has been found so far. The corrected SFFS algorithm is given in Fig. 4.5 — however, the original version can be obtained by removing the underlined parts.

Somol et al. also present more sophisticated versions of SFFS and SBFS. They call the approach adaptive floating search (AFS), where the algorithm is allowed to switch to the generalized versions GSFS(g) and GSBS(g) when it is close to the desired number of selected variables. According to their results, the sets selected are moderately better when compared to those chosen by the non-adaptive floating search methods, at the expense of a rather heavily increased computational load.

While the floating search has been found to give results superior to the non-backtracking algorithms (Pudil et al., 1994, Jain and Zongker, 1997, Kudo and Sklansky, 2000 — however, see also Reunanen, 2003), one potential problem in the use of even the non-adaptive version is that there is no way to know in advance for how long the algorithm is going to run. If the dataset is such that no backtracking is done, then the time complexity is comparable to those of SFS and SBS, but the result is no better. If there is lots of backtracking, then more time is taken, but there is a good chance that the result will be better as well. On the other hand, the algorithm can be stopped at any time: because it maintains the set of best variable sets, good results are usually obtained even if the algorithm has not finished yet.

```
function SFFS(n, J)              Returns a set of variable subsets of different sizes (B)
begin         n
    S := (0,...,0);             Start with an empty set
    k := 0;
    B := ∅;                     Initialize the set of best variable sets found
    while k < n                 Repeat until the set of all variables is reached
        R := ∅;                 Initialize the set of evaluations of different branches
        for each {j | S_j = 0}  Repeat for each possible branch
            S' := S;
            S'_j := 1;          Add the jth variable
            R(j) := J(S');      Evaluate the branch
        end;
        k := k + 1;
        j := argmin R(·);       Find the best branch
        if R(j) ≥ J(B(k))       Was this branch the best of its size found so far?
            S := B(k);          If no, abruptly switch to the best one
        else
            S_j := 1;           If yes, take the branch
            B(k) := S;          Store the newly found subset
            t := 1;             This is reset when backtracking is to be stopped
            while k > 2 ∧ t = 1 Backtrack until o better subsets are found
                R := ∅;         Initialize the set of evaluations of different branches
                for each {j | S_j = 1} Repeat for each possible branch
                    S' := S;
                    S'_j := 0;      Prune the jth variable
                    R(j) := J(S');  Evaluate the branch
                end;
                j := argmin R(·);   Find the best branch
                if R(j) < J(B(k-1)) Was a better subset of size k-1 found?
                    k := k - 1;     If yes, backtrack
                    S_j := 0;
                    B(k) := S;      Store the newly found subset
                else
                    t := 0;         If no, stop backtracking
                end;
            end;
        end;
    end;
    return B;
end;
```

Fig. 4.5. Sequential forward floating selection algorithm; the fix by Somol et al. (1999) is pointed out by underlining the lines to be added

4.4.5 Oscillating Search

Another recent method by Somol and Pudil (2000) is that of oscillating search (OS). An oscillating search is initialized by running for example SFS or SFFS to yield a good guess for the best variable subset of the desired size d. The actual OS steps consist of what Somol and Pudil call "down-swings" and "up-swings". During a down-swing, the algorithm searches for subsets that are smaller than d. Likewise, an up-swing step searches for bigger subsets. Thus, the sizes of the sets that are evaluated oscillate around the desired size d. The amplitude of the oscillation is reset to one if an improvement in the results is seen during the swings, and increased otherwise. The motivation behind the oscillation is that previous sequential algorithms spend lots of computing power evaluating sets whose sizes may be far from what is desired. In OS, more time is spent in those parts of the search space where the user really wants to pick the result from.

4.4.6 Compound Operators

With sequential pruning and many irrelevant variables, the search will spend a lot of time discovering the irrelevance of these variables over and over again. Because only one of them can be discarded at a time, and at the next step the exclusion of each variable needs to be considered again, there will be lots of processing. If it were possible to utilize the information that the variables seem to be irrelevant, then one would be able to immediately drop most of such variables.

In the context of backward selection, the compound operator approach (Kohavi and Sommerfield, 1995, Kohavi and John, 1997) works as follows: At a given step, each variable in the current subset is considered for pruning. Without the compound operators, the candidate whose removal gives the best results is excluded, and a new step can be started. In the compound operator method, the removal of that best candidate is combined to the removal of the second best candidate, which yields the first compound operator. More of them can be generated by adding the removal of the next best candidates as well. This can continue as long as the application of these operators does not degrade the results. In Fig. 4.6, the use of compound operators is described in an algorithmic form.

4.5 Stochastic Search

The search algorithms described so far contain no random component in addition to that potentially incurred by the evaluation of a variable set. Given a particular initialization, they should always return the same subsets, provided that the evaluation function evaluates the same variable set repeatedly in the same way. This sounds like a nice property for a search algorithm, but

function SBS-C(n, J) *Returns a set of variable subsets of different sizes (B)*

begin

  $S := (\overbrace{1, \ldots, 1}^{n})$;   *Start with the full set of all n variables*

  $k := n$;

  $B := \emptyset$;   *Initialize the set of best variable sets found*

  while $k > 1$ *Repeat for as long as there are branches to compare*

    $R := \emptyset$;   *Initialize the set of evaluations of different branches*

    for each $\{j \mid S_j = 1\}$ *Repeat for each possible branch*

      $S' := S$;   *Copy the variable set*

      $S'_j := 0$;   *Prune the jth variable*

      $R(j) := J(S')$;   *Evaluate the branch*

    end;

    $c := 0$;   *Initialize the compound operator index*

    while $c < \infty$ *Continue until no improvement, or no variables left*

      $j := \arg\min R(\cdot)$;   *Find the best remaining operator*

      $R(j) := \infty$;   *Make the operator unavailable for further use*

      $S' := S$;

      $S'_j := 0$;   *Apply the operator*

      if $c > 0$ *Was this the basic, non-compound operation?*

        if $J(S') > J(S)$ *If not, then check the benefits*

          $c := \infty$;   *If the results degraded, then stop*

        end;

      end;

      if $c < \infty$ *Should the operator be applied?*

        $k := k - 1$;   *If yes, then apply*

        $S := S'$;

        $B(k) := S$;   *Store the new subset*

        $c := c + 1$;   *Update the compound operator index*

        if $k < 2$ *Check that there are still variables left*

          $c := \infty$;   *If not, then stop*

        end;

      end;

    end;

  end;

  return B;

end;

Fig. 4.6. Sequential backward selection using compound operators

unfortunately the variable set that is returned can be very sensitive to the particular dataset, and it can substantially change even if only one sample is excluded. Hence, the determinism of these algorithms may not be such an important property in many applications.

Several researchers have used stochastic optimization algorithms for variable selection. This section starts by describing two important approaches that are to some extent similar, yet fundamentally different: simulated annealing, and genetic algorithms. They were first suggested for variable selection by Siedlecki and Sklansky (1988). Finally, the section is concluded by Sect. 4.5.3, which shortly describes a recent randomized method more tailor-made for variable selection.

4.5.1 Simulated Annealing

Simulated annealing (SA) is a search method that is based on how physical matter cools down and freezes, ending up in a crystal structure that minimizes the energy of the body. Kirkpatrick et al. (1983) pointed out the analogy between this minimization process of the nature and the general search for the minimum of an abstract system, such as a function.

In SA, the search starts with an initial, potentially random, variable subset in a high "temperature". At each step, a small random change is introduced to the subset. If the change results in a better subset, it is accepted. If the result is worse than the previous solution, the change is accepted with a probability that depends on the temperature: in a high temperature, an adverse change is more likely to be accepted than in a low temperature. As steps are completed, the temperature is declined every now and then — more often when no improvements can be found. Thus, the search will not get stuck in a local optimum in the beginning of the search process when the temperature is high. Still, it is able to find the exact local optimum in the end of the search, because of the low temperature that does not allow deteriorating changes to be made anymore.

A simple example of simulated annealing is given in Fig. 4.7. It could be extended by allowing larger than one-bit changes, by changing how the temperature decreases, by changing how the temperature and a change in the cost function affect the acceptance probability, or by having bookkeeping for the best subset of each size that is found, instead of just the current set S.

4.5.2 Genetic Algorithms

Genetic algorithms (GAs) constitute another family of stochastic optimization algorithms (for a comprehensive introduction, see, e.g., Michalewicz, 1992). While SA is inspired by physics, the motivation for GAs comes from biological evolution, where the best individuals have a higher probability of survival. An algorithmic difference between SA and GAs is that while SA only keeps

```
function SA(n, J, T₀, T₁, m, v)  Returns the best variable subset found (S)
begin
    k := rand-int(1, n);          Get a random subset size k ∈ [1, n] to start with
    S := rand-subset(k);          Generate a random variable subset of size k
    T := T₀;                      Initialize the temperature
    while T ≥ T₁                   Repeat until cold enough
        i := 0;                   Initialize the counter for non-beneficial evaluations
        while i < m               Stay in the current temperature for a while
            S' := S;              Make a copy of the current subset
            j := rand-int(1, n);  Get a random variable index j ∈ [1, n]
            S'ⱼ := 1 − S'ⱼ;        Flip the jth bit
            ΔJ := J(S') − J(S);   Compute the change in cost function
            if ΔJ < 0             Was the change beneficial?
                S := S';          If yes, move there
            else                  If not, . . .
                r := rand-real(0, 1);  Get a random real value r ∈ [0, 1]
                if r < exp(−ΔJ/T)      Depending on the temperature and the randomness,
                    S := S';           consider changing to the new set anyway
            end;
            i := i + 1;           No improvement: increase the counter
        end;
    end;
    T := T × v;                   Decrease the temperature: v ∈]0, 1[
    end;
    return S;
end;
```

Fig. 4.7. An example of variable selection based on simulated annealing

one variable subset in memory, GAs maintain a set of them. In the GA terminology, the solution vectors are usually called chromosomes and a set of chromosomes is called a population. The biological vocabulary is further exploited by defining genetic operations like mutation or crossover. In mutation, a random bit or several bits of the chromosome are flipped to yield a new chromosome. In crossover, an offspring of two chromosomes is obtained by cutting both at some random position and swapping the tails. A new population is typically formed by retaining some of the chromosomes in the old population and composing new chromosomes by applying genetic operations on the old chromosomes. The better a chromosome is, the higher is its probability of being selected to the new population, or as a parent in a genetic operation.

Siedlecki and Sklansky (1989) reported the first results when compared to "classical" methods like many of those reviewed in Sects. 4.3 and 4.4. In their comparison article, Kudo and Sklansky (2000) explicitly recommend that GAs should be used for large-scale problems with more than fifty candidate variables. They also describe a practical implementation of GAs for variable selection.

4.5.3 Rapid Randomized Pruning

While SA and GAs are known as general-purpose strategies for hard optimization problems, randomized methods tailored more specifically for variable selection have also been suggested. The idea of rapidly removing variables that seem to be irrelevant was discussed in the context of compound operators in Sect. 4.4.6. A recent approach by Stracuzzi and Utgoff (2004) is based on a similar idea, but seeks further performance benefits by adding a random component into the process.

Concentrating on problems where the proportion of relevant variables is small, Stracuzzi and Utgoff compute the probability that an important variable is included in a randomly selected subset of size k that is considered for immediate pruning. The value of k is chosen to enable at once the removal of several variables that all still have a high probability of being irrelevant. If the error estimate goes up as a consequence of this exclusion, then we suppose that one or more of the pruned variables was indeed relevant. Thus, we cancel the removal, and pick a new random subset. This can go on until too many consecutive trials fail, which we take as a sign of all the remaining candidates being actually relevant with respect to the prediction task.

4.6 On the Different Levels of Testing

In a variable selection process, essentially two major types of accuracy testing are involved: the tests performed to guide the search, and those carried out to obtain estimates for the final performance of the selected subsets. Tests of the former kind will be called validation in what follows, whereas the latter case shall be called just testing. The failure to make the distinction between these two may lead to making invalid conclusions based on the results of the tests.

4.6.1 Do Not Use the Guiding Validation Results After the Search

The direction of the search is in all the discussed algorithms determined by the scores output by the evaluation function $J(\cdot)$. The assumption underlying the whole construction is that these scores are correlated with the ultimate accuracy that the final predictor will be able to attain, using the corresponding subsets. With filter-type criteria, this correlation may sometimes be question-able, as the specific predictor architecture is not at all used for the evaluation.

On the other hand, even the accuracy on a separate data set as in cross-validation has a non-zero variance. When a myriad of such validation results are compared to each other, the best score is no longer a useful estimate for the accuracy of the winner model. Jensen and Cohen (2000) have given a highly illustrative example of this perhaps counter-intuitive fact. In variable selection, using the best validation score as an estimate for the true expected accuracy is bound to give results that are often just wrong (Reunanen, 2004). *Overfitting* was discussed in Chap. 2 — this is the same, but on a higher level: even the validation scores overfit, when validation is done extensively enough.

To obtain useful estimates of the expected final performance, one therefore needs to have, in addition to the validation done during the search, also some testing afterwards. This will lead to holding some of the data completely out during the search and using it to test the subset(s) selected when the search has completed, or even to an outer loop of cross-validation if execution time is of no essence.

4.6.2 Guidance for the Search Without Validation

On the other hand, it is also possible to omit the validation completely during the selection process, and to use some more straightforward and quicker-to-compute estimate for guiding the search. Rapid guidance is enabled, for in-stance, by the filter-type criteria such as those discussed in Chapter 3. How-ever, it is also possible to take the ultimate predictor architecture into account: for instance, Ng (1998) has proposed using the *training error* of the predictor as the guide for the search.

Once several subsets — typically of different sizes — are thus selected, they can all be tested utilizing the held-out test data, allowing one to choose the best-performing subset. Of course, then the test score of the winner should not be used as an unbiased estimate of its accuracy anymore, because that very score was used to select the subset in the first place.

4.7 The Best Strategy?

A practitioner does not necessarily have to know all the algorithms reviewed in this chapter. Rather, he or she would just like to know which algorithm is the best, and to use that method only. To serve this purpose, empirical comparisons between the different algorithms have been published (Jain and Zongker, 1997, Kudo and Sklansky, 2000). They show clear results that the floating search methods (SFFS and SBFS, Sect. 4.4.4) are able to outperform the straightforward sequential methods (SFS and SBS, Sect. 4.3) in almost any problem. Moreover, Kudo and Sklansky recommend the use of genetic algorithms (Sect. 4.5.2) for any large-scale selection problem with more than 50 variables.

However, it seems that some of such comparisons, for instance the one by Kudo and Sklansky (2000), have used as the final scores to be compared the same estimates (for example due to cross-validation) that were used to guide the search. Unfortunately, such scores should not be used to assess the final performance, as was pointed out in Sect. 4.6. A search algorithm evaluating a large number of subsets will usually find sets that seem to be better than those found by a method that terminates quickly. Alas, it may turn out that these seemingly better subsets are not at all better when making predictions for new, previously unseen samples (Reunanen, 2003).

The discrepancy between the evaluation score and the ultimate performance makes a desperate search for the "optimal" subset (Sect. 4.2) somewhat questionable (Reunanen, 2004), if not even a little old-fashioned. In the future, approaches like those described in Sects. 4.4.6 and 4.5.3 that seek to evaluate a minimal number of subsets may be the ones to look for, especially when the number of candidate variables to try is very high.

Unfortunately, not many recommendations with a valid experimental foundation for the selection of the subset search algorithm seem to exist currently. The practitioner is urged to try a couple of strategies with different complexities; the selection of the methods to try can be based on the size and nature of the datasets at hand. The important thing is that the algorithms are compared properly, which includes testing with data that is not used during the search process.

References

D. W. Aha and R. L. Bankert. A comparative evaluation of sequential feature selection algorithms. In D. Fisher and J.-H. Lenz, editors, *Artificial Intelligence and Statistics V*, pages 199–206. Springer-Verlag, 1996.

T. M. Cover and J. M. van Campenhout. On the possible orderings in the measurement selection problem. *IEEE Transactions on Systems, Man and Cybernetics*, 7(9):657–661, 1977.

I. Guyon and A. Elisseeff. An introduction to variable and feature selection. *Journal of Machine Learning Research*, 3:1157–1182, 2003.

A. K. Jain and D. Zongker. Feature selection: Evaluation, application, and small sample performance. *IEEE Transactions on Pattern Analysis and Machine Intelligence*, 19(2):153–158, 1997.

D. D. Jensen and P. R. Cohen. Multiple comparisons in induction algorithms. *Machine Learning*, 38(3):309–338, 2000.

S. Kirkpatrick, C. D. Gelatt, and M. P. Vecchi. Optimization by simulated annealing. *Science*, 220(4598):671–680, 1983.

J. Kittler. Feature set search algorithms. In C. H. Chen, editor, *Pattern Recognition and Signal Processing*, pages 41–60. Sijthoff & Noordhoff, 1978.

R. Kohavi and G. John. Wrappers for feature subset selection. *Artificial Intelligence*, 97(1–2):273–324, 1997.

R. Kohavi and D. Sommerfield. Feature subset selection using the wrapper model: Overfitting and dynamic search space topology. In *Proc. of the 1st Int. Conf. on Knowledge Discovery and Data Mining (KDD-95)*, pages 192–197, Montreal, Canada, 1995.

M. Kudo and J. Sklansky. Comparison of algorithms that select features for pattern classifiers. *Pattern Recognition*, 33(1):25–41, 2000.

T. Marill and D. M. Green. On the effectiveness of receptors in recognition systems. *IEEE Transactions on Information Theory*, 9(1):11–17, 1963.

Z. Michalewicz. *Genetic Algorithms + Data Structures = Evolution Programs*. Springer-Verlag, 1992.

P. M. Narendra and K. Fukunaga. A branch and bound algorithm for feature subset selection. *IEEE Transactions on Computers*, 26(9):917–922, 1977.

A. Y. Ng. On feature selection: Learning with exponentially many irrelevant features as training examples. In *Proc. of the 15th Int. Conf. on Machine Learning (ICML-98)*, pages 404–412, Madison, WI, USA, 1998.

P. Pudil, J. Novovičová, and J. Kittler. Floating search methods in feature selection. *Pattern Recognition Letters*, 15(11):1119–1125, 1994.

J. Reunanen. Overfitting in making comparisons between variable selection methods. *Journal of Machine Learning Research*, 3:1371–1382, 2003.

J. Reunanen. A pitfall in determining the optimal feature subset size. In *Proc. of the 4th Int. Workshop on Pattern Recognition in Information Systems (PRIS 2004)*, pages 176–185, Porto, Portugal, 2004.

W. Siedlecki and J. Sklansky. On automatic feature selection. *International Journal of Pattern Recognition and Artificial Intelligence*, 2(2):197–220, 1988.

W. Siedlecki and J. Sklansky. A note on genetic algorithms for large-scale feature selection. *Pattern Recognition Letters*, 10(5):335–347, 1989.

P. Somol and P. Pudil. Oscillating search algorithms for feature selection. In *Proc. of the 15th Int. Conf. on Pattern Recognition (ICPR'2000)*, pages 406–409, Barcelona, Spain, 2000.

P. Somol, P. Pudil, and J. Kittler. Fast branch & bound algorithms for optimal feature selection. *IEEE Transactions on Pattern Analysis and Machine Intelligence*, 26(7):900–921, 2004.

P. Somol, P. Pudil, J. Novovičová, and P. Paclík. Adaptive floating search methods in feature selection. *Pattern Recognition Letters*, 20(11–13):1157–1163, 1999.

S. D. Stearns. On selecting features for pattern classifiers. In *Proc. of the 3rd Int. Joint Conf. on Pattern Recognition*, pages 71–75, Coronado, CA, USA, 1976.

D. J. Stracuzzi and P. E. Utgoff. Randomized variable elimination. *Journal of Machine Learning Research*, 5:1331–1362, 2004.

A. W. Whitney. A direct method of nonparametric measurement selection. *IEEE Transactions on Computers*, 20(9):1100–1103, 1971.

B. Yu and B. Yuan. A more efficient branch and bound algorithm for feature selection. *Pattern Recognition*, 6(26):883–889, 1993.

Chapter 5

Embedded Methods

Thomas Navin Lal[1], Olivier Chapelle[1], Jason Weston[2], and André Elisseeff[3]

[1] Max Planck Institute for Biological Cybernetics, Tübingen, Germany
{navin.lal, olivier.chapelle}@tuebingen.mpg.de
[2] NEC Research, Princeton, U.S.A. jasonw@nec-labs.com
[3] IBM Research, Zürich, Switzerland ael@zurich.ibm.com

5.1 Introduction

Although many embedded feature selection methods have been introduced during the last few years, a unifying theoretical framework has not been developed to date. We start this chapter by defining such a framework which we think is general enough to cover many embedded methods. We will then discuss embedded methods based on *how* they solve the feature selection problem.

Embedded methods differ from other feature selection methods in the way feature selection and learning interact. Filter methods do not incorporate learning. Wrapper methods use a learning machine to measure the quality of subsets of features without incorporating knowledge about the specific structure of the classification or regression function, and can therefore be combined with any learning machine. In contrast to filter and wrapper approaches embedded methods do not separate the learning from the feature selection part — the structure of the class of functions under consideration plays a crucial role. For example, Weston et al. (2000) measure the importance of a feature using a bound that is valid for Support Vector Machines only (Section 5.3.1) - thus it is not possible to use this method with, for example, decision trees.

Feature selection can be understood as finding the feature subset of a certain size that leads to the largest possible generalization or equivalently to minimal risk. Every subset of features is modeled by a vector $\boldsymbol{\sigma} \in \{0,1\}^n$ of indicator variables, $\sigma_i := 1$ indicating that a feature is present in a subset and $\sigma_i := 0$ indicating that that feature is absent ($i = 1, \cdots, n$). Given a parameterized family of classification or regression functions[4] $f : \Lambda \times \mathbb{R}^n \to \mathbb{R}, (\boldsymbol{\alpha}, \mathbf{x}) \mapsto f(\boldsymbol{\alpha}, \mathbf{x})$ we try to find a vector of indicator variables $\boldsymbol{\sigma}^* \in \{0,1\}^n$ and an $\boldsymbol{\alpha}^* \in \Lambda$ that minimize the expected risk

---

[4]For example, in the case of a linear Support Vector Machine, the vector $\boldsymbol{\alpha}$ codes the weight vector $\mathbf{w} \in \mathbb{R}^n$ and the offset $b \in \mathbb{R}$ of the hyperplane.

T.N. Lal et al.: *Embedded Methods*, StudFuzz **207**, 137–165 (2006)
www.springerlink.com

$$R(\boldsymbol{\alpha}, \boldsymbol{\sigma}) = \int L[f(\boldsymbol{\alpha}, \boldsymbol{\sigma} \odot \mathbf{x}), y] \, dP(\mathbf{x}, y), \tag{5.1}$$

where $\odot$ denotes the entrywise product (Hadamard product), L is a loss function and P is a measure on the domain of the training data (X, Y). In some cases we will also have the additional constraint $s(\boldsymbol{\sigma}) \leq \sigma_0$, where $s : [0, 1]^n \to \mathbb{R}^+$ measures the sparsity of a given indicator variable $\boldsymbol{\sigma}$. For example, s could be defined to be $s(\boldsymbol{\sigma}) := l_0(\boldsymbol{\sigma}) \leq \sigma_0$, that is to bound the zero "norm" $l_0(\boldsymbol{\sigma})$ - which counts the number of nonzero entries in $\boldsymbol{\sigma}$ - from above by some number σ_0.[5]

The wrapper approach to approximate minimizers of (5.1) can be formulated in the following way

$$\min_{\boldsymbol{\sigma} \in \{0,1\}^n} G(f^*, \boldsymbol{\sigma}, X, Y) \quad \text{s.t.} \quad \begin{cases} s(\boldsymbol{\sigma}) \leq \sigma_0 \\ f^* = \widetilde{T}(\mathcal{F}, \boldsymbol{\sigma}, X, Y), \end{cases} \tag{5.2}$$

where $\mathcal{F} \subset \mathbb{R}^{\mathbb{R}^n}$ denotes the family of classification or regression functions. Given such a family $\mathcal{F}$, a fixed indicator vector $\boldsymbol{\sigma}$ and the training data (X, Y) the output of the function $\widetilde{T}$ is a classifying or regression function f^* trained on the data X using the feature subset defined by $\boldsymbol{\sigma}$. The function G measures the performance of a trained classifier $f^*(\boldsymbol{\sigma})$ on the training data for a given $\boldsymbol{\sigma}$. It is very important to understand that — although we write $G(f^*, \cdot, \cdot, \cdot)$ to denote that G depends on the classifying or regression function f^* — the function G does not depend on the structure of f^*; G can only access f^* as a black box, for example in a cross-validation scheme. Moreover, G does not depend on the specific learner $\widetilde{T}$. In other words $\widetilde{T}$ could be any off-the-shelf classification algorithm and G guides the search through the space of feature subsets.[6]

If we allow G to depend on the learner $\widetilde{T}$ and on parameters of f^* we get the following formulation:

$$\min_{\boldsymbol{\sigma} \in \{0,1\}^n} G(\boldsymbol{\alpha}^*, \widetilde{T}, \boldsymbol{\sigma}, X, Y) \quad \text{s.t.} \quad \begin{cases} s(\boldsymbol{\sigma}) \leq \sigma_0 \\ \boldsymbol{\alpha}^* = \widetilde{T}(\boldsymbol{\sigma}, X, Y). \end{cases} \tag{5.3}$$

To emphasize that G can access the structure of the classifying or regression function we use the notation $f(\alpha^*, \cdot)$ instead of f^*. In this formulation the function G can use information about the learner and the class of functions on which it operates. Thus, G could evaluate, e.g., a bound on the expected

---

[5]Please note that $l_0(\cdot)$ is not a norm, since $l_0(c\boldsymbol{\sigma}) \neq |c| \, l_0(\boldsymbol{\sigma})$ for $\boldsymbol{\sigma} \neq 0, |c| \notin \{0, 1\}$.

[6]A function is no more than a complete collection of all input-output pairs. Thus one could argue that having f^* as a black box is equivalent to having f^* as an analytical expression. That is true in theory, but for many applications it does not hold true, for example for reasons of feasibility.

risk valid for the specific choice of $\tilde{T}$ and the classifying function $f(\alpha^*, \cdot)$ (see Section 5.3.1). To simplify the notation we will also write $G(\alpha^*, \sigma, X, Y)$ whenever $\tilde{T}$ and $\mathcal{F}(\Lambda)$ are defined in the context. We define methods of type (5.3) as embedded methods.

Some embedded methods do not make use of a model selection criterion to evaluate a specific subset of features. Instead, they directly use the learner $\tilde{T}$. Assuming that many learning methods $\tilde{T}$ can be formulated as an optimization problem[7]:

$$\alpha^* = \operatorname*{argmin}_{\alpha \in \Lambda} T(\alpha, \sigma, X, Y) = \tilde{T}(\sigma, X, Y)$$

we can rewrite the minimization problem (5.3) for the special case of $G = T$ as

$$\min_{\alpha \in \Lambda, \sigma \in \{0,1\}^n} T(\alpha, \sigma, X, Y) \quad \text{s.t.} \quad s(\sigma) \leq \sigma_0. \tag{5.4}$$

Unfortunately, both minimization problems (5.3) and (5.4) are hard to solve. Existing embedded methods approximate solutions of the minimization problem. In this chapter we discuss embedded methods according to *how* they solve problem (5.3) or (5.4):

1. Methods that iteratively add or remove features from the data to greedily approximate a solution of minimization problem (5.3) or (5.4) are discussed in Section 5.2.
2. Methods of the second type relax the integrality restriction of $\sigma \in \{0,1\}^n$ and minimize G over the compact set $[0,1]^n$. In this case we refer to $\sigma \in [0,1]^n$ as scaling factors instead of indicator variables (Chapelle, 2002). Section 5.3 is devoted to these methods.
3. If T and s are convex functions and if we assume $\sigma \in [0,1]^n$, problem (5.4) can be converted into a problem of the form

$$\min_{\alpha \in \Lambda, \sigma \in [0,1]^n} T(\alpha, \sigma, X, Y) + \lambda s(\sigma). \tag{5.5}$$

More specifically, let T and s be strictly convex functions and let (α^*, σ^*) be a solution of problem (5.4) for a given $\sigma_0 > 0$. Then there exists a unique $\lambda > 0$ such that (α^*, σ^*) solves (5.5). Furthermore if (α^*, σ^*) solves (5.5) for a given $\lambda > 0$ then there exists one and only one σ_0 such that (α^*, σ^*) solves (5.4). The focus of Section 5.4 is on methods that can be formulated as a minimization problem of type (5.5).

5.2 Forward-Backward Methods

In this section we discuss methods that iteratively add or remove features from the data to greedily approximate a solution of the minimization problem

---

[7]For better readability, we assume that the function has one and only one minimum. However, this is not the case for all algorithms.

(5.3). These methods can be grouped into three categories. *Forward selection methods:* these methods start with one or a few features selected according to a method-specific selection criteria. More features are iteratively added until a stopping criterion is met. *Backward elimination methods:* methods of this type start with all features and iteratively remove one feature or bunches of features. *Nested methods:* during an iteration features can be added as well as removed from the data.

5.2.1 Sensitivity of the Output

Several methods that rank features according to their influence on the regression or discriminant function f have been proposed. These methods make use of a variety of criteria. For instance, one can measure the sensitivity of the output on the point $\mathbf{x}_k$ with respect to the i-th feature by[8]

$$t_{i,k} := \left.\frac{\partial f}{\partial x^i}\right|_{\mathbf{x}=\mathbf{x}_k}, \tag{5.6}$$

and the final criterion for that feature could be the ℓ_1 or ℓ_2 norm of the vector $\mathbf{t}_i$.

The underlying assumption of methods based on derivatives of the classifying function is the following: if changing the value of a feature does not result in a major change of the regression or classification function, the feature is not important. This approach can be used for a wide range of regression or classification functions. For example, the squared derivative of a linear model $f_{w,b}(x) = b + \mathbf{x} \cdot \mathbf{w}$ w.r.t. feature i yields $t_{i,k} = w_i^2$, $\forall k$. This quantity is used as a ranking criterion in Recursive Feature Elimination (RFE) (Guyon et al., 2002). The sensitivity of the output has also been used for neural networks. A good introduction can be found in Leray and Gallinari (1999). The authors also discuss methods that include second derivatives of the regression function (second order methods) as well as methods based on the regression function itself (zero-order methods). The sensitivity criterion can be used by methods of all three categories: forward, backward and nested.

In the following paragraph we develop a link between the minimization problem (5.4) of embedded methods and methods that use the sensitivity of the classifying function as a measure for feature relevance. The optimization of (5.4) is done over Λ and $\{0,1\}^n$. Since this minimization problem is difficult to solve, many methods fix $\boldsymbol{\sigma}$ and optimize over Λ, fix $\boldsymbol{\alpha}$ and optimize over Σ in an alternating fashion. In many cases, like RFE and Optimal Brain Damage, the optimization over Λ is a standard learning method, like SVM (RFE) or neural networks (OBD). Oftentimes, the full calculation of $\min_{\boldsymbol{\sigma}\in\{0,1\}^n} G(\boldsymbol{\sigma},\boldsymbol{\alpha})$ (for a fixed $\boldsymbol{\alpha}$) is too expensive since it might involve setting one σ_i at a turn to zero, training the learner and evaluating the result.

---

[8]Notation: $\mathbf{x}_k$ denotes the kth training point, $x_{k,i}$ the ith feature of training point $\mathbf{x}_k$, and x^i denotes the ith feature of training point $\mathbf{x}$.

Instead, methods that are based on the sensitivity of the classifying (or regression) function or weight based methods (Section 5.2.3) use the gradient of G with respect to an indicator variable to approximate the expensive calculation of the minimum.

Suppose that we want to learn a parameterized function $f(\boldsymbol{\alpha}, \mathbf{x})$. The vector $\boldsymbol{\alpha}$ can, for example, be the weights of a neural network. We would like to find the weight vector $\boldsymbol{\alpha}$ which minimizes a regularized functional

$$T(\boldsymbol{\alpha}) = \sum_{k=1}^{m} L(f(\boldsymbol{\alpha}, \mathbf{x}_k), y_k) + \Omega(\boldsymbol{\alpha}). \tag{5.7}$$

Here L is a loss function and Ω is a regularization term such as $\lambda ||\boldsymbol{\alpha}||^2$.

Let us introduce a scaling factor $\sigma_i \in [0, 1]$ for every input component i. The functional T now also depends on the vector of scaling factors $\boldsymbol{\sigma}$ and we write it as $T(\boldsymbol{\alpha}, \boldsymbol{\sigma})$. Let G be defined as

$$G(\boldsymbol{\sigma}) = \min_{\boldsymbol{\alpha}} T(\boldsymbol{\alpha}, \boldsymbol{\sigma}) = \min_{\boldsymbol{\alpha}} \sum_{k=1}^{m} L(f(\boldsymbol{\alpha}, \boldsymbol{\sigma} \odot \mathbf{x}_k), y_k) + \Omega(\boldsymbol{\alpha}).$$

The removal of feature p corresponds to changing the p-th component of the scaling vector $\boldsymbol{\sigma} := (1, \cdots, 1)$ to 0. We will refer to this vector as $\mathbf{1}_0(p)$. A greedy backward method would start with all features, i.e. $\boldsymbol{\sigma} = \mathbf{1}$. As said earlier, for every $p \in \{1, ..., n\}$ it would have to calculate $G(\mathbf{1}_0(p))$, remove the feature p that minimizes this value and continue with the remaining features. However, the value $G(\mathbf{1}_0(p))$ can be approximated by the value of $G(\mathbf{1})$ and the gradient of T at the minimal point:

$$G(\mathbf{1}_0(p)) \approx G(\mathbf{1}) - \left.\frac{\partial G(\boldsymbol{\sigma})}{\partial \sigma_p}\right|_{\boldsymbol{\sigma}=\mathbf{1}} \qquad (p = 1, ..., n). \tag{5.8}$$

Let us write $G(\boldsymbol{\sigma}) = T(\boldsymbol{\alpha}^*(\boldsymbol{\sigma}), \boldsymbol{\sigma})$, where $\boldsymbol{\alpha}^*(\boldsymbol{\sigma})$ denotes the parameter vector which minimizes T (for a given $\boldsymbol{\sigma}$). Then,

$$\frac{\partial G(\boldsymbol{\sigma})}{\partial \sigma_p} = \sum_j \frac{\partial \alpha_j^*(\boldsymbol{\sigma})}{\partial \sigma_p} \overbrace{\left.\frac{\partial T(\boldsymbol{\alpha}, \boldsymbol{\sigma})}{\partial \alpha_j}\right|_{\boldsymbol{\alpha}=\boldsymbol{\alpha}^*(\boldsymbol{\sigma})}}^{= 0} + \left.\frac{\partial T(\boldsymbol{\alpha}, \boldsymbol{\sigma})}{\partial \sigma_p}\right|_{\boldsymbol{\alpha}=\boldsymbol{\alpha}^*(\boldsymbol{\sigma})} \tag{5.9}$$

$(p = 1, ..., n)$ and with equation (5.8) we obtain

$$G(\mathbf{1}_0(p)) \approx G(\mathbf{1}) - \left.\frac{\partial T(\boldsymbol{\alpha}, \boldsymbol{\sigma})}{\partial \sigma_p}\right|_{(\boldsymbol{\alpha}, \boldsymbol{\sigma})=(\boldsymbol{\alpha}^*(\mathbf{1}), \mathbf{1})} \qquad (p = 1, ..., n). \tag{5.10}$$

For every $p \in \{1, ..., n\}$ the gradient of T can be expressed as

$$\left[\frac{\partial T(\boldsymbol{\alpha}, \boldsymbol{\sigma})}{\partial \sigma_p}\bigg|_{\sigma=1}\right]^2 \overset{\text{(eq.5.7)}}{=} \left[\sum_{k=1}^{m} L'(f(\boldsymbol{\alpha}, \mathbf{x}_k), y_k) \, \mathbf{x}_{k,p} \, \frac{\partial f}{\partial x^p}(\boldsymbol{\alpha}, \mathbf{x}_k)\right]^2$$

$$\overset{\text{(Cauchy-Schwarz)}}{\leq} \sum_{k=1}^{m} [L'(f(\boldsymbol{\alpha}, \mathbf{x}_k), y_k) \, \mathbf{x}_{k,p}]^2 \sum_{k=1}^{m} \left[\frac{\partial f}{\partial x^p}(\boldsymbol{\alpha}, \mathbf{x}_k)\right]^2$$

$$\leq \max_{k=1,\dots,m} L'(f(\boldsymbol{\alpha}, \mathbf{x}_k), y_k)^2$$

$$\cdot \sum_{k=1}^{m}(\mathbf{x}_{k,p})^2 \sum_{k=1}^{m} \left[\frac{\partial f}{\partial x^p}(\boldsymbol{\alpha}, \mathbf{x}_k)\right]^2$$

$$\overset{(*)}{=} C \sum_{k=1}^{m} \left[\frac{\partial f}{\partial x^p}(\boldsymbol{\alpha}, \mathbf{x}_k)\right]^2 \propto ||\mathbf{t}_p||_2^2,$$

where $L'(\cdot, \cdot)$ denotes the derivative of L with respect to its first variable, C is a constant independent of p and $\mathbf{t}_p$ has been defined in (5.6). For equality $(*)$ to hold, we assume that each input has zero mean and unit variance. This inequality and the approximation (5.10) link the definition (5.4) to methods based on the sensitivity of the output: The removal of a feature does not change the value of the objective G significantly if the classifying function f is not sensitive to that feature.

5.2.2 Forward Selection

In this section we describe the forward selection methods *Gram-Schmidt Orthogonalization*, *decision trees* and the *Grafting*-method. The three approaches iteratively increase the number of features to construct a target function.

Forward Selection with Least Squares

If S is a set of features, let X_S denotes the submatrix of the design matrix where only the features in S are included: $X_S := (\mathbf{x}_{k,p})_{k \in 1,\dots,m, i \in S}$. When doing standard regression with the subset S, the residuals on the training points are given by

$$P_S Y \quad \text{with} \quad P_S := I - X_S^\top (X_S X_S^\top)^{-1} X_S. \tag{5.11}$$

Note that P_S is a projection matrix, i.e. $P_S^2 = P_S$.

The classical forward selection for least square finds greedily the components which minimize the residual errors:

1. Start with $Y = (y_1 \dots y_m)^\top$ and $S = \emptyset$.
2. Find the component i such that $||P_{\{i\}}Y||^2 = Y^\top P_{\{i\}} Y$ is minimal.
3. Add i to S

4. Recompute the residuals Y with (5.11)
5. Stop or go back to 2

More recent algorithms based on this idea include Gram-Schmidt Orthogonalization described below and *Least Angle Regression* (LARS) (Efron et al., 2004).

Grafting

For fixed $\lambda_0, \lambda_1, \lambda_2 > 0$, Perkins et al. (2003) suggested minimizing the function

$$C(\boldsymbol{\alpha}) = \frac{1}{m} \sum_{k=1}^{m} L(f(\boldsymbol{\alpha}, \mathbf{x}_k), y_k) + \lambda_2 \|\boldsymbol{\alpha}\|_2 + \lambda_1 \|\boldsymbol{\alpha}\|_1 + \lambda_0 \, l_0(\boldsymbol{\alpha}) \qquad (5.12)$$

over the set of parameters Λ that defines the family of regression or classification functions $\mathcal{F} = \{f(\alpha, \cdot) \mid \alpha \in \Lambda\}$.[9] We restrict our analysis to linear models, i.e. $\mathcal{F} := \{f : \mathbb{R}^n \to \mathbb{R} \mid f(\mathbf{x}) = \mathbf{w} \cdot \mathbf{x} + b, \mathbf{w} \in \mathbb{R}^n, \ b \in \mathbb{R}\}$. In this case w_i is associated with feature i $(i = 1, \cdots, n)$. Therefore requiring the l_0 as well as the l_1 norm of $\mathbf{w}$ to be small results in reducing the number of features used. Note that for families of non-linear functions this approach does not necessarily result in a feature selection method. Since the objective C has the structure of minimizing a loss function plus a sparsity term it will also be mentioned in Section 5.4.

Perkins et al. (2003) solve problem (5.12) in a greedy forward way. In every iteration the working set of parameters is extended by one α_i and the newly obtained objective function is minimized over the enlarged working set. The selection criterion for new parameters is $|\partial C/\partial \alpha_i|$. Thus, the α_i is selected which yields the maximum decrease of the objective function C after one gradient step. Furthermore, the authors discuss possible stopping criteria.

Note that Grafting is an embedded method, because the structure of the target functions f (namely linear functions) is important for the feature selection part.

Decision Trees

Decision trees can be used to learn discrete valued functions. The most prominent approaches include CART (Breiman et al., 1984), ID3 (Quinlan, 1986) and C4.5 (Quinlan, 1993). Decision trees are iteratively built by recursively partitioning the data depending on the value of a specific feature. The "splitting" feature is chosen according to its importance for the classification task. A widely used criterion for the importance of a feature is the mutual information between feature i and the outputs Y (Duda et al., 2001, Section 8.3.2):

---

[9]The original optimization problem is slightly more complex. It includes weighted Minkowski norms.

$$MI(X^i, Y) = H(Y) - H(Y|X^i),$$

where H is the entropy (Cover and Thomas, 1991, Chapter 2) and $X^i :=$ $(x_{1,i}, \cdots, x_{m,i})$. In case of a continuous input space, the components of vector X_i are replaced by binary values corresponding to decisions of the type $x_{k,i} \leq$ *threshold*. Another choice is to split the data in such a way that the resulting tree has minimal empirical error.

In many cases, only a subset of the features is needed to fully explain the data. Thus, feature selection is implicitly built into the algorithm and therefore decision tree learning can be understood as an embedded method.

Gram-Schmidt Orthogonalization

Gram-Schmidt Orthogonalization (e.g. Chen et al., 1989) uses the angle of a feature to the target as an evaluation criterion to measure the importance of a feature for classification or regression purposes:

$$\cos(X^i, Y) = \frac{\langle X^i, Y \rangle^2}{\|X^i\|^2 \|Y\|^2}, \qquad (i = 1, ..., n)$$

In an iterative procedure the angle of every feature to the target is computed and the feature maximizing this quantity is selected. The remaining features are mapped to the null subspace of the previously selected features and the next iteration starts. During each iteration the least-squares solution of a linear-in-its-parameters model is computed based on the already ranked features. Thus the selection method is an embedded method for a linear least-square predictor. It is similar in spirit to the partial least squares (PLS) algorithm (Wold, 1975), but instead of constructing features that are linear combinations of the input variables, the Gram-Schmidt variable selection method selects variables.

Rivals and Personnaz (2003) suggested using polynomials as baseline functions. In a first step all monomials of degree smaller than an a priori fixed number d are ranked using Gram-Schmidt Orthogonalization. In a second step the highest ranked monomials are used to construct a polynomial. To reduce the complexity of the constructed polynomial one monomial at a time is removed from the polynomial in an iterative procedure until a stopping criterion is met. For large data sets the authors use the Fisher test as a stopping criterion; for small data sets the leave-one-out error is used since the statistical requirements for the hypothesis test are not satisfied. Both criteria are based on the mean squared error. The paper also describes how the ranked monomials can be used to select a subset of features as inputs to a neural network. Since this approach has more the flavor of a filter method, we do not discuss the details here.

(Stoppiglia et al., 2003) introduced a stopping criterion for the iteration during Gram-Schmidt Orthogonalization by adding a random feature to the set of features. The idea is that features ranked higher than the random feature carry information about the target concept whereas features ranked lower than the random feature are irrelevant. See also Chapter 2 for more details on Gram-Schmidt Orthogonalization, probe methods and the Fisher test.

5.2.3 Backward Elimination

Starting with all features, backward elimination methods iteratively remove features from the data according to a selection criterion until a stopping criterion is met. In this section we first discuss methods that derive the selection criterion from the analysis of the weight vector of the classifying or regression function. In the second part we briefly mention a method that uses a learning bound as a selection criterion.

Weight Based Analysis

In this section we review algorithms which choose features according to the weights given to those features by a classifier. The motivation for this is based on the idea that the value of a feature is measured by the change in expected value of error when removing it. Given this assumption, these methods approximate this goal by using the weights given by the classifier itself.

Recursive Feature Elimination

Recursive Feature Elimination (RFE) is a recently proposed feature selection algorithm described in Guyon et al. (2002). The method, given that one wishes to employ only $\sigma_0 < n$ input dimensions in the final decision rule, attempts to find the best subset of size σ_0 by a kind of greedy backward selection. It operates by trying to choose the σ_0 features which lead to the largest margin of class separation, using an SVM classifier (see Chapter 1). This combinatorial problem is solved in a greedy fashion at each iteration of training by removing the input dimension that decreases the margin the least until only σ_0 input dimensions remain.

Algorithm 1: Recursive Feature Elimination (RFE) in the linear case.

1: **repeat**
2:    Find $\mathbf{w}$ and b by training a linear SVM.
3:    Remove the feature with the smallest value $|w_i|$.
4: **until** σ_0 features remain.

The algorithm can be accelerated by removing more than one feature in step 2. RFE has shown good performance on problems of gene selection for microarray data (Guyon et al., 2002, Weston et al., 2003, Rakotomamonjy, 2003). In such data there are thousands of features, and the authors usually remove half of the features in each step.

One can also generalize the algorithm to the nonlinear case (Guyon et al., 2002). For SVMs the margin is inversely proportional to the value $W^2(\alpha) := \sum \alpha_k \alpha_l y_k y_l k(\mathbf{x}_k, \mathbf{x}_l)(= \|\mathbf{w}\|^2)$. The algorithm thus tries to remove features which keep this quantity small[10]. This leads to the following iterative procedure:

Algorithm 2: Recursive Feature Elimination (RFE) in the nonlinear case.

  1: **repeat**
  2:    Train a SVM resulting in a vector α and scalar b.
  3:    Given the solution α, calculate for each feature p:

$$W_{(-p)}^2(\alpha) = \sum \alpha_k \alpha_l y_k y_l k(\mathbf{x}_k^{-p}, \mathbf{x}_l^{-p})$$

     (where $\mathbf{x}_k^{-p}$ means training point k with feature p removed).
  4:    Remove the feature with smallest value of $|W^2(\alpha) - W_{(-p)}^2(\alpha)|$.
  5: **until** σ_0 features remain.

If the classifier is a linear one, this algorithm corresponds to removing the smallest corresponding value of $|w_i|$ in each iteration. The nonlinear formulation helps to understand how RFE actually works. Given the criterion of small W^2 (large margin) as a measure of generalization ability of a classifier, one could at this point simply choose one of the usual search procedures (hill climbing, backward elimination, etc.). However, even if we were to choose backward elimination, we would still have to train $\frac{1}{2}(n^2 + n - \sigma_0^2 + \sigma_0)$ SVMs (i.e. on each iteration we have to train as many SVMs as there are features) which could be prohibitive for problems of high input dimension. Instead, the trick in RFE is to estimate in each iteration the change in W^2 by only considering the change in the kernel k resulting from removing one feature, and assuming the vector α stays fixed.

Note that RFE has been designed for two-class problems although a multi-class version can be derived easily for a one-against-the-rest approach (Weston et al., 2003). The idea is then to remove the features that lead to the smallest value of $\sum_{c=1}^{Q} |W_c^2(\alpha^c) - W_{c,(-p)}^2(\alpha^c)|$ where $W_c^2(\alpha^c)$ is the inverse margin for the machine discriminating class c from all the others. This would lead to removing the *same* features from all the hyperplanes that build up the multiclass classifier, thus coupling the classifiers. This is useful if the same features

---

  [10] Although the margin has been shown to be related to the generalization ability by theoretical bounds for SVMs (Vapnik, 1998), if you change the feature space by removing features, bounds on the test error usually also involve another term controlling this change, e.g. the bound $R^2 \|\mathbf{w}\|^2$ in equation (5.16) also includes the radius R of the sphere around the data in feature space. However, if the data is suitably normalized, it appears that RFE works well.

are relevant for discriminating between different classes, if this is not the case it may not be a good idea to couple the hyperplanes in this way.

Clearly, it is easy to generalize RFE to other domains as well, such as other classifications algorithms or regression. In Zhu and Hastie (2003) the authors show how to adapt RFE to penalized logistic regression to give performance similar to SVMs, but with the added benefit of being able to output probability of correctness, rather than just predictions. They demonstrate that RFE gives performance superior to a univariate correlation score feature selector using this classifier.

Finally, in Lal et al. (2004) the authors show how to treat groups of features with RFE in the domain of Brain Computer Interfaces (BCI). In this problem there are several features for each EEG channel, and one is interested in finding which channels are relevant for a particular task. Therefore the authors modify RFE to remove the least important *channel* on each iteration by measuring the importance of a channel with $\sum_{i \in C_j} |w_i|$ for each channel (set of features) C_j.

RFE-Perceptron

In Gentile (2004), the authors propose an algorithm similar to RFE but for (approximate) large margin perceptrons using the p-norm (Grove et al., 1997) (ALMA$_p$). For $p = 2$ the algorithm is similar to an SVM and for $p = \infty$ the algorithm behaves more like an multiplicative update algorithm such as Winnow (Kivinen and Warmuth, 1995). They adapt the p-norm according to the number of features so when the feature size is shrunk sufficiently their algorithm behaves more like an SVM (they report that using the full feature set Winnow generally outperforms an SVM). The authors then propose a feature selection method that removes the features with the smallest weights until w fails to satisfy: $\sum_i |w_i| \geq 1 - \alpha(1 - \alpha)\gamma^{p/(p-1)}$ where $\alpha \in (0, 1]$ is a hyperparameter controlling the degree of accuracy and γ is the margin obtained from running ALMA$_p$. This differentiates it from RFE which requires setting the number of features to remove in advance (see Chapter 2 for statistical tests and cross-validation). The authors report this automatic choice results in a small number of chosen features, and generalization ability often close to the optimum number of features for RFE chosen using the test set.

Optimal Brain Damage

Optimal Brain Damage (OBD) (LeCun et al., 1990) is a technique for pruning weights in a Neural Network. The method works by, after having chosen a reasonable network architecture and training that network, first computing the second derivatives h_{ii} for each weight w_i. It then sorts the weights by saliency, defined as $h_{ii}w_i^2/2$ and deletes r weights with low saliency, and then repeats. Essentially, this measures the change in error when a weight is decreased. Thus OBD removes weights which do not appear to influence the training error. The authors demonstrate good performance on an Optical Character

Recognition (OCR) problem. Starting with a highly constrained and sparsely connected network with 2,578 weights, trained on 9300 examples, they were able to delete more than 1000 weights without reducing training or testing error.

Motivation for Weight Based Approaches

In this section we motivate weight-based approaches based on the minimization problem (5.4) for embedded methods. Suppose that the function f is a typical neural network:

$$f(\mathbf{w}, \mathbf{x}) = \tilde{f}(\ldots, w_{iq} x^i, \ldots, \tilde{\mathbf{w}}),$$

where the vector $\mathbf{w}$ of weights has been split between the weights w_{iq} of the first layer and the rest of the parameters $\tilde{\mathbf{w}}$. The connection of the i-th neuron of the input layer and the q-th neuron of the hidden layer is weighted by w_{iq} ($i \in \{1, \cdots, n\}, q \in \{1, \cdots, Q\}, Q \in \mathbb{N}$).

For a fixed $\boldsymbol{\sigma} = \mathbf{1}$, let $\mathbf{w}^*$ be a minimizer of $T(\mathbf{w}, \boldsymbol{\sigma}) = \sum_k L(f(\mathbf{w}, \boldsymbol{\sigma} \odot \mathbf{x}_k), y_k) + \lambda \Omega(\mathbf{w})$. With equation (5.9) we get for $p = 1, ..., n$:

$$\left. \frac{\partial G(\boldsymbol{\sigma})}{\partial \sigma_p} \right|_{\boldsymbol{\sigma}=1} = \left. \frac{\partial T(\mathbf{w}, \boldsymbol{\sigma})}{\partial \sigma_p} \right|_{(\mathbf{w}, \boldsymbol{\sigma})=(\mathbf{w}^*, 1)}$$

$$= \sum_{k=1}^{m} L'(f(\mathbf{w}^*, \mathbf{x}_k), y_k) \, x_k^p \sum_{q=1}^{Q} w_{pq}^* \left. \frac{\partial \tilde{f}(\ldots, t_{pq}, \ldots, \tilde{\mathbf{w}}^*)}{\partial t_{pq}} \right|_{t_{pq}=w_{pq}^* x_p} \quad (5.13)$$

On the other hand, since the weight vector $\mathbf{w}^*$ is a minimizer of $T(\mathbf{w}, \mathbf{1})$, we also have

$$0 = \left. \frac{\partial T(\mathbf{w}, \boldsymbol{\sigma})}{\partial w_{pq}} \right|_{(\mathbf{w}, \boldsymbol{\sigma})=(\mathbf{w}^*, 1)}$$

$$= \sum_{k=1}^{m} L'(f(\mathbf{w}^*, \mathbf{x}_k), y_k) \, x_k^p \left. \frac{\partial \tilde{f}(\ldots, t_{pq}, \ldots, \tilde{\mathbf{w}}^*)}{\partial t_{pq}} \right|_{t_{pq}=w_{pq}^* x_p} + \left. \frac{\partial \lambda \Omega(\mathbf{w})}{\partial w_{pq}} \right|_{\mathbf{w}=\mathbf{w}^*} .$$

By multiplying this last equation by w_{pq}, summing over q and combining it with (5.13), we finally get

$$\left. \frac{\partial G(\boldsymbol{\sigma})}{\partial \sigma_p} \right|_{\boldsymbol{\sigma}=1} = -\sum_{q=1}^{Q} w_{pq}^* \left. \frac{\partial \lambda \Omega(\mathbf{w})}{\partial w_{pq}} \right|_{\mathbf{w}=\mathbf{w}^*} .$$

If the regularization term Ω is quadratic, this last equation yields

$$\left. \frac{\partial G(\boldsymbol{\sigma})}{\partial \sigma_p} \right|_{\boldsymbol{\sigma}=1} = -2\lambda \sum_{q=1}^{Q} \left(w_{pq}^* \right)^2 .$$

This equation suggests to remove the component such that the l_2 norm of the weights associated with this input is small. More generally, weight based methods remove the component with the smallest influence on G after one gradient step.

Bounds for Support Vector Machines

Rakotomamonjy (2003) reports a greedy backward feature Rakotomamonjy (2003) reports a greedy backward feature selection method similar to RFE. Two bounds on the leave-one-out error LOO of a trained *hard margin* SVM classifier are used as ranking criterion: the radius margin bound (Vapnik, 1998):

$$LOO \leq 4R^2 \|\mathbf{w}\|^2, \tag{5.14}$$

where R denotes the radius of the smallest sphere that contains the training data[11], and the span-estimate (Vapnik and Chapelle, 2000):

$$LOO \leq \sum_k \alpha_k^* S_k^2, \qquad S_k^2 = \frac{1}{(\tilde{\mathbf{K}}^{-1})_{kk}}, \qquad \tilde{\mathbf{K}} = \begin{pmatrix} \mathbf{K}_{SV} & \mathbf{1} \\ \mathbf{1}^\top & 0 \end{pmatrix},$$

where $\tilde{\mathbf{K}}$ is the "augmented" kernel matrix computed only on the support vectors. Such SVM having kernel matrix $\mathbf{K}$ is trained with squared slack variables. Based on these bounds, two types of ranking criteria for features are introduced:

(i) Feature i is removed from the training data. An SVM is trained and the leave-one-out error LOO^{-i} is computed. The same procedure is repeated for the remaining features. The features are ranked according to the values of LOO^{-i} (the feature i with the highest value of LOO^{-i} is ranked first). This approach might be computationally expensive but it can be approximated in the following way: An SVM is trained using all features. Let $\boldsymbol{\alpha}^*$ be the solution. To measure the importance of feature i, the i-th component is removed from the kernel matrix. The bound $\widetilde{LOO}^{-i}$ is calculated using the reduced kernel matrix and $\boldsymbol{\alpha}^*$. The features are ranked according to the values of $\widetilde{LOO}^{-i}$. The assumption that would justify this approach is that the solution vector $\boldsymbol{\alpha}_{-i}^*$ of an SVM trained on the training data where the i-th feature was removed can be approximated by the solution $\boldsymbol{\alpha}^*$.

(ii) Instead of ranking features according to how their removal changes the value of the bound, the author also suggested to use the sensitivity of the bound with respect to a feature as a ranking criterion. In other words, a feature i is ranked according to the absolute value of the derivative of the bound with respect to this feature (the feature yielding the highest value is ranked first).

Like RFE, the method starts with a trained SVM that uses all features. The features are ranked using one of the two approaches (i) or (ii). The least ranked feature is removed from the data and the next iteration starts. This method is closely related to the work of Weston et al. (2000): Instead of iteratively

---

[11] Please note that R was previously also used for the risk.

removing features, scaling factors are introduced and optimized by a gradient descent (see also Section 5.3.1).

The method described by (Rakotomamonjy, 2003) can be interpreted as an approximation to the optimization problem (5.3). The functional $\widetilde{T}$ is the Support Vector Machine and the functional G is either the radius/margin bound or the leave-one-out estimate. The minimization over Σ is approximated in a greedy backward way without the guarantee of finding a global minimum. In (i), the criterion G is explicitly computed after a tentative feature has been removed, whereas in (ii), the gradient $\partial G/\partial\sigma_i$ is calculated to measure the importance of that feature.

5.3 Optimization of Scaling Factors

5.3.1 Scaling Factors for SVM

One method of using SVMs for feature selection is choosing the scaling factors which minimize a bound, as proposed by the authors of Weston et al. (2000). The idea is the following. Feature selection is performed by scaling the input parameters by a vector $\sigma \in [0,1]^n$. Larger values of σ_i indicate more useful features. Thus the problem is now one of choosing the best kernel of the form:

$$k_\sigma(\mathbf{x}, \mathbf{x}') = k(\sigma \odot \mathbf{x}, \sigma \odot \mathbf{x}'),\qquad(5.15)$$

where $\odot$ is element-wise multiplication, i.e. we wish to find the optimal parameters σ. This can be achieved by choosing these hyperparameters via a generalization bound.

Gradient Descent on the R^2w^2 Bound

Taking the expectation on both sides of inequality (5.14) and using the fact that the leave-one-out error is an unbiased estimator of the generalization error (Luntz and Brailovsky, 1996) one gets the following bound for hard margin SVMs,[12]

$$EP_{err} \leq \frac{4}{m} E\left\{R^2\|\mathbf{w}\|^2\right\},\qquad(5.16)$$

if the training data of size m belong to a sphere of size R and are separable with margin $1/\|\mathbf{w}\|^2$ (both in the feature space). Here, the expectation is taken over sets of training data of size m. Although other bounds exist, this one is probably the simplest to implement. A study of using other bounds for the same purpose is conducted in (Chapelle et al., 2002). The values of σ can be found by minimizing such a bound by gradient descent, providing

---

[12]For soft margin SVMs, the simplest way to still make this apply is to add a ridge to kernel matrix (Chapelle et al., 2002).

the kernel itself is differentiable. This can be solved by iteratively training an SVM, updating $\boldsymbol{\sigma}$ (the kernel) according to the gradient, and retraining until convergence. An implementation is available at (Spider, 2004). This method has been successfully used for visual classification (face detection, pedestrian detection) (Weston et al., 2000) and for gene selection for microarray data (Weston et al., 2000, 2003).

After optimization of the scaling factors, most of them will hopefully be small and one can explicitly discard some components and optimize the scaling factors for the remaining components. There are 3 different ways of discarding irrelevant components:

- The most straightforward is to remove the components which have small scaling factors. This is somehow similar to the RFE algorithm (see Section 5.2.3, as well as the end of this section).
- A more expensive strategy is to explicitly remove one feature, retrain an SVM and see by how much the radius margin bound increases (cf. Chapelle (2002), Rakotomamonjy (2003) and Section 5.2.3). Add this feature back to the data and remove the next feature and so on.
- To avoid such expensive retrainings, (Chapelle, 2002) suggest to use second order information and, similarly to OBD (LeCun et al., 1990), remove the features p which have a small value of

$$
\sigma_p^2 \frac{\partial R^2 \|\mathbf{w}\|^2}{\partial \sigma_p^2}.
$$

One could easily use the same trick for classifiers other than SVMs if a bound for that classifier is readily available. Alternatively one could use a validation set instead, this route is explored in Chapelle et al. (2002).

Other Criteria

In the previous section, the hyperparameters $\boldsymbol{\sigma}$ were optimized by gradient descent on $R^2\mathbf{w}^2$. Note that other model selection criteria are possible such as the span bound (Vapnik and Chapelle, 2000) or a validation error. Experimental results of such approaches are presented in Chapelle et al. (2002), but from a practical point of view, these quantities are more difficult to optimize because they are highly non-convex. In equation (5.16), one can use the variance instead of the radius, i.e.

$$
R^2 \approx \frac{1}{m} \sum_{k=1}^{m} \left(\Phi(\mathbf{x}_k) - \frac{1}{m} \sum_{l=1}^{m} \Phi(\mathbf{x}_l) \right)^2 = \frac{1}{m} \sum_{k=1}^{m} k(\mathbf{x}_k, \mathbf{x}_k) - \frac{1}{m^2} \sum_{k,l=1}^{m} k(\mathbf{x}_k, \mathbf{x}_l).
$$

$$(5.17)$$

The first advantage in using the variance instead of the radius is that it is easier to compute. But a second advantage is that it is less sensitive to outliers and that it is theoretically sound (Bousquet, 2002).

A similar approach to the one described in the previous section is to consider the scaling factors $\boldsymbol{\sigma}$ not anymore as hyperparameters but as parameters of the learning algorithm (Grandvalet and Canu, 2003). A scalar parameter σ_0 controls the norm of $\boldsymbol{\sigma} \geq 0$:

$$\min_{\sigma} \max_{\alpha} \sum_{k} \alpha_k - \frac{1}{2} \sum_{k,l} \alpha_k \alpha_l y_k y_l k_{\sigma}(\mathbf{x}_k, \mathbf{x}_l)$$

under constraints

$$0 \leq \alpha_k \leq C, \quad \sum \alpha_k y_k = 0 \quad \text{and} \quad ||\boldsymbol{\sigma}||_p = \sigma_0, \tag{5.18}$$

with k_{σ} defined in (5.15). The closer the hyperparameter p is to 0, the sparser the solution, but also the more difficult the optimization.

Linear SVMs

We now study the special case of optimizing the scaling factors for a linear kernel,

$$k_{\sigma}(\mathbf{x}_k, \mathbf{x}_l) = \sum_{i=1}^{n} \sigma_i^2 x_{k,i} x_{l,i}.$$

Assume that the data are centered and that each component is rescaled such that it has unit variance. Then the variance criterion (5.17), as an approximation of the radius, gives

$$R^2 = \frac{1}{m} \sum_{i=1}^{n} \sum_{k=1}^{m} \sigma_i^2 (x_{k,i})^2 = \sum_{i=1}^{n} \sigma_i^2.$$

Since for hard-margin SVMs the maximum of the dual is equal to $\mathbf{w}^2/2$, the optimization of the radius-margin criterion (5.16) can be rewritten as the maximization of the margin under constant radius[13],

$$\min_{\sigma} \max_{\alpha} \sum_{k} \alpha_k - \frac{1}{2} \sum_{k,l=1}^{m} \alpha_k \alpha_l y_k y_l \sum_{i=1}^{n} \sigma_i^2 x_{k,i} x_{l,i} \tag{5.19}$$

under constraints

$$\alpha_k \geq 0, \quad \sum \alpha_k y_k = 0 \quad \text{and} \quad \sum \sigma_i^2 = 1.$$

Note that in the linear case, we recover the optimization procedure proposed in Grandvalet and Canu (2003) with $p = 2$ in (5.18).

---

[13]It is indeed possible to fix the radius since multiplying the vector $\boldsymbol{\sigma}$ by a scalar λ would result in a multiplication of R^2 by λ^2 and a weight vector $\tilde{w}$ with norm $||w||/\lambda$ and thus would not affect the bound.

As explained in the beginning of this section, it is possible to find the optimal scaling factors $\boldsymbol{\sigma}$ by gradient descent. Moreover, it turns out that (5.19) is a convex optimization problem in σ_i^2 (as a pointwise maximum of linear functions) and is thus easy to solve. However, when the number of variables is larger than the number of training points, it might be advantageous to take the dual of (5.19) and solve the following optimization problem:

$$\max_{\alpha} \sum_{k=1}^{m} \alpha_k$$

under constraints

$$\alpha_k \geq 0, \quad \sum \alpha_k y_k = 0$$

$$-1 \leq \sum_{k=1}^{m} \alpha_k y_k x_{k,i} \leq 1, \quad 1 \leq i \leq n.$$

Let μ_i^+ and μ_i^- be the Lagrange multipliers associated with the two sets of constraints above. Then, one can recover the scaling parameters σ_i by $\sigma_i^2 = \mu_i^+ + \mu_i^-$, which is typically a sparse vector. A similar approach was proposed by Peleg and Meir (2004).

Please note that although this kind of approach is appealing, this might lead to overfitting in the case where there are more dimensions than training samples.

Link with RFE

When the input components are normalized to have variance 1, the derivative of the radius/margin bound with respect to the square of the scaling factors is

$$\frac{\partial R^2 \mathbf{w}^2}{\partial \sigma_p^2} = -R^2 w_p^2 + \mathbf{w}^2.$$

¿From this point of view, RFE amounts to making one gradient step and removing the components with the smallest scaling factors.

5.3.2 Automatic Relevance Determination

Automatic Relevance Determination has first been introduced in the context of Neural Networks (MacKay, 1994, Neal, 1996). In this section, we follow the Sparse Bayesian learning framework described in Tipping (2001). Further details as well as application to gene expression data can be found in Li et al. (2002).

In a probabilistic framework, a model of the likelihood of the data is chosen $P(\mathbf{y}|\mathbf{w})$ as well as a prior on the weight vector, $P(\mathbf{w})$. To predict the output of a test point $\mathbf{x}$, the average of $f_{\mathbf{w}}(\mathbf{x})$ over the posterior distribution $P(\mathbf{w}|\mathbf{y})$ is computed. If this integral is too complicated to estimate, a standard solution is to predict using the function $f_{\mathbf{w}_{\mathrm{MAP}}}$, where $\mathbf{w}_{\mathrm{MAP}}$ is the vector of parameters called the Maximum a Posteriori (MAP), i.e.

$$\mathbf{w}_{\mathrm{MAP}} = \underset{\mathbf{w}}{\mathrm{argmax}}\, P(\mathbf{w}|\mathbf{y}) = \underset{\mathbf{w}}{\mathrm{argmin}}\, -\log P(\mathbf{y}|\mathbf{w}) - \log P(\mathbf{w}). \qquad (5.20)$$

The right hand side of the previous equation can be interpreted as the minimization of a regularized risk, the first term being the empirical risk and the second term being the regularization term.

One way to get a sparse vector $\mathbf{w}_{\mathrm{MAP}}$ is to introduce several hyperparameters β_i controlling the variance of the w_i (Tipping, 2001),

$$P(\mathbf{w}|\boldsymbol{\beta}) = \prod_{i=1}^{n} \sqrt{\frac{\beta_i}{2\pi}} \exp(-\beta_i w_i^2/2).$$

The vector $\boldsymbol{\beta}$ is learned by *maximum likelihood type II*, i.e. by finding the vector $\boldsymbol{\beta}$ which maximizes (assuming a (improper) flat hyperprior on $\boldsymbol{\beta}$),

$$P(\mathbf{y}|\boldsymbol{\beta}) = \int P(\mathbf{y}|\mathbf{w})P(\mathbf{w}|\boldsymbol{\beta})d\mathbf{w}. \qquad (5.21)$$

Using a Laplace approximation to compute the above integral and setting the derivative of (5.21) to 0, in can be shown that at the optimum the vector $\boldsymbol{\beta}$ satisfies (MacKay, 1994, Tipping, 2001)

$$\beta_i = \frac{1 - \beta_i H_{ii}^{-1}}{(w_i)_{\mathrm{MAP}}^2}, \qquad (5.22)$$

where H is the Hessian of the log posterior around its maximum, $H = -\nabla^2 \log P(\mathbf{w}|\mathbf{y})|_{\mathbf{w}_{\mathrm{MAP}}}$.

An iterative scheme is adopted: compute $\mathbf{w}_{\mathrm{MAP}}$ with (5.20) and update $\boldsymbol{\beta}$ using (5.22).

- When a component is not useful for the learning task, the maximization of the marginalized likelihood (5.21) over the corresponding β_i will result in $\beta_i \to \infty$ and thus $w_i \to 0$, effectively pruning the weight w_i.
- From this point of view, $\beta_i^{-1/2}$ is the Bayesian equivalent of the scaling factors σ_i introduced in Section 5.3.1.
- The prior on the weights w_i can be computed as

$$P(w_i) = \int P(w_i|\beta_i)P(\beta_i)d\beta_i.$$

When the hyperprior on β_i is a Gamma distribution, this gives a Student-t distribution on $P(w_i)$ (Tipping, 2001): it is sharply peaked around 0 and thus tends to set the weights w_i at 0, leading to a sparse solution.

5.3.3 Variable Scaling: Extension to Maximum Entropy Discrimination

The Maximum Entropy Discrimination (MED) framework has been introduced in Jaakkola et al. (1999). It is a probabilistic model in which one does not learn parameters of a model, but distributions over them. Those distributions are found by minimizing the KL divergence with respect to a prior distribution while taking into account constraints given by the labeled examples. For classification, it turns out that the optimization problem solved in Jaakkola et al. (1999) is very similar to the SVM one.

Feature selection can be easily integrated in this framework (Jebara and Jaakkola, 2000). For this purpose, one has to specify a prior probability p_0 that a feature is active.

If w_i would be the weight associated with a given feature for a linear model (as found by a linear SVM for instance), then the expectation of this weight taking into account this sparse prior is modified as follows Jebara and Jaakkola (2000),

$$\frac{w_i}{1 + \frac{1-p_0}{p_0} \exp(-w_i^2)}.$$

This has the effect of discarding the components for which

$$w_i^2 \ll \log \frac{1 - p_0}{p_0}.$$

For this reason, even though the feature selection is done in a complete different framework than a standard SVM, this algorithm turns out to be similar to RFE in the sense that it ignores features whose weights are smaller than a threshold. The MED framework was recently extended to the case where multiple inter-related learning tasks are jointly solved (Jebara, 2004).

5.3.4 Joint Classifier and Feature Optimization (JCFO)

The Joint Classifier and Feature Optimization (JCFO) algorithm (Krishnapuram et al., 2004) is similar to sparse Gaussian Process for classification (see also Seeger (2000), Williams and Barber (1998)) with an ARD prior (Section 5.3.2). It considers classifiers of the form:

$$\sum_{k=1}^{m} \alpha_k k(\boldsymbol{\sigma} \odot \mathbf{x}, \boldsymbol{\sigma} \odot \mathbf{x}_k) + b.$$

The aim of this classifier is to find a function with a sparse vector $\boldsymbol{\alpha}$ *and* a sparse vector of scaling factors $\boldsymbol{\sigma}$. This is achieved by using a Laplacian prior on $\boldsymbol{\alpha}$ and $\boldsymbol{\sigma}$. The inference in this probabilistic model is done by a type of EM algorithm which uses a a conjugate gradient descent on $\boldsymbol{\sigma}$ during the M-step. Note that, as in Section 5.3.1, this gradient descent can be avoided in the case of a linear kernel.

5.4 Sparsity Term

In the case of linear models, indicator variables are not necessary as feature selection can be enforced on the parameters of the model directly. This is generally done by adding a *sparsity term* to the objective function that the model minimizes. To link minimization problem (5.4) of the introductory section to methods that use a sparsity term for feature selection purposes, we consider linear decision functions of the form $f(\mathbf{w}, \mathbf{x}) = \mathbf{w} \cdot \mathbf{x}$, indicator variables $\boldsymbol{\sigma} \in \{0,1\}^n$ and we make use of the following lemma,

Lemma: *If the learning criterion T can be expressed as $T(\mathbf{w}, \boldsymbol{\sigma}) = \sum L(f(\mathbf{w} \odot \boldsymbol{\sigma}, \mathbf{x}_k), y_k) + \Omega(\mathbf{w})$ and Ω is component-wise minimized at 0, then*

$$\min_{\mathbf{w},\, l_0(\boldsymbol{\sigma})=\sigma_0} T(\mathbf{w}, \boldsymbol{\sigma}) = \min_{l_0(\mathbf{w})=\sigma_0} T(\mathbf{w}, \mathbf{1}), \tag{5.23}$$

where the zero "norm" l_0 is defined as $l_0(\boldsymbol{\sigma}) := cardinality(\{i \in \{1, \cdots, n\} : \sigma_i \neq 0\})$.

Proof: Let $\mathbf{w}^*$ and $\boldsymbol{\sigma}^*$ minimize the left hand side of (5.23). Then $T(\mathbf{w}^*, \boldsymbol{\sigma}^*) \geq T(\mathbf{w}^* \odot \boldsymbol{\sigma}^*, \mathbf{1})$ because setting one of the component of $\mathbf{w}$ to 0 will decrease Ω by assumption. Thus, the left hand side of (5.23) is greater or equal than its right hand side. On the other hand, if $\mathbf{w}^*$ is the minimizer of the right hand side, defining $\sigma_i = 1_{w_i^* \neq 0}$ shows the other inequality.

In other words, in the linear case, one can ignore the scaling factors $\boldsymbol{\sigma}$ and directly find a sparse vector $\mathbf{w}$. The following section presents in more detail how this is implemented for several algorithms. For the sake of simplicity, we will assume that the target is in $\{-1, 1\}$. When generalizations to more complex problems (regression or multi-class) are possible, it will be stated clearly.

5.4.1 Feature Selection as an Optimization Problem

Most linear models that we consider can be understood as the result of the following minimization:

$$\min_{\mathbf{w},b} \frac{1}{m} \sum_{k=1}^{m} L(\mathbf{w} \cdot \mathbf{x}_k + b, y_k) + C\Omega(\mathbf{w}),$$

where $L(f(\mathbf{x}_k), y_k)$ measures the loss of a function $f(\mathbf{x}) = (\mathbf{w} \cdot \mathbf{x} + b)$ on the training point $(\mathbf{x}_k, y_k)$, $\Omega(\mathbf{w}) : \mathbb{R}^n \to \mathbb{R}_+$ is a penalizing term and C is a trade-off coefficient balancing the empirical error with this penalizing term. Examples of empirical errors are:

1. *The ℓ_1 hinge loss*:

$$\ell_{\text{hinge}}(\mathbf{w} \cdot \mathbf{x} + b, y) := |1 - y(\mathbf{w} \cdot \mathbf{x} + b)|_+,$$

where $|z|_+ = z$ if $z > 0$ and $|z|_+ = 0$ otherwise.

2. *The ℓ_2 loss*:
$$\ell_2\left(\mathbf{w}\cdot\mathbf{x}+b,y\right):=\left(\mathbf{w}\cdot\mathbf{x}+b-y\right)^2.$$

3. *The Logistic loss*:

$$\ell_{\text{Logistic}}\left(\mathbf{w}\cdot\mathbf{x}+b,y\right):=\log(1+e^{-y(\mathbf{w}\cdot\mathbf{x}+b)}).$$

This loss, usually used in logistic regression, is based on the following generalized linear model: $\log\left(\frac{P(y=1|\mathbf{w},b)}{1-P(y=1|\mathbf{w},b)}\right)=\mathbf{w}\cdot\mathbf{x}+b$.

The penalizing terms that we will consider here will be of two types:

1. *The ℓ_0 norm*:
$$\Omega(\mathbf{w})=\ell_0(\mathbf{w})$$

representing the number of non-zero coordinates of $\mathbf{w}$.
2. *The ℓ_1 norm*:
$$\Omega(\mathbf{w})=\sum_{i=1}^{n}|w_i|.$$

Table 5.1 introduces the possible combinations of loss and sparsity along with the corresponding algorithms.

Table 5.1. Methods that have been designed to enforce feature selection during the training of a linear model. A cross means that such combination has not been considered so far. The number in brackets refers to the section number.

Loss \ Sparsity	ℓ_0	ℓ_1
hinge	FSV [5.4.3]	ℓ_1 SVM [5.4.2]
ℓ_2	multiplicative update [5.4.3]	LASSO [5.4.4]
Logistic	×	Generalized Lasso [5.4.4]

5.4.2 ℓ_1 Support Vector Machine

The ℓ_1 Support Vector Machine (ℓ_1-SVM) of Bradley and Mangasarian (1998) (Mangasarian, 1968) for classification solves the following optimization problem:

$$\min_{\mathbf{w},b}\sum_{i=1}^{n}|w_i|+C\sum_{k=1}^{m}\xi_k$$

subject to: $\xi_k\geq 0$ and $y_k\left(\mathbf{w}\cdot\mathbf{x}_k+b\right)\geq 1-\xi_k$. The main difference compared to a classical SVM is the replacement of the quadratic regularization term $\|\mathbf{w}\|_2^2$ by the ℓ_1 norm $\sum_i |w_i|$. This slight change in the regularization term induces a big difference in the final outcome of the optimization. This is due to the strict convexity shape of the quadratic norm. Assume that two linear models parameterized by $\mathbf{w}_1$ and $\mathbf{w}_2$ are consistent on the training set. Assume furthermore that $\mathbf{w}_1$ uses only the first half of the features and $\mathbf{w}_2$ the second half (the input space is built from redundant features). Then any linear combination: $\mathbf{w} = (1 - \lambda)\mathbf{w}_1 + \lambda\mathbf{w}_2$ ($\lambda \in (0,1)$) will have a smaller ℓ_2 norm than $\mathbf{w}_1$ or $\mathbf{w}_2$. This implies - in this particular case - that choosing a vector $\mathbf{w}$ with more features induces a strictly smaller ℓ_2 norm. This shows that the SVM tends to return a model that uses many redundant features. It is by the way one of the strengths of the SVM to distribute the classification decision among many redundant features, making it more robust to the overall noise.[14] In the context of feature selection, such a property might be a drawback. The introduction of the ℓ_1 norm tends to remove this property by giving the same value of the regularization term to all the $\mathbf{w} \in [\mathbf{w}_1, \mathbf{w}_2]$. This lets other factors (like the minimization of the empirical error) choose the right model and hence choose a sparse model if it decreases the empirical error. Note that the trade-off parameter C can be used to balance the amount of sparsity relative to the empirical error. A small C will lead to a sparse linear model but whose empirical error might be larger as with a large C.

The ℓ_1 SVM was applied as a feature selection method by different authors. Bradley and Mangasarian (1998) introduced this version of the SVM more like a general classification technique but they noted the ability of this method to return sparse linear models. (Fung and Mangasarian, 2003) exploited this property to perform feature selection introducing a new optimization technique. Note that the ℓ_1 SVM can be defined for regression as well. (Bi et al., 2003) use this approach in the context of drug design.

5.4.3 Concave Minimization

In the case of linear models, feature selection can be understood as the following optimization problem:

$$\min_{\mathbf{w}, b} \ell_0(\mathbf{w})$$

subject to: $y_k (\mathbf{w} \cdot \mathbf{x}_k + b) \geq 0$. Said differently, feature selection is interpreted as finding a $\mathbf{w}$ with as few non zero coordinates as possible such that the derived linear model is consistent on the training set. This problem is known to be NP-hard (Amaldi and Kann, 1998) and hence cannot be solved directly. In this section, we present two attempts to approximately optimize it. Both approximations are based on replacing ℓ_0 by a smooth function whose gradient

---

[14]If each redundant feature is perturbed by the same zero mean random noise, adding these features would reduce the overall influence of this noise to the output of the SVM.

can be computed and can be used to perform a gradient descent. Note that the above problem does not match the goal of feature selection in machine learning. The latter is indeed interested in improving generalization error. In the current set-up, we are just interested in finding the smallest number of features consistent with the training set. This can obviously lead to overfitting when the number of training samples is much smaller than the number of dimensions.[15] This supports the idea that exact minimization of the ℓ_0 norm is not desired and it motivates approximations that push the solution towards low capacity systems for which the generalization error is better controlled. Such systems might be, for instance, large margin classifiers. The iterative nature of the methods that we describe below make it possible as well to use early stopping. The latter technique computes an estimate of the generalization error on a validation set and stops the procedure when the error estimate is too high.

Feature Selection Concave (FSV)

Bradley and Mangasarian (1998) propose to approximate the function $\ell_0(\mathbf{w})$ as:

$$\ell_0(\mathbf{w}) \approx \sum_{i=1}^{n} 1 - \exp(-\alpha|w_i|).$$

The coefficient α controls the steepness of the function and its closeness to $\ell_0(\mathbf{w})$. In their paper, Bradley and Mangasarian suggest to take $\alpha = 5$ as a first guess. Note that this function is not differentiable directly but a constrained gradient descent can be applied without any difficulty. Algorithm 3 presents the method. Although this algorithm is presented here for separable datasets, it is described in the original paper for non-separable datasets. The errors are then computed using the hinge loss.

Multiplicative Update

Weston et al. (2003) use a slightly different function. They replace the ℓ_0 norm by:

$$\ell_0(\mathbf{w}) \leftrightarrow \sum_{i=1}^{n} \log\left(\epsilon + |w_i|\right).$$

Although not a good approximation of ℓ_0, Weston *et al.* argue that its minimum is close to the minimum of the ℓ_0 norm. The interest of taking such an approximation is that it leads directly to an iterative scheme whose basic step is a classical SVM optimization problem. The multiplicative update can therefore be implemented very quickly based on any SVM optimization code. Algorithm 4 describes the approach.

---

[15]Having a small number of samples increases the chance of having a noisy feature completely correlated with the output target.

Algorithm 3: **F**eature **S**election Concave (FSV)

Require: α: controls the steepness of the objective function
1: Start with $\mathbf{v}^0$
2: cont=true; t=0;
3: **while** (cont==true) **do**
4:     Let $\mathbf{v}^*$ be the solution of the Linear Program: {Find descent direction}

$$\min_{\mathbf{v}} \sum_{k=1}^{n} \alpha e^{-\alpha v_k^t} \left(v_k - v_k^t \right)$$

    subject to: $y_k \left(\mathbf{w} \cdot \mathbf{x}_k + b \right) \geq 1, \; -v_k \leq w_k \leq v_k$
5:     $\mathbf{v}^{t+1} = \mathbf{v}^*$;
6:     **if** $(\mathbf{v}^{t+1} == \mathbf{v}^t)$ **then**
7:         cont=false; {If nothing changes, stop}
8:     **end if**
9:     t=t+1;
10: **end while**

Algorithm 4: Multiplicative update

1: Start with $\sigma^0 = (1, ..., 1) \in \mathbb{R}^n$
2: cont=true; t=0;
3: **while** (cont==true) **do**
4:     Let $\mathbf{w}^*$ be the minimum of: {SVM optimization}
    $\min_{\mathbf{w}} \|\mathbf{w}\|_2^2$
    subject to: $y_k \left(\mathbf{w} \cdot (\sigma^t \odot \mathbf{x}_k) + b \right) \geq 1$
5:     $\sigma^{t+1} = \sigma^t \odot \mathbf{w}^*$; {$\odot$ is the component-wise multiplication}
6:     **if** $(\sigma^{t+1} == \sigma^t)$ **then**
7:         cont=false; {If nothing changes, stop}
8:     **end if**
9:     t=t+1;
10: **end while**

As for FSV, the multiplicative update can be extended to non separable datasets. The errors are then measured with a quadratic loss. Both FSV and multiplicative update can be generalized to regression problems. We refer the reader to the original papers for more details.

5.4.4 LASSO

The LASSO technique (Least Absolute Shrinkage and Selection Operator) (Tibshirani, 1996) is very similar in its spirit to the ℓ_1 SVM. It minimizes the following problem:

$$\min_{\mathbf{w},b} \sum_{k=1}^{m} \left(\mathbf{w} \cdot \mathbf{x}_k - y_k \right)^2$$

subject to: $\sum_{i=1}^{n} |w_i| \leq \sigma_0$. The use of the ℓ_1 norm constraint on the parameter leads to a sparse model as in the case of the ℓ_1 SVM. It can be used for regression and classification. Recently, the LASSO technique has been generalized to handle classification problem with a more adequate loss. Roth (2003) defined what is called *generalized LASSO* as:

$$\min_{\mathbf{w},b} \sum_{k=1}^{m} \log(1 + e^{-y_k(\mathbf{w}\cdot\mathbf{x}_k+b)})$$

subject to: $\sum_{i=1}^{n} |w_i| \leq \sigma_0$. Although the problem is convex, it is not quadratic nor linear. To solve it, Roth suggests to use an Iterative Reweighed Least Square scheme. The generalized LASSO has the advantage of producing sparse models whose outputs can be interpreted as probabilities.

5.4.5 Other Methods

This section is by no means an exhaustive survey of all machine learning techniques that can be understood as minimizing an empirical error plus a sparsity term. Such an interpretation encompasses too many methods to discuss them in this chapter. We have instead presented examples of which we believe that they cover a wide range of approaches. Other methods, like sparse kernel Fisher discriminant (Mika et al., 2000), the grafting method (Perkins et al., 2003) (see also Section 5.2.2) or the Potential Support Vector Machine (Hochreiter and Obermayer, 2004) can be understood - in a certain sense - as minimizing an empirical error plus a sparsity term. In fact, any machine learning technique involving a linear model can be extended to implement feature selection by adding a sparsity term. We have shown two of those sparsity terms in this section (namely the ℓ_1 and the ℓ_0 norm). We believe they can be combined with most objective functions and optimized using the methods described, or variants.

5.5 Discussions and Conclusions

The introduction of this chapter provides a theoretical framework which unifies many embedded feature selection methods that were introduced during the last few years. This framework is built upon the concept of scaling factors. We discussed embedded methods along *how* they approximate the proposed optimization problem:

- Explicit removal or addition of features - the scaling factors are optimized over the discrete set $\{0,1\}^n$ in a greedy iteration,
- Optimization of scaling factors over the compact interval $[0,1]^n$, and
- Linear approaches, that directly enforce sparsity of the model parameters.

From the literature that covers embedded methods it is neither possible to infer a ranking that reflects the relative abilities of the methods nor is it possible to state which methods work best in which scenario. These conclusions could be drawn only from a systematic comparison - which clearly is beyond the scope of this chapter.

Every family of feature selection methods (filter, wrapper and embedded) has its own advantages and drawbacks. In general, filter methods are fast, since they do not incorporate learning. Most wrapper methods are slower than filter methods, since they typically need to evaluate a cross-validation scheme at every iteration. Whenever the function that measures the quality of a scaling factor can be evaluated faster than a cross-validation error estimation procedure, we expect embedded methods to be faster than wrapper approaches. Embedded methods tend to have higher capacity than filter methods and are therefore more likely to overfit. We thus expect filter methods to perform better if only small amounts of training data are available. Embedded methods will eventually outperform filter methods as the number of training points increase.

Acknowledgements

We would like to thank Bernhard Schölkopf, N. Jeremey Hill and Michael Schröder for their help with this work. This work was supported in part by the IST Programme of the European Community, under the PASCAL Network of Excellence, IST-2002-506778. T.N.L. was supported by a grant from the Studienstiftung des deutschen Volkes.

References

E. Amaldi and V. Kann. On the Approximability of Minimizing non zero Variables or Unsatisfied Relations in Linear Systems. *Theoretical Computer Science*, 209: 237–260, 1998.

J. Bi, K. Bennett, M. Embrechts, C. Breneman, and M. Song. Dimensionality Reduction via Sparse Support Vector Machines. *Journal of Machine Learning Research*, 3:1229–1243, 2003.

O. Bousquet. *Concentration Inequalities and Empirical Processes Theory Applied to the Analysis of Learning Algorithms*. PhD thesis, École Polytechnique, 2002.

P. S. Bradley and O. L. Mangasarian. Feature Selection via Concave Minimization and Support Vector Machines. In *Proc. 15th International Conf. on Machine Learning*, pages 82–90. Morgan Kaufmann, San Francisco, CA, 1998.

L. Breiman, J.H. Friedman, R.A. Olshen, and C.J. Stone. *Classification and Regression Trees*. Wadsworth and Brooks, 1984.

O. Chapelle. *Support Vector Machines: Induction Principles, Adaptive Tuning and Prior Knowledge*. PhD thesis, LIP6, Paris, 2002.

O. Chapelle, V. Vapnik, O. Bousquet, and S. Mukherjee. Choosing Multiple Parameters for Support Vector Machines. *Machine Learning*, 46(1-3):131–159, 2002.

S. Chen, S.A. Billings, and W. Luo. Orthogonal Least Squares and Their Application to Non-linear System Identification. *International Journal of Control*, 50:1873–1896, 1989.

T. Cover and J. Thomas. *Elements of Information Theory.* Wiley and Sons, USA, 1991.

R.O. Duda, P.E. Hart, and D.G. Stork. *Pattern Classification.* John Wiley and Sons, New York, USA, second edition, 2001.

B. Efron, T. Hastie, I. Johnstone, and R. Tibshirani. Least angle regression. *Annals of Statistics*, 32(2):407–499, 2004.

G. Fung and O. L. Mangasarian. A Feature Selection Newton Method for Support Vector Machine Classification. *Computational Optimization and Aplications*, pages 1–18, 2003.

C. Gentile. Fast Feature Selection from Microarray Expression Data via Multiplicative Large Margin Algorithms. In Sebastian Thrun, Lawrence Saul, and Bernhard Schölkopf, editors, *Advances in Neural Information Processing Systems 16.* MIT Press, Cambridge, MA, 2004.

Y. Grandvalet and S. Canu. Adaptive Scaling for Feature Selection in SVMs. In S. Thrun S. Becker and K. Obermayer, editors, *Advances in Neural Information Processing Systems*, volume 15, Cambridge, MA, USA, 2003. MIT Press.

A. J. Grove, N. Littlestone, and D. Schuurmans. General Convergence Results for Linear Discriminant Updates. In *Computational Learing Theory*, pages 171–183, 1997.

I. Guyon, J. Weston, S. Barnhill, and V. Vapnik. Gene Selection for Cancer Classification using Support Vector Machines. *Machine Learning*, 46:389–422, January 2002.

S. Hochreiter and K. Obermayer. Gene Selection for Microarray Data. In B. Schölkopf, K. Tsuda, and J.-P. Vert, editors, *Kernel Methods in Computational Biology.* MIT Press, Cambridge, Massachusetts, 2004.

T. Jaakkola, M. Meila, and T. Jebara. Maximum Entropy Discrimination. Technical Report AITR-1668, Massachusetts Institute of Technology, Artificial Intelligence Laboratory, 1999.

T. Jebara. Multi-Task Feature and Kernel Selection For SVMs. In *Proceedings of the 21st International Conference on Machine Learning (ICML)*, 2004.

T. Jebara and T. Jaakkola. Feature Selection and Dualities in Maximum Entropy Discrimination. In *Proceedings of the 16th Conference on Uncertainty in Artificial Intelligence*, 2000.

J. Kivinen and M. Warmuth. The Perceptron Algorithm vs. Winnow: Linear vs. Logarithmic Mistake Bounds when few Input Variables are Relevant. In *Proceedings of the eighth annual conference on Computational learning theory*, pages 289–296, New York, USA, 1995. ACM Press.

B. Krishnapuram, L. Carin, and A. Hartemink. Gene Expression Analysis: Joint Feature Selection and Classifier Design. In B. Schölkopf, K. Tsuda, and J.-P. Vert, editors, *Kernel Methods in Computational Biology.* MIT Press, Cambridge, MA, 2004.

T.N. Lal, M. Schröder, T. Hinterberger, J. Weston, M. Bogdan, N. Birbaumer, and B. Schölkopf. Support Vector Channel Selection in BCI. *IEEE Transactions on Biomedical Engineering. Special Issue on Brain-Computer Interfaces*, 51(6):1003–1010, June 2004.

Y. LeCun, J. Denker, S. Solla, R. E. Howard, and L. D. Jackel. Optimal Brain Damage. In D. S. Touretzky, editor, *Advances in Neural Information Processing Systems II*, San Mateo, CA, 1990. Morgan Kauffman.

P. Leray and P. Gallinari. Feature Selection with Neural Networks. *Behaviormetrika*, 26(1), 1999.

Y. Li, C. Campbell, and M. Tipping. Bayesian Automatic Relevance Determination Algorithms for Classifying Gene Expression Data. *Bioinformatics*, 18(10):1332–1339, 2002.

A. Luntz and V. Brailovsky. On the Estimation of Characters Obtained in Statistical Procedure of Recognition. *Technicheskaya Kibernetica*, 1996.

D. J. C. MacKay. Bayesian non-linear modelling for the prediction competition. *ASHRAE Transactions*, 100(2):1053–1062, 1994.

O. Mangasarian. Multisurface method of pattern separation. *IEEE Transactions on Information Theory*, 14(6):801 – 807, 1968.

S. Mika, G. Rätsch, and K.-R. Müller. A Mathematical Programming Approach to the Kernel Fisher Algorithm. In S.A. Solla, T.K. Leen, and K.-R. Müller, editors, *Advances in Neural Information Processing Systems*, pages 591–597, Cambridge, MA, USA, 2000. MIT Press.

R. Neal. *Bayesian Learning for Neural Networks*, volume 118 of *Lecture Notes in Statistics*. Springer, 1996.

D. Peleg and R. Meir. A feature selection algorithm based on the global minimization of a generalization error bound. In *NIPS 18*, 2004.

S. Perkins, K. Lacker, and J. Theiler. Grafting: Fast, Incremental Feature Selection by Gradient Descent in Function Space. *Journal of Machine Learning Research*, 3:1333–1356, 2003.

J.R. Quinlan. *C4.5: Programs for Machine Learning*. Morgan Kaufmann, 1993.

J.R. Quinlan. Induction of Decision Trees. *Machine Learning*, 1(1):81–106, 1986.

A. Rakotomamonjy. Variable Selection Using SVM-based Criteria. *Journal of Machine Learning Research*, 3:1357–1370, 2003.

I. Rivals and L. Personnaz. MLPs (Mono-Layer Polynomials and Multi-Layer Perceptrons) for Nonlinear Modeling. *Journal of Machine Learning Research*, 3: 1383–1398, 2003.

V. Roth. The Generalized LASSO. *IEEE Transactions on Neural Networks*, 2003.

M. Seeger. Bayesian Model Selection for Support Vector Machines, Gaussian Processes and Other Kernel Classifiers. In S.A. Solla, T.K. Leen, and K.-R. Müller, editors, *Advances in Neural Information Processing Systems*, volume 12, Cambridge, MA, USA, 2000. MIT Press.

Spider. Machine Learning Toolbox http://www.kyb.tuebingen.mpg.de/bs/people/spider/, 2004.

H. Stoppiglia, G. Dreyfus, R. Dubois, and Y. Oussar. Ranking a Random Feature for Variable and Feature Selection. *Journal of Machine Learning Research*, 3: 1399–1414, 2003.

R. Tibshirani. Regression Shrinkage and Selection via the Lasso. *Journal of the Royal Statistical Society. Series B(Methodological)*, 58(1):267–288, 1996.

M. E. Tipping. Sparse Bayesian Learning and the Relevance Vector Machine. *Journal of Machine Learning Research*, 1:211–244, 2001.

V. Vapnik and O. Chapelle. Bounds on Error Expectation for Support Vector Machines. *Neural Computation*, 12(9), 2000.

V. N. Vapnik. *Statistical Learning Theory.* John Wiley and Sons, New York, USA, 1998.

J. Weston, S. Mukherjee, O. Chapelle, M. Pontil, T. Poggio, and V. Vapnik. Feature selection for SVMs. In S.A. Solla, T.K. Leen, and K.-R. Müller, editors, *Advances in Neural Information Processing Systems*, volume 12, pages 526–532, Cambridge, MA, USA, 2000. MIT Press.

J. Weston, A. Elisseeff, B. Schölkopf, and M. Tipping. Use of the Zero-Norm with Linear Models and Kernel Methods. *Journal of Machine Learning Research*, 3: 1439–1461, March 2003.

C. Williams and D. Barber. Bayesian Classification with Gaussian Processes. *IEEE Transactions on Pattern Analysis and Machine Intelligence*, 12(20), 1998.

H. Wold. Soft modeling by latent variables; the nonlinear iterative partial least squares approach. In J. Gani, editor, *Perspectives in Probability and Statistics, Papers in Honours of M.S. Bartlett*, London, 1975. Academic Press.

J. Zhu and T. Hastie. Classification of Gene Microarrays by Penalized Logistic Regression. *Biostatistics*, 5(3):427–443, 2003.

Chapter 6

Information-Theoretic Methods

Kari Torkkola[1]

Motorola, Intelligent Systems Lab, Tempe, AZ, USA
Kari.Torkkola@motorola.com

6.1 Introduction

Shannon's seminal work on information theory provided the conceptual framework for communication through noisy channels (Shannon, 1948). This work, quantifying the information content of coded messages, established the basis for all current systems aiming to transmit information through any medium.

By using information theory, variable selection and feature construction can be viewed as coding and distortion problems. Variables or features can be understood as a "noisy channel" that conveys information about the message. The aim would be to select or to construct features that provide as much information as possible about the "message". The message usually comes in the form of another variable which could be the class label or the target for prediction or regression.

This chapter gives a brief tutorial to the use of information-theoretic concepts as components of various variable selection and feature construction methods. We begin by introducing basic concepts in information theory, mainly entropy, conditional entropy, and mutual information. Criteria for variable selection are then presented, concentrating on classification problems. Relationships between the Bayes error rate and mutual information are reviewed. Evaluation and optimization of these criteria are also discussed. Besides variable selection, this chapter also briefly touches feature construction, distance metric construction, and distributional clustering using information theory.

6.2 What is Relevance?

6.2.1 Defining Variable Selection and Feature Construction

Variable selection generally must contain two major components. There needs to be a criterion, which, given a set of variables, evaluates joint relevance of

K. Torkkola: *Information-Theoretic Methods*, StudFuzz **207**, 167–185 (2006)
www.springerlink.com

the set. The second component is a search mechanism that adds or removes variables to the current set (Chapter 4). It may also be that the criterion only evaluates the relevance of a single variable, or a small number of variables at a time. This may lead to different search strategies.

Feature construction problems may be divided in two categories depending upon whether the constructed features are a continuous or a discrete-valued function of the original variables. In the former case the relevance criterion should be expressible as a function of the training set, and it should be differentiable with respect to constructed features, which in turn should be differentiable with respect to parameters of the construction function. This would make it possible to perform numerical optimization of the criterion. The latter case leads to a similar search problem as in variable selection.

Both the selection and the construction problems require a definition of "relevance". Once defined, we can formalize both as follows. We denote the original variables by $X \in R^d$, and the selected subset by Φ. Here Φ could also denote features after construction, $\Phi_i = g(\boldsymbol{x}_i, \boldsymbol{\theta})$, where $\boldsymbol{\theta}$ denotes parameters of the construction function or the indices of selected variables, and $\boldsymbol{x}_i$ is the ith data point. The whole variable selection or feature construction process can then be written as

$$\boldsymbol{\theta}^* = \operatorname*{argmax}_{\boldsymbol{\theta}}[I(Y; \boldsymbol{\Phi}(X, \boldsymbol{\theta}))], \qquad (6.1)$$

subject to some constraints, such as the dimension of $\boldsymbol{\Phi}$, $d_\Phi < d$. Without constraints or regularization the optimal $\boldsymbol{\Phi}$ would be equal to the original variables, of course. I denotes a function that evaluates the relevance between some variable of interest Y, such as a class label, and a representation Φ. This chapter is mostly concerned with using *mutual information* as the function.

6.2.2 The Bayes Error is Relevant for Classification

In classification, relevance is closely related to discrimination. An optimal criterion should reflect the Bayes risk in the selected or transformed variable space (Chapter 1). The Bayes risk is defined in terms of a problem specific loss function as $e_{bayes}(X) = E_x[L(y, \hat{y})]$, where $\hat{y}$ denotes the estimated variable and y the true variable value. In the simplest case of 0/1-loss for classification this can be written as the Bayes error

$$e_{bayes}(X) = E_x[Pr(y \neq \hat{y})] = \int_x p(\boldsymbol{x}) \left(1 - \max_i(p(y_i|\boldsymbol{x}))\right) d\boldsymbol{x}, \qquad (6.2)$$

where $\boldsymbol{x}$ denotes the variable vector in the selected or transformed space and y_i denotes the class label. Note that the direct use of this criterion would require the full knowledge of posterior probability density functions of classes $p(y_i|\boldsymbol{x})$. Estimating the Bayes error in practice would thus require estimating posterior probability density functions of classes, and numerical integration of a nonlinear function of those, which is difficult given only a training data set.

Another noteworthy point is that the Bayes error is the lower bound on attainable error rate given perfect knowledge of the classes (and thus a perfect classifier). In practical problems, the aim is to train a predictor using a limited data set in order to minimize the generalization error of the predictor, i.e., the error on unseen future data drawn from the same joint distribution $p_{X,Y}$ as the training data. However, for a training data set of limited size and for a given predictor, and a given loss function, two problems with the same Bayes risk may result in very different generalization errors. This is why the predictor and the loss function should ideally also be included in the variable selection process (Kohavi and John, 1997, Tsamardinos and Aliferis, 2003).

There is a wide spectrum of other class separability measures mainly for feature construction problems, that could be roughly categorized from more heuristic to more principled measures as follows:

1. *Sums of distances between data points of different classes.* The distance metric between data points could be Euclidean, for example.
2. *Nonlinear functions of the distances or sums of the distances.*
3. *Probabilistic measures based on class conditional densities.* These measures may make an approximation to class conditional densities followed by some distance measure between densities. For example, this distance measure could be Battacharyya distance or divergence (Devijver and Kittler, 1982, Guorong et al., 1996, Saon and Padmanabhan, 2001). Some of these probabilistic measures can be shown to bound the Bayes error (Devijver and Kittler, 1982). A Gaussian assumption usually needs to be made about the class-conditional densities to make numerical optimization tractable. Equal class covariance assumption, although restrictive, leads to the well known Linear Discriminant Analysis (LDA), which has an analytic solution (Chapter 1). Some measures allow non-parametric estimation of the class conditional densities.
4. *The Bayes error.*

A thorough overview of all these measures is given by Devijver and Kittler (1982).

We now introduce some basic concepts in information theory, and discuss how they are related to class separability and the Bayes error. Later we move into actually using information-theoretic concepts in variable selection.

6.3 Information Theory

6.3.1 Elementary Concepts

Assume a continuous random variable $X \in R^d$ representing available variables or observations[1], and a discrete-valued random variable Y representing the class labels.

---

[1] If X is discrete, integrals in the following equations can be replaced by sums.

The uncertainty or entropy in drawing one sample of Y at random is expressed in terms of class prior probabilities. According to Shannon's definition

$$H(Y) = E_y[\log_2 \frac{1}{p(y)}] = -\sum_y p(y) \log_2(p(y)).$$ (6.3)

If all values of Y have equal probabilities the entropy is at maximum, decreasing as the "unevenness" of $p(Y)$ increases. Entropy can also be written for a continuous variable as

$$H(X) = E_x[\log_2 \frac{1}{p(x)}] = -\int_x p(x) \log_2(p(x)) dx.$$ (6.4)

Whereas (6.3) gives the absolute entropy of a discrete variable, (6.4) gives a differential entropy. It is only meaningful to compare entropies between two distributions rather than to look at the absolute values.

After having made an observation of a variable vector x, the uncertainty of the class identity is defined in terms of the conditional density $p(y|x)$ and it is accordingly called conditional entropy or equivocation (the average ambiguity of the received signal)

$$H(Y|X) = \int_x p(x) \left(-\sum_y p(y|x) \log_2(p(y|x)) \right) dx.$$ (6.5)

The expression in parenthesis is the entropy of Y for a particular value x. $H(Y|X)$ is thus the expectation of class entropy over all possible values of x.

The amount by which the class uncertainty is reduced after having observed the variable vector x is called the mutual information between X and Y. It can be written as

$$I(Y, X) = H(Y) - H(Y|X)$$ (6.6)

$$= \sum_y \int_x p(y, x) \log_2 \frac{p(y, x)}{p(y)p(x)} dx$$ (6.7)

Equation (6.7) can be obtained from (6.6) by using the identities $p(y, x) = p(y|x)p(x)$ and $p(y) = \int_x p(y, x) dx$. Entropy and mutual information are often illustrated in the form of a Venn diagram as in Fig. 6.1.

Mutual information measures dependence between variables, in this case between Y and X. It vanishes if and only if $p(y, x) = p(y)p(x)$, that is, when the joint density of Y and X can be factored as a product of marginal densities. This is the condition for independence. Mutual information can also be seen as the Kullback-Leibler divergence measure between $p(y, x)$ and its factored form $p(y)p(x)$. In general, for two densities $p_1(x)$ and $p_2(x)$, the Kullback-Leibler divergence $D_{KL}(p_1||p_2)$ is defined as

$$D_{KL}(p_1||p_2) = \int_x p_1(x) \log_2 \frac{p_1(x)}{p_2(x)} dx.$$ (6.8)

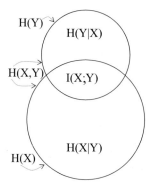

Fig. 6.1. Relationships between the entropies of two dependent variables X and Y. $H(X)$ and $H(Y)$ are each represented by a circle. Joint entropy $H(X,Y)$ consists of the union of the circles, and mutual information $I(X;Y)$ the intersection of the circles. Thus $H(X,Y) = H(X) + H(Y) - I(X;Y)$.

6.3.2 Channel Coding Theorem and Rate-Distortion Theorem

Shannon's work shows that the mutual information (MI) given by (6.7) is the solution to several important problems in communication. The two most well known results are the channel coding theorem and the rate-distortion theorem.

Shannon showed that the rate of transmission of information through a channel with input X and output Y' is $R = H(X) - H(X|Y') = I(X, Y')$. The capacity of this particular (fixed) channel is defined as the maximum rate over all possible input distributions, $C = \max_{p(X)} R$. Maximizing the rate means thus choosing an input distribution that matches the channel since we cannot have an effect on the channel itself. The maximization is possibly carried out under some constraints, such as fixed power or efficiency of the channel.

Variable selection and feature construction have the following analogy to the communication setting. Channel input now consists of the available variables X, which can be thought of as a result of source coding. By some process the real source Y is now represented (encoded) as X. In contrast to the communication setting, now the channel input distribution X is fixed, but we can modify how the input is communicated to the receiver by the channel either by selecting a subset of available variables or by constructing new features. Both can be represented by modeling the channel output as $\Phi = g(X, \theta)$ where g denotes some selection or construction function, and θ represents some tunable parameters. In Shannon's case θ was fixed but X was subject to change. Now the "channel" capacity can be represented as $C = \max_\theta R$ subject to some constraints, such as keeping the dimensionality of the new feature representation as a small constant. Maximizing MI between Y and Φ thus produces a representation that provides maximal information about Y.

The rate-distortion theorem is concerned with finding the simplest representation (in terms of bits/sec) to a continuous source signal within a given tolerable upper limit of distortion. Such a signal representation could then be transmitted through a channel without wasting the channel capacity. In this case the solution for a given distortion D is the representation Φ that minimizes the rate $R(D) = \min_{E(d) \leq D} I(X, \Phi)$. This rate-distortion function alone is not as relevant to the feature selection problem as the channel capacity. However, a combination of the two results in a loss function that does not require setting constraints to the dimensionality of the representation, rather it emerges as the solution (Tishby et al., 1999). We discuss briefly this *information bottleneck* in Section 6.7.

6.4 Information-Theoretic Criteria for Variable Selection

The variable selection problem can now be viewed as a communication problem. The aim would be to construct the "channel" by selecting a variable or a set of variables X that maximizes the information transmitted about the message Y as measured by mutual information $I(X, Y)$. According to previous treatment, this determines the best rate of information about the message Y. We discuss next how MI bounds the Bayes error in classification problems, and how MI is related to other criteria, such as maximum likelihood estimation.

6.4.1 Optimality of MI in Classification

Since the Bayes error is the ultimate criterion for any procedure related to discrimination, any proxy criterion such as mutual information should be related to the Bayes error.

Antos et al. (1999) have shown that no Bayes error estimate can be trusted for all data distributions, not even if the sample size tends to infinity. Even though there are consistent classification rules (such as k-nn), their convergence to the Bayes error may be arbitrarily slow. It is thus not possible *ever* to make claims about universally superior feature extraction method (for classification).

A similar notion is discussed by Feder et al. (1992) and Feder and Merhav (1994). The Bayes error is not uniquely determined by conditional entropy, but the latter provides bounds for the former. Although intuitively thinking, the higher the uncertainty of a variable, the more difficult the prediction of a variable would appear to be, this is not always the case. Two random variables with the same conditional entropy may have different Bayes error rates, within the following bounds.

An upper bound on the Bayes error

$$e_{bayes}(X) \leq \frac{1}{2}H(Y|X) = \frac{1}{2}(H(Y) - I(Y, X)) \tag{6.9}$$

was obtained by Hellman and Raviv (1970) for the binary case and Feder and Merhav (1994) for the general case. A lower bound on the error also involving conditional entropy or mutual information is given by Fano's inequality (Fano, 1961)

$$e_{bayes}(X) \geq 1 - \frac{I(Y,X) + \log 2}{\log(|Y|)}, \tag{6.10}$$

where $|Y|$ refers to the cardinality of Y. Extensions to this bound have been presented by Han and Verdú (1994). Both bounds are minimized when the mutual information between Y and X is maximized, or when $H(Y|X)$ is minimized. This will serve as one justification of using $I(Y,X)$ as a proxy to the Bayes error. These bounds are relatively tight, in the sense that both inequalities can be obtained with equality. Every point in the boundary can be attained (depending on $p(y|\boldsymbol{x})$), as discussed by Feder and Merhav (1994).

6.4.2 Alternative Views to Mutual Information and to Conditional Entropy

Besides the Bayes error bounds, usage of MI or conditional entropy in variable selection or feature construction can also be justified from several different points of view. We enumerate some of them here.

1. *Entropy of the posteriors.* In order to make classifying each point $\boldsymbol{x}$ in the variable (or feature) space as certain as possible without ambiguity, the posterior probability of just one class $p(y_i|\boldsymbol{x})$ should dominate, and $p(y_j|\boldsymbol{x})$, $j \neq i$, should be close to zero. This can be conveniently quantified by entropy of the class label distribution. For a given $\boldsymbol{x}$, the class label entropy $-\sum_i p(y_i|\boldsymbol{x}) \log_2(p(y_i|\boldsymbol{x}))$ should be as small as possible, and thus one should minimize this over the entire space leading to minimizing $E_{\boldsymbol{x}}[-\sum_i p(y_i|\boldsymbol{x}) \log_2(p(y_i|\boldsymbol{x}))]$, which, in fact, is the definition of $H(Y|X)$.
2. *Divergence between class conditional densities and marginal density.* For independent x, y, we can write $p(x,y) = p(x)p(y)$, whereas in general, $p(x,y) = p(x|y)p(y)$. One should thus attempt to maximize a distance measure between $p(x)$ and $p(x|y)$ in order to find x maximally dependent on y, the labels. This can also be viewed as making each class conditional density $p(x|y)$ stand out from the marginal density $p(x)$. The KL-divergence

$$D_{KL}(p(\boldsymbol{x}|y)||p(y)) = \int_{\boldsymbol{x}} p(\boldsymbol{x}|y) \log_2 \frac{p(\boldsymbol{x}|y)}{p(\boldsymbol{x})} d\boldsymbol{x} \tag{6.11}$$

between the two should be maximized over all classes. Taking the expectation with respect to y gives

$$E_y[D_{KL}(p(\boldsymbol{x}|y)||p(y))] = \sum_y p(y) \int_{\boldsymbol{x}} p(\boldsymbol{x}|y) \log_2 \frac{1}{p(\boldsymbol{x})} p(\boldsymbol{x}|y) d\boldsymbol{x}$$

$$= \sum_y \int_{\boldsymbol{x}} p(\boldsymbol{x},y) \log_2 \frac{p(\boldsymbol{x},y)}{p(\boldsymbol{x})p(y)} d\boldsymbol{x}, \tag{6.12}$$

which can be seen to be the expression for mutual information between $\boldsymbol{x}$ and y.

3. *Maximum likelihood.* Conditional entropy is also related to maximum likelihood parameter estimation (Peltonen and Kaski, 2005). Log-Likelihood of a generative parametric model of the posterior densities of classes equals

$$L(\boldsymbol{\theta}) = E_{\boldsymbol{x},y}[\log_2 \hat{p}(y|\boldsymbol{x};\boldsymbol{\theta})] = \sum_y \int_{\boldsymbol{x}} p(\boldsymbol{x},y) \log_2 \hat{p}(y|\boldsymbol{x};\boldsymbol{\theta}) d\boldsymbol{x}$$

$$= \sum_y \int_{\boldsymbol{x}} p(\boldsymbol{x})p(y|\boldsymbol{x}) \log_2 \hat{p}(y|\boldsymbol{x};\boldsymbol{\theta}) d\boldsymbol{x}$$

$$= \int_{\boldsymbol{x}} p(\boldsymbol{x}) \left(\sum_y p(y|\boldsymbol{x}) \log_2 \hat{p}(y|\boldsymbol{x};\boldsymbol{\theta}) \right) d\boldsymbol{x}$$

$$= \int_{\boldsymbol{x}} p(\boldsymbol{x}) \left(\sum_y p(y|\boldsymbol{x}) \log_2 p(y|\boldsymbol{x}) - D_{KL}(p(y|\boldsymbol{x})||\hat{p}(y|\boldsymbol{x};\boldsymbol{\theta})) \right) d\boldsymbol{x}$$

$$= -H(Y|X) - E_{\boldsymbol{x}}[D_{KL}(p(y|\boldsymbol{x})||\hat{p}(y|\boldsymbol{x};\boldsymbol{\theta}))]. \tag{6.13}$$

The true data generating model is denoted by $p(y|\boldsymbol{x})$ and the generative parametric model by $\hat{p}(y|\boldsymbol{x};\boldsymbol{\theta})$. Thus, maximizing the likelihood of a generative model is equivalent to minimization of the sum of the conditional entropy and the discrepancy between the model and the true data generator. In the large sample limit with consistent density estimators the latter term vanishes.

6.4.3 An Example in Variable Ranking using Maximum Mutual Information

Mutual information is widely used in decision tree construction to rank variables one at a time (Breiman et al., 1984, Quinlan, 1993). In this context, and in some others, for example, in text classification, it has been called *Information Gain* (IG). We describe first variable ranking in decision trees, after which the setting is changed to a typical variable selection problem.

Given a set of training data at a node in a decision tree, the aim is to choose one variable and a test on values of the variable, such that the chosen test best separates the classes. IG is defined as the reduction of class label entropy when the class distribution of training data in a branch of the decision tree is compared to average entropies of the new s_k partitions (new branches of the tree) constructed using a test on values of chosen variable k.

$$I_{\text{gain}}(k) = H(\hat{p}) - \sum_{i=1}^{s_k} \frac{n_i}{n} H(\hat{p}_{ik}) \tag{6.14}$$

Here $\hat{p}$ denotes the class label distribution of a set of data in a node of a decision tree before making a split, $\hat{p}_{ik}$ denotes the label class distribution in the ith partition, and n_i denotes the number of examples in partition i. Entropies are usually estimated using simply observed counts as in (6.16). Comparing to (6.6), information gain can be seen to be essentially the same expression as the mutual information between variable k and the class label. If variable k is continuous, the split is made into two partitions, finding a threshold that maximizes IG. Continuous and discrete variables can thus be compared on the same basis. Note that this corresponds to discretizing the continuous variable in two levels for the current split.

However, IG favors variables that result in splitting the node into a large number of trivial partitions. This is a problem with discrete variables that have a large number of levels and little training data. As a remedy, the information gain ratio (IGR) has been suggested, which is defined as the IG normalized by the entropy of the split itself (Quinlan, 1993):

$$I_{\text{gain ratio}}(k) = I_{\text{gain}}(k)/H(\hat{p}_k) \qquad (6.15)$$

where

$$H(\hat{p}_k) = -\sum_{i=1}^{s_k} \frac{n_i}{n} \log_2 \frac{n_i}{n}. \qquad (6.16)$$

This leaves only the effect of class entropy within new nodes, not in the number of new nodes.

Instead of IG, the Gini-index has also been used in decision tree construction for CART (Breiman et al., 1984). For two classes, the Gini-index replaces $\sum_{i=1}^2 p_i \log_2 p_i$ by $2 \sum_{i=1}^2 (1 - p_i)p_i$. The behavior of the two functions is very similar.

In a typical variable selection problem, one wishes to evaluate (6.14) for each variable X_k and perhaps pick the highest ranking one to the current working set of variables (Section 6.4.5). Equation (6.14) can now be written as

$$
\begin{aligned}
I(Y, X_k) &= H(Y) - H(Y|X_k) \\
&= H(Y) - \sum_{i=1}^{s_k} p(x_{ki})H(Y|x_{ki}) \\
&= \sum_{j=1}^{m} p(y_j) \log_2 p(y_j) - \sum_{i=1}^{s_k} p(x_{ki}) \sum_{j=1}^{m} p(y_j|x_{ki}) \log_2 p(y_j|x_{ki}).
\end{aligned}
\qquad (6.17)
$$

Here x_{ki} denotes the ith discrete value of X_k, $p(x_{ki})$ denotes its probability of occurrence in X_k, and Y denotes the class labels. In practice, the probabilities have been replaced by empirical frequencies, which unfortunately introduces a bias (Section 6.4.4). IG has been often used for variable ranking in text analysis and information retrieval applications (Yang and Pedersen, 1997, Forman, 2003).

6.4.4 Evaluation or Optimization of MI in Practice

MI is defined in terms of full knowledge of the joint density. In practice this is not available. Between two variables the non-parametric histogram approach can be used (Battiti, 1994), but in higher dimensions any amount of data is too sparse to bin. Thus the practical estimation of MI from data based on (6.7) is difficult. In higher dimensions one might have to resort to simple parametric class density estimates (such as Gaussians) and plug them into the definition of MI.

Since we might most often be interested in evaluating the conditional entropy or MI, and MI can be expressed as a difference between two entropies, the estimation of entropy is a key issue. The simplest way is the maximum likelihood estimate based on histograms (such as (6.16)). This estimate is known to have a negative bias that can be corrected to some extent by the so-called Miller-Madow bias correction. This consists of adding $(\hat{m}-1)/2N$ to the estimate, where $\hat{m}$ denotes an estimate of the number of bins with nonzero probability, and N is the number of observations. As shown by Paninski (2003), this cannot be done in many practical cases, such as when the number of bins is close to the number of observations.

Bayesian techniques can be used if some information about the underlying probability density function is available in terms of a prior (Wolpert and Wolf, 1995, Zaffalon and Hutter, 2002). The same applies to the estimation of MI.

At this point it is worthwhile noting that stochastic optimization does not necessarily require estimating the underlying quantity but only its gradient with respect to parameters to be optimized. This is applicable in feature construction problems where the MI is a function of the constructed features, which in turn are a function of a parameter to be optimized to maximize the MI (Section 6.5).

6.4.5 Usage of MI in Variable Selection

The purpose of variable selection is twofold. First, to remove irrelevant variables and, second, to remove redundant variables.

Simple variable ranking by any criterion, including information-theoretic criteria, is sub-optimal. This may be good enough in many cases, though. It is clear that addressing redundancy requires evaluating a *joint* measure of relevancy, such as MI. Measuring irrelevancy cannot be done one variable at a time, either. It is not difficult to devise cases where the single most relevant variable according to any criterion is not among the joint two most relevant variables.

Thus greedy variable selection based on ranking may not work very well when there are dependencies among relevant variables. Because the evaluation of a joint criterion is often hard or impossible (See Section 6.4.4), except for a small number of variables (such as two), there are a number of practical approaches that remove the overlap or redundancies in the selected variable set by evaluating a *pairwise* criterion. One such approach is MIFS (mutual information based feature selection) (Battiti, 1994) and variants. MIFS adds that variable $\hat{X}$ to the current set of chosen variables Φ, which maximizes

$$I(Y,\hat{X}) - \beta \sum_{X_j \in \Phi} I(\hat{X}, X_j). \qquad (6.18)$$

The first term attempts to maximize the relevancy of $\hat{X}$ to Y, and the second term attempts to minimize the redundancy between $\hat{X}$ and the already chosen set Φ. The balance between maximizing relevance and minimizing redundancy is determined by parameter β.

MIFS
1: Set $\hat{X} = \operatorname{argmax}_{X_i} I(Y, X_i)$;
    set $\Phi \leftarrow \{\hat{X}\}$;
    set $F \leftarrow \{X_1, ..., X_N\} \setminus \{\hat{X}\}$.
2: For all pairs (i,j), $X_i \in F$ and $X_j \in \Phi$
    evaluate and save $I(X_i, X_j)$ unless already saved.
3: Set $\hat{X} = \operatorname{argmax}_{X_i} \left[I(Y, X_i) - \beta \sum_{X_j \in \Phi} I(X_i, X_j) \right]$;
    set $\Phi \leftarrow \Phi \cup \{\hat{X}\}$;
    set $F \leftarrow F \setminus \{\hat{X}\}$,
    and repeat from step 2 until $|\Phi|$ is desired.

This is, of course, only a heuristic approximation to evaluating the full joint criterion, which is nearly impossible to estimate when d_Φ is high. The same basic strategy with minor variations has also been adopted in other works (Yu and Liu, 2003, Vasconcelos, 2003, Fleuret, 2004). In some special cases a modification of this approximation can be shown to be equal to optimizing the true joint criterion (Vasconcelos, 2003).

6.4.6 Usefulness of Measures Other Than Shannon's

Shannon derived the entropy measure axiomatically and showed, for example, that no other measure would fulfill all the axioms. However, if the aim is not to calculate an absolute value of the entropy or divergence, but rather to find a distribution that minimizes/maximizes the entropy or divergence, the axioms used in deriving the measure can be relaxed and still the result of the optimization is the same distribution (Kapur, 1994). One example is the Renyi entropy (Renyi, 1961, Kapur, 1994), which is defined for a discrete variable Y and for a continuous variable X as

$$H_\alpha(Y) = \frac{1}{1-\alpha} \log_2 \sum_y p(y)^\alpha; \qquad H_\alpha(X) = \frac{1}{1-\alpha} \log_2 \int_x p(\boldsymbol{x})^\alpha d\boldsymbol{x},$$

$$(6.19)$$

where $\alpha > 0$, $\alpha \neq 1$, and $\lim_{\alpha \to 1} H_\alpha = H$. This is a parametric family that has the Shannon entropy as one member of the family ($\alpha \to 1$), but could exhibit other desirable properties with some other parameter values. For example, quadratic Renyi entropy is straightforward to estimate from a set of samples using the Parzen window approach (Principe et al., 2000). The Renyi entropy has been earlier used in feature construction and image registration (Principe et al., 2000, Hero et al., 2001).

6.4.7 Markov Blankets

A Markov Blanket (MB) of a joint probability distribution of a target variable is defined as the minimal union of all variables that make the target independent of all other variables (strongly relevant features) (Koller and Sahami, 1996). Thus if we have a variable set $\boldsymbol{X}$, the Markov blanket of a target variable Y, denoted by $M(\boldsymbol{X}, Y)$ is the smallest subset of $\boldsymbol{X}$ such that Y is independent of the rest of the variables $\boldsymbol{X} \setminus M(\boldsymbol{X}, Y)$. Thus the Markov blanket minimizes $D_{KL}[p(Y|M(\boldsymbol{X}, Y))||p(Y|\boldsymbol{X})]$. All information to estimate the probability distribution of Y is contained in $M(\boldsymbol{X}, Y)$. However, for a 0/1-loss function, only the most probable classification is needed, and thus a MB may contain unnecessary variables. Redundancy between the features must be dealt with other methods.

Since MB does not contain irrelevant variables, finding the MB might be useful as the first step in feature selection in applications where the data has a large number of irrelevant variables. Such applications are common in biology. Gene expression analysis is one an example (Aliferis et al., 2003, Xing et al., 2001).

Inducing the MB may take time exponential in the size of the MB if inferring a full dependency network is attempted. To overcome this limitation, fast alternatives have been developed recently (Aliferis et al., 2003, Tsamardinos et al., 2003). Extracting the MB from a decision tree after training it for classification has been shown to outperform methods that directly attempt to infer the dependency network (Frey et al., 2003). This approach is very similar to extracting important variables from a random forest after having trained it to a classification task first (Breiman, 2001, Tuv, 2005).

6.5 MI for Feature Construction

As a criterion for feature construction, MI is as valid as for variable selection. Given a set of training data $\{\boldsymbol{x}_i, y_i\}$ as samples of a continuous-valued random variable X, $\boldsymbol{x}_i \in R^d$, and class labels as samples of a discrete-valued random variable Y, $y_i \in \{1, 2, ..., m_y\}, i \in \{1, ..., m\}$, the objective is to find a transformation (or its parameters $\boldsymbol{\theta}$) to $\boldsymbol{\Phi}_i \in R^{d_\Phi}, d_\Phi < d$ such that $\boldsymbol{\Phi}_i = g(\boldsymbol{\theta}, \boldsymbol{x}_i)$ maximizes $I(Y, \Phi)$, the mutual information (MI) between transformed data Φ and class labels Y (6.1). The procedure is depicted in Fig. 6.2. The crux is of course to express I as a function of the data set, $I(\{\boldsymbol{\Phi}_i, y_i\})$, in a differentiable form. Once that is done, we can perform gradient ascent on I as follows

$$\boldsymbol{\theta}_{t+1} = \boldsymbol{\theta}_t + \eta \frac{\partial I}{\partial \boldsymbol{\theta}} = \boldsymbol{\theta}_t + \eta \sum_{i=1}^{m} \frac{\partial I}{\partial \boldsymbol{\Phi}_i} \frac{\partial \boldsymbol{\Phi}_i}{\partial \boldsymbol{\theta}}. \tag{6.20}$$

In its simplest form $\boldsymbol{\Phi}_i = S \boldsymbol{x}_i$, where S is a $d_\Phi \times d$ projection matrix. In the general case $\boldsymbol{\Phi}_i$ could be any nonlinear parametric transform of $\boldsymbol{x}_i$, such as a neural network.

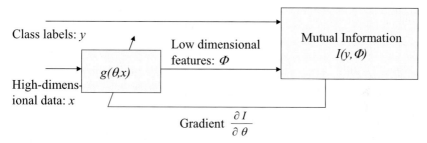

Fig. 6.2.   Learning feature transforms by maximizing the mutual information between class labels and transformed features.

No combinatorial optimization is required as in the search of the best variable combination. Instead, a continuous function optimization problem needs to be solved. Evaluating the actual MI and its full gradient can be hard to do in the general multi-class case without resorting to modeling the classes as single Gaussians. This is exactly the direction some research has taken (Guorong et al., 1996, Saon and Padmanabhan, 2001).

Another alternative is to approximate the MI by a criterion that allows computational shortcuts. One example is to replace the KL-divergence between the joint density and the product of marginals by a quadratic divergence which is simpler to differentiate and to evaluate in a non-parametric fashion (Devijver and Kittler, 1982, Aladjem, 1998, Principe et al., 2000, Torkkola, 2003).

Optimization can also be done using the stochastic gradient, which means that the actual function may not need to be evaluated accurately but only its gradient with respect to a single data point (Torkkola, 2003, Peltonen and Kaski, 2005).

6.6 Information Theory in Learning Distance Metrics

Variable selection and feature construction can be seen as two particular ways to ignore irrelevant information in the data and to make relevant information more explicit. In continuous feature spaces there exists another alternative for the same purpose: learning a distance metric that reflects the relevance of the target variable.

Both selection and (linear) feature construction can be expressed as $\boldsymbol{\Phi}_i = S\boldsymbol{x}_i$ where S is a $d_{\Phi} \times d$ projection or selection matrix. These selected or constructed features implicitly define a global Euclidean metric between two original data samples $\boldsymbol{x}$ and $\boldsymbol{x}'$ as

$$d_A^2(\boldsymbol{x}, \boldsymbol{x}') = (\boldsymbol{x} - \boldsymbol{x}')^T S^T S(\boldsymbol{x} - \boldsymbol{x}') = (\boldsymbol{x} - \boldsymbol{x}')^T A(\boldsymbol{x} - \boldsymbol{x}'). \qquad (6.21)$$

Now, A is fixed throughout the original feature space since variable selection and feature construction are typically global in the sense that the selected or constructed features are then used the same way everywhere in the sample space. Whatever criterion for selection or construction is used, the attempt is to find such a matrix A that minimizes or maximizes its expected value over the whole sample space. As A is a result of global optimization, it may not be optimal locally. A particular feature may be more relevant in one location in the sample space whereas another feature may be more relevant elsewhere.

In a more general case the matrix A depends on the location of the data point:

$$d_A^2(\boldsymbol{x}, \boldsymbol{x} + d\boldsymbol{x}) = d\boldsymbol{x}^T A(\boldsymbol{x})d\boldsymbol{x}. \qquad (6.22)$$

It is possible to *learn a metric* in the original space of variables that varies according to location, as the relevance to an auxiliary variable Y also varies according to the location. The distance between two nearby points in the feature space should reflect the difference in the conditional distributions of the auxiliary variable $p(y|\boldsymbol{x})$ as follows:

$$d_J^2(\boldsymbol{x}, \boldsymbol{x} + d\boldsymbol{x}) = D_{KL}[p(y|\boldsymbol{x})||p(y|\boldsymbol{x} + d\boldsymbol{x})] = \frac{1}{2}d\boldsymbol{x}^T J(\boldsymbol{x})d\boldsymbol{x}. \qquad (6.23)$$

$J(\boldsymbol{x})$ equals the Fisher information matrix, which is derived from the Kullback-Leibler divergences between the conditional auxiliary variable at two different locations, $\boldsymbol{x}$ and $\boldsymbol{x} + d\boldsymbol{x}$. It scales the local distance between the two locations (Kaski and Sinkkonen, 2004, Peltonen et al., 2004). Rather than learning or selecting features that remain fixed throughout the feature space, this results in learning local metrics that enhance information relevant to the task. Distances between distant points are defined as minimum path integrals of (6.23). The metric then needs to be embedded into, for example, a clustering algorithm resulting in "semi-supervised" clustering that reflects the auxiliary variable (Peltonen et al., 2004).

6.7 Information Bottleneck and variants

The information bottleneck method (IB) takes a rate-distortion approach to feature construction by finding a representation Φ of original variables X so as to minimize the loss function

$$\mathcal{L}(p(\phi|x)) = I(X, \Phi) - \beta I(\Phi, Y). \tag{6.24}$$

The loss function is a compromise between trying to preserve as much information as possible about a "relevant" variable Y while at the same time attempting to form as compact a representation as possible by minimizing $I(X, \Phi)$ (Tishby et al., 1999). The representation Φ can be seen as a bottleneck that extracts relevant information about Y from X.

The solution to the minimization problem is a set of iterative equations. One of the equations calculates Φ using the KL-divergence between $p(Y|X)$ and $p(Y|\Phi)$. Viewing this KL-divergence as an appropriate distortion measure between conditional distributions, the algorithm can be seen as very similar in structure to the EM algorithm. Indeed, it has been shown recently that both the IB and the EM algorithms are special cases of more general Bregman divergence optimization (Banerjee et al., 2004).

In its original form IB constructed clusters of original variables as the new representation Φ (Tishby et al., 1999). For example, if the variables X represent co-occurrence data, such as whether a particular word occurred in document x_i or not, IB finds a reduced number of word clusters by grouping together words such that the new co-occurrence data Φ provides as much information as possible about the document class or identity. This is a very natural representation in the text domain. Extensions exist for multivariate cases, for continuous variables, and for feature construction, in which case Φ would be a function of the original variables (Globerson and Tishby, 2003).

6.8 Discussion

Information theory provides a principled way of addressing many of the problems in variable selection and feature construction. Shannon's seminal work showed how mutual information provides a measure of the maximum transmission rate of information through a channel. This chapter presented an analogy to variable selection and feature construction with mutual information as the criterion to provide maximal information about the variable of interest.

Even though this chapter has elaborated more on classification, selecting or constructing features that maximize the mutual information between the variable of interest Y and the features Φ has a wider applicability. Such features constitute a maximally informative statistic of Y, which is a generalization of the concept of a sufficient statistic (Wolf and George, 1999). Any inference strategy based on Y may be replaced by a strategy based on a sufficient statistic. However, a sufficient statistic of a low dimensionality may not necessarily exist, whereas a maximally informative statistic exists.

Pearson's chi-squared (χ^2) test is commonly used in statistics to test whether two observed discrete distributions are produced by the same underlying distribution (or to test whether two distributions are independent). In feature selection literature this statistic is also called the strength of association (Chapter 3). The test statistic is $\chi^2 = \sum_i (O_i - E_i)^2 / E_i$, where O_i

are the observed counts and E_i are the expected counts. However, Pearson developed the χ^2 test only as a computationally convenient approximation[2] to the log-likelihood ratio test or the $\mathcal{G}$ test because calculating likelihood ratios was laborious at the time. The statistic is $\mathcal{G} = 2\sum_i O_i \log(O_i/E_i)$. Now, this can be seen as the Kullback-Leibler divergence between the two distributions.

In supervised learning one typically minimizes the Mean Squared Error (MSE) between the learner output and the target variable. This is equivalent to minimizing a second order statistic, the energy or the variance of the error. If the pdf of the error signal is Gaussian this is naturally enough. However, this may not be the case. In order to transfer all information from the target variable to the learner, one should constrain the statistics of all orders, not just the second. One way to do this is to minimize the error entropy as discussed by Erdogmus et al. (2002, 2003). Likewise, mutual information has a connection to the Canonical Correlation Analysis (CCA), a method that makes use of second order cross-statistic. CCA attempts to maximize the correlation between linear transforms of two different sets of variables. Assuming Gaussian input variables, maximizing the mutual information between the outputs is equivalent to CCA (Becker, 1996, Chechik et al., 2005). Again, if the Gaussianity does not hold, and if the transform is anything more than a linear transform, maximizing the MI provides a more general solution.

Unlike in communications applications, in statistical inference from data, the underlying true densities appearing in the definitions of information-theoretic criteria are generally not available. The application of these criteria hinges then on estimating them accurately enough from available data. As discussed by Paninski (2003), no unbiased estimator for entropy or mutual information exists. However, recent work towards lower bias estimators holds some promise even for small data sets (Paninski, 2003, Kraskov et al., 2003). Furthermore, applying the information-theoretic criteria optimally would entail estimating joint criteria of multiple variables, which is a hard problem by itself. Current work has approached the problem either by avoiding it, by devising strategies where evaluation of pairwise MI suffices, or by approximating it by parametric estimation of the underlying densities or by replacing the definition of MI by an approximation that is easier to compute. Future work needs to characterize which approach is appropriate for what kind of data.

References

M.E. Aladjem. Nonparametric discriminant analysis via recursive optimization of Patrick-Fisher distance. *IEEE Transactions on Systems, Man, and Cybernetics*, 28(2):292–299, April 1998.

C. Aliferis, I. Tsamardinos, and A. Statnikov. HITON, a novel Markov blanket algorithm for optimal variable selection. In *Proceedings of the 2003 American Medical*

---

[2]The approximation is not valid for small observed counts.

Informatics Association (AMIA) Annual Symposium, pages 21–25, Washington, DC, USA, November 8-12 2003.

A. Antos, L. Devroye, and L. Gyorfi. Lower bounds for Bayes error estimation. *IEEE Transactions on PAMI*, 21(7):643–645, July 1999.

A. Banerjee, I. Dhillon, J. Ghosh, and S. Merugu. An information theoretic analysis of maximum likelihood mixture estimation for exponential families. In *Proc. International Conference on Machine Learning (ICML)*, pages 57–64, Banff, Canada, July 2004.

R. Battiti. Using mutual information for selecting features in supervised neural net learning. *Neural Networks*, 5(4):537–550, July 1994.

S. Becker. Mutual information maximization: Models of cortical self-organization. *Network: Computation in Neural Systems*, 7(1), February 1996.

L. Breiman. Random forests. *Machine Learning*, 45(1):5–32, 2001.

L. Breiman, J.F. Friedman, R.A. Olshen, and P.J. Stone. *Classification and regression trees*. Wadsworth International Group, Belmont, CA, 1984.

G. Chechik, A. Globerson, N. Tishby, and Y. Weiss. Information bottleneck for gaussian variables. *Journal of Machine Learning Research*, 6:168–188, 2005.

P.A. Devijver and J. Kittler. *Pattern recognition: A statistical approach*. Prentice Hall, London, 1982.

D. Erdogmus, K.E. Hild, and J.C. Principe. Online entropy manipulation: Stochastic information gradient. *IEEE Signal Processing Letters*, 10:242–245, 2003.

D. Erdogmus, J.C. Principe, and K.E. Hild. Beyond second order statistics for learning: A pairwise interaction model for entropy estimation. *Natural Computing*, 1: 85–108, 2002.

R.M. Fano. *Transmission of Information: A Statistical theory of Communications*. Wiley, New York, 1961.

M. Feder and N. Merhav. Relations between entropy and error probability. *IEEE Trans. on Information Theory*, 40:259–266, 1994.

M. Feder, N. Merhav, and M. Gutman. Universal prediction of individual sequences. *IEEE Trans. on Information Theory*, 38:1258–1270, 1992.

F. Fleuret. Fast binary feature selection with conditional mutual information. *Journal of Machine Learning Research*, 5:1531–1555, 2004.

G. Forman. An extensive empirical study of feature selection metrics for text classification. *Journal of Machine Learning Research*, 3:1289–1305, March 2003.

L. Frey, D. Fisher, I. Tsamardinos, C. Aliferis, and A. Statnikov. Identifying Markov blankets with decision tree induction. In *Proc. of IEEE Conference on Data Mining*, Melbourne, FL, USA, Nov. 19-22 2003.

A. Globerson and N. Tishby. Sufficient dimensionality reduction. *Journal of Machine Learning Research*, 3:1307–1331, 2003.

X. Guorong, C. Peiqi, and W. Minhui. Bhattacharyya distance feature selection. In *Proceedings of the 13th International Conference on Pattern Recognition*, volume 2, pages 195 – 199. IEEE, 25-29 Aug. 1996.

T. S. Han and S. Verdú. Generalizing the fano inequality. *IEEE Trans. on Information Theory*, 40(4):1147–1157, July 1994.

M.E. Hellman and J. Raviv. Probability of error, equivocation and the Chernoff bound. *IEEE Transactions on Information Theory*, 16:368–372, 1970.

A.O. Hero, B. Ma, O. Michel, and J. Gorman. Alpha-divergence for classification, indexing and retrieval. Technical Report CSPL-328, University of Michigan Ann Arbor, Communications and Signal Processing Laboratory, May 2001.

J.N. Kapur. *Measures of information and their applications*. Wiley, New Delhi, India, 1994.

S. Kaski and J. Sinkkonen. Principle of learning metrics for data analysis. *Journal of VLSI Signal Processing, special issue on Machine Learning for Signal Processing*, 37:177–188, 2004.

R. Kohavi and G.H. John. Wrappers for feature subset selection. *Artificial Intelligence*, 97:273–324, 1997.

D. Koller and M. Sahami. Toward optimal feature selection. In *Proceedings of ICML-96, 13th International Conference on Machine Learning*, pages 284–292, Bari, Italy, 1996.

A. Kraskov, H. Stögbauer, and P. Grassberger. Estimating mutual information. e-print arXiv.org/cond-mat/0305641, 2003.

L. Paninski. Estimation of entropy and mutual information. *Neural Computation*, 15:1191–1253, 2003.

J. Peltonen and S. Kaski. Discriminative components of data. *IEEE Transactions on Neural Networks*, 2005.

J. Peltonen, A. Klami, and S. Kaski. Improved learning of Riemannian learning metrics for exploratory analysis. *Neural Networks*, 17:1087–1100, 2004.

J.C. Principe, J.W. Fisher III, and D. Xu. Information theoretic learning. In Simon Haykin, editor, *Unsupervised Adaptive Filtering*. Wiley, New York, NY, 2000.

J. R. Quinlan. *C4.5: Programs for Machine Learning*. Morgan Kaufmann, San Mateo, CA, 1993.

A. Renyi. On measures of entropy and information. In *Proceedings of the Fourth Berkeley Symposium on Mathematical Statistics and Probability*, pages 547–561. University of California Press, 1961.

G. Saon and M. Padmanabhan. Minimum Bayes error feature selection for continuous speech recognition. In T. K. Leen, T. G. Dietterich, and V. Tresp, editors, *Advances in Neural Information Processing Systems 13 (Proc. NIPS'00)*, pages 800–806. MIT Press, 2001.

C. Shannon. A mathematical theory of communication. *The Bell System Technical Journal*, 27:379–423, 623–656, July, October 1948.

N. Tishby, F. Pereira, and W. Bialek. The information bottleneck method. In *Proceedings of the 37-th Annual Allerton Conference on Communication, Control and Computing*, pages 368–377, 1999.

K. Torkkola. Feature extraction by non-parametric mutual information maximization. *Journal of Machine Learning Research*, 3:1415–1438, March 2003.

I. Tsamardinos, C. Aliferis, and A. Statnikov. Algorithms for large scale Markov blanket discovery. In *The 16th International FLAIRS Conference*, St. Augustine, Florida, USA, 2003.

I. Tsamardinos and C.F. Aliferis. Towards principled feature selection: Relevancy, filters and wrappers. In *Proceedings of the Workshop on Artificial Intelligence and Statistics*, 2003.

E. Tuv. Feature selection and ensemble learning. In I. Guyon, S. Gunn, M. Nikravesh, and L. Zadeh, editors, *Feature Extraction, Foundations and Applications*. Springer, New York, 2005.

N. Vasconcelos. Feature selection by maximum marginal diversity: optimality and implications for visual recognition. In *Proc. IEEE Conf on CVPR*, pages 762–772, Madison, WI, USA, 2003.

D.R. Wolf and E.I. George. Maximally informative statistics. In José M. Bernardo, editor, *Bayesian Methods in the Sciences*. Real Academia de Ciencias, Madrid, Spain, 1999.

D.H. Wolpert and D.R. Wolf. Estimating functions of distributions from a finite set of samples. *Phys. Rev. E*, 52(6):6841–6854, 1995.

E.P. Xing, M.I. Jordan, and R.M. Karp. Feature selection for high-dimensional genomic microarray data. In *Proc. 18th International Conf. on Machine Learning*, pages 601–608. Morgan Kaufmann, San Francisco, CA, 2001.

Y. Yang and J.O. Pedersen. A comparative study on feature selection in text categorization. In *Proc. 14th International Conference on Machine Learning*, pages 412–420. Morgan Kaufmann, 1997.

L. Yu and H. Liu. Feature selection for high-dimensional data: A fast correlation-based filter solution. In *ICML'03*, Washington, D.C., 2003.

M. Zaffalon and M. Hutter. Robust feature selection by mutual information distributions. In *Proceedings of the 18th Conference on Uncertainty in Artificial Intelligence*, pages 577–584, San Francisco, 2002. Morgan Kaufmann.

Chapter 7

Ensemble Learning

Eugene Tuv*

Intel, `eugene.tuv@intel.com`

7.1 Introduction

Supervised ensemble methods construct a set of base learners (experts) and
use their weighted outcome to predict new data. Numerous empirical studies confirm that ensemble methods often outperform any single base learner
(Freund and Schapire, 1996, Bauer and Kohavi, 1999, Dietterich, 2000b). The
improvement is intuitively clear when a base algorithm is unstable. In an
unstable algorithm small changes in the training data lead to large changes
in the resulting base learner (such as for decision tree, neural network, etc).
Recently, a series of theoretical developments (Bousquet and Elisseeff, 2000,
Poggio et al., 2002, Mukherjee et al., 2003, Poggio et al., 2004) also confirmed
the fundamental role of stability for generalization (ability to perform well on
the unseen data) of any learning engine. Given a multivariate learning algorithm, model selection and feature selection are closely related problems (the
latter is a special case of the former). Thus, it is sensible that model-based
feature selection methods (wrappers, embedded) would benefit from the regularization effect provided by ensemble aggregation. This is especially true for
the fast, greedy and unstable learners often used for feature evaluation.

In this chapter we demonstrate two types of model-based variable scoring
and filtering: embedded and sensitivity based. In both cases the ensemble aggregation plays a key role in *robust* variable relevance estimation. We briefly
review ensemble methods in section 7.2. We distinguish between two main
types of ensembles: parallel and serial, and describe in detail two representative techniques: bagging and boosting. In section 7.3 we define importance
metrics, and illustrate their properties on a simulated data. The vote aggregation in the context of the Bayesian framework is described in section 7.4.
Even though throughout the chapter we used a binary decision tree as a base
learner, the methods discussed are generic. The generalization is explicitly
pointed out in section 7.3.

E. Tuv: *Ensemble Learning*, StudFuzz **207**, 187–204 (2006)
`www.springerlink.com`   © Springer-Verlag Berlin Heidelberg 2006

7.2 Overview of Ensemble Methods

Ensemble methods combine outputs from multiple base learners to improve
the performance of the resulting committee. There are two primary approaches
to ensemble construction: parallel and serial.

A parallel ensemble combines independently constructed accurate and diverse base learners. That is, an individual base learner needs to have an error
rate better than a random guess, and different base learners should make
different errors on new data. It is intuitive then, that an ensemble of such
base learners will outperform any single of its components since diverse errors
will cancel out. More formal treatments of ensembles of diverse and accurate base learners were provided by Hansen and Salamon (1990), Amit and
Geman (1997). Parallel ensembles are variance-reduction techniques, and in
most cases, they are applied to unstable, high-variance algorithms (like trees).
Although, Valentini and Dietterich (2003) showed that ensembles of low-bias
support vector machines (SVMs) often outperform a single, best-tuned, canonical SVM (Boser et al., 1992).

In serial ensembles, every new expert that is constructed relies on previously built experts in such a way that the resulting weighted combination of
base learners forms an accurate learning engine. A serial ensemble algorithm
is often more complex, but it is targeted to reduce both bias and variance,
and can show excellent performance.

7.2.1 Parallel Ensembles

Many methods were developed to impose diversity in the ensemble construction process. Bagging (Bootstrap Aggregation) trains each base learner on
a different bootstrap sample drawn from the data (Breiman, 1996). Other
methods generate diversity by injecting randomness into the learning algorithm (e.g. randomly initialized weights in neural networks), manipulating
the input features, manipulating the output targets, etc. More comprehensive
overviews of ensemble methods were presented by Dietterich (2000a), Valentini and Masulli (2002).

Random Forest (Breiman, 2001) is an improved bagging method that extends the "random subspace" method (Ho 1998). It grows a forest of random
trees on bagged samples showing excellent results comparable with the best
known classifiers. Random Forest (RF) does not overfit, and can be summarized as follows:

1. a number n is specified much smaller than the total number of variables
   N (typically $n \sim \sqrt{N}$)
2. each tree of maximum depth is grown on a bootstrap sample of the training
   set
3. at each node, n variables are selected at random out of the N
4. the split used is the best split on these n variables

Note that for every tree grown in RF, about one-third of the cases are out-of-bag (out of the bootstrap sample). The out-of-bag (OOB) samples can serve as a test set for the tree grown on the non-OOB data. We discuss in section 7.3.1 how OOB samples can be used for variable scoring.

7.2.2 Serial Ensembles

A serial ensemble results in an additive model built by a forward-stagewise algorithm. The *Adaboost* algorithm was introduced by Freund and Schapire (1996). At every step of ensemble construction the boosting scheme adds a new base learner that is forced (by reweighting the training data) to concentrate on the training observations that are misclassified by the previous sequence. Boosting showed dramatic improvement in accuracy even with very weak base learners (like decision stumps). In the numerous studies, Adaboost showed remarkable performance on a variety of problems, except for datasets with high levels of noise in the target variable.

   Breiman (1997), Friedman et al. (2000) showed that the Adaboost algorithm is a form of gradient optimization in functional space, and is equivalent to the following forward-stagewise, additive algorithm with the exponential loss function $\Psi(y, F(\mathbf{x}) = exp(-yF(\mathbf{x}))$ and base learner family $\{b(\mathbf{x}, \gamma)\}_\gamma$

Algorithm 5: Forward stagewise boosting

   1. Initialize $F(\mathbf{x})=0$.
   2. For l=1 to L
      (a) Compute $(\beta_l, \gamma_l) = arg\,min_{\beta, \gamma} \sum_{i=1}^m \Psi(y_i, F_{l-1}(x_i) + \beta b(x_i, \gamma))$
      (b) Set $F_l(\mathbf{x}) = F_{l-1}(\mathbf{x}) + \beta_l b(\mathbf{x}, \gamma_l)$
   3. Output $F(\mathbf{x}) = F_L(\mathbf{x}) = \sum_{l=1}^L \beta_l b(\mathbf{x}, \gamma_l)$

A *Gradient Tree Boosting (GTB)* learning machine (Friedman, 1999a,b) uses a numeric approximation to solve (2a) in the Algorithm 5 with robust loss functions like L_1 loss: $\Psi(F(\mathbf{x}), y) = |F(\mathbf{x}) - y|$ or Huber's loss function

$$: \Psi(F(\mathbf{x}), y) = \begin{cases} |y - F(\mathbf{x})|^2, & \text{for } |F(\mathbf{x}) - y| \leq \delta \\ \delta(|y - F(\mathbf{x})| - \delta/2), \text{otherwise} \end{cases} \text{ where } \delta = \alpha^{th} -$$

quantile$|F(\mathbf{x}) - y|$, usually $\alpha = 0.1$

GTB uses gradient descent in functional space for numerical optimization, and at every iteration l of GTB a new base learner (a shallow tree in this case) T_l is fitted to the generalized residuals with respect to a loss function Ψ

$$-\left[\frac{\partial \Psi(y_i, F(x_i))}{\partial F(x_i)}\right]_{F=F_{l-1}} \qquad (7.1)$$

This provides terminal regions $R_{jl}, j = 1, 2, ..., J_l$. The corresponding constants γ_{jl} are solutions to

$$\gamma_{jl} = arg \min_{\gamma} \sum_{x_i \in R_{jl}} \Psi(y_i, F_{l-1}(x_i) + \gamma) \qquad (7.2)$$

and updated $F_l(\mathbf{x})$ in $(2b)$ of Algorithm 5 is given by

$$F_l(\mathbf{x}) = F_{l-1}(\mathbf{x}) + \sum_{j=1}^{J_l} \gamma_{jl} I(\mathbf{x} \in R_{jl}) \qquad (7.3)$$

In a K-class classification problem with response y taking K unordered values $\{c_1, ..., c_K\}$ GTB builds a multinomial logistic model to estimate the class conditional probabilities $p_k(x) = Pr(y = c_k|x)$, for $k = 1, 2, ..., K$ and

$$p_k(x) = \frac{e^{F_k(x)}}{\sum_{l=1}^{K} e^{F_l(x)}} \qquad (7.4)$$

with the $\sum_{l=1}^{K} F_l(x) = 0$.

Penalized, stagewise formulation and feature selection

Friedman (1999a) also showed empirically that an introduction of shrinkage factor $0 < \nu < 1$ in $(2b)$ step of Algorithm 5 could dramatically improve the performance of the resulting ensemble

$$F_l(\mathbf{x}) = F_{l-1}(\mathbf{x}) + \nu \cdot \beta_l b(\mathbf{x}, \gamma_l) \qquad (7.5)$$

There is a strong connection between GTB's additive tree expansion with a shrinkage strategy (7.5) and a "lasso" penalized linear regression on all possible (J-region) trees (Tibshirani, 1996) (7.6),(7.7).

$$\hat{F}(\mathbf{x}) = \sum \hat{a}_k T_k(\mathbf{x}) \qquad (7.6)$$

where

$$\hat{a}_k = arg \min_{a_k} \sum_{i=1}^{m} \Psi(y_i, \sum a_k T_k(\mathbf{x})) + \lambda \sum |a_k| \qquad (7.7)$$

They produce very similar solutions as the shrinkage parameter in (7.5) becomes arbitrary small ($\nu \to 0$) with the number of trees in (7.6) $\sim 1/\lambda$. It was first suggested by Hastie et al. (2001), and rigorously derived by Efron et al. (2004).

Both, the commonly used L_2 penalty $\sum a_k^2$ (common for ridge regression, support vector machines), and the "lasso" penalty in (7.7) penalize larger absolute values of the coefficients a_m. However, the L_2 penalty discourages variation in absolute values, and tends to produce coefficients similar in absolute values, whereas the L_1 penalty is not concerned with variance in coefficients, and tends to produce *sparse* solutions with more variability in the absolute value of coefficients, with many of them being zero. Thus, GTB with the shrinkage strategy carries out feature selection in the form of selecting a small subset of relevant features (simple functions of inputs - shallow trees) out of all possible trees.

Hybrid GTB and RF (GTB-FS)

For data sets with very large number of variables standard GTB with an exhaustive search over all variables could be computationally prohibitory expensive. Random Forests, on the other hand, are comprised from trees of maximum depth, and they are challenging to build for very large sample sizes. For massive datasets (in both dimensions) Borisov et al. (2005) proposed a hybrid (GTB and RF) ensemble method, GTB-FS, that is at least as accurate as both of them, and incomparably faster. GTB-FS is built with shallow trees using a small subset of variables at every step, and capable of handling huge (in both dimensions) datasets. The main idea is to select a small sample of features at every step of the ensemble construction. The sampling distribution is modified at every iteration to promote features more relevant to the learning task (generalized residuals from the previous iteration). A measure of relevance could be approximated by variable importance evaluated over a historical window of prior iterations. The averaging window could be moving, weighted by distance in time, shrinking in time, etc.. Naturally, the sampling strategy is closely tied to GTB's regularization coefficient. Sampling weights could be initialized using prior knowledge or data (from an initial run of a single tree or a simple univariate measure of relevance, etc.), or set to equal.

7.3 Variable Selection and Ranking with Tree Ensembles

In this section we give the formal definitions for embedded and sensitivity-based variable scoring metrics. We explore several feature selection aspects on a simulated data comparing different importance metrics and ensemble types.

7.3.1 Relative Variables Importance Metrics

A decision tree partitions the $\mathbf{X}$ space into a set of disjoint regions, and assigns a response value to each corresponding region. It uses a greedy, top-down recursive partitioning strategy. At every step a decision tree uses exhaustive search by trying all combinations of variables and split points to achieve the maximum reduction in impurity. Therefore, the tree construction process itself can be considered as a type of variable selection (a kind of forward selection embedded algorithm, see also (Chapter 5), and the impurity reduction due to a split on a specific variable could indicate the relative importance of that variable to the tree model. Note, that this relative importance is multivariate-model based, and is different from the relevance measured by standard, univariate filter methods (Chapter 3).

For a single decision tree a measure of variable importance is proposed in (Breiman et al., 1984):

$$VI(x_i, T) = \sum_{t \in T} \Delta I(x_i, t) \tag{7.8}$$

where $\Delta I(x_i, t) = I(t) - p_L I(t_L) - p_R I(t_R)$ is the decrease in impurity due to an actual (or potential) split on variable x_i at a node t of the optimally pruned tree T. p_L, p_R are the proportions of cases sent to the left(or right) by x_i. Node impurity $I(t)$ for regression is defined as $\frac{1}{N(t)} \sum_{s \in t}(y_s - \bar{y})^2$ where the sum and mean are taken over all observations s in node t, and $N(t)$ is the number of observations in node t. For classification $I(t) = Gini(t)$ where $Gini(t)$ is the Gini index of node t:

$$Gini(t) = \sum_{i \neq j} p_i^t p_j^t \tag{7.9}$$

and p_i^t is the proportion of observations in t whose response label equals i ($y = i$) and i, j run through all response class numbers. The Gini index is in the same family of functions as $entropy - \sum_i p_i^t log(p_i^t)$, and measures node impurity. It is zero when t has observations only from one class, and is maximum when classes are perfectly mixed.

An important question remains for tree based models: how to rank variables that were masked by others with slightly higher splitting scores, but could provide as accurate a model if used instead.

One of key features of CART (Breiman et al., 1984) is a notion of surrogate splits. In CART methodology, the surrogate splits is used mainly to handle missing values, detect masking, and assess variable-importance ranking. The predictive association of a surrogate variable x_s for the best splitter x^* at a tree node t is defined through the probability that x_s predicts the action of x^* correctly and this is estimated as:

$$p(x_s, x^*) = p_L(x_s, x^*) + p_R(x_s, x^*)$$

where $p_L(x_s, x^*)$ and $p_R(x_s, x^*)$ define the estimated probabilities that both x_s and x^* send a case in t left (right). The predictive measure of association $\lambda(x^*|x_s)$ between x_s and x^* is defined as

$$\lambda(x^*|x_s) = \frac{min(p_L, p_R) - (1 - p(x_s, x^*))}{min(p_L, p_R)} \tag{7.10}$$

It measures the relative reduction in error due to using x_s to predict x^* ($1 - p(x_s, x^*)$) as compared with the "naïve" rule that matches the action with $max(p_L, p_R)$ (with error $min(p_L, p_R)$). If $\lambda(x^*|x_s) < 0$ then x_s is disregarded as a surrogate for x^*.

The sum in (7.8) is taken over all internal tree nodes where x_i is a primary splitter or a surrogate variable ($\lambda(x^*|x_i) > 0$ for a primary splitter x^*). Often a variable that does not appear in a tree will still be ranked high on the variable importance list constructed using surrogate variables.

In the following example the Iris classification solution (Fisher, 1936) is given by a single tree that used only one variable to split nodes. Relative variable ranking takes into account surrogate variables, and Figure 7.1 depicts an adequate relative variable importance pareto for all four variables.

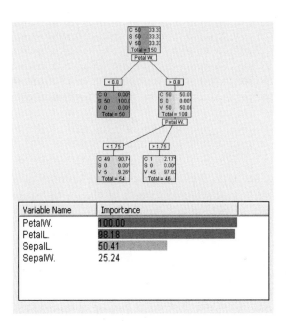

Variable Name	Importance
PetalW.	100.00
PetalL.	98.18
SepalL.	50.41
SepalW.	25.24

Fig. 7.1. Iris flower data, three species are classified: setosa (S), versicolor (C), and virginica (V) using four predictors: sepal length, sepal width, petal length, and petal width. Relative variable ranking calculation takes into account the surrogate variables. Even though the decision tree uses only one variable (petal width), an adequate relative importance is presented for all four variables.

For the stochastic tree ensembles (GTB, RF, GTB-FS) of M trees this importance measure is easily generalized. It is simply averaged over the trees

$$M(x_i) = \frac{1}{M}\sum_{j=1}^{M} VI(x_i, T_j) \tag{7.11}$$

The regularization effect of averaging makes this measure more reliable, and because of the stochastic nature of the ensembles (a slightly different version of the data is used for each expert for both RF / GTB) the masking issue is not a problem (especially for the independently built RF). Therefore, in (7.11) the sum is evaluated only over internal nodes where the variable of interest is the primary splitter. For a classification problem, GTB and GTB-FS build separate models $f_k(x)$ to predict each of the K classes

$$F_k(x) = \sum_{j=1}^{M} T_{kj}(x)$$

In this case (7.11) generalizes to

$$M(x_i, k) = \frac{1}{M} \sum_{j=1}^{M} M(x_i, T_{kj})$$

and represents the relevance of X_i to separate class k from the rest of them. The overall relevance of X_i can be obtained by averaging over all classes

$$M(x_i) = \frac{1}{K} \sum_{k=1}^{K} M(x_i, k)$$

The matrix $\{M_{ik} := M(x_i, k)\}$ could be used in number of ways. One could average the matrix over a selected subsets of classes to estimate the variable importance for that subset. In the same way, one could determine what class a chosen subset of variables separates best. For a parallel ensemble (RF) it is not obvious how to extract embedded importance ranking by class.

It is clear that the relative variable importance approach naturally generalizes beyond trees to any unstable scoring mechanism. Relevance scores are calculated over multiple permutations of data, variables, or both with a fixed scoring method and averaged. The score can be obtained from a base learner or a simple univariate score. If one is only interested in a simple variable ranking, then ranks from heterogeneous ranking methods could be combined too.

Breiman (2001) proposed a *sensitivity* based measure of variable relevance evaluated by a Random Forest. For a classification problem it is summarized as follows:

- Classify the OOB cases (those out of the bootstrap sample) and count the number of votes cast for the correct class in every tree grown in the forest
- Randomly permute the values of variable n in the OOB cases and classify these cases down the tree
- Subtract the number of votes for the correct class in the variable-n-permuted OOB data from the untouched OOB data
- Average this number over all trees in the forest to obtain the raw importance score for variable n.

Clearly, this measure of relevance could be calculated by class and overall. For regression, the same method is used, but variable importance is measured by the residual sum of squares.

Breiman (2002) also noticed that the correlations of these scores between trees are quite low, and therefore one can compute the standard errors in the standard way, divide the raw score by its standard error to get a z-score, and assign a significance level to the z-score assuming normality. (The null hypothesis tested is that the mean score is zero, against the one-sided alternative that the mean score is positive).

Again, the "sensitivity" criterion described above can be easily generalized to any *cross-validated committees* (Parmanto et al., 1996) of any learning machine. Each variable can be scored by the change in the cross-validation error after a rerun of the same learning algorithm on the same data with the variable of interest noised-out (randomly permuted). Significant positive change in the error would indicate that the variable is important (at least to a chosen learning scheme).

In classical statistical modeling the importance of each predictor variable is proportional to its contribution to the variation reduction of the target variable. For example, in multiple linear regression, the *partial R^2* gives the incremental predictive power of each additional input variable measured by the marginal contribution of an input variable when all others are already included in the model. To illustrate how the methods described above rank variables we generated 5000 samples from a simple linear model $z = x_1 + 2x_2 + ... + 10x_{10} + \epsilon$ with independent x_i. For this example an input's *partial $R^2 \sim$* the square of the corresponding coefficient.

Figure (7.2) shows the relative variable ranking (compared to the maximum) for this simple linear model calculated using three methods: RF-ERR, RF-MSE, GTB. Here, RF-ERR represents the sensitivity-based measure evaluated by Random Forest. RF-MSE and GTB are the impurity reduction measures (7.11) calculated by RF and GTB ensembles, correspondingly. To compare these scores to the "true" linear coefficient we took the square root of all scores. All methods rank variables correctly. GTB gives relative scores very close to the theoretical. RF-MSE also does very well but slightly overestimates the importance of the least significant predictors. RF-ERR overestimates the importance of all but the top few predictors.

7.3.2 Variable Importance Ranking, Filtering, and Compactness

In this section we look at the problem of variable-relevance scoring from several slightly different angles: variable ranking (where interest is in the relative relevance score to the target for all input variables), variable filtering (where we are more interested in separation of irrelevant inputs), best compact subset (where we want to find a small subset of independent variables with the most predictive power). We use a simple metric to measure compactness of variable scoring: the proportion of top ranked variables covering all relevant predictors.

To study these properties of the different variable importance metrics defined above, we simulated a dataset that conceptually represents an important class of problems often dealt with in the industrial applications: a large number of variables with a small number relevant. Among the relevant variables only a few are highly influential, and the rest are the weak predictors. Input variables are often correlated.

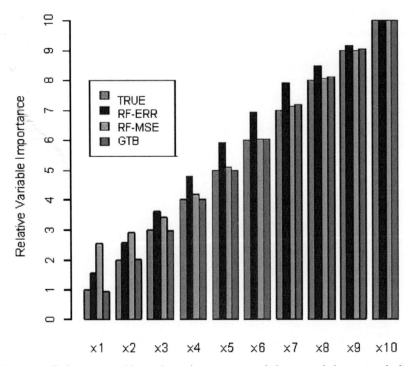

Fig. 7.2. Relative variable ranking (square root of the scores) for a simple linear model $z = x_1 + 2x_2 + \ldots + 10x_{10} + \epsilon$ computed by the sensitivity based measure RF-ERR, and the impurity-reduction measures calculated by Random Forest (RF-MSE) and Gradient Tree Boosting (GTB) ensembles.

The data generated had one numeric response, and 203 input variables: $x_1, \ldots, x_{100}$ are highly correlated with one another, and are reasonably predictive of the response ($R^2 \sim 0.5$); a, b, and c are independent variables that are much weaker predictors ($R^2 \sim 0.1$); $y_1, \ldots, y_{100}$ are i.i.d. $N(0,1)$ noise variables. The actual response variable was generated using $z = x_1 + a + b + c + \epsilon)$, where $\epsilon \sim N(0,1)$. Therefore, the best compactness score for our artificial data is $4/203 \sim 0.2\%$

Consider first the impurity-reduction measures embedded in the tree based learners. For a single tree (7.8) and its generalization for ensembles (7.11) this measure strongly depends on the complexity (and the generalizability) of the model. In the high-dimensional problems even with no predictive relationship, a single tree would have no problem finding split variables/points. The relative variable ranking would convey no useful information since the optimal tree corresponds to a root node. The generalization error would grow with every split and that would indicate the anomaly in this analysis.

Figure 7.3 shows the variable importance results of fitting a single tree to the full model $z \sim a + b + c + x_1 + x_2 + \ldots + x_{100} + y_1 + y_2 + \ldots + y_{100}$.

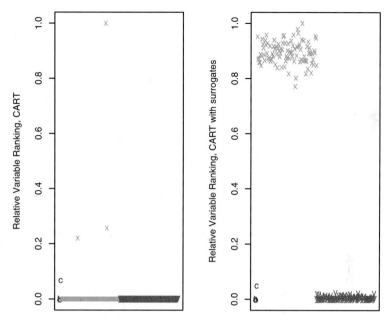

Fig. 7.3. Relative variable ranking computed by CART using impurity reduction scores. The left graph corresponds to the optimally-pruned tree with no surrogate variables included. The right graph corresponds to the same tree, but the scores are computed for all variables that satisfy the surrogate condition $\lambda(x^*|x_s) < 0$ defined by (7.10). In both cases the single tree failed to detect the importance of the weak predictors a, b.

The relative variable ranking is calculated from the optimally-pruned single tree using the impurity-reduction scores. The left graph corresponds to the variable importance evaluated over a single tree with no surrogate variables included. The right graph corresponds to the same tree, but the scores are computed for all variables that satisfy the surrogate condition $\lambda(x^*|x_s) < 0$ defined by (7.10). In both cases the single tree failed to detect the importance of the weak independent predictors a, b (both have relevance scores 0). The surrogate method correctly ranked all correlated x_s similarly high on the importance list.

Figure 7.4 shows the relative variable ranking computed by Gradient Tree Boosting (GTB) using impurity-reduction scores. The left graph corresponds to the standard (tuned) GTB, and the right graph corresponds to GTB with dynamic feature selection (GTB-FS). For GTB-FS 50 variables are selected at every split for every expert-tree in the ensemble. Both provide similar accuracy (with GTB-FS slightly better), but GTB-FS is four times faster. Both produce very compact importance scores ($\sim 2\%$), with the bulk of the variables well separated from a small group of relevant champions. Standard GBT produced an accurate relative ranking within relevant variables while

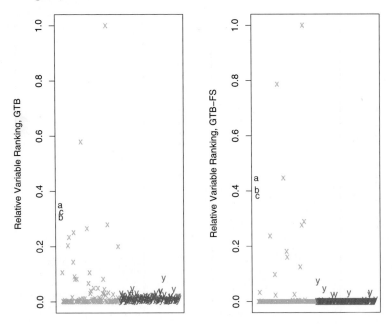

Fig. 7.4. Relative variable ranking computed by Gradient Tree Boosting (GTB) using impurity-reduction scores. The left graph corresponds to the standard (tuned) GTB, and the right graph corresponds to GTB with dynamic-feature selection (GTB-FS). For GTB-FS 50 variables are selected at every split for every expert-tree, and the variable sampling distribution is modified at every iteration to promote features that are more relevant to its learning task from the generalized residuals from the previous iteration

GTB-FS upweighted scores for weak predictors a, b, c. In general, in the presence of correlated predictors a sequential ensemble like GTB (that acts similar to a forward, stepwise procedure) will produce a more compact representation of relevant variables than a parallel ensemble where every expert sees practically the same copy of the data.

Figure 7.5 shows the relative variable ranking computed by RF using the impurity-reduction measure (7.11). The left graph corresponds to the best tuned RF, while the right graph corresponds to a very fast RF with only one variable selected at random at every split. There is no dramatic difference in generalization error between two forests, but there is noticeable difference between the two variable rankings. The fast RF gave the least compact representation of important variables with noise variables y_s having minimum relative importance around 15%, and weak predictors a, b, c scoring close to noise. But it seemed to be perfect tool to assess variable masking. The tuned RF gave accurate relative scores between the top strong predictor $x(100\%)$, weak predictors a, b, c ($\sim 40\%$), and noise variables y_s ($< 5\%$). In terms of the

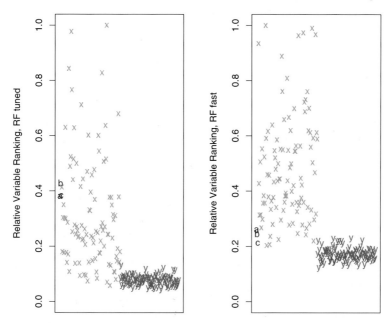

Fig. 7.5. Relative variable ranking computed by RF using the impurity-reduction scores. The left graph corresponds to the best tuned RF, and the right graph corresponds to a very fast RF with only one variable selected at random at every split.

best subset selection the compactness score of the tuned RF is $\sim 15\%$, while it is close to 50% for the fast one.

Figure 7.6 shows the sensitivity-based variable relevance translated to the z scores generated by RF. Solid and dotted lines represent 5% and 1% significance levels, respectively. Weak, but *independent* predictors a, b, c show higher relevance than strong, but highly *correlated* predictors x_s. The incorrect relative ranking among the relevant predictors is due to the fact that there is no predictive "substitutes" for independent variables a, b, c, whereas the predictive power of x_s is uniformly distributed among the 100 variables by RF. Noise variables y_s are seen to have statistically insignificant relevance to the response. Notice also that the compactness score for this measure is very good too $\sim 2\%$. Also, unlike embedded impurity reduction ranking, the "honest" sensitivity measure provides a clear statistical cutoff point between noise and useful features. Thus, if the additional computational burden is not a concern, the sensitivity-based relevance measure seems to do a fine job for variable filtering (correct ranking order could still be a challenge). It proved to be a reliable noise filtering mechanism on many real datasets too.

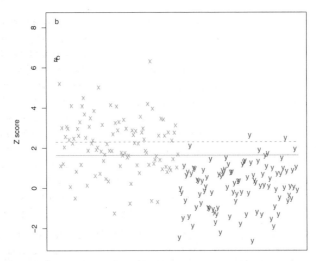

Fig. 7.6. RF sensitivity-based variable relevance translated to the z scores. The solid and dotted lines represent 5%/1% significance-level performance. Weak, but *independent* predictors a, b, c show higher relevance than strong, but highly *correlated* predictors x_s. Noise variables y_s show insignificant relevance to the response.

7.4 Bayesian Voting

In Bayesian learning (Gelman et al., 1995) a sampling model $P(y|x, \theta)$ given the parameters θ is defined, and the prior distribution $P(\theta)$ that reflects our knowledge about the parameters is specified also. After observing the data D the posterior distribution

$$P(\theta|D) = \frac{P(D|\theta)P(\theta)}{\int P(D|\theta)P(\theta)} \tag{7.12}$$

represents the updated knowledge about the model parameters, and is used for the exact inference (prediction)

$$P(y|x_{new}, D) = \int P(x_{new}|\theta) \cdot P(\theta|D)d\theta \tag{7.13}$$

The integral in (7.13) is often difficult to evaluate, and numeric approximations are employed that use Markov Chain Monte Carlo Methods (MCMC) to sample from the posterior distribution. Samples are generated from a Markov chain with a stationary distribution being the desired posterior. There is a number of MCMC procedures including Gibbs sampling and the Metropolis-Hastings algorithm for sampling from the posterior. MCMC methods were discussed by Gillks et al. (1996).

A Gibbs classifier provides a simple approximation to the Bayesian optimal classifier. It draws a single sample from the posterior distribution for the

parameter θ, and then classifies to the class label corresponding to that parameter. Ng and Jordan (2001) proposed an ensemble of Gibbs classifiers called
Voting Gibbs (VG) that draws m samples from the posterior distribution and
takes a majority vote to make the final prediction. They showed that the relative error of Voting Gibbs compared to the Bayesian optimal classifiers decays
at the rate of $O(1/m)$. More interestingly in the context of this chapter, it was
shown that there is a particular choice of priors that make the VG algorithm
highly resistant to irrelevant features. Moreover, they showed that the VG
algorithm with a particular misspecified prior has sample complexity that is
logarithmic in the number of irrelevant features.

Multi-Layer Perceptrons such as Bayesian Neural Networks carryout internal feature selection through the Automatic Relevance Determination (ARD)
mechanism (MacKay, 1994, Neal, 1996). ARD uses hierarchical hyperpriors
on the input weights. Conditional on the values of these hyperparameters, the
input weights have independent Gaussian prior distributions with standard
deviations given by the corresponding hyperparameters. If a hyperparameter specifies a small standard deviation, the effect of the corresponding input
variable is likely to be small. Conversely, a larger standard deviation indicates
a likely strong effect. The posterior distribution of these hyperparameters reveals an input's relevance after observing the training data.

A popular trend in Bayesian model (variable) selection is to estimate posterior model probabilities using *variable dimension* MCMC methods (Green,
1995, Stephens, 2000). These methods allow jumps between models with different dimensionality in the parameters, and therefore sampling includes posterior distributions of input combinations. The variable dimension MCMCs
visit models according to their posterior probabilities, and hence the most
probable models are visited in a finite sample. Then the posterior probabilities of the models-input combinations are estimated based on the number of
visits for each model, and the input variables could be ranked based on the
marginal distributions (Vehtari and Lampinen, 2002).

7.5 Discussions

We described ensemble-based variable ranking and filtering methods that
proved to be very effective (the top second and third entry at this feature
selection challenge used Random Forest for variable selection). Despite many
obvious advantages in using tree as a base learner, there are some issues with
using trees for the feature ranking. Trees tend to split on variables with more
distinct values. This effect is more pronounced for categorical predictors with
many levels. It often makes a less relevant (or completely irrelevant) input
variable more "attractive" to split on only because it has high cardinality.
Clearly it could have a strong effect on embedded impurity reduction measures (7.8),(7.11). It is unlikely that sensitivity based ranking is completely
immune to this problem either. To alleviate this problem some cardinality ad-

justments could be made either during the tree construction process or when the variable scoring is evaluated over the ensemble. An example of such an adjustment is the information gain ratio used in some decision tree algorithms. Another sensible solution would be to factor in the generalizability of a potential or actual split on a hold-out portion of the data. That would be especially appropriate for ensembles with dynamic feature selection GTB-FS described in section 7.2.2 .

As was discussed in section 7.4, the Bayesian framework provides feature selection as an internal mechanism making it practically insensitive to a potentially very large number of irrelevant inputs. If the variable ranking is of interest, simple and very efficient, embedded-ensemble-based ranking described in section 7.3.2 clearly would be a preferred choice compared to a computationally challenging MCMC procedure. The sensitivity-based approach outlined in section 7.3.2 also would be a preferred approach if one needs to separate relevant features from the noise. However, if one is interested in a small model with the same predictive accuracy as a full model, Bayesian *variable dimension* MCMC methods provide a ranking of subsets of inputs giving a natural way to find smaller combinations of potentially useful input variables. The other ensemble-based methods presented in the chapter (sequential to a lesser degree) have to use some heuristic procedure (like backward elimination) to find a good small-subset model. However, it is questionable how capable Bayesian best-subset selection is when the underling model has a set of weak but relevant predictors and many important and highly correlated inputs - MCMC chain is more likely to visit combinations of inputs with many relevant and redundant inputs. It seems reasonable though that for the best-subset problem a sequential ensemble could use a penalization strategy to prevent redundant variables from entering the model by penalizing the algorithm for adding new variables.

It is often desirable to understand effects of interactions, interdependencies (such as masking, etc.) between input variables. In the Bayesian framework input dependencies could be examined through correlations between ARD values. Also posterior joint probabilities of inputs could provide insights on joint predictive effects and dependencies of input variables.

A relatively straightforward and generic approach to rank input interactions could be devised using ensembles of trees, specifically the serial GTB ensemble described in the section 7.3.2. It is applicable to practically any supervised settings (including mixed type data). Assume that the ANOVA decomposition of the target function $f(\mathbf{x})$ is

$$f(\mathbf{x}) = \sum_j f_j(\mathbf{x}_j) + \sum_{ij} f_{ij}(\mathbf{x}_i, \mathbf{x}_j) + \sum_{ijk} f_{ijk}(\mathbf{x}_i, \mathbf{x}_j, \mathbf{x}_k) + ... \qquad (7.14)$$

The first term in the decomposition represents *main* effects, the second term has the second-order interactions, etc. In the tree-based ensembles the order of interactions in the model is controlled by the tree depth (size) (Hastie et al., 2001). Tree stumps limit an ensemble model to main effects only, trees with at most three terminal nodes allow second order interactions in the model, etc. Suppose we would like to rank the predictive importance of second-order interactions. Given the additive nature of a decomposition it is easy to remove the main effects by building a model with an ensemble of stumps and taking residuals. The next step would be building a model for the residuals with an ensemble of trees of size at most three, and ranking pairs of distinct variables participating in any consecutive split in terms of impurity reduction score averaged over the ensemble.

It is also not difficult to imagine a scheme to evaluate masking effects for input variables over an ensemble using predictive measure of association (7.10).

References

Y. Amit and D. Geman. Shape quantization and recognition with randomized trees. *Neural Computation*, 9(7):1545–1588, 1997.

E. Bauer and R. Kohavi. An empirical comparison of voting classification algorithms: Bagging, boosting, and variants. *Machine Learning*, 36:525–536, 1999.

A. Borisov, V. Eruhimov, and E. Tuv. *Feature Extraction, Foundations and Applications*, chapter Dynamic soft feature selection for tree-based ensembles. Springer, 2005.

B. Boser, I. Guyon, and V. Vapnik. A training algorithm for optimal margin classifiers. In *Fifth Annual Workshop on Computational Learning Theory*, pages 144–152, Pittsburgh, 1992.

O. Bousquet and A. Elisseeff. Algorithmic stability and generalization performance. In *Advances in Neural Information Processing Systems 13*, pages 196–202, 2000. URL citeseer.nj.nec.com/bousquet01algorithmic.html.

L. Breiman. Bagging predictors. *Machine Learning*, 24:123–140, 1996.

L. Breiman. Random forests. *Machine Learning*, 45(1):5–32, 2001.

L. Breiman. *Manual On Setting Up, Using, And Understanding Random Forests V3.1*, 2002.

L. Breiman, J.H. Friedman, R.A. Olshen, and C.J. Stone. *Classification and Regression Trees*. Wadsworth and Brooks, Monterey, CA, 1984.

Leo Breiman. Arcing the edge. Technical Report 486, Statistics Department, University of California at Berkeley, 1997.

T.G. Dietterich. Ensemble methods in machine learning. In *Multiple Classifier Systems. First International Workshop*, volume 1857. Springer-Verlag, 2000a.

T.G. Dietterich. An experimental comparison of three methods for constructing ensembles of decision trees: Bagging, boosting, and randomization. *Machine Learning*, 40(2):139–157, 2000b. available at ftp://ftp.cs.orst.edu/pub/tgd/papers/tr-randomized-c4.ps.gz.

B. Efron, T. Hastie, I. Johnstone, and R. Tibshirani. Least angle regression. *The Annals of Statistics*, 32(2):407–451, 2004.

R. Fisher. The use of multiple measurements in taxonomic problems. *Annals of Eugenics*, 7(II):179–188, 1936.

Y. Freund and R.E. Schapire. Experiments with a new boosting algorithm. In *Machine Learning: Proceedings of Thirteenth International Conference*, pages 148–156, 1996.

J. Friedman. Greedy function approximation: a gradient boosting machine, 1999a. IMS 1999 Reitz Lecture, February 24, 1999, Dept. of Statistics, Stanford University.

J. Friedman. Stochastic gradient boosting. Technical report, Dept. of Statistics, Stanford University, 1999b.

J. Friedman, T. Hastie, and R. Tibshirani. Additive logistic regression: A statistical view of boosting. *The Annals of Statistics*, 28:832–844, 2000.

A. Gelman, J. Carlin, H. Stern, and D. Rubin. *Bayesian Data Analysis*. Chapman and Hall, 1995.

W.R. Gillks, S. Richardson, and D.J. Spiegelhalter. *Markov Chain Monte Carlo in practice*. Chapman and Hall, 1996.

P. Green. Reversible jump markov chain monte carlo computation and bayesian model determination. *Biometrika*, 82(4):711–732, 1995.

L.K. Hansen and P. Salamon. Neural network ensembles. *IEEE Trans. Pattern Analysis and Machine Intelligence*, 12(10):993–1001, 1990.

T. Hastie, R. Tibshirani, and J. Friedman. *The Elements of Statistical Learning: Data Mining, Inference, and Prediction*. Springer, 2001.

D. J. C. MacKay. Bayesian non-linear modelling for the prediction competition. *ASHRAE Transactions: Symposia*, OR-94-17-1, 1994.

S. Mukherjee, P. Niyogi, T. Poggio, and R. Rifkin. Statistical learning: Stability is sufficient for generalization and necessary and sufficient for consistency of empirical risk minimization. AI Memo 2002-024, MIT, 2003.

R. Neal. *Bayesian Learning for Neural Networks*. Springer-Verlag, 1996.

A.Y. Ng and M.I. Jordan. Convergence rates of the voting gibbs classifier, with application to bayesian feature selection. In *ICML 2001*, pages 377–384, 2001.

B. Parmanto, P.W. Munro, and H.R. Doyle. Improving committee diagnosis with resampling techniques. In D.S. Touretzky, M.C. Mozer, and M.E. Hasselmo, editors, *Advances in Neural Information Processing Systems*, volume 8, pages 882–888. The MIT Press, 1996.

T. Poggio, R. Rifkin, S. Mukherjee, and P. Niyogi. General conditions for predictivity in learning theory. *Nature*, 428:419–422, 2004.

T. Poggio, R. Rifkin, S. Mukherjee, and A. Rakhlin. Bagging regularizes. AI Memo 2002-003, MIT, 2002.

M. Stephens. Bayesian analysis of mixtures with an unknown number of components an alternative to reversible jump methods. *The Annals of Statistics*, 28(1):40–74, 2000.

R. Tibshirani. Regression shrinkage and selection via lasso. *J. Royal Statist. Soc.*, 58:267–288, 1996.

G. Valentini and T. Dietterich. Low bias bagged support vector machines. In *ICML 2003*, pages 752–759, 2003.

G. Valentini and F. Masulli. Ensembles of learning machines. In M. Marinaro and R. Tagliaferri, editors, *Neural Nets WIRN Vietri-02*, Lecture Notes in Computer Sciences. Springer-Verlag, Heidelberg, 2002.

A. Vehtari and J. Lampinen. Bayesian input variable selection using posterior probabilities and expected utilities. Technical Report Report B31, Laboratory of Computational Engineering, Helsinki University of Technology, 2002.

Chapter 8

Fuzzy Neural Networks

Madan M. Gupta[1], Noriyasu Homma[2], and Zeng-Guang Hou[3]

[1] Intelligent Systems Research Laboratory, College of Engineering, University of Saskatchewan, 57 Campus Drive, Saskatoon, Saskatchewan, Canada S7N 5A9 guptam@sask.usask.ca

[2] School of Health Sciences, Faculty of Medicine, Tohoku University, 2-1 Seiryo-machi, Aoba-ku, Sendai, Japan 980-8575 homma@abe.ecei.tohoku.ac.jp

[3] Institute of Automation, The Chinese Academy of Sciences, P. O. Box 2728, Beijing, P.R. CHINA 100080 zengguang.hou@mail.ia.ac.cn

8.1 Introduction

The theory of *fuzzy logic*, founded by Zadeh (1965), deals with the linguistic notion of graded membership, unlike the computational functions of the digital computer with bivalent propositions. Since mentation and cognitive functions of brains are based on *relative grades* of information acquired by the natural (biological) sensory systems, fuzzy logic has been used as a powerful tool for modeling human thinking and cognition (Gupta and Sinha, 1999, Gupta et al., 2003). The *perceptions* and *actions* of the cognitive process thus act on the graded information associated with fuzzy concepts, fuzzy judgment, fuzzy reasoning, and cognition. The most successful domain of fuzzy logic has been in the field of feedback control of various physical and chemical processes such as temperature, electric current, flow of liquid/gas, and the motion of machines (Gupta, 1994, Rao and Gupta, 1994, Sun and Jang, 1993, Gupta and Kaufmann, 1988, Kiszka et al., 1985, Berenji and Langari, 1992, Lee, 1990a,b). Fuzzy logic principles can also be applied to other areas. For example, these fuzzy principles have been used in the area such as fuzzy knowledge–based systems that use fuzzy IF–THEN rules, *fuzzy software engineering*, which may incorporate fuzziness in data and programs, and fuzzy database systems in the field of medicine, economics, and management problems. It is exciting to note that some consumer electronic and automotive industry products in the current market have used technology based on fuzzy logic, and the performance of these products has significantly improved (Al-Holou et al., 2002, Eichfeld et al., 1996).

Conventional forms of fuzzy systems have low capabilities for learning and adaptation. Fuzzy mathematics provides an inference mechanism for approximate reasoning under cognitive uncertainty, while neural networks offer advantages such as learning and adaptation, generalization, approximation

M.M. Gupta: *Fuzzy Neural Networks*, StudFuzz **207**, 205–233 (2006)
www.springerlink.com

and fault tolerance. These networks are also capable of dealing with computational complexity, nonlinearity, and uncertainty. The integration of these two fields, *fuzzy logic and neural networks*, has given birth to an innovative technological field called *fuzzy neural networks* (FNNs) (Qi and Gupta, 1991, 1992a,b, Rao and Gupta, 1994, Jin et al., 1995). Extensive studies have indicated that FNNs, with the unique capabilities of dealing with numerical data, and linguistic knowledge and information, have the potential of capturing the attributes of these two fascinating fields—fuzzy logic and neural networks—into a single capsule, *fuzzy neural networks*. In view of the robust capabilities of FNNs, it is believed that they posses a great potential as emulation machines for a variety of behaviors associated with human cognition and intelligence (Gupta and Sinha, 1999).

Although much progress has been made in the field of fuzzy neural networks (FNNs), there are no universally accepted models of FNNs so far. FNNs can be defined as distributed parallel information processing schemes that employ neuronlike processing unit with learning capabilities and fuzzy operations for dealing with fuzzy signals[§]. Among such FNNs, two main classes of FNNs have been studied extensively, and have been proved to have robust capabilities for processing fuzzy information for specified tasks. The first category of FNNs has fuzzy triangular inputs and outputs, and it implements a mapping from a fuzzy input set to a fuzzy output set, and has the potential for realizing fuzzy logic functions on a compact fuzzy set. The other class of FNNs deals with crisp input and output signals. However, the internal structure of this type of FNN contains many fuzzy operations and approximate reasoning using the rule-based knowledge framework. It can be expected that this type of FNNs could implement fuzzy systems for real-world applications. Studies on the first class of FNNs can be traced back to 1974 (Lee and Lee, 1974), when the concepts of fuzzy sets into neural networks were introduced for the generalization of the McCulloch–Pitts (Mc-P) model by using intermediate values between zero and one. Various types of fuzzy neurons were developed using the notions of standard fuzzy arithmetic and fuzzy logic such as t-norm, t-conorm, and fuzzy implications (Hayashi and Buckley, 1993c,b,a, 1994a, Pedrycz, 1991, 1993). Some applications of this class of FNNs have been reported (Sun and Jang, 1993, Kosko, 1992, Wang, 1993). Important contributions have also been made on the universal approximation capabilities of fuzzy systems that can be expressed in the form of FNNs, and genetic algorithms have also been used in the learning schemes of FNNs (Hayashi and Buckley, 1994b, Sun and Jang, 1990, Jang, 1992, Kosko, 1994, Pedrycz, 1995, Mendel and Wang, 1992, 1993, Wang, 1993).

The objective of this chapter is to provide an overview of the basic principles, mathematical descriptions, and the state-of-the-art developments of FNNs. It contains four sections. In Section 8.2 the foundations of fuzzy sets

---

[§]This definition is very general and it may cover other approaches combining fuzzy logic and neural networks (Gupta and Sinha, 1999)

and systems are briefly reviewed in order to provide the necessary mathematical background. The basic definitions of fuzzy neurons with fuzzy input signals and weights are introduced in Section 8.3. Both the structures and learning mechanisms of hybrid fuzzy neural networks (HFNNs) are studied in Section 8.4. Function approximation capabilities of FNNs are also discussed in this section. The material presented in this chapter not only provides an overview of the existing results but also presents some state-of-the-art new achievements and open problems in the field of fuzzy neural computing.

8.2 Fuzzy Sets and Systems: An Overview

Fuzzy set theory is a generalization of conventional set theory and was introduced by Zadeh in 1965 (Zadeh, 1965, 1972, 1973). It provides a mathematical tool for dealing with linguistic variables associated with natural languages. Some introductory definitions of fuzzy sets, fuzzy logic, and fuzzy systems are reviewed in this section. Systematic descriptions of these topics can be found in several texts (Zadeh and Bellman, 1977, Prade and Dubois, 1980, Gupta and Kaufmann, 1985, 1988). A central notion of fuzzy set theory, as described in the following sections, is that it is permissible for elements to be only partial elements of a set rather than full membership.

8.2.1 Some Preliminaries

A "fuzzy" set is defined as a set whose boundary is not sharp. Let $X = \{x\}$ be a conventional set with generic elements x. A fuzzy set A is characterized by a membership function $\mu_A(x)$ defined on X, a set of ordered pairs $A = \{x, \mu_A(x)\}, x \in X$, where $\mu_A(x)$ is the grade of membership of x in A, and is defined as

$$\mu_A : X \to [0,1] \tag{8.1}$$

Thus, a fuzzy set A in X can also be represented as

$$A = \{(x, \mu_A(x)) : x \in X\} \tag{8.2}$$

The set X may be either a discrete set with discrete elements or a continuous set with continuous elements. For instance, $X = \{1, 2, 3, \ldots, 35\}$ is a discrete set, and $X = \Re^+ = [0, +\infty)$ is a continuous set. In this case, alternative ways of expressing a fuzzy set A in a discrete set $X = \{x_1, x_2, \ldots, x_m\}$ are

$$A = \{(x, \mu_A(x)) : x \in X\} = \{\mu_A(x_1)/x_1, \mu_A(x_2)/x_2, \ldots, \mu_A(x_m)/x_m\}$$
$$= \sum_{i=1}^{m} \mu_A(x_i)/x_i = \sum_{x_i \in X} \mu_A(x_i)/x_i$$

Similarly, a fuzzy set A of a continuous set X is represented by

Table 8.1. Fuzzy set–theoretic definitions and operations

Inclusion:	$A \subset B$ implies that $\mu_A(x) \leq \mu_B(x), \quad \forall x \in \boldsymbol{X};$
Intersection:	$A \cap B$, an intersection of A and B, implies that
	$\mu_{A \cap B}(x) = \min[\mu_A(x), \mu_B(x)] = \mu_A(x) \wedge \mu_B(x)$
	$= A \textbf{ AND } B, \quad \forall x \in \boldsymbol{X}$
Union:	$A \cup B$, a union of A and B, implies that
	$\mu_{A \cup B}(x) = \max[\mu_A(x), \mu_B(x)] = \mu_A(x) \vee \mu_B(x)$
	$= A \textbf{ OR } B, \quad \forall x \in \boldsymbol{X}$
Complement:	$\overline{A}$, a complement of A, implies that
	$\mu_{\overline{A}} = 1 - \mu_A(x) = \textbf{NOT } A, \quad \forall x \in \boldsymbol{X};$

$$A = \int_{\boldsymbol{X}} \mu_A(x)/x$$

where the signs $\sum$ and $\int$ do not mean conventional summation and integration, and "/" is only a marker between the membership $\mu_A(x_i)$ and its element x_i and does not represent division.

A fuzzy set is said to be a *normal fuzzy set* if and only if

$$\max_{x \in \boldsymbol{X}} \mu_A(x) = 1$$

Assume that A and B are two normal fuzzy sets defined on $\boldsymbol{X}$ with membership functions $\mu_A(x)$ and $\mu_B(x)$, $x \in \boldsymbol{X}$. The set-theoretic definitions and operations such as inclusion ($\subset$), intersection ($\cap$), union ($\cup$), and the complement of the two fuzzy sets are defined as follows:

(i) The *intersection* of fuzzy sets A and B corresponds to the connective "**AND**." Thus, $A \cap B = A \textbf{ AND } B$.

(ii) The *union* of fuzzy sets A and B corresponds to the connective "**OR**." Thus, $A \cup B = A \textbf{ OR } B$.

(iii) The operation of *complementation* corresponds to the negation **NOT**. Thus, $\overline{A} = \textbf{NOT } A$.

Fuzzy set operations are summarized in Table 8.1.

Given two sets A and B as shown in Fig. 8.1a, the logic operations listed above are shown in Figs. 8.1b–8.1d. An example is also given below.

Example 1. Assume $\boldsymbol{X} = \{a, b, c, d, e\}$. Let

$$A = \{0.5/a, \ 0.9/b, \ 0.7/c, \ 0.6/d, \ 1/e\}$$

and

$$B = \{0.7/a, \ 1/b, \ 0.8/c, \ 0.5/d, \ 0/e\}$$

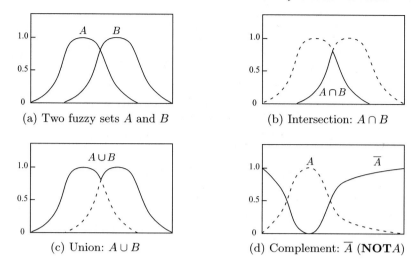

Fig. 8.1. Some logic operations on fuzzy sets.

Then

$$A \cap B = \{0.5/a, \ 0.9/b, \ 0.7/c, \ 0.5/d, \ 0/e\}$$

and

$$A \cup B = \{0.7/a, \ 1/b, \ 0.8/c, \ 0.6/d, \ 1/e\}$$

and

$$\overline{A} = \{0.5/a, \ 0.1/b, \ 0.3/c, \ 0.4/d, \ 0/e\} \qquad\qquad \blacksquare$$

Some other operations of two-fuzzy sets are defined as follows:

(i) The product of two fuzzy sets A and B, written as $A \cdot B$, is defined as

$$\mu_{A \cdot B} = \mu_A(x) \cdot \mu_B(x), \quad \forall x \in X \qquad (8.3)$$

(ii) The algebraic sum of two fuzzy sets A and B, written as $A \oplus B$, is defined as

$$\mu_{A \oplus B} = \mu_A(x) + \mu_B(x) - \mu_A(x) \cdot \mu_B(x), \quad \forall x \in X \qquad (8.4)$$

(iii) A fuzzy relation R between the two (nonfuzzy) sets X and Y is a fuzzy set in the Cartesian product $X \times Y$; that is, $R \subset X \times Y$. Hence, the fuzzy relation R is defined as

$$R = \{\mu_R(x, y)/(x, y)\}, \quad \forall (x, y) \in X \times Y \qquad (8.5)$$

For example, a fuzzy relation R between two sets $X = \{x_1, x_2, \ldots, x_m\}$ and $Y = \{y_1, y_2, \ldots, y_n\}$ can be represented by

$$R = \sum_{i=1,j=1}^{i=m,j=n} \mu_R(x_i, y_j)/(x_i, y_j)$$

or, a matrix expression of the relation R is

$$R = \begin{bmatrix} \mu_R(x_1, y_1) & \mu_R(x_1, y_2) & \cdots & \mu_R(x_1, y_n) \\ \mu_R(x_2, y_1) & \mu_R(x_2, y_2) & \cdots & \mu_R(x_2, y_n) \\ \vdots & \vdots & \ddots & \vdots \\ \mu_R(x_m, y_1) & \mu_R(x_m, y_2) & \cdots & \mu_R(x_m, y_n) \end{bmatrix}$$

(iv) The max–min composition of two fuzzy relations $R \subset X \times Y$ and $S \subset Y \times Z$, written as $R \circ S$, is defined as a fuzzy relation $R \circ S \subset X \times Z$ such that

$$\mu_{R \circ S}(x, z) = \max_{y \in Y}(\mu_R(x, y) \wedge \mu_S(y, z)) \tag{8.6}$$

for each $x \in X, z \in Z$, where $\wedge = \min$.

(v) The Cartesian product of two fuzzy sets $A \subset X$ and $B \subset Y$, written as $A \times B$, is defined as a fuzzy set in $X \times Y$, such that

$$\mu_{A \times B}(x, y) = \mu_A(x) \wedge \mu_B(y) \tag{8.7}$$

for each $x \in X$ and $y \in Y$.

8.2.2 Fuzzy Membership Functions (FMFs)

The definitions of *fuzzy membership functions* (FMFs) of fuzzy sets play an important role in fuzzy set theory and its applications. The following are several types of fuzzy membership functions on a one-dimensional continuous space, as illustrated in Fig. 8.2, which are either continuous, or discontinuous in terms of a finite number of switching points:

(i) Triangular function:

$$\mu(x, a, b, c) = \max\left(\min\left(\frac{x-a}{b-a}, \frac{c-x}{c-b}\right), 0\right), \quad a \neq b \text{ and } c \neq b \tag{8.8}$$

(ii) Trapezoidal function:

$$\mu(x, a, b, c, d) = \max\left(\min\left(\frac{x-a}{b-a}, 1, \frac{d-x}{d-c}\right), 0\right) \tag{8.9}$$

$$a \neq b \text{ and } c \neq d$$

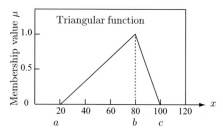
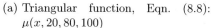

(a) Triangular function, Eqn. (8.8): $\mu(x, 20, 80, 100)$

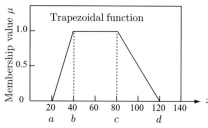

(b) Trapezoidal function, Eqn. (8.9): $\mu(x, 20, 40, 80, 120)$

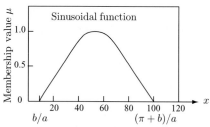

(c) Sinusoidal function, Eqn. (8.10): $\mu(x, \pi/90, \pi/9)$

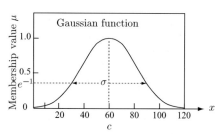

(d) Gaussian function, Eqn. (8.11): $\mu(x, 30, 60)$

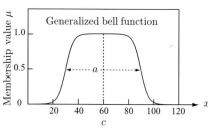

(e) Generalized bell function, Eqn. (8.12): $\mu(x, 5, 35, 85, 115)$

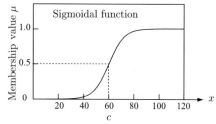

(f) Sigmoidal function, Eqn. (8.13): $\mu(x, 1, 60)$

Fig. 8.2. Examples of some fuzzy membership functions (FMFs), Eqns. (8.8)–(8.13).

(iii) Sinusoidal function:

$$\mu(x, a, b) = \begin{cases} \sin(ax - b), & \text{if } \dfrac{b}{a} \le x \le \dfrac{\pi + b}{a}, \ a \neq 0 \\ 0, & \text{otherwise} \end{cases} \tag{8.10}$$

(iv) Gaussian function:

$$\mu(x, \sigma, c) = e^{-[(x-c)/\sigma]^2} \tag{8.11}$$

where c is a center parameter for controlling the center position of $\mu(x, \sigma, c)$ and σ is a parameter for defining the width of $\mu(x, \sigma, c)$.

(v) Generalized bell function:

$$\mu(x, a, b, c) = \frac{1}{1 + \left|\frac{x-c}{a}\right|^{2b}}, \quad \text{with} \quad b > 0 \tag{8.12}$$

(vi) Sigmoidal function:

$$\mu(x, a, c) = \frac{1}{1 + \exp(-a(x - c))} \tag{8.13}$$

where the parameter c determines the position of $\mu(x, a, c)|_{x=c} = 0.5$.

It should be noted that an FMF contains a set of parameters that define the shape of the membership function. Usually, these parameters can be predetermined by human experience, knowledge, or known data. However, in fuzzy-neural systems they can be adapted online according to the specified environment in order to achieve the optimal performance.

Since the early 1970s, because of the simplicity in their formulations and computational efficiency, both triangular and trapezoid functions have been used extensively as FMFs in fuzzy logical systems (Gupta and Kaufmann, 1985, 1988). However, these two types of FMFs consist of straight line segments, and are not smooth at the switching points, which are determined by the preselected parameters. This raises some difficulties for fuzzy neural computing. Some studies have indicated that continuous and differentiable FMFs such as Gaussian functions, sigmoidal functions, and sinusoidal functions are good candidates for fuzzy neural computing (Sun and Jang, 1993, Jin et al., 1994, 1995).

8.2.3 Fuzzy Systems

A fuzzy system with a basic configuration as depicted in Fig. 8.3 has four principal elements: *fuzzifier, fuzzy rule base, fuzzy inference engine,* and *defuzzifier.* Without the loss of generality, we will consider here multiinput single-output fuzzy systems: $S \subset \Re^n \to \Re$, where S is a compact set.

In such a fuzzy system, the *fuzzifier* deals with a mapping from the input space $S \subset \Re^n$ to the fuzzy sets defined in S, which are characterized by a membership function $\mu_F : S \to [0,1]$, and is labeled by a linguistic variable F such as "small," "medium," "large," or "very large." The most commonly used fuzzifier is a singleton fuzzifier, which is defined as follows:

$$\boxed{\begin{aligned} &x \in S \to \text{fuzzy set } A_x \subset S \text{ with } \mu_{A_x}(x) = 1, \text{ and} \\ &\mu_{A_x}(x') = 0 \text{ for } x' \in S \text{ and } x' \neq x \end{aligned}} \tag{8.14}$$

Thus, by defining a membership function of the input, the fuzzifier changes the range of crisp values of input variables into a corresponding universe of

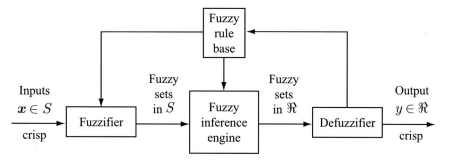

Fig. 8.3. A schematic representation of a fuzzy system.

discourse, and converts nonfuzzy (crisp) input data into suitable linguistic values.

The *fuzzy rule base* consists of a set of linguistic rules of the following form: "IF a set of conditions are satisfied, THEN a set of consequences are inferred."

In other words, a fuzzy rule base is a collection of IF–THEN values. Moreover, we consider in this chapter a fuzzy rule base having M rules of the following forms

$$\boxed{\begin{aligned} R_j(j = 1, 2, \ldots, M): & \text{ IF } x_1 \text{ is } A_1^j, \textbf{ AND } x_2 \text{ is } A_2^j, \\ \textbf{AND}, \ldots, & \textbf{ AND } x_n \text{ is } A_n^j, \text{ THEN } y \text{ is } B^j. \end{aligned}} \qquad (8.15)$$

where $x_i(i = 1, 2, \ldots, n)$ are the input variables to the fuzzy system, y is the output variable of the fuzzy system, and A_i^j and B^j are the linguistic variables characterized by the fuzzy membership functions $\mu_{A_i^j}$ and μ_{B^j}, respectively. In practical applications, the rules can be extracted from either numerical data or human knowledge for the problem of concern. A simple example of three rules for the single-input and single-output case can be given as $R_j(j = 1, 2, 3)$: IF x is A^j THEN y is B^j, where A^j and B^j are characterized by trapezoidal membership functions. A^1 and B^1 are labeled by the linguistic variable "small," A^2 and B^2 are "medium," and A^3 and B^3 are "large," respectively, as illustrated in Fig. 8.4.

Each rule R_j can be viewed as a fuzzy implication

$$A_1^j \times \cdots \times A_n^j \to B^j$$

which is a fuzzy set in $S \times \Re$ with

$$\mu_{A_1^j \times \cdots \times A_n^j \to B^j}(x_1, \ldots, x_n, y) = \mu_{A_1^j}(x_1) \otimes \cdots \otimes \mu_{B^j}(y) \qquad (8.16)$$

for $x \in S$ and $y \in \Re$. The most commonly used operations for $\otimes$ are product and min operations defined as

$$\text{Product operation: } \mu_{A_1^j}(x_1) \otimes \mu_{A_2^j}(x_2) = \left[\mu_{A_1^j}(x_1) \cdot \mu_{A_2^j}(x_2) \right] \qquad (8.17)$$

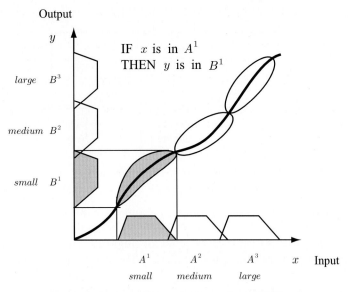

Fig. 8.4. An example of the fuzzy IF–THEN rule.

Min operation: $\mu_{A_1^j}(x_1) \otimes \mu_{A_2^j}(x_2) = \min\left[\mu_{A_1^j}(x_1), \mu_{A_2^j}(x_2)\right]$ (8.18)

The *fuzzy inference engine* is a decisionmaking logic that uses the fuzzy rules provided by the fuzzy rule base to implement a mapping from the fuzzy sets in the input space S to the fuzzy sets in the output space $\Re$. The efficiency of a fuzzy inference engine greatly depends on the knowledge base of the system considered. Let A_x be an arbitrary fuzzy set in S. Then each R_j of the fuzzy rule base creates a fuzzy $A_x \circ R_j$ in $\Re$ based on the sup–star composition:

$$\mu_{A_x \circ R_j} = \sup_{\boldsymbol{x}' \in S}\left[\mu_{A_x}(\boldsymbol{x}') \otimes \mu_{A_1^j \times \cdots \times A_n^j \to B^j}(x_1', \dots, x_n', z)\right]$$

$$= \sup_{\boldsymbol{x}' \in S}\left[\mu_{A_x}(\boldsymbol{x}') \otimes \mu_{A_1^j}(x_1') \otimes \cdots \otimes \mu_{A_n^j}(x_n') \otimes \mu_{B^j}(z)\right] \quad (8.19)$$

The *defuzzifier* provides a mapping from the fuzzy sets in $\Re$ to crisp points in $\Re$. The following centroid defuzzifier, which performs a mapping from the fuzzy set $A_x \circ R_j (j = 1, 2, \dots, M)$ in $\Re$ to a crisp point $y \in \Re$, is the most commonly used method (Mendel 1995), and is defined as follows:

$$y = \frac{\displaystyle\sum_{j=1}^{M} c_j \mu_{A_x \circ R_j}(c_j)}{\displaystyle\sum_{j=1}^{M} \mu_{A_x \circ R_j}(c_j)} \quad (8.20)$$

where c_j is the point in $\Re$ at which $\mu_{B^j}(c_j)$ achieves the maximum value $\mu_{B^j}(c_j) = 1$.

Next, if one assumes that $\otimes$ is a product operation (product inference), then for $\mu_{A_x}(\boldsymbol{x}) = 1$ and $\mu_{A_x}(\boldsymbol{x}') = 0$ for all $\boldsymbol{x}' \in S$ with $\boldsymbol{x}' \neq \boldsymbol{x}$, replacing $\otimes$ in Eqn. (8.19) with the conventional product yields

$$
\mu_{A_x \circ R_j}(c_j) = \sup_{\boldsymbol{x}' \in S} \left[\mu_{A_x}(\boldsymbol{x}')\, \mu_{A_1^j}(x_1') \cdots \mu_{A_n^j}(x_n')\, \mu_{B^j}(c_j) \right]
$$

$$
= \prod_{i=1}^{n} \mu_{A_i^j}(x_i) \tag{8.21}
$$

Thus, the analytical relationship between the crisp input $\boldsymbol{x}$ and the crisp output y is

$$
y = \frac{\displaystyle\sum_{j=1}^{M} c_j \left(\prod_{i=1}^{n} \mu_{A_i^j}(x_i) \right)}{\displaystyle\sum_{j=1}^{M} \left(\prod_{i=1}^{n} \mu_{A_i^j}(x_i) \right)} \tag{8.22}
$$

Other types of defuzzifiers, such as a maximum defuzzifier, mean of maxima defuzzifier, and height defuzzifier, can also be applied to form the mapping from the crisp input $\boldsymbol{x}$ to the crisp output $y \in \Re$.

It is worth to note that from Eqn. (8.22), one may get a Gaussian kernel by using the Gaussian membership function given in Eqn. (8.11). However, a fundamental difference between the kernel method and the fuzzy system is that the former needs to center the kernels on sample data in general, while the latter does not need. Also, the fuzzy membership function does not express a probability of the input, but a possibility of the input. Thus, it is not necessary that the sum or the integration of the membership function for all the input values is 1.

8.3 Building Fuzzy Neurons (FNs) Using Fuzzy Arithmetic and Fuzzy Logic Operations

Integrating the basic mathematics of fuzzy logic discussed in the previous section and the basic structure of neurons (Gupta et al., 2003), some models of fuzzy neurons (FNs) are introduced in this section.

8.3.1 Definition of Fuzzy Neurons

When we consider fuzzy uncertainties within neural units, the inputs and/or the weights of a neuron can be expressed in terms of their membership functions, and several types of *fuzzy neurons* (FNs) based on fuzzy arithmetic and logic operations can be defined. According to the nature of neural inputs and weights (fuzzy or nonfuzzy), we define the following three types of fuzzy neurons with fuzzy operations:

(i) FN_1 has nonfuzzy neural inputs but fuzzy synaptic weights;
(ii) FN_2 has fuzzy neural inputs and nonfuzzy synaptic weights;
(iii) FN_3 has fuzzy neural inputs and fuzzy synaptic weights.

Restricting the synaptic weights to fuzzy quantities may avoid deformation of fuzzy input signals in fuzzy neural computation. Since FN_1 and FN_2 may be considered as special cases of FN_3, emphasis will be devoted only to FN_3, which will be simply referred to as an FN in the following discussion.

The mathematical operations involved in a conventional neuron are

(i) The weighting of the neural inputs with synaptic weights;
(ii) The aggregation of these weighted neural inputs;
(iii) The nonlinear operation on this aggregation.

The mathematical operations in fuzzy neural networks can be carried out using either fuzzy arithmetic operations or fuzzy logic operations. In this section, we briefly describe fuzzy neurons first using fuzzy arithmetic operations and then using fuzzy logic operations.

Fuzzy Arithmetic–Based Fuzzy Neurons

The weighting of fuzzy neural inputs using the synaptic weights can be expressed by fuzzy multiplication, and the aggregation operation of weighted neural inputs by fuzzy addition, and these modifications lead to a fuzzy neural architecture. On the basis of fuzzy arithmetic operations, the mathematical expression of such an FN is given by the following equation

$$y = \sigma\left((+)_{i=0}^{n} w_i(\cdot)x_i \right), \quad x_0 = 1 \qquad (8.23)$$

where σ is a neural activation function, $(+)$ and $(\cdot)$ respectively are the fuzzy addition and fuzzy multiplication operators, and w_0 is the threshold.

Fuzzy neural inputs and fuzzy synaptic weights are defined on an n-dimensional hypercube in terms of their membership functions x_i and w_i, and are as follows:

$$\left.\begin{array}{l} \boldsymbol{x} = [x_0\ x_1\ x_2\ \cdots\ x_n]^T \in [0,1]^{(n+1)}, \quad x_0 = 1 \\ \boldsymbol{w} = [w_0\ w_1\ w_2\ \cdots\ w_n]^T \in [0,1]^{(n+1)} \end{array}\right\} \qquad (8.24)$$

Fuzzy Logic–Based Fuzzy Neurons

Alternatively, fuzzy logic operations, using **OR**, **AND**, and **NOT**, or their generalized versions, can be employed to perform fuzzy neural operations. In this case, fuzzy logic operations can be expressed by the following two neural models

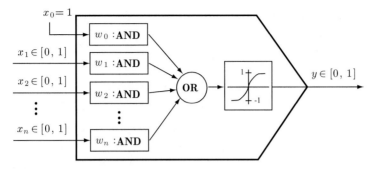

Fig. 8.5. OR–AND-type fuzzy neuron.

(i) **OR**–**AND**-*type fuzzy neuron (Fig. 8.5):*

This type of fuzzy neuron is described by

$$y = \sigma(\ \mathbf{OR}_{i=0}^{n}(w_i\ \mathbf{AND}\ x_i)\ )$$
$$= \sigma(\ \mathbf{OR}(w_0\ \mathbf{AND}\ x_0,\ w_1\ \mathbf{AND}\ x_1,\ \ldots, w_n\ \mathbf{AND}\ x_n)\ ) \qquad (8.25)$$

A schematic representation of this neuron is shown in Fig. 8.5. This **OR**–**AND** fuzzy operations–based neuron is similar to that of the conventional type of neurons (Gupta et al., 2003)

(ii) **AND**–**OR**-*type fuzzy neuron (Fig. 8.6):*

This type of fuzzy neuron is shown in Fig. 8.6, and is described by

$$y = \sigma(\ \mathbf{AND}_{i=0}^{n}(w_i\ \mathbf{OR}\ x_i)\ )$$
$$= \sigma(\ \mathbf{AND}(w_0\ \mathbf{OR}\ x_0,\ w_1\ \mathbf{OR}\ x_1,\ \ldots, w_n\ \mathbf{OR}\ x_n)\ ) \qquad (8.26)$$

This **AND**–**OR**-type of fuzzy neuron is similar to that of the radial basis function (RBF) neurons, and is useful for pattern recognition and other decisionmaking problems. However, only the **OR**–**AND**-type of fuzzy neurons is explored in the following discussions.

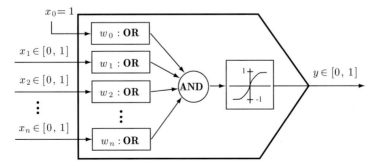

Fig. 8.6. AND–OR-type fuzzy neuron.

8.3.2 Utilization of T and S Operators

Definition and Properties

The $\mathbf{T}$ operator (t-norm) and $\mathbf{S}$ operator (t-conorm), which are generalized $\mathbf{AND}$ and $\mathbf{OR}$ operations, respectively, can be employed effectively to deal with the fuzzy operations given in Eqns. (8.25) and (8.26). Let $x_1, x_2 \in [0,1]$ be two triangular fuzzy numbers. The $\mathbf{T}$ operator $\mathbf{T} : [0,1] \times [0,1] \to [0,1]$ represents the generalized $\mathbf{AND}$ operation, and is defined as

$$(x_1 \ \mathbf{AND} \ x_2) \overset{\triangle}{=} (x_1 \ \mathbf{T} \ x_2) = \mathbf{T}(x_1, x_2) \tag{8.27}$$

Similarly, the $\mathbf{S}$ operator $\mathbf{S} : [0,1] \times [0,1] \to [0,1]$ represents the generalized $\mathbf{OR}$ operation, and is defined as

$$(x_1 \ \mathbf{OR} \ x_2) \overset{\triangle}{=} (x_1 \ \mathbf{S} \ x_2) = \mathbf{S}(x_1, x_2) \tag{8.28}$$

In fact, a $\mathbf{T}$ operator (t-norm) is a nonlinear mapping from $[0,1] \times [0,1]$ onto $[0,1]$. For three fuzzy numbers x, y, and $z \in [0,1]$ the $\mathbf{T}$ operator satisfies the following properties

(i) $\mathbf{T}(x, y) = \mathbf{T}(y, x)$ (commutativity)
(ii) $\mathbf{T}(\mathbf{T}(x, y), z) = \mathbf{T}(x, \mathbf{T}(y, z))$ (associativity)
(iii) $\mathbf{T}(x_1, y_1) \geq \mathbf{T}(x_2, y_2)$ if $x_1 \geq x_2$ and $y_1 \geq y_2$ (monotonicity)
(iv) $\mathbf{T}(x, 1) = x$ (boundary condition)

An $\mathbf{S}$ operator (t-conorm) is also a nonlinear mapping from $[0,1] \times [0,1]$ onto $[0,1]$ that differs from a $\mathbf{T}$ operator only in the property (iv), the boundary condition. For the $\mathbf{S}$ operator, the boundary conditions are

$$\mathbf{S}(x, 0) = x \tag{8.29}$$

Some additional properties of the $\mathbf{T}$ and $\mathbf{S}$ operators are

$$\left. \begin{array}{l} \mathbf{T}(0, 0) = 0, \ \mathbf{T}(1, 1) = 1 \\ \mathbf{S}(0, 0) = 0, \ \mathbf{S}(1, 1) = 1 \end{array} \right\} \tag{8.30}$$

Also, using the $\mathbf{T}$ and $\mathbf{S}$ operators, De Morgan's theorems are stated as follows:

$$\mathbf{T}(x_1, x_2) = 1 - \mathbf{S}(1 - x_1, 1 - x_2) \tag{8.31}$$

and

$$\mathbf{S}(x_1, x_2) = 1 - \mathbf{T}(1 - x_1, 1 - x_2) \tag{8.32}$$

Indeed, negation $\mathbf{N}$ on $x_1 \in [0,1]$ is defined as a mapping

$$\mathbf{N}(x_1) = 1 - x_1 \tag{8.33}$$

which implies $\mathbf{N}(0) = 1, \mathbf{N}(1) = 0$, and $\mathbf{N}(\mathbf{N}(x)) = x$.

Table 8.2. T and **S** operators on fuzzy variables x and $y \in [0,1]$

No.	$\mathbf{T}\,(x,y)$: **AND** operation	$\mathbf{S}\,(x,y)$: **OR** operation	$\mathbf{N}\,(x)$
1	$\min(x,y)$	$\max(x,y)$	$1-x$
2	xy	$x+y-xy$	$1-x$
3	$\max(x+y-1,0)$	$\min(x+y,1)$	$1-x$

Fuzzy Logic Neuronal Equations

By means of the **T** and **S** operators just discussed, the input–output function $y = f(x_1, x_2, \ldots, x_n)$ of the **OR–AND** fuzzy neuron defined in Eqn. (8.25) can be represented further as

$$u = \mathbf{S}_{i=0}^{n}\left[(w_i\,\mathbf{T}\,x_i)\right]$$
$$= \mathbf{S}_{i=0}^{n}\left[\mathbf{T}(w_i, x_i)\right]$$
$$= \mathbf{S}\left[\mathbf{T}(w_0, x_0),\ \mathbf{T}(w_1, x_1),\ \ldots, \mathbf{T}(w_n, x_n)\right] \in [0,1], \quad x_0 = 1 \quad (8.34)$$

and

$$y = \sigma(u) \in [0,1], \quad \text{for} \quad u \geq 0 \tag{8.35}$$

where $u \in [0,1]$ is an intermediate variable that is introduced to simplify the mathematical expression of such a fuzzy neural operation. It can be noted that even if a bipolar activation function $\sigma(\cdot) \in [-1,1]$ is employed in Eqn. (8.35), the output y, which is also a fuzzy quantity in terms of the membership grade, is always located in the unit interval $[0,1]$ because $u \geq 0$.

There are many alternative ways to define the expressions for the **T** and **S** operators. However, for simplicity, only the three types of **T** and **S** operators proposed previously are summarized in Table 8.2. Since in fuzzy neural computing, the operations of the **T** and **S** operators defined in Table 8.2 are often on more than two fuzzy variables, the generalized versions of **T** and **S** operators given in Table 8.2 are provided in Table 8.3 for dealing with n fuzzy variables $x_1, x_2, \ldots, x_n \in [0,1]$.

According to the three definitions of the **T** and **S** operators given in Tables 8.2 and 8.3, we now give the mathematical expressions for three different types of **OR–AND** fuzzy neurons.

Type I (min–max fuzzy neuron):

The operational equation for this type of min–max FN is obtained using the first type of **T** and **S** operators given in Table 8.3 as follows:

$$u = \max_{0 \leq i \leq n} \left(\min(w_i, x_i)\right)$$
$$= \max\left(\min(w_0, x_0), \min(w_1, x_1), \ldots, \min(w_n, x_n)\right) \tag{8.36}$$

Table 8.3. T and **S** operators for n fuzzy variables $x_1, x_2, \ldots, x_n \in [0,1]$

No.	$\mathbf{T}(x_1, x_2, \ldots, x_n)$ (**AND** operation)	$\mathbf{S}(x_1, x_2, \ldots, x_n)$ (**OR** operation)
1	$\min(x_1, x_2, \ldots, x_n)$	$\max(x_1, x_2, \ldots, x_n)$
2	$\prod_{i=1}^{n} x_i$	$\sum_{i=1}^{n} x_i - \sum_{j=1}^{n} \sum_{1 \le i_1 < \cdots < i_j \le n} x_{i_1} \cdots x_{i_j}$ or equivalently $\begin{cases} v_1 = x_1 \\ v_i = v_{i-1} + x_i - x_i v_{i-1}, \quad i = 2, 3, \ldots, n \\ x_n + v_{n-1} - x_n v_{n-1} \end{cases}$
3	$\begin{cases} v_1 = x_1 \\ v_i = \max(v_{i-1} + x_i - 1, 0), \quad i = 2, 3, \ldots, n \\ \max(v_{n-1} + x_n - 1, 0) \end{cases}$	$\min(x_1 + x_2 + \cdots + x_n, \ 1)$

and

$$y = \sigma(u)$$

Type II (product–sum fuzzy neuron):

The product–sum fuzzy neuron is of the second type and is expressed by the following recursive formulations

$$\left. \begin{aligned} v_0 &= w_0 x_0 \\ v_i &= w_i x_i + v_{i-1} - w_i x_i v_{i-1}, \quad i = 1, 2, 3, \ldots, n \\ u &= v_n \end{aligned} \right\} \tag{8.37}$$

and

$$y = \sigma(u) = \sigma(v_n) \tag{8.38}$$

or equivalently

$$\left. \begin{aligned} v_i &= w_i x_i, \quad i = 0, 1, 2, \ldots, n \\ u &= \sum_{i=0}^{n} v_i - \sum_{j=1}^{n} \sum_{0 \le i_1 < i_2 < \cdots < i_j \le n} v_{i_1} v_{i_2} \cdots v_{i_j} \end{aligned} \right\} \tag{8.39}$$

and

$$y = \sigma(u)$$

For instance, when $n = 2$, Eqn. (8.37) becomes

$$
\left.
\begin{aligned}
v_0 &= w_0 x_0, \, x_0 = 1 \\
v_1 &= w_1 x_1 + w_0 x_0 - w_1 x_1 w_0 x_0 \\
v_2 &= w_2 x_2 + w_1 x_1 + w_0 x_0 - w_1 x_1 w_0 x_0 \\
&\quad - w_2 x_2 w_1 x_1 - w_2 x_2 w_0 x_0 - w_2 x_2 w_1 x_1 w_0 x_0 \\
u &= v_2
\end{aligned}
\right\}
$$

and

$$y = \sigma(u) \tag{8.40}$$

Type III (max–min fuzzy neuron):

The third type of fuzzy neuron is the max–min fuzzy neuron, which is obtained by using the third type of $\mathbf{T}$ and $\mathbf{S}$ operators given in Table 8.3.

$$
\begin{aligned}
u &= \min \Big(\max(w_0 + x_0 - 1, 0) + \max(w_1 + x_1 - 1, 0) + \cdots \\
&\qquad + \max(w_n + x_n - 1, 0), 1 \Big) \\
&= \min \left(\sum_{i=0}^{n} \max(w_i + x_i - 1, 0), 1 \right)
\end{aligned}
\tag{8.41}
$$

Noting that since for x_1 and $x_2 \in [0,1]$

$$\max(x_1 + x_2 - 1, 0) = \max(x_1 + x_2, 1) - 1$$

Eqn. (8.41) can equivalently be expressed as follows:

$$
\begin{aligned}
u &= \min \Big(\max(w_0 + x_0, 1) - 1 + \max(w_1 + x_1, 1) - 1 + \cdots \\
&\qquad + \max(w_n + x_n, 1) - 1, 1 \Big) \\
&= \min \left(\sum_{i=0}^{n} \max(w_i + x_i, 1) - n, 1 \right)
\end{aligned}
\tag{8.42}
$$

and

$$y = \sigma(u) \tag{8.43}$$

Example 2. Consider a fuzzy neuron with four inputs x_1, x_2, x_3, and x_4 as shown in Fig. 8.7. Let the nonlinear activation function be a sigmoidal function $\sigma(u) = \tanh(u)$. Using the three types of fuzzy neural operations just discussed above, the output of the fuzzy neuron can be obtained as given in Table 8.4.
∎

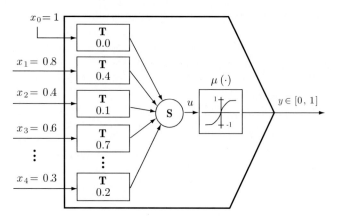

Fig. 8.7. Example 2: the fuzzy neuron.

Table 8.4. Example 2: output of the fuzzy neurons

Type	u	Output $y = \tanh(u)$
I: min–max	0.6	0.8090
II: product–sum	0.6441	0.8478
III: max–min	0.5	0.7071

8.4 Hybrid Fuzzy Neural Networks (HFNNs)

The three types of fuzzy neural units discussed in the previous section can be used to form a class of fuzzy neural networks (FNNs). These FNNs can be used for nonlinear approximating mappings from the input hypercube $[0,1]^n$ to the output hypercube $[0,1]^m$ in a fuzzy logic–based format. Since these FN models are built by using the standard fuzzy logic, the networks formed by these FNs are termed *regular fuzzy neural networks* (RFNNs).

The universal approximation capability of neural networks is one of the promising advantages for their applications to areas such as identification, control, and pattern recognition. How well does a neural network approximate an unknown function? This is an important question that is asked about all types of neural networks, including multilayered feedforward neural networks (MFNNs) and dynamic neural networks (DNNs). However, it has been proved that the RFNNs are not universal approximators (Gupta et al., 2003).

To modify the fuzzy operations in the RFNN so that the universal approximation capability of the fuzzy neural networks is ensured, some new structures of fuzzy neurons are discussed in this section.

8.4.1 Difference-Measure-Based Two-Layered HFNNs

A new architecture of a hybrid fuzzy neural network (HFNN) is shown in Fig. 8.8, where the HFNN has multiple inputs and a single output, and consists of an input neural node, one hidden fuzzy neural layer, and an output neural layer.

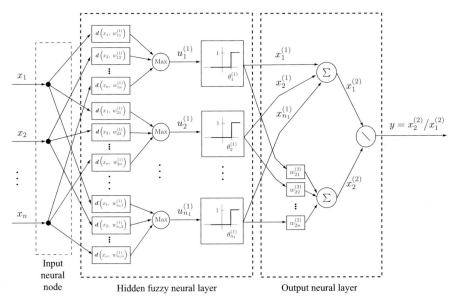

Fig. 8.8. A difference-measure-based two-layered hybrid fuzzy neural network (HFNN).

In the HFNN, all of the neural inputs are distributed to all the neurons in the hidden fuzzy neural layer. In such fuzzy neural computing, it is proposed that this operation can be replaced by a *difference measure* of the input signal $x_i \in [0,1]$ and the weight $w_i \in [0,1]$ defined by

$$\boldsymbol{d}(x_i, w_i) \begin{cases} = 0, & \text{if} \quad x_i = w_i \\ > 0, & \text{otherwise} \end{cases} \tag{8.44}$$

Thus, the output of the neurons in the hidden fuzzy neural layer can be obtained as follows:

$$\begin{aligned} u_i^{(1)} &= \max\left(\left[\boldsymbol{d}(x_1, w_1), \boldsymbol{d}(x_2, w_2), \ldots, \boldsymbol{d}(x_n, w_n)\right]\right) \\ &= \max_{i=1}^{n}\left(\ \boldsymbol{d}(x_i, w_i)\ \right) \end{aligned} \tag{8.45}$$

and

$$x_i^{(i)} = \begin{cases} 1, & \text{if } u_i^{(1)} > \theta_i^{(1)} \\ 0, & \text{otherwise} \end{cases} \tag{8.46}$$

where $\theta_i^{(1)}$ is a threshold associated with the neuron FN(1,i).

There are only two neurons in the output layer. The output of the first neuron in the output layer is simply the summation of the outputs of all the neurons in the first layer:

$$x_1^{(2)} = \sum_{j=1}^{n_1} x_j^{(1)} \tag{8.47}$$

The output of the second neuron in the output layer is a weighted summation of the form

$$x_2^{(2)} = \sum_{i=1}^{n_1} w_{2j}^{(2)} x_j^{(1)} \tag{8.48}$$

Finally, the output of the network is defined as

$$\begin{cases} \dfrac{x_2^{(2)}}{x_1^{(2)}}, & \text{if } x_1^{(2)} > 0 \\ 0, & \text{otherwise} \end{cases} \tag{8.49}$$

In this definition all the weights and thresholds are assumed to be triangle fuzzy numbers. Thus, the operation from the hidden layer to the output layer deals with a *centroid defuzzifier*, where the input signals $x_1^{(1)}, x_2^{(1)}, \ldots, x_{n_1}^{(1)}$ are unipolar binary signals (Hayashi and Buckley, 1993a, 1994a). These unipolar binary signals can easily be extended to bipolar binary signals as well as to some modified neural architectures and different types of fuzzy numbers.

It has been proved that the HFNN described above is a universal approximator (Hayashi and Buckley, 1993a). However, no related learning algorithm has been reported to carry out such an approximation.

It is to be noted that the input–output mapping of such an HFNN is discontinuous since the max operation and hard-limiting functions are used in forming the input–output mapping. This may cause some difficulties with the learning phase.

8.4.2 Fuzzy Neurons and Hybrid Fuzzy Neural Networks (HFNNs)

A conventional neuron involves a somatic operation, which is a confluence or similarity measure operation between the input signals and the corresponding synaptic weights (Gupta et al., 2003) In the FN model discussed in Section 8.3, this operation was replaced by the **T** operator. This operation can also be replaced with a difference measure on the neural input signal $x_i \in [0,1]$ and the synaptic weight $w_i \in [0,1]$. This difference measure operation, as discussed in Section 8.4.1 on two-layered HFNNs and as seen in Eqn. (8.44), is denoted by $\boldsymbol{d}(x_i, w_i)$, which satisfies $0 \leq \boldsymbol{d}(x_i, w_i) \leq 1$. For example, $\boldsymbol{d}(x_i, w_i)$ can be selected as

$$\boldsymbol{d}(x_i, w_i) = |x_i, w_i|^n \tag{8.50}$$

In the following discussion, we assume that $\boldsymbol{d}(x_i, w_i)$ is of the quadratic form

$$\boldsymbol{d}(x_i, w_i) = (x_i - w_i)^2 \tag{8.51}$$

This difference measure operation is then used to replace the **T** operator in the fuzzy neuron introduced in Section 8.3. As shown in Fig. 8.9, all these difference measures between the inputs and the associate weights can be combined by means of the standard **S** operator.

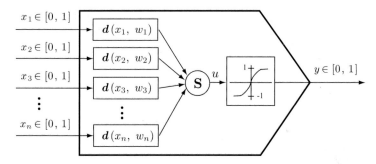

Fig. 8.9. A fuzzy neuron with the difference measure and **S** operator, Eqns. (8.52) and (8.53).

Thus, the output of such a fuzzy neuron is given by

$$
\begin{aligned}
u &= \mathbf{S}_{i=1}^n \left(\, \boldsymbol{d}(x_i, w_i) \, \right) \\
&= \mathbf{S} \left(\, \boldsymbol{d}(x_1, w_1), \boldsymbol{d}(x_2, w_2), \ldots, \boldsymbol{d}(x_n, w_n) \, \right)
\end{aligned}
\tag{8.52}
$$

and, finally, the neural output is given by

$$y = \sigma(u) \tag{8.53}$$

Obviously, this type of fuzzy neuron is no longer monotonic in terms of its fuzzy inputs $x_1, x_2, \ldots, x_n$.

Using this type of fuzzy neuron, a two-layered fuzzy neural network, the *hybrid fuzzy neural network*, can easily be formed. The operational equations of such an HFNN are as follows:

$$\text{FN}(1, i) : \begin{cases} u_1^{(1)} = \boldsymbol{S}_{\ell=1}^n \left(\boldsymbol{d} \left(w_{i\ell}^{(1)}, x_\ell \right) \right), & i = 1, 2, \ldots, p \\[2mm] x_1^{(1)} = \sigma \left(u_i^{(1)} \right), & i = 1, 2, \ldots, p \end{cases} \qquad (8.54)$$

$$\text{FN}(2, j) : \begin{cases} u_j^{(2)} = \boldsymbol{S}_{q=1}^p \left(\boldsymbol{d} \left(w_{jq}^{(2)}, x_q^{(1)} \right) \right), & i = 1, 2, \ldots, m \\[2mm] y_j = \sigma \left(u_j^{(2)} \right), & i = 1, 2, \ldots, m \end{cases} \qquad (8.55)$$

It seems that this HFNN has a capability for approximation of functions, but no strict mathematical proof is currently available as to its universal approximation for this network. The following example will show that the HFNN is capable of solving the **XOR** problem.

Example 3. From this example, it will be seen that a two-variable binary **XOR** function $y = x_1 \oplus x_2$ can be implemented by a difference-measure-based two-layered HFNN with two neurons in the hidden layer and one in the output layer, as shown in Fig. 8.10. In fact, if the **S** operator is selected as a max operation, the input–output equation of such a network can be obtained as

$$u_1^{(1)} = \max\{|x_1 - 1|, \; x_2\}$$
$$u_2^{(1)} = \max\{x_1, \; |x_2 - 1|\}$$
$$y = \max \left\{ \left| u_1^{(1)} - 1 \right|, \; \left| u_2^{(1)} - 1 \right| \right\}$$

The input–output relationship for this HFNN is given in Table 8.5, which clearly shows an **XOR** operation of this network.

It is to be noted that in this example, although we have used the **S**-operation as a max operation, other types of the **S**-operation, as tabulated in Table 8.2, can also be used. ∎

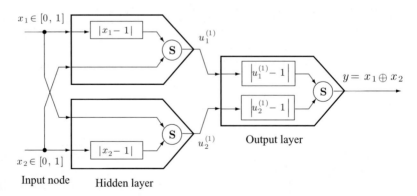

Fig. 8.10. Example 3: a difference-measure-based HFNN with two neurons in the hidden layer and one neuron in the output layer for the implementation of a two-variable **XOR** function, $y = x_1 \oplus x_2$.

Table 8.5. Example 3, fuzzy **XOR** operation

x_1	x_2	$y = x_1 \oplus x_2$
0	0	0
0	1	1
1	0	1
1	1	0

In addition to the realization of fuzzy **XOR** operation by HFNNs, an another architecture, called *fuzzy basis function networks* (FBFNs), may be introduced to analyze universal approximation capabilities of fuzzy systems (Gupta et al., 2003).

Compared to the fuzzy system that has a fuzzifier, a fuzzy rule base, a fuzzy inference engine, and a defuzzifier discussed in Section 8.2.3 and illustrated in Fig. 8.3, the fuzzy neural network (FNN) structures discussed so far dealt only with the fuzzy input signals for a specified task as shown in Fig. 8.11. On the other hand, an FBFN is used to express an *entire* fuzzy system in the context of forming a desired input–output mapping function for crisp signals, as illustrated in Fig. 8.12.

It has been indicated that if a Gaussian membership function is applied, the fuzzy system is functionally equivalent to a modified Gaussian network. Thus, well-known results for the Gaussian network such as online and offline learning algorithms, and universal approximation capabilities might be employed directly in the design and analysis of fuzzy systems (Gupta et al., 2003)

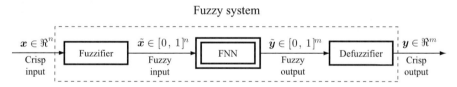

Fig. 8.11. A fuzzy neural network (FNN) as a component of a fuzzy system.

Fig. 8.12. A fuzzy neural network (FNN) implements an entire fuzzy system.

8.4.3 Backpropagation Learning for Hybrid Fuzzy Neural Networks

The learning procedure for the free parameters in such a neural network is considered on the basis of the elements of the set of the training patterns. Given a set of desired data pairs $(\boldsymbol{x}(k), \boldsymbol{y}(k))$, an error index is defined as

$$E(k) = \tfrac{1}{2}\sum_{j=1}^{m}\left[y_j^d(k) - y_j(k)\right]^2 = \tfrac{1}{2}\sum_{j=1}^{m} e_j^2(k)$$

For a two-layered hybrid fuzzy neural network (HFNN) with the operational equations, Eqns. (8.54) and (8.55), the fuzzy backpropagation (FBP) algorithm is given by the following updating formulations

$$w_{i\ell}^{(1)}(k+1) = sat\left(w_{i\ell}^{(1)}(k) + \Delta w_{i\ell}^{(1)}(k)\right)$$
$$i = 1, 2, \dots, p; \quad \ell = 1, 2, \dots, n \tag{8.56}$$
$$w_{jq}^{(2)}(k+1) = sat\left(w_{jq}^{(2)}(k) + \Delta w_{jq}^{(2)}(k)\right)$$
$$j = 1, 2, \dots, m; \quad q = 1, 2, \dots, p \tag{8.57}$$

where $sat(\cdot)$ is a unipolar saturating function defined as

$$sat(x) = \begin{cases} 1, & \text{if} \quad x > 1 \\ x, & \text{if} \quad 0 \le x \le 1 \\ 0, & \text{if} \quad x < 0 \end{cases} \tag{8.58}$$

and

$$\Delta w_{i\ell}^{(1)} = -\eta_1 \frac{\partial E}{\partial w_{i\ell}^{(1)}} \tag{8.59}$$
$$\Delta w_{jq}^{(2)} = -\eta_2 \frac{\partial E}{\partial w_{jq}^{(2)}} \tag{8.60}$$

In order to derive these updating formulations, the concept of the error partial derivatives δs is introduced below. For such a two-layered fuzzy neural structure these intermediate variables are denoted as

$$\delta_i^{(1)} \triangleq -\frac{\partial E}{\partial u_i^{(1)}}, \quad i = 1, 2, \dots, p \tag{8.61}$$
$$\delta_j^{(2)} \triangleq -\frac{\partial E}{\partial u_j^{(2)}}, \quad j = 1, 2, \dots, m \tag{8.62}$$

where $\delta_i^{(1)}$ is the partial derivative of FN(1, i) and $\delta_j^{(2)}$ is that of FN(2, j). Therefore, Eqns. (8.59) and (8.60) can be represented as

$$\Delta w_{i\ell}^{(1)} = -\eta_1 \frac{\partial E}{\partial u_i^{(1)}} \frac{\partial u_i^{(1)}}{\partial w_{i\ell}^{(1)}} = -\eta_1 \delta_i^{(1)} \frac{\partial u_i^{(1)}}{\partial w_{i\ell}^{(1)}} \tag{8.63}$$

$$\Delta w_{jq}^{(2)} = -\eta_2 \frac{\partial E}{\partial u_j^{(2)}} \frac{\partial u_j^{(2)}}{\partial w_{jq}^{(2)}} = -\eta_2 \delta_j^{(2)} \frac{\partial u_j^{(2)}}{\partial w_{jq}^{(2)}} \tag{8.64}$$

It is seen that the definition of these partial derivatives not only keeps the derivation simple but also plays an important role in the final learning formulations. It is easy to derive the δs for the output neurons as follows:

$$\delta_j^{(2)} = e_j \sigma' \left(\delta_j^{(2)} \right) \tag{8.65}$$

For simplicity, assume that the **S** operator is defined as a max operation in the following derivation. Then

$$\frac{\partial u_j^{(2)}}{\partial w_{jq}^{(2)}} = \begin{cases} \left(w_{jq}^{(2)} - x_q^{(1)} \right), & \text{if } u_j^{(2)} = \left(w_{jq}^{(2)} - x_q^{(1)} \right)^2 \\ 0, & \text{otherwise} \end{cases} \tag{8.66}$$

Thus, the updating formulations for the weights in the output layer are obtained as follows:

$$w_{jq}^{(2)}(k+1) = \begin{cases} sat \left(w_{jq}^{(2)}(k) + \eta_2 \delta_j^{(2)} \left(w_{jq}^{(2)}(k) - x_q^{(1)}(k) \right) \right), \\ \qquad \text{if} \quad u_j^{(2)} = \left(w_{jq}^{(2)} - x_q^{(1)} \right)^2 \\ w_{jq}^{(2)}(k), \qquad \text{otherwise} \end{cases} \tag{8.67}$$

The next task is to derive the updating formulations for the weights associated with the hidden neurons. To do so, we first deal with the δs associated with the hidden neurons. By means of the chain law, Eqn. (8.61) can be represented as

$$\delta_i^{(1)} = -\frac{\partial E}{\partial u_i^{(1)}} = -\sum_{j=1}^{m} \frac{\partial E}{\partial u_j^{(2)}} \frac{\partial u_j^{(2)}}{\partial u_i^{(1)}} = \sum_{j=1}^{m} \delta_j^{(2)} \frac{\partial u_j^{(2)}}{\partial u_i^{(1)}}$$

$$= \sum_{j=1}^{m} \delta_j^{(2)} \frac{\partial u_j^{(2)}}{\partial x_i^{(1)}} \frac{\partial x_i^{(1)}}{\partial u_i^{(1)}} = \sum_{j=1}^{m} \delta_j^{(2)} \sigma' \left(u_i^{(1)} \right) \frac{\partial u_j^{(2)}}{\partial x_i^{(1)}} \tag{8.68}$$

Noting the relationship between $u_j^{(2)}$ and $x_i^{(1)}$ given by Eqn. (8.66), the following partial derivative formulations can be obtained:

$$\frac{\partial u_j^{(2)}}{\partial x_i^{(1)}} = \begin{cases} \left(x_i^{(1)} - w_{ji}^{(2)} \right), & \text{if } u_j^{(2)} = \left(w_{ji}^{(2)} - x_i^{(1)} \right)^2 \\ 0, & \text{otherwise} \end{cases} \tag{8.69}$$

Furthermore

$$\frac{\partial u_i^{(1)}}{\partial w_{i\ell}^{(1)}} = \begin{cases} \left(w_{i\ell}^{(1)} - x_i \right), & \text{if } u_i^{(1)} = \left(w_{i\ell}^{(1)} - x_i \right)^2 \\ \\ 0, & \text{otherwise} \end{cases} \tag{8.70}$$

In this case, the weight updating formulations are obtained as follows

$$w_{i\ell}^{(1)}(k+1) = \begin{cases} sat\left(w_{i\ell}^{(1)}(k) + \eta_1 \delta_i^{(1)} \left(w_{i\ell}^{(1)}(k) - x_\ell(k) \right) \right), \\ \qquad\qquad \text{if } u_i^{(1)} = \left(w_{i\ell}^{(1)} - x_\ell \right)^2 \\ \\ w_{i\ell}^{(1)}(k), & \text{otherwise} \end{cases} \tag{8.71}$$

The fuzzy backpropagation (FBP) algorithm obtained above has a two-way information transfer. First the input fuzzy signals are calculated in the feedforward path and then the error signals that are used for updating the process are produced in the backward path. In other words, the input signals are processed starting from the input layer to the output layer. The error signals are calculated in the output layer and then propagated to the lower neural layers. The term *backpropagation* is used here to reflect this interesting fact.

8.5 Concluding Remarks

Fuzzy neural networks (FNNs) incorporate both neural networks and fuzzy mathematics. A neural network is a computational network that has some special characteristics such as learning, adaptation, and generalization. On the other hand, fuzzy mathematics has the capacity for processing the approximate reasoning and knowledge based information by using fuzzy logic operations. FNNs retain the advantages of both of these two structures and are capable of dealing with both numerically expressed and knowledge-based information. In practice, the learning and adaptation mechanisms of FNNs can enhance the approximate reasoning power of fuzzy systems.

Neural structures employing fuzzy logic operations, such as t-norm, t-conorm, and fuzzy implications, can be used for classification, approximation, and rule generation. Although the various definitions of t-norm and t-conorm could give different mathematical descriptions for a network mechanism, the final results of the mapping realized by the network are quite similar. This suggests that more attention to this type of FNN should be placed on hybrid fuzzy neural networks (HFNNs), which may have functional approximation capability. Also, fuzzy backpropagation (FBP) learning algorithms and genetic algorithms can be applied effectively to tune the parameters in such a fuzzy network using the data or online sensor measurements. On the other hand, fuzzy basis function networks (FBFNs) can be used to express fuzzy systems such that the learning and adaptation capabilities are easily enhanced for adapting both system parameters and membership functions. Both the gradient descent technique-based online learning schemes and clustering, and the generalized inverse approaches-based offline approaches can be employed in the learning of FBFNs to perform tasks such as modeling, control, and pattern recognition.

The purpose of this chapter is to help the reader learn not only the existing results in the field but also the state-of-the-art achievements. The topics studied in this chapter cover definition, structure, mathematical models, and learning and adaptation mechanisms of FNNs. The materials reported here form a basis for applications such as fuzzy modeling and control, pattern recognition, and fuzzy neural reasoning. Behind the foundations presented in this chapter, the advanced topics such as fuzzy genetic algorithms, dynamic fuzzy neural structures, and real-time implementations of FNNs have also been studied extensively since the mid-1990s. However, these topics are not discussed in this chapter. An extensive list of references in some literature (Gupta et al., 2003) will help the readers explore this field in more detail.

References

N. Al-Holou, T. Lahdhiri, J.D. Sung, J. Weaver, and F. Al-Abbas. Sliding mode neural network inference fuzzy logic control for active suspension systems. *IEEE Trans. Fuzzy Systems*, 10(2):234–246, 2002.

H.R. Berenji and R. Langari. *Handbook of intelligent control*, chapter Fuzzy Logic in Control Engineering, pages 93–140. Van Nostrand, New York, 1992.

H. Eichfeld, T. Kunemund, and M. Menke. A 12b general-purpose fuzzy logic controller chip. *IEEE Trans. Fuzzy Systems*, 4(4):460–475, 1996.

M.M. Gupta. *Neuro Control*, chapter Fuzzy Logic and Neural Networks. IEEE Press, New York, 1994.

M.M. Gupta, L. Jin, and N. Homma. *Static and Dynamic Neural Networks – From Foundamentals to Advanced Theory*. IEEE Press & Wiley, Hoboken, NJ, 2003.

M.M. Gupta and A. Kaufmann. *Introduction to Fuzzy Arithmetic*. Van Nostrand, New York, 1985.

M.M. Gupta and A. Kaufmann. *Fuzzy Mathematical Models in Engineering and Management Science*. North Holland, Amsterdam, 1988.

M.M. Gupta and N.K. Sinha. *Soft Computing and Intelligent Systems – Theory and Applications*. Academic Press, New York, 1999.

Y. Hayashi and J.J. Buckley. Can fuzzy neural nets approximate continuous fuzzy functions. *Fuzzy Sets and Systems*, 61(1):43–52, 1993a.

Y. Hayashi and J.J. Buckley. Hybrid neural nets can be fuzzy controllers and fuzzy expert systems. *Fuzzy Sets and Systems*, 60(2):135–142, 1993b.

Y. Hayashi and J.J. Buckley. Numerical relationships between neural networks, continuous functions and fuzzy systems. *Fuzzy Sets and Systems*, 60(1):1–8, 1993c.

Y. Hayashi and J.J. Buckley. Fuzzy genetic algorithm and applications. *Fuzzy Sets and Systems*, 61(2):129–136, 1994a.

Y. Hayashi and J.J. Buckley. Fuzzy neural networks: A survey. *Fuzzy Sets and Systems*, 66(1):1–13, 1994b.

J.S.R. Jang. Self-learning fuzzy controllers based on temporal back-propagation. *IEEE Trans. Neural Networks*, 3(5):714–723, 1992.

L. Jin, M.M. Gupta, and P.N. Nikiforuk. *Intelligent Control Systems*, chapter Approximation Capabilities of Feedforward and Recurrent Neural Networks, pages 234–264. IEEE Press, New York, 1994.

L. Jin, M.M. Gupta, and P.N. Nikiforuk. *Fuzzy Logic and Intelligent Control*, chapter Neural Networks and Fuzzy Basis Functions for Functional Approximation, pages 17–68. Kluwer Academic Publishers, 1995.

J. Kiszka, M.M. Gupta, and P.N. Nikiforuk. Energetistic stability of fuzzy dynamic systems. *IEEE Trans. Syst., Man. Cybernetics*, 15(5):783–792, 1985.

B. Kosko. *Neural Networks and Fuzzy Systems*. Prentice Hall, Englewood Cliffs, 1992.

B. Kosko. Fuzzy systems as universal approximators. *IEEE Trans. Computers*, 43 (11):1329–1333, 1994.

C.C. Lee. Fuzzy logic in control systems: Fuzzy logic controller, part i. *IEEE Trans. Syst., Man. Cybernetics*, 20(2):404–418, 1990a.

C.C. Lee. Fuzzy logic in control systems: Fuzzy logic controller, part ii. *IEEE Trans. Syst., Man. Cybernetics*, 20(2):419–435, 1990b.

E.T. Lee and S.C. Lee. Fuzzy sets and neural networks. *J. Cybernetics*, 4(2):83–101, 1974.

J.M. Mendel and L.X. Wang. Generating fuzzy rules from numerical data, with applications. *IEEE Trans. Syst., Man. Cybernetics*, 22(6):1414–1472, 1992.

J.M. Mendel and L.X. Wang. Fuzzy adaptive filters, with application to nonlinear channel equalization. *IEEE Trans. Fuzzy Systems*, 1(3):161–170, 1993.

W. Pedrycz. A referential scheme of fuzzy decision making and its neural network structure. *IEEE Trans. Syst., Man. Cybernetics*, 21(6):1593–1604, 1991.

W. Pedrycz. Fuzzy neural networks and neurocomputations. *Fuzzy Sets and Systems*, 56(1):1–28, 1993.

W. Pedrycz. Genetic algorithms for learning in fuzzy relational structures. *Fuzzy Sets and Systems*, 69(1):37–52, 1995.

H. Prade and D. Dubois. *Fuzzy Sets and Systems: Theory And Applications*. Academic, Orlando FL, 1980.

J. Qi and M.M. Gupta. On fuzzy neuron models. In *IJCNN*, pages 1431–1456, 1991.

J. Qi and M.M. Gupta. *Fuzzy logic for the management of uncertainty*, chapter On Fuzzy Neuron Models, pages 479–491. Wiley, New York, 1992a.

J. Qi and M.M. Gupta. Theory of t-norms and fuzzy inference method. *Fuzzy Sets and Systems*, 40:431–450, 1992b.

D.H. Rao and M.M. Gupta. On the principles of fuzzy neural networks. *Fuzzy Sets and Systems*, 61(1):1–18, 1994.

C.T. Sun and J.S.R. Jang. Neuro-fuzzy modeling and control. *Proc. IEEE*, 83(3): 378–406, 1990.

C.T. Sun and J.S.R. Jang. Functional equivalence between radial basis function networks and fuzzy inference systems. *IEEE Trans. Neural Networks*, 4(1):156–159, 1993.

L.X. Wang. Stable adaptive fuzzy control of nonlinear systems. *IEEE Trans. Fuzzy Systems*, 1(2):146–155, 1993.

L.A. Zadeh. Fuzzy sets. *Information and Control*, 8:338–353, 1965.

L.A. Zadeh. A rational for fuzzy control. *J. Dyn. Syst. Meas. Contr*, 34:3–4, 1972.

L.A. Zadeh. Outline of a new approach to the analysis of complex systems and decision processes. *IEEE Trans. Syst., Man. Cybernetics*, 3:28–44, 1973.

L.A. Zadeh and R.E. Bellman. *Modern uses of multiple-valued logic*, chapter Local and Fuzzy Logics, pages 103–165. Reidel, Dordrecht, Netherlands, 1977.

Part II

Feature Selection Challenge

Chapter 9

Design and Analysis of the NIPS2003 Challenge

Isabelle Guyon[1], Steve Gunn[2], Asa Ben Hur[3], and Gideon Dror[4]

[1] ClopiNet, 955 Creston Rd., Berkeley, CA 94708, USA. isabelle@clopinet.com
[2] School of Electronics and Computer Science, University of Southampton, Southampton, United Kingdom. s.r.gunn@ecs.soton.ac.uk
[3] Department of Genome Sciences, University of Washington, 1705 NE Pacific St., Seattle WA 98195, USA. asa@gs.washington.edu
[4] The Academic College of Tel-Aviv-Yaffo, 4 Antokolski St., Tel-Aviv 64044, ISRAEL. gideon@mta.ac.il

Summary. We organized in 2003 a benchmark of feature selection methods, whose results are summarized and analyzed in this chapter. The top ranking entrants of the competition describe their methods and results in more detail in the following chapters. We provided participants with five datasets from different application domains and called for classification results using a minimal number of features. Participants were asked to make on-line submissions on two test sets: a validation set and a "final" test set, with performance on the validation set being presented immediately to the participant and performance on the final test set presented at the end of the competition. The competition took place over a period of 13 weeks and attracted 78 research groups. In total 1863 entries were made on the validation sets during the development period and 135 entries on all test sets for the final competition. The winners used a combination of Bayesian neural networks with ARD priors and Dirichlet diffusion trees. Other top entries used a variety of methods for feature selection, which combined filters and/or wrapper or embedded methods using Random Forests, kernel methods, neural networks as classification engine. The classification engines most often used after feature selection are regularized kernel methods, including SVMs. The results of the benchmark (including the predictions made by the participants and the features they selected) and the scoring software are publicly available. The benchmark is available at http://www.nipsfsc.ecs.soton.ac.uk/ for post-challenge submissions to stimulate further research.

9.1 Introduction

Section I provided the reader with a variety of tools to address feature selection problems. Section II puts these tools to work in the context of a benchmark of classification problems. The reader has the opportunity of comparing the relative strengths and weaknesses of the methods and trying to match or outperform the results obtained by the competitors, since the benchmark is

I. Guyon et al.: *Design and Analysis of the NIPS2003 Challenge*, StudFuzz **207**, 237–263 (2006)
www.springerlink.com   © Springer-Verlag Berlin Heidelberg 2006

still open for submissions. The results of the competition were reported and discussed at the NIPS 2003 workshops (Guyon et al., 2005, to appear) and are presented here in more details.

Many data repositories including the well known UCI Machine Learning repository (Murphy and Aha, 1994), and dozens of other sites (Kazakov et al.) provide other opportunities for testing new ideas. Yet, the proliferation of datasets combined with the creativity of researchers in designing experiments makes it difficult to compare one method with another (LaLoudouana and Tarare, 2002). Our benchmark provides a framework that is limited in scope (two-class classification problems), but facilitates making comparisons and deriving conclusions. We summarize the main findings:

- **Feature selection can be performed effectively.** Even though some of the best entries use all the features, there is always another entry close in performance, which uses only a small fraction of the original features. Using a scoring method that favors small feature sets when performance is not statistically significantly different, 7 out of the 10 winning entries use feature selection.

- **Eliminating meaningless features is not critical.** By design, our datasets include many irrelevant "distracters" features: We have purposely added a large fraction of features whose values are distributed similarly to the real features but carry no information about classification task. In contrast with redundant features, which may not be needed to improve accuracy but carry information, those distracters are "pure noise" and can only hurt classification performance. Surprisingly, some of the best entries of the benchmark use all the features.

- **Filter methods are powerful.** For many years, filter methods have dominated feature selection for computational reasons. It was understood that wrapper and embedded methods are more powerful, but too computationally expensive. Some of the top ranking entries use one or several filters as their only selection strategy. A filter as simple as the Pearson correlation coefficient proves to be very effective, even though it does not remove feature redundancy and therefore yields unnecessarily large feature subsets. Other entrants combined filters and embedded methods to further reduce the feature set and eliminate redundancies.

- **Embedded methods are preferred over wrappers.** Wrapper methods were not used by any of the top entrants. Some of the top entrants used embedded method to add or eliminate features in the process of learning. For instance, Bayesians use an Automatic Relevance Determination (ARD) prior (MacKay, 1994, Neal, 1996) as a form of backward elimination. Users of ensembles of tree classifiers (like Random Forests) performed forward selection or feature ranking. Only very few of the top entrants used search strategies more sophisticated than forward selection or backward elimination.

- **Non-linear classifiers do not necessarily overfit.** Several challenge datasets included a very large number of features (up to 100,000) and only a few hundred examples. Therefore, only methods that avoid overfitting can succeed in such adverse aspect ratios. Not surprisingly, the winning entries use as classifies either ensemble methods or strongly regularized classifiers. More surprisingly, non-linear classifiers often outperform linear classifiers. Hence, with adequate regularization, non-linear classifiers do not overfit the data, even when the number of features exceeds the number of examples by orders of magnitude.
- **Some methods are both effective and versatile.** Although the challenge datasets were all restricted to be two class problems, they were selected to have a wide range of characteristics: the features were binary or continuous, they were both sparse and non-sparse, the classes were balanced in number of examples or strongly unbalanced, the number or training examples ranged from 100 to 6000, the number of features ranged from 500 to 100,000, and the ratio of number of examples to number of features ranged from 8:1000 to 4:1. By asking participants to enter results on all five datasets, we could identify several versatile methods, including the Bayesian neural networks and regularized kernel methods, like regularized least squares and SVMs. Some methods perform better for specific problems. For instance the Dirichlet Diffusion Tree method proved efficient in exploiting the clustered structure of the ARCENE data that was obtained by merging data from several sources.
- **Unsupervised dimensionality reduction works.** Principal Component Analysis was successfully used by several researchers to reduce the dimension of input space down to a few ten features, without any knowledge of the class labels. This was not harmful to the prediction performances and greatly reduced the computational load of the learning machines.

In Section 9.2, we provide some details on the benchmark design. The datasets and their formatting is further described in Appendix C. In Section 9.3 we analyze the benchmark results. The complete result tables are presented in Appendix D. Fact sheets summarizing the characteristic of the top ranking entries are found in Appendix C. We also provide supplementary information on our web site: `http://clopinet.com/isabelle/Projects/NIPS2003/analysis.html`.

9.2 Benchmark Design

9.2.1 Synopsis

We formatted five datasets (Tables 9.1 and 9.2) from various application domains in a standard text format. All datasets are two-class classification problems. The data are split into three subsets: a training set, a validation set,

Table 9.1. NIPS 2003 challenge datasets. We show for the various datasets their type (dense, sparse, or sparse binary), and number of features, training examples, validation examples, and test examples. All problems are two-class classification problems.

Dataset	Domain	Type	# Feat.	# Tr.	# Val.	# Te.
ARCENE	Mass Spec.	Dense	10000	100	100	700
DEXTER	Text categorization	Sparse	20000	300	300	2000
DOROTHEA	Drug discov.	Sp. bin.	100000	800	350	800
GISETTE	Digit recog.	Dense	5000	6000	1000	6500
MADELON	Artificial data	Dense	500	2000	600	1800

and a test set. All three subsets were made available at the beginning of the benchmark, on September 8, 2003. The class labels for the validation set and the test set were withheld. The identity of the datasets and of the features (some of which were random features artificially generated) were kept secret. The participants could submit prediction results on the validation set and get their performance results and ranking on-line for a period of 12 weeks. On December 1st, 2003, the participants had to turn in their results on the test set. The validation set labels were released at that time. On December 8th, 2003, the participants could make submissions of test set predictions, after having trained on both the training and the validation set.

9.2.2 Challenge Format

We gave to our benchmark the format of a challenge to stimulate participation. We made available a web-based automatic system to submit prediction results and get immediate feed-back, inspired by the system of the NIPS2000 and NIPS2001 unlabelled data competitions (S. Kremer et al., 2000-2001). However, unlike what had been done for the unlabelled data competitions, the subset of the data called "validation set" was used during the development period, and the separate "test set" was used for final scoring.

During development participants could submit validation results on any of the five datasets proposed (not necessarily all). But at the challenge deadline results on all five independent test sets had to be returned. This avoided a common problem of "multiple track" benchmarks in which hardly any conclusion can be drawn because too few participants enter each track.

To promote collaboration between researchers, reduce the level of anxiety, and let people explore various strategies (e.g. "pure" methods and "hybrids"), we allowed participating groups to submit a total of five final entries on December 1st and five entries on December 8th.

Our format was very successful: it attracted 78 research groups who competed for 13 weeks and made a total of 1863 entries. Twenty groups were eligible for being ranked on December 1st (56 submissions), and 16 groups on December 8th (36 submissions). Our feature selection benchmark residing on the web site http://www.nipsfsc.ecs.soton.ac.uk/results/ remains active as a resource for researchers in feature selection.

9.2.3 Choice of the Theme and the Datasets

As of 1997, when a special issue on relevance including several papers on variable and feature selection was published (Blum and Langley, 1997, Kohavi and John, 1997), few domains explored used more than 40 features[5]. The situation has changed considerably in the past few years: in the 2003 special issue edited by one of the authors for JMLR (Guyon and Elisseeff, 2003), most papers explore domains with hundreds to tens of thousands of variables or features. The applications are driving this effort: bioinformatics, cheminformatics, text processing, speech processing, and machine vision provide machine learning with problems in very high dimensional spaces, but often with relatively few examples (hundreds).

Performing feature selection may have various motivations. Like for other space dimensionality reduction methods, the goals of feature selection include: compressing data, improving prediction performance, and visualizing data. But, the goals of feature selection also include: data understanding, reducing the number of measurements, reducing the storage requirements, and reducing training and utilization times.

The necessary computing power to handle large datasets is now available in simple laptops, so there is a proliferation of solutions proposed for such problems. Yet, there does not seem to be an emerging unity, be it from the standpoint of experimental design, algorithms, or theoretical analysis. We formatted five datasets for the purpose of benchmarking variable selection algorithms in a controlled manner (Guyon, 2003) (see Table 9.1). The datasets were chosen to span a variety of domains: cancer prediction from mass-spectrometry data, handwritten digit recognition, text classification, and prediction of molecular activity, used in drug discovery. One dataset is artificial. We chose datasets that had sufficiently many examples to create a large enough test set to obtain statistically significant results. The input variables are continuous or binary, sparse or dense.

To prevent researchers familiar with the datasets to have an advantage, we concealed the identity of the datasets during the benchmark and we disguised the datasets as explained in the next section.

---

[5]In this paper, we do not make a distinction between features and variables. The benchmark addresses the problem of selecting input variables. Those may actually be features derived from the original variables using a preprocessing step.

9.2.4 Data Preparation

The details of data preparation can be found in Appendix B. The data preparation was designed to facilitate classification (by extracting features), disguise the datasets and compress them, facilitate performance assessment (by adding meaningless features), and facilitate importing data into various platforms. We examine each aspect of our design and summarize some statistics in Table 9.2 showing the variety of the resulting tasks.

Table 9.2. Range of task difficulty. We show for the various datasets their sparseness (fraction of zero values), whether the features are binary or continuous, the "aspect ratio" of the data that is the ratio of the number of features to the number of training examples, the fraction of probes in the original feature set, and the number of clusters per class.

Dataset	Sparseness	Binary	#Feat/#Tr	Fprobe	Cluster
ARCENE	0.50	No	100	0.30	> 3
DEXTER	0.995	No	67	0.503	-
DOROTHEA	0.99	Yes	125	0.50	-
GISETTE	0.87	Almost	0.83	0.50	1-2?
MADELON	< 0.01	No	0.25	0.96	16

Dataset Profiles

We briefly describe the datasets we are using:

- **Arcene:** ARCENE is a biomedical application. The task is to separate cancer samples from normal samples, using mass-spectra of blood serum. Three datasets from two sources (NCI and EVMS) were merged (E. F. Petricoin III et al., 2002, Adam Bao-Ling et al., 2002). All used the same SELDI instrument. Two datasets included prostate cancer samples and one included ovarian cancer samples. The NCI ovarian cancer data consist of 253 spectra (162 cancer, 91 control), and 15154 features. The NCI prostate cancer data consist of 322 spectra (69 cancer, 253 control), and 15154 features. The EVMS prostate cancer data consist of 652 spectra from 326 samples (167 cancer, 159 control), and 48538 features. The spectra preprocessing included mass-alignment and baseline removal. The resulting features are normalized spectral intensities for the various mass over charge values.

- **Dexter:** DEXTER is a text categorization application. The task is to identify texts about "corporate acquisitions". The data were originally collected and labeled by Carnegie Group, Inc. and Reuters, Ltd. in the course of developing the CONSTRUE text categorization system. The particular subset of texts used was assembled and preprocessed by Thorsten Joachims (Joachims, 1998). The data contains 1300 texts about corporate acquisitions and 1300 texts about other topics, and 9947 features representing frequencies of occurrence of word stems in text.

- **Dorothea:** DOROTHEA is a drug discovery application. The task is to identify compounds that bind to a target site on thrombin, a key receptor in blood clotting. The data were provided by DuPont Pharmaceuticals Research Laboratories for the KDD Cup 2001. It includes 2543 compounds tested for their ability to bind to thrombin; 192 active (bind well); the rest inactive. The 139,351 binary features describe three-dimensional properties of the molecule.

- **Gisette:** GISETTE is a handwritten digit recognition application. The task is to separate two confusable classes: four and nine. The data were originally compiled by NIST and are made available by Yann Le Cun (LeCun). The dataset includes 13500 digits size-normalized and centered in a fixed-size image of dimension 28x28. We selected one difficult two-class separate (separating the digits "four" from "nine") and added to the pixels features that are products of pixels in the region of the image that is most informative.

- **Madelon:** MADELON is an artificial task inspired by Perkins et al. (Perkins et al., 2003). Gaussian clusters are positioned on the vertices of a hypercube and labelled randomly. We used a hypercube in five dimensions with thirty-two clusters of points (16 per class). Five redundant features were added, obtained by multiplying the useful features by a random matrix. Some of the previously defined features were repeated to create 10 more features. The other 480 features are drawn from a Gaussian distribution and labelled randomly. The data were distorted by adding noise, flipping labels, shifting and rescaling. The program of data generation is provided in Appendix B (Guyon, 2003).

Data Modifications and Formatting

The datasets were modified by introducing a number of meaningless features called *probes*. Such probes have a function in performance assessment: a good feature selection algorithm should eliminate most of the probes. For each dataset we chose probes according to a distribution that resembles that of the real features. For ARCENE, the probes were obtained by randomly permuting the intensity values (across all examples) in the least informative regions of the spectra. For DEXTER, we added random probes drawn according to Zipf's law. For DOROTHEA, we ranked the features according to their correlation to the target and permuted the values of the 50000 least informative across all examples. For GISETTE, we created additional features that were products of pixels and permuted their values across examples. For MADELON, the probes were drawn from a Gaussian distribution. Details are found in Appendix B.

The number of features ended up being very different from the original one, making the datasets more difficult to recognize. To further disguise the dataset, we sometimes eliminated a few patterns to round the number of examples. We randomized the order of the features and the patterns. Finally, non-binary features were shifted, scaled and rounded to obtain 1000 integers in the range 0 to 999, for ease of encoding them in ASCII and subsequently compressing them. Although the formats are trivial to read, we also provided sample Matlab code to read the file.

Performance assessment was facilitated by the randomization of examples (patterns). This ensured that training, validation, and test set would be similarly distributed. We ran preliminary classification experiments to estimate the best error rate that could be obtained. We reserved a test set of sufficient size to provide statistically significant results using a simple scheme based on well-known error bounds (Guyon et al., 1998): the test set size should be 100/E, where E is the error rate of the best classifier. We verified *a posteriori* that our estimates were good: According to the McNemar test, there were always at least a pair of entries with results statistically significantly different in the top ten entries (in spite of the high correlation between entries due to the fact that each group could submit 5 entries).

9.2.5 Result Format and Performance Assessment

We asked the participants to return files containing the class predictions for training, validation, and test sets for all five tasks proposed, and the list of features used. Optionally, classification confidence values could be provided. During the development period, submissions with validation set results on one or more tasks were accepted. But a final entry had to be a full submission.

The performances were assessed with several metrics:

- BER: The balanced error rate, that is the average of the error rate of the positive class and the error rate of the negative class. This metric was used because some datasets (particularly DOROTHEA) are unbalanced.
- AUC: Area under the ROC curve. The ROC curve is obtained by varying a threshold on the discriminant values (outputs) of the classifier. The curve represents the fraction of true positive as a function of the fraction of false negative. For classifiers with binary outputs, BER=1-AUC.
- Ffeat: Fraction of features selected.
- Fprobe: Fraction of probes found in the feature set selected.

We ranked the participants with the test set results using a score combining BER, Ffeat and Fprobe as follows:

- Make pairwise comparisons between classifiers for each dataset:
  - Use the McNemar test to determine whether classifier A is better than classifier B according to the BER with 5% risk. Obtain a score of 1 (better), 0 (don't know) or 1 (worse).
  - If the score is 0, break the tie with Ffeat if the relative difference is larger than 5%.
  - If the score is still 0, break the tie with Fprobe.
- The overall score for each dataset is the sum of the pairwise comparison scores (normalized by the maximum achievable score, that is the number of submissions minus one).
- The global score is the average of the dataset scores.

With this scheme, we obtain positive or negative scores. Even a score of zero is good, because out of the 75, only 20 self-selected participants decided to make final submissions on December 1st and 16 on December 8th. The scheme favors accuracy over compactness of the feature set. Still some top ranking entries have very compact feature sets. One advantage of our scheme is that it has only 2 parameters (the risk of the test and the threshold for Ffeat) to which scoring is relatively insensitive. One disadvantage is that the scoring changes when new classification entries are made in the challenge. However, we found that the five top ranking groups are consistently ranked at the top and in the same order under changes of the set of submissions.

9.3 Challenge Results

9.3.1 Participant Performances

The winners of the benchmark (both December 1st and 8th) are Radford Neal and Jianguo Zhang, with a combination of Bayesian neural networks and Dirichlet diffusion trees. Their achievements are significant since they win on the overall ranking with respect to our scoring metric, and also with respect to the balanced error rate (BER), the area under the ROC curve (AUC), and they have the smallest feature set among the top entries that have performance not statistically significantly worse. They are also the top entrants December 1st for ARCENE and DEXTER and December 1st and 8th for DOROTHEA.

The scores of the best entries are shown in Table 9.3 and are further commented upon in Section 9.3.2. The full result tables are found in Appendix D.

BER Distribution

In Figure 9.1 we show the distribution of the BER performance obtained throughout the development period on the validation set. We plotted results on the validation set because many more entries were made on the validation set only. This allows us to compare the difficulty of the five tasks of the

Table 9.3. Challenge results. We show the top entries sorted by their score, the balanced error rate in percent (BER) and corresponding rank in parenthesis, the area under the ROC curve times 100 (AUC) and corresponding rank in parenthesis, the percentage of features used (Ffeat), and the percentage of probes in the features selected (Fprob). The entries emphasized were submitted for verification (see text.) The columns "Chapter" refers to the chapter number and the corresponding "fact sheets" summarizing the methods found in the book appendices.

(a) **December 1st 2003 challenge results.**

Method	Group	Chapter	Score	BER (Rk)	AUC (Rk)	Ffeat	Fprob
BayesNN-DFT	Neal/Zhang	10	88	6.84 (1)	97.22 (1)	80.3	47.77
BayesNN-DFT	Neal/Zhang	10	86.18	6.87 (2)	97.21 (2)	80.3	47.77
BayesNN-small	*Neal*	10	*68.73*	*8.20 (3)*	*96.12 (5)*	*4.74*	*2.91*
BayesNN-large	Neal	10	59.64	8.21 (4)	96.36 (3)	60.3	28.51
RF+RLSC	*Torkkola/Tuv*	11	*59.27*	*9.07 (7)*	*90.93 (29)*	*22.54*	*17.53*
final2	Chen	12	52	9.31 (9)	90.69 (31)	24.91	11.98
SVMBased3	*Zhili/Li*	13	*41.82*	*9.21 (8)*	*93.60 (16)*	*29.51*	*21.72*
SVMBased4	Zhili/Li	13	41.09	9.40 (10)	93.41 (18)	29.51	21.72
final1	*Chen*	12	*40.36*	*10.38 (23)*	*89.62 (34)*	*6.23*	*6.1*
transSVM2	Zhili	13	36	9.60 (13)	93.21 (20)	29.51	21.72
myBestValid	Zhili	13	36	9.60 (14)	93.21 (21)	29.51	21.72
TransSVM1	Zhili	13	36	9.60 (15)	93.21 (22)	29.51	21.72
BayesNN-E	Neal	10	29.45	8.43 (5)	96.30 (4)	96.75	56.67
Collection2	*Saffari*	14	*28*	*10.03 (20)*	*89.97 (32)*	*7.71*	*10.6*
Collection1	Saffari	14	20.73	10.06 (21)	89.94 (33)	32.26	25.5

(b) **December 8th 2003 challenge results.**

Method	Group	Chapter	Score	BER (Rk)	AUC (Rk)	Ffeat	Fprob
BayesNN-DFT	Neal/Zhang	10	71.43	6.48 (1)	97.20 (1)	80.3	47.77
BayesNN-large	Neal	10	66.29	7.27 (3)	96.98 (3)	60.3	28.51
BayesNN-small	Neal	10	61.14	7.13 (2)	97.08 (2)	4.74	2.91
final_2-3	Chen	12	49.14	7.91 (8)	91.45 (25)	24.91	9.91
BayesNN-large	Neal	10	49.14	7.83 (5)	96.78 (4)	60.3	28.51
final2-2	Chen	12	40	8.80 (17)	89.84 (29)	24.62	6.68
Ghostminer1	Ghostminer	23	37.14	7.89 (7)	92.11 (21)	80.6	36.05
RF+RLSC	Torkkola/Tuv	11	35.43	8.04 (9)	91.96 (22)	22.38	17.52
Ghostminer2	Ghostminer	23	35.43	7.86 (6)	92.14 (20)	80.6	36.05
RF+RLSC	Torkkola/Tuv	11	34.29	8.23 (12)	91.77 (23)	22.38	17.52
FS+SVM	Lal	20	31.43	8.99 (19)	91.01 (27)	20.91	17.28
Ghostminer3	Ghostminer	23	26.29	8.24 (13)	91.76 (24)	80.6	36.05
CBAMethod3E	CBA group	22	21.14	8.14 (10)	96.62 (5)	12.78	0.06
CBAMethod3E	CBA group	22	21.14	8.14 (11)	96.62 (6)	12.78	0.06
Nameless	Navot/Bachr.	17	12	7.78 (4)	96.43 (9)	32.28	16.22

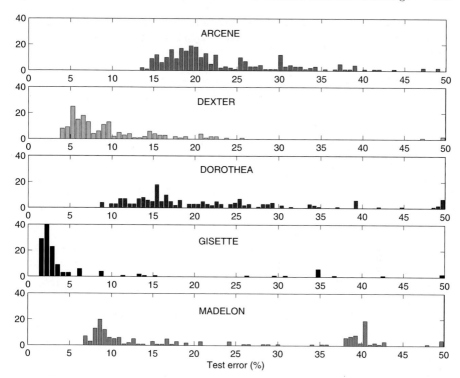

Fig. 9.1. Distribution of validation set balanced error rate (BER).

benchmark. The GISETTE dataset is the easiest one to learn and has also the smallest variance among participants. Next come DEXTER. The ARCENE and DOROTHEA datasets are harder and show a lot of variance among participants. This could be attributed to particular difficulties of these datasets: ARCENE was obtained by grouping data from several sources and DOROTHEA has strongly unbalanced classes. Finally, the MADELON datasets has a bimodal performance distribution. This property of the disdribution may be traced to the fact that MADELON is a very non-linear problem requiring feature selection techniques that do not select features for their individual predictive power. The worst performing methods may have failed to select the informative features or may have used linear classifiers.

Error Bars

The test set sizes were computed to obtain reasonable error bars on the prediction errors, with the empirical formula $M \cdot E = 100$, where M is the number of test examples and E is the error rate (Guyon et al., 1998). Not knowing in advance what the best error rate would be, we made rough estimates. A

Table 9.4. Error bars on the best entries. We show the BER of the best entries (in percent) and the corresponding error bars (standard error for the binomial distribution).

Dataset	Test set size M	Best BER	$M \cdot BER$	Error bar
ARCENE	700	10.73 (Dec. 8, Chen)	75.11	1.17
DEXTER	2000	3.30 (Dec. 8, Lal)	66	0.40
DOROTHEA	800	8.54 (Dec. 1, Neal)	68.32	0.99
GISETTE	6500	1.26 (Dec. 8, Neal)	81.90	0.14
MADELON	1800	6.22 (Dec. 8, Neal)	111.96	0.57

posteriori, we can verify that the test set sizes chosen give us acceptable error bars. For simplicity, we treat the balanced error rate (BER) as a regular error rate and calculate error bars assuming i.i.d. errors using the Binomial distribution. This gives us a standard error of $\sqrt{(BER(100 - BER))/M}$ (in percent), which we use as error bar. For every data set, such error bars are around 10% of the BER. The value of $M \cdot E$ is also not far from our target value 100 that we anticipated (Table 9.4). Having a large enough number of test examples allowed us making reliable comparisons between classifiers and assessing the significance of the difference in balanced error rates with statistical tests. The McNemar test at 5% risk always uncovered at least one significant difference in the top ten ranking entries for every dataset and for both submissions of December 1st and 8th (see the tables in appendix).

Model Selection

Several participants used their performance on the validation set as an indicator of how well they would perform on the test set. This method of model selection proved to be accurate, since the ranking of the best challengers has changed little when the final performance was computed with the test set. We show in Table 9.5 the validation set error bars computed in the same way we computed the test set error bars. Clearly, the error bar for the ARCENE validation set makes it unfit for model selection. Indeed, we computed the correlation between the validation set BER and test set BER of all the entries submitted during the development period and found a poor correlation for the ARCENE dataset. Neal and Zhang report on their paper having been mislead by the validation set error rate in their ARCENE submission.

Effectiveness of Feature Selection

One of the main question the benchmark attempted to answer is whether feature selection is harmful or useful to prediction performance and, if harmful,

Table 9.5. Use of the validation set for model selection. We show the validation set error bars for the best BER (in percent) and the R^2 of the test BER and the validation BER.

Dataset	Validation set size	Error bar	R^2
ARCENE	100	3.09	81.28
DEXTER	300	1.03	94.37
DOROTHEA	350	1.49	93.11
GISETTE	1000	0.35	99.71
MADELON	600	0.98	98.62

whether one can significantly reduce the number of features without significantly reducing performance. To answer this question, we examined the smallest fraction of features used as a function of the BER obtained by the participants. To assess the significance, we used the error bars (sigma) computed in Table 9.4. Our observations are:

- DEXTER. The winning entries of both dates have the smallest BER and a significantly reduced feature set.
- GISETTE. The winning entries of both dates are within one sigma of the smallest BER entry and also correspond to a significant reduction in the number of features.
- MADELON. Entries were made that are within 2 sigma of the best entry and use only 8 features (the true dimensionality of the problem is 5, only 20 redundant features are relevant).
- ARCENE. The December 8 winners succeeded in reducing significantly the number of features within one sigma of the best performance.
- DOROTHEA. The winning entries use all the features.

We show in Tables 9.6 and 9.7 the statistics of the winning entries. We notice that the reduced feature sets eliminated most of the random probes. In conclusion, even though some classifiers can accommodate large numbers of irrelevant features, it is possible to get rid of those irrelevant features without significantly degrading performance.

Performance Progress in Time

We investigated how the participants progressed over time. In Figure 9.2, we show the evolution of the balanced error rate on the test set. We see that except for a last minute improvement during the final submission, the performance remained steady after about a month, in spite of the fact that the number of entrants kept increasing. The performances on the validation

Table 9.6. December 1st 2003 winners by dataset.

Dataset	Method	Group	Score	BER	AUC	Ffeat	Fprob
ARCENE	BayesNN-DFT	Neal & Zhang	98.18	13.30 (1)	93.48 (1)	100.00	30.00
DEXTER	BayesNN-DFT	Neal & Zhang	96.36	3.90 (1)	99.01 (2)	1.52	12.87
DOROTHEA	BayesNN-DFT	Neal & Zhang	98.18	8.54 (1)	95.92 (2)	100.00	50.00
GISETTE	final 2	Yi-Wei Chen	98.18	1.37 (8)	98.63 (31)	18.26	0.00
MADELON	Bayesian+SVM	Chu Wei	100.00	7.17 (5)	96.95 (7)	1.60	0.00

Table 9.7. December 8th 2003 winners by dataset.

Dataset	Method	Group	Score	BER	AUC	Ffeat	Fprob
ARCENE	BayesNN-small	Radford Neal	94.29	11.86 (7)	95.47 (1)	10.70	1.03
DEXTER	FS+SVM	Navin Lal	100.00	3.30 (1)	96.70 (23)	18.57	42.14
DOROTHEA	BayesNN-DFT	Neal & Zhang	97.14	8.61 (1)	95.92 (2)	100.00	50.00
GISETTE	final2-2	Yi-Wei Chen	97.14	1.35 (7)	98.71 (22)	18.32	0.00
MADELON	Bayesian+SVM	Chu Wei	94.29	7.11 (13)	96.95 (10)	1.60	0.00

set evolved slightly differently. After about 2 months, there seemed to be an improvement for ARCENE and DOROTHEA. But this improvement is not reflected by an improvement on the test set. Therefore, we might trace it to a problem of overfitting.

9.3.2 Methods Employed

We performed a survey of the methods employed among the contributer of this book, who are all among the top ranking participants. Since a wide variety of methods were employed, we grouped them into coarse categories to be able to draw useful conclusions. The results are listed in Table 9.3. Our findings include:

- **Feature selection.** Although the winners and several top ranking challengers use a combination of filters and embedded methods, several high ranking participants obtain good results using only filters, even simple correlation coefficients. The second best entrants use Random Forests (an ensemble of tree classifiers) as a filter. Search strategies are generally unsophisticated (simple feature ranking, forward selection or backward elimination.) Only 2 out of 17 in our survey used a more sophisticated search

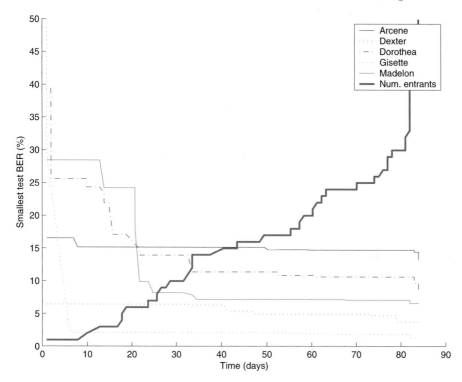

Fig. 9.2. Evolution of the test BER.

strategy. The selection criterion used is usually based on cross-validation. A majority use K-fold, with K between 3 and 10. Two groups (including the winners) made use of "random probes" that are random features purposely introduced to track the fraction of falsely selected features. One group used the area under the ROC curve computed on the training set.

- **Classifier.** Although the winners use a neural network classifier, kernel methods are most popular: 12/17 in the survey. Of the 12 kernel methods employed, 8 are SVMs. In spite of the high risk of overfitting, 8 of the 12 groups using kernel methods found that Gaussian kernels gave them better results than the linear kernel on ARCENE, DEXTER, DOROTHEA, or GISETTE (for MADELON all best ranking groups used a Gaussian kernel).

- **Ensemble methods.** Some groups relied on a committee of classifiers to make the final decision. The techniques to build such committee include sampling from the posterior distribution using MCMC (Neal, 1996) and bagging (Breiman, 1996). The groups that used ensemble methods reported improved accuracy, except Torkkola and Tuv, who found no significant improvement.

Table 9.8. Methods employed by the challengers. The classifiers are grouped
in four categories: N=neural network, K=SVM or other kernel method, T=tree classifiers, O=other. The feature selection engines (Fengine) are grouped in three categories: C=single variable criteria including correlation coefficients, T=tree classifiers
used as a filter, E=Wrapper or embedded methods. The search methods are identified by: E=embedded, R=feature ranking, B=backward elimination, F=forward
selection, S=more elaborated search. Fact sheets summarizing the methods are appended.

Group	Chapter	Classifier	Fengine	Fsearch	Ensemble	Transduction
Neal/Zhang	10	N/O	C/E	E	Yes	Yes
Torkkola/Tuv	11	K	T	R	Yes	No
Chen/Lin	12	K	C/T/E	R/E	No	No
Zhili/Li	13	K	C/E	E	No	Yes
Saffari	14	N	C	R	Yes	No
Borisov et al	15	T	T	E	Yes	No
Rosset/Zhu	16	K	K	E	No	No
Navot/Bachrach	17	K	E	S	No	Yes
Wei Chu et al	18	K	K	R	No	No
Hohreiter et al	19	K	K	R	Yes	No
Lal et al	20	K	C	R	No	No
Embrechts/Bress	21	K	K	B	Yes	No
CBA group	22	K	C	R	No	No
Ghostminer	23	K	C/T	B	Yes	No
SK Lee et al	24	K/O	C	S	No	No
Boullé	25	N/O	C	F	No	No
Lemaire/Clérot	26	N	N	B	No	No

- **Transduction.** Since all the datasets were provided since the beginning
  of the benchmark (validation and test set deprived of their class labels),
  it was possible to make use of the unlabelled data as part of learning
  (sometimes referred to as transduction (Vapnik, 1998)). Only three groups
  took advantage of that, including the winners.
- **Preprocessing.** Centering and scaling the features was the most common
  preprocessing used. Some methods required discretization of the features.
  One group normalized the patterns. Principal Componant Analysis (PCA)
  was used by several groups, including the winners, as a means of constructing features.

9.4 Post-Challenge Verifications

In this section, we provide the results of some experiments conducted after the end of the competition to verify some of the observations made and verify the validity of the entries.

9.4.1 Verification of the Observations

We conducted a limited set of experiments to strengthen the conclusions of the previous section.

In a first set of experiments, we selected fifty entries (five for each dataset for each submission date), which offered reduced feature sets (of the order of 10% of the total number of features or less. We trained an SVM with a choice of preprocessing and hyperparameters and selected the best one by 10-fold cross-validation. The details about our experimental setup are found in Appendix A.

Linear vs. Non-Linear Classifier

We verified empirically the observation from the challenge results that non-linear classifiers perform better than linear classifiers by training an SVM on feature subsets submitted at the benchmark.

For each dataset, across all 10 feature sets (5 for December 1st and 5 for December 8th), for the top five best classifiers obtained by hyperparamenter selection, we computed statistics about the most frequently used kernels and methods of preprocessing.[6] In Table 9.9, we see that the linear SVM rarely shows up in the top five best hyperparameters combinations.

Preprocessing

We used three methods of preprocessings:

- **prep1 = feature standardization:** Each feature was separately standardized, by substracting its mean and dividing by the standard deviation. To avoid dividing by too small values, the standard deviation was augmented by a fudge factor equal to the median of the standard deviations of all features. In case of sparse data, the mean was not substracted.
- **prep2 = pattern normalization:** Each example is divided by its L_2 norm.
- **prep3 = global normalization:** Each example is divided by the mean L_2 norm of examples.

---

[6] Since the performance of the 5th performer for some datasets was equal to that of the 6th 7th and even the 8th, to avoid the arbirtrariness of picking the best 5 by "sort" we picked also the next several equal results.

Table 9.9. SVM experiments on 50 feature subsets from the challenge results. The values shown are the percentages of time a given hyperparameter choice showed up in the five top choices selected by ten-fold cross-validation, out of all hyperparameter settings tried.

		Kernel								Preprocessing		
Dataset	linear	poly2	poly3	poly4	poly5	rbf 0.5	rbf 1	rbf 2	prep1	prep2	prep3
ARCENE	0.00	0.00	38.46	38.46	23.08	0.00	0.00	0.00	46.15	30.77	23.08
DEXTER	13.33	0.00	13.33	33.33	20.00	0.00	20.00	0.00	6.67	93.33	0.00
DOROTHEA	0.00	20.00	20.00	0.00	40.00	20.00	0.00	0.00	0.00	60.00	40.00
GISETTE	0.00	0.00	23.08	23.08	23.08	0.00	15.38	15.38	0.00	100.00	0.00
MADELON	0.00	0.00	0.00	0.00	0.00	20.00	40.00	40.00	100.00	0.00	0.00

Our hyperparameter selection study indicates that, except for MADELON and ARCENE, L_2 normalization (prep2) is quite effective compared to the more popular prep1 (each feature is centered by its means and and scaled by its standard deviation.)

Effectiveness of SVMs

One observation of the challenge is that SVMs (and other kernel methods) are effective classifiers and that they can be combined with a variety of feature selection methods used as "filters". Our hyperparameter selection experiments also provide us with the opportunity to check whether we can get performance that is similar to that of the challengers by training an SVM on the feature sets they declared. This allows us to determine whether the performance is classifier dependent. As can be seen in Figure 9.3, for most datasets, SVMs achieve comparable results as the classifiers used by the challengers. We achieve better results for GISETTE. For DOROTHEA, we do not match the results of the challengers, which can partly be explained by the fact that we did not select the bias properly (our AUC results compare more favorably to the challengers' AUCs.)

Effectiveness of Feature Selection

One observation is that strongly regularized classifiers are immune to a large fraction of irrelevant features. We performed feature selection experiments with two simple feature selection methods:

- A feature ranking method with a correlation coefficient (we used Golub's criterion (T. R. Golub et al., 1999) that is, for each feature, the ratio of the

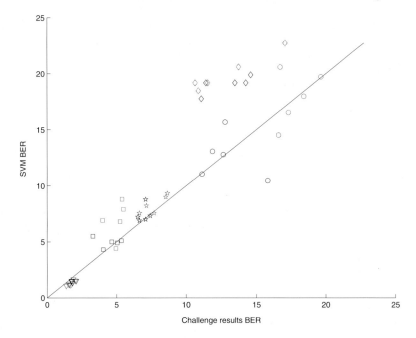

Fig. 9.3. Results of the best SVM in the fifty feature subset study. The symbol shapes represent datasets: ARCENE =circle, DEXTER =square, DOROTHEA =diamond, GISETTE =triangle, MADELON =star. The symbol colors represent the date: Dec. 1st=red, Dec. 8th=black.

distance between the class means over the average of the class standard deviation.)

- An embedded method with backward selection of features using and SVM (RFE-SVM (Guyon et al., 2002), see Chapter 5)).

We ranked the features using the entire training data set. We then selected nested subsets of the top ranking features to train our classifier. We generated numbers of features approximately equally spaces in a logarithmic scale, such that altogether we have 25 values per dataset. For each feature set we conducted a search in hyperparameter space (kernel, normalization scheme, C) using 10-fold cross-validation. The choices of hyperparameters and normalization are the same as in the previous section.

We tried two variants of our method: selecting the best SVM based on 10-fold cross-validation, and voting on the 5 best selected. Based on our experiments, it seems that the voting method does not improve on the best. Our experiments indicate that SVMs perform better without feature selection on the datasets of the challenge and are therefore immune to the presence of meaningless features.

9.4.2 Verification of the Validity of the Entries

Our benchmark design could not hinder participants from "cheating" in the following way. An entrant could "declare" a smaller feature subset than the one used to make predictions. To deter participants from cheating, we warned them that we would be performing a stage of verification. Such verification is described in this section.

Examining the Fraction of Probes

As part of the design of the benchmark, we introduced in the datasets random features (called probes) to monitor the effectiveness of the feature selection methods in discarding irrelevant features. In the result Table 9.3, we see that entrants who selected on average less than 50% of the features have feature sets generally containing less than 20% of probes. But in larger subsets, the fraction of probes can be substantial without altering the classification performance. Thus, a small fraction of probes is an indicator of quality of the feature subset, but a large fraction of probes cannot be used to detect potential breaking of the rules of the challenge.

Specifically, for MADELON, all feature subsets under examination are completely free of probes. Thus the best entrants were very successful at filtering out all the irrelevant features in this artificially generated dataset. For ARCENE, all the subsets have less than 12% of probes, and the subset with 12% of probes selected a total of 44% of the feature. So we have no reason to suspect anything wrong. For GISETTE, all the feature sets investigated have less than 0.5% probes, except for one that has 19% probes, for 10% of features selected. For DOROTHEA, one entry has a high fraction of random probes (pwKPKbca, P-SVM Sepp Hochreiter 60%). For DEXTER, the subsets selected have up to 40% probes. The test presented in the next sections cleared these entries of any suspicions, showing that the corresponding feature selection methods are inefficient at filtering irrelevant features but that the classifiers employed are immune to a large fraction of probes.

Testing New Examples Restricted to the Specified Feature Subset

We performed a more reliable check using MADELON data. We generated additional test data (two additional test sets of the same size as the original test set: 1800 examples each). We selected entries with small feature sets of the five top groups from December 1st (corresponding the the entries emphasized in Table 9.3). We sent to these five groups the new test sets, restricted to the feature subsets of their selected entries, namely each group received the additional test data projected on one of the feature sets it submitted. We asked them to return the class label predictions. We scored the results and compared them to their results on the original test set to detect an eventual significant discrepancy. Specifically, we performed the following calculations:

Table 9.10. Result verification using MADELON **data.** We provided to five entrants additional test data restricted to the feature set they had declared. We adjusted the performances to account for the dataset variability (values shown in parentheses, see text for details). All five entrants pass the test since the adjusted performances on the second and third datasets are not worse than those on the first one, within the computed error bar 0.67.

Entry	Ffeat	Fprob	Test1 BER	Test2 BER	Test3 BER
BayesNN-small	3.4	0	7.72 (8.83)	9.78 (9.30)	9.72 (9.05)
RF+RLSC	3.8	0	6.67 (7.63)	8.28 (7.87)	8.72 (8.12)
SVMbased3	2.6	0	8.56 (9.79)	10.39 (9.88)	9.55 (8.89)
final1	4.8	16.67	6.61 (7.56)	7.72 (7.34)	8.56 (7.97)
Collection2	2	0	9.44 (10.80)	10.78 (10.25)	11.38 (10.60)

- Compute for each test set the mean error rate over the 5 entries checked: μ_1, μ_2, μ_3.
- Compute the overall mean $\mu = (\mu_1 + \mu_2 + \mu_3)/3$.
- Compute corrective factors $\mu/\mu_1, \mu/\mu_2, \mu/\mu_3$.
- Apply the corrective factors to the error rates obtained.
- Compute an error bar as $\sqrt{\mu(1-\mu)/m}$, where m is the size of each test set.

We find that the increases in error rates from the original test set to the new test sets are within the error bar (Table 9.10). Therefore, all five entries checked pass the test satisfactorily. The December 8th entries were not checked.

Comparing with the Performance of a Reference Classifier

We performed another check on a selection of entries with small feature sets (five for each data set and each submission date, i.e. a total of 50). We trained an SVM with the datasets restricted to the selected features (we used SVM-light (Joachims, 1998) with a hyperparameter search, as detailed in Appendix A). We looked for outliers of BER performance that might be indicative that the feature subset submitted was fraudulent. To correct for the fact that our best SVM classifier does not perform on average similarly to the challengers' classifiers, we computed a correction for each dataset k:

$$c_k = < \text{Reference SVM BER} - \text{Corresponding challenge BER} >$$

where the average denoted by the bracket is taken for each dataset over 10 feature sets (5 for each date.) We computed for each challenge entry being checked the following T statistic:

$$T = \frac{\text{Reference SVM BER} - \text{Corresponding challenge BER} - c_k}{\text{stdev}(\text{Difference}_k)}$$

where k is the dataset number and stdev(Difference$_k$) is the standard deviation of (Reference SVM BER−Corresponding challenge BER−c_k), computed separately for each dataset. We show in Figure 9.4 the T statistics as a function of the corresponding challenge BER for the various entries. We also show the limits corresponding to one sided risks of 0.01 (dashed line) and 0.05 (dotted line) to reject to null hypothesis that the T statistic is equal to zero (no significant difference in performance between the entry and our reference method.) No entry is suspicious according to this test. The only entry above the 0.05 line is for MADELON and had a fraction of probes of zero, so it is really sound. N.B. The lines drawn according to the T distributions were computed for 9 degrees of freedom (10 entries minus one.) Using the normal distribution instead of the T distribution yields a dotted line (risk 0.05) at 1.65 and a dashed line (risk 0.01) at 2.33. This does not change our results.

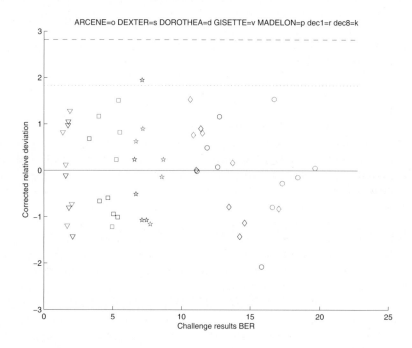

Fig. 9.4. Second verification. Fifty submitted feature sets were tested to see whether their corresponding challenge performance significantly differed from our reference SVM classifier. The symbol shapes represent datasets: ARCENE =circle, DEXTER =square, DOROTHEA =diamond, GISETTE =triangle, MADELON =star. The symbol colors represent the date: Dec. 1st=red, Dec. 8th=black.

9.5 Conclusions and Future Work

The challenge results demonstrate that a significant reduction in the number of features can be achieved without significant performance degradation. A variety of methods are available to perform feature selection and the simplest ones (including feature ranking with correlation coefficients) perform very competitively. Eliminating meaningless features is not critical to obtain good classification performances: Regularized classifier (e.g. Bayesian methods using ARD priors, regularized least-square or SVMs) do not need feature selection to perform well. Further, such regularized methods often attain better performance with non-linear classifiers, overcoming the problem of overfitting despite the large dimensionality of input space (up to 100,000 features) and the relatively small number of training examples (hundreds).

While the challenge helped answering some important questions, it left some questions unanswered and raised new ones:

- *Do ensemble methods work better?* Some participants reported improvements with ensemble methods, others found that the best performing single classifier performed similarly to the ensemble.
- *Does transduction help?* By design, we had shuffled the examples to have a uniform distribution of examples in the training, validation and test sets. We did not expect transductive methods to perform significantly better. Yet some participants reported some benefits of using the unlabeled data of the the validation and test set as part of training.
- *How predominant is the role of hyperparameter selection?* Several participants obtained very different performances with the same learning machine and learning algorithm. Could this be traced to differences in the hyperparamenter selection strategies?
- *Which method of space dimensionality reduction works best?* PCA was used successfully by some of the top entrants. But no comparison was made with other methods, perhaps because PCA is so easy to use, perhaps because it is widespread. Does it work better that the many other methods?
- *Why are wrappers not popular?* Wrappers were not among the methods used by the top entrants. Is it because they are too slow, to complicated or because they do not perform well?
- *Which model selection method works best?* Possible strategies to select the optimum number of features include K-fold cross-validation and the computation of pvalues. The hyperparameters may be selected jointly with the number of features, or before or after feature selection. A variety of strategies were used by the top ranking participants so it is not clear whether there is an optimum way of tackling this problem.

Several pending questions revolve around the problems of model selection and ensemble methods. A new benchmark is in preparation to clarify these issues. See: `http://clopinet.com/isabelle/Projects/modelselect/`.

Acknowledgments

We are very thankful to the data donor institutions without whom nothing would have been possible: the National Cancer Institute (NCI), the Eastern Virginia Medical School (EVMS), the National Institute of Standards and Technology (NIST), DuPont Pharmaceuticals Research Laboratories, and Reuters Ltd. and the Carnegie Group, Inc. We also thank the people who formatted the data and made them available: Thorsten Joachims, Yann Le Cun, and the KDD Cup 2001 organizers. The workshop co-organizers and advisors Masoud Nikravesh, Kristin Bennett, Richard Caruana, and André Elisseeff, are gratefully acknowledged for their help, and advice, in particular with result dissemination.

References

Adam Bao-Ling et al. Serum protein fingerprinting coupled with a pattern-matching algorithm distinguishes prostate cancer from benign prostate hyperplasia and healthy men. *Cancer Research*, 62:3609–3614, July 2002.

A. Blum and P. Langley. Selection of relevant features and examples in machine learning. *Artificial Intelligence*, 97(1-2):245–271, December 1997.

Leo Breiman. Bagging predictors. *Machine Learning*, 24(2):123–140, 1996.

E. F. Petricoin III et al. Use of proteomic patterns in serum to identify ovarian cancer. *The Lancet*, 359, February 2002.

I. Guyon. Design of experiments of the NIPS 2003 variable selection benchmark. *http://www.nipsfsc.ecs.soton.ac.uk/papers/NIPS2003-Datasets.pdf*, 2003.

I. Guyon and A. Elisseeff. An introduction to variable and feature selection. *JMLR*, 3:1157–1182, March 2003.

I. Guyon, J. Makhoul, R. Schwartz, and V. Vapnik. What size test set gives good error rate estimates? *PAMI*, 20(1):52–64, 1998.

I. Guyon, J. Weston, S. Barnhill, and V. Vapnik. Gene selection for cancer classification using support vector machines. *Machine Learning*, 46(1-3):389–422, 2002.

I. Guyon, S. Gunn, A. Ben Hur, and G. Dror. Result analysis of the nips 2003 feature selection challenge. In *Advances in Neural Information Processing Systems 17*, Cambridge, MA, 2005, to appear. MIT Press.

T. Joachims. Making large-scale support vector machine learning practical. In A. Smola B. Schölkopf, C. Burges, editor, *Advances in Kernel Methods: Support Vector Machines*. MIT Press, Cambridge, MA, 1998.

D. Kazakov, L. Popelinsky, and O. Stepankova. MLnet machine learning network on-line information service. In *http://www.mlnet.org*.

R. Kohavi and G. John. Wrappers for feature selection. *Artificial Intelligence*, 97 (1-2):273–324, December 1997.

D. LaLoudouana and M. Bonouliqui Tarare. Data set selection. In *NIPS02 http://www.jmlg.org/papers/laloudouana03.pdf*, 2002.

Y. LeCun. The MNIST database of handwritten digits. *http://yann.lecun.com/exdb/mnist/*.

D. J. C. MacKay. Bayesian non-linear modeling for the energy prediction competition. *ASHRAE Transactions*, 100:1053–1062, 1994.

P. M. Murphy and D. W. Aha. UCI repository of machine learning databases. In *http://www.ics.uci.edu/~mlearn/MLRepository.html*, 1994.

R. M. Neal. *Bayesian Learning for Neural Networks*. Number 118 in Lecture Notes in Statistics. Springer-Verlag, New York, 1996.

S. Perkins, K. Lacker, and James Theiler. Grafting: Fast, incremental feature selection by gradient descent in function space. *JMLR*, 3:1333–1356, March 2003.

S. Kremer et al. NIPS 2000 and 2001 unlabeled data competitions. *http://q.cis.uoguelph.ca/~skremer/Research/NIPS2000/*   *http://q.cis.uoguelph.ca/~skremer/Research/NIPS2001/*, 2000-2001.

T. R. Golub et al. Molecular classification of cancer: Class discovery and class prediction by gene expression monitoring. *Science*, 286:531–537, 1999.

V. Vapnik. *Statistical Learning Theory*. John Wiley and Sons, N.Y., 1998.

A Details About the Fifty Feature Subset Study

We selected 50 feature subsets, five for each dataset both for December 1st and for December 8th entries (see Table 9.11). You can download the corresponding challenge entries, including the subsets themselves from: http://clopinet.com/isabelle/Projects/NIPS2003/Resu_Challenge.zip. We used a soft-margin SVM implemented in SVMlight (Joachims, 1998), with a choice of kernels (linear, polynomial, and Gaussian) and a choice of preprocessings and hyperparameters selected with 10-fold cross-validation. For the December 8th feature sets, we trained on both the training set and the validation set to be in the same conditions as the challengers. Specifically, for each (dataset, feature-set, date) the following settings were tried:

- Eight different kernels:
  - linear,
  - polynomial of degree 2, 3, 4, 5,
  - RBF with width 0.1, 1, 10.
- Three values of C were tried:
  - $C = 0.1, 1, 10$;
- Three preprocessings:
  - prep1: each feature was separately standardized, by substracting its mean and dividing by (std + fudge factor). In case of sparse data, the mean is not substracted.
  - prep2: each example is divided by its L_2 norm.
  - prep3: each example is divided by the mean L_2 norm of examples. On some datasets we also used the raw data. However, for some datasets (e.g. MADELON) the SVM classifiers spent too much time on each trial, so we removed it from the bank of normalization methods.

In total, for each (dataset, feature-set, date), we estimated the performance of the SVM classifier on $8 \times 3 \times 3 = 72$ hyper parameter sets. The

Table 9.11. Fifty entries selected to perform experiments. The balanced error rate (BER) and the area under the ROC curve (AUC) are multiplied by 100. The challenge rank is in parenthesis. We also show the percentage of features used (**Ffeat**), and the percentage of probes in the features selected (**Fprob**). The team are named by one of their member, see the papers for the complete list of authors.

Data set	Entry ID	Method	Team	BER (Rk)	AUC (Rk)	Ffeat	Fprobe
			December 1st				
ARCENE	vqUNMWlP	inf5	Saffari	17.30 (17)	82.70 (38)	5.00	0.00
	VeJsRriv	KPLS	Embrecht	16.71 (12)	83.67 (34)	5.14	8.56
	DZSQTBfy	BayesNN	Neal	16.59 (10)	91.15 (8)	10.70	1.03
	XNexBfhI	multi23	SK Lee	18.41 (20)	81.59 (40)	1.85	0.00
	HEdOWjxU	Mod. R	Ng	19.64 (32)	86.72 (25)	3.60	8.06
DEXTER	DZSQTBfy	BayesNN	Neal	4.00 (4)	99.03 (1)	1.52	12.87
	voCGSrBX	Collection	Saffari	4.95 (11)	95.05 (35)	5.01	36.86
	EoeosKio	IDEAL	Borisov	5.50 (18)	98.35 (10)	1.00	25.50
	UFXqDSHR	RF+RLSC	Torkkola	5.40 (16)	94.60 (38)	2.50	28.40
	XyxUJuLe	Nameless	Navot	5.25 (12)	98.80 (5)	7	43.71
DOROTHEA	aIiushFx	Nameless	Navot	10.86 (6)	92.19 (13)	0.30	0.00
	DZSQTBfy	BayesNN	Neal	10.63 (5)	93.50 (5)	0.50	0.40
	WSRyGqDz	SVMb3	Zhili	11.52 (11)	88.48 (20)	0.50	18.88
	HEdOWjxU	Mod. RF	Ng	13.72 (16)	91.67 (16)	0.40	20.75
	pwKPKbca	P-SVM	Hochreiter	17.06 (36)	90.56 (17)	0.14	60.00
GISETTE	tKBRMGYn	final 1	YW Chen	1.37 (9)	98.63 (32)	18.26	0.00
	DRDFlzUC	transSVM	Zhili	1.58 (11)	99.84 (9)	15.00	0.00
	GbXDjoiG	FS + SVM	Lal	1.69 (16)	98.31 (33)	14.00	0.00
	UFXqDSHR	RF+RLSC	Torkkola	1.89 (19)	98.11 (39)	6.14	0.00
	DZSQTBfy	BayesNN	Neal	2.03 (26)	99.79 (14)	7.58	0.26
MADELON	kPdhkAos	BayesSVM	Chu Wei	7.17 (5)	96.95 (7)	1.60	0.00
	UFXqDSHR	RF+RLSC	Torkkola	6.67 (3)	93.33 (33)	3.80	0.00
	WOzFIref	P-SVM	Hochreiter	8.67 (20)	96.46 (12)	1.40	0.00
	DZSQTBfy	BayesNN	Neal	7.72 (7)	97.11 (4)	3.40	0.00
	DRDFlzUC	transSVM	Zhili	8.56 (13)	95.78 (20)	2.60	0.00
			December 8th				
ARCENE	caBnuxYN	BayesNN	Neal	11.86 (7)	95.47 (1)	10.70	1.03
	UHLVxHZK	RF	Ng	12.63 (10)	93.79 (6)	3.80	0.79
	RpBzOXAE	CBAMet3	CBA	11.12 (4)	94.89 (2)	28.25	0.28
	ZdtHhsyQ	FS+SVM	Lal	12.76 (12)	87.24 (22)	47.00	5.89
	wCttZQak	Nameless	Navot	15.82 (19)	84.18 (29)	44.00	12.48
DEXTER	ZdtHhsyQ	FS+SVM	Lal	3.30 (1)	96.70 (23)	18.57	42.14
	caBnuxYN	BayesNN	Neal	4.05 (7)	99.09 (3)	1.52	12.87
	bMLBEZfq	Sparse	DIMACS	5.05 (14)	94.37 (29)	0.93	6.49
	PVxqJatq	RF+RLSC	Torkkola	4.65 (10)	95.35 (27)	2.50	28.40
	BuVgdqsc	final2	YW Chen	5.35 (15)	96.86 (21)	1.21	2.90
DOROTHEA	wCttZQak	Nameless	Navot	11.40 (7)	93.10 (7)	0.40	0.00
	caBnuxYN	BayesNN	Neal	11.07 (6)	93.42 (6)	0.50	0.40
	untoIUbA	ESNB+NN	Boulle	14.59 (17)	91.50 (13)	0.07	0.00
	UHLVxHZK	RF	Ng	14.24 (16)	91.40 (14)	0.32	4.38
	oIiuFkuN	original	Zhili	13.46 (14)	86.54 (22)	0.50	18.88
GISETTE	RqsEHylt	P-SVM/nu	Hochreiter	1.82 (19)	99.79 (10)	4.00	0.50
	PVxqJatq	RF+RLSC	Torkkola	1.77 (17)	98.23 (29)	6.14	0.00
	MLOvCsXB	P-SVM/nu	Hochreiter	1.75 (15)	99.79 (13)	9.90	19.19
	oIiuFkuN	original	Zhili	1.58 (11)	99.84 (9)	15.00	0.00
	caBnuxYN	BayesNN	Neal	2.09 (25)	99.78 (17)	7.58	0.26
MADELON	PDDChpVk	Bayesian	Chu Wei	7.11 (13)	96.95 (10)	1.60	0.00
	caBnuxYN	BayesNN	Neal	6.56 (4)	97.62 (3)	3.40	0.00
	BuVgdqsc	final2	YW Chen	7.11 (12)	92.89 (25)	3.20	0.00
	jMqjOeOo	RF+RLSC	Torkkola	6.67 (6)	93.33 (22)	3.80	0.00
	NiMXNqvY	GhostMiner	Team	7.44 (14)	92.56 (26)	3.00	0.00

estimation was performed using 10-fold cross validation on the training set (for December 8 submissions, the 10-fold cross validation was performed on the training+validation set).

Biasing. Within the 10 fold cross validation we first trained the predictor on the whole training set and ran it on the test set part of the data split, then chose a bias such the ratio of positive to negative examples corresponds to that of the training set. This bias is then subtracted from the scores of predictor when ran on each fold.

Code availability. The datasets and the Matlab code of this study can be dowloaded from our web site at `http://clopinet.com/isabelle/Projects/NIPS2003/download`, file `matlab_batch_may04.zip` as well as the results of our best classifiers selected by CV (file `best_reference_svm.zip`) and the results of the vote of our top 5 best classifiers selected by CV (file `vote_reference_svm.zip`). For questions, please contact Gideon Dror.

Chapter 10

High Dimensional Classification with Bayesian Neural Networks and Dirichlet Diffusion Trees

Radford M. Neal[1] and Jianguo Zhang[2]

[1] Dept. of Statistics and Dept. of Computer Science, University of Toronto
radford@stat.utoronto.ca, http://www.cs.utoronto.ca/~radford/
[2] Dept. of Statistics, University of Toronto
jianguo@stat.utoronto.ca

Our winning entry in the NIPS 2003 challenge was a hybrid, in which our predictions for the five data sets were made using different methods of classification, or, for the Madelon data set, by averaging the predictions produced using two methods. However, two aspects of our approach were the same for all data sets:

- We reduced the number of features used for classification to no more than a few hundred, either by selecting a subset of features using simple univariate significance tests, or by performing a global dimensionality reduction using Principal Component Analysis (PCA).
- We then applied a classification method based on Bayesian learning, using an Automatic Relevance Determination (ARD) prior that allows the model to determine which of these features are most relevant.

Selecting features with univariate tests is a simple example of the "filter" approach discussed in Chapter 3. Reducing dimensionality using a method that looks at all (or most) features is an alternative to feature selection in problems where the cost of measuring many features for new items is not an issue. We used PCA because it is simple, and feasible for even the largest of the five data sets.

Two types of classifiers were used in our winning entry. Bayesian neural network learning with computation by Markov chain Monte Carlo (MCMC) is a well-developed methodology, which has performed well in past benchmark comparisons (Neal, 1996, 1998, Rasmussen, 1996, Lampinen and Vehtari, 2001). Dirichlet diffusion trees (Neal, 2001, 2003) are a new Bayesian approach to density modeling and hierarchical clustering. We experimented with using the trees produced by a Dirichlet diffusion tree model (using MCMC) as the basis for a classifier, with excellent results on two of the data sets. As allowed by the challenge rules, we constructed these trees using both the training

R.M. Neal and J. Zhang: *High Dimensional Classification with Bayesian Neural Networks and Dirichlet Diffusion Trees*, StudFuzz **207**, 265–296 (2006)
www.springerlink.com

data and the unlabelled data in the validation and test sets. This approach —
sometimes referred to as "transduction" or "semi-supervised learning" — tries
to exploit the possible connection between the distribution of features and the
relationship of the class to these features. We also used the available unlabelled
data in a more minor way when finding the principal components looked at
by some of the neural network models.

The ARD prior used in both the neural network and Dirichlet diffusion
tree models is an example of (soft) feature selection using an "embedded"
approach, as discussed in Chapter 5. The number of features in four of the
five data sets is too large for ARD to be used directly, due to computational
costs, which is why some feature selection or dimensionality reduction was
done first. (With 500 features, the Madelon data set can be handled directly,
though the computations are a bit arduous.) However, by using ARD, we can
afford to keep more features, or reduce dimensionality less, than if we used a
classification method that could not learn to ignore some features. This lessens
the chance that useful information will be discarded by the fairly simple-
minded feature selection and dimensionality reduction methods we used.

In this chapter, we describe the methods used in our winning entry, and
discuss why these methods performed as they did. We also present refined and
simplified versions of the methods using Bayesian neural networks with PCA
and variable selection, for which scripts have been made publicly available (at
`www.cs.utoronto.ca/~radford/`). The scripts use the R language (available
at `www.r-project.org`) for preprocessing, and the neural network module
of R. M. Neal's Software for Flexible Bayesian Modeling (available at `www.
cs.utoronto.ca/~radford/`) for fitting Bayesian models using MCMC. To
begin, however, we discuss some general issues regarding Bayesian and non-
Bayesian learning methods, and their implications for feature selection.

10.1 Bayesian Models vs. Learning Machines

Chapter 1 of this book discusses classification in terms of "learning machines".
Such a machine may be visualized as a box with inputs that can be set to
the features describing some item, and which then produces an output that
is the machine's prediction for that item's class. The box has various knobs,
whose settings affect how this prediction is done. Learning is seen as a process
of fiddling with these knobs, in light of how well the box predicts the classes
of items in a known training set, so as to produce good predictions for the
classes of future items. This view of learning leads one to choose machines
of adequate but limited complexity, providing a suitable trade-off of "bias"
and "variance", perhaps estimated by means of cross-validation. One way of
limiting complexity is to choose a small set of features to input to the machine.

The Bayesian approach to classification is quite different. Rather than
a process of adjusting knobs on a machine, learning is seen as probabilistic
inference, in which we predict the unknown class of a new item using the

conditional probability distribution for this class given the known features for this item, and the known features and classes for the items in the training set. Using X to denote sets of features and Y to denote classes, we will write this as $P(Y_{\text{new}} \mid X_{\text{new}}, X_{\text{train}}, Y_{\text{train}})$. Such predictive distributions are usually expressed in terms of a *model* for the conditional probability of the class of an item given its features. The model usually has *parameters*, θ, whose values are unknown, and which are therefore assigned a probability distribution, $P(\theta)$, known as the *prior* distribution. The choice of both the model and the prior should reflect our knowledge of the problem before seeing the training set. When we then condition on the known features and classes of items in the training set, we obtain the *posterior* distribution for the model parameters:

$$P(\theta \mid X_{\text{train}}, Y_{\text{train}}) \;=\; \frac{P(\theta)\,P(Y_{\text{train}} \mid X_{\text{train}}, \theta)}{\int P(\theta)\,P(Y_{\text{train}} \mid X_{\text{train}}, \theta)\,d\theta} \tag{10.1}$$

This equation assumes that the distribution of the feature values, X, contains no information about how items should be classified, as modeled using θ. In a Bayesian approach to semi-supervised learning, this would not be assumed, and hence additional factors of $P(X \mid \theta)$ would appear. In our approach to the challenge problems, however, the semi-supervised aspects were instead done in an *ad hoc* way.

Using this posterior distribution, the predictive distribution for the class of a new item can be expressed in terms of an integral over the parameters, which averages the conditional probabilities for the class of the new item given its features, as defined by the model, with respect to the posterior distribution of the model parameters:

$$P(Y_{\text{new}} \mid X_{\text{new}}, X_{\text{train}}, Y_{\text{train}}) \;=\; \int P(Y_{\text{new}} \mid X_{\text{new}}, \theta)\,P(\theta \mid X_{\text{train}}, Y_{\text{train}})\,d\theta \tag{10.2}$$

The integral above is usually analytically intractable. Since θ is generally high-dimensional, a standard approach is to approximate it using Monte Carlo methods, after obtaining a sample of many values for θ drawn from the posterior distribution by simulating a suitable Markov chain. Reviewing such Markov chain Monte Carlo methods is beyond the scope of this chapter. Introductions to MCMC methods include (Neal, 1993) and (Liu, 2001).

These two approaches to classification correspond to the "frequentist" and "Bayesian" frameworks for statistics, whose respective merits have long been debated (see Bernardo and Smith, 1994). One difference is that the Bayesian approach provides, in theory, a unique solution to the classification problem — the one derived from the model and prior that express our prior beliefs about the situation — whereas the "learning machine" approach inevitably leaves some choices to be made arbitrarily, or by applying prior knowledge informally. For high-dimensional classification problems, however, a purely Bayesian approach may sometimes be impractical, because the intellectual effort required to formalize our prior knowledge of a complex situation and the computational effort of computing integrals over a parameter space of enormous dimension may both be prohibitive. Use of *ad hoc* frequentist methods may then be necessary in order to reduce the problem to a manageable size.

The need for such compromises should not make one lose sight of the fundamentally different nature of Bayesian methods. In a pure Bayesian approach to classification, we would *not* remove some features from the model, unless we really had certain prior knowledge that they are irrelevant — the best prediction for the class of a new item is found using a model that looks at *all* features that might possibly be relevant. In situations where we wish to eliminate some features in order to avoid the costs of measuring these features when classifying new items, we would quantify these costs, and treat the selection of a subset of features as a decision problem trading off the cost of measurement with the cost of sub-optimal classification. The optimal solution to this decision problem would base the choice of a prediction rule that looks at a subset of features on evaluations made using a predictive model that looks at all features. When the cost of measuring features for new items is not an issue, we would eliminate features from a Bayesian model only to reduce the effort involved in training and prediction, not because including all features would reduce the performance of a properly formulated and implemented model.

This may seem surprising from the "learning machine" viewpoint. Won't a model in which the number of features, and hence also the number of parameters, is much greater than the number of training items "overfit" the data, and make bad predictions for new items? The crucial point to keep in mind here is that the parameters of a Bayesian model do *not* correspond to the "knobs" that are adjusted to make a learning machine predict better. Predictions based on a Bayesian model are made using equation (10.2). Seen as a function of X_{new}, with X_{train} and Y_{train} fixed, this prediction function is generally *not* expressible as $P(Y_{\text{new}} \mid X_{\text{new}}, \theta)$ for some fixed value of θ. When the training set is small, the predictive distribution for Y_{new} will tend to be "simple", because it will be found by integrating $P(Y_{\text{new}} \mid X_{\text{new}}, \theta)$ with respect to a fairly broad posterior distribution for θ, which will cause the complexities of $P(Y_{\text{new}} \mid X_{\text{new}}, \theta)$ for individual θ values to be averaged away. With a larger training set, the posterior distribution of θ will be more concentrated, and the predictive distribution may be more complex, if a complex distribution is actually favoured by the data. No explicit adjustment of the model on the basis of the size of the training set is necessary in order to achieve this effect.

The practical import of this is that when selecting a subset of features, or otherwise reducing dimensionality, our goal is merely to reduce the number of features to the point where using a Bayesian model is intellectually and computationally feasible. There is no need to reduce the number of features to an even smaller number to avoid "overfitting". An appropriate Bayesian model, using an appropriate prior, will be able to learn to what extent each of the remaining features are relevant to the task of predicting the class. If there is a cost to measuring features in the future, their number can be reduced further after fitting this model by seeing which features turned out to play a substantial role in predictions. In practice, when models are imperfectly specified and implemented, it is possible that a model using a small number of features may sometimes do better than a less-than-perfect model using a larger number of features, but there is no reason to expect this in general.

10.2 Selecting Features with Univariate Tests

One way to reduce the number of variables used in the model to what can be handled computationally is to select a subset of features that appear to be associated with the class. There are many ways of doing this, as discussed in Chapter 3. We used a fairly simple approach, in which we looked at only one feature at a time. There is no guarantee that this will work — it could be that no single feature contains any information about the class, but that that some pairs (or larger subsets) of features are informative when looked at jointly. However, it may be that for most real problems, features that are relevant to predicting the class are at least somewhat relevant on their own. If so, we will be able to select the relevant features with univariate tests of association with the class. We can then use a model that looks only at these features, and that can explore possible interactions among them.

One criterion for identifying relevant features is the sample correlation of a feature with the class, numerically encoded in some way (e.g., as -1 and $+1$). The sample correlation of $x_1, \ldots, x_m$ with $y_1, \ldots, y_m$ is defined as

$$r_{xy} = \frac{(1/m)\sum_{i=1}^{m}(x_i - \bar{x})(y_i - \bar{y})}{s_x\, s_y} \tag{10.3}$$

where $\bar{x}$ and s_x are the sample mean and standard deviation of $x_1, \ldots, x_m$ and $\bar{y}$ and s_y are the sample mean and standard deviation of $y_1, \ldots, y_m$.

The correlation of a feature with the class will seldom be exactly zero, which raises the question of how big r_{xy} must be for us to consider that x is a good candidate for a relevant feature. A frequentist approach to answering this question is to compute the probability that a value for r_{xy} as big or bigger than the actual value would arise if the feature actually has no association with the class. If this "p-value" is quite small, we might consider it unlikely that there is no association. We can find the p-value using a *permutation test* (Cox and Hinkley, 1974, Section 6.2), based on the idea that if there is no real association, the class labels might just as well have been matched up with the features in a completely different way.

One way to compute this permutation test p-value is as follows:

$$p = 2 \min\left(\frac{1}{m!}\sum_{\pi} I\big(r_{xy_\pi} \geq r_{xy}\big),\ \frac{1}{m!}\sum_{\pi} I\big(r_{xy_\pi} \leq r_{xy}\big)\right) \tag{10.4}$$

where the sums are over all $m!$ possible permutations of $y_1, \ldots, y_m$, with y_π denoting the class labels as permuted by π. The indicator functions have the value one if the correlation of x with y_π is at least as high (or at least as low) as the actual correlation. We take the minimum so that the test is sensitive to either positive or negative correlations. The multiplication by 2 ensures that p has (approximately) a uniform distribution over $(0, 1)$ when there is no real association, which is the conventional property of p-values. Since $m!$ is generally very large, we will in practice approximate p using averages of the indicator functions over a few thousand randomly chosen permutations.

In the special case where $x_1, \ldots, x_m$ are binary, p can be computed exactly using hypergeometric probabilities. It may be useful to convert some non-binary features to binary by thresholding them at some value, and then testing for an association of the class with this binary feature. This might reveal an association that is not apparent when looking at the original feature.

One might also look at the correlation of the class with some other transformation of a feature, such as its square or square root. There are endless such possibilities, however. Rather than try them all, we can instead transform the feature values to ranks — i.e., if x_i is greater than or equal to k of the other x_j, it is replaced by the value k — and then find the correlation of the class with these ranks. The result (sometimes called the "Spearman correlation") is invariant to any monotonic transformation of the original feature, and hence can detect any monotonic relationship with the class.

Nonmonotonic relationships can be detected using a *runs test* (Gibbons, 1985, Chapter 3). Let π be a permutation for which $x_{\pi_i} \leq x_{\pi_{i+1}}$ for all i. If the class is unrelated to x, the corresponding permutation of the class values, $y_{\pi_1}, \ldots, y_{\pi_m}$, should not differ systematically from a random permutation. We can test this by counting how often y_{π_i} equals $y_{\pi_{i+1}}$, and then computing a p-value as the probability of a count this big or bigger arising when the y values are randomly permuted. If there are ties among the x_i, there will be more than one permutation π that orders them, so we average the counts over all such permutations (or in practice, a random sample of them).

The tests described above are of increasing generality, testing for linear, monotonic, and arbitrary nonmonotonic relationships. The increase in generality comes at the cost of power, however — a linear relationship that produces a small p-value in a permutation test based on correlation with the original feature value might not produce such a small p-value when the test is based on correlation with the ranks, or when a runs test is used. Accordingly, it may be desirable to select features using more than one test, perhaps with different p-value thresholds for each.

Finally, one should note that an optimistic bias can result from selecting a subset of features using significance tests and then using this subset in a model as if they were the only features. The predictive probabilities produced using such a model may be more certain (i.e., closer to zero or one) than is actually justified. In the extreme, one might think the class is highly predictable when in fact none of the features are informative — the selected features having appeared to be associated with the class only by chance. One check for such problems is to see whether the p-values found for all the features using one of the procedures above appear to be distributed uniformly over $(0, 1)$, as would be expected if none of them is actually associated with the class. For all the challenge data sets, the distribution of p-values is far from uniform, indicating that at least some features are informative.

10.3 Reducing Dimensionality Using PCA

Rather than select a manageably small subset of the original features to use in our model, we can instead create a small set of new variables, which we hope will capture the important information in the full data set. Of the many such methods for "dimensionality reduction", we used the simplest, known as Principal Components Analysis (PCA), in which the new variables are linear combinations of the original feature, chosen to capture as much of the original variance as possible. There is no guarantee that a small number of principal components with the highest variance will contain the information needed for classification — conceivably, the important information could be contained in linear combinations of variables whose variance is low rather than high. In practice, however, the first few principal components usually contain at least some useful information.

Let X be the matrix of feature values, with m rows representing cases, and n columns of feature values. (This notation is consistent with other chapters in this book, but note that the universal convention in statistics is that the number of features is p and the size of the training set is n.) The principal component vectors are the eigenvectors of the $n \times n$ matrix $X^T X$, ordered by decreasing magnitude of the corresponding eigenvalue. When there are more features than cases (i.e., $n > m$), only the m eigenvectors with non-zero eigenvalues are meaningful. In this case, the principal component vectors can more efficiently be computed from the $m \times m$ matrix XX^T, using the fact that if v is an eigenvector of XX^T then $X^T v$ is an eigenvector of $X^T X$, with the same eigenvalue. Once the principal component vectors have been found, the n feature values for a case are reduced to k values by finding the projections of the feature vector on the first k principal component vectors.

Before finding the principal components, the matrix of feature values, X, may be modified by subtracting the mean from each feature value, and perhaps dividing by its standard deviation. However, if zero is a value of special significance, one might decide not to subtract the mean, in order to preserve this meaning. Similarly, dividing by the standard deviations might be harmful if the standard deviations of the features are a good guide to their relevance.

10.4 Bayesian Logistic Regression

Before discussing the methods used for our winning entry, it is worthwhile discussing some much simpler methods based on *logistic regression*, which also give very good results on some of the challenge data sets.

The linear logistic regression model expresses the probability distribution for the class of item i, given the features for that item, as follows:

$$P(Y_i = 1 \,|\, X_i = x_i) \;=\; \left[1 + \exp\left(-\left(\alpha + \sum_{j=1}^{n} \beta_j x_{ij} \right) \right) \right]^{-1} \quad (10.5)$$

Maximum likelihood is sometimes used to select values for the parameters α and β that best fit the data, but when the classes can be linearly separated, the maximum likelihood estimates will be both infinite and non-unique. This situation will arise whenever the number of features (n) is greater than the number of items in the training set (m). Bayesian methods for logistic regression have no problem in this situation, however, since the Bayesian predictions are obtained by averaging (10.5) with respect to the posterior distribution for the parameters, rather than selecting some particular estimate.

To do Bayesian logistic regression, we must first select a prior distribution for the parameters. The intercept parameter, α, can be given a rather broad prior distribution, since the data will suffice to constrain this parameter in any case. The prior used for the β_j parameters is more crucial. Perhaps the simplest option is to give β a multivariate Gaussian prior with mean zero and diagonal covariance $\sigma^2 I_n$, where I_n is the n by n identity matrix. The value of σ will determine to what extent the model thinks the class can be predicted from the features — if σ is zero, for instance, the features will have no effect on the predictions. Often, we would have little prior idea as to how predictable the class is, and would therefore not be in a position to fix σ to any specific value. Instead, we can treat σ as an additional "hyperparameter", to which we give a fairly broad prior distribution, allowing the degree of predictability to be inferred from the training data. It is convenient for this prior to take the form of a gamma distribution for $1/\sigma^2$, since this form of prior is "conjugate", which allows some operations to be done analytically. Extremely vague (or improper) priors must be avoided, since the limit $\sigma \to 0$ will not be completely excluded by the data, and in the case of linearly separable data, $\sigma \to \infty$ will also be compatible with the data. A prior that gives substantial probability to values of σ ranging over a few orders of magnitude is often appropriate.

This prior distribution for β is spherically symmetric. If R is any orthogonal matrix (e.g., a rotation), transforming the X_i to $X_i' = R X_i$ and β to $\beta' = R\beta$ will have no effect on the probabilities in (10.5), since $\beta'^T X_i' = \beta^T X_i$. Furthermore, with a spherically symmetric prior, β' will have the same prior density as β. Applying any such transformation to the data will therefore have no effect on the final predictions from a Bayesian model with such a prior.

This invariance to orthogonal transformations has two important implications. First, we see that logistic regression with a spherically symmetric prior cannot be seen as performing any sort of feature selection, since the original coordinate system, in terms of which the features are defined, has no significance. Second, when the number of features (n) is greater than number of items in the training set (m), we can exploit the model's invariance under orthogonal transformations to reduce the computation time needed for the MCMC runs, by effectively reducing the number of features looked at from n to m.

We do this by finding an orthogonal transformation, R, for which all but m of the components of the transformed features, RX_i, are zero in all m training cases. Such a transformation always exists. One convenient choice is the transformation that projects X_i onto the m principal components found from the training cases, and projects the portion of X_i normal to the space of these principal components onto some set of $n-m$ additional orthogonal directions. For the training cases, the projections in these $n-m$ other directions will all be zero, so that X'_{ij} will be zero for $j > m$. Clearly, one then need only compute the first m terms of the sum in (10.5). For a test case, one can imagine that its component normal to the space of principal components is aligned with one of the additional directions, so that all but $m+1$ of the components in the transformed vector are zero. The final result is that only $\beta_1, \ldots, \beta_{m+1}$ need be represented, which produces a large reduction in computation time when, for example, $m = 100$ and $n = 10000$, as in the Arcene dataset.

Despite its possibly excessive simplicity, this model sometimes produces quite good results on problems such as classifying tumors based on gene expression data from DNA microarrays. (We have found this to be true, for example, on the data from Golub et al., 1999.) Although such problems have a great many features, they may nevertheless be rather easy, if a large number of these features carry information about the class. Computational cost can be further reduced by looking only at the projections onto the first k principal components, with $k < m$. This sometimes has a negligible effect, when the variance in the remaining directions is very small, and when the effect is not negligible, it is sometimes beneficial. Results on the challenge datasets using such a model with $k = 40$ are reported in Section 10.7.

Bayesian logistic regression can also be done using a prior that expresses the belief that some of the features are likely to be much more relevant to predicting the class than others — i.e., that some of the β_j parameters are likely to be much bigger than others. As for the simpler prior described above, it will usually be appropriate to include a hyperparameter, σ, in the model, that captures the overall scale of the β_j parameters, and hence the overall degree to which the class can be predicted from the features. Conditional on this hyperparameter, we can let the β_j be independent under the prior. To allow some of the β_j to be much larger than others, we can let the prior for β_j given σ be a heavy-tailed distribution (e.g., a t distribution with two degrees of freedom), with mean zero, and width parameter σ. If only a few features are relevant, the posterior distribution for σ will be concentrated on values close to zero, and most of the β_j will therefore also be close to zero. However, due to the heavy tails of the prior, a few of the β_j can still take on large values, allowing a few of the features to influence the class probabilities.

This model is equivalent to one in which each β_j has a Gaussian prior distribution, but with individual variances, s_j^2, that are variable hyperparameters. If $1/s_j^2$ is given a gamma distribution, the resulting effective prior for β_j (integrating over s_j) will be a t distribution. This view of the model is related to the ARD priors for neural networks discussed in the next section.

When the β_j are given non-Gaussian priors, the model is no longer spherically symmetrical. Accordingly, the number of features can no longer be reduced from n to m by applying an orthogonal transformation. The posterior distribution of the parameters may also have many modes, in which different sets of features are regarded as relevant. For these reasons, MCMC computations become time-consuming when the number of features exceeds a thousand or so. Section 10.7 reports results using logistic regression models with heavy-tailed priors applied to the challenge datasets after reducing dimensionality using PCA, or selecting a subset of features using univariate significance tests.

10.5 Bayesian Neural Network Models

In the logistic regression models of the previous section, the features of an item influence class probabilities only through a linear combination of the feature values. For many problems, this is not an adequate model. One way of extending logistic regression to allow for non-linear relationships is to model the class probability function using a *neural network* with *hidden units*, also known as a *multilayer perceptron* network.

Many architectures for such networks are possible. The neural network methods we used for the feature selection challenge used two "layers" of hidden units, producing a model that can be written as follows:

$$P(Y_i = 1 \,|\, X_i = x_i) \; = \; \left[1 + \exp(-f(x_i))\right]^{-1} \qquad (10.6)$$

$$f(x_i) \; = \; c + \sum_{l=1}^{H} w_\ell h_\ell(x_i) \qquad (10.7)$$

$$h_\ell(x_i) \; = \; \tanh\left(b_\ell + \sum_{k=1}^{G} v_{k\ell} g_k(x_i)\right) \qquad (10.8)$$

$$g_k(x_i) \; = \; \tanh\left(a_k + \sum_{j=1}^{n} u_{jk} x_{ij}\right) \qquad (10.9)$$

This network is illustrated in Figure 10.1. A large class of functions, f, can be represented by such a network as we vary the network's parameters — the "biases" a_k, b_ℓ, and c and the "weights" u_{jk}, $v_{k\ell}$, and w_ℓ — provided that G and H, the numbers of hidden units in the two layers, are fairly large.

The traditional methods for training such networks (Rumelhart, Hinton, and Williams, 1986), based on maximum likelihood estimation of the parameters, have been successful for many problems, but difficulties with local maxima and with overfitting can arise. These problems can be avoided in a

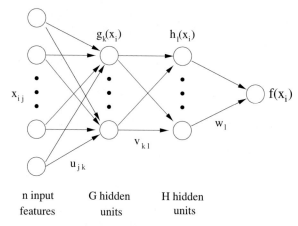

Fig. 10.1. A neural network with two hidden layers. The circles on the left represent the feature values for a particular item. The circles in the middle represent the values of hidden units, which are computed by applying the tanh function to a linear combination of the values to which they are connected by the arrows. The circle on the right represents the function value computed by the network, which is used to define the class probabilities.

Bayesian approach using Markov chain Monte Carlo, in which predictions are based not on one network, but on the average class probabilities from many networks, in which the parameters are drawn at random from their posterior distribution, as simulated using a Markov chain. Full details of this approach can be found in (Neal, 1996). Here, we can only briefly discuss some issues relating to feature selection.

Analogously with the logistic regression models discussed in Section 10.4, the computation needed for Bayesian neural network models with a large number of features can be greatly reduced if the prior distribution for parameters is spherically symmetrical. This will be the case if the prior distribution for the input-to-hidden weights, u_{jk}, is a multivariate Gaussian with mean zero and covariance matrix σI_n, where σ may be a hyperparameter that is given a higher-level prior distribution. By applying an orthogonal transformation to the feature space, the m training items can be changed so that $n - m$ of the features are always zero. Consequently, these features and their associated weights, u_{jk}, need not be represented, since features that are always zero will have no effect in equation (10.9).

Such models sometimes work well, but for many problems, it is important to learn that some features are more relevant to predicting the class than others. This can be done using an "Automatic Relevance Determination" (ARD) prior (MacKay, 1994, Neal, 1996) for the input-to-hidden weights, u_{jk}. In an ARD prior, each feature is associated with a hyperparameter, σ_j, that expresses how relevant that feature is. Conditional on these hyperparameters, the u_{jk} have a multivariate Gaussian distribution with mean zero and diagonal covariance matrix, with the variance of u_{jk} being σ_j^2. The σ_j hyperparameters are themselves given a fairly broad prior distribution, which may depend on a yet higher-level hyperparameter. The posterior distribution for σ_j will tend to concentrate on small values if feature j is not useful for predicting the class, given the other features that are present. Random associations between the class and this feature will then have little detrimental influence on the predictions made by the model, since the weights, u_{jk}, for that feature will be small.

ARD models with tens of thousands of features are not presently feasible, for two reasons. First, a simple prior distribution (e.g., based on a gamma distribution) for the σ_j may not be adequate for such a large number of features. It might instead be necessary to use a more elaborate prior that allowed for a larger class of distributions for these relevance hyperparameters. Whether this is actually necessary is not known, however, because of the second reason — simulating a Markov chain that adequately samples from the posterior distribution over tens of thousands of relevance hyperparameters would take an infeasibly large amount of computation time. For these reasons, dimensionality reduction or preliminary selection of features, as described in Sections 10.2 and 10.3, is necessary to reduce the number of features to around a thousand at most, after which an ARD model can discover which of these features are most relevant.

10.6 Dirichlet Diffusion Tree Models

For two of the challenge datasets, we tried out a method that hierarchically clusters the training and test items, and then classifies the unlabelled items on the basis of their positions in the hierarchy. This is a "semi-supervised" approach, since the unlabelled data is used during training.

The hierarchical clustering was based on the "Dirichlet diffusion tree" model of Neal (2001, 2003). This is a Bayesian clustering method, which produces not a single tree, but a set of trees drawn from the posterior distribution of trees. We applied this Dirichlet diffusion tree model to a set of principal component values derived from all the features. Hyperparameters analogous to those used in ARD priors allowed the model to learn which of these principal components were involved in the clusters, and which instead had distributions that could be modeled independently of the others.

A Dirichlet diffusion tree has the data items as terminal (leaf) nodes. The way in which these items are grouped in a hierarchy is analogous to how species are arranged in a phylogenetic tree to represent their evolutionary relationships. The model can be defined by a random procedure for generating both a dataset and a tree that describes the relationships underlying the data. This procedure is based on a Gaussian diffusion process, also known as Brownian motion, which is described in terms of a fictitious time variable, which varies from $t = 0$ to $t = 1$.

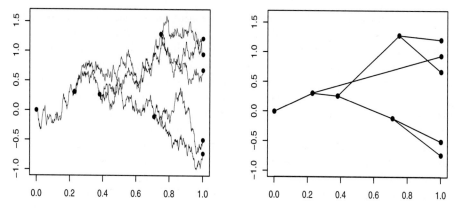

Fig. 10.2. A Dirichlet diffusion tree with five items. Diffusion time is on the horizontal axis. The vertical axis represent the value of a single feature. The plot on the left shows the tree with details of the Brownian diffusion paths included. The plot on the right shows the same tree with details of the paths suppressed between divergence points, as is done in practice when performing computations. The diffusion standard deviation used was $\sigma = 1$; the divergence function was $a(t) = 1/(1-t)$.

The process is illustrated for a single feature, x, in Figure 10.2. The first data point (which could be any of the five shown) is generated by starting a diffusion process with $x = 0$ at $t = 0$. As t increases, x diffuses, with the variance of $x(t + \epsilon) - x(t)$ being equal to $\sigma^2 \epsilon$, where σ is a hyperparameter giving the diffusion standard deviation for the feature. The value of x when $t = 1$ gives the value of the feature for this first data point. The second data point is generated in the same fashion, except that the diffusion path for this data point initially follows the *same* diffusion path as the first data point. At a random time, however, the path of the second data point diverges from that of the first, and thereafter proceeds independently.

The distribution of the time when the path to the second point diverges is determined by a "divergence function", $a(t)$. In the example, $a(t) = 1/(1-t)$. If the path has not diverged by time t, the probability of divergence within the next small time interval ϵ is $a(t)\epsilon$. If $\int_0^1 a(t)dt$ is infinite, divergence is guaranteed to occur before time $t = 1$, and hence the second data point will not be the same as the first.

Later data points are generated in analogous fashion, each following previous paths for a while, but at some time diverging and then proceeding independently. When, before divergence, the path to the new point comes to a branch, where previous paths diverged, the new path takes a branch chosen randomly with probabilities proportional to the number of times each branch was taken previously. A similar "reinforcement" mechanism modifies the divergence probability — the probability of divergence occurring between time t and $t + \epsilon$ is $a(t)\epsilon/d$, where d is the number of previous data points that followed the current path.

As shown by Neal (2001, 2003), even though this procedure produces data points sequentially, the order of the data points produced is not actually significant — the distribution over datasets is "exchangeable", meaning that the probability density for a dataset is not changed when the items are permuted.

When there is more than one feature, the values for different features are produced by independent diffusions, but with the same tree structure (and divergence times) being used for all features. It is often useful to model the observed values of the features as being the values of the terminal nodes in the tree plus Gaussian "noise". By letting both the standard deviation of this noise and the diffusion standard deviation be hyperparameters specific to each feature, we can express the idea that the variation of some of the features may be mostly "noise" that is unrelated to the other feature values, while the variation of other features is mostly related to the underlying tree (and hence to other such features that also depend on the tree), with little "noise".

Given some particular dataset, we can try to find the posterior distribution over trees that might have been produced along with this dataset, if the data were the result of the Dirichlet diffusion tree process. Since this posterior distribution is complex, we will need to apply Markov chain Monte Carlo methods, as described by Neal (2003). The Markov chain can sample not just for the underlying tree, but also for the noise and diffusion standard deviation hyperparameters, and for hyperparameters that determine the divergence function.

Once we have a sample of trees from the posterior distribution, we might use them in various ways to classify unlabelled items. One class of methods use only the tree structure, without paying attention to the divergence times of the branch points in the tree. Another class of methods use the divergence times in the tree to define a measure of distance between items. Similar methods for classification based on trees have been developed independently by Kemp, Griffiths, Stromsten, and Tenenbaum (2004).

The simplest method using only the structure of the tree predicts the class of an unlabelled item based on the nearest neighboring items in the tree that have labels. The probability that the unlabelled item has class +1 based on this tree is then estimated as the proportion of these labelled nearest neighbors that have class +1. For the tree in Figure 10.3, for example, this procedure will predict that items B, C, and E have class +1 with probability 1, since in each case there is only a single nearest labelled item, and it is of class +1. Item H, however, has four nearest labelled items, of which two are labelled +1 and two are labelled −1, and is therefore predicted to be of class +1 with probability $2/4 = 0.5$. These probabilities based on a single tree are averaged for all the trees in the sample from the posterior to get the final predictive probabilities.

We can try to make better use of the unlabelled data by classifying unlabelled items in succession, using the newly added labels when classifying later items. At each stage, we find one of the unlabeled items that is closest to a labeled item, as measured by divergence time (though the final result actual

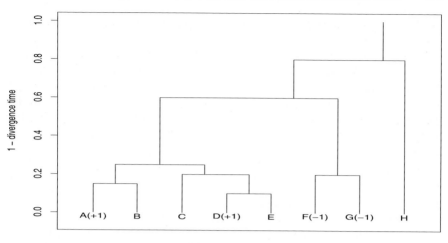

Fig. 10.3. A tree with labelled and unlabelled items. This dendrogram shows a single tree with both labelled (training) and unlabelled (test) items at the leaves. Labels for the unlabelled items can be predicted in various ways, as described in the text.

depends only on the tree structure, not the divergence times), and assign a label to this unlabelled item based on the labeled items nearest it, including items that were assigned labels in previous stages. For the tree in Figure 10.3, we would start with item E, and predict that it is in class +1 with probability 1. Next we will look at items B and C, and again predict that they are in class +1 with probability 1. So far, the results are the same as for the previous method. When we come to item H, however, we will predict that it is in class +1 with probability 5/7, since five of its seven nearest neighbors are now labelled as being in class +1, a result that is different from the previous method's prediction of equal probabilities for class +1 and −1. When some of the nearest neighbors are not classified either way with probability 1, we treat them as contributing the appropriate fraction to the counts for both classes. As before, the final predictive probabilities are obtained by averaging over all trees in the sample from the posterior distribution.

The second class of methods measures the distance between two items by one minus the time at which the paths to the two items diverged. In the tree in Figure 10.3, for example, items B and D have distance of approximately 0.25, and items C and G have distance of about 0.6.

These distances can be averaged over a sample of trees drawn from the posterior distribution to obtain a final distance measure, which can be used to classify the test data by various methods. The simplest is to classify based on the k nearest neighbors according to this distance measure. We also tried using this distance to define a covariance function for a Gaussian distribution of latent values representing the logits of the class probabilities for each item (Neal, 1999).

Alternatively, we can classify unlabelled items by a weighted average in which training items nearer the test item are given more weight. Suppose the divergence time in a particular tree between items j and k is t_{jk} (i.e., their distance in this tree is $1 - t_{jk}$). For an unlabelled test observation i, we let the weight of the labelled observation j be $e^{rt_{ij}} - 1$, which gives the following probability that the unknown label, Y_i, of the test item is $+1$:

$$P(Y_i = 1) \; = \; \sum_{j \,:\, y_j = +1} \left(e^{rt_{ij}} - 1\right) \Big/ \sum_{j} \left(e^{rt_{ij}} - 1\right) \qquad (10.10)$$

Here, j indexes the training observations. The value of r to use for each tree can be chosen by leave-one-out cross validation. The final predictions are made by averaging the class probabilities found using each of the trees drawn from the posterior distribution.

From our preliminary experiments, it is unclear which of the above somewhat *ad hoc* methods is best. Further work is needed to evaluate these methods, and to develop methods that have clearer justifications.

10.7 Methods and Results for the Challenge Data Sets

In this section, we present and discuss the results obtained by our winning challenge submission, BayesNN-DFT-combo, and by two other submissions, BayesNN-small and BayesNN-large, which came second and third (with almost equal overall balanced error rates).

The BayesNN-small and BayesNN-large entries used Bayesian neural network models (see Section 10.5). BayesNN-large used as many features as seemed desirable in terms of predictive performance. BayesNN-small used as small a set of features as seemed possible without seriously affecting predictive performance. In both cases, balanced error rate on the validation set was the primary criterion for estimating performance on the test set. For some datasets, a small set of features was judged best, in which case the BayesNN-small and BayesNN-large methods were the same. BayesNN-DFT-combo was the same as BayesNN-large except for the Arcene dataset, for which a Dirichlet diffusion tree method was used, and the Madelon dataset, for which predictions from Dirichlet diffusion tree and neural network methods were averaged. Two entries were made for each method — one in which only the training data was used, and one in which both training and validation data was used. The second submission for each method was the same as the first, except for the use of additional data, and sometimes a different length MCMC run.

In these original challenge submissions, various *ad hoc* methods were employed, which seemed in our judgement to be desirable in light of the specifics of the datasets, but which make it difficult to say exactly what was responsible for the good performance. For some of the original challenge submissions, the Markov chain Monte Carlo runs were allowed to continue for several days, which might or might not have been essential, but which seemed advisable when the aim was to achieve the best predictive performance. (For models of the sort we used, it is difficult to tell for sure whether an MCMC run has truly converged.) To better understand the methods involved, we have therefore included results obtained using simplified versions of the logistic regression and neural network methods, implemented using scripts that are available on the web, at `www.cs.utoronto.ca/~radford/`. There are six simplified methods, which use various feature selection or dimensionality reduction methods, and either linear logistic regression or neural network models.

Three of these simplified methods use the first 40 principal components rather than individual features (see Section 10.3). These principal components were based on the subset of features that were non-zero in at least four of the training cases. The New-Bayes-lrg-pc method applies linear logistic regression with a spherically-symmetric Gaussian prior to the values of the first 40 principal components. As discussed in Section 10.4, this may be similar to applying logistic regression with a spherically-symmetric Gaussian prior to all the original features. The New-Bayes-lr-pc method instead uses a prior in which (conditional on a single high-level width hyperparameter) the regression coefficients for the 40 principal components are independent and have a t distribution with two degrees of freedom as their prior. This heavy-tailed prior allows some of the principal components to have much more influence than others. The New-Bayes-nn-pc method applies a neural network model with two hidden layers, with 20 and 8 hidden units, to the 40 principal components, using an ARD prior that again allows some of the principal components to have more influence than others.

The other three simplified methods use a subset of the features. A subset was first selected based on the results of the univariate significance tests discussed in Section 10.2, with the number of features chosen by hand, based on what appeared to be adequate. The New-Bayes-lr-sel method uses these features in a linear logistic regression model with a heavy-tailed prior that allows some features to be much more important than others. The New-Bayes-nn-sel method uses these features in a neural network model with two hidden layers, with 20 and 8 hidden units, with an ARD prior that allows some features to have more influence. A smaller subset of features was then found by looking at the posterior distribution of the relevance of each feature in the Bayes-new-lr-sel and Bayes-new-nn-sel models, with the number of features to retain again being chosen by hand. This smaller subset was used in the New-Bayes-nn-red method, which was otherwise the same as New-Bayes-nn-sel.

For each dataset, the method from among these six that had the smallest validation error when fit using only the training data was selected as the method to use for that dataset in the New-Bayes-large method. The best from among New-Bayes-lr-sel, New-Bayes-nn-sel, and New-Bayes-nn-red was selected as the method to use for the New-Bayes-small method.

As for the original challenge entries, two submissions were made for each method, one using only the training data, and one using both training and validation data. The submissions were otherwise identical. In particular, feature subsets were chosen once, based solely on the training data (apart from some final model selection based on validation error). The same subset was then used for the second submission, without redoing the feature selection using the additional data in the validation set.

The neural network methods used in the original challenge submissions (the BayesNN methods) were generally similar to the corresponding New-Bayes methods, but differed in many details, some specific to particular datasets. The BayesNN methods based on principal components used the first 50 components. The use of only 40 components for the New-Bayes methods was aimed at reducing the computational burden for people reproducing the results. Similarly, the first hidden layer for the BayesNN methods usually had 25 units, rather than the 20 units for the New-Bayes methods. For BayesNN, features were often transformed in ways that increased their correlation with the class before being used, but this was not done for New-Bayes, for the sake of simplicity. Many of these differences might be expected to result in the New-Bayes methods having somewhat worse performance than the original submissions, though some refinements and possibly-different human judgements could produce the reverse effect instead.

The remainder of this section details how these methods were applied to each of the five challenge datasets, and discusses the resulting performance. Tables 1 through 5 below summarize the performance of each method in terms of "balanced error rate" (which averages the error rate for both classes) when trained using the training data alone, and when using the training plus the validation data. These tables also show the number of features used by each method, and the performance of the best of the original challenge entries (the best is sometimes different for the two training scenarios). Note that although the source for each dataset is briefly described below, this information was not available at the time of the original BayesNN submissions, and was deliberately not used when designing the New-Bayes methods.

10.7.1 The Arcene Dataset

The Arcene dataset is derived from mass spectrograms of blood serum used as a way of distinguishing people with cancer (class $+1$) from people without cancer (class -1). The data comes from three sources, two for prostate cancer and one for ovarian cancer. There were 10000 features. The feature values were thresholded so that around half of the values were zero. Approximately 44% of the cases were positive. Accordingly, in order to minimize the posterior expected loss under the balanced error rate criterion, test cases were predicted to be in class $+1$ if the probability of class $+1$ was greater than 0.44.

For the New-Bayes methods, features that are non-zero in fewer than four training cases were eliminated from consideration. The first 40 principal com-

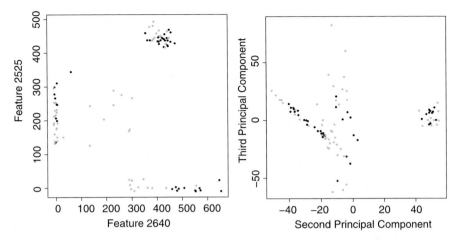

Fig. 10.4. Scatterplots of Arcene training data. The plot on the left shows two informative features, with values jittered by ±8 to reduce overlap. The plot on the right shows two informative principal components. Black dots indicate items in class +1, gray dots items in class −1.

ponents were found after centering and scaling the remaining 9656 features, and used for New-Bayes-lrg-pc, New-Bayes-lr-pc, and New-Bayes-nn-pc. A large number of the features were relevant according to one or more of the univariate significance tests of Section 10.2. Using fairly stringent criteria of p-values less than 0.0008, 0.002, and 0.002 for the permutation test on correlation with ranks, the hypergeometric test, and the runs test, a subset of 484 features was chosen for use in New-Bayes-lr-sel and New-Bayes-nn-sel. A smaller subset of 100 features was chosen based on the ARD hyperparameter values in the MCMC run for New-Bayes-nn-sel. This smaller subset was used by New-Bayes-nn-red.

As can be seen in Figure 10.4, although both individual features and the principal component values contain some information about the class, the relationship is sometimes rather complex. We would therefore expect non-linear models such as neural networks to do better than linear logistic regression. This is confirmed by the results in Table 10.1. The best performance on the test cases was achieved by New-Bayes-nn-red, but since this method had the worst error rate on the validation set, it was not chosen for New-Bayes-small or New-bayes-large, which instead used the methods of New-Bayes-nn-sel and New-Bayes-nn-pc. Error on the small validation set (100 cases) was clearly not a very good guide for model selection.

The time required (on a 3GHz Pentium 4) for the New-Bayes MCMC runs (on the training data only) ranged from 2 minutes for New-Bayes-lrg-pc to 7.5 hours for New-Bayes-nn-sel.

The original BayesNN-large and BayesNN-small submissions were similar to New-Bayes-large and New-Bayes-small. One difference is that BayesNN-

Table 10.1. Arcene results (10000 features, 100 train, 100 validation, 700 test).

Method	Number of Features	Training Data Only Valid. Error	Test Error	Train+Valid. Test Error
New-Bayes-lrg-pc	9656	.1372	.1954	.1687
New-Bayes-lr-pc	9656	.1461	.1979	.1725
New-Bayes-lr-sel	484	.1891	.1877	.1424
New-Bayes-nn-pc/large	9656	.1282	.1699	.1469
New-Bayes-nn-sel/small	484	.1778	.1607	.1218
New-Bayes-nn-red	100	.2029	.1599	.0720
BayesNN-small	1070	.1575	.1659	.1186
BayesNN-large	9656	.1396	.1919	.1588
BayesNN-DFT-combo	10000	.0722	.1330	.1225
Best Original Entry				
BayesNN-DFT-combo	10000	.0722	.1330	
final2-2 (Yi-Wei Chen)	10000			.1073

large looked at 50 principal components, found after a power transformation of each feature (with power of 1/2, 1, or 2) chosen to maximize correlation with the class. Based on validation error rates, such transformations were not used for BayesNN-small.

For the winning BayesNN-DFT-combo entry, we decided on the basis of validation error rates (which in this instance proved to be a good guide) not to use one of the neural network methods for this dataset, but instead to use Dirichlet diffusion tree methods, as described in Section 10.6.

The Dirichlet diffusion tree models were applied to principal component values for all the cases (training, validation, and test), producing a sample of trees drawn from the posterior distribution. Feature values were always centred before principal components were taken. We tried both rescaling each feature to have standard deviation of one before taking principal components, and leaving the features with their original scales. We also tried rescaling the resulting principal component values to have standard deviation one, and leaving them with their original scales. The combinations of these choices produce four possible methods.

For all four methods, we used a prior in which each feature had hyperparameters for the diffusion standard deviation and the noise standard deviation, which allows the model to decide that some features are mostly "noise", and should not influence the clustering. The models we used were based on the first 100 principal components, because we found that when models with more principal components were fitted, most components beyond these first 100 had noise standard deviations that were greater than their diffusion standard deviation (at least when features were not rescaled).

We found that the various classification methods described in Section 10.6 produced quite similar results. In our challenge entry, we used the method of classifying unlabelled items in succession, using the newly added labels when classifying later items. We found that the validation error rates when using rescaled features were higher than when features were not rescaled. However, we decided nevertheless to base our predictions on the average probabilities from all four methods (i.e., rescaling features or not, and rescaling principal components or not). This produced smaller validation error than any of the individual methods.

Figure 10.5 shows one tree from the posterior distribution of a Dirichlet diffusion model (with only training cases shown). The early split into three subtrees is evident, which likely corresponds to the three sources of the data that were merged to produced this dataset. The labels for the leaves of the tree are clearly related to the clustering, providing a good basis for classification.

As seen in Table 10.1, the BayesNN-DFT-combo entry using Dirichlet diffusion trees had the best performance of any entry when using the training data only. However, when both training and validation data was used, both the original BayesNN-small method and the New-Bayes-small method did better than BayesNN-DFT-combo, and New-Bayes-nn-red did even better. A possible explanation for this is that the Dirichlet diffusion tree method is very good at accommodating the three very distinct clusters underlying this dataset, allowing it to obtain good results even with only the 100 training cases. Once a total of 200 cases (training plus validation) are used, however, the neural network models seem to be more capable.

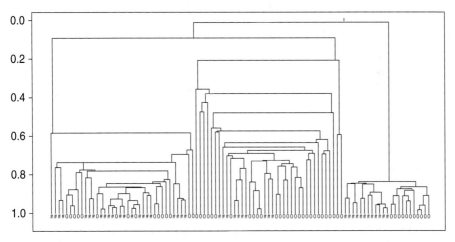

Fig. 10.5. A dendrogram of a tree drawn from the posterior distribution of a Dirichlet diffusion tree model for the Arcene data (based on un-rescaled principal components found from un-rescaled features). Only training cases are shown here, though the tree was constructed based on all the cases. Class +1 is labelled with "#"; class −1 with "0". The vertical axis shows divergence time in the tree.

10.7.2 The Gisette Dataset

The Gisette dataset is derived from images of handwritten examples of the digits "4" and "9". The digits were scaled to a uniform size and centred in a 28-pixel by 28-pixel field. Gisette has 5000 features, which include the pixel values, thresholded to produce a substantial fraction of zeros, many products of pairs of values for randomly-selected pixels, and many "probe" features unrelated to the class. Note that both thresholding and the introduction of random products are probably undesirable when using a flexible model such as a neural network. There are equal numbers of cases in the two classes.

For the New-Bayes methods, a few features were eliminated from consideration because they were non-zero in fewer than four training cases. The first 40 principal components of the remaining 4724 features were found (with scaling and centering) using only the training data (to save time and memory), and used for New-Bayes-lrg-pc, New-Bayes-lr-pc, and New-Bayes-nn-pc. According to the univariate significance tests of Section 10.2, a large number of features were related to the class — e.g., for 2021 features, the p-value for the hypergeometric significance test applied to the binary form of the feature (zero/non-zero) is less than 0.05. A subset of features of size 1135 with the smallest p-values was chosen based on the need to reduce the number of features for computational reasons, and used for New-Bayes-lr-sel and New-Bayes-nn-sel. A smaller subset of 326 features was chosen based on the ARD hyperparameter values in the MCMC run for New-Bayes-nn-sel, and on the magnitude of the regression coefficients in the MCMC run for New-Bayes-lr-sel. This smaller subset was used for New-Bayes-nn-red. Based on results on the validation set the New-Bayes-nn-pc method was used for New-Bayes-large, and the New-Bayes-nn-sel method was used for New-Bayes-small.

As can be seen in Table 10.2, the best performance was achieved by New-Bayes-nn-pc. Figure 10.6 shows how some of the principal components used by this method relate to the class. The advantage that a non-linear neural network model will have over linear logistic regression can be seen in the left plot of this figure.

The time required (on a 3GHz Pentium 4) for the New-Bayes MCMC runs (on the training data only) ranged from 3 hours for New-Bayes-lrg-pc to 25 hours for New-Bayes-nn-sel.

The best original challenge submission was BayesNN-large (also used for BayesNN-DFT-combo), which is similar in both method and performance to New-Bayes-nn-pc.

10.7.3 The Dexter Dataset

The Dexter dataset is derived from data on the frequency of words in documents, with class +1 consisting of documents about corporate acquisitions, and class −1 consisting of documents about other topics. There are equal

Table 10.2. Gisette results (5000 features, 6000 train, 1000 validation, 6500 test).

Method	Number of Features	Training Data Only Valid. Error	Training Data Only Test Error	Train+Valid. Test Error
New-Bayes-lrg-pc	4724	.0300	.0295	.0298
New-Bayes-lr-pc	4724	.0320	.0295	.0295
New-Bayes-lr-sel	1135	.0190	.0186	.0175
New-Bayes-nn-pc/large	4724	.0170	.0134	.0125
New-Bayes-nn-sel/small	1135	.0190	.0188	.0168
New-Bayes-nn-red	326	.0200	.0203	.0186
BayesNN-small	379	.0250	.0203	.0209
BayesNN-large/combo	5000	.0160	.0129	.0126
Best Original Entry BayesNN-large/combo	5000	.0160	.0129	.0126

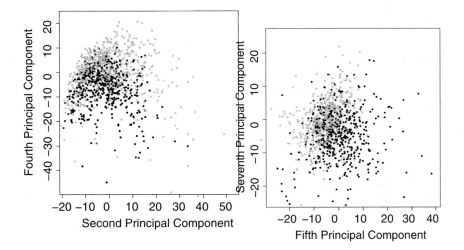

Fig. 10.6. Scatterplots of principal components from Gisette training data. Only 1000 randomly chosen training cases are shown, with the classes distinguished by black and gray dots.

numbers of cases in the two classes. The 20000 features for Dexter are mostly zero; the non-zero values are integers from 1 to 999.

The New-Bayes methods looked only at the 1458 features that are non-zero in four or more of the 300 training cases. Only these features were used when finding the first 40 principal components (without centering or scaling, using all the data), and when selecting a smaller subset of features. Many of the individual features are obviously informative, as are a number of the principal components, as can be seen in Figure 10.7.

The New-Bayes-lr-sel and New-Bayes-nn-sel methods looked at the subset of 298 features for which the p-value from at least one of three univariate significance tests was less than 0.05, a threshold chosen since it produced

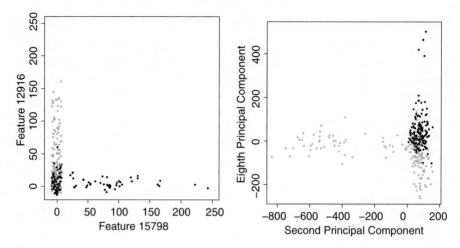

Fig. 10.7. Scatterplots of Dexter training data. The plot on the left shows two informative features, with values jittered by ±8 to reduce overlap. The plot on the right shows two informative principal components. Black dots indicate items in class +1, gray dots items in class −1.

what seemed to be a reasonable number of features. The three tests applied were a permutation test based on the correlation of the class with the ranked data, a hypergeometric significance test with the features converted to binary form according to whether they were zero or non-zero, and a runs test sensitive to non-monotic relationships. Given knowledge of the source of this data (not available at the time of the original challenge), the runs test would seem inappropriate, since it is unlikely that the frequencies of any words would be non-monotonically related to the class.

A smaller subset of 100 features was selected by looking at the relevance hyperparameters from the MCMC run done for the New-Bayes-nn-sel method. Looking instead at the magnitudes of weights from the MCMC run for New-Bayes-lr-sel would have produced mostly the same features. The somewhat arbitrary decision to choose 100 features was motivated simply by a desire to reduce the number substantially from 298, without drastically affecting predictive performance. The New-Bayes-nn-red method using this reduced feature set had the lowest validation error of the six New-Bayes methods, and was therefore chosen as the method to use for both New-Bayes-small and New-Bayes-large.

As can be seen from Table 10.3, the performance of the New-Bayes methods that looked at the 298 selected features was better than that of the methods that used principal components, especially when only the 300 training cases were used. The best performance from among the New-Bayes methods was obtained by New-Bayes-lr-sel. The difference in error rate between New-Bayes-lr-sel and the next best method, New-Bayes-nn-sel, is statistically significant for the results using training data only (where the methods pro-

Table 10.3. Dexter results (20000 features, 300 train, 300 validation, 2000 test).

Method	Number of Features	Training Data Only Valid. Error	Test Error	Train+Valid. Test Error
New-Bayes-lrg-pc	1458	.0800	.0770	.0505
New-Bayes-lr-pc	1458	.0900	.0705	.0475
New-Bayes-lr-sel	298	.0767	.0360	.0455
New-Bayes-nn-pc	1458	.0900	.0740	.0505
New-Bayes-nn-sel	298	.0800	.0510	.0505
New-Bayes-nn-red/small/large	100	.0733	.0550	.0555
BayesNN-small/large/combo	303	.0533	.0390	.0405
Best Original Entry				
BayesNN-small/large/combo	303	.0533	.0390	
FS+SVM (Navin Lal)	3714			.0330

Note: The best original entry using training data was actually BayesNN-large, but it used the same method with 303 features as BayesNN-small. Due to a glitch, the result submitted for BayesNN-large was slightly different (and slightly better) than that for BayesNN-small. I have listed BayesNN-small as best since it essentially was.

duced different predictions for 52 test cases), but not for the results using training and validation data (where the methods differed for 40 test cases). The New-Bayes-nn-red method, which had the best validation error rate, was worse than New-Bayes-nn-sel, though the difference is not statistically significant.

The New-Bayes-nn-red method does achieve reasonably good performance using a quite small set of features. Excluding some preliminary BayesNN submissions that used methods similar to New-Bayes-nn-red, the best method in the original challenge that used less than 150 features was CBAMethod3E, which used 119 features and achieved an error rate of 0.0600 using both training and validation data.

One puzzling aspect of the results is that, except for the methods using principal components, using both training and validation data did not help performance, and indeed seemed sometimes to have hurt it. (The biggest drop in performance, for New-Bayes-lr-sel, results from differences in predictions for 63 test cases, giving a p-value of 0.023.)

The time required (on a 3GHz Pentium 4) for the New-Bayes MCMC runs (on the training data only) ranged from 10 minutes for New-Bayes-lrg to 14 hours for New-Bayes-nn-sel.

In the original BayesNN-small method (also used for BayesNN-large and BayesNN-DFT-combo), various power transformations of the features were considered, with the result that most features were transformed by taking the square root, as this improved the correlation with the class. This may explain why the performance of BayesNN-small is better than that of New-Bayes-small. With knowledge of the data source, a transformation such as square root seems natural, since it leads to more emphasis on whether or not a word is used at all, with less emphasis on exactly how many times it is used.

10.7.4 The Dorothea Dataset

The Dorothea dataset was derived from data on whether or not small organic molecules bind to thrombin. Class +1 consists of molecules that do bind, and class −1 of molecules that do not bind. The 100000 features are binary indicators of the presence or absence of various features in the molecule's three-dimensional structure. The features are very sparse, with less than 1% of the values being one. Only a few (9.75%) of the molecules are in class +1. Accordingly, to minimize the balanced error rate, test cases were predicted to be in class +1 if the probability of class +1 was greater than 0.0975.

The New-Bayes methods looked only at the 55230 features that are non-zero in four or more of the 800 training cases. These features were used when finding the first 40 principal components (without centering or scaling, using all the data), and when selecting a smaller subset of 683 features (using hypergeometric significance tests with p-value threshold of 0.001). This subset of 683 features was used for New-Bayes-lr-sel and New-Bayes-nn-sel, and a smaller subset of 72 features was then found using the hyperparameter values in the MCMC runs for these methods. On the basis of the balanced error rates on the validation cases, the New-Bayes-nn-pc method was chosen for New-Bayes-large, and the New-Bayes-nn-sel method was chosen for New-Bayes-small.

As seen in Table 10.4, the New-Bayes-nn-pc and New-Bayes-nn-sel methods did better than the other New-Bayes methods (with little difference between them). The smaller subset used by New-Bayes-nn-red appears to have been too small (or not chosen well). New-Bayes-lr-pc and New-Bayes-lr-sel did almost as well as the neural network methods, showing that this dataset exhibits only a slight (though not negligible) degree of non-linearity, as can be seen also from Figure 10.8.

As was the case with Dexter, there is a puzzling tendency for the methods to perform slightly worse rather than better when using both training and validation data.

The time required (on a 3GHz Pentium 4) for the New-Bayes MCMC runs (on the training data only) ranged from 26 minutes for New-Bayes-lrg-pc to 21 hours for New-Bayes-nn-sel.

The BayesNN-small method from the original challenge was similer to New-Bayes-nn-sel. However, BayesNN-large, whose performance is better than that of any other method, differed in several ways from New-Bayes-large. Like New-Bayes-large, it was based on a neural network with two hidden layers. Like New-Bayes-large, it also looked at principal components, but differed in using the first 50, found from all features that are non-zero in two or more cases (training, validation, or test). In addition, BayesNN-large looked at 60 of the original features, selected on the basis of ARD hyperparameters from other neural network runs. Finally, the neural network for BayesNN-large had three additional inputs, equal to the logs of one plus the numbers of singleton, doubleton, and tripleton features that are non-zero in each case. A singleton feature is one that is non-zero in exactly one of the cases (training,

Table 10.4. Dorothea results (100000 features, 800 train, 350 validation, 800 test).

Method	Number of Features	Training Data Only Valid. Error	Test Error	Train+Valid. Test Error
New-Bayes-lrg-pc	55230	.1330	.1396	.1367
New-Bayes-lr-pc	55230	.1314	.1176	.1226
New-Bayes-lr-sel	683	.1321	.1126	.1193
New-Bayes-nn-pc/large	55230	.1131	.1166	.1046
New-Bayes-nn-sel/small	683	.1253	.1110	.1121
New-Bayes-nn-red	72	.1830	.2053	.1956
BayesNN-small	500	.1110	.1063	.1107
BayesNN-large/combo	100000	.0547	.0854	.0861
Best Original Entry BayesNN-large/combo	100000	.0547	.0854	.0861

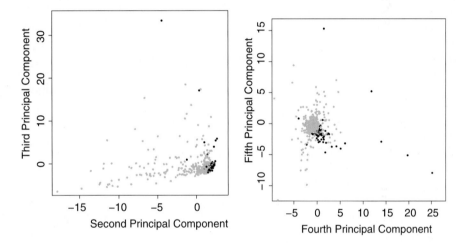

Fig. 10.8. Scatterplots of Dorothea training data, showing four informative principal components. Black dots indicate items in class +1, gray dots items in class −1.

validation, or test). There are 10896 singleton features. A doubleton feature is one that is non-zero in exactly two of the cases. There are 7199 doubleton features. Similarly, there are 5661 tripleton features. There are 12 training cases in which one or more of the singleton features is non-zero. (In fact, in all these cases, at least 36 singleton features are non-zero.) Every one of these 12 training cases is in class +1. The presense of singleton features therefore appears to be highly informative, though the reason for this is obscure, even with some knowledge of the source of the data. It is likely that BayesNN-large performs better than New-Bayes-large principally because it was provided with this information.

10.7.5 The Madelon Dataset

The Madelon dataset was artificially generated by placing 16 clusters of points from class +1 and 16 clusters of points from class −1 on the vertices of a five-dimensional hypercube, and then using various transformations of the five coordinates of these points as features (along with some additional useless features). Extra noise was added to features as well, and a few (1%) of the class labels were flipped at random. There were equal numbers of cases in the two classes. The 500 features had values ranging from 0 to 999, but 99% of the feature values were between 400 and 600.

With 500 features and 2000 training cases, the Madelon dataset is just small enough that it is feasible to use neural network models that look at all features. For uniformity with the other datasets, however, the New-Bayes-lr-sel and New-Bayes-nn-sel methods looked at a subset of features chosen according to univariate significance tests. The New-Bayes-nn-red method uses a smaller set of features selected based on the ARD hyperparameters from the MCMC runs for New-Bayes-nn-sel along with those for an MCMC run of a neural network model that looked at all the features. Methods looking at principal components (found with centering and scaling) were also run.

As can be seen in Figure 10.9, some of the features, as well as some of the principal components, are highly informative, but their relationship to the class is not linear. This explains the poor performance of methods using linear logistic regression (see Table 10.5). The smallest validation error rate was achieved by New-Bayes-nn-red, which was therefore chosen for both New-Bayes-small and New-Bayes-large. New-Bayes-nn-red did prove the best of the New-Bayes methods when using only the training data, and was close

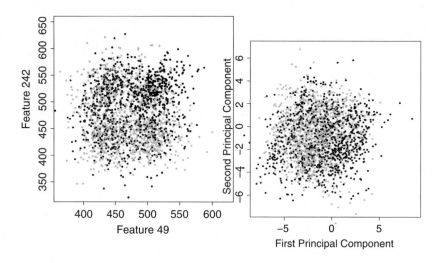

Fig. 10.9. Scatterplots of Madelon training data. The plot on the left shows two informative features; that on the right shows two informative principal components. Black dots indicate items in class +1, gray dots items in class −1.

Table 10.5. Madelon results (500 features, 2000 train, 600 validation, 1800 test).

Method	Number of Features	Training Data Only Valid. Error	Test Error	Train+Valid. Test Error
New-Bayes-lrg-pc	500	.4117	.3994	.3944
New-Bayes-lr-pc	500	.4100	.3922	.3961
New-Bayes-lr-sel	38	.4233	.4133	.3983
New-Bayes-nn-pc	500	.1333	.1239	.1300
New-Bayes-nn-sel	38	.1383	.1144	.0950
New-Bayes-nn-red/small/large	21	.0917	.0883	.0994
BayesNN-small/large	17	.0733	.0772	.0656
BayesNN-DFT-combo	500	.0677	.0717	.0622
Best Original Entry				
final1 (Yi-Wei Chen)	24	.0750	.0661	
BayesNN-DFT-combo	500			.0622

to the best when using both training and validation data, despite a puzzling worsening in performance with the addition of the validation data (perhaps explainable by poor MCMC convergence).

The computer time (on a 3GHz Pentium 4) for the MCMC runs done for the New-Bayes methods (on the training data only) ranged from one hour for New-Bayes-lrg-sel to 36 hours for the MCMC run with the neural network model looking at all 500 features.

The original BayesNN-small method (which was also used for BayesNN-large) was similar to New-Bayes-nn-red. However, ARD hyperparameters from a larger set of MCMC runs were used, including one run of a Gaussian process classification model (Neal, 1999) using all 500 features. This produced better results, perhaps because of a better selection of features.

For the winning BayesNN-DFT-combo entry, we averaged the predictive probabilities from BayesNN-small and from a Dirichlet diffusion tree model, since the validation error rates for these two methods were quite close. The Dirichlet diffusion trees were fit to the first six principal components, since preliminary runs indicated that only the first five principal components were important (one more was kept out of a sense of caution). Principal components were found after the features were centred, but not rescaled, since rescaling produced a larger validation error rate. Classification was done by successively assigning unlabelled items to classes, using the newly added class labels when classifying later items.

The performance of this BayesNN-DFT-combo entry was the best of any method when using both training and validation data. When using only training data, it was close to the best.

10.8 Conclusions

We have shown that Bayesian neural network models can perform very well on high-dimensional classification problems with up to 100000 features by first reducing dimensionality to no more than about a thousand features, and then applying ARD priors that can automatically determine the relevance of the remaining features. For some problems, reducing dimensionality using PCA (applied to most or all features) worked best. For others, selecting a subset of features using univariate significance tests worked better. A further reduction in the number of features based on the values of the ARD hyperparameters sometimes improved performance further, or at least did not reduce performance much. Classification using Dirichlet diffusion trees also performed very well on some problems, and is a promising area for further research.

A further advantage of Bayesian methods not discussed earlier is that they produce good indications of uncertainty. This is evidenced by the very good performance of the best of our methods in terms of the area under the ROC curve (AUC), which was often substantially greater than that of other methods, even when these methods had a similar balanced error rate.

From a Bayesian standpoint, the methods we used are a compromise, in which computational cost forces the use of somewhat *ad hoc* methods for dimensionality reduction before a Bayesian model is applied. With improvements in computer hardware and in algorithms for Bayesian computation, we may hope to reduce the extent to which dimensionality needs to be reduced for computational reasons. Problems in which the number of features is too great for them all to be used will probably always exist, however. Accepting this, we might nevertheless hope to avoid the optimistic bias that can arise when we select a subset of features and then use the same data used for selection when finding the posterior distribution for a model looking only at this subset of features. This is a topic on which we are currently working.

Another area in which further research is needed is selection of a model (or models) from among those that have been tried. For the New-Bayes methods (and mostly for the original challenge submissions as well), we made this choice based on performance on the validation set, but this did not always work well (especially for Arcene). According to Bayesian theory, models should be compared on the basis of their "marginal likelihood" (also known as the "evidence"), but this approach also has problems, notably a great sensitivity to prior specifications, and exceptional computational difficulties.

A final lesson can be learned from the fact that, for several of the datasets, our original challenge submissions performed better than the similar but more systematic methods used for the New-Bayes methods. Even when no knowledge of the nature of the datasets was available, exploratory data analysis (e.g., looking at plots) was able to suggest various modifications specific to each dataset, which do appear to have been mostly beneficial. The diversity of possible classification methods is such that no automatic procedure is yet able to completely replace human judgement. We expect this to be even more true when prior information regarding the nature of the dataset is available.

Acknowledgements

This research was supported by the Natural Sciences and Engineering Research Council of Canada. The first author holds a Canada Research Chair in Statistics and Machine Learning.

References

Bernardo, J. M. and Smith, A. F. M. (1994) *Bayesian Theory*, Wiley.

Cox, D. R. and Hinkley, D. V. (1974) *Theoretical Statistics*, Chapman and Hall/CRC.

Gibbons, J. D. (1985) *Nonparametric Statistical Inference*, Marcel Dekker.

Golub, T. R., Slonim, D. K., Tamahyo, P., Huard, C., Gaasenbeek, M., Mesirov, J. P., Coller, H., Loh, M. L., Downing, J. R., Caligiuri, M. A., Bloomfield, C. D., Lander, E. S. (1999), "Molecular classification of cancer: class discovery and class prediction by gene expression monitoring", *Science*, vol. 286, pp. 531–537.

Kemp, C., Griffiths, T. L. Stromsten, S., and Tenenbaum, J. B. (2004) "Semisupervised learning with trees", in S. Thrun, *et al.* (editors) *Advances in Neural Information Processing Systems 16*.

Lampinen, J. and Vehtari, A. (2001) "Bayesian approach for neural networks — review and case studies", *Neural Networks*, vol. 14, pp. 257-274.

Liu, J. S. (2001) *Monte Carlo Strategies in Scientific Computing*, Springer-Verlag.

MacKay, D. J. C. (1994) "Bayesian non-linear modeling for the energy prediction competition", *ASHRAE Transactions*, vol. 100, pt. 2, pp. 1053-1062.

Neal, R. M. (1993) *Probabilistic Inference Using Markov Chain Monte Carlo Methods*, Technical Report CRG-TR-93-1, Department of Computer Science, University of Toronto, 144 pages. Available from `http://www.cs.utoronto.ca/~radford/`.

Neal, R. M. (1996) *Bayesian Learning for Neural Networks*, Lecture Notes in Statistics No. 118, Springer-Verlag.

Neal, R. M. (1998) "Assessing relevance determination methods using DELVE", in C. M. Bishop (editor) *Neural Networks and Machine Learning*, pp. 97-129, Springer-Verlag.

Neal, R. M. (1999) "Regression and classification using Gaussian process priors" (with discussion), in J. M. Bernardo, *et al* (editors) *Bayesian Statistics 6*, Oxford University Press, pp. 475-501.

Neal, R. M. (2001) "Defining priors for distributions using Dirichlet diffusion trees", Technical Report No. 0104, Dept. of Statistics, University of Toronto, 25 pages.

Neal, R. M. (2003) "Density modeling and clustering using Dirichlet diffusion trees", in J. M. Bernardo, *et al.* (editors) *Bayesian Statistics 7*, pp. 619-629, Oxford University Press.

Rasmussen, C. E. (1996) *Evaluation of Gaussian Processes and Other Methods for Non-linear Regression*, PhD Thesis, Dept. of Computer Science, University of Toronto.

Rumelhart, D. E., Hinton, G. E., and Williams, R. J. (1986) "Learning internal representations by error propagation", in D. E. Rumelhart and J. L. McClelland (editors) *Parallel Distributed Processing: Explorations in the Microstructure of Cognition, Volume 1: Foundations*, Cambridge, Massachusetts: MIT Press.

Chapter 11

Ensembles of Regularized Least Squares Classifiers for High-Dimensional Problems

Kari Torkkola[1] and Eugene Tuv[2]

[1] Motorola, Intelligent Systems Lab, Tempe, AZ, USA,
`Kari.Torkkola@motorola.com`
[2] Intel, Analysis and Control Technology, Chandler, AZ, USA,
`eugene.tuv@intel.com`

Summary. It has been recently pointed out that the Regularized Least Squares Classifier (RLSC), continues to be a viable option for binary classification problems. We apply RLSC to the datasets of the NIPS 2003 Feature Selection Challenge using Gaussian kernels. Since RLSC is sensitive to noise variables, ensemble-based variable filtering is applied first. RLSC makes use of the best-ranked variables only. We compare the performance of a stochastic ensemble of RLSCs to a single best RLSC. Our results indicate that in terms of classification error rate the two are similar on the challenge data. However, especially with large data sets, ensembles could provide other advantages that we list.

11.1 Introduction

Regularized least-squares regression and classification dates back to the work of Tikhonov and Arsenin (1977), and to ridge regression (Hoerl and Kennard, 1970). It has been re-advocated and revived recently by Poggio and Smale (2003), Cucker and Smale (2001), Rifkin (2002), Poggio et al. (2002). The Regularized Least Squares Classifier (RLSC) is an old combination of a quadratic loss function combined with regularization in reproducing kernel Hilbert space, leading to a solution of a simple linear system. In many cases in the work cited above, this simple RLSC appears to equal or exceed the performance of modern developments in machine learning such as support vector machines (SVM) (Boser et al., 1992, Schölkopf and Smola, 2002).

This simplicity of the RLSC approach is a major thread in this paper. We verify the above mentioned findings using the NIPS 2003 Feature Selection Challenge data sets which all define binary classification problems.

The combination of RLSC with Gaussian kernels and the usual choice of spherical covariances gives an equal weight to every component of the feature vector. This poses a problem if a large proportion of the features consists of noise. With the datasets of the challenge this is exactly the case. In order to succeed in these circumstances, noise variables need to be removed or weighted

K. Torkkola and E. Tuv: *Ensembles of Regularized Least Squares Classifiers for High-Dimensional Problems*, StudFuzz **207**, 297–313 (2006)
`www.springerlink.com`

down. We apply *ensemble-based variable filtering* to remove noise variables. A Random Forest (RF) is trained for the classification task, and an importance measure for each variable is derived from the forest (Breiman, 2001). Only the highest ranking variables are then passed to RLSC. We chose Random Forests (RF) for this task for several reasons. RF can handle huge numbers of variables easily. A global relative variable importance measure can be derived as a by-product from the Gini-index used in the forest construction with no extra computation involved. In order to compare RF variable selection and RF classification to our procedure of RF variable selection and RLSC classification, see the results of Ng and Breiman (2005).

In this paper we study empirically how a stochastic ensemble of RLSCs with minimum regularization compares to a single optimized RLSC. Our motivation to do this is the well known fact that ensembles of simple weak learners are known to produce stable models that often significantly outperform an optimally tuned single base learner (Breiman, 1996, Freund and Schapire, 1995, Breiman, 2001). A further advantage of ensembles is the possibility of parallelization. Using much smaller sample sizes to train each expert of an ensemble could be faster than training a single learner using a huge data set.

For an ensemble to be effective, the individual experts need to have low bias and the errors they make should be uncorrelated (Bousquet and Elisseeff, 2000, Breiman, 2001). Using minimum regularization with RLSC reduces the bias of the learner making it a good candidate for ensemble methods. Diversity of the learners can be accomplished by training base learners using independent random samples of the training data.

The structure of this paper is as follows. We begin by briefly describing the RLSC, theory behind it, and its connections to support vector machines. We discuss ensembles, especially ensembles of RLSCs and the interplay of regularization and bias in ensembles. The scheme for variable filtering using ensembles of trees is discussed next, after which we describe experimentation with the challenge data sets. We discuss our findings regarding ensembles of RLSCs, and conclude by touching upon several possible future directions.

11.2 Regularized Least-Squares Classification (RLSC)

In supervised learning the training data $(\boldsymbol{x}_i, y_i)_{i=1}^{m}$ are used to construct a function $f : X \to Y$ that predicts or generalizes well. To measure goodness of the learned function $f(\boldsymbol{x})$ a loss function $L(f(\boldsymbol{x}), y)$ is needed. Some commonly used loss functions for regression are as follows:

- Square loss or L_2: $L(f(\boldsymbol{x}), y) = (f(\boldsymbol{x}) - y)^2$ (the most common),
- Absolute value, or L_1 loss: $L(f(\boldsymbol{x}), y) = |f(\boldsymbol{x}) - y|$,
- Huber's loss function : $\begin{cases} |y - f(\boldsymbol{x})|^2, & \text{for } |f(\boldsymbol{x}) - y| \leq \delta \\ \delta(|y - f(\boldsymbol{x})| - \delta/2), & \text{otherwise} \end{cases}$

- Vapnik's ϵ-insensitive loss: $L(f(\boldsymbol{x}), y) = (|f(\boldsymbol{x}) - y| - \epsilon)_+$, where $(\cdot)_+ = 0$ if the argument is negative.

Examples of loss functions for binary classification $(-1, +1)$ are

- Misclassification: $L(f(\boldsymbol{x}), y) = I(\text{sign}(f(\boldsymbol{x})) \neq y)$
- Exponential (Adaboost): $L(f(\boldsymbol{x}), y) = \exp(-yf(\boldsymbol{x}))$
- Hinge loss (implicitly introduced by Vapnik) in binary SVM classification: $L(f(\boldsymbol{x}), y) = (1 - yf(\boldsymbol{x}))_+$
- Binomial deviance (logistic regression): $L(f(\boldsymbol{x}), y) = \log(1 + \exp(-2yf(\boldsymbol{x})))$
- Squared error: $L(f(\boldsymbol{x}), y) = (1 - yf(\boldsymbol{x}))^2$

Given a loss function, the goal of learning is to find an approximation function $f(\boldsymbol{x})$ that minimizes the expected risk, or the generalization error

$$E_{P(\boldsymbol{x},y)}L(f(\boldsymbol{x}), y) \qquad (11.1)$$

where $P(\boldsymbol{x}, y)$ is the unknown joint distribution of future observations $(\boldsymbol{x}, y)$.

Given a finite sample from the (X,Y) domain this problem is ill-posed. The regularization approach of Tikhonov and Arsenin (1977) restores well-posedness (existence, uniqueness, and stability) to the empirical loss by restricting the hypothesis space, the functional space of possible solutions:

$$\hat{f} = \underset{f \in H}{\text{argmin}} \, \frac{1}{m} \sum_{i=1}^{m} L(f(\boldsymbol{x}_i), y_i) + \gamma \|f\|_k^2 \qquad (11.2)$$

The hypothesis space H here is a Reproducing Kernel Hilbert Space (RKHS) defined by a kernel k. γ is a positive regularization parameter.

The mathematical foundations for this framework as well as a key algorithm to solve (11.2) are derived elegantly by Poggio and Smale (2003) for the square loss function. The algorithm can be summarized as follows:

(1.) Start with the data $(\boldsymbol{x}_i, y_i)_{i=1}^m$.
(2.) Choose a symmetric , positive definite kernel, such as

$$k(\boldsymbol{x}, \boldsymbol{x}') = e^{-\frac{\|\boldsymbol{x} - \boldsymbol{x}'\|^2}{2\sigma^2}}. \qquad (11.3)$$

(3.) The representer theorem (Schölkopf and Smola, 2002) states that the solution to (11.2) is

$$f(\boldsymbol{x}) = \sum_{i=1}^{m} \alpha_i k(\boldsymbol{x}_i, \boldsymbol{x}). \qquad (11.4)$$

For functions of form (11.4) the norm in (11.2) is $\|f\|_k^2 = \boldsymbol{\alpha}^T K \boldsymbol{\alpha}$, where $K_{ij} = k(\boldsymbol{x}_i, \boldsymbol{x}_j)$. It is now easy to insert the square loss function, the norm, and (11.4) in (11.2), and differentiate w.r.t $\boldsymbol{\alpha}$. This leads to solving

$$(m\gamma I + K)\boldsymbol{\alpha} = \mathbf{y} \qquad (11.5)$$

representing well-posed linear ridge regression (Hoerl and Kennard, 1970).

The generalization ability of this solution, as well choosing the regularization parameter γ were studied by Cucker and Smale (2001, 2003).

Thus, the regularized least-squares algorithm solves a simple well defined linear problem. The solution is a linear kernel expansion of the same form as the one given by support vector machines (SVM.) Note also that SVM formulation naturally fits in the regularization framework (11.2). Inserting the SVM hinge loss function $L(f(\boldsymbol{x}), y) = (1 - yf(\boldsymbol{x}))_+$ to (11.2) leads to solving a quadratic optimization problem instead of a linear solution.

RLSC with the quadratic loss function, that is more common for regression, has also proven to be very effective in binary classification problems (Suykens and Vandervalle, 1999, Rifkin, 2002).

11.3 Model Averaging and Regularization

11.3.1 Stability

Generalization ability of a learned function is closely related to its stability. Stability of the solution could be loosely defined as continuous dependence on the data. A stable solution changes very little for small changes in data. A comprehensive treatment of this connection can be found in (Bousquet and Elisseeff, 2000).

Furthermore, it is well known that bagging (bootstrap aggregation) can dramatically reduce the variance of unstable learners providing some regularization effect (Breiman, 1996). Bagged ensembles do not overfit, and they are limited by the learning power of base learners. Key to the performance is a low bias of the base learner, and low correlation between base learners.

Evgeniou (2000) experimented with ensembles of SVMs. He used a few datasets from the UCI benchmark collection tuning all parameters separately for both a single SVM and for an ensemble of SVMs to achieve the best performance. He found that both perform similarly. He also found that generalization bounds for ensembles are tighter than for a single machine.

Poggio et al. (2002) studied the relationship between stability and bagging. They showed that there is a bagging scheme, where each expert is trained on a disjoint subset of the training data, providing strong stability to ensembles of non-strongly stable experts, and therefore providing the same order of convergence for the generalization error as Tikhonov regularization. Thus, at least asymptotically, bagging strongly stable experts would not improve generalization ability of the individual member, since regularization would provide the same exact effect.

11.3.2 Ensembles of RLSCs

Since the sizes of the challenge datasets are relatively small, we compare simple stochastic aggregation of LSCs to the best individually trained RLSC.

We are looking for diverse low biased experts. For RLSC, bias is controlled by the regularization parameter, and by the σ in case of Gaussian kernel. Instead of bootstrap sampling from training data which imposes a fixed sampling strategy, we found that often much smaller sample sizes improve performance, but in general it is data set specific.

Combining the outputs of the experts in an ensemble can be done in several ways. We performed majority voting over the outputs of the experts. In binary classification this is equivalent to averaging the discretized (+1,-1) predictions of the experts. In our experiments this performed better than averaging the actual numeric expert outputs before applying their decision function (sign).

11.4 Variable Filtering with Tree-Based Ensembles

For most datasets from the challenge we noticed a significant improvement in accuracy when only a small (but important) fraction of the original variables was used in kernel construction.

We used fast exploratory tree-based models for variable filtering. One of many important properties of CART (Breiman et al., 1984) is its embedded ability to select important variables during tree construction using greedy recursive partition, where impurity reduction is maximized at every step. Impurity measures the unevenness of class distribution. If data of a node in a decision tree belongs to a single class it is pure, or, impurity equals zero. If every class is equally probable, impurity is at the maximum. Entropy is obviously one such measure. CART uses the Gini index $I_G(t) = \sum_{q \neq r} \hat{p}_{tq} \hat{p}_{tr}$, where $\hat{p}_{tq}$ is the probability of class q estimated from the data at node t. Variable importance then can be defined as a sum over the tree nodes

$$M(x_i, T) = \sum_{t \in T} \Delta I_G(x_i, t), \qquad (11.6)$$

where $\Delta I_G(x_i, t)$ is the decrease in impurity due to an actual or potential split on variable x_i at a node t of the optimally pruned tree T. The sum in (11.6) is taken over all internal tree nodes where x_i was a primary splitter or a surrogate variable. Consequently, no additional effort is needed for its calculation.

Two recent advances in tree ensembles, Multivariate Adaptive Regression Trees, or MART (Friedman, 1999a,b) and Random Forests, or RF (Breiman, 2001), inherit all nice properties of a single tree, and provide a more reliable estimate of this value, as the importance measure is averaged over the N_T trees in the ensemble

$$M(x_i) = \frac{1}{N_T} \sum_{k=1}^{N_T} M(x_i, T_k). \qquad (11.7)$$

In both cases the trees have the same depth. Thus there is no need to normalize individual tree importances. The maximum of $M(x_i)$ over the variables is scaled to be 100.

MART builds shallow trees using all variables, and hence, can handle large datasets with a moderate number of variables. RF builds maximal trees but chooses a small random subset of variables at every split, and easily handles thousands of variables in datasets of moderate size. For datasets massive in both dimensions a hybrid scheme with shallow trees and dynamic variable selection has been shown to have at least the same accuracy but to be much faster than either MART or RF (Chapter 15).

Note that the index of variable importance defined in the above measures is the global contribution of a variable to the learned model. It is not just a univariate response-predictor relationship.

For the NIPS challenge we used RF to select important variables. A forest was grown using the training data until there was no improvement in the generalization error. Typically, this limit was around 100 trees. As an individual tree is grown, a random sample of the variables is drawn, out of which the best split is chosen, instead of considering all of the variables. The size of this sample was typically $\sqrt{N}$, N being the number of original variables. A fast Intel in-house implementation of the RF was used. Breiman's original Fortran implementation can be used for the same purpose.

11.5 Experiments with Challenge Data Sets

As discussed in the chapter presenting the challenge, the data sets were divided in three parts: the actual training data set, a validation set, and the test set. In the first phase of the challenge, only the labels of the training set were provided. Preliminary submissions were evaluated using the validation data, final submissions using the test data.

In the second phase of the challenge the validation data set labels were also provided for a period of one week. We did not repeat feature selection experiments by pooling labeled validation data together with the training data. We used the same feature sets and classifier structures that were chosen in the first phase, and only retrained those using the combined training and validation data sets. Since this was suboptimal for the second phase of the challenge, we only report experiments that were performed for the first phase.

11.5.1 Variable Selection Experiments

Initial experimentation was performed to determine whether variable selection was necessary at all. We trained ensembles of RLSCs for three smallest data sets, Madelon, Dexter, and Arcene. Results are given in Table 11.1 together with later selection experiments. The last column shows error rates using all variables, and the middle column shows these initial selection experiments.

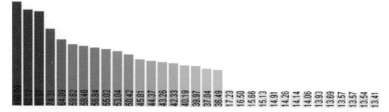

Fig. 11.1. The importance of the top 33 out of 500 variables of Madelon derived from a training set of 2000 cases in 500 trees. Variable importance has a clear cut-off point at 19 variables. Compare to Fig 11.2.

The numbers are averages of ten-fold cross validation (CV), by which we mean dividing the training data set in ten equal sized parts, training using nine parts, testing using the remaining part, repeating by leaving a different part aside, and averaging the ten tests.

These results appearer to indicate that RLSC is sensitive to noise variables in data, and that variable selection based on importances derived from Random Forests improves cross-validation results.

For the rest of the experiments, we adopted the following variable selection procedure. Importances are calculated by a random forest as described in Sec. 11.4. Variable importances $M(x_j)$ are then sorted in descending order, and differences $\Delta M(x_j) = M(x_j) - M(x_{j+1})$ are calculated. Sorting the differences in descending order produces a list of candidates for cut-off points. Figure 11.1 shows a clear example of such a cut-off point. For comparison, and to demonstate the stability of the selection method, Figure 11.2 plots the same importances derived from the validation set, which is smaller. Variables and their order are the same (not shown), only the cut-off point is not as clear as with a larger amount of data.

For each data set, the smallest possible variable set as indicated by a cut-off point was tried first. If the results were unsatisfactory (not competitive with other entrants), variable selection by the next cut-off point was tried, and so on, until satisfactory results were obtained. The maximum number of variables considered was about 500. We display these results in Table 11.1. Full cross-validation was thus not done over the whole possible range of the number of selected variables.

The variable set was thereafter fixed to the one that produced the smallest cross-validation error, with two exceptions: Contrary to other data sets, on Arcene the error rate using the validation set did not follow cross-validation error but was the smallest when all variables were used. Arcene is evidently such a small data set that variable selection and classifier training both using the 100 training samples, will overfit. The second exception is Dexter, which gave the best results using 500 variables ranked by maximum mutual information with the class labels (Torkkola, 2003), (Chapter 6).

Fig. 11.2. Importances derived from the 600 cases in the validation set for the top 33 out of 500 variables of Madelon. The top 19 variables are the same as those derived from the training data set (Figure 11.1), but the cut-off point is not as clear.

Table 11.1. Experimentation with the number of selected variables. Ten-fold cross-valdation errors on training data using a number of best ranked variables. See text for Arcene and Dexter.

Data set	Selection			All variables
	11	**19**	25	500
MADELON	0.1295	**0.1071**	0.1497	0.254
	49	109	432	20000
DEXTER	0.0767	0.0738	0.033	0.324
	13	87	434	10000
ARCENE	0.267	0.0406	0.0430	0.1307
	44	156	**307**	5000
GISETTE	0.0453	0.0247	**0.0218**	-
			284	1000000
DOROTHEA			**0.0280**	-

At this point we also studied preprocessing. In every case, the mean is subtracted first. In addition, we experimented with [1] standardization of individual variables, and [2] weighting variables by their importance. Due to lack of space these experiments are not tabulated, but the decisions are summarized in table 11.2.

Once the variable set and the preprocessing is fixed, the next task is to optimize the classifier parameters.

11.5.2 Parameters of a Single RLSC and an Ensemble of RLSCs

An individual RLSC has two parameters that need to be determined by cross-validation. These are

Table 11.2. Variable selection, standardization, and variable weighting decisions.

Data set	Original variables	Selected variables	Selection method	Stand- ardize?	Weight- ing?
MADELON	500	19	RF	yes	no
DEXTER	20000	500	MMI	yes	by MI
ARCENE	10000	10000	none	no	no
GISETTE	5000	307	RF	no	no
DOROTHEA	100000	284	RF	no	no

(1.) *Kernel parameters.* After a brief experimentation with polynomial ker- nels, we adopted Gaussian kernels. The only parameter is the kernel width σ^2, the range of which was determined as a function of the average squared distance between samples in the training set, d_{av}^2. This was ap- proximated as the average distance between 1000 randomly chosen sam- ple pairs. Since the optimal width depends on the data, the kernel width was varied typically within $d_{av}^2/30 < \sigma^2 < 10 d_{av}^2$ in cross-validation ex- periments.

(2.) *Regularization.* For a single RLSC, regularization is critical in order not to overfit. The choice was again made by cross-validation, and appears to be very data dependent.

A stochastic ensemble of RLSCs has two additional parameters. We discuss each in turn.

(1.) *Kernel parameters.* To make the best possible use of an ensemble, individ- ual classifiers should overfit in order to reduce the bias of the ensemble. Variance of the individual classifiers is averaged out by combining the individual outputs. In general, this leads to the optimum kernel width be- ing narrower than the optimum width for a single RLSC. As with a single RLSC, the optimum width needs to be determined by cross-validation.

(2.) *Regularization.* The quest for low bias also typically leads to regularizing RLSCs minimally in order to guarantee the existence of the solution to equation (11.5) in badly conditioned cases.

(3.) *Ensemble size.* Not too much attention needs to be paid to the size of an ensemble N_T as long as it is reasonably large. If the training data set or the fraction sampled from it is small, determining the weights of an RLSC is computationally so light that the ensemble size can just be kept as a relatively large number, typically $N_T = 200$. Increasing the size did not have a detrimental effect in our experiments.

(4.) *Fraction of training data sampled for individual RLSCs.* A typical bag- ging procedure would sample the training data with replacement (boot- strap sample), that is, the same data point could occur in the sample multiple times (Breiman, 1996). With decision trees these duplicated samples contribute to the statistics of classes in the nodes. The case is different with RLSC. Duplication of the same sample would only lead to a duplication of the same kernel function in the classifier. This does not change the solution and gains nothing for the classifier. Thus, contrary to typical bagging, we sampled the training data without replacement for each RLSC. The optimal fraction f_{td} of training data to sample is mainly a function of the size of the data set, and the nature of the data set. A larger data set generally can do with a smaller fraction of the data sampled for each RLSC. Ensembles with a variable f_{td} were not experimented with.

To determine the parameters for the single best RLSC, we have thus a two-dimensional parameter space in which to run the cross-validation experiments. For an ensemble, although the problem seems to be a search in a four-dimensional parameter space, it is in fact two-dimensional at most. Regularization coefficient and ensemble size can be dropped from the parameter search, since the former needs to be very close to zero, and the latter needs to be a number around 100-200. Ensemble performance does not seem to be terribly sensitive to a proportion of data sampled for each base learner either. It is also noteworthy that leave-one-out (LOO) cross-validation can be done analytically as described by Rifkin (2002). If f_S denotes the function obtained using all training data, and f_{S^i} the function when data sample i is removed, the LOO value is

$$y_i - f_{S^i}(\boldsymbol{x}_i) = \frac{y_i - f_S(\boldsymbol{x}_i)}{1 - G_{ii}}, \qquad (11.8)$$

where $G = (K + m\gamma I)^{-1}$. Rifkin (2002) also shows that the number of LOO errors can be bound by

$$|\boldsymbol{x}_i : y_i f(\boldsymbol{x}_i) \leq \frac{k(\boldsymbol{x}_i, \boldsymbol{x}_i)}{k(\boldsymbol{x}_i, \boldsymbol{x}_i) + m\gamma}|. \qquad (11.9)$$

However, we used empirical 10-fold cross-validation in the experimentation.

11.5.3 Classifier Parameter Optimization Experiments

Kernel width and regularization parameter search comparing a single RLSC to an ensemble of RLSCs on Madelon is presented in Fig. 11.3. This is a very typical plot for a single RLSC. For each kernel width there is a narrow range of optimal regularization parameters. There is also a relatively narrow range of kernel widths around the optimal one.

Optimal kernel width is different for a single RLSC and for an ensemble. With every challenge data set, the optimal ensemble kernel was narrower than that of a single RLSC. One illustration is given in Fig. 11.4.

There are two further observations we can make about ensembles of RLSCs on the basis of these two plots: [1] Tuning of the regularization parameter can

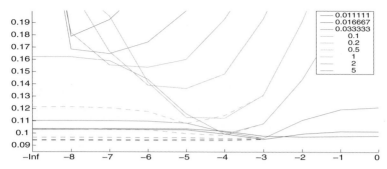

Fig. 11.3. Cross-validation experimentation in order to find the optimal combination of kernel width and regularization parameter for the Madelon data set. The vertical axis is the 10-fold cross-validation error rate on training data, the horizontal axis is $\log_{10}(\gamma)$, and each curve corresponds to a specific kernel width. The legend displays the multiplier to $d_{av}^2 = 37.5$. A single RLSC is denoted by solid lines, and dashed lines denote an ensemble of 200 RLSCs, each trained using a random sample of 40% of the training data. The color coding remains the same.

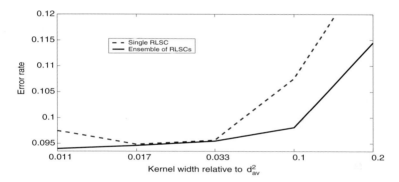

Fig. 11.4. Dashed line: Cross-validation error of a single RLSC as a function of the kernel width. The solid line plots the same for an ensemble. The regularization parameter was kept fixed.

be eliminated unless ill-conditioning requires some positive value. [2] Not much can be gained in terms of reducing the error rate using the ensemble vs. a single optimized RLSC.

The size of the ensemble does not appear to matter much, as long as it is above a certain limit (100-200 experts.) As an example we plot the cross-validation accuracy on Gisette as a function of the ensemble size and the fraction of the data sampled for each expert (Figure 11.5.)

In general, the trend in the challenge data sets appears to be that the larger the training data set, the smaller f_{td}, the fraction of data to sample for each expert in the ensemble, can be. This is advantageous regarding the computational complexity of an ensemble. Solving the linear system directly using all training data is $O(N^3)$, where N is the training set size. There is a

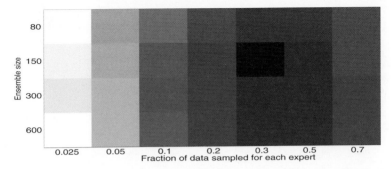

Fig. 11.5. Cross-validation experimentation to illustrate the optimal combination of the ensemble size against the fraction of data sampled for Gisette. Dark squares indicate low error rate.

limit to N (even with current computers) which makes solving the full system if not impossible, at least very inconvenient. Solving instead 200 systems of size $N/10$ can be much more feasible if $O(N^3)$ passes some resource limit but $O((N/10)^3)$ does not. A further advantage of ensembles is that they can be easily parallelized.

We present classification error rates both for CV and for the validation set in Table 11.3. Corresponding ensemble parameters are shown in Table 11.4. Even though there is no difference in CV/validation error rates between using a single RLSC or an ensemble of LSCs, we decided to make our entry to the challenge using ensembles. This was motivated by a combination of two facts, [1] small sample sizes in some of the data sets, and [2] because of the reduced sensitivity of an ensemble to parameter choices. Results of the challenge entry on the test set are presented in Table 11.5 together with the results of the winner.

Our overall score was 5th of the entries in the challenge and 2nd of the entrants. To what extent this success is due to the RF variable selection and to what extent due to the classifier, is unclear. However, on Arcene no variable selection was used. The results are thus fully attributable to the classifier, and appear to be competitive with any other classifier in the challenge. On Madelon, RF picked the correct variable set with no probes. On Gisette no probes were chosen either. But with Dexter, Dorothea, and Gisette we stopped increasing the number of selected variables too early. Since this process includes some heuristic parameters (Sec. 11.5.1), cross-validation experimentation should be run over a wider range of expected number of variables than what was done in this work.

Table 11.3. Error rates (both using 10-fold cross validation and the validation data set) after optimizing σ^2 and γ for a single RLSC, and σ^2 and the fraction of data sampled for each RLSC in an ensemble of 200 classifiers.

Data set	Single RLSC, cross-validation	Single RLSC, validation set	Ensemble, cross-validation	Ensemble, validation set
MADELON	0.0949	0.0700	0.0927	0.0667
DEXTER	0.0287	0.0633	0.0300	0.0633
ARCENE	0.1133	0.1331	0.1128	0.1331
GISETTE	0.0165	0.0210	0.0202	0.0210
DOROTHEA	0.0270	0.1183	0.0284	0.1183

Table 11.4. Ensemble parameter choices used in the challenge. Kernel width here denotes a multiplier to d_{av}^2.

Data set	Ensemble size	Data fraction	Regularization	Kernel width
MADELON	200	0.4	0	1/30
DEXTER	200	0.7	1e-4	1/2.5
ARCENE	200	0.8	0	1/10
GISETTE	200	0.3	0	1
DOROTHEA	30	0.6	0	1/10

Table 11.5. NIPS 2003 challenge results on the test data for ensembles of RLSCs.

Dec. 1st Dataset	Our best challenge entry					The winning challenge entry					
	Score	BER	AUC	Feat	Probe	Score	BER	AUC	Feat	Probe	Test
OVERALL	59.27	9.07	90.93	22.54	17.53	88.00	6.84	97.22	80.3	47.8	0.6
ARCENE	81.82	15.14	84.86	100.0	30.0	98.18	13.30	93.48	100.0	30.0	0
DEXTER	49.09	5.40	94.60	2.5	28.4	96.36	3.90	99.01	1.5	12.9	1
DOROTHEA	23.64	16.23	83.77	0.28	29.23	98.18	8.54	95.92	100.0	50.0	1
GISETTE	45.45	1.89	98.11	6.14	0	98.18	1.37	98.63	18.3	0.0	1
MADELON	96.36	6.67	93.33	3.8	0	100.00	7.17	96.95	1.6	0.0	0

11.5.4 Summary of the Approach

We summarize here the procedure used in our submission to the challenge.

Variable ranking by Random Forest:

(1.) Compute an importance index for each variable for each tree in the process of growing the tree as the decrease of impurity to an actual or potential split on the variable at a node (eq. 11.6).

(2.) Grow around 100 trees (until no more generalization improvement.) Tree variability is obtained by bootsrap sampling training data, and by drawing at random $\sqrt{N}$ variables at each node and choosing the best split on these variables.

(3.) Compute the average importance index over all trees in the forest (eq. 11.7).

The optimum number of features is selected using ten-fold CV by trying subsets of variables corresponding to "cut-off points" (Sec. 11.5.1).

Training RLSC ensembles:

(1.) Train multiple RLSCs by subsampling a fraction f_{td} of the training data, without replacement (computationally efficient because one does not need to recompute the kernel matrix.) Fix the regularization parameter to a vanishingly small value. The ensemble size should be large (around 200) since there is no risk of overfitting.

(2.) Optimize the kernel width and the fraction f_{td}.

(3.) Vote on the classifier decisions with weight 1.

11.6 Future Directions

We outline here three major directions that we feel do have a chance of improving or extending the work presented in this paper.

11.6.1 Diversification

The ensemble construction process is computationally light because the whole kernel matrix K for all available training data can be pre-computed, and sampled from, to construct the individual classifiers of the ensemble.

However, only one source of diversification is provided to the experts, the random sample of training data points. For ensemble methods to work optimally, the experts should be as diverse as possible.

We believe that a kernel width that varies across the experts randomly over a range that is derived from sample distances, could improve ensemble performance. Combination of this with ensemble postprocessing discussed later would lead to elimination of the kernel width selection procedure altogether. Also the whole kernel matrix would not have to be re-calculated for each sampled kernel width. Only a function of an existing kernel matrix needs to be calculated.

Another possibility for diversification is to use random feature selection, in addition to random training sample selection. Unfortunately, unlike in Random Forests, a computational cost is incurred, since the kernel needs to be explicitly re-evaluated for each different random feature combination.

11.6.2 Supervised Kernels

We describe an approach in this paper that consists of two disjoint systems, Random Forests for variable selection, and RLSC for the actual classification. Even though the two systems nicely complement each other, RF providing fast embedded variable selection and RLSC providing highly capable base learners to compensate for the lack of smoothness of the trees of an RF, an integrated approach would be desirable. We describe an idea towards such a system.

RF could act as one type of supervised kernel generator using the pairwise similarities between cases. Similarity for a single tree between two cases could be defined as the total number of common parent nodes, normalized by the level of the deepest case, and summed up for the ensemble. Minimum number of common parents to define nonzero similarity is another parameter that could be used like width in Gaussian kernels.

Figure 11.6 illustrates the difference between a Gaussian kernel and the proposed supervised kernel.

An advantage of the method is that it works for any type of data, numeric, categorical, or mixed, even for data with missing values. This is because the base learners of the Random Forest can tolerate these.

A further advantage is that explicit variable selection is bypassed altogether. Important variables will become used in the trees of the forest, and they thus participate implicitly in the evaluation of the kernel.

11.6.3 Ensemble Postprocessing

A well known avenue to improve the accuracy of an ensemble is to replace the simple averaging of individual experts by a weighting scheme. Instead of giving equal weight to each expert, the outputs of more reliable experts are weighted up. Linear regression can be applied to learn these weights. To avoid overfitting, the training material for this regression should be produced by passing only those samples through an expert, that did not participate in construction of the particular expert (out-of-bag samples.) Since the outputs of the experts are correlated (even though the aim is to minimize this correlation), Partial Least Squares or Principal Components Regression could well be used to find aggregation weights.

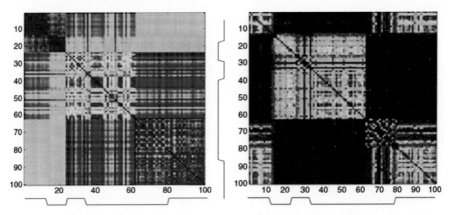

Fig. 11.6. Gaussian kernel compared to a supervised kernel using the Arcene dataset. Left side depicts the 100×100 Gaussian kernel matrix of the dataset clustered in three clusters. Each cluster has samples from both classes. Class identity of a sample is depicted as the graphs below and between the kernel matrix images. For ideal classification purposes, the kernel matrix should reflect the similarity within a class and dissimilarity between classes. This can be seen on the right side of the figure, where the proposed supervised kernel has split the first cluster (top left corner) into the two classes nicely. Splits on the second and third clusters are not that clean but still visible, and much more so than what can be seen in the Gaussian kernel.

11.7 Conclusion

We have described a relatively simple approach to the combined problem of variable selection and classifier design as a combination of random forests and regularized least squares classifiers. As applied to the NIPS 2003 Feature Selection Challenge, this approach ranked the 5th of entries and the 2nd of entrants in the initial part of the challenge.

Even though individual RLSC could achieve comparable accuracy through careful tuning over parameter space, we found that ensembles are easier to train, and are less sensitive to parameter changes. The generalization error stabilizes after a relatively small number of base learners is added to the ensemble (less than 200.) Naturally it does not hurt to add more experts as there is no overfitting. The optimal kernel width for an ensemble is consistently smaller than that of the best single RLSC, and the range of optimal values is wider. It is consistent with known results on superior performance of ensembles consisting of low bias and high variance base learners.

The random sample of data used to build each individual learner was relatively small. The fact that the larger the data set, the smaller the sample size needed, makes parallel ensemble construction much more efficient than constructing a single RLSC using all of the data.

References

B. Boser, I. Guyon, and V. Vapnik. A training algorithm for optimal margin classifiers. In *Proc. COLT*, pages 144–152, 1992.

O. Bousquet and A. Elisseeff. Algorithmic stability and generalization performance. In *Proc. NIPS*, pages 196–202, 2000.

L. Breiman. Random forests. *Machine Learning*, 45(1):5–32, 2001.

L. Breiman. Bagging predictors. *Machine Learning*, 24(2):123–140, 1996.

L. Breiman, J.H. Friedman, R.A. Olshen, and C.J. Stone. *Classification and Regression Trees*. CRC Press, 1984.

F. Cucker and S. Smale. On the mathematial foundations of learning. *Bulletin of the American Mathematical Society*, 89(1):1–49, 2001.

F. Cucker and S. Smale. Best choices for regularization parameters in learning theory: on the bias-variance problem. *Foundations of Computational Mathematics*, 2(4):413–428, 2003.

T. Evgeniou. *Learning with Kernel Machine Architectures*. PhD thesis, Massachusetts Institute of Technology, EECS, July 2000.

Y. Freund and R. E. Schapire. A decision-theoretic generalization of on-line learning and an application to boosting. In *European Conference on Computational Learning Theory*, pages 23–37, 1995.

J.H. Friedman. Greedy function approximation: a gradient boosting machine. Technical report, Dept. of Statistics, Stanford University, 1999a.

J.H. Friedman. Stochastic gradient boosting. Technical report, Dept. of Statistics, Stanford University, 1999b.

A. Hoerl and R. Kennard. Ridge regression; biased estimation for nonorthogonal problems. *Technometrics*, 12(3):55–67, 1970.

V. Ng and L. Breiman. Random forest variable selection. In I. Guyon, S. Gunn, M. Nikravesh, and L. Zadeh, editors, *Feature Extraction, Foundations and Applications*. Springer, New York, 2005. This volume.

T. Poggio and S. Smale. The mathematics of learning: Dealing with data. *Notices of the American Mathematical Society (AMS)*, 50(5):537–544, 2003.

T. Poggio, R. Rifkin, S. Mukherjee, and A. Rakhlin. Bagging regularizes. CBCL Paper 214, Massachusetts Institute of Technology, Cambridge, MA, February 2002. AI Memo #2002-003.

R. Rifkin. *Everything Old Is New Again: A Fresh Look at Historical Approaches in Machine Learning*. PhD thesis, MIT, 2002.

B. Schölkopf and A. Smola. *Learning with Kernels*. MIT Press, 2002.

J.A.K. Suykens and J. Vandervalle. Least squares support vector machines. *Neural Processing Letters*, 9(3):293–300, June 1999.

A.N. Tikhonov and V.Y. Arsenin. *Solutions of Ill-posed Problems*. W.H.Wingston, Washington, D.C., 1977.

K. Torkkola. Feature extraction by non-parametric mutual information maximization. *Journal of Machine Learning Research*, 3:1415–1438, March 2003.

Chapter 12

Combining SVMs with Various Feature Selection Strategies

Yi-Wei Chen and Chih-Jen Lin

Department of Computer Science, National Taiwan University, Taipei 106, Taiwan
b88052@csie.ntu.edu.tw,cjlin@csie.ntu.edu.tw

Summary. This article investigates the performance of combining support vector machines (SVM) and various feature selection strategies. Some of them are filter-type approaches: general feature selection methods independent of SVM, and some are wrapper-type methods: modifications of SVM which can be used to select features. We apply these strategies while participating to the NIPS 2003 Feature Selection Challenge and rank third as a group.

12.1 Introduction

Support Vector Machine (SVM) (Boser et al., 1992, Cortes and Vapnik, 1995) is an effective classification method, but it does not directly obtain the feature importance. In this article we combine SVM with various feature selection strategies and investigate their performance. Some of them are "filters": general feature selection methods independent of SVM. That is, these methods select important features first and then SVM is applied for classification. On the other hand, some are wrapper-type methods: modifications of SVM which choose important features as well as conduct training/testing. We apply these strategies while participating to the NIPS 2003 Feature Selection Challenge. Overall we rank third as a group and are the winner of one data set.

In the NIPS 2003 Feature Selection Challenge, the main judging criterion is the balanced error rate (BER). Its definition is:

$$\text{BER} \equiv \frac{1}{2}\left(\frac{\#\text{ positive instances predicted wrong}}{\#\text{ positive instances}} + \frac{\#\text{ negative instances predicted wrong}}{\#\text{ negative instances}}\right). \tag{12.1}$$

For example, assume a test data set contains 90 positive and 10 negative instances. If all the instances are predicted as positive, then BER is 50% since the first term of (12.1) is 0/90 but the second is 10/10. There are other

Y.-W. Chen and C.-J. Lin: *Combining SVMs with Various Feature Selection Strategies*, StudFuzz **207**, 315–324 (2006)
www.springerlink.com   © Springer-Verlag Berlin Heidelberg 2006

judging criteria such as the number of features and probes, but throughout the competition we focus on how to get the smallest BER.

This article is organized as follows. In Section 2 we introduce support vector classification. Section 3 discusses various feature selection strategies. In Section 4, we show the experimental results during the development period of the competition. In Section 5, the final competition results are listed. Finally, we have discussion and conclusions in Section 6. All competition data sets are available at `http://clopinet.com/isabelle/Projects/NIPS2003/`.

12.2 Support Vector Classification

Recently, support vector machines (SVMs) have been a promising tool for data classification. Its basic idea is to map data into a high dimensional space and find a separating hyperplane with the maximal margin. Given training vectors $\mathbf{x}_k \in R^n, k = 1, \ldots, m$ in two classes, and a vector of labels $\mathbf{y} \in R^m$ such that $y_k \in \{1, -1\}$, SVM solves a quadratic optimization problem:

$$\min_{\mathbf{w}, b, \xi} \frac{1}{2}\mathbf{w}^T\mathbf{w} + C\sum_{k=1}^{m} \xi_k , \qquad (12.2)$$

$$\text{subject to } y_k(\mathbf{w}^T\phi(\mathbf{x}_k) + b) \geq 1 - \xi_k,$$

$$\xi_k \geq 0, k = 1, \ldots, m,$$

where training data are mapped to a higher dimensional space by the function ϕ, and C is a penalty parameter on the training error. For any testing instance $\mathbf{x}$, the decision function (predictor) is

$$f(\mathbf{x}) = \text{sgn}\left(\mathbf{w}^T\phi(\mathbf{x}) + b\right).$$

Practically, we need only $k(\mathbf{x}, \mathbf{x}') = \phi(\mathbf{x})^T\phi(\mathbf{x}')$, the kernel function, to train the SVM. The RBF kernel is used in our experiments:

$$k(\mathbf{x}, \mathbf{x}') = \exp(-\gamma\|\mathbf{x} - \mathbf{x}'\|^2) . \qquad (12.3)$$

With the RBF kernel (12.3), there are two parameters to be determined in the SVM model: C and γ. To get good generalization ability, we conduct a validation process to decide parameters. The procedure is as the following:

(1.) Consider a grid space of (C, γ) with $\log_2 C \in \{-5, -3, \ldots, 15\}$ and $\log_2 \gamma \in \{-15, -13, \ldots, 3\}$.
(2.) For each hyperparameter pair (C, γ) in the search space, conduct 5-fold cross validation on the training set.
(3.) Choose the parameter (C, γ) that leads to the lowest CV balanced error rate.
(4.) Use the best parameter to create a model as the predictor.

12.3 Feature Selection Strategies

In this Section, we discuss feature selection strategies tried during the competition. We name each method to be like "A + B," where A is a filter to select features and B is a classifier or a wrapper. If a method is "A + B + C," then there are two filters A and B.

12.3.1 No Selection: Direct Use of SVM

The first strategy is to directly use SVM without feature selection. Thus, the procedure in Section 12.2 is considered.

12.3.2 F-score for Feature Selection: F-score + SVM

F-score is a simple technique which measures the discrimination of two sets of real numbers. Given training vectors $x_k, k = 1, \ldots, m$, if the number of positive and negative instances are n_+ and n_-, respectively, then the F-score of the ith feature is defined as:

$$F(i) \equiv \frac{\left(\bar{\mathbf{x}}_i^{(+)} - \bar{\mathbf{x}}_i\right)^2 + \left(\bar{\mathbf{x}}_i^{(-)} - \bar{\mathbf{x}}_i\right)^2}{\frac{1}{n_+ - 1} \sum_{k=1}^{n_+} \left(x_{k,i}^{(+)} - \bar{\mathbf{x}}_i^{(+)}\right)^2 + \frac{1}{n_- - 1} \sum_{k=1}^{n_-} \left(x_{k,i}^{(-)} - \bar{\mathbf{x}}_i^{(-)}\right)^2}, \quad (12.4)$$

where $\bar{\mathbf{x}}_i, \bar{\mathbf{x}}_i^{(+)}, \bar{\mathbf{x}}_i^{(-)}$ are the average of the ith feature of the whole, positive, and negative data sets, respectively; $x_{k,i}^{(+)}$ is the ith feature of the kth positive instance, and $x_{k,i}^{(-)}$ is the ith feature of the kth negative instance. The numerator indicates the discrimination between the positive and negative sets, and the denominator indicates the one within each of the two sets. The larger the F-score is, the more likely this feature is more discriminative. Therefore, we use this score as a feature selection criterion.

A disadvantage of F-score is that it does not reveal mutual information among features. Consider one simple example in Figure 12.1. Both features of this data have low F-scores as in (12.4) the denominator (the sum of variances of the positive and negative sets) is much larger than the numerator.

Despite this disadvantage, F-score is simple and generally quite effective. We select features with high F-scores and then apply SVM for training/prediction. The procedure is summarized below:

(1.) Calculate F-score of every feature.
(2.) Pick some possible thresholds by human eye to cut low and high F-scores.
(3.) For each threshold, do the following
    a) Drop features with F-score below this threshold.
    b) Randomly split the training data into X_{train} and X_{valid}.

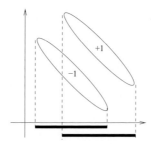

Fig. 12.1. A two-feature example showing the disadvantage of F-score

    c) Let X_{train} be the new training data. Use the SVM procedure in Section 12.2 to obtain a predictor; use the predictor to predict X_{valid}.
    d) Repeat the steps above five times, and then calculate the average validation error.
(4.) Choose the threshold with the lowest average validation error.
(5.) Drop features with F-score below the selected threshold. Then apply the SVM procedure in Section 12.2.

In the above procedure, possible thresholds are identified by human eye. For data sets in this competition, there is a quite clear gap between high and lower scores (see Figure 12.2, which will be described in Section 12.4). We can automate this step by, for example, gradually adding high-F-score features, until the validation accuracy decreases.

12.3.3 F-score and Random Forest for Feature Selection: F-score + RF + SVM

Random Forest (RF) is a classification method, but it also provides feature importance (Breiman, 2001). Its basic idea is as follows: A forest contains many decision trees, each of which is constructed from instances with randomly sampled features. The prediction is by a majority vote of decision trees. To obtain a feature importance criterion, first we split the training sets to two parts. By training with the first and predicting the second we obtain an accuracy value. For the jth feature, we randomly permute its values in the second set and obtain another accuracy. The difference between the two numbers can indicate the importance of the jth feature.

In practice, the RF code we used cannot handle too many features. Thus, before using RF to select features, we obtain a subset of features using F-score selection first. This approach is thus called "F-score + RF + SVM" and is summarized below:

(1.) F-score
    a) Consider the subset of features obtained in Section 12.3.2.
(2.) RF

a) Initialize the RF working data set to include all training instances with the subset of features selected from Step 1. Use RF to obtain the rank of features.
b) Use RF as a predictor and conduct 5-fold CV on the working set.
c) Update the working set by removing half of the features which are least important and go to Step 2b.
   Stop if the number of features is small.
d) Among various feature subsets chosen above, select one with the lowest CV error.

(3.) SVM
a) Apply the SVM procedure in Section 12.2 on the training data with the selected features.

Note that the rank of features is obtained at Step 2a and is not updated throughout iterations. An earlier study on using RF for feature selection is (Svetnik et al., 2004).

12.3.4 Random Forest and RM-bound SVM for Feature Selection: RF + RM-SVM

Chapelle et al. (2002) directly use SVM to conduct feature selection. They consider the RBF kernel with feature-wise scaling factors:

$$k(\mathbf{x}, \mathbf{x}') = \exp\left(-\sum_{i=1}^{n} \gamma_i (x_i - x_i')^2\right) . \tag{12.5}$$

By minimizing an estimation of generalization errors, which is a function of $\gamma_1, \ldots, \gamma_n$, we can have feature importance. Leave-one-out (loo) errors are such an estimation and are bounded by a smoother function (Vapnik, 1998):

$$\text{loo} \le 4\|\tilde{\mathbf{w}}\|^2 \tilde{R}^2 . \tag{12.6}$$

We refer to this upper bound as the radius margin (RM) bound. Here, $\tilde{\mathbf{w}}^T \equiv [\mathbf{w}^T \ \sqrt{C}\boldsymbol{\xi}^T]$ and $(\mathbf{w}, \boldsymbol{\xi})$ is the optimal solution of the L2-SVM:

$$\min_{\mathbf{w}, b, \boldsymbol{\xi}} \frac{1}{2}\mathbf{w}^T\mathbf{w} + \frac{C}{2}\sum_{k=1}^{m} \xi_k^2 ,$$

under the same constraints of (12.2); $\tilde{R}$ is the radius of the smallest sphere containing all $[\phi(\mathbf{x}_k)^T \ \mathbf{e}_k^T/\sqrt{C}]$, $k = 1, \ldots, m$, where $\mathbf{e}_k$ is a zero vector except the kth component is one.

We minimize the bound $4\|\tilde{\mathbf{w}}\|^2 \tilde{R}^2$ with respect to C and $\gamma_1, \ldots, \gamma_n$ via a gradient-based method. Using these parameters, an SVM model can be built for future prediction. Therefore we call this machine an RM-bound SVM. When the number of features is large, minimizing the RM bound is time consuming. Thus, we apply this technique only on the problem MADELON, which has 500 features. To further reduce the computational burden, we use RF to pre-select important features. Thus, this method is referred to as "RF + RM-SVM."

12.4 Experimental Results

In the experiment, we use LIBSVM[1] (Chang and Lin, 2001) for SVM classi-fication. For feature selection methods, we use the randomForest (Liaw and Wiener, 2002) package[2] in software R for RF and modify the implementation in (Chung et al., 2003) for the RM-bound SVM[3]. Before doing experiments, data sets are scaled. With training, validation, and testing data together, we scale each feature to [0, 1]. Except scaling, there is no other data preprocessing.

In the development period, only labels of training sets are known. An on-line judge returns BER of what competitors predict about validation sets, but labels of validation sets and even information of testing sets are kept unknown.

We mainly focus on three feature selection strategies discussed in Sections 3.1-3.3: SVM, F-score + SVM, and F-score + RF + SVM. For RF + RM-SVM, due to the large number of features, we only apply it on MADELON. The RF procedure in Section 12.3.3 selects 16 features, and then RM-SVM is used. In all experiments we focused on getting the smallest BER.

For the strategy F-score + RF + SVM, after the initial selection by F-score, we found that RF retains all features. That is, by comparing cross-validation BER using different subsets of features, the one with all features is the best. Hence, F+RF+SVM is in fact the same as F+SVM for all the five data sets. Since our validation accuracy of DOROTHEA is not as good as that by some participants, we consider a heuristic by submitting results via the top 100, 200, and 300 features from RF. The BERs of the validation set are 0.1431, 0.1251, and 0.1498, respectively. Therefore, we consider "F-score + RF top 200 + SVM" for DOROTHEA.

Table 12.1 presents the BER on validation data sets by different feature selection strategies. It shows that no method is the best on all data sets.

In Table 12.2 we list the CV BER on the training set. Results of the first three problems are quite different from those in Table 12.1. Due to the small training sets or other reasons, CV does not accurately indicate the future performance.

In Table 12.3, the first row indicates the threshold of F-score. The second row is the number of selected features which is compared to the total number

---

[1]http://www.csie.ntu.edu.tw/~cjlin/libsvm
[2]http://www.r-project.org/
[3]http://www.csie.ntu.edu.tw/~cjlin/libsvmtools/
[4]Our implementation of RF+RM-SVM is applicable to only MADELON, which has a smaller number of features.

Table 12.1. Comparison of different methods during the development period: BERs of validation sets (in percentage); bold-faced entries correspond to approaches used to generate our final submission

Dataset	ARCENE	DEXTER	DOROTHEA	GISETTE	MADELON
SVM	13.31	11.67	33.98	2.10	40.17
F+SVM	**21.43**	**8.00**	21.38	**1.80**	13.00
F+RF+SVM	21.43	8.00	**12.51**	1.80	13.00
RF+RM-SVM[4]	–	–	–	–	**7.50**

Table 12.2. CV BER on the training set (in percentage)

Dataset	ARCENE	DEXTER	DOROTHEA	GISETTE	MADELON
SVM	11.04	8.33	39.38	2.08	39.85
F+SVM	9.25	4.00	14.21	1.37	11.60

of features in the third row. Figure 12.2 presents the curve of F-scores against features.

Table 12.3. F-score threshold and the number of features selected in F+SVM

Dataset	ARCENE	DEXTER	DOROTHEA	GISETTE	MADELON
F-score threshold	0.1	0.015	0.05	0.01	0.005
#features selected	661	209	445	913	13
#total features	10000	20000	100000	5000	500

12.5 Competition Results

For each data set, we submit the final result using the method that leads to the best validation accuracy in Table 12.1. A comparison of competition results (ours and winning entries) is in Table 12.4 and 12.5.

For the December 1st submissions, we rank 1st on GISETTE, 3rd on MADE-LON, and 5th on ARCENE. Overall we rank 3rd as a group and our best entry is the 6th, using the criterion of the organizers. For the December 8th submissions, we rank 2nd as a group and our best entry is the 4th.

322   Yi-Wei Chen and Chih-Jen Lin

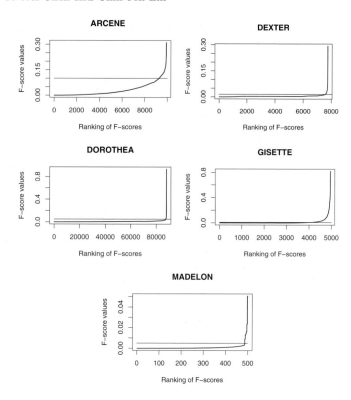

Fig. 12.2. Curves of F-scores against features; features with F-scores below the horizontal line are dropped

12.6 Discussion and Conclusions

Usually SVM suffers from a large number of features, but we find that a direct use of SVM works well on GISETTE and ARCENE. After the competition, we realize that GISETTE comes from an OCR problem MNIST (LeCun et al., 1998), which contains 784 features of gray-level values. Thus, all features are of the same type and tend to be equally important. Our earlier experience indicates that for such problems, SVM can handle a rather large set of features. As the 5,000 features of GISETTE are a combination of the original 784 features, SVM may still work under the same explanation. For ARCENE, we need further investigation to know why direct SVM performs well.

For the data set MADELON, the winner (see Chapter 18) uses a kind of Bayesian SVM (Chu et al., 2003). It is similar to RM-SVM by minimizing a function of feature-wise scaling factors. The main difference is that RM-SVM uses an loo bound, but Bayesian SVM derives a Bayesian evidence function. For this problem Tables 12.4- 12.5 indicate that the two approaches achieve very similar BER. This result seems to indicate a strong relation between

Table 12.4. NIPS 2003 challenge results on December 1st

| Dec. 1st | Our best challenge entry | | | | | The winning challenge entry | | | | | |
Dataset	Score	BER	AUC	Feat	Probe	Score	BER	AUC	Feat	Probe	Test
OVERALL	52.00	9.31	90.69	24.9	12.0	88.00	6.84	97.22	80.3	47.8	0.4
ARCENE	74.55	15.27	84.73	100.0	30.0	98.18	13.30	93.48	100.0	30.0	0
DEXTER	0.00	6.50	93.50	1.0	10.5	96.36	3.90	99.01	1.5	12.9	1
DOROTHEA	-3.64	16.82	83.18	0.5	2.7	98.18	8.54	95.92	100.0	50.0	1
GISETTE	98.18	1.37	98.63	18.3	0.0	98.18	1.37	98.63	18.3	0.0	0
MADELON	90.91	6.61	93.39	4.8	16.7	100.00	7.17	96.95	1.6	0.0	0

Table 12.5. NIPS 2003 challenge results on December 8th

| Dec. 8th | Our best challenge entry | | | | | The winning challenge entry | | | | | |
Dataset	Score	BER	AUC	Feat	Probe	Score	BER	AUC	Feat	Probe	Test
OVERALL	49.14	7.91	91.45	24.9	9.9	88.00	6.84	97.22	80.3	47.8	0.4
ARCENE	68.57	10.73	90.63	100.0	30.0	94.29	11.86	95.47	10.7	1.0	0
DEXTER	22.86	5.35	96.86	1.2	2.9	100.00	3.30	96.70	18.6	42.1	1
DOROTHEA	8.57	15.61	77.56	0.2	0.0	97.14	8.61	95.92	100.0	50.0	1
GISETTE	97.14	1.35	98.71	18.3	0.0	97.14	1.35	98.71	18.3	0.0	0
MADELON	71.43	7.11	92.89	3.2	0.0	94.29	7.11	96.95	1.6	0.0	1

the two methods. Though they are derived from different aspects, it is worth investigating the possible connection.

In conclusion, we have tried several feature selection strategies in this competition. Most of them are independent of the classifier used. This work is a preliminary study to determine what feature selection strategies should be included in an SVM package. In the future, we would like to have a systematic comparison on more data sets.

References

B. Boser, I. Guyon, and V. Vapnik. A training algorithm for optimal margin classifiers. In *Proceedings of the Fifth Annual Workshop on Computational Learning Theory*, pages 144–152, 1992.

Leo Breiman. Random forests. *Machine Learning*, 45(1):5–32, 2001. URL citeseer. nj.nec.com/breiman01random.html.

Chih-Chung Chang and Chih-Jen Lin. *LIBSVM: a library for support vector machines*, 2001. Software available at http://www.csie.ntu.edu.tw/~cjlin/libsvm.

O. Chapelle, V. Vapnik, O. Bousquet, and S. Mukherjee. Choosing multiple parameters for support vector machines. *Machine Learning*, 46:131–159, 2002.

W. Chu, S.S. Keerthi, and C.J. Ong. Bayesian trigonometric support vector classifier. *Neural Computation*, 15(9):2227–2254, 2003.

Kai-Min Chung, Wei-Chun Kao, Chia-Liang Sun, Li-Lun Wang, and Chih-Jen Lin. Radius margin bounds for support vector machines with the RBF kernel. *Neural Computation*, 15:2643–2681, 2003.

C. Cortes and V. Vapnik. Support-vector network. *Machine Learning*, 20:273–297, 1995.

Yann LeCun, L. Bottou, Y. Bengio, and P. Haffner. Gradient-based learning applied to document recognition. *Proceedings of the IEEE*, 86(11):2278–2324, November 1998. MNIST database available at http://yann.lecun.com/exdb/mnist/.

Andy Liaw and Matthew Wiener. Classification and regression by randomForest. *R News*, 2/3:18–22, December 2002. URL http://cran.r-project.org/doc/Rnews/Rnews_2002-3.pdf.

V. Svetnik, A. Liaw, C. Tong, and T. Wang. Application of Breiman's random forest to modeling structure-activity relationships of pharmaceutical molecules. In F. Roli, J. Kittler, and T. Windeatt, editors, *Proceedings of the 5th International Workshopon Multiple Classifier Systems*, Lecture Notes in Computer Science vol. 3077., pages 334–343. Springer, 2004.

Vladimir Vapnik. *Statistical Learning Theory*. Wiley, New York, NY, 1998.

Chapter 13

Feature Selection with Transductive Support Vector Machines

Zhili Wu[1] and Chunhung Li[2]

[1] Department of Computer Science, Hong Kong Baptist University
   vincent@comp.hkbu.edu.hk
[2] Department of Computer Science, Hong Kong Baptist University
   chli@comp.hkbu.edu.hk

Summary. SVM-related feature selection has shown to be effective, while feature selection with transductive SVMs has been less studied. This paper investigates the use of transductive SVMs for feature selection, based on three SVM-related feature selection methods: filtering scores + SVM wrapper, recursive feature elimination (RFE) and multiplicative updates(MU). We show transductive SVMs can be tailored to feature selection by embracing feature scores for feature filtering, or acting as wrappers and embedded feature selectors. We conduct experiments on the feature selection competition tasks to demonstrate the performance of Transductive SVMs in feature selection and classification.

13.1 Introduction

SVMs have been studied to work with different categories of feature selection methods, like wrappers, embedded methods, and filters. For example, the SVM as a wrapper can be used to select features, while the feature set quality is indicated by the performance of the trained SVM (Yu and Cho, 2003). Feature selection can also be conducted during the process of training SVMs. This induces some embedded feature selection methods (Brank et al., 2002, Guyon et al., 2002, Weston et al., 2003, 2000, Guyon et al., 2003, Perkins et al., 2003). SVMs are also useful for filter methods. Although filters often score or rank features without utilizing a learning machine, they can utilize a SVM as the final predictor for classification. Moreover, filters are often integrated as a preprocessing step into wrappers and embedded methods, such that they can help feature selection of wrappers or embedded methods.

The development of SVM-related feature selection methods motivates us to explore the role of Transductive SVMs (TSVMs) in feature selection. Transductive SVMs are a type of transductive learning (Vapnik, 1998) machine. They aim to build SVM models upon both labeled and unlabeled data. Transductive SVMs, as generalized from inductive SVMs, inherit the advantages of

Z. Wu and C. Li: *Feature Selection with Transductive Support Vector Machines*, StudFuzz **207**, 325–341 (2006)
www.springerlink.com

inductive SVMs such as the large margin (Boser et al., 1992) and regularization formulation as well as kernel mapping. Besides, TSVMs are suitable for the tasks we consider here, which are characterized by many available unlabeled data. We thus consider extending SVM-related feature selection to TSVM-related feature selection.

Feature selection with transduction has drawn some attentions (Weston et al., 2002, Wu and Flach, 2002, Jebara and Jaakkola, 2000). Particularly Weston initiated the use of TSVMs for feature selection (Weston et al., 2002). Among many variants of transductive SVMs (Bennett and Demiriz, 1998, Fung and Mangasarian, 2001, Joachims, 1999b), we in this paper try to systematically adapt the 2-norm TSVM powered by the SVMLight (Joachims, 1999b) to feature selection, based on modification of SVM-related feature selection methods.

The paper is organized as follows. In section 2, the formulations of SVMs and TSVMs are briefly reviewed, including some practical issues like implementation and model selection. In section 3, several SVM-related feature selection methods are discussed. In section 4, we explain how to extend them to transductive SVM-related feature selection. In section 5, we give some analysis on the five feature selection competition tasks and a synthetic dataset. Section 6 discusses our approach and concludes this paper.

13.2 SVMs and Transductive SVMs

13.2.1 Support Vector Machines

Given a set of n-dimensional training patterns $\boldsymbol{X} = \{\boldsymbol{x}_i\}_{i=1}^m$ labeled by $\{y_i\}_{i=1}^m$, and their mapping $\{\phi(\boldsymbol{x}_i)\}_{i=1}^m$ via a kernel function $\mathbf{k}(\boldsymbol{x}_i, \boldsymbol{x}_j) = \phi(\boldsymbol{x}_i)^T \phi(\boldsymbol{x}_j)'$, a SVM has the following primal form **P1**.

P1:   Minimize over $(\boldsymbol{w}, b, \xi_1, \ldots, \xi_m)$:

$$||\boldsymbol{w}||_p^p + C \sum_{i=1}^m \xi_i$$

subject to: $\forall_{i=1}^m : y_i(\boldsymbol{w}^T \phi(\boldsymbol{x}_i)' + b) \geq 1 - \xi_i, \xi_i \geq 0$

The SVM predictor for a pattern $\boldsymbol{x}$, as shown below, is determined by the vector inner product between the $\boldsymbol{w}$ and the mapped vector $\phi(\boldsymbol{x})$, plus constant b.

$$y = \text{sgn}(\boldsymbol{w}^T \phi(\boldsymbol{x})' + b).$$

The predictor actually corresponds to a separating hyperplane in the mapped feature space. The prediction for each training pattern $\boldsymbol{x}_i$ is associated with a violation term ξ_i. The C is a user-specified constant to control the penalty to these violation terms.

The parameter p in the above **P1** indicates, which type of norm of $\boldsymbol{w}$ is evaluated. It is usually set to 1 or 2, resulting in the 1-norm(l_1-SVM) and 2-norm SVM(l_2-SVM) respectively. The l_2-SVM has the following dual form **P2**, which is a typical quadratic programming problem.

P2: Maximize over $(\alpha_1, \ldots, \alpha_m)$:

$$J = -\frac{1}{2} \sum_{i,j=1}^{m} \alpha_i \alpha_j y_i y_j \phi(\boldsymbol{x}_i)^T \phi(\boldsymbol{x}_j)' + \sum_{i=1}^{m} \alpha_i$$

subject to: $\sum_{i=1}^{m} \alpha_i y_i = 0$ and $\forall_{i=1}^{m} : 0 \leq \alpha_i \leq C$

The dual form is a convex problem. It can be efficiently solved by sequential minimal optimization (Platt, 1998), or by decomposition into subtasks suitable for standard quadratic programming routines (Joachims, 1999a).

Hyperparameter selection is particularly important for SVMs. The performance of SVM models crucially depends on the type of kernel, kernel parameters, and the constant C. Some theoretical bounds are useful for model selection (Weston et al., 2000). But an easy way to implement hyperparameter selection is to perform a grid search guided by the cross validation accuracy. Though the approach is computationally intensive, it is still tractable in most cases where computational resources are plentiful. Some built-in routines are available (Chang and Lin, 2001) for grid model search.

13.2.2 Transductive Support Vector Machines

The 1-norm and 2-norm transductive SVMs have been discussed in (Bennett and Demiriz, 1998) and (Joachims, 1999b) respectively. Their general setting can be described as **P3**.

P3:   Minimize over $(y_1^*, \ldots, y_k^*, \boldsymbol{w}, b, \xi_1, \ldots, \xi_m, \xi_1^*, \ldots, \xi_k^*)$:

$$||\boldsymbol{w}||_p^p + C \sum_{i=1}^{m} \xi_i + C^* \sum_{j=1}^{k} \xi_j^*$$

subject to: $\forall_{i=1}^{m} : y_i(\boldsymbol{w}^T \phi(\boldsymbol{x}_i)' + b) \geq 1 - \xi_i, \xi_i > 0$

$\forall_{j=1}^{k} : y_j^*(\boldsymbol{w}^T \phi(\boldsymbol{x}_j^*)' + b) \geq 1 - \xi_j^*, \xi_j^* > 0$

where each y_j^* is the unknown label for $\boldsymbol{x}_j^*$, which is one of the k unlabelled patterns.

Compared with the **P1**, the formulation of TSVMs (**P3**) takes the unlabelled data into account, by modelling the violation terms ξ_j^* caused by predicting each unlabelled pattern $\phi(\boldsymbol{x}_j^*)$ into y_j^*. The penalty to these violation terms is controlled by a new constant C^*.

Exactly solving the transductive problem requires searching all possible assignments of $y_1^*, \ldots, y_k^*$ and specifying many terms of ξ^*, which is often

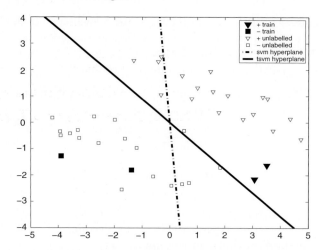

Fig. 13.1. A Toy TSVM Example. The solid line is the decision plane of an inductive SVM, which only train labeled data denoted by filled rectangles and triangles. The TSVM as depicted by the dashed line utilizes the additional unlabeled data

intractable for large datasets. But it is worth mentioning the l_2-TSVM implemented in the SVMLight (Joachims, 1999a). It efficiently approximates the solution by using a form of local search, and is scalable to cases of thousands of data points. We briefly mention the TSVM implementation by SVMLight as follows.

- 1. Build an inductive SVM **P2**, and use it to predict unlabeled data into positive and negative classes as a pre-specified ratio. Initialize the constant C^* to be a small value.
- 2. Train a new inductive SVM **P3** with predicted labels of y_j^*.
- 3. Pick a pair of unlabelled patterns, which are near the separating plane and are associated with positive and negative predicted labels respectively. Try to switch their labels to reduce the objective value of **P3** and rebuild a SVM. And repeat this step till no more pair can be found
- 4. Increase C^* a little, and go back to step 2. If C^* is large enough, stop and report the final model.

The TSVM model is obtained from training a number of inductive SVM models. In time/resource-limited cases, we can stop the training process early and use the incomplete model. The constant C^* can also be set to be small to make the training process short.

13.3 Feature Selection Methods Related with SVMs

13.3.1 Filtering Scores+SVM Wrapper

This approach first filters some features based on scores like Pearson correlation coefficients, Fisher scores or mutual information (Forman, 2003). It is then followed by feature selection using SVM as a wrapper to maximize the SVM performance such as the cross validation accuracy. This wrapper approach selects features in a manner of sequential forward/backward search as ordered by filtering scores.

Different scores may perform differently for different datasets, wrappers or the final predictors. The Fisher score, which is commonly used in SVM-related feature selection, is given by

$$s_t = (\mu_t^+ - \mu_t^-)^2 / ((\sigma_t^+)^2 + (\sigma_t^-)^2)$$

where μ_t^+, σ_t^+ are the mean and standard deviation for the t-th feature in the positive class. Similarly, μ_t^-, σ_t^- for the negative class. But for tasks with very unbalanced positive/negative training patterns, scores taking the unbalance into account might be more suitable. For example, we find a variant of odds ratio (Mladenic and Grobelnik, 1999) more suitable for the Dorothea data. It is calculated by

$$s_t = \exp(tpr_t - fpr_t),$$

where the tpr means the true positive rate: the rate of positive examples containing positive feature values, and fpr means the false positive rate: the rate of negative examples containing positive feature values. They can be more easily calculated according to the (Table 13.1).

Table 13.1. odds ratio

	Feature Value		
	0	1	
Class -1	a	b	$fpr=b/(a+b)$
Class +1	c	d	$tpr=d/(c+d)$

Simple scoring schemes plus the sequential search of wrapper are easy to implement and use. However, simple scoring schemes may not be able to distinguish redundant features, and cannot deal with features nonlinearly correlated with each other. Efforts can be put to invent or modify scoring schemes, to adopt complicated search strategies other than the score-based sequential one, or to use specific performance measures for wrappers. But alternatively some embedded feature selection methods such as the RFE and MU described below can be used.

13.3.2 Recursive Feature Elimination

The RFE approach operates in an iterative manner to eliminate features with small weights approximated from l_2–SVM. At each iteration, the weight of the t-th feature is given by the change of the objective value J in **P2** by leaving this feature out, assuming the set of all $\{\alpha_i\}$ will remain unchanged.

The objective value after leaving the t-th feature out is

$$J'_t = -\frac{1}{2}\sum_{i,j=1}^{m}\alpha_i\alpha_j y_i y_j \phi(\boldsymbol{x}_i \setminus \boldsymbol{x}_{it})^T \phi(\boldsymbol{x}_j \setminus \boldsymbol{x}_{jt})' + \sum_{i=1}^{n}\alpha_i$$

where $(\boldsymbol{x}_i \setminus \boldsymbol{x}_{it})$ denotes the input pattern i with feature t removed.

The weight of the t-th feature can be defined by

$$s_t = \sqrt{|\Delta J_t|} = \sqrt{|J - J'_t|}. \tag{13.1}$$

The following approximation suggested in (Guyon et al., 2003) is easier to calculate.

$$s_t^2 \approx \sum_{i,j=1}^{m}\alpha_i\alpha_j y_i y_j \phi(\boldsymbol{x}_{it})^T \phi(\boldsymbol{x}_{jt})' \tag{13.2}$$

Specifically, the feature weights are identical to the $\boldsymbol{w}$ if the SVM is built upon a linear kernel.

13.3.3 Multiplicative Updates

The multiplicative updates as proposed in (Weston et al., 2003) formulates feature selection as a l_0–SVM, and suggests iteratively multiplying the $\boldsymbol{w}$ specified by a l_1 or l_2 SVM back into each mapped pattern $\phi(\boldsymbol{x})$. Some features then will be deactivated by small $\boldsymbol{w}$ values. The MU method by its formulation is general to feature selection in some implicit feature space. But for kernels like RBF kernels the feature selection in feature space is hard because the $\boldsymbol{w}$ could be infinitely long.

The features we deal with in this paper are limited to those in the input space. To realize multiplicative updates by multiplying feature weights into patterns, we use the technique of RFE to approximate feature weights. This version of MU can be summarized as follows:

1   Let $\mathbf{z}$ be an all-one vector matching the length of input vectors
2   Solve the following SVM variant with $p = 2$:

P4:   Minimize over $(\boldsymbol{w}, b, \xi_1, \dots, \xi_m)$:

$$||\boldsymbol{w}||_p^p + C\sum_{i=1}^{m}\xi_i$$

subject to: $\forall_{i=1}^{m} : y_i(\boldsymbol{w}^T\phi(\mathbf{z} * \boldsymbol{x}_i) + b) \geq 1 - \xi_i, \xi_i > 0$

where $\mathbf{z}*\boldsymbol{x}_i$ denotes the component-wise product between the weight vector $\mathbf{z}$ and the input pattern $\boldsymbol{x}_i$ as given by $\mathbf{z}*\boldsymbol{w} = (\mathbf{z}_1\boldsymbol{x}_{i1}, \mathbf{z}_2\boldsymbol{x}_{i2}\ldots)$. Note the exact formulation of MU actually requires $\mathbf{z}*\boldsymbol{w}$.

3   Let $\boldsymbol{s}$ be the vector of all weights s_t based on (Eq.13.1 or 13.2). Set $\mathbf{z} \leftarrow \mathbf{z}*\boldsymbol{s}$.
4   Back to the step 2 till there are enough iterations/ features.

13.4 Transductive SVM-Related Feature Selection

Extending previous SVM-related feature selection methods to transductive feature selection is straightforward, only requiring replacing SVMs by TSVMs. Specifically, we can devise three TSVM-related feature selection methods: 1) Filtering scores+TSVM by using TSVMs as the wrapper and final predictor; 2) TSVM+RFE by iteratively eliminating features with weights calculated from TSVM models; and 3) TSVM+MU by deactivating features by weights approximated from TSVM models.

To save space, we simply describe the three TSVM-related feature selection approaches as the following general procedures.

1   Preprocess data and calculate filtering scores $\bar{\boldsymbol{s}}$. And optionally further normalize data.
2   Initialize $\mathbf{z}$ as an all-one input vector.
3   Set $\mathbf{z} \leftarrow \mathbf{z}*\bar{\boldsymbol{s}}$. Set part of small entries of $\mathbf{z}$ to zero according to a ratio/threshold, and possibly discretize non-zero $\mathbf{z}$ to 1.
4   Get a (sub-)optimal TSVM as measured by CV accuracy or other bounds,

$$\text{Minimize over} \ (y_1^*, \ldots, y_k^*, \boldsymbol{w}, b, \xi_1, \ldots, \xi_m, \xi_1^*, \ldots, \xi_k^*) :$$

$$\frac{1}{2}||\boldsymbol{w}||_2^2 + C\sum_{i=1}^{m}\xi_i + C^*\sum_{j=1}^{k}\xi_j^*$$

$$\text{subject to:} \quad \forall_{i=1}^{m} : y_i(\boldsymbol{w}^T\phi(\mathbf{z}*\boldsymbol{x}_i)' + b) \geq 1 - \xi_i, \xi_i > 0$$
$$\forall_{j=1}^{k} : y_j^*(\boldsymbol{w}^T\phi(\mathbf{z}*\boldsymbol{x}_j^*)' + b) \geq 1 - \xi_j^*, \xi_j^* > 0$$

5   For RFE or MU approaches, approximate feature weights $\bar{\boldsymbol{s}}$ from the model in step 4 according to (Eq.13.1 or Eq.13.2).
6   Go back to step 3 unless there is an expected number of features/iterations.
7   Output the final predictor and features indicated by large values of $\mathbf{z}$.

The step 3 involves selecting a ratio/number of features according to a threshold cutting the vector $\mathbf{z}$. For filtering scores and the RFE approach, the vector $\mathbf{z}$ is converted to a binary vector. And for the MU approach, part of small entries of $\mathbf{z}$ can be set to zero. And then the $\mathbf{z}*\boldsymbol{x}$ has the effect of pruning or deactivating some features.

The threshold is often set to prune a (fixed) number/ratio of features at each iteration. The quality of remaining features is then measured by the optimality of the TSVM model obtained in step 4. We generally use the cross-validation accuracy as the performance indicator of TSVM models. For a subset of features as selected by choosing a threshold value, we run the model search upon the free parameters like $[C, C^*, \sigma(RBF), d(Poly)]$ and determine the best parameter set that results in the highest cross validation accuracy.

13.5 Experimentation

13.5.1 Overall Performance on the Five Feature Selection Tasks

We only participated in the competition due on 1 Dec, 2003. During which both validation and test data are used as unlabelled data in our experiments. Our best challenge results are given in (Table 13.2). Compared with the winning challenge entry, our approach achieves positive scores for all tasks except the Arcene data. The Arcene data is challenging for us because of the overfitting caused by the nonlinear RBF kernels. For the Arcene data, we actually select features constructed from PCA on the Arcene data, and then train RBF TSVMs, but unfortunately the learning models we constructed does not generalize well. In some sense, it actually reflects the nontrivial importance of the learning machines in feature selection tasks for classification.

And for comparison with other contestant's submissions, we refer readers to Chen & Lin's paper, which has intensively investigated some SVM-related feature selection methods. Since their results even show better scores than ours, together with the fact that the McNemar tests conducted by the competition organizers show no significant difference between these top entries except the winning one, we hereby do not claim any guaranteed superiority of our TSVM-related approach. We mainly intend to show that SVM-related feature selection can be extended to TSVMs. These TSVM-related methods usually have comparable performance, and even show improvement for some tasks like the Dexter data.

For the five challenge tasks, we use various preprocessing techniques and different combination of feature selection procedures, which are briefly summarized in (Table 13.3). And more miscellaneous techniques can be found in the supplementary website [3]. Filtering scores are usually calculated first, and part of features with small values are then discarded by inspecting the score distribution. And then a normalization step is often adopted. For Arcene and Madelon data, we normalize the features to be with zero mean and unit standard deviation. As for Dexter and Dorothea data, we divide each entry of their data matrix by the square root of the production of row sum and column sum. At each iteration of wrapper feature selection, the model parameters are often set to be the same as those used in the final predictor. For Dexter and Madelon data, we also test the RFE and MU embedded approach.

---

[3]http://www.comp.hkbu.edu.hk/~vincent/nipsTransFS.htm

Table 13.2. NIPS 2003 challenge results for Wu & Li

Dec. 1st	Our best challenge entry				The winning challenge entry					
Dataset	Score	BER	AUC	Feat	Probe	Score	BER	AUC	Feat	Probe	Test
Overall	41.82	9.21	93.60	29.51	21.72	88.00	6.84	97.22	80.3	47.8	0.8
Arcene	-41.82	20.01	85.97	100	30	98.18	13.30	93.48	100.0	30.0	1
Dexter	70.91	4.40	97.92	29.47	59.71	96.36	3.90	99.01	1.5	12.9	0
Dorothea	78.18	11.52	88.48	0.5	18.88	98.18	8.54	95.92	100.0	50.0	1
Gisette	56.36	1.58	99.84	15	0	98.18	1.37	98.63	18.3	0.0	1
Madelon	45.45	8.56	95.78	2.6	0	100.00	7.17	96.95	1.6	0.0	1

Table 13.3. Methods Used. RBF: Gaussian radial basis function $\mathbf{k}(\boldsymbol{x}_i, \boldsymbol{x}_j) = \exp\left(-\frac{||\boldsymbol{x}_i - \boldsymbol{x}_j||^2}{\sigma^2}\right)$ where σ is a user-specified scaling factor and $g = 1/\sqrt{2\sigma^2}$. Lin: Linear kernel function $\mathbf{k}(\boldsymbol{x}_i, \boldsymbol{x}'_j) = \boldsymbol{x}_i^T \boldsymbol{x}'_j$. Poly: Polynomial kernel function $\mathbf{k}(\boldsymbol{x}_i, \boldsymbol{x}'_j) = (\boldsymbol{x}_i^T \boldsymbol{x}'_j + 1)^d$ where d is a user-specified positive integer.

	Arcene	Dexter	Dorothea	Gisette	Madelon
Filtering Scores		odds ratio	Fisher	Fisher	Fisher
Normalization	Centered PCA	Divide row and column sum			Mean 0, std 1
Wrapper	TSVM	TSVM	TSVM	SVM	TSVM
Embedded		RFE/MU			MU
Final Predictor	RBF$(C, g = 2^5, 2^{-6})$	Lin	Lin$(\frac{C_+}{C_-} = 20)$	Poly $(d = 2)$	RBF$(C, g = 1)$

In this paper, we present some main observations on dealing with the contest tasks. Specifically, in Sec.13.5.2, we show filtering scores together with TSVM wrappers are simple and useful. And then we present two examples of using the TSVM-RFE and TSVM-MU. At the end of this section, we analyze the performance of different methods in selecting salient features for a synthetic data.

13.5.2 Filtering Scores + TSVM Wrapper

Features relevant to labels are usually given large filtering scores. Even without rigorous criteria, a significant portion of features with small scores can be removed based on the intuition about the score distribution. After the initial feature pruning, and based on the feature ranking by filtering scores, several iterations of backward feature elimination guided by the performance of SVM/TSVM models can result in a more compact feature set.

For example, the Madelon data is artificially created to be very nonlinear and noisy. It often makes RBF SVMs perform as badly as random guess. But the Fisher scores can indicate that a small set of Madelon features are much more relevant to labels than others. The SVM and TSVM models further

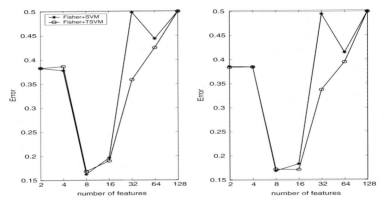

Fig. 13.2. 5-fold CV errors of Madelon training data (Left), and the average testing errors of validation data (Right) by the 5 CV models. (RBF kernel,$\sigma = \sqrt{2}/2, C = 1$)

pinpoint that the range of (8~16) features induces the most accurate and stable prediction, measured by the 5-fold cross validation(CV) errors of training data (Figure 13.2-left) or the mean error of predicting the validation dataset by using the five CV models (Figure 13.2-right).

Another example is the Dexter data(Figure 13.3), which are originated from the Reuters text categorization benchmark. Both the SVM and TSVM wrappers indicate about 30% features with highest Fisher scores lead to prediction improvement. Moreover, the clear performance gap between SVMs and TSVMs actually verifies that TSVMs can achieve better performance than the inductive SVMs for this text data (Joachims, 1999b), even with random features added by competition organizers.

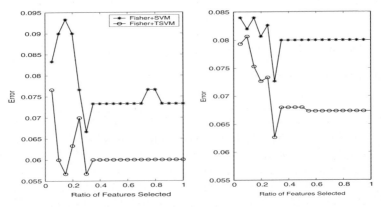

Fig. 13.3. 5-fold CV errors of Dexter training data (Left) and the mean testing errors of validation data (Right). (Linear kernel, $C = 1$)

The Dorothea data is very unbalanced with the ratio of positive and negative patterns in the training set approaching $78/722$. To deal with the unbalance property, we use the odd-ratio score, which favors the positive samples. And for model training, we follow the formulation of introducing an additional C to control the positive and negative patterns differently. Through model selection we find the ratio of 20 runs well. Results in (Figure 13.4) indicate that features ranked by odd-ratio scores are generally better than those based on Fisher scores. And the TSVM approach based on features ranked by odd-ratio has some advantages in constructing wrapper models and the final predictor.

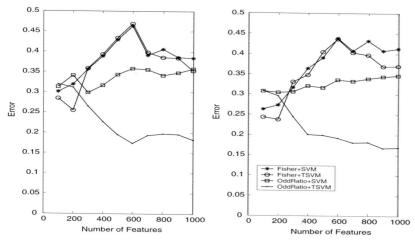

Fig. 13.4. 5-fold CV errors of Dorothea training data (Left) and the mean testing error of validation dataset (Right). (Linear kernel, $C = 1, C_+/C_- = 20$)

13.5.3 Embedded Feature Selection by TSVMs

TSVM-MU and SVM-MU

The experiment is conducted upon the Madelon data and results are shown in (Figure 13.5). The methods tested are the TSVM-MU and the SVM-MU. Thirty-two features with highest Fisher scores are selected as the starting feature set, because we know from a previous experiment(Figure 13.2) that the best generalization ability should lie in this feature subset if the Fisher scores are trustable. The performance of selecting 16,8,4,2 features using the TSVM-MU and SVM-MU are recorded. As shown from the results, both the TSVM-MU and SVM-MU improve the CV accuracy of the training data and the prediction rate of the validation dataset, compared with the Fisher-TSVM and Fisher-SVM methods. It means the multiplicative updates can refine the

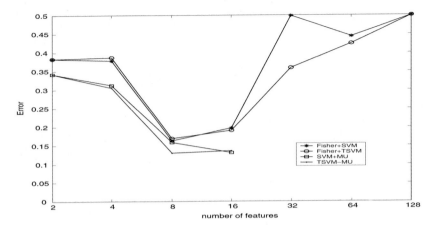

Fig. 13.5. 5-fold CV testing errors of Madelon validation data. (RBF kernel,$\sigma = 1, C = 1$)

features selected by the simple Fisher score. Statistically speaking, the refinement is not significant, especially by considering the cost of approximating the feature weights, which involves operations on the kernel matrix. For the comparison between TSVM-MU and SVM-MU, no strong conclusion can be drawn, though a slight gap can be observed for the two methods.

TSVM-RFE and SVM-RFE

We test the TSVM-RFE and the SVM-RFE on the Dexter data and show results in (Figure 13.6). The observation is that the TSVM-RFE runs better than either the SVM-RFE approach or the Fisher-TSVM. We further submit post-challenges entries to the competition system by taking only the test data as the unlabelled data. The number of features selected ranges from 500 to 6000. As shown in (Table 13.4), the TSVM-RFE has smaller probe ratios and balance error rates than the SVM-RFE. This indicates the TSVM-RFE approach can select better features than the SVM-RFE, and consequently ensures good classification results. These TSVM-RFE submissions also have good ranks compared with other contestants' submissions.

13.5.4 Feature Saliency Check

Without too much information on the features of the data in the competition, we use a toy dataset for feature saliency check and give some discussions on the tradeoff between feature saliency and classification.

We test an artificial dataset similar to the one in (Weston et al., 2003). In our experiments, the size of the test data is fixed at 500. In each of 50 trials, 30 training patterns are generated. The dataset is near linearly separable and has

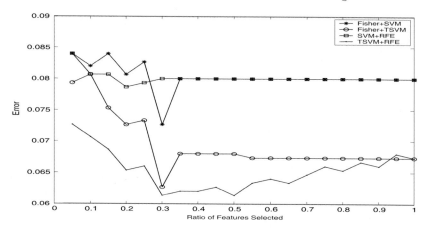

Fig. 13.6. 5-fold CV testing error of Dexter validation data. The 5-fold cross validated pair t-test shows the SVM-RFE and the TSVM-RFE have differences for the feature ratio up to 50% at the confidence rate 95%.(Linear kernel, $C = 1$)

Table 13.4. Comparisons between SVM+RFE and TSVM+RFE

	SVM+RFE			TSVM+RFE		
Feat	BER	AUC	Probe	BER	AUC	Probe
6000	0.0355	99.29	3155	0.0330	98.48	2649
3000	0.0375	99.28	1138	0.0340	98.67	923
500	0.0505	99.17	54	0.0350	98.88	31

100 features. But only six features are relevant and others are random probes. Moreover, among the six relevant features three are redundantly relevant to positive labels, and the other three are redundantly related to negative labels. The most compact size of features thus is two, though the number of relevant features can be up to six. We compare feature selection algorithms when they select 2, 4, 6 features.

We follow the way in (Weston et al., 2003) that uses a linear kernel and a large C ($10^{8 \sim 9}$ in our experiments) to force a hard-margin predictor. To avoid the difficulty in obtaining the hard-margin predictor due to wrongly selecting a high ratio of random probes, we limit the number of epochs of optimizing a SVM QP to be 10^5. If the maximum number of epochs is reached, we stop training the model.

The feature selection performance, as indicated by the average prediction accuracy on test data, is reported in (Table 13.5) and (Table 13.6). Results in the first table are only calculated from models that are converged within 10^5 epochs. Results in the second table are based on all models including those early-stopped.

Table 13.5. Average testing errors of predictions based on converged models. The number in bracket denotes the number of non-converged models within 50 trials

Method	$n_f = 2$	$n_f = 4$	$n_f = 6$
Fisher-SVM	0.2203(42)	0.0968(6)	0.0693
Fisher-TSVM	0.1400(49)	0.0568(19)	0.0538(3)
RFE-SVM	0.0538(15)	0.0412	0.0526
RFE-TSVM	**0.0369**(28)	0.0343(2)	0.0339(1)
MU-SVM	0.0544(6)	0.0420	0.0380
MU-TSVM	0.0459(25)	**0.0328**(2)	**0.0317**

Table 13.6. Average testing errors of predictions based on all models

Method	$n_f = 2$	$n_f = 4$	$n_f = 6$	$n_f = 100$
Fisher-SVM	0.2113	0.1204	0.0693	0.1516
Fisher-TSVM	0.2112	0.1295	0.0639	0.1495
RFE-SVM	0.1062	0.0412	0.0526	0.1516
RFE-TSVM	0.1046	0.0362	0.0348	0.1495
MU-SVM	**0.0705**	0.0420	0.0380	0.1516
MU-TSVM	0.0835	**0.0328**	**0.0317**	0.1495

The feature saliency measure we used is adapted from (Perkins et al., 2003). We define the following saliency score:

$$s = (n_g/n_r) + (n_u/n_f) - 1$$

where n_r is the number of relevant and non-redundant features of the data, and n_g is the number of relevant and non-redundant features selected by feature selection methods. For our task $n_r = 2$, and n_g could be $0, 1, 2$ in each trial. The n_u is the number of relevant features, which excludes random probes but allows redundancy. And n_f is the number of features selected by feature selection methods. We run experiments with n_f set to $2, 4, 6$ respectively. The range of saliency score s thus is $[-1, 1]$. For example, when $n_f = 2$, finding two relevant and non-redundant features gives a maximal score of 1, while finding two random features only results in a minimal score of -1. And the results of saliency scores are shown in Table 13.7.

Results shown in the three tables match some previous observations. All feature selection methods can be used to find some relevant features such that the prediction accuracy is better than training with all features. The TSVM approaches of RFE and MU achieve high accuracy and saliency when the feature size is moderate ($n_f = 4, 6$). But in the extreme case of selecting the

Table 13.7. Feature Saliency. Number in bracket denotes the hit times of selecting at least two non-redundant relevant features within 50 trials

Method	$n_f = 2$	$n_f = 4$	$n_f = 6$
Fisher	0.5100 (1)	0.5800 (23)	0.5367 (37)
RFE-SVM	0.7900 (33)	0.7600 (44)	0.5933 (45)
RFE-TSVM	0.7700 (33)	0.7800 (45)	0.6233 (46)
MU-SVM	**0.8600** (40)	0.7600 (47)	**0.6400** (48)
MU-TSVM	0.8000 (35)	**0.7850** (46)	0.6033 (47)

most compact size of features($n_f = 2$), the MU-SVM ranks the best with the highest feature saliency score (0.86) and success rate of training models (44/50). But note that the smallest classification error is not achieved when the most compact feature sets ($n_f = 2$) are found. It implies that the tasks of feature selection for classification should sacrifice part of feature saliency for a higher classification accuracy.

13.6 Conclusion and Discussion

The paper discusses the use of transductive SVMs for feature selection. It extends three SVM-related feature selection methods to the TSVMs setting. It shows that for some tasks the application of transductive SVMs to feature selection is superior to the application of inductive SVMs in terms of classification accuracy, and feature saliency.

Feature selection incorporated with transduction is a direction to be further explored. Since transduction has become a general learning paradigm containing many transductive algorithms, other transductive algorithms for feature selection are worthy of being studied. More generally, it is interesting to investigate under what circumstance the unlabelled data will (fail to) help feature selection and classification.

For the specific TSVM-based feature selection methods we have dealt with, many issues can be further studied. The computational cost is one of the key concerns. Though the TSVM implementation we use is flexible in training (controlled by the complexity parameter C* and early stopping), it causes great overhead when good TSVM models are needed. Besides, with the rather simple CV indication, the search for the best number of features is also computationally intensive. Efforts for addressing these issues will not only help feature selection, but also contribute to provide efficient TSVM solvers and and help solving challenging model selection problems.

13.7 Acknowledgement

This work is supported by HKBU FRG/01-02/II-67 and the HKBU's High Performance Cluster Computing Center. And the authors want to thank reviewers and editors for their insightful comments and suggestion.

References

K. Bennett and A. Demiriz. Semi-supervised support vector machines. In *NIPS*, 1998.

Bernhard E. Boser, Isabelle Guyon, and Vladimir Vapnik. A training algorithm for optimal margin classifiers. In *Computational Learing Theory*, pages 144–152, 1992.

Janez Brank, Marko Grobelnik, N. Milic-Frayling, and Dunja Mladenic. Feature selection using linear support vector machines. Technical report, 2002.

C.-C. Chang and C.-J. Lin. Libsvm: a library for svms, 2001.

George Forman. An extensive empirical study of feature selection metrics for text classification (kernel machines section). *JMLR*, pages 1289–1305, 2003.

G. Fung and O. Mangasarian. Semi-supervised support vector machines for unlabeled data classification. In *Optimization Methods and Software*, 2001.

Isabelle Guyon, Hans-Marcus Bitter, Zulfikar Ahmed, Michael Brown, and Jonathan Heller. Multivariate non-linear feature selection with kernel multiplicative updates and gram-schmidt relief. In *BISC FLINT-CIBI 2003 workshop*, 2003.

Isabelle Guyon, Jason Weston, Stephen Barnhill, and Vladimir Vapnik. Gene selection for cancer classification using support vector machines. *Machine Learning*, 46(1-3):389–422, 2002.

T. Jebara and T. Jaakkola. Feature selection and dualities in maximum entropy discrimination. In *Uncertainity In Artificial Intellegence*, 2000.

T. Joachims. Making large-scale support vector machine learning practical. In *Advances in Kernel Methods: Support Vector Machines*, 1999a.

Thorsten Joachims. Transductive inference for text classification using support vector machines. In *Proceedings of ICML-99*, pages 200–209, 1999b.

D. Mladenic and M. Grobelnik. Feature selection for unbalanced class distribution and naive bayes. In *ICML*, 1999.

Simon Perkins, Kevin Lacker, and James Theiler. Grafting: Fast, incremental feature selection by gradient descent in function space (kernel machines section). *JMLR*, pages 1333–1356, 2003.

J. Platt. Sequential minimal optimization: A fast algorithm for training support vector machines, 1998.

V. Vapnik. *Statistical Learning Theory*. Wiley, 1998.

J. Weston, F. Perez-Cruz, O. Bousquet, O. Chapelle, A. Elisseeff, and B. Schoelkopf. Feature selection and transduction for prediction of molecular bioactivity for drug design. *Bioinformatics*, 2002.

"Jason Weston, André Elisseeff, Bernhard Schölkopf, and Mike Tipping". Use of the zero-norm with linear models and kernel methods. *JMLR*, 2003.

Jason Weston, Sayan Mukherjee, Olivier Chapelle, Massimiliano Pontil, Tomaso Poggio, and Vladimir Vapnik. Feature selection for SVMs. In *NIPS*, pages 668–674, 2000.

S. Wu and P. A. Flach. Feature selection with labelled and unlabelled data. In *ECML/PKDD'02*, 2002.

E. Yu and S. Cho. Ga-svm wrapper approach for feature subset selection in keystroke dynamics identity verification. In *INNS-IEEE International Joint Conference on Neural Networks*, 2003.

Chapter 14

Variable Selection using Correlation and Single Variable Classifier Methods: Applications

Amir Reza Saffari Azar Alamdari

Electrical Engineering Department, Sahand University of Technology, Mellat Blvd., Tabriz, Iran. amir@ymer.org

Summary. Correlation and single variable classifier methods are very simple algorithms to select a subset of variables in a dimension reduction problem, which utilize some measures to detect relevancy of a single variable to the target classes without considering the predictor properties to be used. In this paper, along with the description of correlation and single variable classifier ranking methods, the application of these algorithms to the NIPS 2003 Feature Selection Challenge problems is also presented. The results show that these methods can be used as one of primary, computational cost efficient, and easy to implement techniques which have good performance especially when variable space is very large. Also, it has been shown that in all cases using an ensemble averaging predictor would result in a better performance, compared to a single stand-alone predictor.

14.1 Introduction

Variable and feature selection have become one of the most important topics in the machine learning field, especially for those applications with very large variable spaces. Examples vary from image processing, internet texts processing to gene expression array analysis, and in all of these cases handling the large amount of datasets is the major problem.

Any method used to select some variables in a dataset, resulting in a dimension reduction, is called a variable selection method, which is the main theme of this book. These methods vary from filter methods to more complex wrappers and embedded algorithms. Filter methods are one of the simplest techniques for variable selection problems, and they can be used as an independent or primary dimension reduction tool before applying more complex methods. Most of filter methods utilize a measure of how a single variable could be useful independently from other variables and from the classifier which, is to be used. So the main step is to apply this measure to each individual variable and then select those with the highest values as the best variables, assuming that this measure provides higher values for better variables. Correlation and single variable classifier are two examples of filter algorithms.

A.R.S.A. Alamdari: *Variable Selection using Correlation and Single Variable Classifier Methods: Applications*, StudFuzz **207**, 343–358 (2006)
www.springerlink.com   © Springer-Verlag Berlin Heidelberg 2006

In Section 14.2, there is a brief introduction to correlation and single variable classifier methods. Details about the mathematical description and concepts of these methods are not included in this section and unfamiliar readers can refer to Chapter 3 in this book for more details. Section 14.3 is an introduction to ensemble averaging methods used as the main predictors in this work, and in Section 14.4, the results and comparisons of applied methods on 5 different datasets of NIPS 2003 Feature Selection Challenge are shown. There is also a conclusion section discussing the results.

14.2 Introduction to Correlation and Single Variable Classifier Methods

Since Chapter 3 covers filter methods in details, this section contains only a short introduction to the correlation and single variable classifier feature ranking algorithms. Consider a classification problem with two classes, λ_1 and λ_2 represented by $+1$ and -1 respectively. Let $X = \{\mathbf{x}_k | \mathbf{x}_k = (x_{k1}, x_{k2}, \ldots, x_{kn})^T \in \mathbf{R}^n, k = 1, 2, \ldots, m\}$ be the set of m input examples and $Y = \{y_k | y_k \in \{+1, -1\}, k = 1, 2, \ldots, m\}$ be the set of corresponding output labels. If $\mathbf{x}^i = (x_{1i}, x_{2i}, \ldots, x_{mi})^T$ denotes the ith variable vector for $i = 1, 2, \ldots, n$ and $\mathbf{y} = (y_1, y_2, \ldots, y_m)^T$ represents the output vector, then the correlation scoring function is given below (Guyon and Elisseeff, 2003):

$$C(i) = \frac{(\mathbf{x}^i - \mu_i)^T (\mathbf{y} - \mu_y)}{\|\mathbf{x}^i - \mu_i\| \times \|\mathbf{y} - \mu_y\|} = \frac{\sum_{k=1}^{m}(x_{ki} - \mu_i)(y_k - \mu_y)}{\sqrt{\sum_{k=1}^{m}(x_{ki} - \mu_i)^2 \sum_{k=1}^{m}(y_k - \mu_y)^2}} \quad (14.1)$$

where μ_i and μ_y are the expectation values for the variable vector $\mathbf{x}^i$ and the output labels vector $\mathbf{y}$, respectively and $\|.\|$ denotes Euclidean norm. It is clear that this function calculates cosine of the angle between the variable and target vector for each variable. In other words, higher absolute value of correlation indicates higher linear correlation between that variable and target.

Single variable classifier (Guyon and Elisseeff, 2003) is a measure of how a single variable can predict output labels without using other variables. In other words, a single variable classifier method constructs a predictor using only the given variable and then measures its correct prediction rate (the number of correct predictions over the total number of examples) on the set of given examples as the corresponding single variable classifier value. The cross-validation technique can be used to estimate the prediction rate, if there is no validation set. Because this method needs a predictor and a validation algorithm, there exists no explicit equation indicating the single variable classifier values.

There is a very simple way to calculate the single variable classifier quantities. This method is used for all experiments in the application section. First of all, for each variable i, a class dependent variable set is constructed:

$X^{i,1} = \{x_{ki}|y_k = 1\}$ and $X^{i,-1} = \{x_{ki}|y_k = -1\}$. Let μ_i^1 and μ_i^{-1} be the expectation values of the $X^{i,1}$ and $X^{i,-1}$ sets, respectively. These values are the concentration point of each class on the *ith* variable axis. The following equation provides a simple predictor based only on the *ith* variable:

$$y = sign((x_i - \frac{\mu_i^1 + \mu_i^{-1}}{2})(\mu_i^1 - \mu_i^{-1})), \quad x_i \in \mathbf{R} \qquad (14.2)$$

where y is the estimated output label, x_i is the input value from *ith* variable, and the $sign(x)$ gives the sign of its input as $+1$ for $x \geq 0$ and -1 for $x < 0$. The first term inside the sign function, determines the distance and the direction of the input variable from the threshold point, $\frac{\mu_i^1 + \mu_i^{-1}}{2}$, and the second term determines the corresponding output class label due to the direction.

Assuming that the test set has the same statistical characteristics (i.e. means and variances), the correct prediction rate of this predictor on the training set can be used to determine the single variable classifier value for each of the variables, and there is no need to do cross-validation operations.

14.2.1 Characteristics of the Correlation and Single Variable Classifier

There are some characteristics of these methods which should be pointed out before proceeding to the applications. The main advantage of using these methods are their simplicity and hence, computational time efficiency. Other methods, which use search methods in possible subsets of variable space, need much more computation time when compared to filter methods. So, if there is a time or computation constraint, one can use these methods. In addition to the simplicity, these methods can also suggest how much class distributions are nonlinear or subjected to noise. In most cases, those variables with nonlinear correlation to the output labels, result in a low value and this can be used to identify them easily. Very noisy variables also can be thought as a highly nonlinear variable. As a result, the scoring functions described above give lower values for both of the noisy and nonlinear variables and it is not possible to distinguish between them using only these methods.

To gain more insight, consider a classification problem with two input variables, shown in Figure 14.1. Both variables are drawn from a normal distribution with different mean values set to (0,0) and (0,3) for class 1 and class 2, respectively. The standard deviations for both classes are equal to 1. The dataset plot is shown in upper right section together with axes interchanged in the lower left to simplify the understanding of images. Also, the histograms of each class distribution are shown in upper left for the vertical axis and in the lower right for the horizontal axis. The total number of examples is 500 for each class. On each axis, the correlation and single variable classifier values are printed. These values are calculated using the methods described

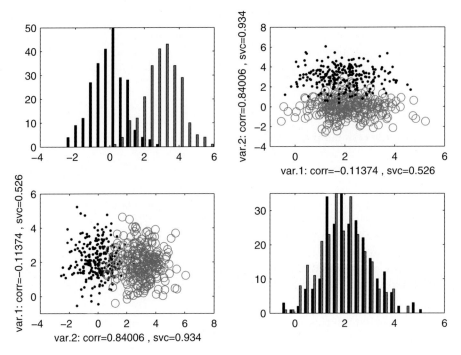

Fig. 14.1. A simple two variable classification problem: var.1 is a pure noise variable, var.2 is a linearly correlated one.

in previous section. As shown in Fig.1, regardless of the class labels, first variable is pure noise. The correlation value for a noisy variable is very low and the single variable classifier value is about 0.5, indicating that the prediction using this variable is the same as randomly choosing target labels. The second variable is a linearly correlated variable with the target labels, resulting in high values. For a nonlinear problem, consider Figure 14.2 which is the famous XOR classification problem. This time each variable has no prediction power when used individually, but can classify the classes when used with the other one. As shown in Figure 14.2, the class distribution on each axis is the same, similar to the situation in noisy variables, and both correlation and single variable classifier values are very low. Summarizing the examples, correlation and single variable classifier methods can distinguish clearly between a highly noisy variable and one with linear correlation to target values, and they can be used to filter out highly noisy variables. But in nonlinear problems these methods are less applicable and would conflict between noisy and nonlinear variables. Another disadvantage of these methods is the lack of redundancy check in the selected variable subset. In other words, if there were some correlated or similar variables, which carry the same information, these methods would select all of them, because there is no check to exclude the similar variables.

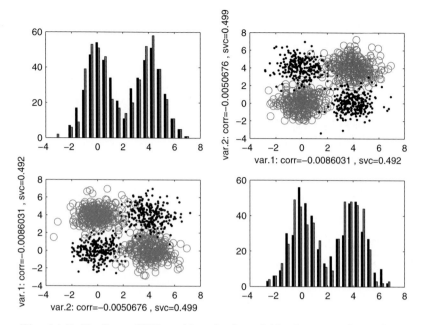

Fig. 14.2. Nonlinear XOR problem: both variables have very low values.

14.3 Ensemble Averaging

Ensemble averaging is a simple method to obtain a powerful predictor using a committee of weaker predictors (Haykin, 1999). The general configuration is shown in Figure 14.3 which illustrates some different experts or predictors sharing the same input, in which the individual outputs are combined to produce an overall output. The main hypothesis is that a combination of differently trained predictors can help to improve the prediction performance with increasing the accuracy and confidence of any decision. This is useful especially when the performance of any individual predictor is not satisfactory whether because of variable space complexity, over-fitting, or insufficient number of training examples comparing to the input space dimensionality.

There are several ways to combine outputs of individual predictors. The first one is to vote over different decisions of experts about a given input. This is called the voting system: each expert provides its final decision as a class label, and then the class label with the higher number of votes is selected as final output of the system.

If the output of each predictor before decision is a confidence value, , then another way to combine outputs is to average the predictor confidence for a given input, and then select the class with higherest confidence value as final decision. This scheme is a bit different from the previous system, because in voting each predictor shares the same right to select a class label, but in

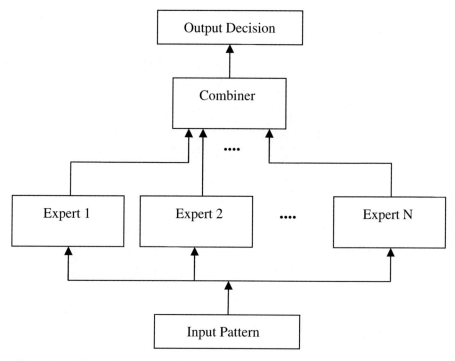

Fig. 14.3. General structure of an ensemble averaging predictor using N experts.

confidence averaging those with less confidence values have lower effect on the final decision than those with higher confidence values.

For example, consider a classification problem in which both of its class examples are drawn from a normal distribution with mean values of (0,0) for class 1 and (1,1) for class 2 with standard deviations equal to 1, as shown in Figure 14.4. The individual prediction error of 9 MLP neural networks with different initial weights is shown in Table 14.1. Here the values are prediction errors on 20000 unseen test examples. All networks have 2 tangent sigmoid neurons in hidden layer and are trained using scaled conjugate gradient (SCG) on 4000 examples.

The average prediction error of each network is about 0.4795 which is a bit better than a random guess, and this is due to the high overlap and conflict of class distributions. Using a voting method, the overall prediction error turns to be 0.2395 which shows a 0.2400 improvement. This is why in most cases ensemble averaging can convert a group of weak predictors to a stronger one easily. Even using 2 MLPs in the committee would result in a 0.2303 improvement. Additional MLPs are presented here to show that the bad performance of each MLP is not due to the learning processes. The effect of more committee members on the overall improvement is not so much here, but might be important in more difficult problems. Note that the second row

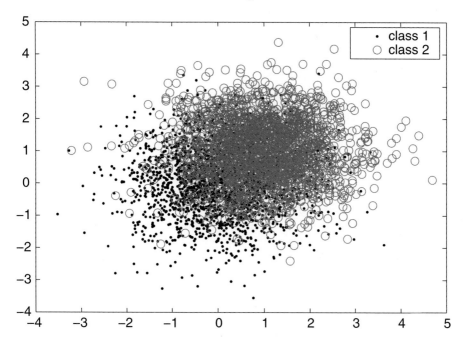

Fig. 14.4. Dataset used to train neural networks in ensemble averaging example.

in Table 14.1 shows the ensemble prediction error due to the addition of each
MLP to the committee.

Since the original distribution is normal, the Bayesian optimum estimation
of the class labels can be carried out easily. For each test example, its distance
from the mean points of each class can be used to predict the output label.
Using this method, the test prediction error is 0.2396. Again this shows that
ensemble averaging method can improve the prediction performance of a set of
weak learners to a near Bayesian optimum predictor. The cost of this process
is just training more weak predictors, which in most of cases is not too high
(according to computation time).

Table 14.1. Prediction error of individual neural networks, the first row, and the
prediction error of the committee according to the number of members in the ensemble, the second row.

Network No.	1	2	3	4	5	6	7	8	9
	0.4799	0.4799	0.4784	0.4806	0.4796	0.4783	0.4805	0.4790	0.4794
Member Num.	1	2	3	4	5	6	7	8	9
	0.4799	0.2492	0.2392	0.2425	0.2401	0.2424	0.2392	0.2403	0.2395

For more information about ensemble methods and other committee machines, refer to Chapter 1 and Chapter 7 in this book and also (Haykin, 1999, Bauer and Kohavi, 1999, Freund and Schapire, 1996, Opitz and Maclin, 1999). Note that this section is not related explicitly to the variable selection issues.

14.4 Applications to NIPS 2003 Feature Selection Challenge

This section contains applications of the discussed methods to the NIPS 2003 Feature Selection Challenge. The main goal in this challenge was to reduce the variable space as much as possible while improving the performance of predictors as higher as possible. There were five different datasets with different size of variable spaces ranging from 500 to 100,000. The number of training examples was also different and in some cases was very low with respect to the space dimensionality. In addition, some pure noisy variables were included in the datasets as random probes to measure the quality of variable selection methods.

The results of the correlation and single variable classifier analysis for each dataset are shown in Figure 14.5. Values are sorted in descending manner according to the correlation values. Since the descending order of variables for the correlation and single variable classifier values are not the same, there are some irregularities in the single variable classifier plots. Note that the logarithmic scale is used for the horizontal axis for more clarity on first parts of the plot.

Before proceeding to the applications sections, it is useful to explain the common overall procedures applied to the challenge datasets in this work. The dataset specific information will be given in next subsections. There are three different but not independent processes to solve the problem of each dataset: variable selection, preprocessing, and classification. The followings are the summarized steps for these three basic tasks:

(1.) First of all, constant variables, which their values do not change over the training set, are detected and removed from the dataset.
(2.) The variables are normalized to have zero mean values and also to fit in the $[-1, 1]$ range, except the DOROTHEA (see DOROTHEA subsection).
(3.) For each dataset, using a k-fold cross-validation (k depends on the dataset), a MLP neural network with one hidden layer is trained to estimate the number of neurons in the hidden layer.
(4.) The correlation and single variable classifier values are calculated and sorted for each variable in the dataset, as shown in Figure 14.5.
(5.) The first estimation for the number of good variables in each dataset is computed using a simple cross-validation method for the MLP predictor in step 2. Since an online validation test was provided through the challenge website, these numbers were optimized in next steps to be consistent with the actual preprocessing and also predictors.

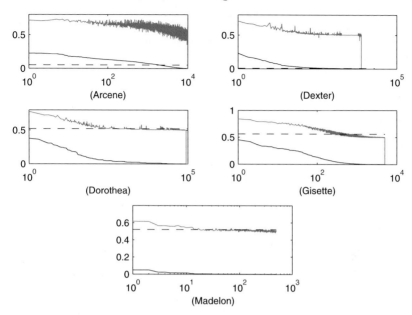

Fig. 14.5. Correlation and single variable classifier values plot for 5 challenge datasets. Correlation values are plotted with black lines while single variable classifier values are in grey. The dashed horizontal line indicates the threshold.

(6.) 25 MLP networks with different randomly chosen initial weights are trained on the selected subset using SCG algorithm. The transfer function of each neuron is selected to be tangent sigmoid for all predictors. The number of neurons in the hidden layer is selected on the basis of the experimental results of the variable selection step, but is tuned manually according to the online validation tests.

(7.) After the training, those networks with acceptable training error performances are selected as committee members (because in some cases the networks are stuck in a local minima during the training sessions). This selection procedure is carried out by filtering out low performance networks using a threshold on the training error.

(8.) For validation/test class prediction, the output values of the committee networks are averaged to give the overall confidence about the class labels. The sign of this confidence value gives the final predicted class label.

(9.) The necessity of a linear PCA (Haykin, 1999) preprocessing method usage is also determined for each dataset by applying the PCA to the selected subset of variables and then comparing the validation classification results to the non-preprocessing system.

(10.) These procedures are applied for both correlation and single variable classifier ranking methods in each dataset, and then one with higher validation performance (lower classification error) and also lower number of variables is selected as the basic algorithm for the variable selection in that dataset.

(11.) Using online validation utility, the number of variables and also the number of neurons in the hidden layer of MLPs are tuned manually to give the best result.

The following subsections have the detailed information about the application of the described methods on each dataset specifically. More information about this competition, the results, and the descriptions of the datasets can be found in the following website:
`http://clopinet.com/isabelle/Projects/NIPS2003/#challenge`.

14.4.1 ARCENE

This is the first dataset with a high number of variables (10000) and relatively low number of examples (100). The correlation values are sorted and those with values higher than 0.05 are selected, which is about 20% of all the variables, see Figure 14.5.a.

The correlation analysis shows that in comparison to other datasets discussed below, the numbers of variables with relatively good correlation values are high in ARCENE. As a result, it seems that this dataset consists of many linearly correlated parts with less contributed noise. The fraction of random probes included in the selected variables is 2.92% which again shows that correlation analysis is good for noisy variables detection and removal.

A linear PCA is applied to the selected subset and the components with low contribution to overall variance are removed. Then 25 MLP networks with 5 hidden neurons are trained on the resulting dataset, as discussed in previous section. It is useful to note that because of the very low number of examples, all networks are subject to over-fitting. The average prediction error for single networks on the unseen validation set is 0.2199. Using a committee the prediction error lowers to 0.1437 which shows a 0.0762 improvement. This result is expected for the cases with low number of examples and hence low generalization. The prediction error of ensemble on unseen test examples is 0.1924.

14.4.2 DEXTER

The second dataset is DEXTER with again unbalanced number of variables (20000) and examples (300). The correlation values are sorted and those with higher values than 0.0065 are selected which is about 5% of overall variables, see Figure 14.5.b.

Note that there are many variables with fixed values in this and others datasets. Since using these variables gains no power in prediction algorithm, they can be easily filtered out. These variables consist about 60% of overall variables in this dataset. There are also many variables with low correlation values. This indicates a highly nonlinear or a noisy problem compared to the previous dataset. Another fact that suggests this issue, is seen from the number of selected variables (5%) with very low threshold value of 0.0065 which is very close to the correlation values of pure noisy variables. As a result, the fraction of random probes included in the selected variables is 36.86%, which is very high.

There is no preprocessing for this dataset, except the normalization applied in first steps. 25 MLP networks with 2 hidden neurons are trained on the resulting dataset. The prediction error average for single networks on the unseen validation set is 0.0821, where using a committee improves the prediction error to 0.0700. The ensemble prediction error on the unseen test examples is 0.0495.

14.4.3 DOROTHEA

DOROTHEA is the third dataset. Its variable are all binary values with a very high dimensional input space (100000) and relatively low number of examples (800). Also this dataset is highly unbalanced according to the number of positive and negative examples, where positive examples consist only 10% of all the examples. The single variable classifier values are sorted and those with higher values than 0.52 are selected, which represents about 1.25% of variables, see Figure 14.5.c.

Fig.5.c together with the number of selected variables (1.25%) with low threshold value of 0.52 for single variable classifier shows that this problem has again many nonlinear or noisy parts. The fraction of random probes included in the selected variables is 13.22%, indicating that lowering the threshold value results in a higher number of noise variables to be included in the selected set.

In preprocessing step, every binary value of zero in dataset is converted to -1. 25 MLP networks with 2 hidden neurons are trained on the resulting dataset. Since the number of negative examples is much higher than positive ones, each network tends to predict more negative. The main performance measure of this competition was balanced error rate (BER), which calculates the average of the false detections according to the number of positive and negative examples by:

$$BER = 0.5(\frac{F_p}{N_p} + \frac{F_n}{N_n}) \tag{14.3}$$

where N_p and N_n are the total number of positive and negative examples, respectively, and F_p and F_n are the number of false detections of the positive and negative examples, respectively. As a result, the risk of an erroneous prediction for both classes is not equal and a risk minimization (Bishop, 1997) scenario must be used. In this way, the decision boundary, which is zero for other datasets, is shifted toward -0.7. This results in the prediction of negative label if the confidence were higher than -0.7. So, only the examples for which the predictor is more confident about them are detected as negative. The -0.7 bias value is calculated first with a cross-validation method and then optimized with online validation tests manually.

The prediction error average for single networks on the unseen validation set is 0.1643. The committee has prediction error of 0.1020 and shows a 0.0623 improvement, which is again expected because of the low number of examples, especially positive ones. The ensemble prediction error on the unseen test set is 0.1393.

14.4.4 GISETTE

The fourth dataset is GISETTE with a balanced number of variables (5000) and examples (6000). The single variable classifier values are sorted and those with higher values than 0.56 are selected which is about 10% of all the variables, see Figure 14.5.d.

single variable classifier analysis shows that this example is not much non-linear or subjected to noise, because the number of variables with good values is high. The fraction of random probes included in the selected variables is zero, indicating very good performance in noisy variables removal.

A linear PCA is applied and the components with low contribution to the overall variance are removed. Then 25 MLP networks with 3 hidden neurons are trained on the resulting dataset. Because of the relatively high number of examples according to the difficulty of the problem, it is expected that the performance of a committee and individual members would be close. Prediction error average for single networks on the unseen validation set is 0.0309. Using a committee, the prediction error only improves by 0.0019 and becomes 0.0290. The ensemble prediction error on the unseen test set is 0.0258.

14.4.5 MADELON

The last dataset is MADELON with (2000) number of examples and (500) variables. The single variable classifier values are sorted and those with higher values than 0.55 are selected, which is about 2% of variables, see Figure 14.5.e. This dataset is a highly nonlinear classification problem as seen from single variable classifier values. The fraction of random probes included in the selected variables is zero. Since this dataset is a high dimensional XOR problem, it is a matter of chance to get none of the random probes in the selected subset and this is not an indication of the power of this method.

There is no preprocessing for this dataset, except the primary normalization. 25 MLP networks with 25 hidden neurons are trained on resulting dataset. The number of neurons in hidden layer is more than other cases because of nonlinearity of class distributions. Prediction error average for single networks on unseen validation set is 0.1309 and combining them into a committee, prediction error improves by 0.0292 and reaches 0.1017. The ensemble prediction error on the unseen test set is 0.0944.

14.5 Conclusion

In this paper, the correlation and SVC based variable selection methods were introduced and applied to the NIPS 2003 Feature Selection Challenge. There was also a brief introduction to ensemble averaging methods and it was shown how a committee of weak predictors could be converted to a stronger one.

The overall performance of applied methods to 5 different datasets of challenge is shown in Table 14.2 together with the best winning entry of the challenge. Table 14.3 shows the improvements obtained by using a committee instead of a single MLP network for the validation sets of the challenge datasets.

Table 14.2. NIPS 2003 challenge results for Collection2.

Dec. 1st	Our best challenge entry					The winning challenge entry					
Dataset	Score	BER	AUC	Feat	Probe	Score	BER	AUC	Feat	Probe	Test
OVERALL	28.00	10.03	89.97	7.71	10.60	88.00	6.84	97.22	80.3	47.8	1
ARCENE	25.45	19.24	80.76	20.18	2.92	98.18	13.30	93.48	100.0	30.0	1
DEXTER	63.64	4.95	95.05	5.01	36.86	96.36	3.90	99.01	1.5	12.9	1
DOROTHEA	32.73	13.93	86.07	1.25	13.22	98.18	8.54	95.92	100.0	50.0	1
GISETTE	-23.64	2.58	97.42	10.10	0	98.18	1.37	98.63	18.3	0.0	1
MADELON	41.82	9.44	90.56	2	0	100.00	7.17	96.95	1.6	0.0	1

Table 14.3. Improvements obtained by using a committee instead of a single MLP network on the validation set.

OVERALL	ARCENE	DEXTER	DOROTHEA	GISETTE	MADELON
3.63	7.62	1.21	6.23	0.19	2.29

Summarizing the results, the correlation and single variable classifier are very simple, easy to implement, and computationally efficient algorithms, which have relatively good performance compared to other complex methods. These methods are very useful when the variable space dimension is large and other methods using exhaustive search in subset of possible variables need much more computations. On a Pentium IV, 2.4GHz PC with 512MB RAM running Microsoft Windows 2000 Professional, all computations for variable selection using MATLAB 6.5 finished in less than 15 minutes for both correlation and single variable classifier values of all 5 datasets. This is quite great performance if one considers the large size of the challenge datasets.

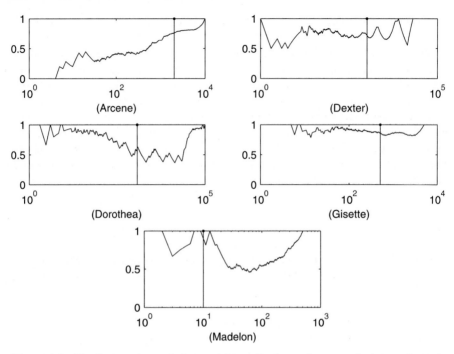

Fig. 14.6. Similarity plots of the variable selection using correlation and single variable classifier methods on challenge datasets. Note that the vertical solid line indicates the number of selected variables for each dataset in the competition.

A simple comparison between the correlation and single variable classifier ranking methods is given in Figure 14.6. Let S_{COR}^N and S_{SVC}^N be the subsets of the original dataset with N selected variables according to their rankings using the correlation and single variable classifier, respectively. In this case the vertical axis of Figure 14.6 shows the fraction of the total number of common elements in these two sets per set sizes, i.e. $N_c = \frac{F(S_{COR}^N \cap S_{SVC}^N)}{N}$, where $F(.)$ returns the number of elements of the input set argument. In other words, this figure shows the similarity in the selected variable subsets according to the correlation and single variable classifier methods.

Table 14.4. The average rate of the common variables using correlation and single variable classifier for the challenge datasets, first row. Second row presents this rate for the number of selected variables in the application section.

	OVERALL	ARCENE	DEXTER	DOROTHEA	GISETTE	MADELON
Average	0.7950	0.8013	0.7796	0.7959	0.8712	0.7269
Application	0.7698	0.7661	0.7193	0.6042	0.8594	0.9000

As it is obvious from this figure, the correlation and single variable classifier share most of the best variables (left parts of the plots) in all of the datasets, except ARCENE. In other words, linear correlation might result in a good single variable classifier score and vice versa. Table 14.4 shows the average of these plots in first row, together with the rate of the common variable in the selected subset of variables for the application section discussed earlier. Another interesting issue is the relation between the rate of the random probes and the rate of the common variables in the subsets of datasets used in the applications. For GISETTE and MADELON the total number of selected random probes was zero. Table 14.4 shows that the common variable rate for these two datasets are also higher, comparing to other datasets. The DEXTER and DOROTHEA had the worst performance in filtering out the random probes, and the rates of common variables for these two sets are also lower than others. In other words, as the filtering system starts to select the random probes, the difference between the correlation and single variable classifier grows higher. Note that these results and analysis are only experimental and have no theoretical basis and the relation between these ranking and filtering methods might be a case of future study.

It is obvious that these simple ranking methods are not the best ones to choose a subset of variables, especially in nonlinear classification problems, in which one has to consider a couple of variables together to understand the underlying distribution. But it is useful to note that these methods can be used to guess the nonlinearity degree of the problem and on the other hand filter out very noisy variables. As a result, these can be used as a primary analysis and selection tools in very large variable spaces, comparing to methods and results obtained by other challenge participants.

Another point is the benefits of using a simple ensemble averaging method over single predictors, especially in situations where generalization is not satisfactory, due to the complexity of the problem, or low number of training examples. Results show a 3.63% improvement in overall performance using an ensemble averaging scenario over single predictors. Training 25 neural networks for each dataset take 5-30 minutes on average depending on the size of the dataset. This is fast enough to be implemented in order to improve prediction performance especially when the numbers of training examples are low.

The overall results can be found in the challenge results website under Collection2 method name. Also, you can visit the following link for some MATLAB programs used by the author and additional information for this challenge: http://www.ymer.org/research/variable.htm.

References

E. Bauer and R. Kohavi. An empirical comparison of voting classification algorithms: bagging, boosting, and variants. *Machine Learning*, 36:105–142, 1999.

C. M. Bishop. *Neural networks for pattern recognition*. Clarendon, Oxford, 1997.

Y. Freund and R. E. Schapire. Experiments with a new boosting algorithm. In *Proceedings of the Thirteen International Conference of Machine Learning*, pages 148–156, 1996.

I. Guyon and A. Elisseeff. An introduction to variable and feature selection. *Machine Learning Research*, 3:1157–1182, 2003.

S. Haykin. *Neural networks: a comprehensive foundation*. Prentice-Hall, New Jersy, 1999.

D. Opitz and R. Maclin. Popular ensemble methods: an empirical study. *Artificial Intelligence Research*, 11:169–198, 1999.

Chapter 15

Tree-Based Ensembles with Dynamic Soft Feature Selection

Alexander Borisov, Victor Eruhimov and Eugene Tuv

Intel,
alexander.borisov@intel.com, victor.eruhimov@intel.com, eugene.
tuv@intel.com

Summary. Tree-based ensembles have been proven to be among the most accurate and versatile state-of-the-art learning machines. The best known are MART (gradient tree boosting) and RF (Random Forest.) Usage of such ensembles in supervised problems with a very high dimensional input space can be challenging. Modelling with MART becomes computationally infeasible, and RF can produce low quality models when only a small subset of predictors is relevant. We propose an importance based sampling scheme where only a small sample of variables is selected at every step of ensemble construction. The sampling distribution is modified at every iteration to promote variables more relevant to the target. Experiments show that this method gives MART a very substantial performance boost with at least the same level of accuracy. It also adds a bias correction element to RF for very noisy problems. MART with dynamic feature selection produced very competitive results at the NIPS-2003 feature selection challenge.

15.1 Background

It is difficult to overestimate the influence of decision trees in general and CART (Breiman et al., 1984) in particular on machine and statistical learning. CART has practically all the properties of a universal learner: fast, works with mixed-type data, elegantly handles missing data, invariant to monotone transformations of the input variables (and therefore resistent to outliers in input space.) Another key advantage of CART is an embedded ability to select important variables during tree construction. Because it is a greedy, recursive algorithm, where impurity reduction is maximized through an exhaustive search at every split, CART is resistant to irrelevant inputs. Moreover, it is easy to define an adequate measure of variable importance such that its computation requires practically no extra work (Breiman et al., 1984):

$$VI(x_i, T) = \sum_{t \in T} \Delta I(x_i, t) \qquad (15.1)$$

A. Borisov et al.: *Tree-Based Ensembles with Dynamic Soft Feature Selection*, StudFuzz **207**, 359–374 (2006)
www.springerlink.com

where $\Delta I(x_i, t) = I(t) - p_L I(t_L) - p_R I(t_R)$ is the decrease in impurity due to an actual (or potential) split on variable x_i at a node t of the optimally pruned tree T. Node impurity $I(t)$ for regression is defined as $\sum_{s \in t} (y_s - \bar{y})^2$ where the sum and mean are taken over all observations s in node t. For classification $I(t) = Gini(t)$, where $N(t)$ is the number of observations in node t, and $Gini(t)$ is the Gini index of node t:

$$Gini(t) = \sum_{i \neq j} p_i^t p_j^t \tag{15.2}$$

where p_i^t is the proportions of observations in t whose response label equals i ($y = i$) and i, j run through all response class numbers. The sum in Equation 15.1 is taken over all internal tree nodes where x_i was a primary splitter or a surrogate variable.

The main limitation of CART is its relatively low accuracy. In response, there was an explosive development of model averaging methods (Breiman, 1996, Freund and Schapire, 1996, Ho, 1998, Breiman, 1998, Dietterich, 2000) during the last decade that resulted in series of very accurate tree-based ensembles. An overview of ensemble methods is given in Chapter 7. Two of the most recent advances in tree ensembles - MART (gradient tree boosting) (Friedman, 1999a,b) and Tree Forests (Ho, 1998, Breiman, 2001) have been proven to be among the most accurate and versatile state-of-the-art learning machines. MART is a serial ensemble where every new expert constructed relies on previously built experts. At every iteration l of MART a new tree T_l is fitted to the generalized residuals with respect to a loss function Ψ

$$-\left[\frac{\partial \Psi(y_i, F(x_i))}{\partial F(x_i)} \right]_{F=F_{l-1}} \tag{15.3}$$

giving terminal regions $R_{jl}, j = 1, 2, ..., J_l$. The corresponding constants γ_{jl} are solutions

$$\gamma_{jl} = \arg \min_{\gamma} \sum_{x_i \in R_{jl}} L(y_i, F_{l-1}(x_i) + \gamma) \tag{15.4}$$

and

$$F_l(\mathbf{x}) = F_{l-1}(\mathbf{x}) + \nu \cdot \sum_{j=1}^{J_l} \gamma_{jl} I(\mathbf{x} \in R_{jl}) \tag{15.5}$$

where $0 < \nu < 1$ is a regularization parameter (learning rate.) The solution is given by

$$\hat{F}(\mathbf{x}) = F_L(\mathbf{x}), \tag{15.6}$$

where the size of the ensemble L is chosen to avoid overfitting (usually by monitoring validation errors.)

RF builds trees independently of each other on a bootstrap sample of the training data, and predicts by a majority vote (or an average in regression.)

At each tree split, a random, small subset of the variables participate in split selection (typically the square root of total number of variables.)

MART and RF inherit all nice properties of a single tree, and also provide (as a byproduct) more reliable estimate of the variable importance. The importance measure Equation 15.1 is averaged over the trees in the ensemble

$$VI(x_i) = \frac{1}{L} \sum_{l=0}^{L} VI(x_i, T_l) \qquad (15.7)$$

MART builds shallow trees using all variables (on a subsample of the training data), and hence, it can handle large datasets with a moderate number of inputs. Very high dimensional data (NIPS-2003 challenge) is extremely challenging for MART. RF builds maximum-sized trees on a bootstrap sample of the training data but chooses a small, random subset of variables at every split. Therefore, it can easily handle thousands of variables in datasets of moderate size, but its performance could be significantly degraded when only a small fraction of inputs is relevant. Thus, for massive and possibly noisy problems some kind of hybrid scheme is very tempting. Such a hybrid ensemble with shallow trees and dynamic variable selection is the primary goal of this study.

15.2 Dynamic Feature Selection

The main idea in this work is to select a small sample of features at every step of the ensemble construction. The sampling distribution is modified at every iteration to promote more relevant features. Throughout this paper we assume that the result of an algorithm's *iteration* is a new base learner (tree) added to an ensemble. Different strategies are considered for serial (MART) and parallel (RF) ensembles.

For MART the proposed strategy attempts to change the sampling distribution for every new expert by up-weighting variables that are more relevant to its learning task (based on the generalized residuals from previous iterations.) A measure of relevance can be generated by using variable importance evaluated over a historical window of prior iterations. This window could be moving, weighted by distance in time, shrinking in time, etc. Naturally, the sampling strategy is closely tied to the MART's regularization coefficient.

For the parallel ensemble (RF), variable relevance to a specific iteration is not an issue. Thus, a reasonable strategy is to minimize the sampling weights for noisy variables while maintaining the sampling probabilities for relevant variables dominant and preferably balanced (to preserve diversity.) Sampling weights could be initialized using prior knowledge, data (such as an initial run of a single tree, for a simple univariate measure of relevance, etc), or set to equal values. For the NIPS challenge we used a simpler strategy to exclude irrelevant variables from the model construction (MART), and at the same time ensuring that less important but relevant variables have a chance to be selected. We operated under the assumption that only a small fraction of variables was relevant (given the huge dimensionality of the datasets.) We chose the variable sampling size large enough to allow all relevant variables to enter the model after an initial "burn-in" period. The sampling weight evaluation window was set to maximum (all prior iterations were used.)

15.2.1 MART and RF in the Presence of Noise

Since MART exhaustively searches through all inputs for the best split, it should be more resistent to noise than RF. The following simulation study illustrates this point. Figure 15.1 presents the comparison of MART and RF predictive accuracy on 200 artificial multi-class problems with 50 inputs (most of which are irrelevant to the target.) The data sets were created using the artificial data generator described in Section 15.3.1. We set the number of variables sampled by RF to 9 (larger values did not improve predictive accuracy.) MART was built with 500 trees of depth 5, learning rate set to 0.1.

One can see that in the presence of noise MART is consistently more accurate (but much slower) than RF, and the difference increases when the fraction of relevant variables shrinks.

15.2.2 MART with Dynamic Feature Selection

For problems in high dimensional input space it would be desirable to apply the variable-sampling technique used by RF. Obviously, such a hybrid scheme would dramatically decrease the computational complexity of MART. However, as shown in the Section 15.3, the uniform sampling used in the parallel ensemble (RF) could cause a significant performance degradation for sequentially boosted trees (MART.)

Instead the variable-sampling probability is modified at every iteration to reflect variable importance learned from the previous iterations (as a byproduct of tree construction) Equation 15.7. Specifically for the NIPS-2003 submission (with datasets with up to 100K inputs) we applied the following sampling scheme:

a small, fixed subset S from a set M of all input variables is sampled with replacement but only distinct elements are kept (sampling without replacement is more appropriate here but it is more computationally expensive.) The sampling probability for a variable i at iteration l is proportional to the corresponding weight

$$p(x_i, l) = w(x_i, l) / \sum_j w(x_j, l) \tag{15.8}$$

Weights have two components

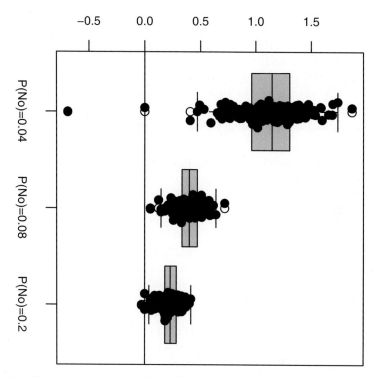

Fig. 15.1. Distribution (log scale) of RF error rates relative to MART on a seven-class problem for different proportions $P(N_0)$ of variables relevant to the response

$$w(x_i, l) = I(x_i, l) + SVI(x_i, l), \tag{15.9}$$

where $SVI(x_i, l) = \sum_{j=1}^{l} VI(x_i, j)$, $VI(x_i, j)$ is the squared-influence of the i-th variable in the j_{th} tree as defined by Equation 15.1; $I(x_i, l)$ is a contribution of the initial influence $I(x_i, 0)$ for the i_{th} variable at the l_{th} iteration. We used the exponentially decreasing in time initial influences

$$I(x_i, l) = I(x_i, 0) \cdot (1 - S/M)^{\alpha l}, \tag{15.10}$$

where α is an adjustable parameter controlling how fast initial weights decrease (empirically chosen in range 0.5-2.) Here, $I(x_i, 0)$ represents prior knowledge about the variable influences, and this governs the sampling weights for a number of initial iterations. For the NIPS challenge we used $I(x_i, 0)$ equal to the sum of variances of log-odds for both response classes.

It is obvious that $I(x_i, l)$ decreases and $SVI(x_i, l)$ grows with the number of iterations l. Therefore, for sufficiently low initial influences, the learned variable importance will dominate the sampling weights after a number of iterations. Note that sampling with replacement (versus without) reduces the computational time for the NIPS data sets up to 5 times. However, it poses additional challenges related to potential "overweighting" effect for a small group of influential variables preventing other relevant variables from entering the model. This effect could be controlled by a gradual transition from initial to learned importance based upon the weights.

Here is the formal description of the dynamic feature weighting algorithm.

MART-WS algorithm for gradient tree boosting with dynamic feature selection

(1.) Set $I(x_i, 0)$ for $i = 1, \ldots, n$ to the initial response deviation. Initialize $w(x_i, 0) = I(x_i, 0)$. Set current residuals (responses) to output variable values. Set $SVI(x_i, 0) = 0$.

(2.) Fit the next MART tree to the current pseudo residuals (15.3), using $p(x_i, l) = w(x_i, l)/\sum_i w(x_i, l)$ as the selection weights. At each tree node, a small fixed number $n_0 << n$ of variables is selected with replacement using selection probabilities $p(x_i, l)$ and the best split is searched only amongst this subset. l is the current iteration number.

(3.) Calculate variable importance $VI(x_i, l)$ on the i-th variable as in (15.1)

(4.) Calculate $SVI(x_i, l+1) = SVI(x_i, l) + VI(x_i, l)$. Update variable weights as $w(x_i, l+1) = I(x_i, 0) \cdot (1 - S/M)^{\alpha \cdot (l+1)} + SVI(x_i, l+1)$.

(5.) Update residuals with the difference between the predicted values and the old residuals.

(6.) Return to step 2 if the maximum iteration number is not exceeded ($l < l_{max}$).

It is also important to note that MART with this dynamic variable selection strategy normally requires a smaller learning/regularization rate, and more iterations might be needed to reach optimal performance. Still it is a very small price to pay for a huge performance gain compared to the standard MART (up to 100 times on NIPS data.)

Example 1 As an illustration consider a simple example with two simulated data sets of 3000 samples consisting of $N = 100$ and $N = 400$ inputs, $x_{i=1,..,N}$ distributed uniformly in $(0,1)$ with response $y = x_1$. A 3/2 partition was used for training/test. For a classification setting y is discretized to 5 classes.

Figure 15.2 shows that MART with uniform feature selection needs to sample almost all variables to match the standard MART accuracy. MART with weighted variable sampling achieves good performance with few selected variables for both regression and classification, and obviously, with dramatic gain in speed.

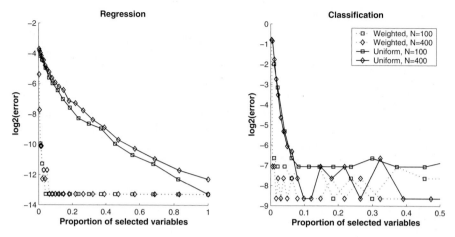

Fig. 15.2. MART performance with uniform and weighted sampling on **Example 1** data with one relevant input, and $N - 1$ irrelevant.

15.2.3 Application of Weighted Sampling to RF

A similar sampling strategy could be applied to RF construction. In the case of a parallel ensemble (RF) each new expert is trained on a bootstrapped version of the same data (whereas residuals from the previous iterations in MART.) The goal is to prevent irrelevant variables from entering the model, and at the same time to allow all relevant inputs to participate at every iteration of the model construction.

The proposed strategy is to adjust sampling weights (15.9) in such a way that the *expected change* in weights is independent from the previous iterations:

$$w(x_i, l) = w(x_i, l - 1) + VI(x_i, l)/p(x_i, l)^{(S)} \tag{15.11}$$

where $p(x_i, l)^{(S)}$ is a probability for the variable i to enter subset S of split candidates at iteration l. This probability can be estimated as $1 - (1 - p(x_i, l))^{|S|}$. Here $p(x_i, l)$ is the sampling probability of variable i at iteration l. Note that there is a considerable difference between MART and RF in terms of $SVI(x_i, l)$ dependence on l. Since MART learns each next tree in the spirit of gradient descent, $SVI(x_i, l)$ tends to a constant with the growth of l. RF learns each tree independently so $SVI(x_i, l)$ is not bounded above but grows like $O(l)$. This is the main reason why the importance weight strategies for MART and RF have to be different. This difference is accounted for by the factor $p(x_i, l)$ in (15.11) and by constant initial influences $w(x_i, 0) = C \cdot I(x_i, 0)$ (we set C to values in the range from 0.1-10.)

Figure 15.3 shows relative performance of standard RF and RF with the dynamic sampling on data from **Example 1**. It is clear that weighted sampling significantly improves accuracy in regression. In classification, there is an improvement for noisier data ($N=400$.) Both schemes provide comparable accuracy in classification when the number of selected features is large enough.

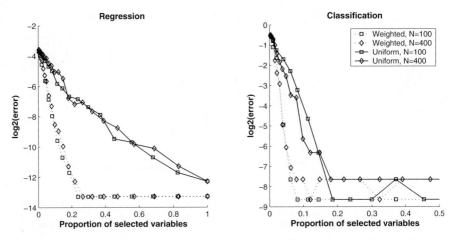

Fig. 15.3. RF performance with uniform (standard) and weighted sampling on **Example 1** data (section 15.2.2) with one relevant input, and $N - 1$ irrelevant.

However, RF with weighted variable sampling achieves the best predictive accuracy with half as many variables per split as does standard RF.

15.2.4 Relative Importance of Input Variables for Ensembles with Weighted Sampling

We did a small simulation to study how relative variable importance estimated from the final model changes when the weighted feature selection is used. As an example, we considered the linear model $y = \sum_{i=1}^{4} a_i x_i + \varepsilon$, where $a_i = i$, and ε is small $N(0, 0.5)$ noise. Independent noise variables x_i, $i = 5, \ldots 50$ were added. The joint distribution for all $\mathbf{x} \sim N(\mathbf{0}, \mathbf{I})$. Here, the expected variable importance for $x_{i=1,2,3,4}$ as measured by the squared-influence (15.7) is a_i^2. MART and RF with uniform (-US) and weighted (-WS) sampling were trained, and variable importances averaged over 100 replications. Table 15.1 has the summarized relative squared-influence results for the 4 original variables and an extra row with the maximum relative squared-influence achieved by one of the 50 noise variables. Variables with the highest influence were assigned score 1, and importances of the rest were scaled correspondingly (for each algorithm.)

From Table 15.1 one can see that the relative influence of noise variables learned by both MART and RF is consistently (with tight variance) and significantly smaller when the weighted variable selection (WS) mechanism was used (more so for RF, up to 8 times.) All the methods resulted in relative variable importances consistent with what was expected for the four relevant inputs.

Table 15.1. Input variables relative average squared-influences (and standard deviations) for a linear model learned by MART/RF with uniform and weighted (US/WS) variable sampling. The first column shows the "true" variable relative squared-influences. The last row shows the maximum relative importance of noise variables. Noise is much better separated from the relevant variables with the weighted scheme.

$True RelativeVI_i$	MART-WS	MART-US	MART	RF	RF-WS
0.063	0.050(0.007)	0.076(0.007)	0.073(0.006)	0.088(0.01)	0.036(0.005)
0.25	0.285(0.015)	0.284(0.02)	0.268(0.013)	0.284(0.04)	0.215(0.014)
0.563	0.610(0.025)	0.597(0.027)	0.595(0.029)	0.610(0.07)	0.562(0.025)
1	1.000(0)	1.000(0)	1.000(0)	1.000(0)	1.000(0)
max noise	0.003(5e-4)	0.022(0.002)	0.007(0.001)	0.044(0.003)	0.008(6e-4)

15.3 Experimental results

Denote by MART-WS MART with weighted variable selection, MART-US as MART with uniform variable selection, RF-WS as RF with weighted variable selection. We conducted the following experiments:
1) Compared MART, MART-WS, MART-US, RF and RF-WS on 9 UCI repository and artificial classification/regression data sets.
2) Explored how MART-WS, RF-WS and RF predictive accuracy depends on the proportion of relevant variables and the number of variables sampled.
3) Explored different variable sampling weights $w_i(t)$ initialization strategy. Error was reported as an average test error over multiple runs with a 3/2 train/test partition.

15.3.1 Artificial Data Generator

For generating diverse functional dependencies we use Freidman's random function generator (Freidman 1999a.) Each target function takes the form

$$f(\mathbf{x}) = \sum_{l=1}^{L} a_i g_l(\mathbf{x})$$

where L is an adjustable parameter (50 in our experiments), a_i are random coefficients taken from a uniform distribution $U[-1,1]$, and each $g_l(x)$ is a random function of a randomly selected subset of size n_l, of N_0 input variables. More precisely, each $g_l(x)$ is a multi-dimensional Gaussian function of a selected feature subset with random mean vector and random covariance matrix. The size of each subset is itself taken to be a random number $[1.5 + r]$, where r is drawn from an exponential distribution with mean equal to 2. Thus, the expected number of inputs for each of $g_l(x)$ is between 3 and 4, but there are some higher order interactions if L is large enough. All x_i are sampled from the standard normal distribution $N(\mathbf{0}, \mathbf{I})$. For classification problems we generated $K = 7$ class labels by thresholding each target at corresponding quantiles. Each data set has 5000 observations.

15.3.2 UCI Repository Data Sets

In the first series of experiments we ran all five algorithms (MART, MART-WS, MART-US, RF, RF-WS) against 9 UCI repository classification problems. Parameters were chosen based on an initial exploratory 500 iteration run for each algorithm. Test errors were averaged over 200 runs.

Table 15.2. Misclassification rates for MART, MART-WS, MART-US, RF, RF-WS on UCI repository classification data sets.

Dataset	Samples	Variables	MART-WS	MART-US	MART	RF-WS	RF
connect	7000	43	0.2129	0.2150	0.2062	0.2298	0.2319
dna	1999	181	0.0349	0.0339	0.0354	0.0496	0.0428
letter	20000	17	0.0381	0.0403	0.0404	0.0435	0.0441
musk	6598	167	0.0123	0.0125	0.0124	0.0254	0.0252
diabetes	768	9	0.2283	0.2306	0.2301	0.2268	0.2265
satellite	6435	37	0.0789	0.0785	0.0799	0.0854	0.0853
segment	2310	20	0.0203	0.0213	0.0189	0.023	0.0238
shuttle	10000	10	0.0017	0.0013	0.0011	0.0016	0.0016
spectro	531	94	0.4518	0.4608	0.4791	0.4277	0.4350

The results summarized in Table 15.2 show that MART-WS and MART-US maintain the same predictive accuracy as MART, being 3-15 times faster (with the number of variables sampled $S \sim \frac{1}{2\sqrt{|M|}}$.) However, as shown in the next section, MART-US's relative performance could be worse in the presence of noise. RF-WS is not superior to ordinary RF on the UCI repository data. We tried different variable weights initialization for MART/RF-WS without an evident difference in accuracy.

15.3.3 NIPS Challenge Data

For the NIPS challenge submission we used MART-WS. No data preprocessing was done. Table 15.3 shows cross-validation errors for the challenge datasets for RF, RF-WS, MART, MART-WS. In three cases (where we were able to run the standard MART) MART-WS significantly outperforms MART. It is also nearly 100 times faster. In 3 out of 5 cases RF with feature selection does better than RF.

Table 15.3. NIPS challenge datasets cross-validation errors for best tuned MART,MART-WS,RF,RF-WS. For two datasets it was computationally infeasible to tune the standard MART (error value in the table replaced by "-")

Data	Variables	Observations	Best	RF	RF-WS	MART	MART-WS
GISETTE	5000	6000	0.0126	0.0265	0.026	0.028	0.016
MADELON	500	2000	0.0622	0.270	0.148	0.205	0.1256
ARCENE	10000	100	0.072	0.19	0.13	0.23	0.14
DEXTER	20000	300	0.033	0.08	0.066	-	0.08
DOROTHEA	100000	800	0.0854	0.057	0.056	-	0.0892

Table 15.4. NIPS 2003 challenge, results for MART-WS.

	MART − WS best entry				The winning entry			
Dataset	BER	AUC	Feat	Probe	BER	AUC	Feat	Probe
GISETTE	0.0189	0.9985	12.00	0.00	0.0126	0.9992	94.48	50.40
ARCENE	0.1304	0.9481	100.00	30.00	0.0720	0.9811	1.00	0.0
DEXTER	0.0625	0.9870	0.5	6.00	0.0330	0.9670	18.57	42.14
DOROTHEA	0.0892	0.9480	100.00	50.00	0.0854	0.9592	100.00	50.00
MADELON	0.1256	0.9473	100.00	96.00	0.0622	0.9807	100.00	96.00
OVERALL	0.0853	0.9658	62.50	36.40	0.0648	0.9720	80.30	47.77

The Table 15.4 presents a comparison of our method (MART-WS) results on all five NIPS challenge data sets. All results are reported on the hidden test set.

The results where percent of selected features is less than 100% were obtained using two runs.The first run was to select important variables, the second one to build the model. But in general this does not allow us to build a better model - one can adjust the importance learning rate instead.

15.3.4 Artificial Data – Classification

Using the data set generator (Section 15.3.1) 200 data sets of size 5000 with $N_0 = 2, 4, 10$ significant inputs, and $50 - N_0$ noise inputs were generated from the uniform (0,1) distribution. Test error is calculated from a single run for each algorithm and each of 200 sets. The number of selected features for

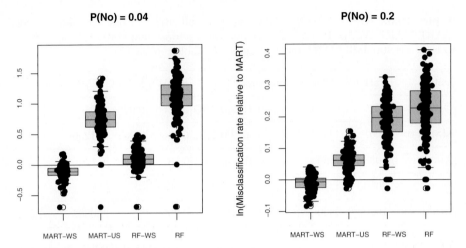

Fig. 15.4. Distribution (log scale) of relative (to MART) multiclass misclassification rate for MART-WS/US, RF, RF-WS for different proportions $P(N_0)$ of relevant variables N_0.

MART-US, MART-WS and RF-WS was taken to be $S = 8$. Figure 15.4 shows (on log scale) distributions of error rates relative to MART for MART/RF-WS/US for two different proportions of relevant variables $P(N_0)$=0.04,0.2.

From Figure 15.4 it is evident that weighted feature selection (MART-WS) gives better results than uniform (MART-US), and the difference is more prominent when a proportion of relevant features is smaller. In the latter case RF-WS gives substantial improvement over standard RF also. Notice that there is a slight improvement (3-5%) in predictive accuracy with MART-WS over standard MART.

Next, we tried to evaluate MART/RF-WS performance dependency on the number of variables sampled and the effect of weights initialization. We used 50 data sets from 200 described in the beginning of the section, and ran all five algorithms 20 times, with different fractions of variables sampled. Figure 15.6 compares errors rates for all classifiers averaged over all 50 data sets and 20 replication runs for different numbers of selected variables and N_0 values. Figure 15.6 shows results on the same data sets with initial variable weights estimated using variable importance (15.1) from a single CART run using all inputs.

Figure 15.5 and 15.6 show that the weighted feature selection for RF is effective only when the proportion of relevant predictors is small. A better estimation of the initial variable weights may help improving the predictive performance or reducing the number of features without altering performance.

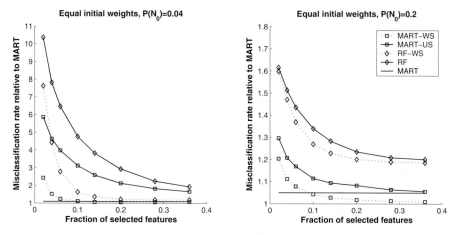

Fig. 15.5. Relative (to MART) misclassification rate for MART-WS/US,RF, RF-WS as a function of feature sampling rate for different fractions $P(N_0) = 0.04, 0.2$ of relevant variables N_0

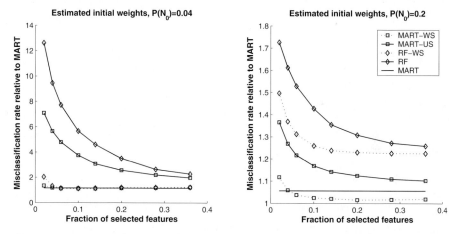

Fig. 15.6. Relative (to MART) misclassification rate for MART-WS/US, RF, RF-WS as a function of feature sampling rate for different proportions $P(N_0) = 0.04, 0.2$ of relevant variables N_0. Initial variable weights were estimated using single CART run.

15.3.5 Artificial Data Sets – Regression

We used the same 50 datasets from Section 15.3.4 but responses were not discretized this time. The number of selected features for MART-US, MART-WS and RF-WS was also taken to be $S = 8$. The number of iterations was 1000, and the learning rate $= 0.01$. Figure 15.7 and 15.8 illustrate the dependency of test error on the number of variables selected per split and the proportion of relevant inputs. In all cases MART-WS outperforms MART/MART-US,

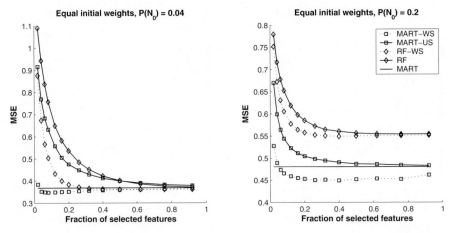

Fig. 15.7. Regression MSE for MART-WS/US, RF, RF-WS as a function of feature sampling rate for different proportions $P(N_0) = 0.04, 0.2$ of relevant variables N_0

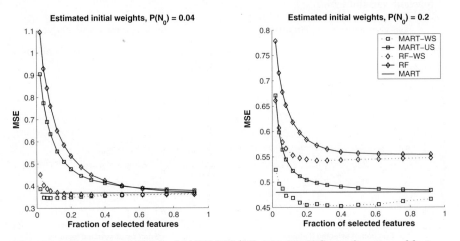

Fig. 15.8. Regression MSE for MART-WS/US, RF, RF-WS as a function of feature sampling rate for different proportions $P(N_0) = 0.04, 0.2$ of relevant variables N_0. Initial variable weights were estimated using a single CART run.

with no effect of weight initialization on accuracy. However, by using CART to estimate the initial variable weights far fewer iterations were needed to reach the optimal predictive accuracy for MART-WS; also, RF-WS needed a smaller fraction of variables sampled (especially in the case of fewer relevant variables.)

15.4 Summary

In this paper we introduced an embedded, dynamic soft variable selection method for tree ensembles. The core idea is to select a small fixed sample of variables at every tree construction step in an ensemble. The sampling distribution is modified at every iteration to up-weight more relevant features based on dynamically learned importances. Different strategies are proposed for serial (MART) and parallel (RF) ensembles.

The considered scheme for MART (MART-WS) tested on the UCI repository and artificial datasets demonstrated a substantial reduction in computational complexity without loss of accuracy (and in noisy data it often outperformed MART.) For very high dimensional problems MART-WS (NIPS-2003 challenge) is up to 100 times faster than MART, and its overall results achieved were ranked among the several best in the feature selection challenge. Simple uniform sampling of variables showed significant degradation in accuracy for MART in the presence of noise.

Thus, this fast, hybrid scheme - stage-wise stochastic boosting of shallow random trees built on a small intelligently sampled subset of variables - can be applied to noisy, massive (in both dimensions) regression and classification problems with predictive power comparable to the best known learning engines. Based on our experience, this combination of speed, accuracy, and applicability makes MART-WS the best *universal* learner available.

A similar weighted feature selection scheme was applied to RF. The resulting procedure (RF-WS) showed substantial improvement in accuracy for regression models and some improvement in classification when the proportion of relevant features was small. However, individual experts in RF-WS are not independent, and therefore, it loses the attractive computational parallelism and limits the interpretative power of RF (Breiman, 2003).

It was also demonstrated that estimating the initial sampling weights could result in a simpler (and faster) model, but not necessarily a more accurate one.

Several potential improvements could be considered. The proposed method implies that in the presence of noise there is a "burn-in" period in ensemble construction, and therefore some kind of postprocessing scheme to assign an appropriate weight for each expert in the ensemble could be useful. Some preliminary experiments showed that regularized regression (we used ridge regression) applied to the outputs of individual trees treated as new features (with the original response) on a validation/test portion of the data could give a noticeable improvement in accuracy.

The effect of "overweighting" needs to be investigated for problems where a sizeable subset of dominant variables (often collinear) will prevent relevant (but less important) variables from entering the model. A more adaptive strategy to adjust feature sampling weights could help.

References

L. Breiman. Bagging predictors. *Machine Learning*, 24(2):123–140, 1996.

L. Breiman. Arcing classifiers. *The Annals of Statistics*, 3:801–849, 1998.

L. Breiman. Random forests. *Machine Learning*, 45(1):5–32, 2001.

L. Breiman. Two-eyed algorithms and problems. In *Proc. 14th European Conference on Machine Learning - ECML '03*, volume 2873. Springer-Verlag, 2003.

L. Breiman, J. Friedman, R. Olshen, and C. Stone. *Classification and Regression Trees*. Wadsworth, Belmont, MA, 1984.

T.G. Dietterich. An experimental comparison of three methods for constructing ensembles of decision trees: Bagging, boosting, and randomization. *Machine Learning*, 40(2):139–157, 2000.

Y. Freund and R. Schapire. Experiments with a new boosting algorithm. In *Proceedings of the Thirteenth International Conference on Machine Learning*. Morgan Kauffman, San Francisco, 1996.

J.H. Friedman. Greedy function approximation: a gradient boosting machine. Technical report, Dept. of Statistics, Stanford University, 1999a.

J.H. Friedman. Stochastic gradient boosting. Technical report, Dept. of Statistics, Stanford University, 1999b.

T.K. Ho. The random subspace method for constructing decision forests. *IEEE Transactions on Pattern Analysis and Machine Intelligence*, 20(8):832–844, 1998.

Chapter 16

Sparse, Flexible and Efficient Modeling using L_1 Regularization

Saharon Rosset[1] and Ji Zhu[2]

[1] IBM T.J. Watson Research Center, P.O. Box 218, Yorktown Heights, NY 10598
   srosset@us.ibm.com
[2] Department of Statistics, University of Michigan, Ann Arbor, MI 48109-1092
   jizhu@umich.edu

Summary. We consider the generic regularized optimization problem $\hat{w}(\lambda) = \arg\min_w \sum_{k=1}^{m} L(y_k, x_k^T w) + \lambda J(w)$. We derive a general characterization of the properties of (loss L, penalty J) pairs which give piecewise linear coefficient paths. Such pairs allow us to efficiently generate the full regularized coefficient paths. We illustrate how we can use our results to build robust, efficient and adaptable modeling tools.

16.1 Introduction

With the advent of modern high technologies, much of the data we regularly encounter nowadays are:

- High-dimensional, i.e. have a large number of variables or features, possibly many more than observations ($n \gg m$ or $N \gg m$). Biological data, in particular gene microarrays, and commercial databases used in data mining are two obvious examples.
- Noisy and dirty, with mis-specified or mis-measured data, low signal-to-noise ratio or both. Commercial data bases are inherently problematic because of the manual nature of data collection. Scientific and biological datasets may well suffer from low signal-to-noise ratio or from noisy measurements as well.
- Contain many irrelevant variables or features.

Blindly fitting models to such data is guaranteed to give badly over-fitted and useless models. Thus, methods for controlling model complexity (also known as regularization) are essential in modern data analysis.

In this chapter we discuss some of the statistical and computational considerations in defining regularization approaches that lead to useful and practical modeling tools.

In a standard supervised problem, we are given a set of training data (x_1, y_1), (x_2, y_2), ..., (x_m, y_m), where $x_k = (x_{k,1}, x_{k,2}, \ldots, x_{k,n})^T \in \mathbb{R}^n$ is

S. Rosset and J. Zhu: *Sparse, Flexible and Efficient Modeling using L_1 Regularization*, Stud-
Fuzz **207**, 375–394 (2006)
www.springerlink.com

the input, y_k is the output. In regression problems, $y_k \in \mathbb{R}$ is a quantitative continuous variable; while in two-class classification problems, $y_k \in \{1, -1\}$ is a qualitative binary categorical variable. The goal is to build a prediction model $\hat{f}(\boldsymbol{x})$ using the training data, so later on when we see a *new* input $\boldsymbol{x}$, we can either forecast a real response $\hat{f}(\boldsymbol{x})$ if it is a regression problem, or assign a class label $sign\left[\hat{f}(\boldsymbol{x})\right]$ if it is a classification problem.

The generic regularized model fitting problem can be written as:

$$\hat{f} = \arg\min_f \sum_{k=1}^m L(y_k, f(\boldsymbol{x}_k)) + \lambda J(f), \tag{16.1}$$

where $L(y, f)$ is a loss function that describes the goodness of fit of our model to the training data; $J(f)$ is a penalty functional that describes the complexity of the model; λ is a regularization parameter that controls the balance between the loss and the penalty.

For simplicity, in this chapter, we concentrate on the linear model $f(\boldsymbol{x}) = \boldsymbol{x}^T \boldsymbol{w} + b$. Obviously, we can include more flexible models in the linear model framework by enlarging the original input $\boldsymbol{x}$ to higher dimensional bases, e.g. high order polynomials, kernel bases or wavelet bases. To make solving the optimization problems feasible, we consider convex loss functions only. We also only consider the L_q-norm of the coefficient vector ($J(f) = \|\boldsymbol{w}\|_q, q \geq 1$) as the penalty functional. For notational simplicity, we also omit the offset parameter b.

Thus, the regularized optimization problem consists of three components:

- The empirical loss function. It typically corresponds to a likelihood under an assumed error model. For example, squared error loss implicitly assumes an iid Gaussian error distribution, while the logistic log-likelihood loss assumes the data is Bernoulli.
- The complexity penalty. It can be viewed as representing a log-prior over the model parameters. For example, an L_2-norm penalty corresponds to a Gaussian prior, while an L_1-norm penalty corresponds to a double exponential prior.
- The regularization parameter. It balances the loss and the penalty. From a Bayesian perspective, it can be viewed as setting the strength of the prior.

We address statistical and computational issues related to these three components. In section 16.2 we compare the statistical properties of the L_1-norm penalty and the L_2-norm penalty, and propose a *bet on sparseness* principle. In section 16.3 we discuss the favorable computational properties of the L_1-norm penalty. We show that under certain circumstances, the paths of regularized solutions have a *piecewise linear* property which allows us to develop a new approach to solving L_1-regularized problems and efficiently generate the regularized solutions for *all* values of the regularization parameter. In section 16.4 we combine statistical considerations in selecting loss functions with our computational results to suggest a new classification tool. Finally, in section

16.5 we briefly review the performance of our tools on the feature selection challenge datasets.

16.1.1 Some Specific Examples of Regularized Modeling Approaches

Many of the commonly used methods for data mining, machine learning and statistical modeling can be described as exact or approximate regularized optimization approaches.

The obvious examples from the statistics literature are explicit regularized linear regression approaches, such as ridge regression (Hoerl and Kennard, 1970) and the Lasso (Tibshirani, 1996). Both of these use squared error loss, but they differ in the penalty they impose on the coefficient vector $\boldsymbol{w}$ describing the fitted model:

$$\text{Ridge:} \ \hat{\boldsymbol{w}}(\lambda) = \arg\min_{\boldsymbol{w}} \sum_{k=1}^{m} (y_k - \boldsymbol{x}_k^T \boldsymbol{w})^2 + \lambda \|\boldsymbol{w}\|_2^2, \tag{16.2}$$

$$\text{Lasso:} \ \hat{\boldsymbol{w}}(\lambda) = \arg\min_{\boldsymbol{w}} \sum_{k=1}^{m} (y_k - \boldsymbol{x}_k^T \boldsymbol{w})^2 + \lambda \|\boldsymbol{w}\|_1. \tag{16.3}$$

Another example from the statistics literature is the penalized logistic regression model (Wahba, 1990) for classification, which is widely used in medical decisions and credit scoring:

$$\hat{\boldsymbol{w}}(\lambda) = \arg\min_{\boldsymbol{w}} \sum_{k=1}^{m} \log(1 + e^{-y_k \boldsymbol{x}_k^T \boldsymbol{w}}) + \lambda \|\boldsymbol{w}\|_2^2, \tag{16.4}$$

where $L(y, f) = \log(1 + e^{-yf})$ is the negative binomial log-likelihood.

Many "modern" methods for machine learning and signal processing can also be cast in the framework of regularized optimization. For example, the regularized support vector machine (Vapnik, 1995) uses the hinge loss function and the L_2-norm penalty[3]:

$$\hat{\boldsymbol{w}}(\lambda) = \arg\min_{\boldsymbol{w}} \sum_{k=1}^{} (1 - y_k \boldsymbol{x}_k^T \boldsymbol{w})_+ + \lambda \|\boldsymbol{w}\|_2^2, \tag{16.5}$$

where $(\cdot)_+$ is the positive part of the argument.

AdaBoost (Freund and Schapire, 1996) is a popular and highly successful method for iteratively building an additive model from a dictionary of "weak learners". In (Rosset et al., 2004) we show that the AdaBoost algorithm *approximately* follows the path of the L_1-norm regularized solutions to the exponential loss function e^{-yf} as the regularization parameter λ decreases.

---

[3]This representation differs from the "standard" optimization representation of the regularized SVM, however it is mathematically equivalent to it.

16.1.2 An Illustrative Example

Before delving into any technical details, we use the following simple simulated example (discussed in more detail in (Rosset and Zhu, 2003)) to illustrate the importance of all three components of the regularized optimization problem for building good prediction models:

- An appropriate loss function that either matches data error distribution or is robust enough to account for lack of knowledge of this distribution.
- An appropriate regularization scheme.
- Selection of an appropriate value for the regularization parameter.

Take $m = 100$ observations and $n = 80$ variables, where all $x_{k,i}$ are i.i.d $N(0,1)$ and the true model is:

$$y_k = 10 \cdot x_{k,1} + \epsilon_k \tag{16.6}$$

$$\epsilon_k \overset{iid}{\sim} 0.9 \cdot N(0,1) + 0.1 \cdot N(0,100) \tag{16.7}$$

So the true model only depends on the first variable, and the normality of residuals, implicitly assumed by using squared error loss, is violated.

We compare three parametric regularized fitting problems on this data:

$$\text{Ridge: } \hat{w}(\lambda) = \arg\min_{w} \sum_{k=1}^{m} (y_k - x_k^T w)^2 + \lambda \|w\|_2^2 \tag{16.8}$$

$$\text{Lasso: } \hat{w}(\lambda) = \arg\min_{w} \sum_{k=1}^{m} (y_k - x_k^T w)^2 + \lambda \|w\|_1 \tag{16.9}$$

$$\text{Huberized Lasso: } \hat{w}(\lambda) = \arg\min_{w} \sum_{k=1}^{m} L_H(y_k, x_k^T w) + \lambda \|w\|_1 \tag{16.10}$$

where

$$L_H(y, x^T w) = \begin{cases} (y - x^T w)^2 & \text{if } |y - x^T w| \leq 1 \\ 2|y - x^T w| - 1 & \text{otherwise} \end{cases}$$

Figure 16.1 shows the solution coefficient paths $\hat{w}(\lambda)$ for the ridge (left), the Lasso (middle) and the Huberized Lasso (right)[4]. Since the true model is *sparse* with only one non-zero coefficient, the L_1-norm regularization is appropriate here (see section 16.2 for details). Since the error distribution has a long tail, a robust loss function (like Huber's loss L_H) is appropriate. Thus, both Ridge and the Lasso fail in identifying the correct model $E(Y|x) = 10x_1$ while the Huberized Lasso identifies it exactly, *if we choose the appropriate*

---

[4]To follow the tradition of the original Lasso paper (Tibshirani, 1996), we plot $\hat{w}(\lambda)$ as a function of the appropriate norm $\|\hat{w}\|_q$, $q \in \{1,2\}$, rather than the regularization parameter λ. It is easy to show that there is a one to one correspondence between λ and $\|\hat{w}\|_q$.

regularization parameter. That is, the rightmost plot of Figure 16.1, at $\|\hat{\boldsymbol{w}}\|_1 = 10$, recovers exactly the original model from which the data was generated. Figure 16.2 gives the reducible squared error loss of the models along the three regularized paths and illustrates the superiority of the Huberized Lasso.

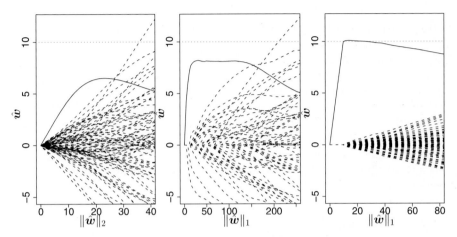

Fig. 16.1. Coefficient paths for ridge (left), Lasso (middle) and Huberized Lasso (right) for data example. The full solid line is for $\hat{\boldsymbol{w}}_1$, and the dashed lines for the noise variables. The true model is $E(Y|\boldsymbol{x}) = 10x_1$.

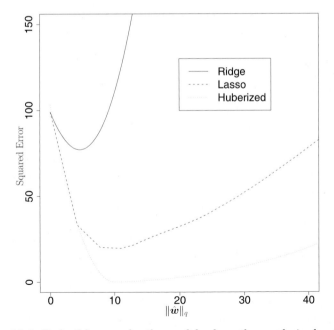

Fig. 16.2. Reducible error for the models along the regularized paths.

Thus we see that in this simple example, the correct combination of loss, penalty and regularization parameter choice allows us to identify the correct model exactly, while failing on any of these three components significantly degrades the performance.

16.2 The L_1-Norm Penalty

In this section we concentrate on the statistical properties of the L_1-norm penalty. We first show that solutions to L_1-regularized problems have a *sparseness* property, i.e. that they cannot have too many non-zero coefficients. Then we show that in the case that L_1-norm penalty is appropriate from a Bayesian perspective, i.e. when the true model is sparse, the problem is "easy" to solve in terms of model complexity. On the other hand, spherically symmetric problems where the L_2-norm penalty is appropriate are "hard" to solve from that perspective. These two properties establish the usefulness of the L_1-norm penalty from a statistical perspective.

16.2.1 Sparseness

A canonical example that uses the L_1-norm penalty is the Lasso for regression problems (16.9). The L_1-norm penalty shrinks the fitted coefficients $\hat{w}$ towards zero. It is well known that this shrinkage has the effect of controlling the variance of $\hat{w}$, hence possibly improves the fitted model's prediction accuracy. Another property of the L_1-norm penalty is that because of the L_1 nature of the penalty, making λ sufficiently large will cause some of the coefficients $\hat{w}_i$'s to be *exactly zero*, i.e. *sparse solution*. Hence, the L_1-norm penalty has an inherent variable/feature selection property. We illustrate the concept of sparseness of $\hat{w}(\lambda)$ with a simple example. We generate 10 training data of two classes. Each data point contains five variables: $x_1, x_2, \ldots, x_5$, but the true classification boundary, $x_1 + x_2 = 0$, only depends on the first two variables. We compare the fitted coefficient paths for the L_1-norm support vector machine and the standard L_2-norm support vector machine. In Figure 16.3, the two solid paths are for x_1 and x_2 (or $\hat{w}_1$ and $\hat{w}_2$); the dashed lines are for the irrelevant noise variables. As we can see in the left panel, when $\|\hat{w}\|_1 \leq 0.7$, only the relevant variables have non-zero fitted coefficients, while the noise variables have zero coefficients. Thus when the regularization parameter varies, the L_1-norm penalty does a kind of continuous variable selection. This is not the case for the standard L_2-norm penalty (right panel).

From a Bayesian perspective, the L_1-norm penalty corresponds to putting a double-exponential log-prior on the coefficient vector describing the model. Thus, the equal-penalty contours in the n-dimensional Euclidean space spanned by the model's coefficients are hyper-diamonds, as illustrated in left panel of Figure 16.4, compared to hyper-spheres for the L_2-norm penalty and hyper-cubes for the L_∞-norm penalty.

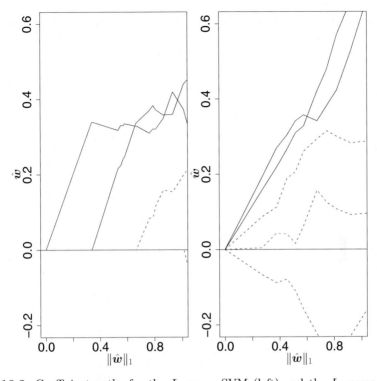

Fig. 16.3. Coefficient paths for the L_1-norm SVM (left) and the L_2-norm SVM (right) for simulated data example. The solid lines correspond to the relevant variables, and the dashed lines correspond to the noise variables.

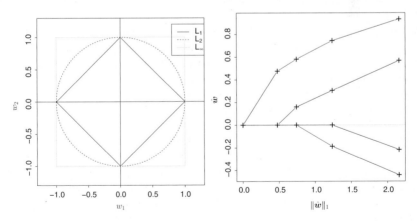

Fig. 16.4. Left panel: Contours of various L_q penalty functions. Right panel: Piecewise linear solution paths for the Lasso on a 4-variable example.

Observing that a hyper-diamond has the vast majority of its volume in the corners gives us an intuitive sense of why we may expect the L_1-norm penalty to give sparse models. It turns out that this is indeed the case, and in fact the following general result holds:

Theorem 1. *There always exists a solution of (16.1) with $J(f) = \|\boldsymbol{w}\|_1$ that has at most m non-zero coefficients, even if $n > m$.*

This theorem establishes the existence of a sparse solution. We omit the proof of the theorem here and refer the readers to (Rosset et al., 2004) for details. With a little more work (again, see (Rosset et al., 2004) for details), under some additional mild conditions, we can also show that either (a) any solution of (16.1) with $J(f) = \|\boldsymbol{w}\|_1$ has at most m non-zero coefficients or (b) the sparse solution is unique.

16.2.2 Bet on Sparseness

Based on the statistical property of the L_1-norm penalty, we quote the *bet on sparseness* principle for high-dimensional problems from (Friedman et al., 2004). Suppose we have 100 data points and our model is a linear combination of 10,000 variables. Suppose the true coefficients of these variables are "randomly drawn" from a Gaussian prior. Then we know that, in a Bayesian sense, the best predictor would be a ridge regression; that is, we should use an L_2-norm rather than an L_1-norm penalty when fitting the coefficients. On the other hand, if there are only a small number of non-zero true coefficients, the L_1-norm penalty will work better. We think of this as a *sparse scenario*, while the first case (Gaussian coefficients) as non-sparse. Note however that in the first scenario, although the L_2-norm penalty is best, neither method does very well since there is too little data from which to estimate 10,000 non-zero coefficients. This is the *curse of dimensionality* taking its toll. In the sparse setting, we can potentially do well with the L_1-norm penalty, since the number of non-zero coefficients is small. The L_2-norm penalty fails again.

In other words, use of the L_1-norm penalty follows what we call the *bet on sparseness* principle for high-dimensional problems:

> *Use a procedure that does well in sparse problems, since no procedure does well in non-sparse problems.*

Figure 16.5 illustrates this point in the context of logistic regression. The details are given in the caption. Note that we are not using the training data to select λ, but rather are reporting the best possible behavior for each method in the different scenarios. The L_2-norm penalty performs poorly everywhere. The L_1-norm penalty performs reasonably well in the only two situations where it can (sparse coefficients). As expected the performance gets worse as the noise level increases (notice that the maximum noise level is 0.25), and as the model becomes less sparse.

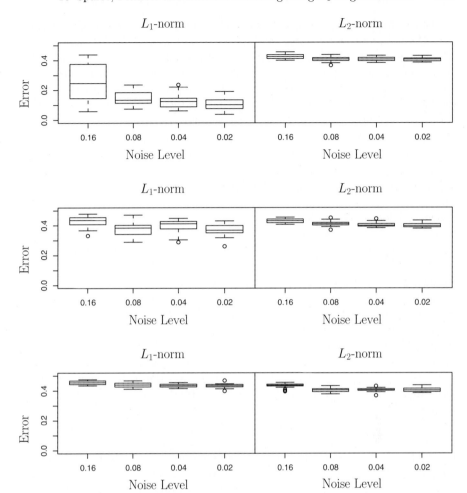

Fig. 16.5. Simulations that show the superiority of the L_1-norm penalty over L_2-norm in classification. Each run has 50 observations with 300 independent Gaussian variables. In the top row only 3 coefficients are non-zero, generated from a Gaussian. In the middle row, 30 are non-zero, and the last row all 300 coefficients are non-zero, generated from a Gaussian distribution. In each case the coefficients are scaled to give different noise levels (we define noise level as $E\left[p(\boldsymbol{x})(1 - p(\boldsymbol{x}))\right]$). L_1-norm penalty is used in the left column, L_2-norm penalty in the right. In both cases we used a series of 100 values of λ, and picked the value that minimized the theoretical misclassification error. In the figures we report the best misclassification error, displayed as box-plots over 20 realizations for each combination.

These empirical results are supported by a large body of theoretical results (e.g. (Donoho and Johnstone, 1994), (Donoho et al., 1992), (Donoho et al., 1995)) that support the superiority of L_1-norm estimation in sparse settings. One recent interesting result is by (Ng, 2004), which shows that when using the L_1-norm penalty, the sample complexity, i.e., the number of training examples required to learn "well", grows only *logarithmically* in the number of irrelevant variables; while the L_2-norm penalty has a worst case sample complexity that grows at least *linearly* in the number of irrelevant variables.

16.3 Piecewise Linear Solution Paths

To get a good fitted model $\hat{f}(\boldsymbol{x})$ that performs well on future data, we also need to select an appropriate regularization parameter λ. In practice, people usually pre-specify a finite set of values for the regularization parameter that covers a wide range, then either use a separate validation dataset or use cross-validation to select a value for the regularization parameter that gives the best performance among the given set. In this section, we concentrate our attention on (loss L, penalty J) pairings where the solution path $\hat{\boldsymbol{w}}(\lambda)$ is *piecewise linear* as a function of λ, i.e. $\exists \lambda_0 = 0 < \lambda_1 < \ldots < \lambda_T = \infty$ and $\boldsymbol{\gamma}_0, \boldsymbol{\gamma}_1, \ldots, \boldsymbol{\gamma}_{T-1} \in \mathbb{R}^n$ such that $\hat{\boldsymbol{w}}(\lambda) = \hat{\boldsymbol{w}}(\lambda_t) + (\lambda - \lambda_t)\boldsymbol{\gamma}_t$ where t is such that $\lambda_t \leq \lambda \leq \lambda_{t+1}$. Such models are attractive because they allow us to generate the whole regularized path $\hat{\boldsymbol{w}}(\lambda)$, $0 \leq \lambda \leq \infty$ simply by calculating the directions $\boldsymbol{\gamma}_1, \ldots, \boldsymbol{\gamma}_{T-1}$, hence help us understand how the solution changes with λ and facilitate the adaptive selection of the regularization parameter λ.

Again, a canonical example is the Lasso (16.9). Recently (Efron et al., 2004) have shown that the piecewise linear coefficient paths property holds for the Lasso. Their results show that the number of linear pieces in the Lasso path is approximately the number of the variables in $\boldsymbol{x}$, and the complexity of generating the whole coefficient path, for all values of λ, is approximately equal to one least square calculation on the full sample.

A simple example to illustrate the piecewise linear property can be seen in the right panel of Figure 16.4, where we show the Lasso solution paths for a 4-variable synthetic dataset. The plot shows the Lasso solutions $\hat{\boldsymbol{w}}(\lambda)$ as a function of $\|\hat{\boldsymbol{w}}\|_1$[5]. Each line represents one coefficient and gives its values at the solution for the range of λ values. We observe that between every two "+" signs the lines are straight, i.e. the coefficient paths are piecewise-linear, as a function of λ, and the 1-dimensional curve $\hat{\boldsymbol{w}}(\lambda)$ is piecewise linear in $\mathbb{R}^4$.

The questions which are of interest to us, in trying to understand the piecewise linear property and extend its usefulness beyond the Lasso, are:

---

[5]It is easy to show that if $\hat{\boldsymbol{w}}(\lambda)$ is piecewise linear in λ, then it is also piecewise linear in $\|\hat{\boldsymbol{w}}\|_1$; the opposite is not necessarily true. See (Rosset and Zhu, 2003) for details.

(1.) What is the general characterization of (loss, penalty) pairs which give piecewise linear solution paths?

(2.) For what members of this family can we build efficient algorithms for generating the whole solution path?

(3.) What are statistically interesting members of this family, which define modeling problems that are of interest, and allow us to utilize our efficient algorithms for building new and useful modeling tools?

In this section and the next we attempt to answer these questions. Our results are based on (Rosset and Zhu, 2003), and allow us to:

- Define general conditions to describe which regularized optimization problems yield piecewise linear regularized solution paths;
- Design efficient algorithms to generate the whole regularized solution paths for some members of this family;
- Suggest new and interesting modeling tools which combine robust loss, L_1-norm regularization and efficient computations.

16.3.1 Sufficient Conditions for Piecewise Linearity

We now develop a general criterion for piecewise linear solution paths in the case that the loss and penalty are both twice differentiable. This will serve us as an intuitive guide to identify regularized models where we can expect piecewise linearity. It will also prove useful as a tool in asserting piecewise linearity for non-twice differentiable functions.

For the coefficient paths to be piecewise linear, we require that $\frac{\partial \hat{w}(\lambda)}{\partial \lambda}$ is a *piecewise constant vector* as a function of λ, or in other words that over ranges of λ values:

$$\lim_{\epsilon \to 0} \frac{\hat{w}(\lambda + \epsilon) - \hat{w}(\lambda)}{\epsilon} = \text{constant} \qquad (16.11)$$

To get a condition, let us start by considering only w values where L, J are both twice differentiable, with bounded second derivatives in the relevant region. Throughout this section we are going to use the notation $L(w) = \sum_{k=1}^{m} L(y_k, x_k^T w)$ in the obvious way, i.e. we make the dependence on the data X, Y implicit, since we are dealing with optimization problems in the coefficients w only, and assuming the data is fixed.

With some algebra, which we skip here for brevity, we obtain the following result:

Lemma 1. *For twice differentiable loss and penalty, a sufficient and necessary condition for piecewise linear coefficient paths is that the direction*

$$\frac{\partial \hat{w}(\lambda)}{\partial \lambda} = -(\nabla^2 L(\hat{w}(\lambda)) + \lambda \nabla^2 J(\hat{w}(\lambda)))^{-1} \nabla J(\hat{w}(\lambda)) \qquad (16.12)$$

is a piecewise constant vector in $\mathbb{R}^n$ as a function of λ.

The obvious realistic situation (i.e. not necessarily twice differentiable loss and penalty) in which Lemma 1 holds and we get piecewise linear coefficient paths is:

- L is piecewise quadratic as a function of $\boldsymbol{w}$ along the optimal path $\hat{\boldsymbol{w}}(\lambda)$.
- J is piecewise linear as a function of $\boldsymbol{w}$ along this path.

So we can conclude that the search for regularized problems with piecewise linear coefficient paths should concentrate on losses L which are "almost everywhere" quadratic and penalties J which are "almost everywhere" linear.

16.3.2 The Piecewise Linear "Toolbox"

Based on this characterization of what constitutes piecewise linear solution paths, we can now create a "toolbox" of loss functions and penalty functionals combinations which yield this property.

On the loss side, we require a "piecewise quadratic" function. Thus, our toolbox includes:

- Pure quadratic loss functions: $(y - \boldsymbol{x}^T \boldsymbol{w})^2$ for regression, $(1 - y\boldsymbol{x}^T \boldsymbol{w})_+^2$ for classification;
- Mixture of quadratic and linear pieces, which includes Huber's loss (16.10) for regression, and the new classification loss we introduce in section 16.4;
- Mixture of linear and constant pieces, such as the absolute loss for regression $|y - \boldsymbol{x}^T \boldsymbol{w}|$ and the hinge loss of SVMs (the first term in (16.5)).

On the penalty side, we require a "piecewise linear" function, which in our family of L_q norms, leaves us with:

- The L_1-norm penalty;
- The L_∞-norm penalty $J(\boldsymbol{w}) = \max_i |w_i|$.

Any combination of loss and penalty from these families will thus give us piecewise linear solution paths, for example:

(1.) 1-norm support vector machines. In (Zhu et al., 2003) we present an algorithm for efficient solution of the 1-norm SVM utilizing this property. The standard (2-norm) SVM has a similar "piecewise linear" property, albeit one not covered by the theory presented here. Details, with a resulting efficient algorithm, can be found in (Hastie et al., 2004).

(2.) The Lasso (16.9) and L_1-norm penalized Huber's loss (16.10), which we have presented in section 16.1.2. In fact, the solution paths in Figure 16.1 were generated using the algorithm we present in the next section which takes computational advantage of the piecewise linear property.

16.3.3 An Interesting Family

We now define a family of *almost quadratic* loss functions whose L_1-norm regularized solution paths are piecewise linear — a family we consider to be most useful and practical, because it yields a simple and provably efficient algorithm to generate the whole regularized solution path. We fix the penalty to be the L_1-norm:

$$J(\boldsymbol{w}) = \|\boldsymbol{w}\|_1 = \sum_i |w_i|. \qquad (16.13)$$

And the loss is required to be differentiable and piecewise quadratic in a fixed function of the sample response and the prediction $\boldsymbol{x}^T \boldsymbol{w}$:

$$L(\boldsymbol{w}) = \sum_k L(y_k, \boldsymbol{x}_k^T \boldsymbol{w}) \qquad (16.14)$$
$$L(y, \boldsymbol{x}^T \boldsymbol{w}) = a(r)r^2 + b(r)r + c(r) \qquad (16.15)$$

where $r = r(y, \boldsymbol{x}^T \boldsymbol{w}) = (y - \boldsymbol{x}^T \boldsymbol{w})$ is the residual for regression and $r = r(y, \boldsymbol{x}^T \boldsymbol{w}) = (y \boldsymbol{x}^T \boldsymbol{w})$ is the margin for classification; and $a(\cdot), b(\cdot), c(\cdot)$ are piecewise constant functions, with a finite (usually small) number of pieces, defined so as to make the function L once differentiable *everywhere*.

Some examples from this family are:

- The Lasso: $L(y, \boldsymbol{x}^T \boldsymbol{w}) = (y - \boldsymbol{x}^T \boldsymbol{w})^2$, i.e. $a \equiv 1, b \equiv 0, c \equiv 0$.
- The Huber loss function with *fixed* knot δ (define $r = y - \boldsymbol{x}^T \boldsymbol{w}$ the residual):

$$L(r) = \begin{cases} r^2 & \text{if } |r| \leq \delta \\ 2\delta|r| - \delta^2 & \text{otherwise} \end{cases} \qquad (16.16)$$

- Squared hinge loss for classification (define $r = y\boldsymbol{x}^T \boldsymbol{w}$ the margin):

$$L(r) = \begin{cases} (1 - r)^2 & \text{if } r \leq 1 \\ 0 & \text{otherwise} \end{cases} \qquad (16.17)$$

Theorem 2. *All regularized problems of the form (16.1) using (16.13) and (16.15) (with r being either the residual or the margin) generate piecewise linear solution coefficient paths $\hat{\boldsymbol{w}}(\lambda)$ as the regularization parameter λ varies.*

The proof is based on explicitly writing and investigating the Karush-Kuhn-Tucker (KKT) conditions for these regularized problems, and we omit it for space considerations. The full proof is available in (Rosset and Zhu, 2003).

Based on this theorem and careful examination of the KKT conditions we can derive a generic algorithm to generate coefficient paths for all members of the almost quadratic family of loss functions with L_1-norm penalty. Our algorithm starts at $\hat{\boldsymbol{w}} = 0$ (or, equivalently, $\lambda = \infty$) and follows the linear pieces, while identifying the *direction change events* and re-calculating the direction when they occur. These direction change events occur when:

- A new variable enters the regularized solution ; (i.e. a 0 coefficient becomes non-0)
- A variable in the solution is dropped (i.e. a non-0 coefficient becomes 0)
- A residual crosses a "knot" of non-twice-differentiability in the loss

The Lar-Lasso algorithm (Efron et al., 2004) is a simplified version of our algorithm since *knot crossing* events do not occur in the Lasso (as the loss is twice differentiable). Notice that the algorithm does not calculate λ explicitly.

Algorithm 6: An algorithm for almost quadratic loss with L_1-norm penalty

Initialization $\hat{\boldsymbol{w}} = 0, \mathcal{A} = \arg\max_i (\nabla L(\hat{\boldsymbol{w}}))_i, \boldsymbol{\gamma}_{\mathcal{A}} = -sgn(\nabla L(\hat{\boldsymbol{w}}))_{\mathcal{A}}, \boldsymbol{\gamma}_{\mathcal{A}^C} = 0,$
    where $\mathcal{A}$ contains the indices of the non-zero coefficients.

Main While $(max|\nabla L(\hat{\boldsymbol{w}})| > 0)$

- $d_1 = \min\{d > 0 : |\nabla L(\hat{\boldsymbol{w}} + d\boldsymbol{\gamma})_i| = |\nabla L(\hat{\boldsymbol{w}} + d\boldsymbol{\gamma})_{i'}|, i \notin \mathcal{A}, i' \in \mathcal{A}\}.$
- $d_2 = \min\{d > 0 : (\hat{\boldsymbol{w}} + d\boldsymbol{\gamma})_i = 0, i \in \mathcal{A}\}$ (hit 0).
- $d_3 = \min\{d > 0 : r(y_k, \boldsymbol{x}_k^T(\hat{\boldsymbol{w}} + d\boldsymbol{\gamma}))$ hits a knot, $k = 1, \ldots, m\}.$
- Set $d = \min(d_1, d_2, d_3)$.
- If $d = d_1$ then add variable attaining equality at d to $\mathcal{A}$.
- If $d = d_2$ then remove variable attaining 0 at d from $\mathcal{A}$.
- $\hat{\boldsymbol{w}} \leftarrow \hat{\boldsymbol{w}} + d\boldsymbol{\gamma}$
- $C = \sum_k a(r(y_k, \boldsymbol{x}_k^T\hat{\boldsymbol{w}}))\boldsymbol{x}_{k,\mathcal{A}}\boldsymbol{x}_{k,\mathcal{A}}^T.$
- $\boldsymbol{\gamma}_{\mathcal{A}} = C^{-1}(-sgn(\hat{\boldsymbol{w}}_{\mathcal{A}})).$
- $\boldsymbol{\gamma}_{\mathcal{A}^C} = 0.$

It should be noted that our formulation here of the almost quadratic family with the L_1-norm penalty has ignored the existence of a non-penalized intercept. This has been done for simplicity of exposition, however incorporating a non-penalized intercept into the algorithm is straightforward. The computer implementation we use for numerical examples throughout this chapter allow for the inclusion of a non-penalized intercept.

The computational complexity of this algorithm turns out to be $O(m^2 n)$ both when $m > n$ and when $n > m$, under mild assumptions (see (Rosset and Zhu, 2003) for details). For this computational cost we get the *full* regularized solution path, i.e. $\hat{\boldsymbol{w}}(\lambda), \forall \lambda$.

16.4 A Robust, Efficient and Adaptable Method for Classification

In section 16.1.2, we presented an example using a robust *Huberized* loss function of the almost quadratic family for regression problems. Algorithm 6 was in fact used to generate the regularized solution paths in the middle and right panels of Figure 16.1. We now suggest a new method for classification, which is of statistical interest and allows us to generate the whole regularized path efficiently. We choose a loss function, which is robust, i.e. linear for large margins. We build on the previous sub-section, in particular Algorithm 6, to generate the solution paths to the regularized problems $\hat{\boldsymbol{w}}(\lambda)$ for the L_1-norm regularized version of this fitting problem.

16.4.1 The Huberized Hinge Loss for Classification

For classification we would like to have a loss, which is a function of the margin: $r(y, \boldsymbol{x}^T \boldsymbol{w}) = y\boldsymbol{x}^T \boldsymbol{w}$. This is true of the loss functions typically used for classification:

$$\text{Logistic regression:} \quad L(y, \boldsymbol{x}^T \boldsymbol{w}) = \log(1 + \exp(-y\boldsymbol{x}^T \boldsymbol{w}))$$
$$\text{Support vector machines:} \quad L(y, \boldsymbol{x}^T \boldsymbol{w}) = (1 - y\boldsymbol{x}^T \boldsymbol{w})_+$$
$$\text{Exponential loss (boosting):} \quad L(y, \boldsymbol{x}^T \boldsymbol{w}) = \exp(-y\boldsymbol{x}^T \boldsymbol{w})$$

The properties we would like from our classification loss are:

- We would like it to be almost quadratic, so we can apply Algorithm 6.
- We would like it to be robust, i.e. linear for large absolute value negative margins, so that outliers would have a small effect on the fit. This property is shared by the loss functions used for logistic regression and support vector machines. The squared hinge loss (16.17) and more so the exponential loss are non-robust in this sense.

This leads us to suggesting for classification the Huberized hinge loss, i.e. (16.17) Huberized at $\delta < 1$:

$$L(r) = \begin{cases} (1 - \delta)^2 + 2(1 - \delta)(\delta - r) & \text{if } r \leq \delta \\ (1 - r)^2 & \text{if } \delta < r \leq 1 \\ 0 & \text{otherwise} \end{cases} \qquad (16.18)$$

The left panel of Figure 16.6 compares some of these classification loss functions: the logistic, exponential, squared hinge loss and our suggested loss function (16.18), with $\delta = -1$. The exponential and logistic are scaled up by 4 to make comparison easier. We can see the non-robustness of the squared and more so the exponential in the way they diverge as the margins become negative.

To illustrate the similarity between our loss (16.18) and the logistic loss, and their difference from the squared hinge loss (16.17), consider the following simple simulated example: $\boldsymbol{x} \in \mathbb{R}^2$ with class centers at $(-1, -1)$ (class "-1") and $(1, 1)$ (class "1") with one big outlier at $(30, 100)$, belonging to the class "-1". The Bayes model, ignoring the outlier, is to classify to class "1" iff $x_1 + x_2 > 0$, hence the best fitted model should have $\hat{w}_1 = \hat{w}_2 > 0$. The data and optimal separator can be seen in the right panel of Figure 16.6.

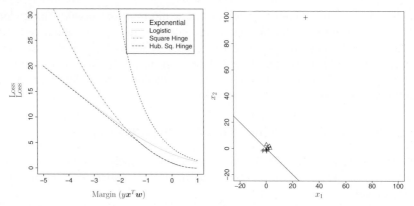

Fig. 16.6. Left: Classification loss functions. Right: Simulated classification data.

Figure 16.7 shows the regularized model paths and misclassification rate for this data using the logistic loss (left), the Huberized hinge loss (16.18) (middle) and the squared hinge loss (16.17) (right). We observe that the logistic and Huberized regularized model paths are similar and they are both less affected by the outlier than the non-Huberized squared loss.

16.5 Results on the NIPS-03 Challenge Datasets

Here we first illustrate our new classification model on the Dexter dataset from NIPS-03 feature selection workshop, then briefly summarize our results on all five challenge datasets. Algorithm 7 contains the general procedure that we follow on all five datasets.

Algorithm 7: General procedure on five challenge datasets

(1.) Pre-selection for each variable, compute the fraction τ of non-zero entries and p-value of the univariate t-statistic; select the variable only if τ is big enough or p is small enough. In the case of Dexter, 1152 variables were pre-selected using this criterion.

(2.) Apply (16.1) on the reduced training dataset. In the case of Dexter, we used the Huberized hinge loss for L and L_1-norm penalty for J.

(3.) Use the validation dataset to select a regularization parameter λ^* that minimizes the balanced error rate.

(4.) Report the model that corresponds to the selected λ^*.

The Dexter dataset contains $m = 300$ training observations, $n = 20,000$ variables and 300 validation observations. Figure 16.8 shows the result. As we can see, the validation error (solid curve) first decreases then increases, as the

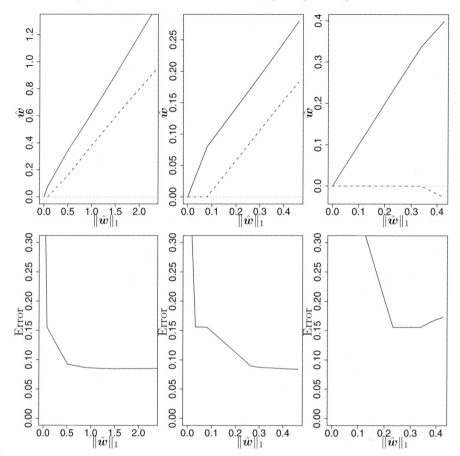

Fig. 16.7. Performance comparison of the logistic loss (left), the Huberized hinge loss (middle) and the squared hinge loss (right). The upper panels are for the regularized solution paths $\hat{w}_1$ (solid) and $\hat{w}_2$ (dashed). The paths for the logistic loss are approximated by solving at pre-fixed 200 different λ values, while the paths for the Huberized hinge loss and the squared hinge loss are solved exactly and efficiently using Algorithm 6. The squared hinge loss gives incorrect sign for $\hat{w}_2$, while the logistic loss and the Huberized hinge loss have similar solution paths. The lower panels show the prediction errors along the solution paths and confirm the relative robustness to the outlier of the logistic loss and the Huberized hinge loss.

regularization parameter λ decreases; while the number of non-zero coefficients or selected variables (dashed curve) *almost* increases monotonically. The best model seems to correspond $\|\hat{\boldsymbol{w}}\|_1 = 0.14$, where the validation error is 0.07 and the number of selected variables is around 120. The entire solution path has 452 linear pieces, and our current un-optimized R implementation takes about 3 minutes to generate the whole solution path on an IBM T30 laptop.

392   Saharon Rosset and Ji Zhu

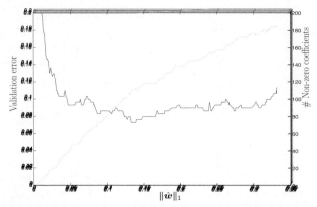

Fig. 16.8. Result on the Dexter dataset: the dotted curve is for the number of non-zero coefficients, and the solid curve is for the validation error.

Table 16.1 summarizes our results on all five challenge datasets and compares them with the best winning results. Due to the space limit, we refer the readers to the fact sheet for details about our procedure and the authors' website for the code[6]. As we can see, except for Gisette, our results are not so satisfactory. Since we are mainly using the original variables as the basis functions, the classification boundaries are simply linear hyper-planes, which are obviously not flexible enough. How to generalize the L_1 regularization to the non-parametric setting will be on our main agenda for future work. For Arcene, Gisette and Madelon, we were able to identify the relevant variables (the fraction of selected "probes" are low); but for Dexter and Dorothea, although we greatly reduced the size of variables (the fraction of selected variables are low), we still kept too many probes. This implies that the L_1-norm regularization may not be penalizing enough in selecting variables. An alternative heavier "penalizer" is the L_0-norm, which corresponds to the best subset selection method in regression (Furnival and Wilson, 1974). However, the L_0-norm is a non-convex functional, which makes the optimization problem computationally difficult, especially when the number of variables is large. We plan to investigate this further, and some discussion of relevant algorithms can be found in chapter Chapter 5.

16.6 Conclusion

In this chapter we have tried to design methods which combine computational and statistical advantages. We emphasize the importance of both appropriate regularization and robust loss functions for successful practical modeling of data. From a statistical perspective, we can consider robustness and regularization as almost independent desirable properties dealing with different issues

---

[6]http://www.stat.lsa.umich.edu/~jizhu/Feature/

Table 16.1. NIPS 2003 challenge results

Dec. 1st	Our best challenge entry				The winning challenge entry					
Dataset	Score	BER	AUC	Feat	Probe	Score	BER	AUC	Feat	Probe	Test
OVERALL	17.82	10.52	92.80	13.99	27.37	88.00	6.84	97.22	80.3	47.8	0.8
ARCENE	0	19.62	88.91	30	5.7	98.18	13.30	93.48	100.0	30.0	1
DEXTER	12.73	6.90	96.28	0.56	44.64	96.36	3.90	99.01	1.5	12.9	1
DOROTHEA	-14.55	15.69	84.51	5.21	76.99	98.18	8.54	95.92	100.0	50.0	1
GISETTE	87.27	1.34	98.26	30	0	98.18	1.37	98.63	18.3	0.0	0
MADELON	3.64	9.06	96.05	4.2	9.52	100.00	7.17	96.95	1.6	0.0	1

in predictive modeling: (a) Robustness mainly protects us against wrong assumptions about our error (or noise) model; (b) Regularization deals mainly with the uncertainty about our model structure by limiting the model space. Thus the main goal of regularization is to make the model estimation problem easier, and match it to the amount of information we have for estimating it, namely our training data. We have also proposed some modeling tools for both regression and classification which are robust (because of the loss function), adaptable (because we can calculate the whole regularized path and choose a good regularization parameter) and efficient (because we can calculate the path with a relatively small computational burden).

References

D. Donoho and I. Johnstone. Ideal spatial adaptation by wavelet shrinkage. *Biometrika*, 81:425–455, 1994.

D. Donoho, I. Johnstone, J. Hoch, and A. Stern. Maximum entropy and the nearly black object. *Journal of Royal Statistical Society B*, 54:41–81, 1992.

D. Donoho, I. Johnstone, G. Kerkyachairan, and D. Picard. Wavelet shrinkage: asymptopia? *Journal of Royal Statistical Society B*, 57:301–337, 1995.

B. Efron, T. Hastie, I. Johnstone, and R. Tibshirani. Least angle regression. *Annals of Statistics*, 32(2), 2004.

Y. Freund and R. Schapire. Experiments with a new boosting algorithm. In *Proceedings of the Thirteenth International Conference on Machine Learning*. Morgan Kauffman, San Francisco, 1996.

J. Friedman, T. Hastie, S. Rosset, R. Tibshirani, and J. Zhu. Discussion of three boosting papers. *Annals of Statistics*, 32(1), 2004. The three papers are by (1) W. Jiang (2) G. Lugosi and N. Vayatis (3) T. Zhang.

G. Furnival and R. Wilson. Regression by leaps and bounds. *Technometrics*, 16: 499–511, 1974.

T. Hastie, S. Rosset, R. Tibshirani, and J. Zhu. The entire regularization path for the support vector machine. *Journal of Machine Learning Research*, 2004. Tentatively accepted.

A. Hoerl and R. Kennard. Ridge regression: Biased estimation for nonorthogonal problems. *Technometrics*, 12(3):55–67, 1970.

A. Ng. Feature selection, l_1 vs. l_2 regularization, and rotational invariance. In *Proceedings of the Twenty-First International Conference on Machine Learning*. Banff, Canada, 2004.

S. Rosset and J. Zhu. Piecewise linear regularized solution paths. Technical Report, Department of Statistics, Stanford University, California, U.S.A., 2003.

S. Rosset, J. Zhu, and T. Hastie. Boosting as a regularized path to a maximum margin classifier. *Journal of Machine Learning Research*, 5:941–973, 2004.

R. Tibshirani. Regression shrinkage and selection via the lasso. *Journal of Royal Statistical Society B*, 58(1), 1996.

V. Vapnik. *The nature of statistical learning*. Springer, 1995.

G. Wahba. *Spline models for observational data*, volume 59 of *CBMS-NSF Regional Conference Series in Applied Mathematics*. SIAM, 1990.

J. Zhu, S. Rosset, T. Hastie, and R. Tibshirani. 1-norm support vector machines. In *Proceedings of Neural Information Processing Systems*, 2003.

Chapter 17

Margin Based Feature Selection and Infogain with Standard Classifiers

Ran Gilad-Bachrach and Amir Navot

The Hebrew University, Jerusalem, Israel
ranb@cs.huji.ac.il, anavot@cs.huji.ac.il

The decision to devote a week or two to playing with the feature selection challenge (FSC) turned into a major effort that took up most of our time a few months. In most cases we used standard algorithms, with obvious modifications for the balanced error measure. Surprisingly enough, the naïve methods we used turned out to be among the best submissions to the FSC.

Using the insights we gained during our participation in the challenge we developed novel feature selection algorithms (*G-flip* and *Simba*), which are based on the large margin principle . We used early versions of these algorithms during the FSC, a detailed description of the mature version of *G-flip* and *Simba* can be found in our contribution to part III of this book (Chapter 29).

17.1 Methods

Basically, we used a mixture of "out-of-the-box", well known methods, with one exception: a novel margin based selection method. We used four stages for each dataset: preprocessing, feature selection, training and finally classification. For each dataset we chose the combination of methods that we found to perform best on the validation dataset. We did not perform exhaustive parameter tuning; instead we tuned the parameters manually using a trial and error procedure. Table 17.1 summaries the methods that were used for each dataset.

17.1.1 Preprocessing

The data from Arcene, Gisette and Madelon were normalized such that the maximum absolute value was set to 1. This was achieved by dividing each value in the instances matrix by the maximum absolute value in the matrix. We refer to this normalization as *abs1*. For Arcene we applied *Principal Component Analysis* (PCA) to the training data and used each Principal Component (PC) as a feature. No preprocessing was applied to Dexter and Dorothea.

R. Gilad-Bachrach and A. Navot: *Margin Based Feature Selection and Infogain with Standard Classifiers*, StudFuzz **207**, 395–401 (2006)
www.springerlink.com

Table 17.1. Summary of methods used.

Dataset	Preprocessing	Feature selection	Classification	Comments
ARCENE	abs1, PCA	margin based	SVM-RBF	$\sigma = 50$, $C = 8$
DEXTER	none	Infogain1	trans. linear SVM	3 trans. rounds of 10%, $C = 10$
DOROTHEA	none	Infogain2	Naïve Bayes	
GISETTE	abs1	Infogain1	Agg. Perceptron	limit=600
MADELON	abs1	margin based	trans. SVM-RBF	$\sigma = 50$, 13 trans. rounds of 10%, $C = 5$

17.1.2 Feature Selection

We used three selection methods. Two of them are based on Mutual Information and log likelihood ratio, and the third is a novel feature selection method that is based on margins:

- **Infogain1:** Features are scored by the mutual information between the feature value and the labels. For non binary data, the values are converted to binary values by comparing to the median. See Chapter 6 and section 5 Chapter 3 for more details about Infogain and feature selection using mutual information in general.
- **Infogain2:** Features are scored by the following value:

$$s_i = (p_i - n_i) \log \left(\frac{p_i}{n_i} \right)$$

where p_i and n_i are estimations of the chance that feature i equals to 1 in the class and outside the class respectively. The data was "binarized" in the same way as for infogain1. We used a simple counts estimations for p_i and n_i, were 0.5 was added to the counts as zero correction.

This method can be considered as variation of infogain (as it involves quantities of the same nature of expectation of log probabilities) or as a natural score for Naïve Bayes classifier (which uses the quantity $\log(p_i/n_i)$ for classification).

Note that this score suppresses the effect of zero values (since the equivalent term that involves the quantities $(1 - p_i)$ and $(1 - n_i)$ is omitted). Thus it is adequate for sparse data were zero values are weak evidence.

- **Margin Based:** In (Crammer et al., 2002), a margin term for Nearest-Neighbor classifiers is presented. The margin of an instance is defined to be half the difference between the distance to the closest point with an alternate label and the closest point with the same label. We selected a subset of features that maximizes the margin on a validation set using a greedy hill climbing procedure. A mature version of this algorithm (*G-flip*) and other margin based feature selection algorithms are described in detail in our contribution to part III of this book. A *Matlab* code is also available.

Infogain1 was used for Gisette and Dexter, Infogain2 for Dorothea and Margin Based for Arcene and Madelon.

17.1.3 Classification

We used SVM (Boser et al., 1992, Vapnik, 1998), transductive SVM (Joachims, 1999), "sleeping" Naïve Bayes (with Good-Turing zero correction (Orlitsky et al., 2003)) and Aggressive Perceptron (Crammer et al., 2003) for classification.

SVM: see section 3.3 of Chapter 1 for detailed description of SVM. We used a tool box by (Cawley, 2000). We applied linear SVM and SVM with a Radial Basis Function (RBF) kernel. In the RBF kernel with a parameter σ, the inner product between two instances x_1 and x_2 is

$$K(x_1, x_2) = e^{-\frac{||x_1 - x_2||^2}{2\sigma^2}}$$

In order to optimize the balanced error rate (BER) we compensated for the different sizes of the two classes as follows. Let R be the ratio between the size of the positive class and the size of the negative class. Then we set the cost of misclassifying an instance to C for positive instances and to CR for negative instances, where C is the standard penalty term of SVM. This was done by setting the parameter ζ of the *train* function in the tool box to 1 for positive instances and to R for negative instances.

Transductive SVM: we started by training an SVM on the training set. Using the classifier obtained we classified the test set. Out of the test set we took a certain percentage of the instances which obtained the largest margin and added it to the training set. We repeated this procedure several times. We denote by *perc* the percentage of the instances from the test set which were added to the train set in each round.

Naïve Bayes: Naïve Bayes classifier classifies an instance x by the rule:

$$\sum_{x_i} \log \frac{p(X_i = x_i | Y = 1)}{p(X_i = x_i | Y = 0)} \gtrless t$$

where t is a threshold. Since we applied it to a sparse binary dataset, we used the "sleeping" version which ignores zero valued features. Thus the classification rule is:

$$\sum_{x_i = 1} \log \frac{p(X_i = 1 | Y = 1)}{p(X_i = 1 | Y = 0)} \gtrless t$$

The threshold was set during training to the value that minimizes the balanced training error. However, the main difficulty is that the probabilities $p(X_i|Y)$ are not known, and should be learned. For this task we used a method reminiscent of Good-Turing estimators. Let $n_y(i)$ be the number of instances x in the training set with label y and $x_i = 1$. The simplest estimator is $P(X_i = 1|Y = y) = \frac{n_y(i)}{n_y}$ where n_y is the size of class y. However, this estimator is inefficient in estimating small probabilities. Therefore we have applied the following correction:

$$P(X_i = 1|Y = y) = \frac{n_y(i) + \frac{1}{\#\{j:n_y(j)=n_y(i)\}}}{n_y}$$

See section 3.1 of Chapter 1 for more details about Naïve Bayes classifiers.

Aggressive Perceptron: The Perceptron algorithms finds a linear separator w by iterating on the training set $\{(x_i, y_i)\}$ and performing the correction $w = w + y_i x_i$ whenever $y_i(w \cdot x_i) < 0$. The aggressive version (Crammer et al., 2003) applies the same correction, but does so whenever $y_i(w \cdot x_i) < limit$ where $limit$ is a predefined positive threshold.

For ARCENE we used SVM with RBF kernel ($\sigma = 0.005$, $C = 8$). For Gisette we used Aggressive Perceptron with a limit set to 600 (i.e. we require that $y(w \cdot x) > 600$ for each (x, y) in the training set). For Dexter we used transductive SVM with a linear kernel, three transduction rounds of 15%, $C = 8$. For Dorothea we used Naïve Bayes with Good-Turing zero correction. For Madelon we used transductive SVM with RBF kernel ($\sigma = 50$), 13 transduction rounds of 10%, $C = 5$.

17.1.4 Implementation

Our implementation was done in *Matlab* . For SVM we used the tool box by (Cawley, 2000). All the rest was written specifically for the challenge. Some of the scripts are provided as supplementary material and can be downloaded from (Gilad-Bachrach and Navot, 2004). Table 17.2 gives the name of the file that implements each method. All the runs were done on a couple of standard PC machines. We did not perform any extensive tuning of parameters, but instead we applied a rough manual tuning.

17.2 Results

Our results for December 1st and December 8th are presented in Table 17.3 and Table 17.4 respectively. According to the average balanced error rate (BER) criterion we rank second as a team, just after the winners. Note that the average BER is the criterion that was used to rank entries during the challenge, and thus we focused on minimizing this term.

Table 17.2. Method Implementation

Purpose	Method	Implementation	Used For
Feature Selection	Infogain1	info_feat_select.m	DEXTER, GISETTE
"	Infogain2	infoSleeping_feat_select.m	DOROTHEA
"	margin-based	distBased_feat_select.m*	MADELON, ARCENE
Classification	SVM	svm_train.m	ARCENE
"	trans. SVM	transsvm_train.m	DEXTER, MADELON
"	Naïve Bayes	naïvebayes_train.m	DOROTHEA
"	Agg. Perceptron	perceptron_train.m	GISETTE

* This function is the original implementation that was used in the challenge. A better, and much faster, implementation can be downloaded from http://www.cs.huji.ac.il/labs/learning/code/feature_selection/. There it appears under the name of *gflip.m*.

For Arcene, we did use feature selection in spite of the fact that we used all the features. We used them to calculate the principal components, but then only 81 principal components were selected using margin-based feature selection. This selection made a vast improvement over using all the principal components.

For Dorothea the accuracy was close to the best using many fewer features than any of the submissions that ranked above us. We only used 0.3% of the features (compared with 100% in the entries above us) and chose no probes (out of 50%).

For Gisette, Madelon and Dexter we were ranked relatively low, but the differences between the different submissions on these datasets are very minor. For example note that our AUC is among the best despite our low rank. This suggests that the ranking on these datasets is statistically insignificant.

17.3 Discussion

The results suggest that besides Radford Neal and Jianguo Zhang, who were the clear winners, many of the other submissions, including ours, performed pretty much the same and can claim second place, depending on the type of ranking. The bottom line is that we achieved good results using the combination of simple feature selection techniques with standard classification methods. Moreover, the computational resources needed to get these results were highly restricted (2 standard desktop PCs).

We also tried to apply 1-Nearest Neighbor (1-NN) instead of SVM with RBF kernel. It turns out that the SVM is only slightly better than the 50+

Table 17.3. NIPS 2003 challenge Dec. 1st results

Dec. 1st	Our best challenge entry					The winning challenge entry					
Dataset	Score	BER	AUC	Feat	Probe	Score	BER	AUC	Feat	Probe	Test
OVERALL	17.45	8.99	95.54	42.18	24.74	88.00	6.84	97.22	80.3	47.8	
ARCENE	18.18	17.20	90.13	100	30	98.18	13.30	93.48	100.0	30.0	
DEXTER	40	5.25	98.8	7	43.71	96.36	3.90	99.01	1.5	12.9	
DOROTHEA	85.45	10.86	92.19	0.3	0	98.18	8.54	95.92	100.0	50.0	
GISETTE	-78.18	3.0	99.63	100	50	98.18	1.37	98.63	18.3	0.0	
MADELON	21.82	8.61	96.97	3.6	0	100.00	7.17	96.95	1.6	0.0	

Table 17.4. NIPS 2003 challenge Dec. 8th results.

Dec. 8th	Our best challenge entry					The winning challenge entry					
Dataset	Score	BER	AUC	Feat	Probe	Score	BER	AUC	Feat	Probe	Test
OVERALL	12	7.78	96.43	32.28	16.22	71.43	6.48	97.2	80.3	47.77	
ARCENE	37.14	12.66	93.37	100	30	94.29	11.86	95.47	10.7	1.03	
DEXTER	11.43	5.0	98.65	7	37.14	100	3.3	96.7	18.57	42.14	
DOROTHEA	68.57	11.4	93.1	0.4	0	97.4	8.61	95.92	100.0	50.0	
GISETTE	-62.86	2.23	99.79	50	13.96	97.14	1.35	98.71	18.32	0.0	
MADELON	5.71	7.61	97.25	4	0	94.29	7.11	96.95	1.6	0.0	

years old nearest neighbor. We also found that the 1-NN is a good predictor for the performance of SVM-RBF. This is useful, since checking the performance of SVM-RBF directly requires the effort of parameter tuning.

Other classification algorithms that we tried such as *Winnow* (Littlestone, 1987) and a few versions of *AdaBoost* (Freund and Schapire, 1997) were inferior to the chosen algorithms.

References

B. Boser, I. Guyon, and V. Vapnik. Optimal margin classifiers. In *In Fifth Annual Workshop on Computational Learning Theory*, pages 144–152, 1992.

G.C. Cawley. Matlab support vector machine toolbox. University of East Anglia, School of Information Systems, Norwich, U.K., 2000. http://theoval.sys.uea.ac.uk/~gcc/svm/toolbox.

K. Crammer, R. Gilad-Bachrach, A. Navot, and N. Tishby. Margin analysis of the lvq algorithm. In *Proceedings of the Fifteenth Annual Conference on Neural Information Processing*, 2002.

K. Crammer, J.S. Kandola, and Y. Singer. Online classification on a budget. In *Proceedings of the Sixteenth Annual Conference on Neural Information Processing*, 2003.

Yoav Freund and Robert E. Schapire. A decision-theoretic generalization of on-line learning and an application to boosting. *Journal of Computer and System Sciences*, 55(1):119–139, 1997.

R. Gilad-Bachrach and A. Navot. MATLAB implementation of FSC submission, 2004. http://www.cs.huji.ac.il/labs/learning/code/fsc/.

T. Joachims. Transductive inference for text classification using support vector machines. In *Proceedings of the Sixteenth International Conference on Machine Learning*, pages 200–209, 1999.

N. Littlestone. Learning quickly when irrelevant attributes abound: A new linear-threshold algorithm. *Machine Learning*, 2(4):285–318, 1987.

A. Orlitsky, N.P. Santhanam, and J. Zhang. Always good turing: Asymptotically optimal probability estimation. *Science*, 302:427–431, 2003.

V. Vapnik. *Statistical Learning Theory*. Wiley, New York, NY, 1998.

Chapter 18

Bayesian Support Vector Machines for Feature Ranking and Selection

Wei Chu[1], S. Sathiya Keerthi[2], Chong Jin Ong[3], and Zoubin Ghahramani[1]

[1] Gatsby Computational Neuroscience Unit, University College London, London, WC1N 3AR, UK. `chuwei@gatsby.ucl.ac.uk`, `zoubin@gatsby.ucl.ac.uk`
[2] Yahoo! Research Lab., Pasadena, CA 91105, USA. `sathiya.keerthi@overture.com`
[3] Department of Mechanical Engineering, National University of Singapore, Singapore, 119260. `mpeongcj@nus.edu.sg`

In this chapter, we develop and evaluate a feature selection algorithm for Bayesian support vector machines. The relevance level of features are represented by ARD (automatic relevance determination) parameters, which are optimized by maximizing the model evidence in the Bayesian framework. The features are ranked in descending order using the optimal ARD values, and then forward selection is carried out to determine the minimal set of relevant features. In the numerical experiments, our approach using ARD for feature ranking can achieve a more compact feature set than standard ranking techniques, along with better generalization performance.

18.1 Introduction

In the classical supervised learning task, we are given a training set of fixed-length feature vectors along with target values, from which to learn a mathematical model that represents the mapping function between the feature vectors and the target values. The model is then used to predict the target for previously unseen instances. The problem of feature selection can be defined as finding relevant features among the original feature vector, with the purpose of increasing the accuracy of the resulting model or reducing the computational load associated with high dimensional problems.

Many approaches have been proposed for feature selection. In general, they can be categorized along two lines as defined by John et al. (1994):

- Filters: the feature selector is independent of a learning algorithm and serves as a pre-processing step to modelling. There are two well-known filter methods, FOCUS and RELIEF. The FOCUS algorithm carries out an exhaustive search of all feature subsets to determine the minimal set

W. Chu et al.: *Bayesian Support Vector Machines for Feature Ranking and Selection*, StudFuzz **207**, 403–418 (2006)
`www.springerlink.com`

of features using a consistency criterion (Almuallim and Dietterich, 1991). RELIEF (Kira and Rendell, 1992) is a randomized algorithm, which attempts to give each feature a weighting indicating its level of relevance to the targets.

• Wrapper: this approach searches through the space of feature subsets using the estimated accuracy from a learning algorithm as the measure of the goodness for a particular feature subset (Langley and Sage, 1994). This method is restricted by the time complexity of the learning algorithm, and when the number of features is large, it may become prohibitively expensive to run.

There are some learning algorithms which have built-in feature selection. Jebara and Jaakkola (2000) formalized a kind of feature weighting in the maximum entropy discrimination framework, and Weston et al. (2001) introduced a method of feature selection for support vector machines by minimizing the bounds on the leave-one-out error. MacKay (1994) and Neal (1996) proposed automatic relevance determination (ARD) as a hierarchical prior over the weights in Bayesian neural networks. The weights connected to an irrelevant input can be automatically punished with a tighter prior in model adaptation, which reduces the influence of such a weight towards zero effectively. In Gaussian processes, the ARD parameters can be directly embedded into the covariance function (Williams and Rasmussen, 1996), which results in a type of feature weighting.

In this paper, we applied Bayesian support vector machines (BSVM) (Chu et al., 2003, 2004) with ARD techniques to select relevant features. BSVM, which is rooted in the probabilistic framework of Gaussian processes, can be regarded as a support vector variant of Gaussian processes. The sparseness in Bayesian computation helps us to tackle relatively large data sets. Bayesian techniques are used to carry out model adaptation. The optimal values of the ARD parameters can be inferred intrinsically in the modelling. Relevance variables are introduced to indicate the relevance level for features. The features can then be ranked from relevant to irrelevant accordingly. In our feature selection algorithm, a forward selection scheme is employed to determine the minimal set of relevant features.

The rest of this paper is organized as follows. Section 18.2 reviews the techniques of BSVM to estimate the optimal values for ARD parameters. Section 18.3 describes a forward selection scheme as post-processing to select the minimal set of relevant features. Section 18.4 presents the results of numerical experiments, followed by the conclusion in Section 18.5.

18.2 Bayesian Framework

As computationally powerful tools for supervised learning, support vector machines (SVMs) were introduced by Boser et al. (1992), and have been widely used in classification and regression problems (Vapnik, 1995). Let us suppose that a data set $\mathcal{D} = \{(\mathbf{x}_i, y_i) | i = 1, \ldots, m\}$ is given for training, where the feature vector $\mathbf{x}_i \in \mathbb{R}^n$ and y_i is the target value. In regression, the target is a real value, while in classification the target is the class label. SVMs map these feature vectors into a high dimensional reproducing kernel Hilbert space (RKHS), where the optimal values of the discriminant function $\{f(\mathbf{x}_i) | i = 1, 2, \ldots, m\}$ can be computed by minimizing a regularized functional. The regularized functional is defined as

$$\min_{f \in \mathrm{RKHS}} \mathcal{R}(f) = \sum_{i=1}^{m} \ell(y_i, f(\mathbf{x}_i)) + \frac{1}{C} \|f\|_{\mathrm{RKHS}}^2, \qquad (18.1)$$

where the regularization parameter C is positive, the stabilizer $\|f\|_{\mathrm{RKHS}}^2$ is a norm in the RKHS and $\sum_{i=1}^{m} \ell(y_i, f(\mathbf{x}_i))$ is the empirical loss term (Evgeniou et al., 1999). For various loss functions, the regularized functional (18.1) can be minimized by solving a convex quadratic programming optimization problem that guarantees a unique global minimum solution. In SVMs for classification (Burges, 1998), hard margin, L_1 soft margin and L_2 soft margin loss functions are widely used. For regression, Smola and Schölkopf (1998) have discussed a lot of common loss functions, such as Laplacian, Huber's, ϵ-insensitive and Gaussian etc.

If we assume that a prior $\mathcal{P}(f) \propto e^{-\frac{1}{C}\|f\|_{\mathrm{RKHS}}^2}$ and a likelihood $\mathcal{P}(\mathcal{D}|f) \propto e^{-\sum_{i=1}^{m} \ell(y_i, f(\mathbf{x}_i))}$, the minimizer of regularized functional (18.1) can be directly interpreted as maximum a posteriori (MAP) estimate of the function f in the RKHS (Evgeniou et al., 1999). Due to the duality between RKHS and stochastic processes (Wahba, 1990), the functions $f(\mathbf{x}_i)$ in the RKHS can also be explained as a family of random variables in a Gaussian process. Gaussian processes have provided a promising non-parametric Bayesian approach to classification problems (Williams and Barber, 1998). The important advantage of Gaussian process models over other non-Bayesian models is the explicit probabilistic formulation. This not only provides probabilistic class prediction but also gives the ability to infer model parameters in a Bayesian framework. We follow the standard Gaussian process to describe a Bayesian framework, in which we impose a Gaussian process prior distribution on these functions and employ particular loss functions in likelihood evaluation. Compared with standard Gaussian processes, the particular loss function results in a different convex programming problem for computing MAP estimates and leads to sparseness in computation.

18.2.1 Prior Probability

The functions in the RKHS (or latent functions) are usually assumed as the realizations of random variables indexed by the input vector $\mathbf{x}_i$ in a stationary zero-mean Gaussian process. The Gaussian process can then be specified by giving the covariance matrix for any finite set of zero-mean random variables

$\{f(\mathbf{x}_i)|i = 1, 2, \ldots, m\}$. The covariance between the outputs corresponding to the inputs $\mathbf{x}_i$ and $\mathbf{x}_j$ can be defined by Mercer kernel functions, such as Gaussian kernel, polynomial kernels and spline kernels (Wahba, 1990). We list two popular covariance functions with ARD parameters in the following:

- ARD Gaussian Kernel: this is a generalization of the popular Gaussian kernel defined as

$$Cov[f(\mathbf{x}_i), f(\mathbf{x}_j)] = \kappa_0 \exp\left(-\frac{1}{2}\sum_{l=1}^{n}\kappa_{a,l}(\mathbf{x}_{i,l} - \mathbf{x}_{j,l})^2\right) + \kappa_b, \quad (18.2)$$

where l is the feature index, $\kappa_0 > 0$, $\kappa_{a,l} > 0$ and $\kappa_b > 0$. κ_0 denotes the average power of $f(\mathbf{x})$ that reflects the noise level, while κ_b corresponds to the variance of the offset in the latent functions.[4] The ARD parameters $\kappa_{a,l}$ $\forall l$ are used to indicate the relevance level of the l-th feature to the targets. Note that a relatively large ARD parameter implies that the associated feature gives more contributions to the modelling, while a feature weighted with a very small ARD parameter implies that this feature is irrelevant to the targets.
- ARD Linear Kernel: this is a type of linear kernel parameterized with ARD parameters defined as

$$Cov[f(\mathbf{x}_i), f(\mathbf{x}_j)] = \sum_{l=1}^{n}\kappa_{a,l}\,\mathbf{x}_{i,l}\,\mathbf{x}_{j,l} + \kappa_b, \quad (18.3)$$

where $\kappa_b > 0$ and $\kappa_{a,l} > 0$.

We collect the parameters in the covariance function (18.2) or (18.3), as θ, the hyperparameter vector. Then, for a given hyperparameter vector θ, the prior probability of the random variables $\{f(\mathbf{x}_i)\}$ is a multivariate Gaussian, which can be simply written as

$$P(\boldsymbol{f}\,|\theta) = \frac{1}{Z_{\boldsymbol{f}}}\exp(-\frac{1}{2}\boldsymbol{f}^T\Sigma^{-1}\boldsymbol{f}), \quad (18.4)$$

where $\boldsymbol{f} = [f(\mathbf{x}_1), f(\mathbf{x}_2), \ldots, f(\mathbf{x}_m)]^T$, $Z_{\boldsymbol{f}} = (2\pi)^{\frac{m}{2}}|\Sigma|^{\frac{1}{2}}$, and Σ is the $m \times m$ covariance matrix whose ij-th element is $Cov[f(\mathbf{x}_i), f(\mathbf{x}_j)]$.

18.2.2 Likelihood

We usually assume that the training data are collected independently. The probability $P(\mathcal{D}|\boldsymbol{f}, \theta)$, known as the likelihood, can be evaluated by:

---

[4]In classification settings, it is possible to insert a "jitter" term in the diagonal entries of the covariance matrix, that could reflect the uncertainty in the corresponding function value.

$$P(\mathcal{D}|\boldsymbol{f},\theta) = \prod_{i=1}^{m} P(y_i|f(\mathbf{x}_i)), \tag{18.5}$$

where $-\ln P(y_i|f(\mathbf{x}_i))$ is usually referred to as the loss function $\ell(y_i, f(\mathbf{x}_i))$.

In regression problems, the discrepancy between the target value y_i and the associated latent function $f(\mathbf{x}_i)$ at the input $\mathbf{x}_i$ is evaluated by a specific noise model. Various loss functions can be used depending on the assumption on the distribution of the additive noise (Chu et al., 2004). In this paper, we focus on binary classification only.

For binary classifier designs, we measure the probability of the class label y_i for a given latent function $f(\mathbf{x}_i)$ at $\mathbf{x}_i$ as the likelihood, which is a conditional probability $P(y_i|f(\mathbf{x}_i))$.[5] The logistic and probit functions are widely used in likelihood evaluation (Williams and Barber, 1998, Neal, 1997b). However these do not result in sparse solutions to the optimization problem. In order to introduce sparseness into this Bayesian framework, Chu et al. (2003) proposed a trigonometric loss function. The trigonometric loss function is defined as

$$\ell_t(y_i, f(\mathbf{x}_i)) = \begin{cases} +\infty & \text{if } y_i \cdot f(\mathbf{x}_i) \in (-\infty, -1]; \\ 2\ln\sec(\frac{\pi}{4}(1 - y_i \cdot f(\mathbf{x}_i))) & \text{if } y_i \cdot f(\mathbf{x}_i) \in (-1, +1); \\ 0 & \text{if } y_i \cdot f(\mathbf{x}_i) \in [+1, +\infty), \end{cases} \tag{18.6}$$

where $\ln$ denotes the natural logarithm and $\sec$ denotes the secant. The trigonometric likelihood function is therefore written as

$$P_t(y_i|f(\mathbf{x}_i)) = \begin{cases} 0 & \text{if } y_i \cdot f(\mathbf{x}_i) \in (-\infty, -1]; \\ \cos^2(\frac{\pi}{4}(1 - y_i \cdot f(\mathbf{x}_i))) & \text{if } y_i \cdot f(\mathbf{x}_i) \in (-1, +1); \\ 1 & \text{if } y_i \cdot f(\mathbf{x}_i) \in [+1, +\infty). \end{cases} \tag{18.7}$$

Note that $P_t(y_i = +1|f(\mathbf{x}_i)) + P_t(y_i = -1|f(\mathbf{x}_i)) = 1$ always holds for any value of $f(\mathbf{x}_i)$, and the trigonometric loss function possesses a flat zero region that is the same as the L_1 and L_2 loss functions in support vector machines.

18.2.3 Posterior Probability

Based on Bayes' theorem, the posterior probability of $\boldsymbol{f}$ can then be written as

$$P(\boldsymbol{f}|\mathcal{D},\theta) = \frac{1}{Z_S} \exp\left(-S(\boldsymbol{f})\right), \tag{18.8}$$

where $S(\boldsymbol{f}) = \frac{1}{2}\boldsymbol{f}^T \Sigma^{-1}\boldsymbol{f} + \sum_{i=1}^{n} \ell(y_i, f(\mathbf{x}_i))$, $\ell(\cdot)$ is the loss function we used and $Z_S = \int \exp(-S(\boldsymbol{f})) \, d\boldsymbol{f}$. Since $P(\boldsymbol{f}|\mathcal{D},\theta) \propto \exp(-S(\boldsymbol{f}))$, the Maximum

---

[5]Here, y_i is a discrete random variable, and the sum of the probabilities for all possible cases of y_i should be equal to 1, i.e. $\sum_{y_i} P(y_i|f(\mathbf{x}_i)) = 1$, which is referred to as the *normalization requirement*.

A Posteriori (MAP) estimate on the values of $\boldsymbol{f}$ is therefore the minimizer of the following optimization problem:

$$\min_{\boldsymbol{f}} S(\boldsymbol{f}) = \frac{1}{2} \boldsymbol{f}^T \Sigma^{-1} \boldsymbol{f} + \sum_{i=1}^{m} \ell(y_i, f(\mathbf{x}_i)). \qquad (18.9)$$

This is a regularized functional. For any differentiable loss function, the solution of the regularized functional $S(\boldsymbol{f})$, is always a linear superposition of covariance functions, one for each data point. This elegant form of a minimizer of (18.9) is also known as the representer theorem (Kimeldorf and Wahba, 1971). A generalized representer theorem can be found in Schölkopf et al. (2001), in which the loss function is merely required to be a strictly monotonically increasing function $\ell : \mathbb{R} \to [0, +\infty)$.

18.2.4 MAP Estimate

Introducing the trigonometric loss function (18.6) into the regularized functional of (18.9), the optimization problem (18.9) can be then restated as the following equivalent optimization problem, which we refer to as the *primal* problem:

$$\min_{\boldsymbol{f}, \boldsymbol{\xi}} \frac{1}{2} \boldsymbol{f}^T \Sigma^{-1} \boldsymbol{f} + 2 \sum_{i=1}^{m} \ln \sec \left(\frac{\pi}{4} \xi_i \right) \qquad (18.10)$$

subject to $y_i \cdot f(\mathbf{x}_i) \geq 1 - \xi_i$ and $0 \leq \xi_i < 2$, $\forall i$. Standard Lagrangian techniques (Fletcher, 1987) are used to derive the *dual* problem. The *dual* problem can be finally simplified as

$$\min_{\alpha} \frac{1}{2} \sum_{i=1}^{m} \sum_{j=1}^{m} (y_i \alpha_i)(y_j \alpha_j) Cov \left[f(\mathbf{x}_i), f(\mathbf{x}_j) \right] - \sum_{i=1}^{m} \alpha_i$$
$$+ \sum_{i=1}^{m} n \left[\frac{4}{\pi} \alpha_i \arctan \left(\frac{2\alpha_i}{\pi} \right) - \ln \left(1 + \left(\frac{2\alpha_i}{\pi} \right)^2 \right) \right] \qquad (18.11)$$

subject to $\alpha_i \geq 0, \forall i$. Refer to Chu et al. (2003) for the derivation details.

The *dual* problem (18.11) is a convex programming problem. The popular SMO algorithm for classical SVMs (Platt, 1999, Keerthi et al., 2001) can be easily adapted to solve the optimization problem (Chu et al., 2003). The MAP estimate on the values of the random variables $\boldsymbol{f}$ can be written in column vector form as

$$\boldsymbol{f}_{\mathrm{MP}} = \Sigma \cdot \boldsymbol{v} \qquad (18.12)$$

where $\boldsymbol{v} = [y_1 \alpha_1, y_2 \alpha_2, \ldots, y_m \alpha_m]^T$. The training samples $(\mathbf{x}_i, y_i)$ associated with non-zero Lagrange multiplier α_i are called *support vectors* (SVs). The other samples associated with zero α_i are not involved in the solution representation and the following Bayesian computation. This property is usually referred to as sparseness, and it reduces the computational cost significantly.

18.2.5 Hyperparameter Inference

The hyperparameter vector θ contains the ARD parameters and other parameters in the covariance function. The optimal values of hyperparameters θ can be inferred by maximizing the posterior probability $\mathcal{P}(\theta|\mathcal{D})$, using $\mathcal{P}(\theta|\mathcal{D}) = \mathcal{P}(\mathcal{D}|\theta)\mathcal{P}(\theta)/\mathcal{P}(\mathcal{D})$. The prior distribution on the hyperparameters $\mathcal{P}(\theta)$ can be specified by domain knowledge. As we typically have little idea about the suitable values of θ before training data are available, we usually assume a flat distribution for $\mathcal{P}(\theta)$, i.e., $\mathcal{P}(\theta)$ is greatly insensitive to the values of θ. Therefore, $\mathcal{P}(\mathcal{D}|\theta)$ (which is known as the evidence of θ) can be used to assign a preference to alternative values of the hyperparameters θ (MacKay, 1992).

The evidence is given by an integral over all $\boldsymbol{f}$: $\mathcal{P}(\mathcal{D}|\theta) = \int \mathcal{P}(\mathcal{D}|\boldsymbol{f},\theta)\mathcal{P}(\boldsymbol{f}|\theta)\,d\boldsymbol{f}$. Using the definitions in (18.4) and (18.5), the evidence can also be written as

$$\mathcal{P}(\mathcal{D}|\theta) = \frac{1}{Z_{\boldsymbol{f}}} \int \exp(-S(\boldsymbol{f}))\,d\boldsymbol{f}, \qquad (18.13)$$

where $S(\boldsymbol{f})$ is defined as in (18.8). Computing the evidence as a high dimensional integral is technically difficult. So far, a variety of approximation techniques have been discussed: Monte Carlo sampling (Neal, 1997a), the MAP approach (Williams and Barber, 1998), bounds on the likelihood (Gibbs, 1997) and mean field approaches (Opper and Winther, 2000, Csató et al., 2000). Recently, Kim and Ghahramani (2003) coupled the Expectation Propagation algorithm (Minka, 2001) with variational methods (Seeger, 1999) for evidence maximization. To maintain the sparseness in Bayesian computation, we apply Laplace approximation at the MAP estimate. The evidence (18.13) could be calculated by an explicit formula, and then hyperparameter inference can be done by gradient-based optimization methods.

The marginalization can be done analytically by considering the Taylor expansion of $S(\boldsymbol{f})$ around its minimum $S(\boldsymbol{f}_{\mathrm{MP}})$, and retaining terms up to the second order. Since the first order derivative with respect to $\boldsymbol{f}$ at the MAP point $\boldsymbol{f}_{\mathrm{MP}}$ is zero, $S(\boldsymbol{f})$ can be written as

$$S(\boldsymbol{f}) \approx S(\boldsymbol{f}_{\mathrm{MP}}) + \frac{1}{2}(\boldsymbol{f} - \boldsymbol{f}_{\mathrm{MP}})^T \left(\left.\frac{\partial^2 S(\boldsymbol{f})}{\partial \boldsymbol{f}\partial \boldsymbol{f}^T}\right|_{\boldsymbol{f}=\boldsymbol{f}_{\mathrm{MP}}} \right) (\boldsymbol{f} - \boldsymbol{f}_{\mathrm{MP}}), \qquad (18.14)$$

where $\frac{\partial^2 S(\boldsymbol{f})}{\partial \boldsymbol{f}\partial \boldsymbol{f}^T} = \Sigma^{-1} + \Lambda$, and Λ is a diagonal matrix coming from the second order derivative of the loss function we used. Introducing (18.14) into (18.13) yields

$$\mathcal{P}(\mathcal{D}|\theta) = \exp(-S(\boldsymbol{f}_{\mathrm{MP}})) \cdot |\mathbf{I} + \Sigma \cdot \Lambda|^{-\frac{1}{2}},$$

where $\mathbf{I}$ is the $m \times m$ identity matrix. Note that, when the trigonometric loss function (18.6) is employed, only a sub-matrix of Σ plays a role in the determinant $|\mathbf{I} + \Sigma \cdot \Lambda|$ due to the sparseness of the diagonal matrix Λ in which only the entries associated with SVs are non-zero. We denote their sub-matrices as Σ_M and Λ_M respectively by keeping their non-zero entries. The MAP estimate of f (18.12) on support vectors can also be simplified as $\Sigma_M \cdot v_M$, where v_M denotes the sub-vector of v by keeping entries associated with SVs. Because of these sparseness properties, the negative log of the evidence can then be simplified as follows

$$-\ln \mathcal{P}(\mathcal{D}|\theta) = \frac{1}{2}v_M^T \cdot \Sigma_M \cdot v_M + 2 \sum_{v \in \text{SVs}} \ln \sec \left(\frac{\pi}{4}\xi_v\right) + \frac{1}{2} \ln |\mathbf{I} + \Sigma_M \cdot \Lambda_M|, \quad (18.15)$$

where $\mathbf{I}$ is the identity matrix with the size of SVs, "$v \in$ SVs" denotes that v is varied over the index set of SVs, and $\xi_v = 1 - y_v \cdot f_{MP}(\mathbf{x}_v)$, $\forall v$. The evidence evaluation is a convenient yardstick for model selection. It is straightforward to consider the posterior distribution $\mathcal{P}(\theta|\mathcal{D}) \propto \mathcal{P}(\mathcal{D}|\theta)\mathcal{P}(\theta)$ by specifying a particular prior distribution $\mathcal{P}(\theta)$. The gradient of $-\ln \mathcal{P}(\theta|\mathcal{D})$ with respect to the variables in the hyperparameter vector θ can be explicitly derived (see Chu et al. (2003) for the detailed derivatives), and then gradient-based optimization methods can be used to find the minimum locally.

The optimization method usually requests evidence evaluation at tens of different θ before the minimum is found. For each θ, a quadratic programming problem should be solved first to find MAP estimate, and then the approximate evidence (18.15) is calculated along with its gradients with respect to the hyperparameters. Due to the sparseness, the quadratic programming problem costs almost the same time as SVMs at the scale about $\mathcal{O}(m^{2.2})$, where m is the size of training data. In gradient evaluations for ARD parameters, the inversion of the matrix Σ_M, corresponding to SVs at the MAP estimate, is required that costs time at $\mathcal{O}(s^3)$ for each feature, where s is the number of SVs that is usually much less than m.

18.3 Post-processing for Feature Selection

The generalization performance of BSVM with ARD techniques are very competitive (Chu et al., 2003). In practical applications, it might be desirable to further select a minimal subset of relevant features for modelling while keeping the accuracy of the resulting model and reducing the computational load. In this section, we describe our method for feature selection based on the techniques described in the previous section. The task of feature selection can be tackled in two steps:

(1.) The original features can be ranked in descending order using the optimal values of the ARD parameters $\{\kappa_a^l\}_{l=1}^m$ we inferred (see Section 18.3.1).
(2.) Then a subset of the top features in the rank is used as the relevant features for modelling. The minimal subset can be determined by the validation performance of the learning machine (see Section 18.3.2).

18.3.1 Feature Ranking

We first introduce a set of relevance variables $\{r^i\}_{i=1}^n$ for the features we are given, which are extracted from the ARD parameters by normalizing them:

$$r^i = \frac{\kappa_{a,i}}{\sum_{j=1}^n \kappa_{a,j}}. \tag{18.16}$$

The relevance variable indicates the relevance level of the feature to the targets, which is independent of the overall scale of ARD parameters.

Since there might be several local minima on the curve of $-\ln \mathcal{P}(\theta|\mathcal{D})$, it is possible that the optimization problem is stuck at local minima in the determination of θ.[6] We may reduce the impact of this problem by minimizing (18.15) several times starting from several different initial states, and simply choosing the one with the highest evidence as our preferred choice for θ. We can also organize these candidates together to represent the evidence distribution that might reduce the uncertainty with respect to the hyperparameters. An approximation scheme is described in the following.

Suppose we started from several different initial states, and reached several minima θ_τ^* of the optimization problem (18.15). We simply assume that the underlying distribution is a superposition of individual distributions with mean θ_τ^*. The underlying distribution $\mathcal{P}(\theta|\mathcal{D})$ is roughly reconstructed as

$$\mathcal{P}(\theta|\mathcal{D}) \approx \sum_{\tau=1}^t w_\tau \mathcal{P}_\tau(\theta; \theta_\tau^*) \tag{18.17}$$

with

$$w_\tau = \frac{\mathcal{P}(\theta_\tau^*|\mathcal{D})}{\sum_{\iota=1}^t \mathcal{P}(\theta_\iota^*|\mathcal{D})}, \tag{18.18}$$

where t is the number of the minima we have discovered by gradient descent methods, and $\mathcal{P}_\tau(\theta; \theta_\tau^*)$ denotes the individual distribution which can be any distribution with the mean θ_τ^*. The values of $\mathcal{P}(\theta_\tau^*|\mathcal{D})$ have been obtained from the functional (18.15) already. In Figure 18.1, we present a simple case as an illustration.

We can evaluate the expected values of the relevance variables based on the approximated $\mathcal{P}(\theta|\mathcal{D})$ (18.17) as follows

$$\hat{r}^i \approx \sum_{\tau=1}^t w_\tau r_\tau^i, \tag{18.19}$$

where r_τ^i is defined as in (18.16) at the minimum θ_τ^*. This is not an approximation designed for the posterior mean of the relevance variables, but helpful to overcome the difficulty caused by the poor minima. Using these values $\{\hat{r}^i\}_{i=1}^d$, we can sort the features in descending order from relevant to irrelevant.

---

[6]Monte Carlo sampling methods can provide a good approximation on the posterior distribution, but might be prohibitively expensive to use for high-dimensional problems.

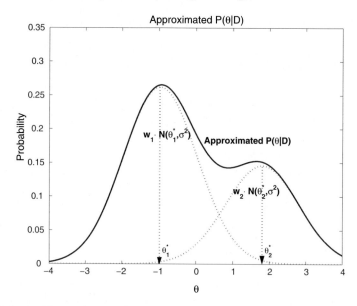

Fig. 18.1. An illustration on the reconstructed curve for the posterior distribution $\mathcal{P}(\theta|\mathcal{D})$. θ_1^* and θ_2^* are the two maxima we found by gradient descent methods. The underlying distribution $\mathcal{P}(\theta|\mathcal{D})$ can be simply approximated as the superposition of two individual distributions (dotted curves) with mean θ_1^* and θ_2^* respectively. The weights w_1 and w_2 for the two distributions are defined as in (18.18). In this graph, Gaussian distributions with same variance were used as individual distribution.

18.3.2 Feature Selection

Given any learning algorithm, we can select a minimal subset of relevant features by carrying out cross validation experiments on progressively larger sets of features, adding one feature at a time ordered by their relevance variables $\hat{r}_i$.[7] Let S denote the set of relevant features being investigated. The features top-ranked according to r^i defined as in (18.19) are added into the set S one by one, and the validation error is calculated. This procedure is repeated as long as adding the next top-ranked feature into S does not increase the validation error significantly. This feature set S is then used along with all the training data for modeling.

18.3.3 Discussion

RELIEF (Kira and Rendell, 1992) also attempts to specify relevance level for features, but the ARD techniques (MacKay, 1994, Neal, 1996) can carry out Bayesian inference systematically. In our approach, the performance of

---

[7]The relevance variables of useless features are usually much less than the average level $1/n$, where n is the total number of features.

Table 18.1. The outline of our algorithm for feature selection.

Ranking	employ a kind of ARD kernel
	randomly select starting points θ_0 for Optimization Package
	while Optimization Package requests evidence/gradient evaluation at θ
	solve (18.11) by convex programming to find the MAP estimate
	evaluate the approximate evidence (18.13)
	calculate the gradients with respect to θ
	return evidence/gradient to the optimization package
	Optimization Package returns the optimal θ^*
	compute the relevance variables defined as in (18.19)
	rank the features in descending order
Selection	initialize validation error to infinity, and $k = 0$
	do
	$k = k + 1$
	use the top k features as input vector to a learning algorithm
	carry out cross validation via grid searching on model parameters
	pick up the best validation error
	while validation error is not increasing significantly
Exit	return the top $k - 1$ features as the minimal subset.

a learning algorithm is used as the criterion to decide the minimal subset; this is analogous to the wrapper approach (Langley and Sage, 1994). The key difference is that we only check the subsets including the top-ranked feature sequentially rather than search through the huge space of feature subsets.

A potential problem may be caused by correlations between features. Such correlations introduce dependencies into the ARD variables. More exactly, at the minima, correlated features may share their ARD values randomly. This makes it difficult to distinguish relevant features based on their ARD values alone. In this case, we suggest a backward elimination process (Guyon et al., 2002). This process begins with the full set and remove the most irrelevant feature one by one. At each step, we carry out inference on the reduced dataset to update their relevance variables. This procedure is computationally expensive, since it requires performing hyperparameter inference m times.

Table 18.2. NIPS 2003 challenge results we submitted. "Score" denotes the score used to rank the results by the organizers (times 100). "BER" denotes balanced error rate (in percent). "AUC" is the area under the ROC curve (times 100). "Feat" is percent of features used. "Probe" is the percent of probes found in the subset selected. "Test" is the result of the comparison with the best entry using the MacNemar test.

Dec. 1st	Our challenge entry					The winning challenge entry					
Dataset	Score	BER	AUC	Feat	Probe	Score	BER	AUC	Feat	Probe	Test
Overall	15.27	9.43	95.70	67.53	38.03	88.00	6.84	97.22	80.3	47.8	0.6
Arcene	78.18	15.17	91.52	100	30	98.18	13.30	93.48	100.0	30.0	0
Dexter	-21.82	6.35	98.57	36.04	60.15	96.36	3.90	99.01	1.5	12.9	1
Dorothea	-25.45	15.47	92.06	100	50	98.18	8.54	95.92	100.0	50.0	1
Gisette	-47.27	2.62	99.67	100	50	98.18	1.37	98.63	18.3	0.0	1
Madelon	100.00	7.17	96.95	1.6	0.0	100.00	7.17	96.95	1.6	0.0	0

18.4 Numerical Experiments

We give an outline of the algorithm for feature selection in Table 18.1.[8] The feature vectors with continuous elements were normalized to have zero mean and unit variance coordinate-wise. The ARD parameters were used for feature weighting. The initial value of the hyperparameters were chosen as $\kappa_0 = 1.0$ and $\kappa_b = 10.0$. We tried ten times starting from different values of κ_a^l to maximize the evidence by gradient descent methods.[9] The ten maxima we found are used to estimate the values of the relevance variables as in (18.19). In the forward feature selection, SVMs with Gaussian kernel, $\exp(-\frac{\kappa}{2}\|x_i - x_j\|^2)$, was used as the learning algorithm. Grid search in the parameter space spanned by the κ in the Gaussian kernel and the regularization factor C as in (18.1), was carried out to locate the best validation output. The primary grid search was done on a 7×7 coarse grid linearly spaced in the region $\{(\log_{10} C, \log_{10} \kappa) | -0.5 \leq \log_{10} C \leq 2.5, -3 \leq \log_{10} \kappa \leq 0\}$, followed by a fine search on a 9×9 uniform grid linearly spaced by 0.1 in the $(\log_{10} C, \log_{10} \kappa)$ space. At each node in this grid, 5-fold cross validation was repeated ten times to reduce the bias in fold generations, and the validation errors were averaged over all the trials.

The NIPS 2003 challenge results we submitted are reported in Table 18.2. On the Madelon data set, we carried out feature selection as described in

---

[8] The source code of Bayesian support vector machines in ANSI C can be accessed at http://guppy.mpe.nus.edu.sg/~chuwei/btsvc.htm.

[9] The number of the minima should be large enough to reproduce the results stably.

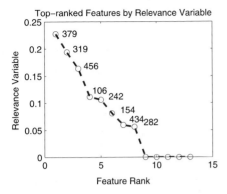
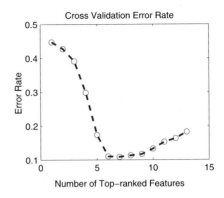

Fig. 18.2. The result on Madelon data set for our algorithm. The values of relevance variables $\hat{r}^i$ for the top 13 features are presented in the left part along with the feature indices, while the validation error rates are plotted in the right part.

Table 18.1. For other datasets, we submitted the predictive results of linear SVMs with L_1 loss function without strict feature selection at that time.[10] The "AUC" performance of our entry is competitive with that of the winning entry.

The Madelon task[11] is to categorize random data into two classes. There are 2000 samples for training and 1800 samples for test. Each sample has 500 features. In Figure 18.2, we present the top 13 features ranked by the estimated values of relevance variables. The performance of 5-fold cross validation (SVMs with L_1 loss function and Gaussian kernel was used) is presented in the right part of Figure 18.2. In the challenge, we selected the top 8 features for modeling, and built up SVMs using ARD Gaussian kernel with fixed relevance variables shown in the left part of Figure 18.2. Cross validation was carried out to decide the optimal regularization factor and the common scale parameter in the Gaussian kernel. The blind test on the 1800 samples got 7.17% error rate. This entry was assigned the highest score by the organizers.

In the next experiment, the popular Fisher score was used in feature ranking for comparison purpose. The Fisher score Chapter 3 is defined as

$$s_i = \frac{|\mu_{i,+} - \mu_{i,-}|}{\sigma_{i,+} + \sigma_{i,-}}, \tag{18.20}$$

where $\mu_{i,+}$ and $\sigma_{i,+}$ are the mean and standard deviation of the i-th feature on the positive samples, while $\mu_{i,-}$ and $\sigma_{i,-}$ are of the negative samples. In Figure 18.3, the top-ranked features by their Fisher score (18.20) are presented in the left graph, and the validation error rate using the top-ranked features incrementally are presented in the right graph. The best validation result is

[10]On the Dexter dataset, we simply removed the features with zero weight in the optimal linear SVMs, and then retrained linear SVMs on the reduced data.
[11]The data set can be found at http://clopinet.com/isabelle/Projects/NIPS2003/.

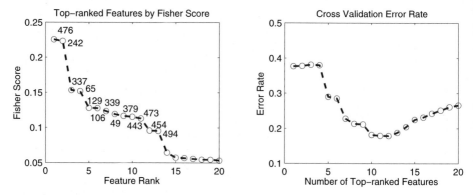

Fig. 18.3. The result on Madelon data set using fisher score for feature ranking. The fisher scores of the 13 top-ranked features are presented in the left part along with the feature indices, while the validation error rates are plotted in the right part.

about 0.18 using 12 features, which is much worse than the result of our algorithm as shown in Figure 18.2.

18.5 Conclusion

In this chapter, we embedded automatic relevance determination in Bayesian support vector machines to evaluate feature relevance. The Bayesian framework provides various computational procedures for hyperparameter inference, which can be used for feature selection. The sparseness property introduced by our loss function (18.6) helps us tackling relatively large data sets. A forward selection method is used to determine the minimal subset of informative features. Overall we have a probabilistic learning algorithm with built-in feature selection that selects informative features automatically. The results of numerical experiments show that this approach can achieve quite compact feature sets, and achieve good generalization performance.

Acknowledgment

This work was supported by the National Institutes of Health and its National Institute of General Medical Sciences division under Grant Number 1 P01 GM63208 (NIH/NIGMS grant title: Tools and Data Resources in Support of Structural Genomics).

References

H. Almuallim and T. G. Dietterich. Learning with many irrelevant features. In *Proc. AAAI-91*, pages 547–552. MIT Press, 1991.

B. Boser, I. Guyon, and V. N. Vapnik. A training algorithm for optimal margin classifier. In *Fifth Annual Workshop on Computational Learning Theory*, pages 144–152, Pittsburgh, USA, 1992.

C. J. C. Burges. A tutorial on support vector machines for pattern recognition. *Data Mining and Knowledge Discovery*, 2(2):121–167, 1998.

W. Chu, S. S. Keerthi, and C. J. Ong. Bayesian trigonometric support vector classifier. *Neural Computation*, 15(9):2227–2254, 2003.

W. Chu, S. S. Keerthi, and C. J. Ong. Bayesian support vector regression using a unified loss function. *IEEE transactions on neural networks*, 15(1):29–44, 2004.

L. Csató, E. Fokoué, M. Opper, B. Schottky, and O. Winther. Efficient approaches to Gaussian process classification. In *Advances in Neural Information Processing Systems*, volume 12, pages 251–257, 2000.

T. Evgeniou, M. Pontil, and T. Poggio. A unified framework for regularization networks and support vector machines. A.I. Memo 1654, MIT, 1999.

R. Fletcher. *Practical methods of optimization.* John Wiley and Sons, 1987.

M. N. Gibbs. *Bayesian Gaussian Processes for Regression and Classification.* Ph.D. thesis, University of Cambridge, 1997.

I. Guyon, J. Weston, and S. Barnhill. Gene selection for cancer classification using support vector machines. *Machine Learning*, 46:389–422, 2002.

T. S. Jebara and T. S. Jaakkola. Feature selection and dualities in maximum entropy discrimination. In *Uncertainty in Artificial Intelligence: Proceedings of the Sixteenth Conference (UAI-2000)*, pages 291–300, San Francisco, CA, 2000. Morgan Kaufmann Publishers.

G. John, R. Kohavi, and K. Pfleger. Irrelevant features and the subset selection problem. In *Proc. ML-94*, pages 121–129. Morgan Kaufmann Publishers, 1994.

S. S. Keerthi, S. K. Shevade, C. Bhattacharyya, and K. R. K. Murthy. Improvements to Platt's SMO algorithm for SVM classifier design. *Neural Computation*, 13 (3):637–649, March 2001.

H. Kim and Z. Ghahramani. The EM-EP algorithm for Gaussian process classification. In *Proc. of the Workshop on Probabilistic Graphical Models for Classification (at ECML)*, 2003.

G. S. Kimeldorf and G. Wahba. Some results on Tchebycheffian spline function. *Journal of Mathematical Analysis and Applications*, 33:82–95, 1971.

K. Kira and L. A. Rendell. The feature selection problem: Traditional methods and a new algorithm. In *Proc. AAAI-92*, pages 129–134. MIT Press, 1992.

P. Langley and S. Sage. Induction of selective Bayesian classifiers. In *Proc. UAI-94*, pages 399–406. Morgan Kaufmann, 1994.

D. J. C. MacKay. A practical Bayesian framework for back propagation networks. *Neural Computation*, 4(3):448–472, 1992.

D. J. C. MacKay. Bayesian methods for backpropagation networks. *Models of Neural Networks III*, pages 211–254, 1994.

T. P. Minka. *A family of algorithm for approximate Bayesian inference.* Ph.D. thesis, Massachusetts Institute of Technology, January 2001.

R. M. Neal. *Bayesian Learning for Neural Networks.* Lecture Notes in Statistics. Springer, 1996.

R. M. Neal. Monte Carlo implementation of Gaussian process models for Bayesian regression and classification. Technical Report No. 9702, Department of Statistics, University of Toronto, 1997a.

R. M. Neal. Regression and classification using Gaussian process priors (with discussion). In J. M. Bernerdo, J. O. Berger, A. P. Dawid, and A. F. M. Smith, editors, *Bayesian Statistics*, volume 6, 1997b.

M. Opper and O. Winther. Gaussian processes for classification: Mean field algorithm. *Neural Computation*, 12(11):2655–2684, 2000.

J. C. Platt. Fast training of support vector machines using sequential minimal optimization. In B. Schölkopf, C. J. C. Burges, and A. J. Smola, editors, *Advances in Kernel Methods - Support Vector Learning*, pages 185–208. MIT Press, 1999.

B. Schölkopf, R. Herbrich, and A. J. Smola. A generalized representer theorem. In *Proc. of the Annual Conference on Computational Learning Theory*, 2001.

M. Seeger. Bayesian model selection for support vector machines, Gaussian processes and other kernel classifiers. In *Advances in Neural Information Processing Systems*, volume 12, 1999.

A. J. Smola and B. Schölkopf. A tutorial on support vector regression. Technical Report NC2-TR-1998-030, GMD First, October 1998.

V. N. Vapnik. *The Nature of Statistical Learning Theory*. New York: Springer-Verlag, 1995.

G. Wahba. *Spline Models for Observational Data*, volume 59 of *CBMS-NSF Regional Conference Series in Applied Mathematics*. SIAM, 1990.

J. Weston, S. Mukherjee, O. Chapelle, M. Pontil, and T. Poggio. Feature selection in SVMs. In Todd Leen, Tom Dietterich, and Volker Tresp, editors, *Advances in Neural Information Processing Systems*, volume 13, 2001. MIT Press.

C. K. I. Williams and D. Barber. Bayesian classification with Gaussian processes. *IEEE Transactions on Pattern Analysis and Machine Intelligence*, 20(12):1342–1351, 1998.

C. K. I. Williams and C. E. Rasmussen. Gaussian processes for regression. In D. S. Touretzky, M. C. Mozer, and M. E. Hasselmo, editors, *Advances in Neural Information Processing Systems*, volume 8, pages 598–604, 1996. MIT Press.

Chapter 19

Nonlinear Feature Selection with the Potential Support Vector Machine

Sepp Hochreiter and Klaus Obermayer

Technische Universität Berlin
Fakultät für Elektrotechnik und Informatik
Franklinstraße 28/29, 10587 Berlin, Germany
hochreit@cs.tu-berlin.de,oby@cs.tu-berlin.de

Summary. We describe the "Potential Support Vector Machine" (P-SVM) which is a new filter method for feature selection. The idea of the P-SVM feature selection is to exchange the role of features and data points in order to construct "support features". The "support features" are the selected features. The P-SVM uses a novel objective function and novel constraints – one constraint for each feature. As with standard SVMs, the objective function represents a complexity or capacity measure whereas the constraints enforce low empirical error. In this contribution we extend the P-SVM in two directions. First, we introduce a parameter which controls the redundancy among the selected features. Secondly, we propose a nonlinear version of the P-SVM feature selection which is based on neural network techniques. Finally, the linear and nonlinear P-SVM feature selection approach is demonstrated on toy data sets and on data sets from the NIPS 2003 feature selection challenge.

19.1 Introduction

Our focus is on the selection of relevant features, that is on the identification of features, which have dependencies with the target value. Feature selection is important (1) to reduce the effect of the "curse of dimensionality" (Bellman, 1961) when predicting the target in a subsequent step, (2) to identify features which allow us understanding the data as well as control or build models of the data generating process, and (3) to reduce costs for future measurements, data analysis, or prediction. An example for item (2) and (3) are gene expression data sets in the medical context (e.g. gene expression patterns of tumors), where selecting few relevant genes may give hints to develop medications and reduce costs through smaller microarrays. Another example is the World Wide Web domain, where selecting relevant hyperlinks corresponds to the identification of hubs and authorities. Regarding items (2) and (3), we investigate feature selection methods, which are not tailored to a certain predictor but are filter methods (cf. Chapter Chapter 3) and lead to compact feature sets.

S. Hochreiter and K. Obermayer: *Nonlinear Feature Selection with the Potential Support Vector Machine*, StudFuzz **207**, 419–438 (2006)
www.springerlink.com

We propose the "Potential Support Vector Machine" (P-SVM, Hochreiter and Obermayer, 2004a) as filter method for feature selection. The P-SVM describes the classification or regression function through complex features vectors (certain directions in input space) rather than through the input vectors as standard support vector machines (SVMs, see Chapter Chapter 1 and Boser et al., 1992, Cortes and Vapnik, 1995, Schölkopf and Smola, 2002, Vapnik, 1998) do. This description imposes no restriction on the chosen function class because it is irrelevant how a function is represented. A feature value is computed by the dot product between the corresponding complex feature vector and an input vector analogous to measurements in physics.

In the following we give an outline of the P-SVM characteristics. (1) The P-SVM avoids redundant information in the selected features as will be discussed in Section 19.3. Redundancy is not only opposed to compact feature sets but may reduce the performance of subsequent model selection methods as shown in (Hochreiter and Obermayer, 2004a) and in Section 19.5.1. For example statistical feature selection approaches (cf. Chapter Chapter 3) suffer from redundant features. (2) The P-SVM has a sparse representation in terms of complex features as SVM-regression has with the ϵ-insensitive loss. (3) The P-SVM assigns feature relevance values, which are Lagrange multipliers for the constraints and, therefore, are easy to interpret. A large absolute value of a Lagrange multiplier is associated with large empirical error if the according complex feature vector is removed from the description. (4) The P-SVM is suited for a large number of features because "sequential minimal optimization" (SMO, Platt, 1999) can be used as solver for the P-SVM optimization problem. Due to the missing equality constraint, for the P-SVM the SMO is faster than for SVMs (Hochreiter and Obermayer, 2004a). (5) The P-SVM is based on a margin-based capacity measure and, therefore, has a theoretical foundation as standard SVMs have.

In this chapter we will first introduce the P-SVM. Then we will extend the basic approach to controlling the redundancy between the selected features. Next, we describe a novel approach, which extends the P-SVM to nonlinear feature selection. As discussed later in Section 19.4, kernelizing is not sufficient to extract the nonlinear relevance of the original features because the nonlinearities, which are investigated are restricted by the kernel. Finally, we apply the generic P-SVM method to the data sets of the NIPS 2003 feature selection challenge. The "nonlinear" variant of the P-SVM feature selection method is tested on the nonlinear MADELON data set.

19.2 The Potential Support Vector Machine

We consider a two class classification task, where we are given the training set of m objects described by input vectors $x_i \in \mathbb{R}^n$ and their binary class labels $y_i \in \{+1, -1\}$. The input vectors and labels are summarized in the matrix $X = (x_1, \ldots, x_m)$ and the vector y. The learning task is to select a classifier

g minimizing a risk functional $\mathrm{R}(g)$, from the set of classifiers

$$g(\boldsymbol{x}) \;=\; \mathrm{sign}\,(\boldsymbol{w} \cdot \boldsymbol{x} \;+\; b\,) \;, \tag{19.1}$$

which are parameterized by the weight vector $\boldsymbol{w}$ and the offset b. The SVM optimization procedure is given by (Vapnik, 1998, Schölkopf and Smola, 2002)

$$\min_{\boldsymbol{w},b} \; \frac{1}{2}\,\|\boldsymbol{w}\|^2 \qquad \text{subject to} \quad y_i\,(\boldsymbol{w} \cdot \boldsymbol{x}_i \;+\; b) \geq 1 \;, \tag{19.2}$$

for linearly separable data. In this SVM formulation the constraints enforce correct classification for a hyperplane in its canonical form whereas the objective function maximizes the margin $\gamma = \|\boldsymbol{w}\|^{-1}$. The margin relates directly to a capacity measure. Let R be the radius of the sphere containing all training data, then the term $\frac{\mathrm{R}}{\gamma}$ is an upper bound for an capacity measure, the VC-dimension (Vapnik, 1998, Schölkopf and Smola, 2002). SVMs are based on the structural risk minimization principle which suggests to select from all classifiers, which correctly classify the training data, the classifier with minimal capacity.

However, the disadvantage of the SVM technique is that it is not scaling invariant, e.g. normalization of the data changes both the support vector solution and the bound $\frac{\mathrm{R}}{\gamma}$ on the capacity. If scaling is justified, we propose to scale the training data such that the margin γ remains constant while R becomes as small as possible. Optimality is achieved when all directions orthogonal the normal vector $\boldsymbol{w}$ of the hyperplane with maximal margin γ are scaled to zero. The new radius is $\tilde{\mathrm{R}} \leq \max_i |\hat{\boldsymbol{w}} \cdot \boldsymbol{x}_i|$, where $\hat{\boldsymbol{w}} := \frac{\boldsymbol{w}}{\|\boldsymbol{w}\|}$. Here we assumed centered data and a centered sphere otherwise an offset allows us to shift the data or the sphere. The new radius is the maximal distance from the origin in a one-dimensional problem. Finally, we suggest to minimize the new objective $\left\|\boldsymbol{X}^\top \boldsymbol{w}\right\|^2$, which is an upper bound on the new capacity measure:

$$\frac{\tilde{\mathrm{R}}^2}{\gamma^2} \;=\; \tilde{\mathrm{R}}^2\,\|\boldsymbol{w}\|^2 \;\leq\; \max_i |\boldsymbol{w} \cdot \boldsymbol{x}_i| \;\leq\; \sum_i (\boldsymbol{w} \cdot \boldsymbol{x}_i)^2 \;=\; \left\|\boldsymbol{X}^\top \boldsymbol{w}\right\|^2 \;. \tag{19.3}$$

The new objective function can also be derived from bounds on the generalization error when using covering numbers because the output range of the training data – which must be covered – is bounded by $2\,\max_i |\boldsymbol{w} \cdot \boldsymbol{x}_i|$. The new objective function corresponds to an implicit sphering (whitening) if the data has zero mean (Hochreiter and Obermayer, 2004c). Most importantly, the solution of eqs. (19.2) with objective function eq. (19.3) is now invariant under linear transformation of the data. Until now we motivated a new objective function. In the following we derive new constraints, which ensure small empirical error.

Definition 1. *A* **complex feature vector** z_j *is a direction in the input space where the feature value $f_{i,j}$ of an input vector $\boldsymbol{x}_i$ is obtained through* $f_{i,j} = \boldsymbol{x}_i \cdot \boldsymbol{z}_j$.

We aim at expressing the constraints, which enforces small empirical error by N complex feature vectors $z_j, 1 \le j \le N$. Complex features and feature values are summarized in the matrices $\boldsymbol{Z} := (z_1, \ldots, z_N)$ and $\boldsymbol{F} = \boldsymbol{X}^\top \boldsymbol{Z}$. The ith feature vector is defined as $\boldsymbol{f}_i = (f_{i,1}, \ldots, f_{i,N}) = \boldsymbol{Z}^\top \boldsymbol{x}_i$. The complex features include Cartesian unit direction, if we set $z_j = \boldsymbol{e}_j$, that is $\boldsymbol{Z} = \boldsymbol{I}$, $\boldsymbol{F} = \boldsymbol{X}^\top$, $f_{i,j} = x_{i,j}$ and $N = n$. In this case we obtain input variable selection. The introduction of complex feature vectors is advantageous for feature construction where a function of $\boldsymbol{Z}$ (e.g. minimal number of directions or statistical independent directions) is optimized and for handling relational data (Hochreiter and Obermayer, 2004c).

We now propose to minimize eq. (19.3) under constraints, which are necessary for the empirical mean squared error $\mathrm{R}_{\mathrm{emp}}\left(g_{\boldsymbol{w},b}\right) = \frac{1}{2\,m} \sum_{i=1}^{m} (\boldsymbol{w} \cdot \boldsymbol{x}_i + b - y_i)^2$ to be minimal ($\nabla_{\boldsymbol{w}} \mathrm{R}_{\mathrm{emp}}\left(g_{\boldsymbol{w},b}\right) = 0$), and we obtain

$$\lim_{t \to 0^+} \frac{\mathrm{R}_{\mathrm{emp}}\left(g_{\boldsymbol{w}+t z_j, b}\right) - \mathrm{R}_{\mathrm{emp}}\left(g_{\boldsymbol{w},b}\right)}{t} = (z_j)^\top \nabla_{\boldsymbol{w}} \mathrm{R}_{\mathrm{emp}}\left(g_{\boldsymbol{w},b}\right) = 0 \quad (19.4)$$

$$\text{and} \quad \frac{\partial \mathrm{R}_{\mathrm{emp}}\left(g_{\boldsymbol{w},b}\right)}{\partial b} = 0 \quad (19.5)$$

for the constraints. The empirical error is a convex function in $(\boldsymbol{w}, b)$ and possesses only one minimum, therefore all constraints can be fulfilled simultaneously. The model selection method which combines both the new objective from eq. (19.3) and the new constraints from eqs. (19.4) is called "Potential Support Vector Machine" (P-SVM). Each complex feature z_j is associated with a constraint in eqs. (19.4).

Our approach enforces minimal empirical error and, therefore, is prone to overfitting. To avoid overfitting, we allow for violation of the constraints, controlled by a hyperparameter ϵ. Standardization (mean subtraction and dividing by the standard deviation) is performed for the feature values $(f_{1,j}, \ldots, f_{m,j})$. We now require, that

$$\left| (z_j)^\top \nabla_{\boldsymbol{w}} \mathrm{R}_{\mathrm{emp}}\left(g_{\boldsymbol{w},b}\right) \right| = \left| \left[\boldsymbol{F}^\top \left(\boldsymbol{X}^\top \boldsymbol{w} - \boldsymbol{y} \right) \right]_j \right| \le \epsilon . \quad (19.6)$$

in analogy to the concept of the ϵ-insensitive loss (Schölkopf and Smola, 2002) for standard SVMs. Hence, absolute constraint values, i.e. directional derivatives, smaller than ϵ are considered to be spurious. Note, that standardization leads to $\boldsymbol{F}^\top \boldsymbol{1} = \boldsymbol{0}$ and the term $\boldsymbol{F}^\top b \, \boldsymbol{1}$ vanishes. The value ϵ correlates with the noise level of the data and is a hyperparameter of model selection. Combining eq. (19.3) and eqs. (19.6) results in the primal P-SVM optimization problem for feature selection:

$$\min_{\boldsymbol{w}} \frac{1}{2} \left\| \boldsymbol{X}^\top \boldsymbol{w} \right\|^2 \quad \text{subject to} \quad \begin{matrix} \boldsymbol{F}^\top \left(\boldsymbol{X}^\top \boldsymbol{w} - \boldsymbol{y} \right) + \epsilon \, \boldsymbol{1} \ge \boldsymbol{0} \\ \boldsymbol{F}^\top \left(\boldsymbol{X}^\top \boldsymbol{w} - \boldsymbol{y} \right) - \epsilon \, \boldsymbol{1} \le \boldsymbol{0} \end{matrix} , \quad (19.7)$$

for which the dual formulation is the

P-SVM feature selection

$$\min_{\alpha^+,\alpha^-} \quad \frac{1}{2}\left(\alpha^+ - \alpha^-\right)^\top \boldsymbol{F}^\top \boldsymbol{F}\left(\alpha^+ - \alpha^-\right) \qquad (19.8)$$

$$- \boldsymbol{y}^\top \boldsymbol{F}\left(\alpha^+ - \alpha^-\right) + \epsilon\,\boldsymbol{1}^\top\left(\alpha^+ + \alpha^-\right)$$

$$\text{subject to} \quad \boldsymbol{0} \le \alpha^+ , \quad \boldsymbol{0} \le \alpha^- .$$

- ϵ: parameter to determine the number of features, large ϵ means few features
- $\alpha_j = \alpha_j^+ - \alpha_j^-$: relevance value for complex feature vector $\boldsymbol{z}_j$, $\alpha_j \ne 0$ means that vector no. j is selected, positive α_j means class 1 indicative vector $\boldsymbol{z}_j$ and negative α_j means class -1 indicative
- $\boldsymbol{F} = \boldsymbol{X}^\top \boldsymbol{Z}$ with data matrix $\boldsymbol{X}$ and the matrix of complex features vectors $\boldsymbol{Z}$ (variable selection: $\boldsymbol{F} = \boldsymbol{X}$)
- $\boldsymbol{y}$: vector of labels

Here α^+ and α^- are the Lagrange multipliers for the constraints (See Hochreiter and Obermayer, 2004a, for the derivation of these equations). Eqs. (19.8) can be solved using a new sequential minimal optimization (SMO) technique (Hochreiter and Obermayer, 2004a). This is important to solve problems with many features because $\boldsymbol{F}^\top \boldsymbol{F}$ is a $N \times N$ matrix, therefore the optimization problem is quadratic in the number of complex features.

Using $\alpha = \alpha^+ - \alpha^-$, the weight vector $\boldsymbol{w}$ and the offset b are given by

$$\boldsymbol{w} = \boldsymbol{Z}\,\alpha = \sum_{j=1}^{N} \alpha_j\,\boldsymbol{z}_j \quad \text{and} \quad b = \frac{1}{m}\sum_{i=1}^{m} y_i . \qquad (19.9)$$

Note, that for feature selection, i.e. $\boldsymbol{Z} = \boldsymbol{I}$, $\boldsymbol{w} = \alpha$ holds, but still we recommend to solve the dual optimization problem because it has only box constraints while the primal has twice as many constraints as input variables.

The classification (or regression) function is the given by

$$g(\boldsymbol{x}) = \boldsymbol{w} \cdot \boldsymbol{x} + b = \sum_{j=1}^{N} \alpha_j\,\boldsymbol{z}_j \cdot \boldsymbol{x} + b = \sum_{j=1}^{N} \alpha_j\,f_j + b . \quad (19.10)$$

Most importantly, the vector $\boldsymbol{w}$ is expressed through a weighted sum of the complex features. Note, that the knowledge of $f_{i,j}$ and labels y_i for all training input vectors $\boldsymbol{x}_i$ is sufficient to select a classifier (see eqs. (19.8)). The complex feature vectors $\boldsymbol{z}_j$ must not be known explicitly. Complex feature vectors corresponding to spurious derivatives (absolute values smaller than ϵ) do not enter $\boldsymbol{w}$ because the corresponding Lagrange multipliers are zero. In particular the term $\epsilon\,\boldsymbol{1}^\top\left(\alpha^+ + \alpha^-\right)$ in the dual eqs. (19.8) leads to a sparse representation of $\boldsymbol{w}$ through complex features and, therefore, to feature selection.

Note, that the P-SVM is basically a classification method. On UCI bench-mark datasets the P-SVM showed comparable to better results than ν-SVMs and C-SVMs (Hochreiter and Obermayer, 2004a). However for classification the constraints are relaxed differently (by slack variables) from the approach presented here.

19.3 P-SVM Discussion and Redundancy Control

19.3.1 Correlation Considerations

In this subsection we focus on feature selection and consider the case of $\boldsymbol{F}^\top \boldsymbol{F} = \boldsymbol{X}\boldsymbol{X}^\top$ for the quadratic term of the optimization problem (19.8). Now it is the empirical covariance matrix of the features. The linear term $\boldsymbol{y}^\top \boldsymbol{X}^\top$ in eqs. (19.8) computes the correlation between features and target. Thus, such features are selected which have large target correlation and are not correlated to other features. Large target correlations result in large negative contributions to the objective function and small mutual feature correlations in small positive contributions. Consequently, highly correlated features are not selected together.

In contrast to statistical methods, the P-SVM selects features not only on the basis of their target correlation. For example, given the values of the left hand side in Table 19.1, the target t is computed from two features f_1 and f_2 as $t = f_1 + f_2$. All values have mean zero and the correlation coefficient between t and f_1 is zero. In this case the P-SVM also selects f_1 because it has negative correlation with f_2. The top ranked feature may not be correlated to the target, e.g. if it contains target-independent information, which can be removed from other features.

Table 19.1. An example for a relevant feature (f_1) without correlation to the target $t = f_1 + f_2$ (left) and an example for an irrelevant feature (f_1) with highest correlation to the target $t = f_2 + f_3$ (right).

f_1	f_2	t	f_1	f_2	f_3	t
-2	3	1	0	-1	0	-1
2	-3	-1	1	1	0	1
-2	1	-1	-1	0	-1	-1
2	-1	1	1	0	1	1

The right hand side of Table 19.1 depicts another situation, where $t = f_2 + f_3$. f_1, the feature which has highest correlation coefficient with the target (0.9 compared to 0.71 of the other features) is not selected because it is correlated to all other features.

19.3.2 Redundancy versus Selecting Random Probes

For the NIPS feature selection challenge we applied the P-SVM technique and found that the P-SVM selected a high percentage of random probes as can be seen in Table 19.5. Random probes are selected because they have by chance a small, random correlation with the target and are not correlated to other selected features. Whereas many features with high target correlation are not selected if they are correlated with other selected features. Avoiding redundancy results in selecting random probes.

In this subsection we extent the P-SVM approach in order to control the redundancy among the selected features. We introduce slack variables in the primal formulation eqs. (19.7) to allow to trade lower correlations in the objective function for errors in the constraints:

$$\min_{\boldsymbol{w}} \quad \frac{1}{2} \|\boldsymbol{X}^\top \boldsymbol{w}\|^2 + C \, \boldsymbol{1}^\top \left(\boldsymbol{\xi}^+ + \boldsymbol{\xi}^-\right) \tag{19.11}$$

$$\text{subject to} \quad \boldsymbol{F}^\top \left(\boldsymbol{X}^\top \boldsymbol{w} - \boldsymbol{y}\right) + \epsilon \, \boldsymbol{1} + \boldsymbol{\xi}^+ \geq \boldsymbol{0}$$

$$\boldsymbol{F}^\top \left(\boldsymbol{X}^\top \boldsymbol{w} - \boldsymbol{y}\right) - \epsilon \, \boldsymbol{1} - \boldsymbol{\xi}^- \leq \boldsymbol{0}, \quad \boldsymbol{0} \leq \boldsymbol{\xi}^+, \boldsymbol{\xi}^- \, .$$

As dual formulation we obtain the following optimization problem.

P-SVM feature selection with redundancy control

$$\min_{\boldsymbol{\alpha}^+, \boldsymbol{\alpha}^-} \quad \frac{1}{2} \left(\boldsymbol{\alpha}^+ - \boldsymbol{\alpha}^-\right)^\top \boldsymbol{F}^\top \boldsymbol{F} \left(\boldsymbol{\alpha}^+ - \boldsymbol{\alpha}^-\right) \tag{19.12}$$

$$- \boldsymbol{y}^\top \boldsymbol{F} \left(\boldsymbol{\alpha}^+ - \boldsymbol{\alpha}^-\right) + \epsilon \, \boldsymbol{1}^\top \left(\boldsymbol{\alpha}^+ + \boldsymbol{\alpha}^-\right)$$

$$\text{subject to} \quad \boldsymbol{0} \leq \boldsymbol{\alpha}^+ \leq C \, \boldsymbol{1}, \; \boldsymbol{0} \leq \boldsymbol{\alpha}^- \leq C \, \boldsymbol{1} \, .$$

- variables as in problem eqs. (19.8)
- C controls redundancy of selected features, small C results in more redundancy

---

The eqs. (19.9) for $\boldsymbol{w}$ and b still hold. The effect of introducing the slack variables can be best seen at the dual problem. Because the α_j are bounded by C, high correlations are lower weighted in the objective function. Consequently, correlated features have a lower positive contribution in the objective function and, therefore, selecting redundant features does not cost as much as in the original P-SVM formulation. The effect is demonstrated at the following two toy experiments.

In the first two class classification experiment six dimension out of 100 are indicative for the class. The class membership was chosen with equal probability (0.5) and with equal probability 0.5 either the first three features were class indicators or the features 4 to 6. If the first three features are class indicators, features a chosen according to $x_{i,j} \sim y_i\,N(j,1)$, $1 \leq j \leq 3$, $x_{i,j} \sim N(0,1)$, $4 \leq j \leq 6$, $x_{i,j} \sim N(0,20)$, $7 \leq j \leq 100$. If features 4 to 6 are class indicators, features a chosen according to $x_{i,j} \sim N(0,1)$, $1 \leq j \leq 3$, $x_{i,j} \sim y_i\,N(j-3,1)$, $4 \leq j \leq 6$, $x_{i,j} \sim N(0,20)$, $7 \leq j \leq 100$. Only the first six feature are class indicators but mutually redundant. Finally, the class labels were switched with probability 0.2. In the experiments ϵ is adjusted to obtain 6 features, i.e. to obtain 6 support vectors. The top part of Table 19.2 shows the result for different values of C. With decreasing C more relevant features are selected because the redundancy weighting is down-scaled.

Table 19.2. Toy example for redundancy control. TOP: Feature ranking where the first 6 features are relevant to predict the class label. With decreasing C more redundant features are selected and, therefore, more relevant features are found (their number is given in column "c"). α values are given in brackets. BOTTOM: 60 relevant features exist. Starting from 4 relevant features (93 % probes) reducing C leads to 40 relevant features (33 % probes).

C	ϵ	c	1.	2.	3.	4.	5.	6.
10	1.5	3	6 (1.78)	2 (1.12)	26(-0.55)	18(-0.44)	3 (0.35)	52(-0.07)
1	2	4	6 (1.00)	2 (0.71)	5 (0.29)	3 (0.13)	18(-0.09)	26(-0.06)
0.5	2	5	2 (0.50)	5 (0.50)	6 (0.50)	3 (0.22)	18(-0.09)	4 (0.04)

C	ϵ	# relevant features	C	ϵ	# relevant features
10	0.65	4	0.1	1.7	23
0.5	1	11	0.05	2	31
0.2	1.45	17	0.003	2	40

In the next experiment we extended the previous experiment by using 940 probes and 60 features (1000 input components), where either the first 30 or features 31 to 60 are indicative for the class label. Indicative features are chosen according to $x_{i,j} \sim N(2,1)$. All other value were as in the previous experiment. The value of ϵ is adjusted to a value that only about 60 features are selected ($\alpha \neq 0$). The bottom part of Table 19.2 shows the result for different values of C. The percentage of probes in the selected variables is 93 % for $C = 10$ and reduces to 33 % for $C = 0.003$.

These simple experiments demonstrated that the original P-SVM selects many random probes because it minimized feature redundancy. Here we controlled this effect by introducing slack variables. Note, that for the NIPS challenge submissions no slack variables were used.

19.3.3 Comparison to Related Methods

1-norm SVMs (Bi et al., 2003). The P-SVM feature selection is related to the 1-norm SVMs because both use a 1-norm sparsity constraint. However the P-SVM contains – in contrast to the 1-norm SVM – a quadratic part. The effect of the quadratic part was demonstrated in Subsection 19.3.1, were we found that important features are selected through correlation with other features. Comparisons can be found in the experiments in Subsection 19.5.1. **Zero-norm SVMs (Weston et al., 2003).** Zero-norm SVMs optimize a different objective than the P-SVM, where the scaling factor of the selected features is no longer important. Scaling factors may, however, be important if different features contain different levels of noise. We compared the P-SVM with zero-norm SVMs in the experiments in Subsection 19.5.1 (only 2 features selected). The zero-norm SVMs select features by successively repeating 1-norm SVMs. The P-SVMs can be extended in a similar way if after standardization features are weighted by their actual importance factors. **LASSO (Tibshirani, 1996).** The LASSO is quite similar to the P-SVM method. In contrast to P-SVM, LASSO does not use the linear term of the dual P-SVM in the objective function but constraints it. P-SVM is derived from an SVM approach, therefore contains a primal and a dual formulation which allows to apply a fast SMO procedure. A major difference between P-SVM and LASSO is that LASSO cannot control the redundancy among the selected features as the P-SVM can with its slack variables as demonstrated in Subsection 19.3.2. Comparisons to LASSO are implicitly contained the NIPS feature selection challenge, where the methods of Saharon Rosset and Ji Zhu are based on the LASSO.

19.4 Nonlinear P-SVM Feature Selection

In this section we extend the P-SVM feature selection approach to assigning relevance values to complex features z_j, where we now consider also nonlinear combinations of the features. To construct new features by nonlinearly combining the original features (Smola et al., 2001, Kramer, 1991, Schölkopf et al., 1997, Oja, 1991, Tishby et al., 1999) by using kernels is possible but not sufficient to extract arbitrary nonlinear dependencies of features. Only those nonlinearities can be detected which are determined by the kernel. A wrong kernel choice does not allow us to extract proper nonlinearities. Therefore, we attempt to construct proper nonlinearities by training multi-layer perceptrons (MLPs). After training we determine the relevance of input variables. Our approach is related to input pruning methods (Hassibi and Stork, 1993, Moody and Utans, 1992) and automatic relevance determination (ARD, MacKay, 1993, Neal, 1996).

For input x the value $y(x)$ is the output function of the MLP and $net_l = w_l \cdot x + b_l$ the net input of the hidden unit l. After training on the training set $\{(x_i, y_i)\}\}$, we set $net_l = 0$ for the forward and backward pass. For training example x_i this leads to a new output of $\tilde{y}_l(x_i)$ and an induced error of $e_l(x_i) = \frac{1}{2}(\tilde{y}_l(x_i) - y(x_i))^2$. The error indicates the relevance of net_l but does not supply a desired value for net_l. However, the gradient descent update signal for net_l supplies a new target value $y_{l,i}$ for $\text{net}_l(\boldsymbol{x}_i)$ and we arrive at the regression task:

$$y_{l,i} = \boldsymbol{w}_l \cdot \boldsymbol{x}_i + b_l , \quad y_{l,i} := -\frac{\partial e_l(\boldsymbol{x}_i)}{\partial \text{net}_l(\boldsymbol{x}_i)} . \tag{19.13}$$

This regression problem is now solved by the P-SVM which selects the relevant input variables for hidden unit l. Fig. 19.1 depicts the regression task. The vectors $\boldsymbol{w}_l$ are now expressed through complex features $\boldsymbol{z}_j$ and allow to assign for each l a relevance value $\alpha_{j,l}$ to $\boldsymbol{z}_j$. Finally, the results for all l are combined and the complex features $\boldsymbol{z}_j$ are ranked by their relevance values $\alpha_{j,l}$, e.g. by the maximal absolute weight or squared weight sum. The pseudo code of the algorithm is shown in Algorithm 8.

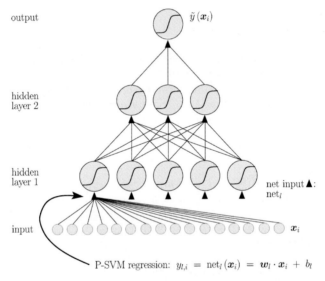

Fig. 19.1. Outline of the nonlinear P-SVM. After MLP training the P-SVM solves the regression task $y_{l,i} = \boldsymbol{w}_l \cdot \boldsymbol{x}_i + b_l$, where $y_{l,i} := -\frac{\partial e_l(\boldsymbol{x}_i)}{\partial \text{net}_l(\boldsymbol{x}_i)}$ for $\boldsymbol{w}_l = \boldsymbol{0}$ and $b_l = 0$. The P-SVM selects the relevant input variables to hidden unit l.

Other targets. The net input $\text{net}_l(\boldsymbol{x}_i)$ as target value instead of $y_{l,i}$ does not take into account that a hidden unit may not be used, may be less used than others, or may have varying influence on the net output over the examples. Setting net_l to other values than 0 (e.g. its mean value) when $y_{l,i}$ is computed works as well as long as saturating regions are avoided. Saturation regions lead to scaling effects through different derivatives at different input regions and reduce the comparability of relevance values of one input variable at different units.

Redundancy. The P-SVM is applied to each unit l, therefore the selected input variables may be redundant after combining the results for each l.

Algorithm 8: Nonlinear P-SVM Feature Selection

BEGIN INITIALIZATION
    training set $\{(\boldsymbol{x}_i, y_i)\}$,
    MLP architecture and activation function,
    MLP training parameters (learning rate),
    MLP learning stop criterion: small error threshold
END INITIALIZATION

BEGIN PROCEDURE
Step 1: perform standardization
Step 2: train an MLP with standard back-propagation until stop criterion
Step 3: {determine feature relevance values for each new feature}
    for all hidden units l in chosen hidden layer **do**
        for $i = 1$ to m **do**
            MLP forward pass with unit l clamped to 0
            MLP backward pass to compute $y_{l,i} = - \frac{\partial e_l(\boldsymbol{x}_i)}{\partial \mathrm{net}_l(\boldsymbol{x}_i)}$
        end for
        hidden layer with the new features (recommended: first hidden layer),
        chose P-SVM parameter ϵ
        solve regression task $\forall_i :\ y_{l,i} = \boldsymbol{w}_l \cdot \boldsymbol{x}_i + b_l$ by the P-SVM method
        and determine relevance values $\alpha_{j,l}$ per new feature
    end for
Step 4: {compute relevance values}
    combine (squared sum or maximal value) all $\alpha_{j,l}$ to determine relevance
    of $\boldsymbol{z}_j$
END PROCEDURE

19.5 Experiments

19.5.1 Linear P-SVM Feature Selection

Weston Data

We consider a 2 class classification task with 600 data points (300 from each class) which is similar to the data set in (Weston et al., 2000) but more difficult. 100 randomly chosen data points are used for feature and model selection. The remaining 500 data points serve as test set. We constructed 2000 input variables from which only the first 20 input variables have dependencies with the class and the remaining 1980 are random probes. For each data point four out of the first 20 input variables are indicative for the class label. The data points are in one of five modes, where the mode determines which input variable is indicative. The modes, which lead to objects groups, are $l = 0, 4, 8, 12, 16$ with associated input variables: $l = 0 \longrightarrow \boldsymbol{x}_{i,1} - \boldsymbol{x}_{i,4}$; $l = 4 \longrightarrow \boldsymbol{x}_{i,5} - \boldsymbol{x}_{i,8}$; $l = 8 \longrightarrow \boldsymbol{x}_{i,9} - \boldsymbol{x}_{i,12}$; $l = 12 \longrightarrow \boldsymbol{x}_{i,13} - \boldsymbol{x}_{i,16}$; $l = 16 \longrightarrow \boldsymbol{x}_{i,17} - \boldsymbol{x}_{i,20}$.

Sepp Hochreiter and Klaus Obermayer

A label from $\{+1, -1\}$ and a mode from $\{0, 4, 8, 12, 16\}$ was randomly and uniformly chosen. Then the four indicative input variables $x_{i,l+\tau}$, $1 \leq \tau \leq 4$, were chosen according to $x_{i,l+\tau} \sim y_i \cdot \mathrm{N}\,(2, 0.5\,\tau)$. Input variables $x_{i,j}$ for $j \neq l + \tau$, $j \leq 20$ were chosen according to $x_{i,j} \sim \mathrm{N}\,(0, 1)$. Finally, for $21 \leq j \leq 2000$ the input variables $x_{i,j}$ were chosen according to $x_{i,j} \sim \mathrm{N}\,(0, 20)$.

We compare the linear P-SVM feature selection technique to the Fisher statistic (Kendall and Stuart, 1977), the Recursive Feature Elimination (RFE) method of Guyon et al. (2002) and the linear R2W2 method (Weston et al., 2000). The experiment is taken from (Hochreiter and Obermayer, 2004b). First we ranked features on the training set, where for RFE the ranking was based on multiple runs. Then we trained a standard C-SVM[1] with the top ranked 5, 10, 15, 20, and 30 input variables. The hyperparameter C was selected from the set $\{0.01, 0.1, 1, 10, 100\}$ through 5-fold cross-validation. Table 19.3 shows the results. The P-SVM method performed best.

The performance of the methods depends on how many modes are represented through the input variables. The results in Table 19.3 must be compared to the classification performance with 20 relevant features (perfect selection), which leads to a fractional error of 0.10, and without feature selection, which leads to a fractional error of 0.38.

Table 19.3. Classification performance for the "Weston" data set. Results are an average over 10 runs on different training and test sets. The values are the fractions of misclassification. The table shows the results using the top ranked 5, 10, 15, 20, and 30 features for the methods: Fisher statistics (Kendall and Stuart, 1977), Recursive Feature Elimination (RFE), R2W2, and the P-SVM. Best results are marked in bold.

Method	number of features				
	5	10	15	20	30
Fisher	0.31	0.28	0.26	0.25	**0.26**
RFE	0.33	0.32	0.32	0.31	0.32
R2W2	0.29	0.28	0.28	0.27	0.27
P-SVM	**0.28**	**0.23**	**0.24**	**0.24**	**0.26**

This benchmark is a very difficult feature selection task because it contains many features but only few of them are indicative, features are indicative for only 1/5 of the data, and features are noisy. It is difficult to extract the few indicative features for all objects groups with the few available examples because it is highly probable that at least some features will be highly correlated to the target.

---

[1] For this experiment we used the Spider-Software, where the C-SVM was easier to use as classifier than the ν-SVM.

Two Best Features Experiment

We compare the P-SVM feature selection technique to the 1-norm and 0-norm support vector machine by performing the benchmark in Weston et al. (2003). The two class classification task has six dimension out of 100 which are indicative for the class. The class membership was chosen with equal probability (0.5) and with probability 0.7 the first three features were class indicators and otherwise the features 4 to 6 are class indicators. For the first case input variables are chosen according to $x_{i,j} \sim y_i \, \mathrm{N}\,(j, 1)$, $1 \leq j \leq 3$, $x_{i,j} \sim \mathrm{N}\,(0, 1)$, $4 \leq j \leq 6$, and $x_{i,j} \sim \mathrm{N}\,(0, 20)$, $7 \leq j \leq 100$. For the second case the input variables are $x_{i,j} \sim \mathrm{N}\,(0, 1)$, $1 \leq j \leq 3$, $x_{i,j} \sim y_i \, \mathrm{N}\,(j - 3, 1)$, $4 \leq j \leq 6$, and $x_{i,j} \sim \mathrm{N}\,(0, 20)$, $7 \leq j \leq 100$. Only the first six input variables are class indicators but mutual redundant. The two top ranked input variables are used for classification. Training and feature selection is performed on 10, 20, and 30 randomly chosen training points and the selected model is tested on additionally 500 test points. The result is an average over 100 trials.

The feature selection methods, which are compared in Weston et al. (2003), are: no feature selection (no FS), 2-norm SVM (largest weights), 1-norm SVM (largest weights), correlation coefficient (CORR), RFE, R2W2, and three approaches to zero-norm feature selection, namely FSV (Bradley et al., 1998, Bradley and Mangasarian, 1998), ℓ_2-AROM, and ℓ_1-AROM (Weston et al., 2003). The correlation coefficient is computed as $(\mu_+ - \mu_-)^2 / (\sigma_+^2 + \sigma_-^2)$, where μ_+ and σ_+ are the mean and the standard deviation of the feature value for the positive class and μ_- and σ_- the according values for the negative class. The authors in (Weston et al., 2003) only mentioned that they used "linear decision rules" while used for the P-SVM a linear ν-SVM with $\nu = 0.3$ as classifier.

Table 19.4 shows the results as an average over 100 trials. The table reports the percentage of test error with the according standard deviation[2] and the number of times that the selected features are relevant and non-redundant. The P-SVM method performs as good as the best methods for 10 and 20 data points but for 30 data points worse than the zero-norm methods and better than other methods. Because the zero-norm approaches solve iteratively one- or two-norm SVM problems, it may be possible to do the same for the P-SVM approach by re-weighting the features by their α-values.

NIPS Challenge

In this section we report the results of the P-SVM method for the NIPS 2003 feature selection challenge. The method and the results are given in the Fact Sheet (see Appendix C and the results at the top of Table 19.5. In order to obtain a compact feature set we applied the P-SVM method without slack

---

[2]Note, that in Weston et al. (2003) the standard deviation of the mean is given, which scales the standard deviation by a factor of 10.

Table 19.4. Comparison of different compact feature set selection methods. The percentage of the test error with its standard deviation in rectangular brackets is given. The number of trials where two relevant non-redundant features are selected is in round brackets. For 10 and 20 point the P-SVM method performs as good as the best methods and for 30 data points the P-SVM performs worse than the zero-norm methods but better than the others. Best results are marked in bold.

Method	10 points	20 points	30 points
no FS	33.8 [std: 6.6] (0)	23.2 [std: 5.6] (0)	16.4 [std: 3.9] (0)
2-norm SVM	26.8 [std:13.9] (3)	16.3 [std: 7.7] (16)	13.4 [std: 4.2] (17)
1-norm SVM	25.9 [std:14.5] (17)	11.0 [std:10.9] (67)	12.1 [std:13.5] (66)
CORR	**23.6** [std:12.9] (9)	15.8 [std: 5.4] (9)	14.3 [std: 3.2] (5)
RFE	30.1 [std:14.5] (10)	11.6 [std:11.0] (64)	8.2 [std: 6.1] (73)
R2W2	26.3 [std:14.1] (14)	9.8 [std: 8.6] (66)	7.8 [std: 6.1] (67)
FSV	24.6 [std:14.9] (17)	9.1 [std: 8.3] (70)	5.9 [std: 5.4] (85)
ℓ_2-AROM	26.7 [std:14.6] (15)	8.8 [std: 9.0] (74)	**5.7** [std: 5.0] (85)
ℓ_1-AROM	25.8 [std:14.9] (20)	8.9 [std: 9.7] (77)	5.9 [std: 5.1] (83)
P-SVM	26.0 [std:13.8] (13)	**8.6** [std: 7.4] (67)	6.9 [std: 9.1] (73)

variables. Therefore, the P-SVM method selects a high percentage of random "probes", i.e. features which are artificially constructed and are not related to the target. Especially prominent is this behavior for the data set ARCENE, where features are highly correlated with each other. This correlation was figured out by a post challenge submission and by the data set description which was made available after the challenge.

We computed the NIPS challenge results for methods with compact feature sets, i.e. methods which based their classification on less than 10 % extracted features. We report only methods, which have a non-negative score to ensure sufficient classification performance. The bottom of Table 19.5 reports the results. The P-SVM method yields good results if compact feature sets are desired. In summary, the P-SVM method has shown good performance as a feature selection method especially for compact feature sets.

19.5.2 Nonlinear P-SVM Feature Selection

Toy Data

In this experiment we constructed two data sets in which the relevant features cannot be found by linear feature selection techniques. We generated 500 data vectors x_i $(1 \leq i \leq 500)$ with 100 input variables $x_{i,j}$ $(1 \leq j \leq 100)$. Each input variable was chosen according to $x_{i,j} \sim N(0,1)$. The attributes y_i of the data vectors x_i were computed from the first two variables by A) $y_i = x_{i,1}^2 + x_{i,2}^2$ and B) $y_i = x_{i,1}x_{i,2}$. We thresholded y_i by $y_i > 1 \Rightarrow y_i = 1$

Table 19.5. NIPS 2003 challenge results for P-SVM. "Score": The score used to rank the results by the organizers (times 100). "BER": Balanced error rate (in percent). "AUC": Area under the ROC curve (times 100). "Feat": Percent of features used. "Probe": Percent of probes found in the subset selected. "Test": Result of the comparison with the best entry using the MacNemar test. TOP: General result table. BOTTOM: Results for compact feature sets with non-negative score. The column "Method" gives the method name. The P-SVM has multiple entries were different weighting of the CV folds is used to select features and hyperparameters. The results are listed according to the percentage of features used.

Dec. 1st	Our best challenge entry					The winning challenge entry					
Dataset	Score	BER	AUC	Feat	Probe	Score	BER	AUC	Feat	Probe	Test
OVERALL	14.18	11.28	93.66	4.6	34.74	88.00	6.84	97.22	80.3	47.8	1
ARCENE	16.36	20.55	87.75	7	61	98.18	13.30	93.48	100	30.0	1
DEXTER	-60	8.70	96.39	2.5	46.6	96.36	3.90	99.01	1.5	12.9	1
DOROTHEA	29.09	16.21	88.00	0.2	29.58	98.18	8.54	95.92	100	50.0	1
GISETTE	18.18	2.06	99.76	12	36.5	98.18	1.37	98.63	18.3	0.0	1
MADELON	67.27	8.89	96.39	1.4	0	100.0	7.17	96.95	1.6	0.0	1

Dec. 1st Method	Feat	Score	BER	AUC	Probe	Test
P-SVM (1)	3.83	0	11.82	93.41	34.6	1
Modified-RF	3.86	6.91	10.46	94.58	9.82	1
P-SVM (2)	4.63	14.18	11.28	93.66	34.74	1
BayesNN-small	4.74	68.73	8.20	96.12	2.91	0.8
final-1	6.23	40.36	10.38	89.62	6.1	0.6
P-SVM (3)	7.38	5.09	12.14	93.46	45.65	1
Collection2	7.71	28	10.03	89.97	10.6	1

and $y_i < -1 \Rightarrow y_i = -1$. For both tasks the correlation coefficient between the target and the relevant input variables is zero. For task A) this follows from the fact that the first and third moments of the zero-mean Gaussian are zero and for task B) if follows from the zero mean of input variables (XOR problem).

We performed 10 trials for each task with the P-SVM nonlinear relevance extraction method. First, a 3-layered multi-layer perceptron (100 inputs, 10 hidden, one output) with sigmoid units in $[-1, 1]$ was trained until the error was 5 % of its initial value. The P-SVM method was applied and features were ranked according to their maximal values of $\alpha_{j,l}$. In all trials the P-SVM ranked the two relevant features $x_{i,1}$ and $x_{i,2}$ on top and produced a clear visible gap between the relevance values of the true relevant features and the remaining features. For comparison we also performed 5 trials with linear P-SVM feature selection on each of both tasks. *The linear version failed to detect the true relevant features.* For comparison we selected input variables with "Optimal Brain Surgeon" (OBS, Hassibi and Stork, 1993) and "Optimal Brain Damage" (OBD, LeCun et al., 1990). We applied OBS and OBD in two ways after the neural network has been trained: first, we computed the saliency values for all weights with OBS and OBD and ranked the features according to their highest values; secondly, we successively deleted weights according to the OBS and OBD procedure and ranked a feature before another feature if at least one input weight is removed later than all the input weights of the other feature. For the latter we retrained the neural network if the error increased more than 10 % since the last training. OBS and OBD lead also to success at this task. This experiment demonstrated that P-SVM nonlinear relevance extraction is able to reliably detect relevant features whereas the linear P-SVM method could not identify relevant features.

NIPS Challenge: Madelon

The data generation procedure of the NIPS feature selection challenge was made public after the challenge. Therefore, we know that the class labels for the data set MADELON were constructed nonlinearly from the input variables. After the challenge, when knowing the data generation process, we computed Pearson's correlation coefficient for each pair of features. That allowed us to extract the 20 relevant features through looking for a set of 20 features, which have high intercorrelation.

For nonlinear feature selection (P-SVM, OBS, OBD) we used 3-layered multi-layer perceptrons (MLPs) with 20 hidden units and 4-layered MLPs with 10 hidden units in each hidden layer. All non-input units have a sigmoid activation function in range $[-1, 1]$. We trained the MLPs with backpropagation until the error was at 5 % of its initial value. Features were ranked by the P-SVM, OBS, and OBD as in Subsection 19.5.2.

The linear P-SVM ranked in 10 runs 13 out of 20 true relevant features at the top. Nonlinear P-SVMs with 3-layered and 4-layered nets ranked in 10 runs always 18 to 20 relevant features at the top, however no gap in relevance values between features and probes was visible. Table 19.6 shows typical results. Increasing the ϵ value produces a gap in the relevance values between the true relevant features and the probes, however fewer true relevant features are ranked at the top (see in Table 19.6). Both the ranking by OBS and OBD through the saliency and through backward elimination lead to inferior results compared to the P-SVM method. On average 3 true relevant features were extracted (Table 19.6 presents typical results). For backward elimination we started by removing a sets of weights (4-layered: 4×500, 3×400, 3×300, 3×200, 2×100, 3×50, 20, $2 \times 10 = 5,090$; 3-layered: 19×500, 200, 100, 2×50, $5 \times 20 = 10,000$). After removing a set we extensively retrained. After removing weight sets, we deleted weights step by step. As seen in previous studies, OBS and OBD tend to keep large weights which result from overfitting (Hochreiter and Schmidhuber, 1997). Only the nonlinear P-SVM was able to rank almost all relevant features at the top. This experiment showed that the nonlinear

Table 19.6. MADELON **nonlinear feature selection examples.** Typical runs for linear (average over four runs with different ϵ) and nonlinear P-SVM, OBS, and OBD. Selected input variables are ordered line-wise and true features are marked boldface. The nonlinear methods are based on a 3-layered and a 4-layered neural network. OBS and OBD ranking uses either the saliency values or successively deleted weights. For the latter a feature is ranked according to when its last weight is deleted. For the P-SVM the ϵ values are given in brackets. The linear P-SVM was not able to find all true relevant features whereas the nonlinear P-SVM finds all of them.

Method	feature ranking									
linear P-SVM	242	476	**337**	**65**	**339**	**454**	**494**	**443**	**49**	**379**
($\epsilon = 3.0$,	**473**	**129**	**106**	431	324	120	425	378	44	11
$2.6, 2.2, 1.8$)	297	56	164	495	121	227	137	283	412	482
nonlinear P-SVM	**452**	**494**	**49**	**319**	**242**	**473**	**443**	**379**	**65**	**456**
(3-layered net)	**106**	**154**	**282**	**29**	**129**	**337**	**339**	**434**	**454**	**476**
($\epsilon = 0.01$)	122	195	223	343	21	402	315	479	409	330
nonlinear P-SVM	**65**	**494**	**242**	**379**	**443**	**454**	**434**	**476**	**129**	**282**
(4-layered net)	**106**	**154**	**473**	**452**	**319**	**339**	**49**	**456**	308	387
($\epsilon = 0.01$)	283	311	139	162	236	457	229	190	16	453
nonlinear P-SVM	**65**	**494**	**242**	**379**	**443**	**476**	**434**	**454**	**106**	**129**
(4-layered net)	**282**	308	387	311	283	139	162	16	236	457
($\epsilon = 0.2$)	229	190	453	35	136	474	359	407	76	336
OBS	62	**49**	169	324	457	424	442	348	302	497
saliency	66	310	61	336	44	299	453	161	212	48
(3-layered net)	78	383	162	5	317	425	197	331	495	153
OBS	425	302	**443**	66	246	**49**	297	497	249	39
saliency	169	164	453	324	166	298	137	11	311	421
(4-layered net)	292	62	433	404	6	310	224	349	**476**	431
OBS	**242**	**49**	**337**	497	324	457	318	50	**154**	62
elimination	162	206	56	299	310	169	348	5	412	128
(3-layered net)	495	27	6	442	415	424	19	47	61	25
OBS	**49**	**443**	283	324	497	297	**319**	164	138	5
elimination	62	249	86	349	246	43	208	6	310	491
(4-layered net)	410	291	298	54	166	302	**476**	457	482	212
OBD	**49**	62	169	457	324	424	442	348	497	310
saliency	61	299	336	**154**	161	78	453	313	293	5
(3-layered net)	495	153	331	292	128	162	121	302	287	411
OBD	425	66	**443**	246	**49**	297	249	169	302	497
saliency	39	453	164	224	324	62	298	349	137	310
(4-layered net)	421	6	43	292	291	58	457	404	166	433

extension of the P-SVM feature selection method can detect relevant features, which are missed with the linear version and also missed by OBS and OBD.

19.6 Conclusion

In the future we intend to investigate how optimization of the complex feature vectors (e.g. to obtain few feature vectors or independent features) can be integrated into our approach. Further we intend to apply the P-SVM method to genomic data (e.g. microsatellites) to identify genetic causes for various diseases (e.g. schizophrenia).

On the NIPS 2003 feature selection challenge data sets we have experimentally shown that the linear P-SVM method is one of the best methods for selecting a compact feature set. The linear P-SVM approach has been generalized to include redundancy control and nonlinearities. Nonlinear P-SVM feature selection does not only extract features, which are missed by its linear version but has the potential to give the features a more appropriate ranking. This property is especially important for data sets, where only few top ranked features control the data generating process.

Acknowledgments. We thank Merlyn Albery-Speyer, Christoph Büscher, Raman Sanyal, Sambu Seo and Peter Wiesing for their help. This work was funded by the DFG (SFB 618), the Monika-Kutzner- and the Anna-Geissler-Stiftung.

References

R. E. Bellman. *Adaptive Control Processes.* Princeton University Press, Princeton, NJ, 1961.

J. Bi, K. Bennett, M. Embrechts, C. Breneman, and M. Song. Dimensionality reduction via sparse support vector machines. *Journal of Machine Learning Research*, 3:1229–1243, 2003. Special Issue on Variable and Feature Selection.

B. E. Boser, I. M. Guyon, and V. Vapnik. A training algorithm for optimal margin classifiers. In D. Haussler, editor, *Proceedings of the 5th Annual ACM Workshop on Computational Learning Theory*, pages 144–152, Pittsburgh, PA, 1992. ACM Press.

P. S. Bradley and O. L. Mangasarian. Feature selection via concave minimization and support vector machines. In J. Shavlik, editor, *Machine Learning Proceedings of the Fifteenth International Conference(ICML '98)*, pages 82–90, San Francisco, California, 1998. Morgan Kaufmann. ftp://ftp.cs.wisc.edu/math-prog/tech-reports/98-03.ps.Z.

P. S. Bradley, O. L. Mangasarian, and W. N. Street. Feature selection via mathematical programming. *INFORMS Journal on Computing*, 10(2):209–217, 1998.

C. Cortes and V. N. Vapnik. Support vector networks. *Machine Learning*, 20: 273–297, 1995.

I. Guyon, J. Weston, S. Barnhill, and V. Vapnik. Gene selection for cancer classification using support vector machines. *Machine Learning*, 46(1-3):389–422, 2002.

B. Hassibi and D. G. Stork. Second order derivatives for network pruning: Optimal brain surgeon. In J. D. Cowan S. J. Hanson and C. L. Giles, editors, *Advances in Neural Information Processing Systems 5*, pages 164–171. San Mateo, CA: Morgan Kaufmann, 1993.

S. Hochreiter and K. Obermayer. Classification, regression, and feature selection on matrix data. Technical Report 2004/2, Technische Universität Berlin, Fakultät für Elektrotechnik und Informatik, 2004a.

S. Hochreiter and K. Obermayer. Gene selection for microarray data. In B. Schölkopf, K. Tsuda, and J.-P. Vert, editors, *Kernel Methods in Computational Biology*, pages 319–355. MIT Press, 2004b.

S. Hochreiter and K. Obermayer. Sphered support vector machine. Technical report, Technische Universität Berlin, Fakultät für Elektrotechnik und Informatik, 2004c.

S. Hochreiter and J. Schmidhuber. Flat minima. *Neural Computation*, 9(1):1–42, 1997.

M. G. Kendall and A. Stuart. *The advanced theory of statistics*. Charles Griffin & Co LTD, 4 edition, 1977.

M. Kramer. Nonlinear principal component analysis using autoassociative neural networks. *AIChE Journal*, 37:233–243, 1991.

Y. LeCun, J. S. Denker, and S. A. Solla. Optimal brain damage. In D. S. Touretzky, editor, *Advances in Neural Information Processing Systems 2*, pages 598–605. San Mateo, CA: Morgan Kaufmann, 1990.

D. J. C. MacKay. Bayesian non-linear modelling for the 1993 energy prediction competition. In G. Heidbreder, editor, *Maximum Entropy and Bayesian Methods, Santa Barbara 1993*. Kluwer, Dordrecht, 1993.

J. E. Moody and J. Utans. Principled architecture selection strategies for neural networks: Application to corporate bond rating prediction. In J. E. Moody, S. J. Hanson, and R. P. Lippman, editors, *Advances in Neural Information Processing Systems 4*, pages 683–690. San Mateo, CA: Morgan Kaufmann, 1992.

R. Neal. *Bayesian Learning for Neural Networks*. Springer Verlag, New York, 1996.

E. Oja. Data compression, feature extraction, and autoassociation in feedforward neural networks. In T. Kohonen, K. Mäkisara, O. Simula, and J. Kangas, editors, *Artificial Neural Networks*, volume 1, pages 737–745. Elsevier Science publishers B.V., North-Holland, 1991.

J. Platt. Fast training of support vector machines using sequential minimal optimization. In B. Schölkopf, C. J. C. Burges, and A. J. Smola, editors, *Advances in Kernel Methods – Support Vector Learning*, pages 185–208, Cambridge, MA, 1999. MIT Press.

B. Schölkopf and A. J. Smola. *Learning with kernels – Support Vector Machines, Reglarization, Optimization, and Beyond*. MIT Press, Cambridge, 2002.

B. Schölkopf, A. J. Smola, and K.-R. Müller. Kernel principal component analysis. In W. Gerstner, A. Germond, M. Hasler, and J.-D. Nicoud, editors, *Artificial Neural Networks – ICANN'97*, pages 583–588, Berlin, 1997. Springer Lecture Notes in Computer Science, Vol. 1327.

A. J. Smola, S. Mika, B. Schölkopf, and R. C. Williamson. Regularized principal manifolds. *Journal of Machine Learning Research*, 1:179–209, 2001. http://www.jmlr.org.

R. Tibshirani. Regression shrinkage and selection via the lasso. *Journal of the Royal Statistical Society, B*, 58(1):267–288, 1996.

N. Tishby, F. C. Pereira, and W. Bialek. The information bottleneck method. In *Proceedings of the 37-th Annual Allerton Conference on Communication, Control and Computing*, pages 368–377, 1999.

V. N. Vapnik. *Statistical Learning Theory*. Adaptive and learning systems for signal processing, communications, and control. Wiley, New York, 1998.

J. Weston, A. Elisseeff, B. Schölkopf, and M. Tipping. Use of the zero-norm with linear models and kernel methods. *Journal of Machine Learning Research*, 3: 1439–1461, 2003. Special Issue on Variable and Feature Selection.

J. Weston, S. Mukherjee, O. Chapelle, M. Pontil, T. Poggio, and V. Vapnik. Feature selection for SVMs. In T. K. Leen, T. G. Dietterich, and V. Tresp, editors, *Advances in Neural Information Processing Systems*, volume 13. MIT Press, Cambridge, MA, 2000.

Chapter 20

Combining a Filter Method with SVMs

Thomas Navin Lal, Olivier Chapelle, and Bernhard Schölkopf

Max-Planck-Institute for Biological Cybernetics, Tübingen, Germany.
navin@tuebingen.mpg.de, olivier.chapelle@tuebingen.mpg.de,
bs@tuebingen.mpg.de

Summary. Our goal for the competition was to evaluate the usefulness of simple machine learning techniques. We decided to use the Fisher criterion (see Chapter 2) as a feature selection method and Support Vector Machines (see Chapter 1) for the classification part. Here we explain how we chose the regularization parameter C of the SVM, how we determined the kernel parameter σ and how we estimated the number of features used for each data set. All analyzes were carried out on the training sets of the competition data. We choose the data set ARCENE as an example to explain the approach step by step.

In our view the point of this competition was the construction of a well performing classifier rather than the systematic analysis of a specific approach. This is why our search for the best classifier was only guided by the described methods and that we deviated from the road map at several occasions. All calculations were done with the software (Spider, 2005).

20.1 The Parameters σ and C of the SVM

For numerical reasons every data point was normalized such that the average l_2-norm is 1: let $\{\mathbf{x}_k, k = 1, ..., m\}$ be the set of training examples. Every $\mathbf{x}$ was divided by $(\frac{1}{m} \sum_k \|\mathbf{x}_k\|^2)^{\frac{1}{2}}$.

For the data sets ARCENE, DEXTER, GISETTE and MADELON a hard margin SVM was calculated. For the unbalanced data set DOROTHEA a soft margin SVM was chosen and the regularization parameter C was obtained by cross validation prior to feature selection. An example of the cross validation error estimates for the data set ARCENE can be found in Figure 20.1. Furthermore, we used a class specific 2-norm penalty by adding a ridge to the kernel matrix (Brown et al., 2000): Let r_1 be the fraction of positive examples in the training data and let r_{-1} be the fraction of negative examples. For each of the two classes we added a different ridge to the kernel matrix k: for positive examples x_i we set $k_{ii} \rightarrow k_{ii} + r_1$ and for negative examples we set $k_{ii} \rightarrow k_{ii} + r_{-1}$. Adding class specific ridges to the diagonal of the kernel ma-

T.N. Lal et al.: *Combining a Filter Method with SVMs*, StudFuzz **207**, 439–445 (2006)
www.springerlink.com

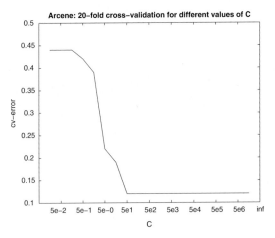

Fig. 20.1. For each value of C (x-axis) the 20-fold cross-validation error (y-axis) is plotted using a linear SVM on data set ARCENE.

trix is equivalent to choosing two different values of C for the different classes (e.g. Schmidt and Gish (1996), Schölkopf and Smola (2002)).

For the data sets DOROTHEA, GISETTE and MADELON we chose a Gaussian kernel. Prior to feature selection, the kernel parameter σ was found by a heuristic: for each k let t_k denote the distance of point x_k to the set formed by the points of the other class. The value of σ was set to the mean of the t_k values[1]. For the remaining two data sets ARCENE and DEXTER we used a linear kernel.

20.2 Feature Ranking

The features were ranked according to their Fisher score (see Chapter 2): For a set $T = \{\mathbf{t}_1, ..., \mathbf{t}_m\} \subset \mathbb{R}^n$ define the mean $\mu_i(T) = \frac{1}{m} \sum_{k=1}^{m} t_{k,i}$ and the variance $V_i(T) = \frac{1}{m} \sum_{k=1}^{m} (t_{k,i} - \mu_i(T))^2$ $(i = 1, ..., n)$. The score R_i of feature i is then given by:

$$R_i(X) = \frac{\left(\mu_j(X^+) - \mu_j(X^-)\right)^2}{V_j(X^+) + V_j(X^-)},$$

with $X^+ := \{\mathbf{x}_k \in X \mid y_k = 1\}$ and X^- similarly. ¿From Figure 20.2 it can be seen that only few of the 10000 features of the data set ARCENE show a high correlation with the labels. However, it is not obvious how many features should be used for classification.

---

[1] In later steps we use an SVM with this σ in a cross validation scheme for further model selection. The calculation of σ involves all labels. Thus, when we test a trained model on a cross validation test set, we make a systematic error, because the label information of the test set is contained in σ. However, we find that the value for σ is not affected much when the test data is removed before the calculation.

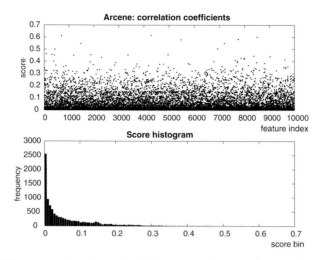

Fig. 20.2. The upper plot shows the Fisher scores (y-axis) for each feature (x-axis) of the data set ARCENE. The lower part of the figure is a histogram of the scores for the same data set. Please note that only few features show a high correlation with the labels.

20.3 Number of Features

The list of ranked features provides an estimate of how valuable a feature is for a given classification task. We were interested in the expected risk of an SVM for any given number N of best features. Provided with these values, we could choose the best number of features, i.e. the number N that minimizes the expected risk.

The expected risk could be estimated by ranking the features using the complete training set. In a second step a cross-validation error estimation can be applied for every number n of best features. However, this approach bears the risk of overfitting since all data is used during the ranking procedure. To avoid this drawback, we proceeded as follows:

For given number N of best features we approximated the expected risk using a ten-fold cross-validation scheme (see Figure 20.3): ten times, the training data were split into a training set which formed 90% of the data and a test set which formed the remaining 10%. The training data were split such that the union of the test sets formed the training data (partition). A training set - test set - pair is called a fold. For each fold we proceeded as follows:

The features were ranked based on the training set. For a given N we restricted the training and the test set examples to the best N ranked features. An SVM was trained on the restricted training set and then tested on the restricted test set of the fold. For each fold we obtained a test error - these errors were averaged over the folds. As a results we got an estimate of the

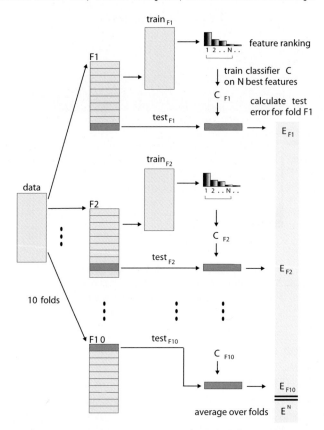

Fig. 20.3. This plot describes the process of calculating an estimate of the expected risk when only using the best N ranked features for classification. First the data is split into 10 train-test folds. The features are ranked on each training set, a classifier is trained using the best N features only and tested on the corresponding test set. The 10 test errors are averaged. Please note that the set of features used by the classifier might vary. Please see also Figure 20.4: every point of the plot is one error estimation for a specific N (from Lal et al. (2004)) .

expected risk for the best N features. We repeated this procedure for different values of N.

Figure 20.4 shows the best N features versus the 10-fold cross-validation errors for the data set ARCENE. Based on this result we used the best 4700 features for the competition.

Once the number N of best features was estimated all data were used, restricted to these best N features and an SVM was trained. To avoid overfitting no further adjustment of C or σ was done. The trained model was then used to predict the labels of the unseen test set examples.

Methods *FS + SVM* used for the competition

Require: data set d, kernel function k, kernel parameter σ

1: normalize data

$$\forall \mathbf{x} \in d \quad \mathbf{x} \rightsquigarrow \mathbf{x}/\sqrt{\frac{1}{|d|}\sum_{\mathbf{y}\in d}\|\mathbf{y}\|^2}$$

2: estimate SVM parameter C via cross-validation on d using all features (for soft margin SVM only).
3: estimate the number n_0 of best features as described in Figure 20.3.
4: use all available data d to rank features according to their Fisher scores.
5: restrict d to the best n_0 ranked features.
6: train an SVM based on the restricted data d using kernel k and regularization parameter C.

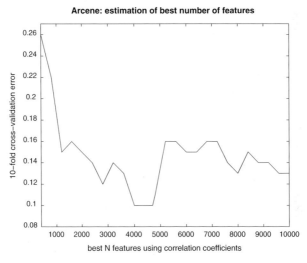

Fig. 20.4. Data set ARCENE: This plot shows an error estimation of the expected risk (y-axis) using only the best N features (x-axis) which are ranked according to their Fisher score. The error estimates were obtained by the cross validation scheme shown in Figure 20.3.

20.4 Summary

We explained how a simple filter method can be combined with SVMs. More specifically we reported how we estimated the number of features to be used for classification and how the kernel parameter as well as the regularization parameter of the SVM can be determined. The performance on the competition data sets can be found in Tables 20.2 and 20.3.

Table 20.1. Parameter values and number of used features for the competition. A value of ∞ for C corresponds to a hard margin SVM.

	C	σ	number used of features
ARCENE	∞	-	4700
DEXTER	∞	-	3714
DOROTHEA	10 + balanced	0.115	1000
GISETTE	∞	0.382	1700
MADELON	∞	0.011	20

Table 20.2. NIPS 2003 challenge results for the December 1st data sets. Comparison of our approach FS + SVM to the best challenge entry.

Dec. 1st Dataset	Our best challenge entry				The winning challenge entry				
	Score	BER	AUC	Feat	Probe	Score	BER	AUC	Feat	Probe
OVERALL	12.73	11.56	88.44	16.91	21.5	88.00	6.84	97.22	80.3	47.8
ARCENE	12.73	18.20	81.80	47.0	13.6	98.18	13.30	93.48	100.0	30.0
DEXTER	85.45	4.20	95.80	18.6	49.8	96.36	3.90	99.01	1.5	12.9
DOROTHEA	-41.82	19.68	80.32	1.0	8.9	98.18	8.54	95.92	100.0	50.0
GISETTE	49.09	1.69	98.31	14.0	0.0	98.18	1.37	98.63	18.3	0.0
MADELON	-41.82	14.06	85.94	4.0	35.0	100.00	7.17	96.95	1.6	0.0

Table 20.3. NIPS 2003 challenge results for the December 8th data sets. Comparison of our approach FS + SVM to the best challenge entry.

Dec. 8th Dataset	Our best challenge entry				The winning challenge entry				
	Score	BER	AUC	Feat	Probe	Score	BER	AUC	Feat	Probe
OVERALL	31.43	8.99	91.01	20.9	17.3	71.43	6.48	97.20	80.3	47.8
ARCENE	65.71	12.76	87.24	47.0	5.9	94.29	11.86	95.47	10.7	1.0
DEXTER	100.00	3.30	96.70	18.6	42.1	100.00	3.30	96.70	18.57	42.1
DOROTHEA	-42.86	16.34	83.66	1.0	3.2	97.14	8.61	95.92	100.0	50.0
GISETTE	82.86	1.31	98.69	34.0	0.2	97.14	1.35	98.71	18.3	0.0
MADELON	-48.57	11.22	88.78	4.0	35.0	94.29	7.11	96.95	1.6	0.0

Acknowledgements

This work was supported in part by the IST Programme of the European Community, under the PASCAL Network of Excellence, IST-2002-506778. T.N.L. was supported by a grant from the Studienstiftung des deutschen Volkes.

References

M.P.S. Brown, W. N. Grundy, D. Lin, N. Cristianini, C.W. Sugnet, T. S Furey, M. Ares Jr., and D. Haussler. Knowledge-based analysis of microarray gene expression data by using support vector machines. *PNAS*, 97(1):262–267, 2000.

T.N. Lal, M. Schröder, T. Hinterberger, J. Weston, M. Bogdan, N. Birbaumer, and B. Schölkopf. Support vector channel selection in BCI. *IEEE Transactions on Biomedical Engineering. Special Issue on Brain-Computer Interfaces*, 51(6): 1003–1010, June 2004.

M. Schmidt and H. Gish. Speaker identification via support vector classifiers. In *Proceedings ICASSP'96*, pages 105–108, Atlanta, GA, 1996.

B. Schölkopf and A. Smola. *Learning with Kernels*. MIT Press, Cambridge, USA, 2002.

Spider. Machine Learning Toolbox http://www.kyb.tuebingen.mpg.de/bs/ people/spider/, 2005.

Chapter 21

Feature Selection via Sensitivity Analysis with Direct Kernel PLS

Mark J. Embrechts[1], Robert A. Bress[1], and Robert H. Kewley[2]

[1] Department of Decision Sciences and Engineering Systems Rensselaer
    Polytechnic Institute, Troy, NY, embrem@rpi.edu
[2] Center for Army Analysis, Washington D.C.

Summary. This chapter introduces Direct Kernel Partial Least Squares (DK-PLS) and feature selection via sensitivity analysis for DK-PLS. The overall feature selection strategy for the five data sets used in the NIPS competition is outlined as well.

21.1 Introduction

A direct kernel method is a general paradigm that extends a linear classification or regression method to a more powerful nonlinear method by introducing a kernel transformation of the data as a preprocessing procedure. The advantages of direct kernel methods in this case are: (i) a transparent explanation for nonlinear modeling, while preserving the structure of the original linear model; (ii) a simple extension of linear methods to nonlinear learning models; (iii) the Mercer condition for the kernel can be relaxed. Disadvantages of direct kernel methods are: (i) the full kernel has to be present in memory; (ii) direct kernel methods are not exactly equivalent to related kernel methods, where the kernel is introduced inherently in the paradigm, rather than explicit in the preprocessing; (iii) direct kernel methods require kernel centering.

This chapter is organized as follows: section 2 gives a brief overview of partial least squares (PLS), section 3 explains direct kernel methods, section 4 describes a procedure for kernel centering, section 5 describes model metrics, section 6 discusses data conditioning and preprocessing, section 7 introduces sensitivity analysis, and section 8 outlines the heuristic feature selection procedure that was applied to the four data sets in the NIPS feature selection challenge.

M.J. Embrechts et al.: *Feature Selection via Sensitivity Analysis with Direct Kernel PLS*,
StudFuzz **207**, 447–462 (2006)
www.springerlink.com

21.2 Partial Least Squares Regression (PLS)

21.2.1 Introduction

Partial Least Squares Regression (PLS) was conceived by the Swedish statistician Herman Wold for econometrics modeling of multivariate time series. The first PLS publication was a sociology application in 1975 (Wold, 1975, Wold et al., 2001). His son, Svante Wold, applied PLS to chemometrics in the early eighties (Wold et al., 2001, Wold, 2001). Currently PLS has become one of the most popular and powerful tools in chemometrics, mainly because of the quality of building models with many variables with collinearity. A decade ago it was not uncommon to have entire Gordon conferences on drug design being dominated by the theory of PLS. The mathematics behind PLS are outlined in Bennett and Embrechts (2003), Wold et al. (2001). The acronym PLS, which originally stood for partial least squares, can be misleading. Therefore Svante Wold proposes that the PLS acronym stands for projection on latent structures, which is more meaningful.

PLS can be viewed as an alternative to principal components regression, but now takes into account the variable dependence. The data are first transformed into a different basis, similar to Principal Component Analysis (PCA), and only a few (the most important) PLS components (or latent variables) are considered for building a regression model (just as in PCA). These PLS components or scores are similar to the principal components in PCA. Unlike PCA, the basis vectors of PLS are not a set of orthonormal eigenvectors of the correlation matrix $X^T X$, where the successive orthogonal directions explain the largest variance in the data, but are actually a set of conjugant gradient vectors to the correlation matrices that span a Krylov space (Ipsen and Meyer, 1998). Just like in PCA the basis vectors can be peeled off from the data matrix X successively in the NIPALS algorithm (Nonlinear Iterative Partial Least Squares), also introduced by Wold (1966).

PLS regression is one of the most powerful data mining tools for large data sets with an overabundance of descriptive features with collinearity. The NIPALS implementation of PLS is elegant and fast. Furthermore, one of the unique characteristics of PLS is the ability to reconstruct logical flow implications and to be able to distinguish cause and consequence relationships from a database (Hulland, 1999).

What makes PLS especially interesting for data mining applications is a recent extension to nonlinear PLS or kernel PLS (Bennett and Embrechts, 2003, Rosipal and Trejo, 2001) which incorporates the kernel transform, similar to support vector machines (Boser et al., 1992, Cristianini and Shawe-Taylor, 2000, Schölkopf and Smola, 2001). Nonlinear Kernel PLS has recently been explained in the context of neural networks (i.e., perceptrons and radial basis functions) (Embrechts and Bennett, 2002).

21.2.2 PLS Algorithm

PLS analysis considers the response vector (or response matrix, in case of multiple responses), typically denoted as $\mathbf{y}$. PLS regression maximizes latent variable correlation with the response vector. Therefore, the first latent variable (which is a linear combination of the original input variables) possesses maximum correlation with the response variable, while remaining conjugate gradient with respect to the correlation matrix to the remaining latent variables. Because the first few partial least squares components, or latent variables, capture most of the correlation with the response, powerful and elegant (linear) prediction models result.

Rosipal and Trejo (2001) modified the PLS algorithm by normalizing the scores rather than the basis vectors in order to make the traditional algorithm ready for the kernel formulation. The original NIPALS algorithm for PLS for one response variable can be modified to the algorithm presented in Algorithm 9 (following Rosipal). Extension to a kernel-based algorithm is now straightforward (Rosipal and Trejo, 2001). The PLS algorithm can be interpreted here as a one-layer linear neural network without activation function and weight vector $\mathbf{w}$. Note that although step 2 of the algorithm calls for a random initialization, the result of the algorithm is not stochastic.

Algorithm 9: Algorithm for PLS for one single response, following Rosipal and Trejo (2001)

(1.) Do for each latent variable

(2.) Randomly initialize an n-dimensional vector $\mathbf{u}$

(3.) Calculate the score, and normalize the m-dimensional vector $\mathbf{t}$

$$\mathbf{t} = X_{nm}^T X_{mn} \mathbf{u}, \quad \mathbf{t} \leftarrow \frac{\mathbf{t}}{||\mathbf{t}||}$$

(4.) Recalibrate $\mathbf{u}$

$$\mathbf{u} = (\mathbf{y}^T \mathbf{t})\mathbf{y}, \quad \mathbf{u} \leftarrow \frac{\mathbf{u}}{||\mathbf{u}||}$$

(5.) Deflate

$$X \leftarrow X - \mathbf{t}\mathbf{t}^T X$$
$$y \leftarrow Y - \mathbf{t}\mathbf{t}^T \mathbf{y}$$

(6.) Store $\mathbf{t}$ in T, store $\mathbf{u}$ in U

(7.) Go to 1 for the next latent variable

(8.) Calculate weights according to

$$\mathbf{w} = X^T U (T^T X X^T U)^{-1} T^T \mathbf{y}$$

21.2.3 Kernel PLS (K-PLS)

The K-PLS method can be reformulated to resemble support vector machines, but it can also be interpreted as a kernel and centering transformation of the descriptor data followed by a regular PLS method (Bennett and Embrechts, 2003). K-PLS was first introduced by Lindgren et al. (1993) in the context of working with linear kernels on data sets with more input descriptors than data, in order to make the PLS modeling more efficient. Early applications of K-PLS were done mainly in this context (Gao and Ren, 1998, 1999, Ren and Gao, 1995). Direct kernel PLS or DK-PLS first considers the kernel as a data transformation, and then applies a regular PLS algorithm. It is important to point out here that DK-PLS is fundamentally a different algorithm than the K-PLS algorithm introduced by Rosipal, because the DK-PLS algorithm operates on the square kernel, KK^T, rather than the original kernel K.

21.3 Regression Models Based on Direct Kernels

The kernel transformation is an elegant way to make a regression model non-linear. The kernel transformation goes back at least to the early 1900's, when Hilbert addressed kernels in the mathematical literature. A kernel is a matrix containing similarity measures in a dataset: either between the data of the dataset itself, or with other data (e.g., support vectors (Boser et al., 1992, Cristianini and Shawe-Taylor, 2000, Schölkopf and Smola, 2001)). A classical use of a kernel is the correlation matrix used for determining the principal components in principal component analysis, where the feature kernel contains linear similarity measures between (centered) attributes. In support vector machines, the kernel entries are similarity measures between data rather than features and these similarity measures are usually nonlinear, unlike the dot product similarity measure that we used before defining a kernel. There are many possible nonlinear similarity measures, but in order to be mathematically tractable the kernel has to satisfy certain conditions, the so-called Mercer conditions (Cristianini and Shawe-Taylor, 2000, Schölkopf and Smola, 2001).

$$
\overleftrightarrow{K}_{mm} =
\begin{bmatrix}
k_{11} & k_{12} & \dots & k_{1m} \\
k_{21} & k_{22} & \dots & k_{2m} \\
& & \dots & \\
k_{m1} & k_{m2} & \dots & k_{mm}
\end{bmatrix}
\tag{21.1}
$$

The expression above, introduces the general structure for the data kernel matrix, $\overleftrightarrow{K}_{mm}$, for m data and n features. The kernel matrix is a symmetrical matrix where each entry contains a (linear or nonlinear) similarity between two data vectors. There are many different possibilities for defining similarity metrics such as the dot product, which is a linear similarity measure and the Radial Basis Function kernel or RBF kernel, which is a nonlinear similarity measure. The RBF kernel is the most widely used nonlinear kernel and the kernel entries are defined by

$$k_{ij} = e^{-\frac{\|\mathbf{x}_j - \mathbf{x}_l\|_2^2}{2\sigma^2}} \qquad (21.2)$$

Note that in the kernel definition above, the kernel entry contains the square of the Euclidean distance (or two-norm) between data points, which is a dissimilarity measure (rather than a similarity), in a negative exponential. The negative exponential also contains a free parameter, σ , which is the Parzen window width for the RBF kernel. The proper choice for selecting the Parzen window is usually determined by an additional tuning, also hyper-tuning, on an external validation set.

The kernel transformation is applied here as a data transformation in a separate pre-processing stage. The features derived for a pattern vector $\mathbf{x}$ of dimension n now consist of a m-dimensional vector containing the similarities between $\mathbf{x}$ and the m training examples. We actually replace the data by a nonlinear data kernel and apply a traditional linear predictive model. Methods where a traditional linear algorithm is used on a nonlinear kernel transform of the data are introduced in this chapter as *direct kernel methods*. The elegance and advantage of such a direct kernel method is that the nonlinear aspects of the problem are captured entirely in the kernel and are transparent to the applied algorithm. If a linear algorithm was used before introducing the kernel transformation, the required mathematical operations remain linear. It is now clear how linear methods such as principal component regression, ridge regression, and partial least squares can be turned into nonlinear direct kernel methods, by using exactly the same algorithm and code: only the data are different, and we operate on the kernel transformation of the data rather than the data themselves.

In order to make out-of-sample predictions on true test data, a similar kernel transformation needs to be applied to the test data. The idea of direct kernel methods is illustrated in Figure 21.1, by showing how any regression model can be applied to kernel-transformed data. One could also represent the kernel transformation in a neural network type of flow diagram and the first hidden layer would now yield the kernel-transformed data, and the weights in the first layer would be just the descriptors of the training data. The second layer contains the weights that can be calculated by applying PLS. When a radial basis function kernel is used, the first layer would be the same as the first layer in a radial basis function neural network, but the weights in the second layer are calculated differently. Note that because of the emphasis on direct kernels as a data transformation we use here the terminology and notation for weights, rather than the familiar alphas in the support vector machine literature. In the next section kernel centering will be discussed. We introduce the symbol in Figure 21.1, indicating multiplication of a centered kernel by a weight vector.

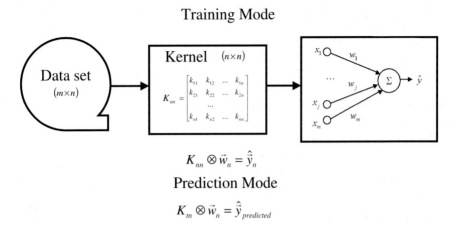

$$K_{nn} \otimes \vec{w}_n = \hat{\vec{y}}_n$$

Prediction Mode

$$K_{tn} \otimes \vec{w}_n = \hat{\vec{y}}_{predicted}$$

Fig. 21.1. Operation schematic for direct kernel methods as a data pre-processing step (the symbol $\otimes$ indicates that the weights should be applied to a centered kernel matrix).

21.4 Dealing with the Bias: Centering the Kernel

There is still one important detail that was overlooked so far, and that is necessary to make direct kernel methods work. By applying weights to the kernel transformed data, there is no constant offset term or bias.

The bias term does not have to be explicitly incorporated in the model when kernel centering is applied. Kernel centering is explained in (Schölkopf and Smola, 2001) which can be extended and summarized in the following recipe: (i) A straightforward way for kernel centering is to subtract the average from each column of the training data kernel, and store these averages for later recall (when centering the test kernel); (ii) A second step for centering the training kernel is going through the newly obtained "vertically" centered kernel again, this time row by row, and subtracting the row average form each horizontal row entry.

The kernel of the test data needs to be centered in a consistent way, following a similar procedure. In this case, the stored column centers from the kernel of the training data will be used for the "vertical" centering part of the kernel of the test data. This vertically centered test kernel is then centered horizontally (i.e., for each row, the average of the vertically centered test kernel is calculated, and each horizontal entry of the vertically centered test kernel is substituted by that entry minus the row average).

The advantage of the kernel-centering algorithm recipe introduced in this section is that it is more general than (Schölkopf and Smola, 2001) because it also applies to rectangular data kernels. The flow chart for pre-processing the data, applying a kernel transform on this data, and centering the kernel for the training data, validation data, and test data is shown in Figure 21.2.

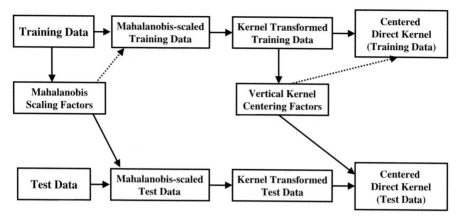

Fig. 21.2. Data pre-processing with kernel centering for direct kernel methods. For a description of the preprocessing and a definition of Mahalanobis scaling, see Section 21.6.

21.5 Metrics for Assessing the Model Quality

An obvious question is the proper assessment for the quality of a model. In the case of a classification problem that would be relatively easy, and one would ultimately present the number of hits and misses in the form of a confusion matrix. For a regression problem, a common way to capture the error is by the Root Mean Square Error index or RMSE, which is defined as the average value of the squared error (either for the training set or the test set). Assuming a response y, and a prediction $\hat{y}$, the RMSE can be expressed as:

$$RMSE = \sqrt{\frac{1}{m} \sum_i (\hat{y}_i - y_i)^2} \qquad (21.3)$$

While the root mean square error is an efficient way to compare the performance of different prediction methods on the same data, it is not an absolute metric in the sense that the RMSE will depend on how the response for the data was scaled. In order to overcome this handicap, additional error measures can be introduced that are less dependent on the scaling and magnitude of the response value. A first metric that will be used for assessing the quality of a trained model is r^2, where r^2 is defined as the correlation coefficient squared between target values and predictions for the response according to:

$$r^2 = \frac{\sum_{i=1}^{m} \left(\hat{y}_i - \bar{\hat{y}}\right)(y_i - \bar{y})}{\sqrt{\sum_{i=1}^{m} \left(\hat{y}_i - \bar{\hat{y}}\right)^2} \sqrt{\sum_{i=1}^{m} (y_i - \bar{y})^2}} \qquad (21.4)$$

where m represents the number of data points in the training set. r^2 takes values between zero and unity, and the higher the r^2 value, the better the model. An obvious drawback of r^2 for assessing the model quality is that r^2 only expresses a linear correlation, indicating how well the predictions follow a line if is plotted as function of y. While one would expect a nearly perfect model when r^2 is unity, this is not always the case. A second and more powerful measure to assess the quality of a trained model is the so-called *Press r squared*, or R^2, often used in chemometric modeling (Johnson and Wichern, 2000), where R^2 is defined as (Golbraikh and Tropsha, 2002):

$$R^2 = 1 - \frac{\sum\limits_{i=1}^{m}(y_i - \hat{y}_i)^2}{\sum\limits_{i=1}^{m}(y_i - \bar{y})^2} \qquad (21.5)$$

We consider R^2 as a more reliable measure than r^2, because it accounts for the residual error as well. The higher the value for R^2, the better the model. Note that in certain cases the R^2 metric can actually be negative. The R^2 metric is commonly used in chemometrics and is generally smaller than r^2. For large datasets, R^2 tends to converge to r^2, and the comparison between r^2 and R^2 for such data often reveals hidden biases.

For assessing the quality of the validation set or a test set, we will introduce similar metrics, q^2 and Q^2, where q^2 and Q^2 are defined as $1 - r^2$ and $1 - R^2$ for the data in the test set. For a model that perfectly predicts on the test data, we now would expect q^2 and Q^2 to be zero. The reason for introducing metrics that are symmetric between the training set and the test set is actually to avoid confusion. Q^2 and q^2 values will always apply to a validation set or a test set; we expect these values to be quite low in order to have a good predictive model. R^2 and r^2 values will always apply to training data, and should be close to unity for a good training model.

An additional metric that can be very useful for classification problems is the area under the ROC curve (Froelicher et al., 2002, Swets et al., 2000). This metric can actually be extended to regression problems as well (Bi and Bennett, 2003), or be used just as is a relative quality indicator for a regression model.

21.6 Data Conditioning and Preprocessing

It is customary in predictive modeling to normalize or Mahalanobis scale all the data first before any further operations. We introduce the term *Mahalanobis scaling* here and prefer it above *normalization* common in the machine learning literature, because normalization has often different meanings for different application domains.[3] By normalizing or Mahalanobis scaling we mean here subtracting the average from each of the descriptive model features or

---

[3]Editor note: This transformation is also referred to as "standardization" in other chapters.

attributes, and dividing each feature entry by the standard deviation. This procedure has to be carried out consistently for training and validation and test data as outlined in Figure 21.2. The division by the standard deviation in Mahalanobis scaling has the obvious advantage that the model is now independent of the metric in which the data were expressed. The centering of the data aspect in Mahalanobis scaling is not always desirable, and generally depends on the nature of the data and the machine learning model. It has been observed by the authors that data centering in large datasets with many sparse binary features can lead to a serious deterioration of the model. Whether or not data should be centered before proceeding with the kernel transform can be determined heuristically by observing the model performance with and without centering on an external validation set.

21.7 Sensitivity Analysis

Sensitivity analysis is a generic and powerful paradigm to select important input features from a predictive model (Kewley and Embrechts, 2000). The underlying hypothesis of sensitivity analysis is that once a model is built, all features are frozen at their average value (i.e., zero in case data were centered first), and then one-by-one the features are tweaked within their allowable range (i.e., usually between -2 and $+2$ in the case of Mahalanobis scaled data with centering). We typically chose 7 settings between $[-2, 2]$, and take the largest difference in response between any of these settings (usually corresponding to the -2 and $+2$ inputs). The features for which the predictions do not vary a lot when they are tweaked are considered less important, and they are slowly pruned out from the input data in a set of successive iterations between model building and feature selection.

To determine the sensitivity of an input feature the sensitivies were averaged over an ensemble of 10 different models where 10% of the training data were left out of the training set in order to obtain more stable results.

The pruning of less relevant features proceeds now in an iterative fashion, where 10% of the least significant features are dropped each time. The feature selection process continues until there is a significant degradation in predictive performance on the validation data. An alternative to determine when to halt the feature pruning process is by introducing a random variable (usually from a uniform distribution) as a gauge feature and to iteratively drop features with sensitivities that are less than the sensitivity of the random variable. Sensitivity analysis for feature selection is a generic wrapper method and can be applied to any machine learning method and is illustrated in Figure 21.3. Feature selection via sensitivity analysis is explained in detailed for neural networks in (Kewley and Embrechts, 2000). This methodology is model independent and can readily be extended to K-PLS, DK-PLS, SVMs and predictive models in general.

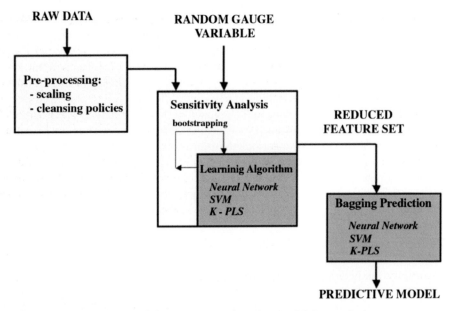

Fig. 21.3. Schematic procedure for sensitivity analysis

21.8 Heuristic Feature Selection Policies for the NIPS Feature Selection Challenge

21.8.1 General Preprocessing and Feature Selection Methodology

Data were preprocessed by centering all the features. It was then decided whether or not to divide the feature entries by their standard deviation based on the performance on the validation set. For partial least squares methods (PLS, K-PLS, or DK-PLS) the response variable is always Mahalanobis scaled. In that case a minority class corresponds to a larger response variable for that class, which is beneficial by putting more emphasis on minority patterns in the least-squares error measure. All predictive models for the NIPS feature selection competition were based on a bagging predictions (Breiman, 1996), where the models were average models obtained from leaving 10% of the data out 100 times. In the case that the data set contained too many patterns for the kernel to fit in the memory (2 GB in this case), a linear PLS model rather than DK-PLS was used during the feature elimination procedure. The final model was then based on a least-squares support vector machine that did not require the full kernel to be in memory (Keerthi and Shevade, 2003). We usually note very little difference between predictions based on K-PLS, DK-PLS, and least squares support vector machines (see next section). Feature elimination proceeds in two stages: a filtering stage and a wrapping stage. The wrapping procedure for feature elimination was based on sensitivity analysis

as described in the previous section. Before proceeding with the wrapping procedure features were eliminated by filtering based on the following recipe:
  Feature elimination by filtering:

(1.) eliminate non-changing features
(2.) eliminate binary features that are mostly zero (99%, but can vary depending on the number of training data)
(3.) inspect features for extreme outliers and drop features that have 6 sigma outliers
(4.) in the case that there are more than 1000 features drop features that show an absolute value of the correlation coefficient less than 0.1.

The software used for model building, filtering, and sensitivity analysis is based on dynamic scripts for the Analyze software [8].

21.8.2 NIPS Feature Selection Challenge

The five feature selection challenge datasets for ARCENE, DEXTER, DOROTHEA, GISETTE and MADELON are summarized in Table 21.1, and the competition results are summarized in Table 21.2 (where AUC stands for area under the ROC curve). The group rank was 11 and the best entry rank 31. Only one submission was made for each dataset, and no second submission was made after the validation data with labels were made available. The features for all

Table 21.1. Description of Feature Selection Challenge Datasets

Dataset	Features	Random Probes	Training Data	Positives
ARCENE	10000	3000	100	44
DEXTER	20000	10053	300	150
DOROTHEA	100000	50000	800	78
GISETTE	5000	2500	6000	3000
MADELON	500	480	2000	1000

Table 21.2. NIPS 2003 challenge results

Dec. 1st Dataset	Our best challenge entry					The winning challenge entry					
	Score	BER	AUC	Feat	Probe	Score	BER	AUC	Feat	Probe	Test
OVERALL	9.82	10.91	89.46	7.17	25.74	88.00	6.84	97.22	80.3	47.8	1
ARCENE	81.82	16.71	83.67	5.14	8.56	98.18	13.30	93.48	100.0	30.0	1
DEXTER	-5.45	6.80	93.47	1.57	28.12	96.36	3.90	99.01	1.5	12.9	1
DOROTHEA	-34.55	19.18	82.10	0.54	92.04	98.18	8.54	95.92	100.0	50.0	1
GISETTE	5.45	2.02	97.92	26.0	0	98.18	1.37	98.63	18.3	0.0	1
MADELON	1.82	9.83	90.16	2.6	0	100.00	7.17	96.95	1.6	0.0	1

datasets are the original inputs and the feature selection procedure with sensitivity analysis was gauged on the validation data: i.e., as soon as there was a relative increase in error on the validation data, the process of iterative pruning with sensitivity analysis was halted. Responses were always normalized. In the case that there was an imbalance between class instances in the training data, a classification cut-off was determined based on cross-validation. While most of the analysis was based on heuristics, the competition results show that sensitivity analysis is a robust feature selection methodology.

ARCENE

For this dataset the data were centered and the 2000 most correlated features were retained. Feature selection was based on a repetitive sensitivity analysis, were during each step 10% of the features were dropped (based on 100 bootstrap models with leave 10% out) using K-PLS with 5 latent variables and $\sigma = 1600$. 514 or 5.14% of the original features were retained based on cross-validation.

DEXTER

The DEXTER data used the original inputs as features, and the inputs that had more than 1% nonzero entries were retained. In a next phase the 1000 most correlated inputs were retained. Feature selection was based on a repetitive sensitivity analysis, were during each step 10% of the features were dropped (based on 100 bootstrap models with leave 10% out) using K-PLS with 5 latent variables and $\sigma = 900$. 205 or 1.57% of the original features were retained based on cross-validation.

DOROTHEA

The DOROTHEA data used the original inputs as features, and the inputs that had more than 3% nonzero entries were retained. The data were centered and in a next phase feature selection was based on a repetitive sensitivity analysis, were during each step 10% of the features were dropped (based on 100 bootstrap models with leave 10% out) using K-PLS with 5 latent variables and $\sigma = 15$. 540 or 0.54% of the original features were retained based on cross-validation.

GISETTE

The data were centered and the 2000 most correlated inputs were retained as features. Feature selection was based on a repetitive sensitivity analysis, were during each step 10% of the features were dropped (based on 10 bootstrap models with leave 10% out) using K-PLS with 5 latent variables and $\sigma = 20$. 1300 or 26% of the original features were retained based on cross-validation.

MADELON

The data were centered and the 2000 most correlated inputs were retained as features. Feature selection was based on a repetitive sensitivity analysis, were during each step 10% of the features were dropped (based on 100 bootstrap models with leave 10% out) using K-PLS with 7 latent variables and $\sigma = 100$. 13 or 2.6% of the original features were retained based on cross-validation. A classification cut-off was determined based on cross-validation.

21.9 Benchmarks

The datasets used for the experiments are shown in Table 21.3 and can be found in the UCI KDD depository. The results are shown in Figure 21.4. Variables and responses are normalized first. All experiments were based on leave 10% out 100 times, except for the Mushroom data which used leave 10% out 10 times. Note that the metrics also include the area under the ROC curve (Froelicher et al., 2002, Swets et al., 2000) for regression problems, even though the ROC curve does not have a clear physical meaning anymore. The RBF kernel was determined by hyper-tuning a K-PLS model. The number of latent variables for the K-PLS and DK-PLS models was 5, except for the Tic Tac Toe, Mushroom and Boston Housing data sets, where 12 latent variables were used.

Table 21.3. Benchmark UCI Datasets

Dataset	Examples	Features	Type
Wine	178	13	3 Class
Cleveland Heart	297	13	2 Class
Tic Tac Toe	958	9	2 Class
Mushroom	8124	14	2 Class
Abalone	4177	8	Regression
Boston Housing	506	13	Regression

The benchmark methods are based on (direct) kernel ridge regression (DKR) (Hoerl and Kennard, 1970), least-squares support vector machines (LS-SVM) (Suykens and Vandewalle, 1999, Suykens et al., 2003), kernel PLS (K-PLS) (Rosipal and Trejo, 2001), direct kernel PLS (DK-PLS), partial least squares (PLS) (Wold et al., 2001), and Support Vector Machines (SVMs) using LibSVM (Wold, 2001). In the above methods we used a RBF kernel, except for PLS, which is a linear method. Metrics are based on % correct, least mean square error, area under the ROC curve, and q^2 and Q^2 as defined in this

Table 21.4. Benchmarks between DKR, LS-SVM, K-PLS, DK-PLS, PLS and SVM on UCI data

(a) Wine Dataset

Method	Correct	LMS	ROC	q^2	Q^2
DKR	99.06	0.176	1	0.052	0.052
LS-SVM	97.88	0.184	1	0.057	0.057
K-PLS	98.42	0.182	1	0.056	0.056
DK-PLS	96.29	0.207	0.999	0.072	0.072
PLS	96.47	0.277	0.991	0.129	0.130
SVM	98.35	0.185	1	0.058	0.058

(b) Cleveland Heart Dataset

Method	Correct	LMS	ROC	q^2	Q^2
DKR	83.07	0.716	0.901	0.510	0.515
LS-SVM	83.31	0.717	0.901	0.510	0.517
K-PLS	83.24	0.731	0.895	0.532	0.536
DK-PLS	83.52	0.727	0.896	0.528	0.530
PLS	83.40	0.725	0.896	0.527	0.529
SVM	82.93	0.730	0.892	0.526	0.536

(c) Tic Tac Toe Dataset

Method	Correct	LMS	ROC	q^2	Q^2
DKR	99.47	0.388	1	0.154	0.165
LS-SVM	91.74	0.595	0.972	0.383	0.390
K-PLS	99.74	0.383	1	0.154	0.162
DK-PLS	88.96	0.640	0.948	0.451	0.452
PLS	67.79	0.930	0.617	0.952	0.956
SVM	99.27	0.399	1	0.163	0.176

(d) Mushroom Dataset

Method	Correct	LMS	ROC	q^2	Q^2
DKR	100.00	0.057	1	0.003	0.003
LS-SVM	99.48	0.269	1	0.070	0.072
K-PLS	100.00	0.060	1	0.004	0.004
DK-PLS	99.08	0.190	1	0.036	0.036
PLS	93.17	0.519	0.968	0.270	0.270
SVM	100.00	0.066	1	0.004	0.004

(e) Abalone Dataset

Method	LMS	ROC	q^2	Q^2
DKR	2.167	[0.872]	0.443	0.444
LS-SVM	2.136	[0.874]	0.444	0.445
K-PLS	2.167	[0.864]	0.457	0.458
DK-PLS	2.206	[0.856]	0.472	0.473
PLS	2.245	[0.862]	0.488	0.488
SVM	2.180	[0.875]	0.445	0.464

(f) Boston Housing Dataset

Method	LMS	ROC	q^2	Q^2
DKR	3.371	[0.952]	0.131	0.139
LS-SVM	3.835	[0.943]	0.173	0.179
K-PLS	3.397	[0.948]	0.123	0.123
DK-PLS	3.761	[0.945]	0.176	0.176
PLS	4.867	[0.931]	0.287	0.289
SVM	3.285	[0.956]	0.131	0.132

chapter. Results are for most cases very similar. The difference between LS-SVMs and the other methods for Tic Tac Toe data can be attributed to the automated heuristic for the selection penalty factor as explained in (Bennett and Embrechts, 2003). The difference between DK-PLS and K-PLS can be entirely attributed to the fact that DK-PLS effectively uses the square of the kernel, rather than the kernel. Only on Tic Tac Toe data are DK-PLS predictions clearly inferior to K-PLS predictions. DK-PLS generally exhibits a slightly inferior performance compared to K-PLS, but is clearly superior to PLS.

21.10 Conclusions

This chapter introduces direct kernel partial least squares and general recipes for feature elimination based on filters and a sensitivity analysis wrapper methodology. Direct kernel methods consider the kernel transformation as a data preprocessing step, and then use a traditional linear regression method for modeling. Direct kernel PLS or DK-PLS is fundamentally different from the original K-PLS introduced by Rosipal. Like all direct kernel methods, kernel centering is an important conditioning step. This chapter introduces a more general kernel centering procedure that applies to rectangular kernels as well as square kernels. The feature selection by sensitivity analysis is an iterative procedure where a small subset of the features are dropped at a time (typically 10 percent). The evaluation of feature sensitivities is based on leave several patterns out several times. It is therefore not uncommon to have actually run through several thousand models during the feature elimination phase before building the final predictive model with selected features. Benchmarking studies compare the performance of DK-PLS with PLS and SVMs. The similarity between PLS and K-PLS on the wine, Cleveland heart, and abalone data suggests that the models are linear. The combination of DK-PLS and iterative feature elimination provides a general-purpose modeling technique that combines the predictive capabilities of nonlinear models with the computational efficiency and transparency of linear models.

Acknowledgements

The author acknowledges the National Science Foundation support of this work (IIS-9979860). The comments from anonymous reviewers and discussions with Kristin Bennett, Robert Bress, Paul Evangelista, Karsten Sternickel, Boleslaw Szymanski and Seppo Ovaska were extremely helpful in preparing this paper.

References

K.P. Bennett and M.J. Embrechts. An optimization perspective on kernel partial least squares regression. In *Advances in learning theory; methods, models and applications*, pages 227–250. IOS Press, Amsterdam, 2003.

J. Bi and K.P. Bennett. Regression error characteristic curves. In *Proceedings of the 20th International Conference on Machine Learning (ICML 2003)*, Washington DC, 2003.

B.E. Boser, I. Guyon, and V. Vapnik. A training algorithm for optimal margin classifiers. In *Fifth Annual Workshop on Computational Learning Theory*, pages 144–152. ACM, 1992.

L. Breiman. Bagging predictors. *Machine Learning*, 22(2):123–140, 1996.

N. Cristianini and J. Shawe-Taylor. *An Introduction to Support Vector Machines and Other Kernel-Based Methods*. Cambridge University Press, 2000.

M.J. Embrechts and K.P. Bennett. Sparse-kernel pls for molecular drug design. In Y. Bengio, editor, *Proceedings of the Learning Workshop*, Snowbird, Utah, 2002.

V. Froelicher, K. Shetler, and E. Ashley. Better decisions through science: exercise testing scores. *Progress in Cardiovascular Diseases*, 44(5):385–414, 2002.

L. Gao and S. Ren. Simultaneous spectrometric determination of manganese, zinc and cobalt by kernel partial least-squares method. *Journal of Automatic Chemistry*, 20(6):179–183, 1998.

L. Gao and S. Ren. Simultaneous spectrophotometric determination of four metals by the kernel partial least squares method. *Chemometrics and Intelligent Laboratory Systems*, 45:87–93, 1999.

A. Golbraikh and A. Tropsha. Beware of q2! *Journal of Molecular Graphics and Modelling*, 20:269–276, 2002.

A.E. Hoerl and R.W. Kennard. Ridge regression: biased estimation for non-orthogonal problems. *Technometrics*, 12:69–82, 1970.

J. Hulland. Use of partial least squares (PLS) in strategic management research: a review of four recent studies. *Strategic Management Journal*, 20:195–204, 1999.

I.C.F. Ipsen and C.D. Meyer. The idea behind krylov methods. *American Mathematical Monthly*, 105:889–899, 1998.

R.A. Johnson and D.A. Wichern. *Applied multivariate statistical analysis (second edition)*. Prentice Hall, 2000.

S.S. Keerthi and S.K. Shevade. SMO algorithm for least squares SVM formulations. *Neural Computation*, 15:487–507, 2003.

R.H. Kewley and M.J. Embrechts. Data strip mining for the virtual design of pharmaceuticals with neural networks. *IEEE Transactions on Neural Networks*, 11(3):668–679, 2000.

F. Lindgren, P. Geladi, and S. Wold. The kernel algorithm for pls. *Journal of Chemometrics*, 7:45–59, 1993.

S. Ren and L. Gao. Simultaneous spectrophotometric determination of copper (II), lead (II) and cadmium (II). *Journal of Automatic Chemistry*, 17(3):115–118, 1995.

R. Rosipal and L. T. Trejo. Kernel partial least squares regression in reproducing kernel hilbert space. *Journal of Machine Learning Research*, 2:97–123, 2001.

B. Schölkopf and A. Smola. *Learning with kernels: Support Vector Machines, regularization, optimization and beyond*. MIT Press, 2001.

J.A.K. Suykens, T. Van Gestel, J. de Brabanter, B. De Moor, and J. Vandewalle. *Least squares support vector machines*. World Scientific Publishing Company, Singapore, 2003.

J.A.K. Suykens and J. Vandewalle. Least-squares support vector machine classifiers. *Neural Processing Letters*, 9(3):293–300, 1999.

J.A. Swets, R.M. Dawes, and J. Monahan. Better decisions through science. *Scientific American*, 283:82–87, 2000.

H. Wold. Estimation of principal components and related models by iterative least squares. In P.R. Krishnaiah, editor, *Multivariate analysis*, pages 391–420. Academic Press, New York, 1966.

H. Wold. Path models with latent variables: the NIPALS approach. In H.M. Balock, editor, *Quantitative sociology: International perspectives on mathematical and statistical model building*, pages 307–357. Academic Press, New York, 1975.

S. Wold. Personal memories of the early PLS development. *Chemometrics and Intelligent Laboratory Systems*, 58:83–84, 2001.

S. Wold, M. Sjölström, and M.L. Erikson. PLS-regression: a basic tool of chemometrics. *Chemometrics and Intelligent Laboratory Systems*, 58:109–130, 2001.

Chapter 22

Information Gain, Correlation and Support Vector Machines

Danny Roobaert, Grigoris Karakoulas, and Nitesh V. Chawla

Customer Behavior Analytics
Retail Risk Management
Canadian Imperial Bank of Commerce (CIBC)
Toronto, Canada
{danny.roobaert,grigoris.karakoulas,nitesh.chawla}@cibc.ca

Summary. We report on our approach, CBAmethod3E, which was submitted to
the NIPS 2003 Feature Selection Challenge on Dec. 8, 2003. Our approach consists
of combining filtering techniques for variable selection, information gain and feature
correlation, with Support Vector Machines for induction. We ranked 13th overall and
ranked 6th as a group. It is worth pointing out that our feature selection method
was very successful in selecting the second smallest set of features among the top-20
submissions, and in identifying almost all probes in the datasets, resulting in the
challenge's best performance on the latter benchmark.

22.1 Introduction

Various machine learning applications, such as our case of financial analytics,
are usually overwhelmed with a large number of features. The task of feature
selection in these applications is to improve a performance criterion such as
accuracy, but often also to minimize the cost associated in producing the features. The NIPS 2003 Feature Selection Challenge offered a great testbed for
evaluating feature selection algorithms on datasets with a very large number
of features as well as relatively few training examples.

Due to the large number of the features in the competition datasets, we
followed the filtering approach to feature selection: selecting features in a single pass first and then applying an inductive algorithm independently. We
chose a filtering approach instead of a wrapper one because of the huge computational costs the latter approach would entail for the datasets under study.
More specifically, we used information gain (Mitchell, 1997) and analysis of
the feature correlation matrix to select features, and applied Support Vector Machines (SVM) (Boser et al., 1992, Cortes and Vapnik, 1995) as the
classification algorithm. Our hypothesis was that by combining those filtering
techniques with SVM we would be able to prune non-relevant features and

D. Roobaert et al.: *Information Gain, Correlation and Support Vector Machines*, StudFuzz
207, 463–470 (2006)
www.springerlink.com   © Springer-Verlag Berlin Heidelberg 2006

learn an SVM classifier that performs at least as good as an SVM classifier learnt on the whole feature set, albeit with a much smaller feature set. The overall method is described in Section 2. Section 3 presents the results and provides empirical evidence for the above hypothesis. Section 4 refers to a few of the alternative techniques for feature selection and induction that we tried. Section 5 concludes the paper with a discussion on lessons learned and future work.

22.2 Description of Approach

We first describe the performance criterion that we aimed to optimize while searching for the best feature subset or parameter tuning in SVM induction. We then present the two filtering techniques for feature selection and briefly describe the specifics of our SVM approach. We report on the approach submitted on Dec. 8. For the Dec. 1 submission, we used an alternative approach (see Section 4) that was abandoned after the Dec. 1 submission because we obtained better performance with the approach described in the following.

22.2.1 Optimization Criterion

For choosing among several algorithms and a range of hyper-parameter settings, the following optimization criterion was followed: Balanced error rate (BER) using random ten-fold cross-validation before Dec. 1 (when the validation labels were not available), and BER on the validation set after Dec. 1 (when the validation labels were available). BER was calculated in the same way as used by the challenge organizers: $BER = \frac{1}{2}\left[\frac{fp}{tn+fp} + \frac{fn}{tp+fn}\right]$, with fp = false positives, tn = true negatives, fn = false negatives and tp = true positives.

22.2.2 Feature Selection

Information Gain

Information gain (IG) measures the amount of information in bits about the class prediction, if the only information available is the presence of a feature and the corresponding class distribution. Concretely, it measures the expected reduction in entropy (uncertainty associated with a random feature) (Mitchell, 1997). Given S_X the set of training examples, $\mathbf{x}_i$ the vector of ith variables in this set, $|S_{\mathbf{x_i}=v}|/|S_X|$ the fraction of examples of the ith variable having value v:

$$IG(S_X, \mathbf{x}_i) = H(S_X) - \sum_{v=values(\mathbf{x}_i)} \frac{|S_{\mathbf{x_i}=v}|}{|S_X|} H(S_{\mathbf{x}_i=v}) \text{ with entropy:}$$

$$H(S) = -p_+(S) \log_2 p_+(S) - p_-(S) \log_2 p_-(S)$$

$p_\pm(S)$ is the probability of a training example in the set S to be of the positive/negative class. We discretized continuous features using information theoretic binning (Fayyad and Irani, 1993).

For each dataset we selected the subset of features with non-zero information gain. We used this filtering technique on all datasets, except the MADELON dataset. For the latter dataset we used a filtering technique based on feature correlation, defined in the next subsection.

Correlation

The feature selection algorithm used on the MADELON dataset starts from the correlation matrix M of the dataset's variables. There are 500 features in this dataset, and we treat the target (class) variable as the 501st variable, such that we measure not only feature redundancy (intra-feature correlation), but also feature relevancy (feature-class correlation). In order to combine redundancy & relevancy information into a single measure, we consider the column-wise (or equivalently row-wise) average absolute correlation $\langle M \rangle_i = \frac{1}{n} \sum_j |M_{ij}|$ and the global average absolute correlation $\langle M \rangle = \frac{1}{n^2} \sum_{i,j} |M_{ij}|$. Plotting the number of column correlations that exceeds a multiple of the global average correlation ($\langle M \rangle_i > t \langle M \rangle$) at different thresholds t, yields Figure 22.1.

As can be observed from Figure 22.1, there is a discontinuity in correlation when varying the threshold t. Most variables have a low correlation, not exceeding about 5 times average correlation. In contrast, there are 20 features that have a high correlation with other features, exceeding 33 times the average correlation. We took these 20 features as input to our model.

The same correlation analysis was performed on the other datasets. However no such distinct discontinuity could be found (i.e. no particular correlation structure could be discovered) and hence we relied on information gain to select variables for those datasets. Note that Information Gain produced 13 features on the MADELON dataset, but the optimization criterion indicated worse generalization performance, and consequently the information gain approach was not pursued on this dataset.

22.2.3 Induction

As induction algorithm, we choose Support Vector Machines (Boser et al., 1992, Cortes and Vapnik, 1995). We used the implementation by Chang and Lin (2001) called LIBSVM. It implements an SVM based on quadratic optimization and an epsilon-insensitive linear loss function. This translates to the following optimization problem in dual variables α:

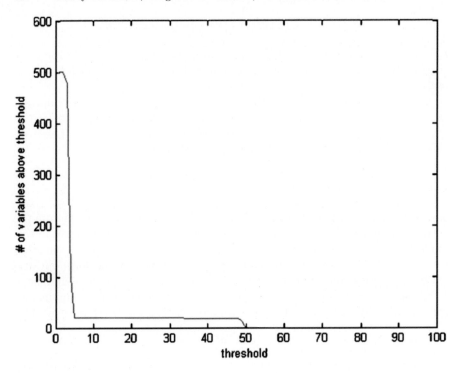

Fig. 22.1. The number of MADELON variables having a column correlation above the threshold

$$\max_{\alpha} \left(\sum_{k=1}^{m} \alpha_k - \frac{1}{2} \sum_{k=1}^{m} \sum_{l=1}^{m} \alpha_k \alpha_l y_k y_l K(\mathbf{x}_k, \mathbf{x}_l) \right) \quad \text{s.t.} \quad \begin{cases} 0 \leq \alpha_k \leq C, \forall k \\ \sum\limits_{k=1}^{m} y_k \alpha_k = 0 \end{cases}$$

where C is the regularization hyper-parameter, $K(\mathbf{x}_k, \mathbf{x}_l)$ the kernel and y_k the target (class) variables. The implementation uses Sequential Minimal Optimization (Platt, 1999) and enhanced heuristics to solve the optimization problem in a fast way. As SVM kernel we used a linear kernel $K(\mathbf{x}_k, \mathbf{x}_l) = \mathbf{x}_k \cdot \mathbf{x}_l$ for all datasets, except for the Madelon dataset where we used an RBF-kernel $K(\mathbf{x}_k, \mathbf{x}_l) = e^{-\gamma \|\mathbf{x}_k - \mathbf{x}_l\|}$. The latter choices were made due to better optimization criterion results in our experiments.

For SVM hyper-parameter optimization (regularization hyper-parameters C and γ in the case of an RBF kernel), we used pattern search (Momma and Bennett, 2002). This technique performs iterative hyper-parameter optimization. Given an initial hyper-parameter setting, upon each iteration, the technique tries a few variant settings (in a certain pattern) of the current hyper-parameter settings and chooses the setting that best improves the performance criterion. If the criterion is not improved, the pattern is applied on a finer scale. If a pre-determined scale is reached, optimization stops.

For the imbalanced dataset DOROTHEA, we applied asymmetrically weighted regularization values C for the positive and the negative class. We used the following heuristic: the C of the minority class was always kept at a factor, $|majorityclass|/|minorityclass|$, higher than the C of the majority class.

22.3 Final Results

Our submission results are shown in Table 22.1. From a performance point of view, we have a performance that is not significantly different from the winner (using McNemar and 5% risk), on two datasets: ARCENE and MADELON. On average, we rank 13th considering individual submissions, and 6th as a group.

Table 22.1. NIPS 2003 challenge results on December 8th

Dec. 8th Dataset	Our best challenge entry[1] Score	BER	AUC	Feat	Probe	The winning challenge entry Score	BER	AUC	Feat	Probe	Test
OVERALL	21.14	8.14	96.62	12.78	0.06	88.00	6.84	97.22	80.3	47.8	0.4
ARCENE	85.71	11.12	94.89	28.25	0.28	94.29	11.86	95.47	10.7	1.0	0
DEXTER	0	6.00	98.47	0.60	0.0	100.00	3.30	96.70	18.6	42.1	1
DOROTHEA	-28.57	15.26	92.34	0.57	0.0	97.14	8.61	95.92	100.0	50.0	1
GISETTE	-2.86	1.60	99.85	30.46	0.0	97.14	1.35	98.71	18.3	0.0	0
MADELON	51.43	8.14	96.62	12.78	0.0	94.29	7.11	96.95	1.6	0.0	1

From a feature selection point of view, we rank 2nd (within the 20 best submission) in minimizing the number of used features, using only 12.78% on average. However we are consistently 1st in identifying probes: on this benchmark, we are the best performer on all datasets.

To show the significance of feature selection in our results, we ran experiments where we ignored the feature selection process altogether, and applied SVMs directly on all features. In Table 22.2, we report the best BER on the validation set of each dataset. These results were obtained using linear SVMs, as in all experiments RBF-kernel SVMs using all features gave worse results compared to linear SVMs. As can be seen from the table, using all features always gave worse performance on the validation set, and hence feature selection was always used.

---

[1] Performance is not statistically different from the winner, using McNemar and 5% risk.

Table 22.2. BER performance on the validation set, using all features versus the described selected features

Dataset	All features	Selected features	Reduction in BER
ARCENE	0.1575	0.1347	-16.87%
DEXTER	0.0867	0.0700	-23.81%
DOROTHEA	0.3398	0.1156	-193.96%
GISETTE	0.0200	0.0180	-11.11%
MADELON	0.4000	0.0700	-471.43%

22.4 Alternative Approaches Pursued

Several other approaches were pursued. All these approaches though gave worse performance (given the optimization criterion) and hence were not used in the final submission. We briefly discuss a few of these approaches, as we are restricted by paper size.

22.4.1 Alternative Feature Selection

We used a linear SVM to remove features. The approach is as follows: we first train a linear SVM (including hyper-parameter optimization) on the full feature set. Then we retain only the features that correspond with the largest weights in the linear function. Finally, we train the final SVM model using these selected features. We experimented with different feature fractions retained, as in general the approach does not specify how to choose the number of features to be retained (or the weight threshold). In Table 22.3, we show a performance comparison at half, the same and double of the size of the feature set finally submitted. We did not try a variant of the above approach called Recursive Feature Elimination (RFE), proposed by (Guyon et al., 2002) due to its prohibitive computational cost.

Table 22.3. BER performance on the validation set, comparing feature selected by LINSVM versus Infogain/Corr

Dataset	Feature Final fraction	LIN SVM feature fraction			InfoGain / Corr. feature
		Half final	Final	Double final	
ARCENE	0.2825	0.1802	0.1843	0.1664	0.1347
DEXTER	0.0059	0.1000	0.1167	0.1200	0.0700
DOROTHEA	0.0057	0.2726	0.3267	0.3283	0.1156
GISETTE	0.3046	0.0260	0.0310	0.0250	0.0180
MADELON	0.0400	0.1133	0.1651	0.2800	0.0700

22.4.2 Combining Feature Selection and Induction

We tried also a linear programming approach to SVM inspired by Bradley and Mangasarian (1998). Here SVM is formulated as a linear optimization problem instead of the typical SVM quadratic optimization. The resulting model only uses a selected number of non-zero weights and hence feature selection is embedded in the induction. Unfortunately the results were not encouraging.

22.5 Discussion and Conclusion

We showed how combining a filtering technique for feature selection with SVM leads to substantial improvement in generalization performance of the SVM models in the five classification datasets of the competition. The improvement is the highest for the datasets Madelon and Dorothea as shown in table 2 above. These results provide evidence that feature selection can help generalization performance of SVMs.

Another lesson learned from our submission is that there is no single best feature selection technique across all five datasets. We experimented with different feature selection techniques and picked the best one for each dataset. Of course, an open question still remains: why exactly these techniques worked well together with Support Vector Machines. A theoretical foundation for the latter is an interesting topic for future work.

Finally, it is worth pointing out that several of the top-20 submissions in the competition relied on using large feature sets for each dataset. This is partly due to the fact that the performance measure for evaluating the results, BER, is a classification performance measure that does not penalize for the number of features used. In most real-world applications (e.g. medical and engineering diagnosis, credit scoring etc.) there is a cost for observing the value of a feature. Hence, in tasks where feature selection is important, such as in this challenge, there is need for a performance measure that can reflect the trade-off of feature and misclassification cost (Turney, 2000, Karakoulas, 1995). In absence of such a measure, our selection of approaches was influenced by this bias. This resulted in the second smallest feature set in the top-20 and the most successful removal of probes in the challenge.

Acknowledgements

Our thanks to Andrew Brown for the Information Gain code and Brian Chambers and Ruslan Salakhutdinov for a helpful hand.

References

B.E. Boser, I. Guyon, and V. Vapnik. A training algorithm for optimal margin classifiers. In *Fifth Annual Workshop on Computational Learning Theory*, pages 144–152. ACM, 1992.

P.S. Bradley and O.L. Mangasarian. Feature selection via concave minimization and support vector machines. In *Proc 15th Int Conf Machine Learning*, pages 82–90, 1998.

C.C. Chang and C.J. Lin. *LIBSVM: a library for support vector machines*, 2001. Software available at http://www.csie.ntu.edu.tw/~cjlin/libsvm.

C. Cortes and V. Vapnik. Support vector networks. *Machine Learning*, 20(3):273–297, 1995.

U. Fayyad and K. Irani. Multi-interval discretization of continuous-valued attributes for classification learning. In *Proc 10th Int Conf Machine Learning*, pages 194–201, 1993.

I. Guyon, J. Weston, S. Barnhill, and V. Vapnik. Gene selection for cancer classification using support vector machines. *Machine Learning*, 46:389–422, 2002.

G. Karakoulas. Cost-effective classification for credit scoring. In *Proc 3rd Int Conf AI Applications on Wall Street*, 1995.

T. Mitchell. *Machine Learning*. McGraw-Hill, New York, 1997.

M. Momma and K.P. Bennett. A pattern search method for model selection of support vector regression. In R. Grossman, J. Han, V. Kumar, H. Mannila, and R. Motwani, editors, *Proceedings of the Second SIAM International Conference on Data Mining*, pages 261–274. SIAM, 2002.

J. Platt. *Fast Training of Support Vector Machines using Sequential Minimal Optimization*, chapter 12, pages 185–208. MIT Press, 1999.

P. Turney. Types of cost in inductive concept learning. In *Workshop cost-sensitive learning, Proc. 17th Int. Conf. Machine Learning*, pages 15–21, 2000.

Chapter 23

Mining for Complex Models Comprising Feature Selection and Classification

Krzysztof Grąbczewski and Norbert Jankowski

Department of Informatics, Nicolaus Copernicus University, Toruń, Poland
kgrabcze@phys.uni.torun.pl, norbert@phys.uni.torun.pl

Summary. Different classification tasks require different learning schemes to be satisfactorily solved. Most real-world datasets can be modeled only by complex structures resulting from deep data exploration with a number of different classification and data transformation methods. The search through the space of complex structures must be augmented with reliable validation strategies. All these techniques were necessary to build accurate models for the five high-dimensional datasets of the NIPS 2003 Feature Selection Challenge. Several feature selection algorithms (e.g. based on variance, correlation coefficient, decision trees) and several classification schemes (e.g. nearest neighbors, Normalized RBF, Support Vector Machines) were used to build complex models which transform the data and then classify. Committees of feature selection models and ensemble classifiers were also very helpful to construct models of high generalization abilities.

23.1 Introduction

Solving classification problems includes both classifiers' learning and relevant preparation of the training data. In numerous domains the stage of data preprocessing can significantly improve the performance of final classification models. A successful data mining system must be able to combine the two analysis stages and take their interaction into account. Each classification method may need differently prepared inputs to build an accurate and reliable model. Therefore we need to search for a robust combination of methods and their parameters.

Using complex model structures implies the necessity of adequate validation. It is very important to validate the whole sequences of transformations and classifiers instead of performing data preprocessing first and then validating the classification algorithms only. Otherwise we are sentenced to overoptimistic results estimates, which do not reflect real generalization abilities.

To build and validate complex models it is very important to have general data analysis tools, which facilitate combining different algorithms and testing their interaction. In the NIPS 2003 Feature Selection Challenge efforts

K. Grąbczewski and N. Jankowski: *Mining for Complex Models Comprising Feature Selection and Classification*, StudFuzz **207**, 471–488 (2006)
www.springerlink.com

we have been supported by our object oriented data mining technology of the GHOSTMINER[1] system. All the algorithms we have used and describe below are components of the package. Recently, we have developed some new functionality of the system to comply with the needs of feature selection, balanced error minimization etc. Thanks to the general, object oriented philosophy of the tool all the components could be easily combined and validated.

It is worth pointing out that all our computations have been run on personal computers including notebooks – thanks to the algorithms no supercomputers or clusters are necessary to obtain interesting results in data mining.

23.2 Fundamental Algorithms

There is no single model architecture, which could be recommended for all the possible applications. To solve different classification problems we need different kinds of models and different data transformation techniques. The search for the final combined model must be based on a set of fundamental algorithms, possibly of different inherent structures and methodologies.

23.2.1 Classification

In our exploration of the datasets we have tested a variety of methods, which implement different ways of cost function minimization. This broadens the search area in the model space. Final models for the five datasets were based on Support Vector Machines, Normalized Radial Basis Functions and Nearest Neighbors approaches. Apart from these we have also tested SSV decision tree, IncNet (Jankowski and Kadirkamanathan, 1997) and Feature Space Mapping (Adamczak et al., 1997) classification algorithms. The SSV is presented here because it was useful for building the feature selection parts of the models.

A special treatment was necessary in the case of the DOROTHEA dataset (and to a lower extent of ARCENE), because the minimization of the standard classification error or MSE leads to completely different models than the optimization of the balanced error rate. The latter is in fact a special case of the classification error defined by a cost matrix, but not all algorithms support it.

Due to the space limitations we are unable to present the methods in full detail. Please refer to the bibliography for more information on the algorithms of interest.

Support Vector Machines (SVMs)

We have used Support Vector Machines for several reasons. One of them is that SVMs optimize margins between class regions. Another one is that with different kernel functions the SVM changes from simple linear model to a

---

[1]GHOSTMINER is a trademark of FQS Poland

nonlinear one. Yet another reason is that the SVM model may be implemented really effectively and can deal with high-dimensional data.

SVMs were proposed initially by Boser et al. (1992) and thoroughly examined by Vapnik (1995, 1998). They are often very accurate and efficient. The idea is applicable to both data classification and function approximation tasks. The statement of the SVM optimization for classification problems may be the following:

$$\min_{\boldsymbol{w},b,\boldsymbol{\xi}} \quad \frac{1}{2}||\boldsymbol{w}||^2 + C\sum_{i=1}^{m}\xi_i \tag{23.1}$$

with constraints:

$$y_i(\boldsymbol{w}^T\boldsymbol{\phi}_i + b) \geq 1 - \xi_i \tag{23.2}$$

$$\xi_i \geq 0, \quad i = 1,\ldots,m \tag{23.3}$$

where m is the number of vectors, $\mathbf{x}_i$ is the ith data vector and $\boldsymbol{y}_i$ is its class label (1 or -1 – the binary classification). The dual problem definition is:

$$\min_{\boldsymbol{\alpha}} \quad \frac{1}{2}\boldsymbol{\alpha}^T\mathbf{Q}\boldsymbol{\alpha} + \mathbf{1}^T\boldsymbol{\alpha} \tag{23.4}$$

with constraints:

$$01\alpha_i 1C, \quad i = 1,\ldots,m \tag{23.5}$$

$$\mathbf{y}^T\boldsymbol{\alpha} = 0 \tag{23.6}$$

where $C > 0$ defines the upper bound for α_i factors, and $\boldsymbol{Q}$ is a matrix defined by:

$$Q_{ij} = y_i y_j k(\boldsymbol{x}_i, \boldsymbol{x}_j). \tag{23.7}$$

The $k(\cdot)$ function is called a kernel and $k(\boldsymbol{x}_i, \boldsymbol{x}_j) \equiv \boldsymbol{\phi}_i^T\boldsymbol{\phi}_j$.

Most often used kernels are: gaussian, linear, sigmoidal and polynomial. With the exception of the linear kernel all the others have some parameters of free choice. Although in our framework we have implemented all the kernels, we recommend only the most useful ones: linear, Gaussian and exponential inverse of distance. In the simplest case $k(\cdot)$ is defined by:

$$k(\boldsymbol{x}, \boldsymbol{x}') = \boldsymbol{x}^T\boldsymbol{x}'. \tag{23.8}$$

To add a nonlinear behavior, the Gaussian kernel can be used instead:

$$k(\boldsymbol{x}, \boldsymbol{x}') = G(\boldsymbol{x}, \boldsymbol{x}'; \sigma) = \exp(-||\boldsymbol{x} - \boldsymbol{x}'||^2/\sigma). \tag{23.9}$$

Interesting results may also be obtained using the following kernel (exponential inverse of distance) which is quite similar to the Gaussian one:

$$k(\boldsymbol{x}, \boldsymbol{x}') = \exp(-||\boldsymbol{x} - \boldsymbol{x}'||/\sigma). \tag{23.10}$$

The main problem with the original definition of SVM was that its learning procedure, the quadratic programming (QP), converged very slowly. Recent years have brought a few novel methods of acceleration of the QP procedure for SVM learning. The most attention deserve the methods proposed by Osuna et al. (1997b), Joachims (1998), Saunders et al. (1998), Platt (1998, 1999) and Chang et al. (2000). Platt's Sequential Minimal Optimization (SMO) algorithm for the QP problems is very fast and provides an analytical solution. Further improvements to the QP procedure were made by Shevade et al. (2000).

The SMO algorithm augmented by the ideas presented in (Shevade et al., 2000) yields very fast and accurate solution of SVM learning problem. We have used such a version of SVM in our research.

The common point of acceleration of the QP procedure is the **decomposition** of $\boldsymbol{\alpha}$ to a working part $(\boldsymbol{\alpha}_B)$ and a fixed part $(\boldsymbol{\alpha}_R)$:

$$\max_{\boldsymbol{\alpha}_B} \quad W(\boldsymbol{\alpha}_B) = (\mathbf{1} - \boldsymbol{Q}_{BR}\boldsymbol{\alpha}_R)^T \boldsymbol{\alpha}_B - \frac{1}{2}\boldsymbol{\alpha}_B^T \boldsymbol{Q}_{BB}\boldsymbol{\alpha}_B \qquad (23.11)$$

with constraints:

$$0 \le \alpha_{B,i} \le C \quad \forall\, i \in B, \qquad (23.12)$$

$$\mathbf{y}_B^T \boldsymbol{\alpha}_B + \mathbf{y}_R^T \boldsymbol{\alpha}_R = 0, \qquad (23.13)$$

where $\begin{bmatrix} \boldsymbol{Q}_{BB} & \boldsymbol{Q}_{BR} \\ \boldsymbol{Q}_{RB} & \boldsymbol{Q}_{RR} \end{bmatrix}$ is a permutation of matrix $\boldsymbol{Q}$.

The decomposition scheme consists of two steps: selection of $\boldsymbol{\alpha}_B$ and optimization of Eq. 23.11. These two steps are repeated as long as the stop criterion is not satisfied. The SMO selects only two $\boldsymbol{\alpha}$ scalars to put them to $\boldsymbol{\alpha}_B$. This is equivalent to the optimization of two (potential) support vectors in a single optimization step. The $\boldsymbol{\alpha}_B$ selection procedure introduced in (Platt, 1998) was optimized by Shevade et al. (2000). Keerthy proposed to optimize Equation 23.11 for the two indices $(B = \{i, j\})$ which violate the KKT conditions Equation 23.2 and Equation 23.3) the most:

$$i = \text{ arg } \max_t(\ \{-\nabla f(\boldsymbol{\alpha})_t \mid y_t = 1 \wedge \alpha_t < C\}\ \cup \qquad (23.14)$$
$$\{\nabla f(\boldsymbol{\alpha})_t \mid y_t = -1 \wedge \alpha_t > 0\}), \qquad (23.15)$$
$$j = \text{ arg } \min_t(\ \{\nabla f(\boldsymbol{\alpha})_t \mid y_t = -1 \wedge \alpha_t < C\}\ \cup$$
$$\{-\nabla f(\boldsymbol{\alpha})_t \mid y_t = 1 \wedge \alpha_t > 0\})$$

where $f(\boldsymbol{\alpha}) = \frac{1}{2}\boldsymbol{\alpha}^T \boldsymbol{Q}\boldsymbol{\alpha} + \mathbf{1}^T\boldsymbol{\alpha}$, and $\nabla f(\cdot)$ defines the gradient. For details on the stopping criterion see (Shevade et al., 2000).

When B consists of two indices the QP optimization defined by Equation 23.11 may be solved analytically. This was proposed in the SMO algorithm (Platt, 1998).

In the case of unbalanced data (large difference of the numbers of representatives of different classes) Osuna et al. (1997a) proposed to use a separate

C parameter for each class. This changes the goal described by Equation 23.1 to:

$$\min_{\boldsymbol{w},b,\boldsymbol{\xi}} \quad \frac{1}{2}||\boldsymbol{w}||^2 + C^+ \sum_{y_i=1} \xi_i + C^- \sum_{y_i=-1} \xi_i \tag{23.16}$$

A method of automatic selection of C (Equation 23.1) and σ (Equation 23.9) parameters can be found in (Jankowski and Grabczewki, 2003).

Normalized Radial Basis Function (NRBF) Networks

The NRBF is a Radial Basis Function network with normalized Gaussian transfer functions. It resembles the concept of Parzen windows. The number of basis functions in the NRBF is equal to the number of vectors in the training dataset. Each basis function is placed exactly at the place defined by given input vector. The NRBF may be seen as lazy learning algorithm because there are no adaptation parameters.

Let $X = \{\boldsymbol{x}_i : i = 1, \ldots, m\}$ be a set of input patterns and $Y = \{y_i : i = 1, \ldots, m\}$ a set of class labels. The final class label (decision) on unseen vector $\boldsymbol{x}$ is computed as a conditional probability of class c given vector $\boldsymbol{x}$:

$$P(c|\mathbf{x}, X, Y) = \sum_{i \in I^c} k(\mathbf{x}; \mathbf{x}_i), \tag{23.17}$$

where $I^c = \{i : \mathbf{x}_i \in X \wedge y_i \in Y \wedge y_i = c\}$ and

$$k(\mathbf{x}; \mathbf{x}_i) = \frac{G(\mathbf{x}, \mathbf{x}_i; \sigma)}{\sum_{j=1}^m G(\mathbf{x}, \mathbf{x}_j; \sigma)}, \tag{23.18}$$

where $G(\mathbf{x}, \mathbf{x}_i; \sigma)$ is the Gaussian kernel (Equation 23.9) with σ parameter. It can be seen that $\sum_{i=1}^K P(i|\mathbf{x}, X, Y) = 1$, where K is the number of classes.

The behavior of NRBF is similar (but not equivalent) to the k nearest neighbors model (see Section 23.2.1) – the classification decision of given vector $\mathbf{x}$ depends on the neighborhood region of $\mathbf{x}$ (on which basis function is the nearest). The biggest difference is that the NRBF decision ($P(c|\mathbf{x}, X, Y)$) changes continuously while for kNN it is discrete.

If the training dataset consists of a large number of vectors the Learning Vector Quantization (Kohonen, 1986) or prototype selection methods (Grochowski and Jankowski, 2004) can be used to reduce the number of vectors appropriately.

k Nearest Neighbors (kNN)

k Nearest Neighbors models were proposed by Cover and Hart (1967) and are designed to classify unseen vectors on the basis of the class labels observed for neighboring reference vectors (typically the training set vectors). The kNN is parameterized by k, the number of nearest neighbors considered during classification. The winner class for a given vector $\mathbf{x}$ may be defined as the majority class within the set $NN(\mathbf{x}; k)$ of its k nearest neighbors.

Typically k is chosen manually. Sub-optimal value of k may be estimated quite effectively via cross-validation based learning – since each fold may estimate a different optimum for k, the sub-optimal value may be estimated by the k for which the average test accuracy (counted for the submodels of the CV based learning) is maximal.

The set $NN(\mathbf{x}; k)$ of *nearest neighbors* depends on the measure used to compute distances between $\mathbf{x}$ and the reference vectors. In most cases the Euclidean measure is used. The Euclidean measure can be simply generalized to the Minkovsky measure:

$$D_M^\alpha(\mathbf{x}, \mathbf{x}') = \sqrt[\alpha]{\sum_{i=1}^{n} |x_i - x_i'|^\alpha} \qquad (23.19)$$

The Euclidean metric corresponds to $\alpha = 2$, which is completely isotropic, and Manhattan metric to $\alpha = 1$.

Sometimes good results can be obtained using the Canberra measure:

$$D_{Ca}(\mathbf{x}, \mathbf{x}') = \sum_{i=1}^{n} \frac{|x_i - x_i'|}{|x_i + x_i'|}. \qquad (23.20)$$

The Chebychev function corresponds to the infinite Minkovsky exponent:

$$D_{Ch}(\mathbf{x}, \mathbf{x}') = \max_{i=1,\dots,n} |x_i - x_i'|. \qquad (23.21)$$

Please note that in the case of symbolic attributes a special metric or a data transformation (see (Grabczewski and Jankowski, 2003)) should be used.

SSV tree

The Separability of Split Value (SSV) criterion is one of the most efficient heuristic used for decision tree construction (Grabczewski and Duch, 1999, 2000). Its basic advantage is that it can be applied to both continuous and discrete features in such a manner that the estimates of *separability* can be compared regardless the substantial difference in types.

The *split* value (or *cut-off point*) is defined differently for continuous and symbolic features. For continuous features it is a real number and for symbolic ones it is a subset of the set of alternative values of the feature. The *left side* (LS) and *right side* (RS) of a split value s of feature f for a given dataset D is defined as:

$$\mathsf{LS}(s, f, D) = \begin{cases} \{x \in D : f(x) < s\} & \text{if } f \text{ is continuous} \\ \{x \in D : f(x) \notin s\} & \text{otherwise} \end{cases} \quad (23.22)$$

$$\mathsf{RS}(s, f, D) = D - \mathsf{LS}(s, f, D)$$

where $f(x)$ is the f's feature value for the data vector x. The definition of the *separability of split value s* is:

$$\mathsf{SSV}(s, f, D) = 2 * \sum_{c \in C} |\mathsf{LS}(s, f, D_c)| * |\mathsf{RS}(s, f, D - D_c)|$$
$$- \sum_{c \in C} \min(|\mathsf{LS}(s, f, D_c)|, |\mathsf{RS}(s, f, D_c)|) \quad (23.23)$$

where C is the set of classes and D_c is the set of data vectors from D assigned to class $c \in C$.

Among all the split values which separate the maximum number of pairs of vectors from different classes the most preferred is the one that separates the smallest number of pairs of vectors belonging to the same class. For every dataset containing vectors, which belong to at least two different classes, for each feature represented in the data by at least two different values, there exists a non-trivial split value with maximum separability. When the feature being examined is continuous and there are several different split values of maximum separability, close to each other, the split value closest to their average is selected. To avoid such ties and to eliminate unnecessary computations, the analysis should be restricted to the split values that are natural for the given dataset (i.e. centered between adjacent feature values that occur in the data vectors). If there are non-maximal (regarding separability) split values between two maxima or if the feature is discrete, then the average is not a reasonable choice – the winner split value should be selected randomly from those of maximum separability.

Decision trees are constructed recursively by searching for best splits among all the splits for all the features. At each stage when the best split is found and the subsets of data resulting from the split are not completely pure (i.e. contain data belonging to more than one class) each of the subsets is analyzed in the same way as the whole data. The decision tree built this way gives maximum possible accuracy (100% if there are no contradictory examples in the data), which usually means that the created model overfits the data. To remedy this a cross validation training is performed to find the optimal parameters for pruning the tree. The optimal pruning produces a tree capable of good generalization of the patterns used in the tree construction process.

Like most decision tree algorithms the SSV based method is independent on the scaling of the feature values, so in particular it is normalization and standardization invariant. The decision borders are perpendicular to the feature space axes and can be described by logical formulae, however in some cases it is a restriction, which limits accuracy. Nevertheless its ability to find informative features can be helpful in feature selection for other classification methods.

The SSV criterion has been successfully used not only for building classification trees, but also for feature selection (Duch et al., 2002, 2003) and data type conversion (from continuous to discrete and in the opposite direction (Grabczewski and Jankowski, 2003)).

23.2.2 Feature Extraction

Providing classifiers with feature spaces, which help obtaining the best possible accuracy is a very complex task. Feature selection and construction play a very important role here. There is no single recipe for good data transformation and no unarguable method to compare the performance of different feature selection algorithms. Moreover each classifier may require different data preparation. To obtain an accurate and stable final model with a particular classifier one must validate a number of data preparation methods.

The term *feature extraction* encompasses both selection and construction of features. Thorough analysis includes testing filters (which are independent on the classifier) and wrappers (which use external classifiers to estimate feature importance). Feature selection strategies either produce a ranking (each feature is assessed separately) or perform *full-featured* selection (select/deselect with respect to the interaction between the features).

CC Based Feature Ranking

The correlation coefficient (CC) is a simple but very robust tool in statistics. It is very helpful also in the task of feature selection. For two random variables X and Y it is defined as

$$\varrho(X,Y) = \frac{E(XY) - E(X)E(Y)}{\sqrt{D^2(X)D^2(Y)}}, \qquad (23.24)$$

where E and D^2 stand for the expected value and variance respectively. $\varrho(X,Y)$ is equal to 0 if X and Y are independent and is equal to 1 when the variables are linearly dependent ($Y = aX + b$).

The correlation coefficient calculated for a feature (treated as a random variable) and the class labels (in fact the integer codes of the labels) is a good measure of feature usefulness for the purpose of classification. The feature list ordered by decreasing absolute values of the CC may serve as feature ranking.

SSV Based Feature Selection

Decision tree algorithms are known to have the ability of detecting the features that are important for classification. Feature selection is inherent there, so they do not need any feature selection at the data preparation phase. Inversely: their capabilities can be used for feature selection.

Feature selection based on the SSV criterion can be designed in different ways. The most efficient (from the computational point of view) one is to create **feature ranking** on the basis of the maximum SSV criterion values calculated for each of the features and for the whole training dataset. The cost is the same as when creating decision stubs (single-split decision trees).

Another way is to create a **single** decision **tree** and *read* feature importance from it. The filter we have used for this type of SSV based feature selection is the algorithm 12.

Feature selection filter based on the SSV criterion

▶ **Input:** A sample X of input patterns and their labels Y (training data)
◀ **Output:** List of features ordered by decreasing importance.

- $T \leftarrow$ the SSV decision tree built for $\langle X, Y \rangle$.
- For each non-final (i.e. which is not a leaf) node N of T,
  $G(N) \leftarrow E(N) - E(N_1) - E(N_2)$, where N_1 and N_2 are the subnodes of N, and $E(N)$ is the number of vectors in X falling into N but incorrectly classified by N.
- $\mathcal{F} \leftarrow$ the set of all the features of the input space.
- $i \leftarrow 0$
- While $\mathcal{F} \neq \emptyset$ do:
  - For each feature $f \in \mathcal{F}$ not used by T define its rank $\mathcal{R}(f) \leftarrow i$. Remove these features from $\mathcal{F}$.
  - Prune T by deleting all the final splits of nodes N for which $G(N)$ is minimal.
  - Prune T by deleting all the final splits of nodes N for which $G(N) = 0$.
  - $i \leftarrow i + 1$
- The result is the list of features in decreasing order of $\mathcal{R}(f)$.

This implements a full-featured filter – the decision tree building algorithm selects the splits locally, i.e. with respect to the splits selected in earlier stages, so that the features occurring in the tree, are complementary. The selection can be done by dropping all the features of rank equal to 0 or by picking a given number of top ranked features.

In some cases the full classification trees use only a small part of the features. It does not allow to select any number of features – the maximum is the number of features used by the tree. To remedy this the Sequential Feature Selection technique (described below) can be used.

The SSV criterion is defined to reflect class separability and has no parameters to adjust it to standard or balanced classification error. Thus we have also used the SSV framework to construct trees with balanced classification error as split eligibility criterion. It was especially useful for the exploration of the DOROTHEA dataset.

Feature Selection wrapper

Wrapper methods use external algorithms and search techniques to determine the best (from the point of view of some particular criterion) values of some parameters. The technique may also be helpful in feature selection. A wrapper method available in the GHOSTMINER package simply adds the features one by one to some initial set of (base) features and estimates the performance of a classifier in so extended feature spaces. If the initial set of features is empty then the classifier is trained on one-dimensional datasets, and the results for all the features are collected – the feature ranking is built according to the accuracies. When started with some base features the method searches for additional features, which extending the preselected set yield a satisfactory improvement in submodel accuracy.

The basic advantage of the wrapper method of feature selection is that its applications are dedicated to some particular models. The major drawback of wrappers is that they require multiple training of their submodels.

Feature Selection Committee

The task of feature selection committees is to combine different feature selection methods and select features which seem attractive from different points of view. Several feature selection models are constructed independently and their selection results collected. The committee selects the features most often selected by its members. If we assign the value of 1 to each selected feature and 0 to not selected then we may sum up the scores obtained from the committee members to get an integer value for each of the features. The committee scores are integer values in the range of 0 to the number of committee members. Setting a threshold value for the scores gives a criterion of final selection. The threshold equal to the committee size selects the features selected by each of the committee members while the value of 1 results in a weak rejection committee. The two border values correspond respectively to the intersection and the sum of the sets of features determined by the members.

Sequential Feature Selection

Full-featured filters select the features providing different (complementary) information about the classification task. They are likely to reject informative features, which although valuable do not introduce anything new to already selected ones. If we want neither simple rankings, which do not reflect features dependencies nor filters that deselect informative but not independent features, the sequential feature selection technique may be of interest. The idea is to select just a number of top ranked features and repeat filtering in the feature space reduced by the selected features. The parameters of this method are the filter algorithm, the number of filter runs and the number of features to select after each step.

The method is helpful in the case of full-featured filters, especially like the ones based on decision trees, which in some circumstances can select only a small number of features. Running them repetitively facilitates selection of any number of features.

Information Theory Based Filters

There is a number of feature filters based on information theory. Unfortunately they usually suffer from the necessity of data discretization by external methods. The equal-width and equal-frequency discretizations are not very robust. Much more interesting results can be obtained with SSV based discretization (Duch et al., 2003, Grabczewski, 2004) but in most cases they are not better than those of SSV feature selection while being more computationally expensive. Some successful methods which employ information gain or mutual information were tried by us on the NIPS FSC datasets. The results were very similar to those obtained with CC or SSV based feature selection. The lack of information theory models inside GHOSTMINER significantly reduced our validation possibilities for these models – this is the major reason why we have not used the methods in our final models.

PCA

The Principal Components Analysis is a well known technique of data transformation. In its standard formulation it finds linear combinations of features which show the directions of largest variance of the data. Viewing the data in two dimensional plots, where the axes are the first and the second principal components is often very informative. Using the largest variance directions as features may significantly lower dimensionality of the space (where most classification models are more effective) without a substantial reduction of information. This feature extraction method constructs valuable features, but it can not be treated as a feature selection technique because all the feature values are still necessary to calculate the new features.

23.3 Fully Operational Complex Models

It can be seen in Section 31.4 that single classification algorithms are often not sufficient to solve a given problem with high accuracy and confidence. It is much more successful to examine different combinations of data transformation and classification models presented in the previous sections (23.2.1

and 23.2.2). Sometimes it is recommended to use even more than one transformation before classification (compare section 23.4.3), but searching for a suitable model sequence and configuration is far from trivial. Sometimes, default parameters (commonly used as starting configuration) are completely inadequate for a specific dataset (compare section 23.4.4). Also, a combination of transformation and classifier useful for one dataset may be useless for another dataset.

It is recommended to search for proper parameters of transformations and classifiers with a meta-learning. Meta-learning should perform internal validation of meta-parameters (the parameters the values of which are searched for), otherwise the learning process is very likely to end up with an overfitted model.

The task of searching for complex models can be viewed as a number of subproblems: testing of base classifiers, testing of feature selection algorithms in the context of their suitability for base classifiers, parameters tuning for models, which seem promising solutions, branching of promising solutions (substitution of parts in the complex models), further search for configuration of complex models, testing of feature selection committees and classification committees. The whole process must switch from one subproblem to another with repetitions and backtracking.

Reliable validation is extremely important in the case of complex models (combinations of transformations and classifiers). If a transformation is supervised (dependent on the class labels) then it is insufficient to validate the classifier – instead, the cross-validation (or other random validation procedure) should run over the whole combination of transformation and classifier. Otherwise the prediction of accuracy and their variance is overoptimistic and falsifies the real generalization possibility. In the case of unsupervised transformations (like PCA, standardization or selection of high variance features), they may be applied before the validation, however if a combination of supervised and unsupervised transformations is used to prepare data for a classification model, then the combination is supervised and as such, must be nested in the validation.

The cross-validation test may be easily changed into a cross-validation committee. It means that all the models built in the cross-validation test can compose a committee. The final decision of the CV committee are based on the voting scheme. To reduce the probability of impasse, it is recommended to use an odd number of CV folds (especially in the case of two–class problems). CV committees have several important advantages: the first is that committee decisions are more stable than those of single models, the second is that the estimated accuracy and variance of the submodels are known directly from the CV, another one is that they avoid the problem of configuration parameters, which although validated may be suitable only for a particular dataset size (the numbers of features, vectors, etc.) and applied to the whole training dataset may produce less successful models.

Table 23.1. NIPS 2003 challenge results.

Dec. 8th	Our best challenge entry					The winning challenge entry					
Dataset	Score	BER	AUC	Feat[2]	Probe[2]	Score	BER	AUC	Feat	Probe	Test
OVERALL	37.14	7.98	92.11	26.8		—	71.43	6.48	97.20	80.3	47.8 0.8
ARCENE	14.29	13.53	86.47	75		—	94.29	11.86	95.47	10.7	1.0 1
DEXTER	71.43	3.50	96.50	40		—	100.00	3.30	96.70	18.6	42.1 0
DOROTHEA	17.14	13.11	86.89	2		—	97.14	8.61	95.92	100.0	50.0 1
GISETTE	57.14	1.31	98.69	14		—	97.14	1.35	98.71	18.3	0.0 0
MADELON	65.71	7.44	92.56	3	0.0	94.29	7.11	96.95	1.6	0.0	1

If for a given dataset we had a number of interesting (complex) models then we would choose the best one according to the following criterion:

$$\text{best-model} = \arg\max_{M} \left[accuracy(M) - \alpha \cdot standard\text{-}deviation(M) \right] \quad (23.25)$$

with the values of α close to 1. The aim is to prefer not only accurate but also confident models.

23.4 Challenge Data Exploration

The challenge datasets differ in many aspects (their size – the number of features and vectors, the source, the representation type, etc.). As a result, the final classification models are also significantly different.

Below, an overview of the most interesting models for the challenge datasets is presented. For each dataset there is a table depicting the structure of our best model and its error rate calculated by the challenge organizers for the test part of the dataset.

The models were submitted to the second part of the contest (December 8th). We assessed the second stage as the most important. Thus, for real, we did not take part in the stage of December 1st. In the final contest we reached the **group rank of 3** and **best entry rank of 7**. Table 23.1 presents a summary of the results.

23.4.1 ARCENE (spectroscopic Data)

ARCENE and DOROTHEA are characterized by high quotient of the number of attributes and the number of input vectors (≈ 100). In such spaces looking for accurate and certain classification is a very hard problem. If a supervised

---

[2]We did not submit the lists of selected features to the contest (except for MADELON). Hence the fractions of features differ from the ones calculated by the organizers and some probe information is missing.

preprocessing is used and then the validation (such as the CV test) performed, the classification accuracy estimates are overoptimistic – real accuracy on unseen data is dramatically higher (the generalization is very poor).

The best model we have found is a CV Committee of combinations of SSV based feature selection and the SVM classifier with linear kernel and class balance. The CV test error was $7.8\% \pm 5\%$.

| CV Committee 9-fold [SSV 7000 → SVM linear] ‖ Test error: 13.5% |

The training set was standardized before feature selection. Quite similar results can be obtained with the correlation coefficient based feature selection. The use of CV Committee increases the CV test accuracy by 1-1.5% It was a surprise, that using SVM with linear kernel and no feature selection the CV test accuracy was only 2% lower (than the best score) with slightly higher variance.

The CV test accuracy goes down (quite monotonically) when SSV or correlation coefficient selection is used to extract less than 7000 features.

It can be observed that SVMs with linear kernel work very well in high-dimensional spaces while with gaussian kernel rather do not.

23.4.2 DEXTER (Corporate Acquisitions)

In the case of DEXTER the first step of the analysis was to remove the zero–variance features. After that the number of features reduced from 20 000 to 11 035.

The DEXTER dataset is sparse. It is important to treat undefined values as zeros, not like a missing value – otherwise the classification is much more difficult. Data standardization also makes the task harder, so we have used the original data. Alternatively, a standardization using the mean and the standard deviation over the whole training set (not *per* feature) can be used. The best model we have found, consists of the correlation coefficient feature selector and SVM with linear kernel. The CV test error was around $4.8\%\pm2\%$.

| CC 8000 → SVM linear ‖ Test error: 3.5% |

Very similar results were obtained using CV Committee of the above combinations (2 more vectors misclassified on the test set). Another model with very similar certainty was a CV Committee of the same combinations but with 5000 features selected.

23.4.3 DOROTHEA (which Compounds Bind to Thrombin)

The DOROTHEA dataset is binary and strongly sparse. As it was already mentioned, it has (over) 100 times more features than vectors.

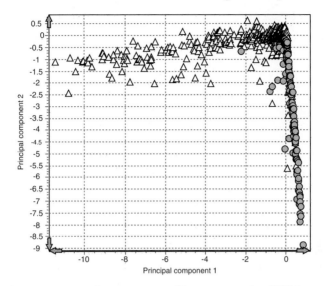

Fig. 23.1. First two principal components of DOROTHEA after SSV feature selection.

In the first step unsupervised feature selection was used. A given feature was selected only if it had a sufficient variance (high variance selection – HVS): in practice, more than p 1s per feature were required. There are no features with high number of 1s in the training data, so the values of the p parameter we have used are: 8, 10 and 11.

After the preprocessing the best combinations of models use both supervised and unsupervised data transformations before final classification. One of the best models starts with selection of features with high variance, next the SSV selects 2000 features (respecting the class balance), then two first principal components are extracted (see Fig. 23.1), and finally the SVM classifier with gaussian kernel is used (also respecting the class balance). The CV test of this model estimated the error of $12.3\% \pm 4.4\%$.

HVS p=11 → SSV 2000+balance → PCA 2 → SVM Gaussian C=50	Test error: 13.1%

Similar results can be obtained with $p = 8$ for the high variance selection or with SSV based selection of 1500 features.

23.4.4 GISETTE (Handwritten Digits)

The GISETTE dataset has nearly balanced numbers of instances and attributes. The major problem of this data was not only to find a proper combination of models but also to tune the parameters of the CC selector and

SVM classifier. Our best model uses 700 features, and gives the CV test error of $1.5\% \pm 0.5\%$:

| CC 700 → SVM Gauss C=1000 bias=0.002 ‖ Test error: 1.31% |

Another interesting model is the combination of correlation coefficient based feature selector (with just 200 features) and kNN with the number of neighbors (k) equal to 5. The CV test error of such configuration is 3.6%, and the standard deviation is smaller than 0.6%. When the correlation coefficient selector is used to select 50 features the CV test error increases to 5%.

23.4.5 MADELON (Random Data)

MADELON is a dataset of its own kind. In comparison to ARCENE and DOROTHEA, it has a small quotient of the numbers of features and instances.

Atypically, to select relevant features the feature selection committee was used. The committee members were a SSV ranking, a SSV single tree (ST) selection and a correlation coefficient selector.

| Selection Committee [SSV, SSV ST, Correlation coefficient] → NRBF ‖ Test error: 7.44% |

The CV test error of the above combination of models is $9\% \pm 0.5\%$ (very stable model). Although the selection committee looks complicated, it selects just 15 features. The validation techniques showed that both selecting more features and reducing the number of features led to a decrease of the accuracy on unseen data. In the cases of MADELON and GISETTE the subtask of finding an accurate and stable classifier was much harder than the feature selection stage.

Slightly worse results can be obtained with kNN model instead of NRBF – the CV test accuracy reduction is close to 1%.

23.5 Conclusions

The challenge efforts of our group brought a number of conclusions which in general are compatible with our earlier experience, however some aspects deserve to be emphasized and some are a surprise:

- It has been confirmed in practice that there is no single architecture (neither learning algorithm nor model combination scheme) of best performance for all the tasks. Although the SVM method was used most often in the final models, its internal structure was not the same each time – some models were based on linear and some on gaussian kernels.

- The models created must be properly validated – all the supervised data preprocessing (like most feature selection methods) must be included in a complex model validated with a CV or similar technique. Otherwise there is a serious danger of overfitting the training data.
- It is advantageous to build committees of models, which proved their generalization abilities in a CV test.
- It is surprising that a feature ranking method as simple as the one based on the correlation coefficient is a very valuable component of complex models.

There is still a lot of work to be done in the area of feature selection and building efficient model combinations. The problem is NP-complete, and the needs are growing due to the bioinformatics and text mining applications. Among other things we should: look for stronger and still effective feature selection algorithms; construct powerful aggregation or aggregation–selection algorithms to help the classifiers by supplying more informative features; develop intelligent meta-learning techniques which could help in automating the search for adequate complex models.

References

R. Adamczak, W. Duch, and N. Jankowski. New developments in the feature space mapping model. In *Third Conference on Neural Networks and Their Applications*, pages 65–70, Kule, Poland, 1997. Polish Neural Networks Society.

B.E. Boser, I. Guyon, and V. Vapnik. A training algorithm for optimal margin classifiers. In *Fifth Annual Workshop on Computational Learning Theory*, pages 144–152. ACM, 1992.

C.-C. Chang, C.-W. Hsu, and C.-J. Lin. The analysis of decomposition methods for support vector machines. *IEEE Transaction on Neural Networks*, 4:1003–1008, 2000.

T.M. Cover and P.E. Hart. Nearest neighbor pattern classification. *IEEE Transactions on Information Theory*, 13(1):21–27, 1967.

W. Duch, J. Biesiada, T. Winiarski, K. Grudziński, and K. Grabczewski. Feature selection based on information theory filters and feature elimination wrapper methods. In *Proceedings of the International Conference on Neural Networks and Soft Computing (ICNNSC 2002)*, Advances in Soft Computing, pages 173–176, Zakopane, 2002. Physica-Verlag (Springer).

W. Duch, T. Winiarski, J. Biesiada, and A. Kachel. Feature selection and ranking filters. In *Artificial Neural Networks and Neural Information Processing – ICANN/ICONIP 2003*, pages 251–254, Istanbul, 2003.

K. Grabczewski. SSV criterion based discretization for Naive Bayes classifiers. In *Proceedings of the 7th International Conference on Artificial Intelligence and Soft Computing*, Zakopane, Poland, June 2004.

K. Grabczewski and W. Duch. A general purpose separability criterion for classification systems. In *Proceedings of the 4th Conference on Neural Networks and Their Applications*, pages 203–208, Zakopane, Poland, June 1999.

K. Grabczewski and W. Duch. The Separability of Split Value criterion. In *Proceedings of the 5th Conference on Neural Networks and Their Applications*, pages 201–208, Zakopane, Poland, June 2000.

K. Grabczewski and N. Jankowski. Transformations of symbolic data for continuous data oriented models. In *Artificial Neural Networks and Neural Information Processing – ICANN/ICONIP 2003*, pages 359–366. Springer, 2003.

M. Grochowski and N. Jankowski. Comparison of instances seletion algorithms II: Algorithms survey. In *Artificial Intelligence and Soft Computing*, pages 598–603, 2004.

N. Jankowski and K. Grabczewki. Toward optimal SVM. In *The Third IASTED International Conference on Artificial Intelligence and Applications*, pages 451–456. ACTA Press, 2003.

N. Jankowski and V. Kadirkamanathan. Statistical control of RBF-like networks for classification. In *7th International Conference on Artificial Neural Networks*, pages 385–390, Lausanne, Switzerland, October 1997. Springer-Verlag.

T. Joachims. *Advances in kernel methods — support vector learning*, chapter Making large-scale SVM learning practical. MIT Press, Cambridge, MA, 1998.

T. Kohonen. Learning vector quantization for pattern recognition. Technical Report TKK-F-A601, Helsinki University of Technology, Espoo, Finland, 1986.

E. Osuna, R. Freund, and F. Girosi. Support vector machines: Training and applications. AI Memo 1602, Massachusetts Institute of Technology, 1997a.

E. Osuna, R. Freund, and F. Girosi. Training support vector machines: An application to face detection. In *CVPR'97*, pages 130–136, New York, NY, 1997b. IEEE.

J. C. Platt. Fast training of support vector machines using sequential minimal optimization. In B. Schölkopf, C. J. C. Burges, and A. J. Smola, editors, *Advances in Kernel Methods - Support Vector Learning*. MIT Press, Cambridge, MA., 1998.

J.C. Platt. Using analytic QP and sparseness to speed training of support vector machines. In M.S. Kearns, S.A. Solla, and D.A. Cohn, editors, *Advances in Neural Information Processing Systems*, volume 11, 1999.

C. Saunders, M.O. Stitson, J. Weston, L. Bottou, B. Schoelkopf, and A. Smola. Support vector machine reference manual. Technical Report CSD-TR-98-03, Royal Holloway, University of London, Egham, UK, 1998.

S.K. Shevade, S.S. Keerthi, C. Bhattacharyya, and K.R.K. Murthy. Improvements to the SMO algorithm for SVM regression. *IEEE Transactions on Neural Networks*, 11:1188–1194, Sept. 2000.

V. Vapnik. *The Nature of Statistical Learning Theory*. Springer-Verlag, New York, 1995.

V. Vapnik. *Statistical Learning Theory*. Wiley, New York, NY, 1998.

Chapter 24

Combining Information-Based Supervised and Unsupervised Feature Selection

Sang-Kyun Lee, Seung-Joon Yi, and Byoung-Tak Zhang

Biointelligence Laboratory
School of Computer Science and Engineering
Seoul National University
Seoul 151-742, Korea
sklee@bi.snu.ac.kr,sjlee@bi.snu.ac.kr,btzhang@bi.snu.ac.kr

The filter is a simple and practical method for feature selection, but it can introduce biases resulting in decreased prediction performance. We propose an enhanced filter method that exploits features from two information-based filtering steps: supervised and unsupervised. By combining the features in these steps we attempt to reduce biases caused by misleading causal relations induced in the supervised selection procedure. When tested with the five datasets given at the NIPS 2003 Feature Extraction Workshop, our approach attained a significant performance, considering the simplicity of the approach. We expect the combined information-based method to be a promising substitute for classical filter methods.

24.1 Introduction

Recent pattern classification studies such as DNA microarray analysis or text mining tend to deal with larger data and it becomes more difficult to manage the high dimensionality of the data. Feature selection can help us improve classification performance or save computational resources by dimension reduction. It also helps us understand the intrinsic properties of the data by extracting meaningful attributes.

The *filter* method is a practical feature selection technique that chooses a subset of highly ranked features with respect to certain scoring criteria (see Chapter 3 for more details). This method is frequently used because of its computational efficiency and statistical robustness against overfitting (Hastie et al., 2001, Guyon and Elisseeff, 2003). However, this method can result in biased features that have negative effects on prediction performance, especially when we try to find the features only informative to the class variable. Our primary objective is to enhance classical supervised filter methods by reducing the effect of such biases.

S.-K. Lee et al.: *Combining Information-Based Supervised and Unsupervised Feature Selection*,
StudFuzz **207**, 489–498 (2006)
www.springerlink.com

In an information theoretical framework, we analyze the dependency between features and the biases caused by misleading causal relations estimated in the supervised feature selection procedure. Based on this analysis, we devise a simple heuristic method to avoid such biases. Finally, we test our method on the five NIPS 2003 Feature Extraction benchmark datasets.

24.2 Methods

The task of feature selection is to find a subset ϕ_M of a set of features ϕ such that $Pr(Y|\phi_M)$ and $Pr(Y|\phi)$ are as close as possible, for the class label Y (Koller and Sahami, 1996). If we find such a subset, the *optimal* feature set, we can use it in place of the whole set of features without losing significant information. We divide the selection procedure into two steps, supervised and unsupervised. In the supervised step we search for the optimal subset, while in the unsupervised step we try to find additional features that are relevant but not selected in the supervised step because of the biases.

24.2.1 Supervised Feature Selection

We can formalize the definition of our optimal feature set by employing the concept of the *Markov blanket*.

Definition 1. *A Markov blanket $\phi_{M(Y)}$ for the class label Y is a set of features, which makes Y be conditionally independent of $\phi - \phi_{M(Y)}$, given $\phi_{M(Y)}$.*

The Markov blanket $\phi_{M(Y)}$ subsumes all the information of the rest of the features with respect to Y, and thus is the optimal subset. To find $\phi_{M(Y)}$ we require some measures to estimate the relevance and the dependence relations of ϕ and Y. We use the idea of *conditional mutual information* (CMI) which is defined as follows (Cover and Thomas, 1991):

$$I(P;Q|R) \triangleq \sum_{p \in P} \sum_{q \in Q} \sum_{r \in R} Pr(p,q,r) \log \frac{Pr(p,q|r)}{Pr(p|r)Pr(q|r)} \qquad (24.1)$$

Conditional mutual information is a measure of the amount of information one variable P contains about another variable Q, given R (to see a more detailed introduction, refer to Chapter 6). Let ϕ_i and ϕ_j be features such that $\phi_i, \phi_j \in \phi$. Based on the CMI we define two metrics, the *significance* and *dependence* as shown in Table 24.1.

Unfortunately, it is often computationally intractable to find the exact Markov blanket of Y. Therefore, instead of finding the exact $\phi_{M(Y)}$, we try to select the features that (1) have considerable information on Y, but (2) are independent of the other features – the two properties of the features of $\phi_{M(Y)}$. Our two metrics are suitable for this purpose: the significance $I(\phi_i; Y|\phi_j)$

Table 24.1. Two metrics for supervised feature selection

Metric	Definition
Significance	$I(\phi_i; Y \vert \phi_j)$
Dependence	$I(\phi_i; \phi_j \vert Y)$

indicates how much information ϕ_i has about Y given ϕ_j; the dependence $I(\phi_i; \phi_j \vert Y)$ evaluates how much information is shared between ϕ_i and ϕ_j given Y, i.e. how they are dependent.

We devise a heuristic multi-objective optimization algorithm to determine a subset of features with high significance and low dependence (see Fig. 24.1). This algorithm first calculates the significance and dependence of each pair of features. Next, it takes out the most significant N_B features and stores them in a bucket B, and then restores the most dependent $N_B/2$ features from the bucket to the original feature pool. It then outputs the features remaining in the bucket and repeats the process until the predefined N_ϕ features are selected.

Procedure SupervisedSelection(ϕ, N_ϕ, N_B)

   ϕ: Set of features

   N_ϕ: Maximum number of features to be selected

   N_B: Size of the bucket

STEP 1. Compute the significance and dependence matrices.

STEP 2. Initialize a bucket B of size N_B containing the features.

STEP 3. // Extract those features with high significance and low dependence //
   Do,
      Sort the significance values in decreasing order.
      Take out the N_B most significant values from ϕ and store in B.
      Restore the $N_B/2$, the most dependent features from B to ϕ.
      Output features in B.
   Until (N_ϕ features are selected).

Fig. 24.1. Outline of the multi-objective optimization algorithm

Note that to satisfy both objectives, significance and dependence, the size of N_B should be carefully chosen ($N_B \geq 2$). For the experiment, we set $N_B = 4$ and the desired maximum number of features N_ϕ to the 50 percent of the total number features in each dataset, as these parameter values showed the best average prediction performance[1]. Moreover, if ϕ_i has high significance values in terms of many other features ϕ_k $(k = 1, \ldots, N)$, the feature ϕ_i can be selected repeatedly in our procedure. Therefore the resulting subset may contain fewer features than specified.

Also note that given $I(\phi_i; Y) = I(\phi_j; Y)$, the value of the significance $I(\phi_i; Y|\phi_p)$ can be higher than $I(\phi_j; Y|\phi_q)$ when there exists a feature ϕ_p which is independent of ϕ_i more than ϕ_q is of ϕ_j. Because we use a forward selection strategy, we can choose features more effectively using this property assuming that there exists a number of sufficiently independent features of ϕ_i and ϕ_j (see Fig. 24.2).

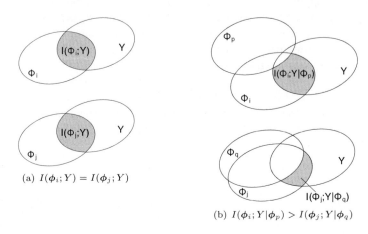

(a) $I(\phi_i; Y) = I(\phi_j; Y)$

(b) $I(\phi_i; Y|\phi_p) > I(\phi_j; Y|\phi_q)$

Fig. 24.2. Conditioning can improve the efficiency of the selection procedure. (a) Let $I(\phi_i; Y) = I(\phi_j; Y)$. Considering only the relevance to Y, we may choose both features ϕ_i and ϕ_j as they are equally favorable. (b) However, let ϕ_p and ϕ_q be the most independent features of ϕ_i and ϕ_j, respectively. If $I(\phi_i; Y|\phi_p) > I(\phi_j; Y|\phi_q)$, the feature ϕ_i is more favorable than ϕ_j because ϕ_i has a more independent buddy ϕ_p and therefore has more chance of being selected when we apply the second criterion, independence.

24.2.2 Unsupervised Feature Selection

One fundamental assumption of the supervised step is that the employed metrics correctly represent the information on the dependence relationship of the variables. This is one of the common assumptions in supervised feature selection approaches using a sort of metrics to estimate the underlying probabilistic distribution. We call this condition *faithfulness* (Spirtes et al., 1993).

---

[1] We also tried $N_B = 2, 4, 8, \ldots, 64$ and $N_\phi = 10, 20, 30, 40, 50$ percent of the total number of features.

Definition 2. *Let G be a causal graph and P a probabilistic distribution implied by G. Then G and P are faithful to one another if and only if every independence relation in P is entailed by the independence relation in G.*

Unfortunately, even if the data are generated faithfully from a model, it is not guaranteed that we can reconstruct the original structure of the model from the data. More specifically, the metrics used to estimate the probability distribution can mislead us to a false image of the original causal relationship (Pearl, 1988, Spirtes et al., 1993, Shipley, 2000). There are two such case:

- **Case 1.** $I(\phi_i; \phi_j | Y) \approx 0$, but $I(\phi_i; \phi_j) \gg 0$.
  In the supervised step, we may select both ϕ_i and ϕ_j as they are independent to each other, given that other conditions are satisfactory (Fig. 3(a)). However, we might have chosen either ϕ_i or ϕ_j, not both, when $I(\phi_i; \phi_j)$ is relatively high (Fig. 3(c)).

- **Case 2.** $I(\phi_i; \phi_j | Y) \gg 0$, but $I(\phi_i; \phi_j) \approx 0$.
  The opposite case. We may not select either ϕ_i or ϕ_j in the supervised step as they are dependent (Fig. 3(c)). However, we might have chosen both features (Fig. 3(d)).

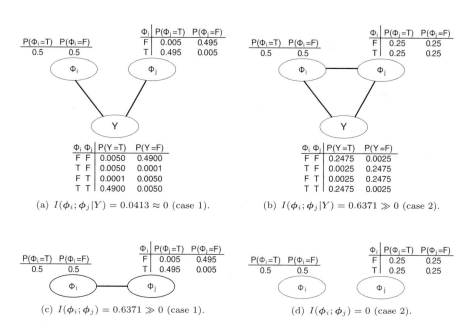

(a) $I(\phi_i; \phi_j | Y) = 0.0413 \approx 0$ (case 1).

(b) $I(\phi_i; \phi_j | Y) = 0.6371 \gg 0$ (case 2).

(c) $I(\phi_i; \phi_j) = 0.6371 \gg 0$ (case 1).

(d) $I(\phi_i; \phi_j) = 0$ (case 2).

Fig. 24.3. Examples of misleading causal relations.

494   Sang-Kyun Lee, Seung-Joon Yi, and Byoung-Tak Zhang

Therefore, we should also consider the value of $I(\phi_i; \phi_j)$ to determine the causal relationship of features. One possible way to perform this task is to calculate $I(\phi_i; \phi_j)$ during the supervised step; but we do not know whether $I(\phi_i; \phi_j|Y)$ or $I(\phi_i; \phi_j)$ shows the true dependence information. We simplify the task by separating it into an unsupervised step which uses only the information in $I(\phi_i; \phi_j)$ and try to find the subset of the most independent features to resolve the second case problem. We ignore the first case problem, however, because it is not serious compared to the second case and requires time-consuming redundancy analysis within the acquired feature subset.

To find the subset with the most independent features, we adopt a hierarchical agglomerative clustering algorithm with average linkage. This is a bottom-up clustering algorithm, which starts from a set of clusters consisting of only one data point, then merges neighboring clusters using a defined distance measure, and finishes with the desired number of clusters (Duda et al., 2001). For the distance measure, the inverse of mutual information between features, $I(\phi_i; \phi_j)^{-1}$, is used. After clustering, the feature ϕ_i with the highest $I(\phi_i; Y)$ value in each cluster is selected as its representative. The number of clusters is set to the 10 percent[2] of the total number of features in each dataset, as it showed the best average prediction performance. Finally, we combine the two resulting feature sets from the previous two steps by the union-set operation.

24.3 Experiments

24.3.1 Datasets

We used five benchmark datasets from the Workshop on Feature Extraction of the NIPS Conference 2003, available at http://clopinet.com/isabelle/Projects/NIPS2003. The properties of each dataset are summarized in Table 24.2. Each dataset is designed to reflect the specific properties of popular research domains.

24.3.2 Preprocessing

First, we discretized all datasets using an EM algorithm that uses the following model. Suppose that the features ϕ_k in each of the 'on' and 'off' states can be modeled by a Gaussian distribution $N(\phi_k|\mu_i, \sigma_i)$, where μ_i and σ_i are the mean and standard deviation ($i = 1, 2$). Given this assumption, the marginal probability of a feature ϕ_k can be modeled by a weighted sum of these

---

[2]We also tried 10,20,30,40,50 percent of the total number of features.

Table 24.2. The properties of the five NIPS datasets

Dataset	Train	Validate	Test	Feature	Research Domain
ARCENE	100	100	700	10,000	Microarray analysis
GISETTE	6,000	1,000	6,500	5,000	OCR(handwritten digits)
DOROTHEA	800	350	800	100,000	Drug response
DEXTER	300	300	2,000	20,000	Text categorization
MADELON	2,000	600	1,800	500	Artificial (XOR pattern)

Gaussian probability functions:

$$P(\phi_k) = w_1 N(\phi_k|\mu_1, \sigma_1) + w_2 N(\phi_k|\mu_2, \sigma_2) \qquad (24.2)$$

This model is called a *univariate mixture model* with two components (Xing, 2002). After finding two Gaussian components that are the best fits for the data, we assigned a binary label to each component. Next we filtered out uninformative features which have a lower mutual information value, $I(\phi_k; Y)$, than the threshold: the threshold was set to the average mutual information value of the randomly permuted feature vectors of each dataset.

24.3.3 Results

We acquired three feature subsets, 'S', 'US', and 'MIXED', from the supervised and unsupervised feature selection and from a combination of these steps. As the test dataset was closed at the time of the experiment, we used validation sets to evaluate the prediction performance of classifiers. We compared the prediction errors using the four feature subsets (S, US, MIXED and SIMPLE) and two classifiers, naïve Bayes (NB) and support vector machine (SVM) (Boser et al., 1992). The SIMPLE feature subset was constructed by applying a classical filter method with information theoretic ranking criteria, to have the same number of features as the MIXED set. We used the classifiers in the Java-based open source data-mining software, WEKA version 3.4 (Witten and Frank, 2000), which is downloadable from http://www.cs.waikato.ac.nz/ml/weka. Because one of our objectives was to investigate the pure performance gain acquired by feature selection, we did not optimize the classifiers specifically for each dataset (the linear kernel and defaults parameters (regularization constant $C = 10$) were used for all datasets).

Table 24.3. NIPS 2003 challenge results.

Dec. 1st	Our best challenge entry					The winning challenge entry					
Dataset	Score	BER	AUC	Feat	Probe	Score	BER	AUC	Feat	Probe	Test
OVERALL	-32.36	18.40	81.60	4.34	31.51	88.00	6.84	97.22	80.3	47.77	1
ARCENE	67.27	18.41	81.59	1.85	0.00	98.18	13.30	93.48	100	30.00	1
DEXTER	-81.82	14.60	85.40	5.09	37.46	96.36	3.90	99.01	1.52	12.87	1
DOROTHEA	10.91	15.26	84.74	0.77	27.68	98.18	8.54	95.92	100	50.00	1
GISETTE	-40.00	2.74	97.26	9.30	0.00	98.18	1.37	98.63	18.26	0.00	1
MADELON	-67.27	38.5	61.50	2.40	91.67	100	7.17	96.95	1.60	0.00	1

Our challenge result of the feature extraction is summarized in Table 24.3. Considering the rank, our best entry (ranked 40th considering individual submissions, and 15th as a group) is not among the best. However, the performance in cases of ARCENE and GISETTE is not significantly different from the winning entry, albeit we did not perform any model optimization. Moreover, we only used less than 5 percent of features in average, while the winners used almost 80 percent. Considering only the entries with less than 5 percent of the total number of features, our method ranked 6th. Our method also showed good overall performance in filtering out probes: especially in cases of ARCENE and GISETTE, we perfected identified the probes. However, the case of MADELON dataset was an exception. The worst performance is also recorded in case of the MADELON. One reason can be found in the characteristic of this dataset which contains artificially constructed nonlinear patterns. The nonlinearity might have a negative effect on our method which is composed of linear procedures. Another possible reason is that we filtered out a large number of features in preprocessing: we might have thrown away discriminating features initially.

The prediction result of the validation data is summarized in Table 24.4. In general, the subsets show lower error rate in order of SIMPLE, US, S, and MIXED. For the naïve-Bayes classifer, the MIXED set shows significant performance gain compared to the S (p-value = 0.0214). For the SVM, we can also identify some performance gain between S and MIXED, in cases of ARCENE, GISETTE and DEXTER datasets. However, the gain is not so significant as the naïve-Bayes case (p-value = 0.2398).

24.4 Conclusions

We have presented an enhanced feature selection method that uses the mutually compensating, combined feature selected from two separate filtering steps based on well-known information theoretic concepts. Our adjustment

Table 24.4. Prediction result of four feature subsets and two classifiers

	naïve-Bayes (Error %)				SVM (Error %)			
Dataset	SIMPLE	US	S	MIXED	SIMPLE	US	S	MIXED
ARCENE	51.00	34.00	31.00	28.00	53.00	35.00	21.00	14.00
GISETTE	49.33	30.60	10.80	7.40	49.30	38.70	3.50	3.10
DOROTHEA	13.38	16.86	6.57	5.43	12.13	12.00	5.43	7.15
DEXTER	45.67	46.67	17.33	13.33	47.00	56.00	20.00	18.67
MADELON	50.45	43.00	38.83	39.00	50.80	41.33	38.83	39.83
p-value[3]	n/a	0.0810	0.0279	0.0214	n/a	0.1397	0.0281	0.2398

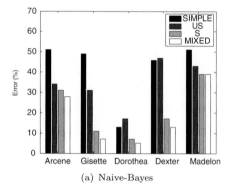

(a) Naive-Bayes

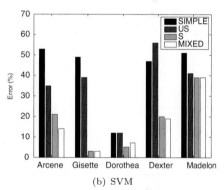

(b) SVM

of selection procedures came as a result of an examination that showed that the biases can be regarded as a consequence of misleading casual relations estimated by certain measures in supervised selection procedures.

The main advantage of our method is that it can expect better accuracy with a computational time similar to the classical filter method. Our experimental results support this aspect showing that our method performs well over the datasets used. However, the algorithms suggested in the filtering steps rely on a set of parameters, which can affect the prediction performance. We empirically determined these parameters, but with a more systematic approach, we may be able to improve the performance. Finally, we expect that devising a fast but more accurate feature combining method may increase the efficiency of the feature selection, which should be the focus of future investigations.

---

[3]The p-values denote the statistical significance of the difference between two feature sets, i.e. (SIMPLE & US), (US & S), and (S & MIXED), which is evaluted by paired t-Test.

References

B.E. Boser, I.M. Guyon, and V.N. Vapnik. A training algorithm for optimal margin classifiers. In *Proceedings of the 5th annual ACM workshop on Computational Learning Theory*, pages 144–152. ACM Press, 1992.

T.M. Cover and J.A. Thomas. *Elements of Information Theory*. John Wiley & Sons, 1991.

R.O. Duda, P.E. Hart, and D.G. Stork. *Pattern Classification*. John Wiley & Sons, second edition, 2001.

I.M. Guyon and A. Elisseeff. An introduction to variable and feature selection. *Journal of Machine Learning Research*, 3:1157–1182, 2003.

T. Hastie, R. Tibshirani, and J. Friedman. *The Elements of Statistical Learning*. Springer, 2001.

D. Koller and M. Sahami. Toward optimal feature selection. In *Proceedings of the Thirteenth International Conference on Machine Learning*, pages 284–292. Morgan Kaufmann, 1996.

J. Pearl. *Probabilistic Reasoning in Intelligent Systems: Networks of Plausible Inference*. Morgan Kaufmann, 1988.

B. Shipley. *Cause and Correlation in Biology: A Users Guide to Path Analysis, Structural Equations and Causal Inference*. Cambridge University Press, 2000.

P. Spirtes, C. Glymour, and R. Scheines. *Causation, prediction, and search*. Springer, 1993.

I.H. Witten and E. Frank. *Data Mining: Practical machine learning tools with Java implementations*. Morgan Kaufmann, 2000.

E.P. Xing. *Feature Selection in Microarray Analysis*, pages 110–131. Kluwer Academic Publishers, 2002.

Chapter 25

An Enhanced Selective Naïve Bayes Method with Optimal Discretization

Marc Boullé

France Telecom R&D, 2 avenue Pierre Marzin, 22307 Lannion Cedex, France
marc.boulle@francetelecom.com

In this chapter, we present an extension of the wrapper approach applied to the predictor. The originality is to use the area under the training lift curve as a criterion of feature set optimality and to preprocess the numeric variables with a new optimal discretization method. The method is experimented on the NIPS 2003 datasets both as a wrapper and as a filter for multi-layer perceptron.

25.1 Introduction

The Naïve Bayes approach is based on the assumption that the variables are independent within each output label, which can harm the performances when violated. In order to better deal with highly correlated variables, the Selective Naïve Bayes approach (Langley and Sage, 1994) uses a greedy forward search to select the variables. The accuracy is evaluated directly on the training set, and the variables are selected as long as they do not degrade the accuracy. For numeric variables, the probability distribution is evaluated according to a Gaussian distribution whose mean and variance are estimated on the training examples.

Although the approach performs quite well on datasets with a reasonable number of variables, it does not scale on very large datasets with hundred or thousands of variables, such as in marketing applications. We propose to enhance the original method by exploiting a new Bayes optimal discretization method called MODL and by evaluating the predictors with a new criterion more sensitive than the accuracy. In the NIPS Challenge, the method is experimented both as a wrapper approach (ESNB) and as a filter for multi-layer perceptron (ESNB+NN). The method is fast, robust and manages to find a good trade-off between the error rate and the number of selected variables. However, the method needs further improvements in order to reach better error rates.

M. Boullé: *An Enhanced Selective Naïve Bayes Method with Optimal Discretization*, StudFuzz
207, 499–507 (2006)
www.springerlink.com

We detail and comment the ESNB method in section 25.2 and focus on the MODL discretization method in section 25.3. We present the results of the method in the NIPS Challenge in section 25.4.

25.2 The Enhanced Selective Naïve Bayes Method

In this section, we describe three enhancements to the Selective Naïve Bayes method: the use of a new discretization method to pre-process numeric variables, the use of the area under the lift curve to evaluate the performance of predictors and a post-processing correction of the predicted output label probabilities. Lift curves summarize the cumulative percent of targets recovered in the top quantiles of a sample (Witten and Franck, 2000).

The evaluation of the probabilities for numeric variables has already been discussed in the literature (Dougherty et al., 1995, Hsu et al., 2002, Yang and Webb, 2002). Experiments have shown that even a simple Equal Width discretization with 10 bins brings superior performances compared to the assumption using a Gaussian distribution. In a selection process, the risk of overfitting the data raises with the number of variables. Slight losses in the quality of the evaluation of the probability distribution of the variables may have cumulated effects and lead to the selection of irrelevant or redundant variables. We propose to use a new supervized discretization method called MODL, which is Bayes optimal. This method is described in section 25.3.

In the wrapper approach, (Kohavi and John, 1997) propose to evaluate the selection process using accuracy with a five-fold cross validation. However, the accuracy criterion suffers from some limits, even when the predictive performance is the only concern. (Provost et al., 1998) propose to use Receiver Operating Analysis (ROC) analysis rather than the accuracy. In marketing applications for example, the lift curves are often used to evaluate predictors. In the context of variable selection, there are other justifications to replace accuracy by another criterion. In case of a skewed distribution of output labels, the accuracy may never be better than the majority accuracy, so that the selection process ends with an empty set of variables. This problem also arises when several consecutive selected variables are necessary to improve the accuracy. In the method proposed in (Langley and Sage, 1994), the selection process is iterated as long as there is no decay in the accuracy. This solution raises new problems, such as the selection of irrelevant variables with no effect on accuracy, or even the selection of redundant variables with either insignificant effect or no effect on accuracy. We propose to use the area under the lift curve, measured directly on the training set, to evaluate whether a new variable should be selected. If we note TP (True Positive), TN (True Negative), FP (False Positive) and FN (False Negative) the four possible outcomes of the confusion matrix, the lift curve is obtained by plotting TP (in %) against $\frac{TP+FP}{TP+FP+TN+FN} \times 100\%$ for each confidence value, starting at (0,1) and ending at (1,0). At each step of the algorithm, the variable which brings the best increase of the area under the lift curve is choosen and the selection process stops as soon as this area does not rise anymore. This allows capturing slight enhancements in the learning process and helps avoiding the selection of redundant variables or probes that have no effect on the lift curve.

The last problem is related to the Naïve Bayes algorithm itself, which is a good rank estimator, but a weak probability estimator (Hand and Yu, 2001). We propose to add a correction to the estimation of the output labels probabilities at the end of the learning process, instead of using the standard 50% probability threshold to predict the output label. For a given probability threshold, we compute the resulting confusion matrix on the training set and score it owing to the chi-square criterion. The higher the chi-square criterion is, the more correlated are the predicted output labels and the true output labels. The best probability threshold is found by evaluating all possible confusion matrices, once the training examples have been sorted by decreasing probability of output label. This corresponds to finding the best point on the lift curve, owing to the maximization of the chi-square criterion of the related confusion matrix.

Altogether, the algorithm can be optimized in $O(n^2 m \log(m))$ time, where n is the number of input variables and m the number of training examples. The pre-processing step needs $O(nm \log(m))$ to discretize all the variables. The forward selection process requires $O(n^2 m \log(m))$ time, owing to the decomposability of the Naïve Bayes formula on the variables. The $O(m \log(m))$ term in the complexity is due to the evaluation of the area under the lift curve, based on the sort of the training examples. The post-processing correction needs $O(m \log(m))$ time by sorting the training examples and evaluating all possible probability thresholds. However, the irrelevant variables can be detected just after the discretization step: they are discretized in a single interval. If n_r is the number of relevant variable and n_s is the number of selected variables at the end of the learning process, the practical complexity of the algorithm is $O(n_r n_s m \log(m))$ time, which is often far below the theoretical complexity when the number of input variables is very high.

Enhanced Selective Naïve Bayes algorithm:

- Initialization
  - Discretize each variable with the MODL discretization method
  - Create an initial empty selected variable set and a set of relevant variables
- Selection process
  Repeat the following steps
  - For each unselected relevant variable
    - Compute the Naïve Bayes predictor with the additional variable, on the basis of the previous best predictor
    - Evaluate the resulting predictor with the area under the lift curve
  - If the evaluation is strictly improved

· Add the best variable to the selected variable set
· Update the best predictor
- Post-processing
  - Find the best decision threshold by maximizing the chi-square criterion of the contingency table

25.3 The MODL Discretization Method

In this section, we present the MODL approach which results in a Bayes optimal evaluation criterion of discretizations and the greedy heuristic used to find near-optimal discretizations.

The objective of the process is to induce a list of intervals that splits the value domain of a numeric input variable. The training examples are described by pairs of values: the numeric input value and the output label. If we sort the training examples according to their input values, we obtain a string S of output labels. We now propose the following formal definition of a discretization model.

Definition: A *standard* discretization model is defined by the following properties:

- the discretization model relies only on the order of the output labels in the string S, without using the values of the input variable;
- the discretization model allows to split the string S into a list of substrings (the intervals);
- in each interval, the distribution of the output labels is defined by the frequencies of the output labels in the interval.

Such a discretization model is called a SDM model.

Notations:
m : number of training examples
J : number of output labels
I : number of intervals
m_i : number of training examples in the interval i
m_{ij} : number of examples with output label j in the interval i
A SDM model is completely defined by the set of parameters $\{I, m_i(1 \leq i \leq I), m_{ij}(1 \leq i \leq I, 1 \leq j \leq J)\}$.

This definition is very general, and most discretization methods rely on SDM models. They first sort the samples according to the variable to discretize (property 1) and try to define a set of intervals by partitioning the string of output labels (property 2). The evaluation criterion is always based on the frequencies of output labels (property 3).

In the Bayesian approach, the best model is found by maximizing the probability $P(model/data)$ of the model given the data. Using the Bayes rule and since the probability $P(data)$ is constant under varying the model, this is equivalent to maximize $P(Model)P(Data/Model)$. We define below a prior which is essentially a uniform prior at each stage of the hierarchy of the SDM model parameters. We also introduce a strong hypothesis of independence of the distributions of the class values. This hypothesis is often assumed (at least implicitely) by many discretization methods, that try to merge similar intervals and separate intervals with significantly different distributions of class values.

Definition: The following distribution prior on SDM models is called the three-stage prior:

- the number of intervals I is uniformly distributed between 1 and m;
- for a given number of intervals I, every division of the string to discretize into I intervals is equiprobable;
- for a given interval, every distribution of output labels in the interval is equiprobable;
- the distributions of the output labels in each interval are independent from each other.

Theorem 1. *(Boullé, 2004b) A SDM model distributed according to the three-stage prior is Bayes optimal for a given set of training examples to discretize if the following criterion is minimal on the set of all SDM models:*

$$\log(m) + \log\binom{m+I-1}{I-1} + \sum_{i=1}^{I} \log\binom{m_i + J - 1}{J-1} + \sum_{i=1}^{I} \log\left(\frac{m_i!}{m_{i,1}!\ldots m_{i,J}!}\right).$$
(25.1)

The first term of the criterion stands for the choice of the number of intervals, the second term for the choice of the bounds of the intervals and the third term for the choice of the output labels distribution in each interval. The last term encodes the probability of the data given the model.

Once the optimality of the evaluation criterion is established, the problem is to design an efficient minimization algorithm. The MODL method uses a greedy bottom-up merge algorithm to perform this optimization, that can be optimized in $O(m \log(m))$ time. This algorithm exploits the additivity of the MODL criterion, memorizes the variations $\Delta value$ of the criterion related to the merges, and keeps these merge evaluations in a maintained sorted list (such as an AVL binary search tree for example).

The method is fully described and experimented in (Boullé, 2004a,b). Compared to other discretization methods, the MODL method obtains better classification performances with fewer intervals. Random variables are discretized with a single interval since this is the most probable discretization model of such variables. The MODL method is thus efficient at detecting probes.

25.4 Results on the NIPS Challenge

In this section, we report the results obtained by the ESNB method on the NIPS 2003 Challenge datasets (Guyon, 2003). Each dataset is divided in 3 sets: training, validation and test. In the first period, only the training sets could be used for training for submission of the *original* challenge entries. In the second period, the training and validation sets were used together for the *validation* entries. We submitted one original challenge entry using the ESNB method. Since the results looked promising, we decided to submit two validation entries, one using the ESNB method and the other using the method as a filter for multi-layer perceptron (ESNB+NN). The ESNB method is fully automatic and does not require any parameter. In the filter approach, we use a non-linear multi-layer perceptron applied on the variables selected by the ESNB method. The multi-layer perceptron is trained using back-propagation with sigmoid activation function, regularized with orthogonal weight decay. The training set of the challenge is used to train and the validation set to stop training. This predictor is trained with a hidden layer containing 1, 5, 10, 25 or 50 neurons, and the best neural architecture is chosen based on the validation set: 50 neurons for the Madelon dataset and 1 neuron for the other datasets. We report in Table 25.1 the results of our ESNB entry by Dec. 1st and in Table 25.2 the results of our ESNB+NN entry by Dec. 8th.

Table 25.1. Challenge results for the ESNB method (Dec. 1st).

Dec. 1st	ESNB challenge entry					The winning challenge entry					
Dataset	Score	BER	AUC	Feat	Probe	Score	BER	AUC	Feat	Probe	Test
OVERALL	-57.82	19.85	85.96	1.02	10.6	88.00	6.84	97.22	80.3	47.77	1
ARCENE	-78.18	31.25	75.93	0.05	40	98.18	13.30	93.48	100.0	30.0	1
DEXTER	-45.45	9.80	96.42	0.17	0	96.36	3.90	99.01	1.52	12.87	1
DOROTHEA	-45.45	21.03	89.43	0.05	1.89	98.18	8.54	95.92	100.0	50.0	1
GISETTE	-56.36	3.12	99.49	3.02	0	98.18	1.37	98.63	18.26	0.0	1
MADELON	-63.64	34.06	68.51	1.8	11.11	100.00	7.17	96.95	1.6	0.0	1

The ESNB method has a low computation time, with on average 5 mn per dataset on a PC 1.7 Mhz. The MODL discretization methods is very efficient at detecting probes and the use of the area under the lift curve as a feature subset selection criterion helps removing redondant variables. This results in small numbers of selected features, on average 1% of the input variables. Compared to the Dec. 1st original entry, the ESNB method is able to exploit the increased number of training examples available in the Dec. 8th experiments. It selects more features (a total of 321 versus 252) while keeping less probes (1 versus 3) and improves the balanced error rate from 19.85% down to 18.25%. The ESNB+NN method largely improves the results of the

Table 25.2. Challenge results for the (ESNB+NN) method (Dec. 8th).

Dec. 8th Dataset	ESNB+NN challenge entry					The winning challenge entry					Test
	Score	BER	AUC	Feat	Probe	Score	BER	AUC	Feat	Probe	
OVERALL	-28	12.42	93.12	1.04		1.43	71.43	6.48	97.20	80.3	47.77	1
ARCENE	-60	22.92	83.78	0.14		7.14	94.29	11.86	95.47	10.7	1.03	1
DEXTER	-25.71	7.20	97.49	0.33		0	100	3.30	96.70	18.57	42.14	1
DOROTHEA	54.29	14.59	91.50	0.07		0	97.14	8.61	95.92	100	50	1
GISETTE	-42.86	2.46	99.64	3.26		0	97.14	1.35	98.71	18.32	0	1
MADELON	-65.71	14.94	93.22	1.4		0	94.29	7.11	96.95	1.6	0	1

ESNB method, especially when the bias of the Naïve Bayes approach is too limiting. Using the ESNB method as a filter approach is thus relevant. In order to evaluate the method both on the balanced error rate and the number of the selected variables, we report the method results and the entry results of all the other participants on a bi-criteria plan displayed in Figure 25.1.

The results of the ESNB methods used as a filter approach are on the Pareto curve in Figure 25.1. Many methods obtain a better error rate, up to twice better than that of the ESNB+NN method, but at the expense of a significantly higher number of selected variables.

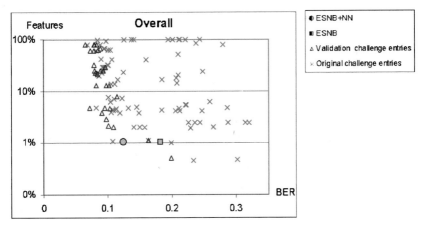

Fig. 25.1. Bi-criteria analysis of the challenge results with the balanced error rate on the x-coordinate and the number of selected variables on the y-coordinate

25.5 Conclusion

The ESNB method is a feature selection method derived from the Naïve Bayes method enclosed in a wrapper approach. It benefits from the use of the Bayes optimal MODL discretization method and from the evaluation of predictor using the area under the lift curve instead of the accuracy. It can be exploited either directly or as a filter approach with a powerful predictor applied on the selected features.

Experiments on the NIPS 2003 datasets show that this fully automatic method is fast, robust, and exhibits results with a good trade-off between error rate and number of selected variables. However, the method suffers from several weaknesses partly related to the bias of the Naïve Bayes approach. In future work, we plan to improve the error rate of the method, with the smallest possible decay in computation time and selected variable number. The wrapper method can be improved by replacing the forward selection algorithm by a more sophisticated search heuristic. Exploiting ensemble methods is a promising direction for selecting a more comprehensive set of selected variables. Although the univariate evaluation of variables is robust owing to the MODL discretization method, the overfitting behaviour resulting from the selection of a set of variables could be reduced by using regularization techniques. A last direction is to decrease the bias of the Naïve Bayes approach by detecting interactions between variables or building new features combining multiple input variables.

Acknowledgements

I am grateful to Vincent Lemaire (author of the ROBELON submission) for providing the multi-layer perceptron software and enabling the ESNB+NN joined submission to the NIPS Challenge. The MODL method is under French patent N 04 00179. Contact the author for conditions of use.

References

M. Boullé. A bayesian approach for supervized discretization. In *Proceedings of the Data Mining 2004 Conference*. WIT Press, 2004a.

M. Boullé. MODL : une méthode quasi-optimale de discrétisation supervisée. Technical Report, NT/FTR&D/84444, France Telecom R&D, Lannion, France, 2004b.

J. Dougherty, R. Kohavi, and M. Sahami. Supervised and unsupervised discretization of continuous features. In *Proceedings 12th International Conference on Machine Learning*, pages 194–202. Morgan Kaufmann, 1995.

I. Guyon. Design of experiments of the NIPS 2003 variable selection benchmark. http://www.nipsfsc.ecs.soton.ac.uk/papers/datasets.pdf, 2003.

D.F. Hand and K.S. Yu. Idiot's bayes-not so stupid after all? *International Statistical Review*, 69(3):385–398, 2001.

C.N. Hsu, H.J. Huang, and T.T. Wong. Implications of the dirichlet assumption for discretization of continuous variables in nave bayesian classifiers. *Machine Learning*, 53(3):235–263, 2002.

R. Kohavi and G.H. John. Wrappers for feature subset selection. *Artificial Intelligence*, 1997.

P. Langley and S. Sage. Induction of selective bayesian classifiers. In *Proceedings 10th Conference on Uncertainty in Artificial Intelligence*. Morgan Kaufman, 1994.

F. Provost, T. Fawcett, and R. Kohavi. The case against accuracy estimation for comparing induction algorithms. In *Proceedings Fifteenth International Conference on Machine Learning*. Morgan Kaufmann, 1998.

I.H. Witten and E. Franck. *Data Mining*. Morgan Kaufmann, 2000.

Y. Yang and G.I. Webb. A comparative study of discretization methods for nave-bayes classifiers. In *Proceedings of the Pacific Rim Knowledge Acquisition Workshop*, 2002.

Chapter 26

An Input Variable Importance Definition based on Empirical Data Probability Distribution

V. Lemaire and F. Clérot

Statistical Information Processing group
France Telecom Research and Development
FTR&D/SUSI/TSI
2 Avenue Pierre Marzin
22307 Lannion cedex FRANCE
vincent.lemaire@rd.francetelecom.com,
fabrice.clerot@rd.francetelecom.com

Summary. We propose in this chapter a new method to score subsets of variables according to their usefulness for a given model. It can be qualified as a variable ranking method 'in the context of other variables'. The method consists in replacing a variable value by another value obtained by randomly choosing a among other values of that variable in the training set. The impact of this change on the output is measured and averaged over all training examples and changes of that variable for a given training example. As a search strategy, backward elimination is used. This method is applicable on every kind of model and on classification or regression task. We assess the efficiency of the method with our results on the NIPS 2003 feature selection challenge.

26.1 Introduction

In this chapter, we describe the ROBELON method used in the NIPS Feature Selection Challenge and its use in variable selection.

The objective of variable selection (Guyon and Elisseeff, 2003) is three-fold: improve the prediction performance of the predictors, provide faster and more cost-effective predictors, and allow a better understanding of the under-lying process that generated the data. Among techniques devoted to variable selection we find filter methods (cf. chapter 3 'Filter methods'), which select variables by ranking them with correlation coefficients, and subset selection methods, which assess subsets of variables according to their usefulness to a given model.

Wrapper methods (cf. chapter 4 'Search strategies' and chapter 5 'Embedded Methods' (Kohavi and John, 1997)) rely on a model as a black box to score subsets of variables according to their usefulness for the modeling task. In practice, one needs to define: (i) how to search the space of all possible

V. Lemaire and F. Clérot: *An Input Variable Importance Definition based on Empirical Data Probability Distribution*, StudFuzz **207**, 509–516 (2006)
www.springerlink.com

variable subsets; (ii) how to assess the prediction performance of a model to guide the search and halt it; (iii) which model to use.

We propose a new method to perform the second point above and to score subsets of variables according to their predictive power for the modeling task. It relies on a definition of the variable importance as measured from the variation of the predictive performance of the model. The method is motivated and described in section 26.2. Its use in variable selection is described in section 26.3. We compare in section 26.4 the performance of the proposed method with other techniques on the NIPS feature selection challenge 2003 and we conclude in section 26.5.

26.2 Analysis of an Input Variable Influence

26.2.1 Motivation and Previous Works

Our motivation is to measure variable importance given a predictive model. The model is considered a perfect black box and the method has to be usable on a very large variety of models for classification (whatever the number of classes) or regression problems.

Since the model is a perfect black box let the model, f, be just an input-output mapping function for an example k : $\mathbf{y^k} = f(\mathbf{x^k}) = f(x_1^k, ..., x_n^k)$.

The 'importance' of a variable for a predictive model is naturally defined in terms of influence on the output when the value of a variable changes. Although natural, this definition hides some pitfalls we briefly discuss below.

The black box model could be a non-linear model for which the variation of the output can be non-monotonous. Hence, the influence of an input variable cannot be evaluated by a local measurement as for example partial derivatives or differences (Réfénes et al., 1994, Moody, 1994, Baxt and White, 1995).

The choice of the input variation range and should depend on the variable: too small a value has the same drawback as the partial derivatives (local information and not well suited for discrete variables), too large a value can be misleading if the function (the model) with respect to an input V is non-monotonous, or periodic.

Recently, Féraud et al. (Féraud and Clérot, 2002) propose a global saliency measurement which seems to answer to the first question. But their definition however does not take into account the true interval of variation of the input variables. They propose to use a prior on the possible values of the input variables. The knowledge needed to define this prior depends on the specificities of the input variable (discrete, positive, bounded, etc). Such individual knowledge is clearly difficult and costly to obtain for databases with a large number of variables. A more automatic way than this 'prior' approach is needed to answer to the second question.

A first step in this direction is given by Breiman in (Breiman, 2001) (paper updated for the version 3.0 of the random forest) where he proposes a method which relies on the distribution of probability of the variable studied. Each example is perturbed by randomly drawing another value of the studied variable among the values spanned by this variable across all examples. The performance of the perturbed set are then compared to the 'intact' set. Ranking variable performance differences allows to rank variable importance. This method allows to automatically determine the possible values of a variable from its probability distribution, even if perturbing every example only once does not explore the influence of the full probability distribution of the variable. Moreover, although (Breiman, 2001) seems to restrict the method to random forests, it can obviously be extended to other models.

The method described in this article combines the definition of the 'variable importance' as given in Féraud et al. (Féraud and Clérot, 2002) ('saliency' in their paper) with an extension of Breiman's idea (Breiman, 2001). This new definition of variable importance both takes into account the probability distribution of the studied variable and the probability distribution of the examples.

26.2.2 Definition of the Variable Importance

The importance of an input variable is a function of examples $\mathbf{x}$ probability distribution and of the probability distribution of the considered variable (V_j). Let us define:

- V_j the variable for which we look for the importance;
- V_{ij} the realization of the variable V_j for the example i;
- $\mathbf{x}_m = (V_{mj})_{j=1...n}$ the example m a vector with n components;
- f the predictive model;
- $P_{V_j}(v)$ the probability distribution of the variable V_j;
- $P_x(u)$ the probability distribution of examples X;
- $f_j(\mathbf{a};b) = f_j(a_1,...,a_n;b) = f(a_1,...,a_{j-1},b,a_{j+1},...,a_n)$ where a_p is the pth component of the vector $\mathbf{a}$.

The importance of the variable V_j is the average of the measured variation of the predictive model output when examples are perturbed according to the probability distribution of the variable V_j. The perturbed output of the model f, for an example $\mathbf{x}_i$, is the model output for this example but having exchanged the jth component of this example with the jth component of another example, k. The measured variation, for the example $\mathbf{x}_i$, is then the difference between the 'true output' $f_j(\mathbf{x}_i;V_{ij})$ and the 'perturbed output' $f_j(\mathbf{x}_i;V_{kj})$ of the model. The importance of the variable V_j is then the average of $|f_j(\mathbf{x}_i;V_{ij}) - f_j(\mathbf{x}_i;V_{kj})|$ on both the examples probability distribution and the probability distribution of the variable V_j. The importance of the variable V_j for the model f is then:

$$S(V_j|f) = \iint P_{V_j}(v)dv P_x(u)du \, |f(u) - f_j(u,v)| \qquad (26.1)$$

26.2.3 Computation

Approximating the distributions by the empirical distributions the computation of the average of $S(V_j|f)$ is:

$$S(V_j|f) = \iint P_{V_j}(v)dv\, P_x(u)du\, |f(u) - f_j(u,v)| \tag{26.2}$$

$$= \int P_x(u)du \int P_{V_j}(v)dv\, |f(u) - f_j(u,v)| \tag{26.3}$$

$$= \frac{1}{m}\sum_{i=1}^{m}\left[\frac{1}{m}\sum_{k=1}^{m}|f_j(\mathbf{x}_i; V_{ij}) - f_j(\mathbf{x}_i; V_{kj})|\right] \tag{26.4}$$

This computation would require to use all the possible values of the variable V_j for all examples available. For m examples in the training set and therefore m possible values of V_j the computation time scales as m^2 and becomes very long for large databases. There are, at least, two faster heuristics to compute $S(V_j|f)$:

(1.) We draw simultaneously $\mathbf{x}_i$ and V_{kj} and compute one realization of $|f_j(\mathbf{x}_i, V_{ij}) - f_j(\mathbf{x}_i, V_{kj})|$. Such realizations are considered as a constant value perturbed by a zero-mean noise; they are successively fed to a Kalman filter to estimate the average (Maybeck, 1979, Welch and Bishop, 2001).

(2.) The empirical variable probability distribution can be approximated using l (arbitrary chosen) representative examples of an ordered statistic (the values of the central individuals of l partiles for example).

$$S(V_j|f) = \frac{1}{m}\sum_{i=1}^{m}\left[\frac{1}{l}\sum_{p=1}^{l}|f_j(\mathbf{x}_i; V_{ij}) - f_j(\mathbf{x}_i; v_p)|\, P(v_p)\right] \tag{26.5}$$

where $P(v_p)$ represent the probability to observe the value v_p. This method is especially useful when V_j takes only discrete values since the inner sum is exact and not an approximation. The computation can also be stopped with a Kalman filter.

A regularization technique has to be applied during the training process and/or a preprocessing to ensure that only not correlated (Moody, 1994) and relevant features will survive after convergence (Burkitt, 1992, Rumelhart et al., 1986).

26.3 Application to Feature Subset Selection

The wrapper methodology offers a simple and powerful way to address the problem of variable selection, regardless the chosen learning machine. The learning machine is considered a perfect black box and the method lends itself to off-the-shelf machine learning software packages. Exhaustive search can only be performed if the number of variables is small and heuristics are otherwise necessary. Among these, backward elimination and 'driven' forward selection which can both rely on the variable importance described above.

In backward elimination one starts with the set of all variables and progressively eliminates the least important variable. The model is re-trained after every selection step. In forward selection, as in (Breiman, 2001), at a first step we train a model with all variables. Then we rank the variables using the method described in this paper. In a second step we train models where variables are progressively incorporated into larger and larger subsets according to their ranks.

Comparison between both methods will be discussed elsewhere. Hereafter we restrict the discussion to backward elimination. We note here that both methods have the appealing property of depending on one parameter only, the degradation of the performance of the model trained with the subset relatively to the best possible performance reached.

To speed up the backward elimination another parameter is added. At each step of the backward elimination we remove all variables with an importance smaller than a very low threshold (10^{-6}). With this implementation the backward elimination method has only two simple parameters, a performance threshold to define the selected subset and an importance threshold to discard variables with 'no' importance.

26.4 Results on the NIPS Feature Selection Challenge

26.4.1 Introduction

The purpose of the NIPS 2003 workshop on feature extraction was to bring together researchers of various application domains to share techniques and methods. Organizers of the challenge formatted a number of datasets for the purpose of benchmarking feature selection algorithms in a controlled manner. The data sets were chosen to span a wide variety of domains. They chose data sets that had sufficiently many examples to create a large enough test set to obtain statistically significant results. The input variables are continuous or binary, sparse or dense. All problems however are two-class classification problems. The similarity of the tasks will allow participants to enter results on all data sets to test the genericity of the algorithms. Each dataset was split in 3 sets: training, validation and test set. All the informations about the challenge, the datasets, the results can be found in this book and on: www.nipsfsc.ecs.soton.ac.uk.

26.4.2 Test Conditions of the Proposed Method

As we wish to investigated the performance of our variable importance measurement, we chose to use a single learning machine for all datasets (no bagging, no Ada-boost, no other bootstrap method): a MLP neural network with 1 hidden layer, tangent hyperbolic activation function and stochastic back-propagation of the squared error as training algorithm.

We added a regularization term active only on directions in weight space which are orthogonal to the training update (Burkitt, 1992). Other methods to measure the sensitivity of the output of neural networks can be found in the chapter 5.2.3 of this book.

In the first period of the challenge, only training sets could be used for training for submission of the *original* challenge entries. For each dataset we split the training set in two sets: a training (70 %) and a validation set (30 %); the validation set of the challenge is then used as a test set. This training set is used to train and the validation set is used to stop training. We made an *original* submission before December first. We named our submissions on the challenge web site ROBELON for RObuts Backward ELimination On Neural network. In the figure 26.1 we call this submission 'ROBELON original entry'.

We made, three months after, a new submission where the training set of the challenge is used to train and the validation set of the challenge is used to stop training. Even if this submission has been made after the 8th december, as we used no information about the databases, this submission is called 'ROBELON validation entry' in the figure 26.1.

In both cases the preprocessing used is only a zero-mean, unit-variance standardization. The variables importance is measured using only the training set. The strategy used to constitute the selected variable subset is the standard backward elimination. The subset of variables was chosen as the smallest subset allowing a performance greater than 95 % of the best performance reached during the selection process.

26.4.3 Comparison with Others Results

'Variable selection' is always somewhat ambiguous when the result is judged from the balanced error rate (BER) only, specially when different learning machines are used, since it is more a matter of balance between the BER and the number of features used rather than a matter of BER only: to prefer a BER=0.1 using 50 % of features to a BER=0.12 using 10 % of the features is mostly a matter of application requirements. In some applications, one would trade some accuracy for less features as, for example, in real-time network applications.

What we expect from a variable selection technique is to adapt itself in such situation by removing as many features as possible. Therefore, what we can expect from the combination of our simple model and our selection technique is to keep a BER reasonably close to the average while using significantly less

Table 26.1. NIPS 2003 challenge results for ROBELON.

Dec. 1st Dataset	Our best challenge entry						The winning challenge entry				
	Score	BER	BER*	AUC	Feat	Probe	Score	BER	AUC	Feat	Probe
Overall	-62.18	16.37	20.1	83.63	1.12	21.47	88.00	6.84	97.22	80.3	47.8
Arcene	-85.45	29.65	20.2	70.35	1.50	60.0	98.18	13.30	93.48	100.0	30.0
Dexter	-49.09	9.70	15.1	90.30	0.61	29.51	96.36	3.90	99.01	1.5	12.9
Dorothea	-56.36	22.24	30.0	77.76	0.07	12.31	98.18	8.54	95.92	100.0	50.0
Gisette	-70.91	3.48	3.5	96.52	1.80	5.56	98.18	1.37	98.63	18.3	0.0
Madelon	-49.09	16.78	31.5	83.22	1.60	0.0	100.00	7.17	96.95	1.6	0.0

BER*: same model using all the variables as a rough guide

features on all datasets. The Table 26.1 presents the detailed original results obtained by the method.

In order to evaluate the method both on the BER and the number of the selected variables, we report the method results and all the challenge entry results on a bi-criteria plan displayed in Figure 26.1.

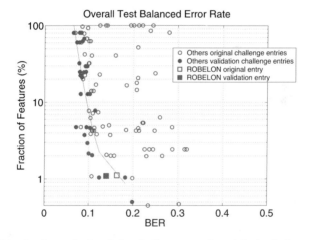

Fig. 26.1. Bi-criteria analysis of the challenge results with the balanced error rate versus the number of selected variables (the line is only a guide for the eyes)

The Table 26.1 and the Figure 26.1 show that restricting ourselves to a simple model with no bootstrap techniques cannot allow us to reach very good BER, particularly on databases as ARCENE where the number of examples is quite small. Several methods obtain a better error rate, up to twice better than the 'ROBELON' method, but at the expense of a significantly higher number of selected variables. Although admittedly not the most adapted for accuracy on some datasets, this simple model indeed reaches a 'reasonable' BER. The proposed method, combined with backward elimination using only one neural network, selects very few variables compared with the other methods. The proposed variable selection technique exhibits the expected behavior by both keeping the BER to a reasonable level (better than the BER with all features, except for ARCENE as already discussed, close to the average result of the challenge) and dramatically reducing the number of features on all datasets. The number of selected probes decreases when the number of the training examples increases.

26.5 Conclusions

We presented a new measure which allows to estimate the importance of each input variable of a model. This measure has no adjustable parameter, is applicable on every kind of model and for classification or regression task.

Experimental results on the NIPS 2003 feature selection challenge show that using this measure coupled with backward elimination allows to reduce considerably the number of input variables with no degradation of the modeling accuracy. Future work will investigate the behavior of this importance measurement when applied to variable selection with bootstrapped models.

References

W. G. Baxt and H. White. Bootstrapping confidence intervals for clinical inputs variable effects in a network trained to identify the presence of acute myocardial infraction. *Neural Computation*, 7:624–638, 1995.

Leo Breiman. Random forest. *Machine Learning*, 45, 2001.

Anthony N. Burkitt. Refined pruning techniques for feed-forward neural networks. *Complex Systems*, 6:479–494, 1992.

Raphael Féraud and Fabrice Clérot. A methodology to explain neural network classification. *Neural Networks*, 15:237–246, 2002.

Isabelle Guyon and André Elisseeff. An introduction to variable and feature selection. *JMLR*, 3(Mar):1157–1182, 2003.

R. Kohavi and G. John. Wrappers for feature subset selection. *Artificial Intelligence*, 97(1-2), 1997.

Peter S. Maybeck. *Stochastic models, estimation, and control*, volume 141 of *Mathematics in Science and Engineering*. 1979.

J. Moody. *Prediction Risk and Architecture Selection for Neural Networks*. From Statistics to Neural Networks-Theory and Pattern Recognition. Springer-Verlag, 1994.

A. N. Réfénes, A. Zapranis, and J. Utans. Stock performance using neural networks: A comparative study with regression models. *Neural Network*, 7:375–388, 1994.

D. E. Rumelhart, G. E. Hinton, and R. J. Williams. Learning internal representations by error propagation. In *Parallel Distributed Processing: Explorations in the Microstructures of Cognition*, volume 1, 1986.

Greg Welch and Gary Bishop. An introduction to the kalman filter. In *SIGGRAPH*, Los Angeles, August 12-17 2001. Course 8.

Part III

New Perspectives in Feature Extraction

Chapter 27

Spectral Dimensionality Reduction

Yoshua Bengio, Olivier Delalleau, Nicolas Le Roux, Jean-François Paiement,
Pascal Vincent, and Marie Ouimet

Département d'Informatique et Recherche Opérationnelle
Centre de Recherches Mathématiques
Université de Montréal
Montréal, Québec, Canada, H3C 3J7
http://www.iro.umontreal.ca/~bengioy

Summary. In this chapter, we study and put under a common framework a number
of non-linear dimensionality reduction methods, such as Locally Linear Embedding,
Isomap, Laplacian eigenmaps and kernel PCA, which are based on performing an
eigen-decomposition (hence the name "spectral"). That framework also includes
classical methods such as PCA and metric multidimensional scaling (MDS). It also
includes the data transformation step used in spectral clustering. We show that in
all of these cases the learning algorithm estimates the principal eigenfunctions of
an operator that depends on the unknown data density and on a kernel that is not
necessarily positive semi-definite. This helps generalizing some of these algorithms
so as to predict an embedding for out-of-sample examples without having to retrain
the model. It also makes it more transparent what these algorithm are minimizing
on the empirical data and gives a corresponding notion of generalization error.

27.1 Introduction

Unsupervised learning algorithms attempt to extract important characteristics of the unknown data distribution from the given examples. High-density
regions are such salient features and they can be described by clustering algorithms (where typically each cluster corresponds to a high density "blob") or
by manifold learning algorithms (which discover high-density low-dimensional
surfaces). A more generic description of the density is given by algorithms that
estimate the density function.

In the context of supervised learning (each example is associated with a
target label) or semi-supervised learning (a few examples are labeled but most
are not), manifold learning algorithms can be used as pre-processing methods
to perform dimensionality reduction. Each input example is then associated
with a low-dimensional representation, which corresponds to its estimated coordinates on the manifold. Since the manifold learning can be done without
using the target labels, it can be applied on all of the input examples (both

Y. Bengio et al.: *Spectral Dimensionality Reduction*, StudFuzz **207**, 519–550 (2006)
www.springerlink.com   © Springer-Verlag Berlin Heidelberg 2006

those labeled and those unlabeled). If there are many unlabeled examples, it
has been shown that they can help to learn a more useful low-dimensional representation of the data (Belkin and Niyogi, 2003b). Dimensionality reduction
is an interesting alternative to feature selection: for instance classical Principal Component Analysis (PCA) is used successfully in Chapter Chapter 10.
Like feature selection it yields a low-dimensional representation, which helps
building lower capacity predictors in order to improve generalization. However, unlike feature selection it may preserve information from all the original
input variables. In fact, if the data really lie on a low-dimensional manifold,
it may preserve almost all of the original information while representing it
in a way that eases learning. For example, manifold learning algorithms such
as those described in this chapter often have the property of "unfolding" the
manifold, i.e. flattening it out, as shown in Figure 27.1. On the other hand,
these techniques being purely unsupervised, they may throw away low variance variations that are highly predictive of the target label, or keep some
with high variance but irrelevant for the classification task at hand. It is still
possible to *combine* dimensionality reduction with a feature extraction algorithm, the latter being applied on the reduced coordinates in order to select
those most appropriate for classification (e.g. in Chapter Chapter 29, a PCA
step is used in experiments for the feature selection challenge). In addition to
being useful as a preprocessing step for supervised or semi-supervised learning, linear and non-linear dimensionality reduction is also often used for data
analysis and visualization, e.g. (Vlachos et al., 2002), since visualizing the projected data (two or three dimensions at a time) can help to better understand
them.

In the last few years, many unsupervised learning algorithms have been
proposed, which share the use of an eigen-decomposition for obtaining a lower-
dimensional embedding of the data that characterizes a non-linear manifold
near which the data would lie: Locally Linear Embedding (LLE) (Roweis
and Saul, 2000), Isomap (Tenenbaum et al., 2000) and Laplacian eigenmaps (Belkin and Niyogi, 2003a). There are also many variants of spectral
clustering (Weiss, 1999, Ng et al., 2002), in which such an embedding is an
intermediate step before obtaining a clustering of the data that can capture
flat, elongated and even curved clusters. The two tasks (manifold learning and
clustering) are linked because the clusters that spectral clustering manages to
capture can be arbitrary curved manifolds (as long as there is enough data to
locally capture the curvature of the manifold): clusters and manifold both are
zones of high density. An interesting advantage of the family of manifold learning algorithms described in this chapter is that they can easily be applied in
the case of non-vectorial data as well as data for which no vectorial representation is available but for which a similarity function between objects can be
computed, as in the MDS (multi-dimensional scaling) algorithms (Torgerson,
1952, Cox and Cox, 1994).

There are of course several dimensionality reduction methods that do not fall in the spectral framework described here, but which may have interesting connections nonetheless. For example, the principal curves algorithms (Hastie and Stuetzle, 1989, Kegl and Krzyzak, 2002) have been introduced based on geometric grounds, mostly for 1-dimensional manifolds. Although they optimize a different type of criterion, their spirit is close to that of LLE and Isomap. Another very interesting family of algorithms is the Self-Organizing Map (Kohonen, 1990). With these algorithms, the low dimensional embedding space is discretized (into topologically organized centers) and one learns the coordinates in the raw high-dimensional space of each of these centers. Another neural network like approach to dimensionality reduction is the auto-associative neural network (Rumelhart et al., 1986, Bourlard and Kamp, 1988, Saund, 1989), in which one trains a multi-layer neural network to predict its input, but forcing the intermediate representation of the hidden units to be a compact code. In section 27.2.7 we discuss in more detail a family of density estimation algorithms that can be written as mixtures of Gaussians with low-rank covariance matrices, having intimate connections with the LLE and Isomap algorithms.

An interesting question that will not be studied further in this chapter is that of selecting the dimensionality of the embedding. This is fundamentally a question of model selection. It could be addressed using traditional model selection methods (such as cross-validation) when the low-dimensional representation is used as input for a supervised learning algorithm. Another approach is that of inferring the dimensionality based on purely unsupervised grounds, using the geometric properties of the empirical data distribution (Kégl, 2003).

27.1.1 Transduction and Induction

The end result of most inductive machine learning algorithms is a function that minimizes the empirical average of a loss criterion (possibly plus regularization). The function can be applied on new points and for such learning algorithms it is clear that the ideal solution is a function that minimizes the expected value of that loss criterion under the unknown true distribution from which the data was sampled. That expected loss is known as the generalization error.

However, such a characterization was missing for spectral embedding algorithms such as metric Multi-Dimensional Scaling (MDS) (Torgerson, 1952, Cox and Cox, 1994), spectral clustering (see (Weiss, 1999) for a review), Laplacian eigenmaps (Belkin and Niyogi, 2003a), Locally Linear Embedding (LLE) (Roweis and Saul, 2000) and Isomap (Tenenbaum et al., 2000), which are used either for dimensionality reduction or for clustering. As such these algorithms are therefore really *transduction* algorithms: any test data for which an embedding is desired must be included in the (unlabeled) "training set" on which the algorithm is applied. For example, if the embedding obtained is used as an input representation for a supervised learning algorithm, the input part of the test examples must be provided at the time of learning the embedding. The basic form of these algorithms does not provide a generic function that can be applied to new points in order to obtain an embedding or a cluster membership, and the notion of generalization error that would be implicitly minimized is not clearly defined either.

As a natural consequence of providing a unifying framework for these algorithms, we provide an answer to these questions. A loss criterion for spectral embedding algorithms can be defined. It is a reconstruction error that depends on pairs of examples. Minimizing its average value yields the eigenvectors that provide the classical output of these algorithms, i.e. the embeddings. Minimizing its expected value over the true underlying distribution yields the eigenfunctions of a linear operator (called L here) that is defined with a similarity function (a kernel, but not necessarily positive semi-definite) and the data-generating density. When the kernel is positive semi-definite and we work with the empirical density there is a direct correspondence between these algorithms and kernel Principal Component Analysis (PCA) (Schölkopf et al., 1998). Our work is also a direct continuation of previous work (Williams and Seeger, 2000) noting that the Nyström formula and the kernel PCA projection (which are equivalent) represent an approximation of the eigenfunctions of the above linear operator. Previous analysis of the convergence of generalization error of kernel PCA (Shawe-Taylor et al., 2002, Shawe-Taylor and Williams, 2003, Zwald et al., 2004) also help to justify the view that these methods are estimating the convergent limit of some eigenvectors (at least when the kernel is positive semi-definite). The eigenvectors can then be turned into estimators of eigenfunctions, which can therefore be applied to new points, turning the spectral embedding algorithms into function induction algorithms. The Nyström formula obtained this way is well known (Baker, 1977), and will be given in eq. 27.2 below. This formula has been used previously for estimating extensions of eigenvectors in Gaussian process regression (Williams and Seeger, 2001), and it was noted (Williams and Seeger, 2000) that it corresponds to the projection of a test point computed with kernel PCA.

In order to extend spectral embedding algorithms such as LLE and Isomap to out-of-sample examples, this chapter defines for these spectral embedding algorithms data-dependent kernels k_m that can be applied outside of the training set. See also the independent work (Ham et al., 2003) for a kernel view of LLE and Isomap, but where the kernels are only applied on the training set.

Obtaining an induced function that can be applied to out-of-sample examples is not only interesting from a theoretical point of view, it is also computationally useful. It allows us to say something about new examples without having to re-do the kernel computations (building the Gram matrix, normalizing it, and computing the principal eigenvectors, which all take at least time quadratic in the number of training examples). The formula proposed requires time linear in the training set size.

Additional contributions of this chapter include a characterization of the empirically estimated eigenfunctions in terms of eigenvectors in the case where the kernel is not positive semi-definite (which is often the case for MDS and Isomap), a convergence theorem linking the Nyström formula to the eigenfunctions of L, as well as experiments on MDS, Isomap, LLE and spectral clustering / Laplacian eigenmaps showing that the Nyström formula for out-of-sample examples is accurate.

27.1.2 Notation

To simplify the presentation, we will consider the vector-space versions of these algorithms, in which we start from a data set $X = (x_1, \ldots, x_m)$ with $x_i \in \mathbb{R}^n$ sampled i.i.d. from an unknown distribution with density $p(\cdot)$. However, the results in this chapter can be readily extended to the case of arbitrary objects, with $p(x)dx$ replaced by $d\mu(x)$ with $\mu(x)$ an appropriate measure, and the only quantity that is required in the algorithms is the similarity or distance between pairs of objects (e.g. similarity $k_m(x_i, x_j)$ below). See for example the treatment of pairwise measurement data for LLE (Saul and Roweis, 2002) and Isomap (Tenenbaum et al., 2000), and for MDS in section 27.2.2.

Below we will use the notation

$$E_x[f(x)] = \int f(x)p(x)dx$$

for averaging over $p(x)$ and

$$\hat{E}_x[f(x)] = \frac{1}{m} \sum_{i=1}^{m} f(x_i)$$

for averaging over the data in X, i.e. over the empirical distribution denoted $\hat{p}(x)$. We will denote kernels with $k_m(x, y)$ or $\tilde{k}(x, y)$, symmetric functions, not always positive semi-definite, that may depend not only on x and y but also on the data X. The spectral embedding algorithms construct an affinity matrix K, either explicitly through

$$K_{ij} = k_m(x_i, x_j) \tag{27.1}$$

or implicitly through a procedure that takes the data X and computes K. We denote by $v_{r,i}$ the i-th coordinate of the r-th eigenvector of K (sorted in order of decreasing eigenvalues), associated with the eigenvalue ℓ_r. With these notations, the Nyström formula discussed above can be written:

$$f_{r,m}(x) = \frac{\sqrt{m}}{\ell_r} \sum_{i=1}^{m} v_{r,i} k_m(x, x_i) \tag{27.2}$$

where $f_{r,m}$ is the r-th Nyström estimator with m samples. We will show in section 27.3 that it estimates the r-th eigenfunction of a linear operator and that it provides an embedding for a new example x.

27.2 Data-Dependent Kernels for Spectral Embedding Algorithms

The first and foremost observation to make is that many spectral embedding algorithms can be cast in a common framework. The spectral embedding algorithms can be seen to build a $(m \times m)$ similarity matrix $\boldsymbol{K}$ (also called the *Gram* matrix)[1] and compute its principal eigenvectors $\boldsymbol{v}_r = (v_{r,1}, \ldots, v_{r,m})^T$ (one entry per exemple) with eigenvalues ℓ_r (sorted by decreasing order). *The embedding associated with the i-th training example is given by the i-th element of the principal eigenvectors, up to some scaling:*

$$\mathbf{P}(\boldsymbol{x}_i) = (v_{1,i}, v_{2,i}, \ldots, v_{N,i})^T \qquad (27.3)$$

where $N \leq m$ is the desired number of embedding coordinates. The scaling factor depends on the algorithm: for instance, in kernel PCA, MDS and Isomap, $v_{r,i}$ is multiplied by $\sqrt{\ell_r}$, and in LLE it is multiplied by $\sqrt{m}$ to obtain the actual embedding coordinates.

In general, we will see that K_{ij} depends not only on $(\boldsymbol{x}_i, \boldsymbol{x}_j)$ but also on the other training examples. Nonetheless, as we show below, it can always be written $K_{ij} = k_m(\boldsymbol{x}_i, \boldsymbol{x}_j)$ where k_m is a "data-dependent" kernel (i.e. it is a function of the m elements of the training set $\boldsymbol{X}$, and not just of its two arguments). In many algorithms a matrix $\tilde{\boldsymbol{K}}$ is first formed from a simpler, often data-independent kernel (such as the Gaussian kernel), and then transformed into $\boldsymbol{K}$. We want to think of the entries of $\boldsymbol{K}$ as being generated by applying k_m to the pairs $(\boldsymbol{x}_i, \boldsymbol{x}_j)$ because this will help us generalize to new examples not in the training set $\boldsymbol{X}$, and it will help us thinking about what happens as m increases.

For each of these methods, by defining a kernel k_m that can be applied outside of the training set, we will be able to generalize the embedding to a new point $\boldsymbol{x}$, via the Nyström formula (eq. 27.2 above, and section 27.3.2). This will only require computations of the form $k_m(\boldsymbol{x}, \boldsymbol{x}_i)$ with $\boldsymbol{x}_i$ a training point.

27.2.1 Kernel Principal Component Analysis

Kernel PCA is an unsupervised manifold learning technique that maps data points to a new space, generally lower-dimensional (but not necessarily). It generalizes the Principal Component Analysis approach to non-linear transformations using the kernel trick (Schölkopf et al., 1996, 1998, Schölkopf et al., 1999). One considers the data mapped into a "feature space", a Hilbert space of possibly infinite dimension such that if $\boldsymbol{x}$ is mapped to $\tilde{\phi}(\boldsymbol{x})$, we have

---

[1]For Laplacian eigenmaps (section 27.2.4) and LLE (section 27.2.6), the matrix $\boldsymbol{K}$ discussed here is not the one defined in the original papers on these algorithms, but a transformation of it to reverse the order of eigenvalues, as we see below.

$\langle \tilde{\phi}(\boldsymbol{x}), \tilde{\phi}(\boldsymbol{y}) \rangle = \tilde{k}(\boldsymbol{x}, \boldsymbol{y})$. Here, $\tilde{k}$ must be a positive (semi)-definite kernel, and is often taken as the Gaussian kernel[2], i.e.

$$\tilde{k}(\boldsymbol{x}, \boldsymbol{y}) = e^{-\frac{\|\boldsymbol{x}-\boldsymbol{y}\|^2}{\sigma^2}}. \tag{27.4}$$

The kernel PCA algorithm consists in performing PCA in the feature space: it implicitly finds the leading eigenvectors and eigenvalues of the covariance of the projection $\tilde{\phi}(\boldsymbol{x})$ of the data. *If the data are centered in feature space* $(\hat{E}_{\boldsymbol{x}}[\tilde{\phi}(\boldsymbol{x})] = 0)$, the (empirical) feature space covariance matrix is $\boldsymbol{C} = \hat{E}_{\boldsymbol{x}}[\tilde{\phi}(\boldsymbol{x})\tilde{\phi}(\boldsymbol{x})^T]$. In general, however, the data are not centered, and we need to define a "centered" mapping

$$\phi_m(\boldsymbol{x}) = \tilde{\phi}(\boldsymbol{x}) - \frac{1}{m}\sum_{i=1}^m \tilde{\phi}(\boldsymbol{x}_i)$$

and an associated *data-dependent* kernel k_m such that $k_m(\boldsymbol{x}, \boldsymbol{y}) = \langle \phi_m(\boldsymbol{x}), \phi_m(\boldsymbol{y}) \rangle$, which rewrites:

$$k_m(\boldsymbol{x}, \boldsymbol{y}) = \tilde{k}(\boldsymbol{x}, \boldsymbol{y}) - \hat{E}_{\boldsymbol{x}'}[\tilde{k}(\boldsymbol{x}', \boldsymbol{y})] - \hat{E}_{\boldsymbol{y}'}[\tilde{k}(\boldsymbol{x}, \boldsymbol{y}')] + \hat{E}_{\boldsymbol{x}',\boldsymbol{y}'}[\tilde{k}(\boldsymbol{x}', \boldsymbol{y}')]. \tag{27.5}$$

The empirical covariance matrix $\boldsymbol{C}$ in "feature space" is thus actually defined by

$$\boldsymbol{C} = \hat{E}_{\boldsymbol{x}}[\phi_m(\boldsymbol{x})\phi_m(\boldsymbol{x})^T] \tag{27.6}$$

with eigenvectors $\boldsymbol{w}_r$ associated with eigenvalues λ_r. As shown in (Schölkopf et al., 1998), this eigen-decomposition of $\boldsymbol{C}$ is related to the one of $\boldsymbol{K}$ (the Gram matrix defined by eq. 27.1) through $\lambda_r = \ell_r/m$ and

$$\boldsymbol{w}_r = \frac{1}{\sqrt{\ell_r}}\sum_{i=1}^m v_{r,i}\phi_m(\boldsymbol{x}_i)$$

where $\boldsymbol{v}_r$ are the eigenvectors of $\boldsymbol{K}$, associated with eigenvalues ℓ_r. As in PCA, one can then obtain the embedding of a training point $\boldsymbol{x}_i$ by the projection of $\phi_m(\boldsymbol{x}_i)$ on the leading eigenvectors $(\boldsymbol{w}_1, \dots, \boldsymbol{w}_N)$ of $\boldsymbol{C}$, which yields exactly the embedding of eq. 27.3, if we multiply $v_{r,i}$ by $\sqrt{\ell_r}$.

Note that, as in PCA, we can also compute the projection $\mathbf{P}(\boldsymbol{x}) = (P_1(\boldsymbol{x}), \dots, P_N(\boldsymbol{x}))^T$ for a new point $\boldsymbol{x}$, which is written

$$P_r(\boldsymbol{x}) = \langle \boldsymbol{w}_r, \phi_m(\boldsymbol{x}) \rangle = \frac{1}{\sqrt{\ell_r}}\sum_{i=1}^m v_{r,i}k_m(\boldsymbol{x}_i, \boldsymbol{x}). \tag{27.7}$$

---

[2]Using the Gaussian kernel is not always a good idea, as seen in section 27.5.1, and other nonlinear kernels, such as polynomial kernels, may be more suited.

This is the key observation that will allow us, in section 27.3.2, to extend to new points the embedding obtained with other spectral algorithms.

27.2.2 Multi-Dimensional Scaling

Metric Multi-Dimensional Scaling (MDS) (Torgerson, 1952, Cox and Cox, 1994) starts from a notion of distance $d(\boldsymbol{x}, \boldsymbol{y})$ that is computed between each pair of training examples to fill a matrix $K_{ij} = d^2(\boldsymbol{x}_i, \boldsymbol{x}_j)$. The idea is to find a low-dimensional embedding of the dataset $\boldsymbol{X}$ that preserves the given distances between training points. To do so, the distances are converted to equivalent dot products using the "double-centering" formula, which makes K_{ij} depend not only on $(\boldsymbol{x}_i, \boldsymbol{x}_j)$ but also on all the other examples:

$$K_{ij} = -\frac{1}{2}\left(\tilde{K}_{ij} - \frac{1}{m}S_i - \frac{1}{m}S_j + \frac{1}{m^2}\sum_k S_k\right) \qquad (27.8)$$

where the S_i are the row sums of $\tilde{\boldsymbol{K}}$:

$$S_i = \sum_{j=1}^{m} \tilde{K}_{ij}. \qquad (27.9)$$

Eq. 27.8 is used to obtain for K_{ij} the centered dot product between $\boldsymbol{x}_i$ and $\boldsymbol{x}_j$ from the pairwise squared distances given by d^2, just as eq. 27.5 yields the centered dot product (in feature space) from the pairwise non-centered dot products given by $\tilde{k}$. The embedding of the example $\boldsymbol{x}_i$ is then given by eq. 27.3 with $v_{r,i}$ multiplied by $\sqrt{\ell_r}$. If d is the Euclidean distance, this is the same embedding as in classical (linear) PCA.

A corresponding data-dependent kernel which generates the matrix $\boldsymbol{K}$ is:

$$k_m(\boldsymbol{x}, \boldsymbol{y}) = -\frac{1}{2}\left(d^2(\boldsymbol{x}, \boldsymbol{y}) - \hat{E}_{\boldsymbol{x}'}[d^2(\boldsymbol{x}', \boldsymbol{y})] - \hat{E}_{\boldsymbol{y}'}[d^2(\boldsymbol{x}, \boldsymbol{y}')] + \hat{E}_{\boldsymbol{x}', \boldsymbol{y}'}[d^2(\boldsymbol{x}', \boldsymbol{y}')]\right). \qquad (27.10)$$

27.2.3 Spectral Clustering

Several variants of spectral clustering have been proposed (Weiss, 1999). They can yield impressively good results where traditional clustering looking for "round blobs" in the data, such as k-means, would fail miserably (see Figure 27.1). It is based on two main steps: first embedding the data points in a space in which clusters are more "obvious" (using the eigenvectors of a Gram matrix), and then applying a classical clustering algorithm such as k-means, e.g. as in (Ng et al., 2002). To construct the spectral clustering affinity matrix $\boldsymbol{K}$, we first apply a data-independent kernel $\tilde{k}$ such as the Gaussian kernel to each pair of examples: $\tilde{K}_{ij} = \tilde{k}(\boldsymbol{x}_i, \boldsymbol{x}_j)$. The matrix $\tilde{K}$ is then normalized, e.g. using "divisive" normalization (Weiss, 1999, Ng et al., 2002)[3] :

$$K_{ij} = \frac{\tilde{K}_{ij}}{\sqrt{S_i S_j}} \tag{27.11}$$

with S_i and S_j defined by eq. 27.9. To obtain N clusters, the first N principal eigenvectors of K are computed and k-means is applied on the embedding coordinates after normalizing each embedding vector to have unit norm: the r-th coordinate of the i-th example is $v_{r,i}/\sqrt{\sum_{l=1}^{N} v_{l,i}^2}$. Note that if one is interested in the embedding prior to normalization, this embedding should be multiplied by $\sqrt{m}$ to be stable (to keep the same order of magnitude) as m varies.

To generalize spectral clustering to out-of-sample points, we will need a kernel that could have generated that matrix K:

$$k_m(\boldsymbol{x}, \boldsymbol{y}) = \frac{1}{m} \frac{\tilde{k}(\boldsymbol{x}, \boldsymbol{y})}{\sqrt{\hat{E}_{\boldsymbol{x}'}[\tilde{k}(\boldsymbol{x}', \boldsymbol{y})]\hat{E}_{\boldsymbol{y}'}[\tilde{k}(\boldsymbol{x}, \boldsymbol{y}')]}} . \tag{27.12}$$

Fig. 27.1. Example of the transformation learned as part of spectral clustering. Input data on the left, transformed data on the right. Gray level and cross/circle drawing are only used to show which points get mapped where: the mapping reveals both the clusters and the internal structure of the two manifolds.

Note that this divisive normalization comes out of the justification of spectral clustering as a relaxed statement of the min-cut problem (Chung, 1997, Spielman and Teng, 1996) (to divide the examples into two groups such as to minimize the sum of the "similarities" between pairs of points straddling the

---

[3]Better embeddings for clustering are usually obtained if we define $S_i = \sum_{j \neq i} \tilde{K}_{ij}$: this alternative normalization can also be cast into the general framework developed here, with a slightly different kernel. Also, one could take the row average instead of the row sum, which seems more natural even if it does not change the embedding.

two groups). The additive normalization performed with kernel PCA (eq. 27.5) makes sense geometrically as a centering in feature space. Both the divisive normalization and the additive normalization procedures make use of a kernel row/column average. It would be interesting to find a similarly pleasing geometric interpretation to the divisive normalization.

27.2.4 Laplacian Eigenmaps

The Laplacian eigenmaps method is a recently proposed dimensionality reduction procedure (Belkin and Niyogi, 2003a) that was found to be very successful for *semi-supervised learning*, where one uses a large unlabeled dataset to learn the manifold structure, thus reducing the dimensionality of labeled data (which can benefit to supervised learning algorithms). Several variants have been proposed by the authors and we focus here on the latest one, but they all share the same spirit.

The Laplacian operator has a natural interpretation as a smoothness functional: we look for an embedding $(\boldsymbol{y}_1, \ldots, \boldsymbol{y}_m)$ of the training points such that $\|\boldsymbol{y}_i - \boldsymbol{y}_j\|^2$ is small when i and j are "near" each other, i.e. when $\tilde{k}(\boldsymbol{x}_i, \boldsymbol{x}_j)$ is large (if $\tilde{k}$ can be interpreted as a similarity function). This corresponds to minimizing $\sum_{ij} \|\boldsymbol{y}_i - \boldsymbol{y}_j\|^2 \tilde{k}(\boldsymbol{x}_i, \boldsymbol{x}_j)$. It has to be done under a norm constraint, and an appropriate one is that, denoting $\boldsymbol{Y}$ the $(m \times N)$ matrix whose i-th row is $\boldsymbol{y}_i$, we force $\boldsymbol{Y}^T \boldsymbol{S} \boldsymbol{Y} = \boldsymbol{I}$, where $\boldsymbol{S}$ is the diagonal matrix with elements S_i from eq. 27.9 (row sums of $\tilde{\boldsymbol{K}}$, possibly ignoring diagonal terms). This norm constraint has the advantage of giving an appropriate weight to examples that are "connected" to more other examples. Rearranging this criterion, the solutions to the constrained optimization problem correspond to the following generalized eigenproblem:

$$(\boldsymbol{S} - \tilde{\boldsymbol{K}})\boldsymbol{z}_r = \sigma_r \boldsymbol{S} \boldsymbol{z}_r \qquad (27.13)$$

with eigenvalues σ_r, and eigenvectors $\boldsymbol{z}_r$ being the columns of $\boldsymbol{Y}$. The solution with smallest (zero) eigenvalue corresponds to the uninteresting solution with constant embedding, so it is discarded. The eigenvalue corresponding to a solution quantifies the above defined smoothness, so we keep the N solutions with smallest non-zero eigenvalues, yielding the desired embedding.

Here, the matrix $\boldsymbol{S} - \tilde{\boldsymbol{K}}$ is the so-called graph Laplacian, and it can be shown (Belkin and Niyogi, 2003a) to be an approximation of the manifold Laplace Beltrami operator, when using the Gaussian kernel or the k-nearest-neighbor kernel for the similarity $\tilde{k}(\cdot, \cdot)$ on the graph. The k-nearest-neighbor kernel is represented by the *symmetric* matrix $\tilde{\boldsymbol{K}}$ whose element (i, j) is 1 if $\boldsymbol{x}_i$ and $\boldsymbol{x}_j$ are *k-nearest-neighbors* ($\boldsymbol{x}_i$ is among the k nearest neighbors of $\boldsymbol{x}_j$ or vice versa) and 0 otherwise. Approximating the Laplace Beltrami operator is motivated by the fact that its eigenfunctions are mappings that optimally preserve the "locality" of data (Belkin and Niyogi, 2003a).

It turns out that the above algorithm results in the same embedding (up to scaling) that is computed with the spectral clustering algorithm from (Shi and Malik, 1997) described in section 27.2.3: as noted in (Weiss, 1999) (Normalization Lemma 1), an equivalent result (up to a component-wise scaling of the embedding) can be obtained by considering the principal eigenvectors v_r of the normalized matrix K defined in eq. 27.11. To fit the common framework for spectral embedding in this chapter, we have used the latter formulation. Therefore, the same data-dependent kernel can be defined as for spectral clustering (eq. 27.12) to generate the matrix K, i.e. spectral clustering just adds a clustering step after a Laplacian eigenmaps dimensionality reduction.

27.2.5 Isomap

Isomap (Tenenbaum et al., 2000) generalizes MDS (section 27.2.2) to nonlinear manifolds. It is based on replacing the Euclidean distance by an empirical approximation of the geodesic distance on the manifold. We define the *geodesic distance $\hat{d}(\cdot, \cdot)$ with respect to a data set X, a distance $d(\cdot, \cdot)$ and a neighborhood k as follows:*

$$\hat{d}(x, y) = \min_{\pi} \sum_{i=1}^{|\pi|} d(\pi_i, \pi_{i+1}) \qquad (27.14)$$

where π is a sequence of points of length $|\pi| = l \geq 2$ with $\pi_1 = x$, $\pi_l = y$, $\pi_i \in X \; \forall i \in \{2, \ldots, l-1\}$ and (π_i, π_{i+1}) are k-nearest-neighbors of each other. The length $|\pi| = l$ is free in the minimization. The Isomap algorithm obtains the normalized matrix K from which the embedding is derived by transforming the raw pairwise distances matrix as follows: (1) compute the matrix $\tilde{K}_{ij} = \hat{d}^2(x_i, x_j)$ of squared geodesic distances with respect to the data X and (2) apply to this matrix the double-centering transformation (eq. 27.8), as for MDS. As in MDS, the embedding of x_i is given by eq. 27.3 with $v_{r,i}$ multiplied by $\sqrt{\ell_r}$. Step (1) can be done in $O(n^3)$ operations very easily (e.g. by Floyd's algorithm), but in (Tenenbaum et al., 2000) it is suggested to use more efficient algorithms exploiting the sparse structure of the neighborhood graph, such as those presented in (Kumar et al., 1994).

There are several ways to define a kernel that generates K and also generalizes out-of-sample. The solution we have chosen simply computes the geodesic distances without involving the out-of-sample point(s) along the geodesic distance sequence (except for the last distance). This is automatically achieved with the above definition of geodesic distance $\hat{d}$, which only uses the training points to find the shortest path between x and y. The double-centering kernel transformation of eq. 27.10 can then be applied to obtain k_m, using the geodesic distance $\hat{d}$ instead of the MDS distance d.

27.2.6 Locally Linear Embedding

The Locally Linear Embedding (LLE) algorithm (Roweis and Saul, 2000) looks for an embedding that preserves the local geometry in the neighborhood of each data point. The idea is to find a low-dimensional representation where the reconstruction of a data point from its neighbors is similar to the one in input space. First, a sparse matrix of local predictive weights W_{ij} is computed, such that $\sum_j W_{ij} = 1$, $W_{ii} = 0$, $W_{ij} = 0$ if $\boldsymbol{x}_j$ is not a k-nearest-neighbor of $\boldsymbol{x}_i$ and $\|(\sum_j W_{ij}\boldsymbol{x}_j) - \boldsymbol{x}_i\|^2$ is minimized. To find those weights, for a given training point $\boldsymbol{x}_i$ with neighbors $(\boldsymbol{y}_{i1}, \ldots, \boldsymbol{y}_{ik})$, a local Gram matrix $\boldsymbol{K}^{(i)}$ is computed, such that $K_{rs}^{(i)} = \langle \boldsymbol{y}_{ir} - \boldsymbol{x}_i, \boldsymbol{y}_{is} - \boldsymbol{x}_i \rangle$. To improve the condition of this Gram matrix (to avoid potential issues when solving the linear system below), it is recommended to add a small multiple of the identity matrix:

$$K_{rs}^{(i)} \leftarrow K_{rs}^{(i)} + \delta_{rs}\frac{\Delta^2}{k}Tr(\boldsymbol{K}^{(i)})$$

with Tr the trace operator, δ the Kronecker symbol, and $\Delta^2 \ll 1$. The weights are then obtained by solving the linear system defined by $\sum_r K_{rs}^{(i)}W_{ir} = 1$ for all s, then rescaling the W_{ir} so that they sum to 1 (Saul and Roweis, 2002).

From the weights W_{ij}, the matrix $\tilde{\boldsymbol{K}} = (\boldsymbol{I} - \boldsymbol{W})^T(\boldsymbol{I} - \boldsymbol{W})$ is formed. The embedding is obtained from the lowest eigenvectors of $\tilde{\boldsymbol{K}}$, except for the eigenvector with the smallest eigenvalue, which is uninteresting because it is proportional to $(1, 1, \ldots, 1)$ (and its eigenvalue is 0). Since we want to select the principal eigenvectors, we define our normalized matrix by $\boldsymbol{K} = c\boldsymbol{I} - \tilde{\boldsymbol{K}}$ (c being any real number) and ignore the top eigenvector (although one could apply an additive normalization to remove the components along the $(1, 1, \ldots, 1)$ direction). The LLE embedding for $\boldsymbol{x}_i$ is then given by eq. 27.3 (multiplied by $\sqrt{m}$), starting at the second eigenvector (since the principal one is constant). If one insists on having a positive semi-definite matrix $\boldsymbol{K}$, one can take for c the largest eigenvalue of $\tilde{\boldsymbol{K}}$ (note that c only changes the eigenvalues additively and has no influence on the embedding of the training set).

In order to define a kernel k_m generating $\boldsymbol{K}$, we first denote by $w(\boldsymbol{x}, \boldsymbol{x}_i)$ the weight of $\boldsymbol{x}_i$ in the reconstruction of any point $\boldsymbol{x} \in \mathbb{R}^n$ by its k nearest neighbors in the training set. This is the same reconstruction as above, i.e. the $w(\boldsymbol{x}, \boldsymbol{x}_i)$ are such that they sum to 1, $\|(\sum_i w(\boldsymbol{x}, \boldsymbol{x}_i)\boldsymbol{x}_i) - \boldsymbol{x}\|^2$ is minimized, and $w(\boldsymbol{x}, \boldsymbol{x}_i) = 0$ if $\boldsymbol{x}_i$ is not in the k nearest neighbors of $\boldsymbol{x}$. If $\boldsymbol{x} = \boldsymbol{x}_j \in \boldsymbol{X}$, we have $w(\boldsymbol{x}, \boldsymbol{x}_i) = \delta_{ij}$. Let us now define a kernel k'_m by $k'_m(\boldsymbol{x}_i, \boldsymbol{x}) = k'_m(\boldsymbol{x}, \boldsymbol{x}_i) = w(\boldsymbol{x}, \boldsymbol{x}_i)$ and $k'_m(\boldsymbol{x}, \boldsymbol{y}) = 0$ when neither $\boldsymbol{x}$ nor $\boldsymbol{y}$ is in the training set $\boldsymbol{X}$. Let k''_m be such that $k''_m(\boldsymbol{x}_i, \boldsymbol{x}_j) = W_{ij} + W_{ji} - \sum_k W_{ki}W_{kj}$ and $k''_m(\boldsymbol{x}, \boldsymbol{y}) = 0$ when either $\boldsymbol{x}$ or $\boldsymbol{y}$ is not in $\boldsymbol{X}$. It can be shown that the kernel $k_m = (c-1)k'_m + k''_m$ is then such that

$$k_m(\boldsymbol{x}_i, \boldsymbol{x}_j) = (c - 1)\delta_{ij} + W_{ij} + W_{ji} - \sum_k W_{ki}W_{kj} = K_{ij}$$

so that it can be used to generate $\boldsymbol{K}$. There could be other ways to obtain a data-dependent kernel for LLE that can be applied out-of-sample: a justification for using this specific kernel will be given in section 27.3.1.

As noted independently in (Ham et al., 2003), LLE can thus be seen as performing kernel PCA with a particular kernel matrix. This identification becomes even more accurate when one notes that getting rid of the constant eigenvector (principal eigenvector of $\boldsymbol{K}$) is equivalent to the centering operation in feature space required for kernel PCA (Ham et al., 2003).

It is interesting to note a recent descendant of Laplacian eigenmaps, Isomap and LLE, called Hessian eigenmaps (Donoho and Grimes, 2003), which considers the limit case of the continuum of the manifold, and replaces the Laplacian in Laplacian eigenmaps by a Hessian. Despite attractive theoretical properties, the Hessian eigenmaps algorithm, being based on estimation of second order derivatives (which is difficult with sparse noisy data), has yet to be applied successfully on real-world high-dimensional data.

27.2.7 Mixtures of Low-Rank Gaussians

Isomap and LLE are two instances of a larger family of unsupervised learning algorithms which characterize the data distribution by a large set of locally linear low-dimensional patches. A simple example of this type of model is a mixture of Gaussians (centered on each example in the standard non-parametric setting) whose covariance matrices are summarized by a few eigenvectors (i.e. principal directions). The mixture of factor analyzers (Ghahramani and Hinton, 1996) is a parametric version of this type of model, in which the EM algorithm is used to estimate the means and the low-rank covariance matrices. A non-parametric version of the mixture of factor analyzers aimed at capturing manifold structure is the Manifold Parzen Windows algorithm (Vincent and Bengio, 2003), which does not require an iterative algorithm for training. With such models, one can obtain a local low-dimensional representation of examples falling near a Gaussian center, but it may be incomparable to the representation obtained for a nearby Gaussian center, because the eigenvectors of the covariance matrices of neighboring Gaussians may not be aligned. In order to perform dimensionality reduction from such models, several algorithms have thus been proposed (Teh and Roweis, 2003, Brand, 2003, Verbeek et al., 2004), which look for a global representation that agrees with each local patch. Although these algorithms do not fall into the "spectral manifold learning" family studied in more detailed in this chapter, they are very close in spirit.

27.2.8 Summary

Algorithm	Description	Connections / Discussion
Kernel PCA	Embedding in a high (possibly infinite) dimensional space (kernel trick) followed by standard PCA.	PCA can be performed using only dot products (i.e. the kernel matrix).
MDS	Distance-preserving embedding in a low-dimensional space.	If distances are obtained from the dot product (or kernel) k, it is equivalent to kernel PCA with the same kernel k.
Spectral Clustering	N-dimensional embedding of the data followed by projection on the unit sphere and any standard clustering algorithm (e.g. N-means), N being the number of clusters.	If the embedding is obtained from the Gaussian kernel as dot product, points nearby have a dot product close to 1 and points far away have a dot product close to 0. Therefore, projecting the points in a space of dimension N will concentrate all points of the same cluster on a single coordinate, each cluster being assigned a different coordinate.
Isomap	Geodesic distance-preserving embedding in a low dimensional space, the geodesic distances being estimated from a neighborhood graph.	Isomap is similar to MDS except it uses (approximate) geodesic distances rather than distances defined from a given kernel.
LLE	Low-dimensional embedding that preserves the linear reconstruction of each point by its neighbors.	LLE is a more "local" algorithm as it only tries to preserve local properties. However, it can be found to be similar to kernel PCA when using a particular kernel matrix. Besides, the removal of the top eigenvector corresponds to the centering of the kernel matrix.
Laplacian eigenmaps	Locality-preserving embedding obtained from the eigenfunctions of the Laplacian operator on a neighborhood graph.	Laplacian eigenmaps is also a "local" algorithm, like LLE. The embedding obtained is the same as spectral clustering (before projecting on the unit sphere): preserving locality is tightly linked to preserving cluster structure. This embedding is motivated by the optimality of the eigenfunctions of the Laplacian on the underlying manifold as locality-preserving mappings.

27.3 Kernel Eigenfunctions for Induction

With the exception of kernel PCA, the spectral manifold learning algorithms presented in section 27.2 do not provide us with an immediate way to obtain the embedding for a new point $x \notin X$. However, for some of them, extensions have already been proposed. We briefly review them in section 27.3.1. In section 27.3.2, we take advantage of the common framework developed in section 27.2: each algorithm being associated with a data-dependent kernel k_m generating a Gram matrix K, we can apply the Nyström formula (eq. 27.2) to obtain the embedding for a new point x.

27.3.1 Extensions to Spectral Embedding Algorithms

For metric MDS, it is suggested in (Gower, 1968) to solve exactly for the coordinates of the new point such that its distances to the training points are the same in the original input space and in the computed embedding, but in general this requires adding a new dimension. Note also that (Williams, 2001) makes a connection between kernel PCA and metric MDS, remarking that kernel PCA is a form of MDS when the kernel is isotropic. In the following, we will pursue this connection in order to obtain out-of-sample embeddings.

A formula has been proposed (de Silva and Tenenbaum, 2003) to approximate Isomap using only a subset of the examples (the "landmark" points) to compute the eigenvectors. Using the notation of this chapter, that formula is

$$e_r(x) = \frac{1}{2\sqrt{\ell_r}} \sum_i v_{r,i}(\hat{E}_{x'}[\hat{d}^2(x', x_i)] - \hat{d}^2(x_i, x)) \qquad (27.15)$$

which is applied to obtain an embedding for the non-landmark examples. One can show (Bengio et al., 2004) that $e_r(x)$ is the Nyström formula when $k_m(x, y)$ is defined as in section 27.2.5. Landmark Isomap is thus equivalent to performing Isomap on the landmark points only and then predicting the embedding of the other points using the Nyström formula, which is the solution we also propose in what follows.

For LLE, with the notations of section 27.2.6, an extension suggested in (Saul and Roweis, 2002) is to take for a new point x the embedding $\mathbf{P}(x) = (P_1(x), \ldots, P_N(x))^T$, where

$$P_r(x) = \sum_{i=1}^m P_r(x_i)w(x, x_i).$$

Interestingly, the same embedding can be obtained from the Nyström formula and the kernel k_m defined in section 27.2.6, when the constant $c \to +\infty$ (Bengio et al., 2004).

27.3.2 From Eigenvectors to Eigenfunctions

From the common framework developed in section 27.2, one can see spectral algorithms as performing a kind of kernel PCA with a specific kernel. (Ng et al., 2002) had already noted the link between kernel PCA and spectral clustering. Recently, (Ham et al., 2003) have also shown how Isomap, LLE and Laplacian eigenmaps can be interpreted as performing a form of kernel PCA.

Here, we propose a similar view, extending the framework to allow negative eigenvalues (which may be the case for Isomap). In addition, those papers did not propose to use this link in order to perform function induction, i.e. obtain an embedding for out-of-sample points. Indeed, since there exists a natural extension to new points for kernel PCA (the projection onto the eigenspaces of the covariance matrix, see eq. 27.7), it is natural to ask whether it makes sense to use such a formula in the more general setting where the kernel may be data-dependent and may have negative eigenvalues.

As noted in (Williams and Seeger, 2000), the kernel PCA projection formula (eq. 27.7) is proportional to the so-called Nyström formula (Baker, 1977, Williams and Seeger, 2000) (eq. 27.2), which has been used successfully to "predict" the value of an eigenvector on a new data point, in order to speed-up kernel methods computations by focusing the heavier computations (the eigen-decomposition) on a subset of examples (Williams and Seeger, 2001). The use of this formula can be justified by considering the convergence of eigenvectors and eigenvalues, as the number of examples increases (Baker, 1977, Koltchinskii, 1998, Koltchinskii and Gin, 2000, Williams and Seeger, 2000). In particular, (Shawe-Taylor et al., 2002, Shawe-Taylor and Williams, 2003, Zwald et al., 2004) give bounds on the kernel PCA convergence error (in the sense of the projection error with respect to the subspace spanned by the eigenvectors), using concentration inequalities.

Based on this kernel PCA convergence results, we conjecture that in the limit, each eigenvector would converge to an eigenfunction for a linear operator (defined below), in the sense that the i-th element of the r-th eigenvector converges to the application of the r-th eigenfunction to x_i. Proposition 2 below formalizes this statement and provides sufficient conditions for such a convergence.

In the following we will assume that the (possibly data-dependent) kernel k_m is *bounded* (i.e. $\exists k_{max}, \forall x, y \ |k_m(x, y)| < k_{max}$) and has a discrete spectrum, i.e. that it can be written as a discrete expansion

$$k_m(x, y) = \sum_{r=1}^{\infty} \alpha_{r,m} \psi_{r,m}(x) \psi_{r,m}(y).$$

Consider the space $\mathcal{H}_p$ of continuous functions f on $\mathbb{R}^n$ that are square integrable as follows:

$$\int f^2(x) p(x) dx < \infty$$

with the data-generating density function $p(x)$. One must note that we actually do not work on functions but on equivalence classes: we say two continuous functions f and g belong to the same equivalence class (with respect to p) if and only if $\int (f(x) - g(x))^2 p(x) dx = 0$ (if p is strictly positive, then each equivalence class contains only one function).

We will assume that k_m converges uniformly in its arguments (in some probabilistic manner, e.g. almost surely or in probability) to its limit k as

$m \to \infty$. We will associate with each k_m a linear operator L_m and with k a linear operator L, such that for any $f \in \mathcal{H}_p$,

$$L_m f = \frac{1}{m} \sum_{i=1}^{m} k_m(\cdot, \boldsymbol{x}_i) f(\boldsymbol{x}_i) \qquad (27.16)$$

and

$$L f = \int k(\cdot, \boldsymbol{y}) f(\boldsymbol{y}) p(\boldsymbol{y}) d\boldsymbol{y} \qquad (27.17)$$

which makes sense because we work in a space of functions defined everywhere. Furthermore, as $k_m(\cdot, \boldsymbol{y})$ and $k(\cdot, \boldsymbol{y})$ are square-integrable in the sense defined above, for each f and each m, the functions $L_m f$ and $L f$ are square-integrable in the sense defined above. We will show that the Nyström formula (eq. 27.2) gives the eigenfunctions of L_m (proposition 1), that their value on the training examples corresponds to the spectral embedding, and that they converge to the eigenfunctions of L (proposition 2). These results will hold even if k_m has negative eigenvalues.

The eigensystems of interest are thus the following:

$$L f_r = \lambda_r f_r \qquad (27.18)$$

and

$$L_m f_{r,m} = \lambda_{r,m} f_{r,m} \qquad (27.19)$$

where (λ_r, f_r) and $(\lambda_{r,m}, f_{r,m})$ are the corresponding eigenvalues and eigenfunctions.

Note that when eq. 27.19 is evaluated only at the $\boldsymbol{x}_i \in \boldsymbol{X}$, the set of equations reduces to the eigensystem

$$\boldsymbol{K} \boldsymbol{v}_r = m \lambda_{r,m} \boldsymbol{v}_r.$$

The following proposition gives a more complete characterization of the eigenfunctions of L_m, even in the case where eigenvalues may be negative. The next two propositions formalize the link already made in (Williams and Seeger, 2000) between the Nyström formula and eigenfunctions of L.

Proposition 1 *L_m has in its image $N \le m$ eigenfunctions of the form:*

$$f_{r,m}(\boldsymbol{x}) = \frac{\sqrt{m}}{\ell_r} \sum_{i=1}^{m} v_{r,i} k_m(\boldsymbol{x}, \boldsymbol{x}_i) \qquad (27.20)$$

with corresponding non-zero eigenvalues $\lambda_{r,m} = \frac{\ell_r}{m}$, where $\boldsymbol{v}_r = (v_{r,1}, \ldots, v_{r,m})^T$ is the r-th eigenvector of the Gram matrix $\boldsymbol{K}$, associated with the eigenvalue ℓ_r.

For $\boldsymbol{x}_i \in \boldsymbol{X}$ these functions coincide with the corresponding eigenvectors, in the sense that $f_{r,m}(\boldsymbol{x}_i) = \sqrt{m} v_{r,i}$.

536   Yoshua Bengio et al.

Proof

First, we show that the $f_{r,m}$ defined by eq. 27.20 coincide with the eigenvectors of K at $x_i \in X$. For $f_{r,m}$ associated with a non-zero eigenvalue,

$$f_{r,m}(x_i) = \frac{\sqrt{m}}{\ell_r} \sum_{j=1}^{m} v_{r,j} k_m(x_i, x_j) = \frac{\sqrt{m}}{\ell_r} \ell_r v_{r,i} = \sqrt{m} v_{r,i}. \qquad (27.21)$$

The v_r being orthonormal the $f_{r,m}$ (for different values of r) are therefore different from each other.

Then for any $x \in \mathbb{R}^n$

$$(L_m f_{r,m})(x) = \frac{1}{m} \sum_{i=1}^{m} k_m(x, x_i) f_{r,m}(x_i) = \frac{1}{\sqrt{m}} \sum_{i=1}^{m} k_m(x, x_i) v_{r,i} = \frac{\ell_r}{m} f_{r,m}(x)$$

$$(27.22)$$

which shows that $f_{r,m}$ is an eigenfunction of L_m with eigenvalue $\lambda_{r,m} = \ell_r/m$. $\square$

Discussion

The previous result shows that the Nyström formula generalizes the spectral embedding outside of the training set. This means the embedding $\mathbf{P}(x) = (P_1(x), \ldots, P_N(x))^T$ for a new point x is given (up to some scaling) by

$$P_r(x) = \frac{f_{r,m}(x)}{\sqrt{m}} = \frac{1}{\ell_r} \sum_{i=1}^{m} v_{r,i} k_m(x, x_i) \qquad (27.23)$$

where the scaling is the same as the one described in section 27.2 (so that the embedding obtained on the training set is coherent with the one obtained from the eigenvectors, thanks to eq. 27.21).

However, there could be many possible generalizations. To justify the use of this particular generalization, the following proposition helps understanding the convergence of these functions as m increases. We would like the out-of-sample embedding predictions obtained with the Nyström formula to be somehow close to the asymptotic embedding (the embedding one would obtain as $m \rightarrow \infty$).

Note also that the convergence of eigenvectors to eigenfunctions shown in (Baker, 1977) applies to data x_i which are deterministically chosen to span a domain, whereas here the x_i form a random sample from an unknown distribution.

Proposition 2 *If $k_m = k$ is bounded and not data-dependent, then the eigenfunctions $f_{r,m}$ of L_m associated with non-zero eigenvalues of multiplicity 1 converge to the corresponding eigenfunctions of L (almost surely, and up to the sign).*

For k_m data-dependent but bounded (almost surely, and independently of m) and converging uniformly to k, if the eigen-decomposition of the Gram *matrix* $(k_m(\boldsymbol{x}_i, \boldsymbol{x}_j))$ *converges*[4] *to the eigen-decomposition of the Gram matrix* $(k(\boldsymbol{x}_i, \boldsymbol{x}_j))$ *then a similar result holds: the eigenfunctions* $f_{r,m}$ *of* L_m *associated with non-zero eigenvalues of multiplicity 1 converge to the corresponding eigenfunctions of* L *(almost surely, and up to the sign).*

Proof

In the following, we will denote by $\hat{f} \in \mathcal{H}_{\hat{p}}$ the restriction of a function $f \in \mathcal{H}_p$ to the training set $\boldsymbol{X} = \{\boldsymbol{x}_1, \ldots, \boldsymbol{x}_m\}$, and by $\hat{L}_m$ the operator in $\mathcal{H}_{\hat{p}}$ defined as in eq. 27.16, which has the same eigenvalues and eigenfunctions as L_m (except the eigenfunctions are restricted to $\boldsymbol{X}$). We start with the case where $k_m = k$. We first take advantage of (Koltchinskii and Gin, 2000), theorem 3.1, that shows that the distance between the eigenvalue spectra of $\hat{L}_m$ and L converges to 0 almost surely. We then use theorem 2.1 from (Koltchinskii, 1998), which is stated as follows. Let k be a symmetric kernel such that $E[\|k(X,X)\|] < +\infty$ and $E[k^2(X,Y)] < +\infty$ (so that the operator L defined by eq. 27.17 is Hilbert-Schmidt and k can be written $k(\boldsymbol{x}, \boldsymbol{y}) = \sum_{i \in I} \mu_i \psi_i(\boldsymbol{x}) \psi_i(\boldsymbol{y})$ with I a discrete set). Suppose that $\mathcal{F}$ is a class of measurable functions such that there exists $F \in \mathcal{H}_p$ verifying $|f(\boldsymbol{x})| \leq F(\boldsymbol{x})$ for all $f \in \mathcal{F}$. Moreover, suppose that for all $i \in I$, $\{f\psi_i : f \in \mathcal{F}\} \in GC(p)$, where $GC(p)$ denotes the set of p-Glivenko-Cantelli classes (see, e.g., (van der Vaart and Wellner, 1996)). Then, for all non-zero eigenvalue λ_r

$$\sup_{f,g \in \mathcal{F}} \left| \langle P_r(\hat{L}_m)\hat{f}, \hat{g} \rangle_{\mathcal{H}_{\hat{p}}} - \langle P_r(L)f, g \rangle_{\mathcal{H}_p} \right| \to 0 \qquad (27.24)$$

almost surely when $m \to +\infty$, with $P_r(L)$ the projection on the r-th eigenspace of L, and $P_r(\hat{L}_m)$, with probability 1 and for m sufficiently large, the projection on the corresponding eigenspace of $\hat{L}_m$ (for more details see (Koltchinskii, 1998)).

Let us consider the r-th eigenspace of L (of dimension 1 because we have considered eigenvalues of multiplicity 1), i.e. the eigenspace spanned by the eigenfunction f_r: the r-th eigenspace of $\hat{L}_m$ is also 1-dimensional, almost surely (because of the convergence of the spectrum), and spanned by $f_{r,m}$. Let $\boldsymbol{x} \in \mathbb{R}^n$ be any point in the input space, and $\mathcal{F} = \{h_{\boldsymbol{x}}\}$ with $h_{\boldsymbol{x}} = k(\boldsymbol{x}, \cdot) \in \mathcal{H}_p$. For any $i \in I$

$$\left| \frac{1}{m} \sum_{j=1}^{m} h_{\boldsymbol{x}}(\boldsymbol{x}_j)\psi_i(\boldsymbol{x}_j) - \int h_{\boldsymbol{x}}(\boldsymbol{y})\psi_i(\boldsymbol{y})p(\boldsymbol{y})d\boldsymbol{y} \right| \to 0$$

almost surely (thanks to the strong law of large numbers), so that $\mathcal{F}$ verifies the hypothesis needed to apply the theorem above. In addition,

$$\langle P_r(L)h_{\boldsymbol{x}}, h_{\boldsymbol{x}} \rangle_{\mathcal{H}_p} = \langle \langle h_{\boldsymbol{x}}, f_r \rangle f_r, h_{\boldsymbol{x}} \rangle = \langle h_{\boldsymbol{x}}, f_r \rangle_{\mathcal{H}_p}^2 = (Lf_r)(\boldsymbol{x})^2 = \lambda_r^2 f_r(\boldsymbol{x})^2$$

---

[4]The convergences should be almost sure, otherwise the result may hold with a different kind of probabilistic convergence, e.g. in probability.

and similarly, using eq. 27.22, we have with probability 1 and for m large enough:

$$\langle P_r(\hat{L}_m)\hat{h}_{\boldsymbol{x}}, \hat{h}_{\boldsymbol{x}}\rangle_{\mathcal{H}_{\hat{p}}} = \langle \hat{h}_{\boldsymbol{x}}, \hat{f}_{r,m}\rangle^2_{\mathcal{H}_{\hat{p}}} = (L_m f_{r,m})(\boldsymbol{x})^2 = \lambda^2_{r,m} f_{r,m}(\boldsymbol{x})^2. \quad (27.25)$$

The conclusion of the theorem thus tells us that

$$\left|\lambda^2_{r,m} f_{r,m}(\boldsymbol{x})^2 - \lambda^2_r f_r(\boldsymbol{x})^2\right| \to 0$$

almost surely. Since we have the convergence of the eigenvalues, this implies

$$\left|f_{r,m}(\boldsymbol{x})^2 - f_r(\boldsymbol{x})^2\right| \to 0 \quad (27.26)$$

almost surely, which shows the (simple) convergence of the eigenfunctions, up to the sign. To get the convergence in $\mathcal{H}_p$, we need to show that $\||f_{r,m}| - |f_r|\|_{\mathcal{H}_p} \to 0$, i.e.

$$\int g_{r,m}(\boldsymbol{x})d\boldsymbol{x} \to 0 \quad (27.27)$$

with $g_{r,m}(\boldsymbol{x}) = (|f_{r,m}(\boldsymbol{x})| - |f_r(\boldsymbol{x})|)^2 p(\boldsymbol{x})$. We will need to show that both $f_{r,m}$ and f_r are bounded (independently of m). Since f_r is an eigenfunction of L, we have $|\lambda_r f_r(\boldsymbol{x})| = |(Lf_r)(\boldsymbol{x})| = |\int k(\boldsymbol{x}, \boldsymbol{y})f_r(\boldsymbol{y})p(\boldsymbol{y})d\boldsymbol{y}| \le c|\int f_r(\boldsymbol{y})p(\boldsymbol{y})d\boldsymbol{y}|$, so that $|f_r(\boldsymbol{x})| \le c'_r$. For $f_{r,m}$, we have

$$|\lambda_{r,m} f_{r,m}(\boldsymbol{x})| = |(L_m f_{r,m}(\boldsymbol{x}))| = \left|\frac{1}{m}\sum_{i=1}^m k(\boldsymbol{x}, \boldsymbol{x}_i)\sqrt{m}v_{r,i}\right| \le \frac{c}{\sqrt{m}}\sum_{i=1}^m |v_{r,i}| \le c$$

because the maximum of $\sum_{i=1}^m |v_{r,i}|$ subject to $\|v_r\|^2 = 1$ is $\sqrt{m}$, so that $|f_{r,m}(\boldsymbol{x})| \le c''_r$ almost surely (thanks to the convergence of $\lambda_{r,m}$). Therefore, we have that (almost surely) $g_{r,m}(\boldsymbol{x}) \le (c'_r + c''_r)^2 p(\boldsymbol{x})$ which is an integrable function, and from eq. 27.26, $g_{r,m}(\boldsymbol{x}) \to 0$ for all $\boldsymbol{x}$. The theorem of dominated convergence can thus be applied, which proves eq. 27.27 is true (almost surely), and there is convergence of the eigenfunctions in $\mathcal{H}_p$.

If k_m is data-dependent but converges, in a way such that the eigen-decomposition of the Gram matrix $(k_m(\boldsymbol{x}_i, \boldsymbol{x}_j))$ converges to the eigen-decomposition of the Gram matrix $(k(\boldsymbol{x}_i, \boldsymbol{x}_j))$, with k the limit of k_m, we want to apply the same reasoning. Because of the convergence of the eigen-decomposition of the Gram matrix, eq. 27.24 still holds. However, eq. 27.25 has to be replaced with a limit, because $h_{\boldsymbol{x}} = k(\boldsymbol{x}, \cdot) \ne k_m(\boldsymbol{x}, \cdot)$. This limit still allows to write eq. 27.26 (possibly with a different form of probabilistic convergence, depending on the convergence of k_m to k), and the same result is obtained. $\quad \square$

Discussion

Kernel PCA has already been shown to be a stable and convergent algorithm (Shawe-Taylor et al., 2002, Shawe-Taylor and Williams, 2003, Zwald et al., 2004). These papers characterize the rate of convergence of the projection error on the subspace spanned by the first N eigenvectors of the feature space covariance matrix. When we perform the PCA or kernel PCA projection on an out-of-sample point, we are taking advantage of the above convergence and stability properties: we trust that a principal eigenvector of the empirical covariance matrix estimates well a corresponding eigenvector of the true covariance matrix. Another justification for applying the Nyström formula outside of the training examples is therefore, as already noted earlier and in (Williams and Seeger, 2000), *in the case where k_m is positive semi-definite,* that it corresponds to the kernel PCA projection (on a corresponding eigenvector of the feature space covariance matrix C).

Clearly, we thus have with the Nyström formula a method to *generalize spectral embedding algorithms to out-of-sample examples,* whereas the original spectral embedding methods only provide the transformed coordinates of training points (i.e. an embedding of the training points). The experiments described in section 27.5 show empirically the good generalization of this out-of-sample embedding. Note however that it is not always clear whether the assumptions needed to apply proposition 2 are verified or not (especially because of the data-dependency of k_m). This proposition mainly gives an intuition of what a spectral embedding technique is doing (estimating eigenfunctions of a linear operator) in the case of ideal convergence.

(Williams and Seeger, 2000) have shown an interesting justification for estimating the eigenfunctions of L. When an unknown function f is to be estimated with an approximation g that is a finite linear combination of basis functions, if f is assumed to come from a zero-mean Gaussian process prior with covariance $E_f[f(x)f(y)] = k(x, y)$, then the best choices of basis functions, in terms of expected squared error, are (up to rotation/scaling) the leading eigenfunctions of the linear operator L defined by eq 27.17.

27.4 Learning Criterion for the Leading Eigenfunctions

Using an expansion into orthonormal bases (e.g. generalized Fourier decomposition in the case where p is continuous), the best approximation of $k(x, y)$ (in the sense of minimizing expected squared error) using only N terms is the expansion that uses the *first N* eigenfunctions of L (in the order of decreasing eigenvalues):

$$\sum_{r=1}^{N} \lambda_r f_r(x) f_r(y) \approx k(x, y).$$

This simple observation allows us to define a *loss criterion for spectral embedding algorithms,* something that was lacking up to now for such algorithms. The limit of this loss converges toward an expected loss whose minimization gives rise to the eigenfunctions of L. One could thus conceivably estimate this generalization error using the average of the loss on a test set. That criterion is simply the kernel reconstruction error

$$R_{k_m}(\boldsymbol{x}_i, \boldsymbol{x}_j) = \left(k_m(\boldsymbol{x}_i, \boldsymbol{x}_j) - \sum_{r=1}^{N} \lambda_{r,m} f_{r,m}(\boldsymbol{x}_i) f_{r,m}(\boldsymbol{x}_j) \right)^2.$$

Proposition 3 *The spectral embedding for a continous kernel k with discrete spectrum is the solution of a sequential minimization problem, iteratively minimizing for $N = 1, 2, \ldots$ the expected value of the loss criterion*

$$R_k(\boldsymbol{x}, \boldsymbol{y}) = \left(k(\boldsymbol{x}, \boldsymbol{y}) - \sum_{r=1}^{N} \lambda_r f_r(\boldsymbol{x}) f_r(\boldsymbol{y}) \right)^2.$$

First, with $\{(f_r, \lambda_r)\}_{k=1}^{N-1}$ already obtained, one gets recursively (λ_N, f_N) by minimizing

$$J_N(\lambda', f') = \int \left(k(\boldsymbol{x}, \boldsymbol{y}) - \lambda' f'(\boldsymbol{x}) f'(\boldsymbol{y}) - \sum_{r=1}^{N-1} \lambda_r f_r(\boldsymbol{x}) f_r(\boldsymbol{y}) \right)^2 p(\boldsymbol{x}) p(\boldsymbol{y}) d\boldsymbol{x} d\boldsymbol{y}$$

(27.28)

where by convention we scale f' such that $\int f'(\boldsymbol{x})^2 p(\boldsymbol{x}) = 1$ (any other scaling can be transferred into λ').

Secondly, if the same hypothesis on k_m as in proposition 2 are verified, the Monte-Carlo average of the criterion R_{k_m}

$$\frac{1}{m^2} \sum_{i=1}^{m} \sum_{j=1}^{m} \left(k_m(\boldsymbol{x}_i, \boldsymbol{x}_j) - \sum_{r=1}^{N} \lambda_{r,m} f_{r,m}(\boldsymbol{x}_i) f_{r,m}(\boldsymbol{x}_j) \right)^2$$

converges in probability to the asymptotic expectation of R_k.

Sketch of proof

The first part of the proposition concerning the sequential minimization of the loss criterion follows from classical linear algebra (Strang, 1980, Kreyszig, 1990). It is an extension of the well-known result stating that the best rank N approximation (for the Frobenius norm) of a symmetric matrix is given by its expansion in terms of its first N eigenvectors. A proof can be found in (Bengio et al., 2003).

To prove the second part, a reasoning similar to the one in the proof of proposition 2 can be done, in order to obtain that (in probability, when $m \to \infty$)

$$\int \left(\lambda_{r,m} f_{r,m}(\boldsymbol{x}) f_{r,m}(\boldsymbol{y}) - \lambda_r f_r(\boldsymbol{x}) f_r(\boldsymbol{y}) \right) p(\boldsymbol{x}) p(\boldsymbol{y}) d\boldsymbol{x} d\boldsymbol{y} \to 0$$

which, combined with the central limit theorem, leads to the desired convergence. $\square$

Discussion

Note that the empirical criterion is indifferent to the value of the solutions $f_{r,m}$ outside of the training set. Therefore, although the Nyström formula gives a possible solution to the empirical criterion, there are other solutions. Remember that the task we consider is that of estimating the eigenfunctions of L, i.e. approximating a similarity function k where it matters according to the unknown density p. Solutions other than the Nyström formula might also converge to the eigenfunctions of L. For example one could use a non-parametric estimator (such as a neural network) to estimate the eigenfunctions. Even if such a solution does not yield the exact eigenvectors on the training examples (i.e. does not yield the lowest possible error on the training set), it might still be a good solution in terms of generalization, in the sense of good approximation of the eigenfunctions of L. It would be interesting to investigate whether the Nyström formula achieves the fastest possible rate of convergence to the eigenfunctions of L.

27.5 Experiments

27.5.1 Toy Data Example

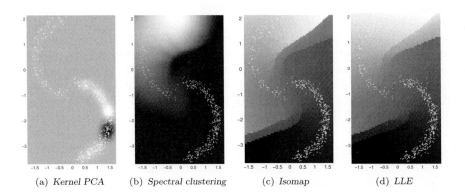

(a) *Kernel PCA*   (b) *Spectral clustering*   (c) *Isomap*   (d) *LLE*

Fig. 27.2. First eigenfunction (gray levels) for various non-linear spectral dimensionality reduction algorithms, on a toy dataset of 500 samples (white dots). Kernel PCA and spectral clustering use the same Gaussian kernel (eq. 27.4) with bandwidth $\sigma = 0.2$, while Isomap and LLE use 20 neighbors.

We first show on a toy dataset what kind of structure can be discovered from the eigenfunctions defined in the previous sections. In Figure 27.2, we display with gray levels the value[5] of the first eigenfunction computed for

---

[5]In the case of spectral clustering, this is the logarithm of the eigenfunction that is displayed, so as to be able to actually see its variations.

kernel PCA, spectral clustering (same as Laplacian eigenmaps), Isomap and LLE, on a toy dataset of 500 examples (remember that the first eigenfunction is proportional to the first coordinate of the projection, as seen in eq. 27.23). This toy dataset is formed of two clusters (white dots) with a similar form, one (top-left) with 100 points, and the other (bottom-right) with 400 points. The clusters are connected (through a low-density region) in such a way that the data lie approximately on a one-dimensional manifold. Our C++ code for performing those spectral dimensionality reduction algorithms can be found in the PLearn library (http://plearn.org).

Although such a toy example does not provide a deep understanding of what these algorithms do, it reveals several characteristics worth pointing out. First of all, kernel PCA should not be used with a Gaussian kernel in order to discover a low-dimensional non-linear manifold. Indeed, one may think kernel PCA does some kind of "local" PCA within neighborhoods of size of the order of the Gaussian kernel's bandwidth, but it is *not* the case. It actually tends to discriminate smaller regions of the data as the bandwidth decreases (and a high bandwidth makes it equivalent to linear PCA). The spectral clustering / Laplacian eigenmaps eigenfunction is more satisfying, in that it obviously reveals the clustered nature of our data (see footnote 5). Note that, even though in this case only one eigenfunction may be enough to discriminate between the two clusters, one should in general compute as many eigenfunctions as desired clusters (because each eigenfunction will tend to map to the same point all clusters but one, e.g. in Figure 2(b) almost all points in the bottom-right cluster are given the same value). As for Isomap and LLE, they give very similar results, showing they correctly captured the underlying one-dimensional non-linear manifold. Although it cannot be seen in this particular example, one should keep in mind that LLE, because it is based only on local computations, does not respect as well as Isomap the global structure of the data.

27.5.2 Discovering Structure in Face Images

Experiments in this section are done over a subset of the database of 698 synthetic face images available at http://isomap.stanford.edu. By selecting only images whose illumination is between 180 and 200, this yields a dataset of 113 examples in 4096 dimensions, which approximately form a 2-dimensional manifold (the two degrees of freedom are the rotation angles of the camera). The Isomap algorithm with 10 nearest neighbors is run on the first 70 examples, while the remaining 43 are projected by the Nyström formula. The embedding thus obtained (Figure 27.3) clearly demonstrates that Isomap captured the intrinsic 2-dimensional manifold, and that the Nyström formula generalizes well. This is a typical example where such a non-linear spectral embedding algorithm can prove very useful for data visualization as well as for dimensionality reduction.

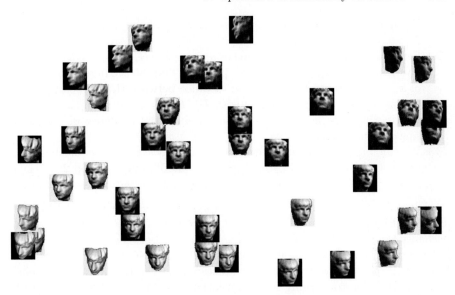

Fig. 27.3. 2-dimensional embedding learned by Isomap from 70 high-dimensional synthetic faces. A few faces from the training set (black background) and from the test set (light gray background) are projected using the Nyström formula. Isomap catpures the intrinsic 2-dimensional structure of the data, each dimension corresponding to the rotation on a different axis (left-right and bottom-top).

27.5.3 Generalization Performance of Function Induction

Here we want to test one aspect of the theoretical results: does the function induction achieved with the Nyström formula work well? We would like to know if the embedding that it predicts on a new point x is close to the embedding that would have been obtained on x if it had been in the training set. However, how do we evaluate the "error" thus obtained? Our idea is to compare it to the variations in embedding that result from small perturbations of the training set (such as replacing a subset of the examples by others from the same distribution).

For this purpose we consider splits of the data in three sets, $X = F \cup R_1 \cup R_2$ and training either with $F \cup R_1$ or $F \cup R_2$, comparing the embeddings on F. For each algorithm described in section 27.2, we apply the following procedure:

(1.) We choose $F \subset X$ with $q = |F|$ samples. The remaining $m - q$ samples in X/F are split into two equal size subsets R_1 and R_2. We train (obtain the eigenvectors) over $F \cup R_1$ and $F \cup R_2$ and we calculate the Euclidean distance between the aligned embeddings obtained for each $x_i \in F$. When eigenvalues are close, the estimated eigenvectors are unstable and can rotate in the subspace they span. Thus we estimate an alignment

(by linear regression) between the two embeddings using the points in
F.

(2.) For each sample $x_i \in F$, we also train over $\{F \cup R_1\}/\{x_i\}$. We apply the
Nyström formula to out-of-sample points to find the predicted embedding
of x_i and calculate the Euclidean distance between this embedding and
the one obtained when training with $F \cup R_1$, i.e. with x_i in the training
set (in this case no alignment is done since the influence of adding a
single point is very limited).

(3.) We calculate the mean **difference** δ (and its standard error) between
the distance obtained in step 1 and the one obtained in step 2 for each
sample $x_i \in F$, and we repeat this experiment for various sizes of F.

The results obtained for MDS, Isomap, spectral clustering and LLE are
shown in Figure 27.4 for different values of $|R_1|/m$ (i.e the fraction of points
exchanged). The vertical axis is δ, the difference between perturbation error and induction error. The horizontal zero line corresponds to no difference
between the embedding error due to induction (vs. transduction) and the embedding error due to training set perturbation. For values of δ above the zero
line, the embedding error due to perturbation is greater than the embedding
error due to out-of-sample prediction. Clearly, the in-sample (transduction)
vs. out-of-sample (induction) difference is of the same order of magnitude as
the change in embedding due to exchanging a small fraction of the data (1 to
5%).

Experiments are done over the same face database as in the previous section, but with the whole set of 698 images. Similar results have been obtained
over other databases such as Ionosphere[6] and Swissroll[7]. Each algorithm generates a two-dimensional embedding of the images, following the experiments
reported for Isomap. The number of neighbors is 10 for Isomap and LLE, and
a Gaussian kernel with a bandwidth of 0.01 is used for spectral clustering /
Laplacian eigenmaps. 95% confidence intervals are drawn beside each mean
difference of error on the figure.

As expected, the mean difference between the two distances is almost
monotonically increasing as the number $|R_1|$ of substituted training samples
grows, mostly because the training set embedding variability increases. We
find in most cases that the out-of-sample error is less than or comparable
to the training set embedding instability when around 2% of the training
examples are substituted randomly.

27.6 Conclusion

Manifold learning and dimensionality reduction are powerful machine learning
tools for which much progress has been achieved in recent years. This chapter

---

[6] http://www.ics.uci.edu/~mlearn/MLSummary.html
[7] http://www.cs.toronto.edu/~roweis/lle/

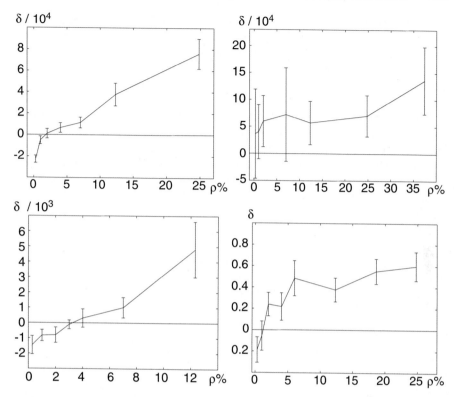

Fig. 27.4. δ (training set variability minus out-of-sample error), w.r.t. ρ (proportion of substituted training samples) on the "Faces" dataset ($m = 698$), obtained with a two-dimensional embedding. Top left: MDS. Top right: spectral clustering or Laplacian eigenmaps. Bottom left: Isomap. Bottom right: LLE. Error bars are 95% confidence intervals. Exchanging about 2% of the training examples has an effect comparable to using the Nyström formula.

sheds light on a family of such algorithms, involving spectral embedding, which are all based on the eigen-decomposition of a similarity matrix.

Spectral embedding algorithms such as spectral clustering, Isomap, LLE, metric MDS, and Laplacian eigenmaps are very interesting dimensionality reduction or clustering methods. However, they lacked up to now a notion of generalization that would allow to easily extend the embedding out-of-sample without again solving an eigensystem. This chapter has shown with various arguments that the well known Nyström formula can be used for this purpose, and that it thus represents the result of a *function induction* process. These arguments also help us to understand that *these methods do essentially the same thing, but with respect to different kernels*: they *estimate the eigenfunctions of a linear operator L* associated with a kernel and with the underlying distribution of the data. This analysis also shows that these methods are *minimizing* *an empirical loss*, and that the solutions toward which they converge are the minimizers of a corresponding expected loss, which thus defines what good generalization should mean, for these methods. It shows that these unsupervised learning algorithms can be extended into function induction algorithms. The Nyström formula is a possible extension but it does not exclude other extensions, which might be better or worse estimators of the eigenfunctions of the asymptotic linear operator L. When the kernels are positive semi-definite, these methods can also be immediately seen as performing kernel PCA. Note that Isomap generally yields a Gram matrix with negative eigenvalues, and users of MDS, spectral clustering or Laplacian eigenmaps may want to use a kernel that is not guaranteed to be positive semi-definite. The analysis in this chapter can still be applied in that case, even though the kernel PCA analogy does not hold anymore. This is important to note because recent work (Laub and Müller, 2003) has shown that the coordinates corresponding to large negative eigenvalues can carry very significant semantics about the underlying objects. In fact, it is proposed in (Laub and Müller, 2003) to perform dimensionality reduction by projecting on the eigenvectors corresponding to the largest eigenvalues **in magnitude** (i.e. irrespective of sign).

In this chapter we have given theorems that provide justification for the Nyström formula in the general case of data-dependent kernels which may not be positive-definite. However, these theorems rely on strong assumptions, which may not hold for particular spectral manifold learning algorithms. To help assess the practical validity of the Nyström formula for predicting the embedding of out-of-sample points, we have performed a series of comparative experiments.

The experiments performed here have shown empirically on several data sets that (i) those spectral embedding algorithms capture different kinds of non-linearity in the data, (ii) they can be useful for both data visualization and dimensionality reduction, and (iii) the predicted out-of-sample embedding is generally not far from the one that would be obtained by including the test point in the training set, the difference being of the same order as the effect of small perturbations of the training set.

An interesting parallel can be drawn between the spectral embedding algorithms and the view of PCA as finding the principal eigenvectors of a matrix obtained from the data. The present chapter parallels for spectral embedding the view of PCA as an estimator of the principal directions of the covariance matrix of the underlying unknown distribution, thus introducing a convenient notion of generalization, relating to an unknown distribution.

Finally, a better understanding of these methods opens the door to new and potentially much more powerful unsupervised learning algorithms. Several directions remain to be explored:

(1.) Using a smoother distribution than the empirical distribution to define the linear operator L_m. Intuitively, a distribution that is closer to the true underlying distribution would have a greater chance of yielding better generalization, in the sense of better estimating eigenfunctions of L. This relates to putting priors on certain parameters of the density, e.g. as in (Rosales and Frey, 2003).

(2.) All of these methods are capturing salient features of the unknown underlying density. Can one use the representation learned through the estimated eigenfunctions in order to construct a good density estimator? Looking at Figure 27.1 suggests that modeling the density in the transformed space (right hand side) should be much easier (e.g. would require fewer Gaussians in a Gaussian mixture) than in the original space.

(3.) These transformations discover abstract structures such as clusters and manifolds. It might be possible to learn even more abstract (and less local) structures, starting from these representations. Ultimately, the goal would be to learn higher-level abstractions on top of lower-level abstractions by iterating the unsupervised learning process in multiple "layers".

Looking for extensions such as these is important because all of the manifold learning algorithms studied here suffer from the following fundamental weakness: they are using mostly the neighbors around each example to capture the local structure of the manifold, i.e. the manifold is seen as a combination of linear patches around each training example. This is very clear in LLE and Isomap, which have a simple geometric interpretation, and it is also clear in non-spectral methods such as Manifold Parzen Windows (Vincent and Bengio, 2003) and other mixtures of factor analyzers (Ghahramani and Hinton, 1996). In low dimension or when the manifold is smooth enough, there may be enough examples locally to characterize the plane tangent to the manifold. However, when the manifold has high curvature with respect to the amount of training data (which can easily be the case, especially with high-dimensional data), it is hopeless to try to capture the local tangent directions based only on local information. Clearly, this is the topic for future work, addressing a fundamental question about generalization in high-dimensional data, and for which the traditional non-parametric approaches may be insufficient.

Acknowledgments

The authors would like to thank Léon Bottou, Christian Léger, Sam Roweis, Yann Le Cun, and Yves Grandvalet for helpful discussions, and the following funding organizations: NSERC, MITACS, IRIS, and the Canada Research Chairs.

References

C.T.H. Baker. *The numerical treatment of integral equations*. Clarendon Press, Oxford, 1977.

M. Belkin and P. Niyogi. Laplacian eigenmaps for dimensionality reduction and data representation. *Neural Computation*, 15(6):1373–1396, 2003a.

M. Belkin and P. Niyogi. Using manifold structure for partially labeled classification. In S. Becker, S. Thrun, and K. Obermayer, editors, *Advances in Neural Information Processing Systems 15*, Cambridge, MA, 2003b. MIT Press.

Y. Bengio, J.F. Paiement, P. Vincent, O. Delalleau, N. Le Roux, and M. Ouimet. Out-of-sample extensions for LLE, Isomap, MDS, Eigenmaps, and Spectral Clustering. In S. Thrun, L. Saul, and B. Schölkopf, editors, *Advances in Neural Information Processing Systems 16*. MIT Press, 2004.

Y. Bengio, P. Vincent, J.-F. Paiement, O. Delalleau, M. Ouimet, and N. Le Roux. Spectral clustering and kernel pca are learning eigenfunctions. Technical Report 1239, Département d'informatique et recherche opérationnelle, Université de Montréal, 2003.

H. Bourlard and Y. Kamp. Auto-association by multilayer perceptrons and singular value decomposition. *Biological Cybernetics*, 59:291–294, 1988.

M. Brand. Charting a manifold. In S. Becker, S. Thrun, and K. Obermayer, editors, *Advances in Neural Information Processing Systems 15*. MIT Press, 2003.

F. Chung. Spectral graph theory. In *CBMS Regional Conference Series*, volume 92. American Mathematical Society, 1997.

T. Cox and M. Cox. *Multidimensional Scaling*. Chapman & Hall, London, 1994.

V. de Silva and J.B. Tenenbaum. Global versus local methods in nonlinear dimensionality reduction. In S. Becker, S. Thrun, and K. Obermayer, editors, *Advances in Neural Information Processing Systems 15*, pages 705–712, Cambridge, MA, 2003. MIT Press.

D.L. Donoho and C. Grimes. Hessian eigenmaps: new locally linear embedding techniques for high-dimensional data. Technical Report 2003-08, Dept. Statistics, Stanford University, 2003.

Z. Ghahramani and G.E. Hinton. The EM algorithm for mixtures of factor analyzers. Technical Report CRG-TR-96-1, Dpt. of Comp. Sci., Univ. of Toronto, 21 1996. URL citeseer.nj.nec.com/ghahramani97em.html.

J.C. Gower. Adding a point to vector diagrams in multivariate analysis. *Biometrika*, 55(3):582–585, 1968.

J. Ham, D.D. Lee, S. Mika, and B. Schölkopf. A kernel view of the dimensionality reduction of manifolds. Technical Report TR-110, Max Planck Institute for Biological Cybernetics, Germany, 2003.

T. Hastie and W. Stuetzle. Principal curves. *Journal of the American Statistical Association*, 84:502–516, 1989.

B. Kegl and A. Krzyzak. Piecewise linear skeletonization using principal curves. *IEEE Transactions on Pattern Analysis and Machine Intelligence*, 24(1):59–74, 2002.

Balázs Kégl. Intrinsic dimension estimation using packing numbers. In S. Becker, S. Thrun, and K. Obermayer, editors, *Advances in Neural Information Processing Systems 15*, pages 681–688. MIT Press, Cambridge, MA, 2003.

T. Kohonen. The self-organizing map. *Proceedings of the IEEE*, 78(9):1464–1480, 1990.

V. Koltchinskii. Asymptotics of spectral projections of some random matrices approximating integral operators. In Eberlein, Hahn, and Talagrand, editors, *Progress in Probability*, volume 43, pages 191–227, Basel, 1998. Birkhauser.

V. Koltchinskii and E. Gin. Random matrix approximation of spectra of integral operators. *Bernoulli*, 6(1):113–167, 2000.

E. Kreyszig. *Introductory Functional Analysis with Applications*. John Wiley & Sons, Inc., New York, NY, 1990.

V. Kumar, A. Grama, A. Gupta, and G. Karypis. *Introduction to Parallel Computing: Design and Analysis of Algorithms*. Benjamin Cummings, Redwood City, CA, 1994.

J. Laub and K.-R. Müller. Feature discovery: unraveling hidden structure in non-metric pairwise data. Technical report, Fraunhofer FIRST.IDA, Germany, 2003.

A. Y. Ng, M. I. Jordan, and Y. Weiss. On spectral clustering: analysis and an algorithm. In T.G. Dietterich, S. Becker, and Z. Ghahramani, editors, *Advances in Neural Information Processing Systems 14*, Cambridge, MA, 2002. MIT Press.

R. Rosales and B. Frey. Learning generative models of affinity matrices. In *Proceedings of the 19th Annual Conference on Uncertainty in Artificial Intelligence (UAI-03)*, pages 485–492, San Francisco, CA, 2003. Morgan Kaufmann Publishers.

S. Roweis and L. Saul. Nonlinear dimensionality reduction by locally linear embedding. *Science*, 290(5500):2323–2326, Dec. 2000.

D.E. Rumelhart, G.E. Hinton, and R.J. Williams. Learning internal representations by error propagation. In D.E. Rumelhart and J.L. McClelland, editors, *Parallel Distributed Processing*, volume 1, chapter 8, pages 318–362. MIT Press, Cambridge, 1986.

L. Saul and S. Roweis. Think globally, fit locally: unsupervised learning of low dimensional manifolds. *Journal of Machine Learning Research*, 4:119–155, 2002.

Eric Saund. Dimensionality-reduction using connectionist networks. *IEEE Transactions on Pattern Analysis and Machine Intelligence*, 11(3):304–314, 1989.

B. Schölkopf, C. J. C. Burges, and A. J. Smola. *Advances in Kernel Methods — Support Vector Learning*. MIT Press, Cambridge, MA, 1999.

B. Schölkopf, A. Smola, and K.-R. Müller. Nonlinear component analysis as a kernel eigenvalue problem. Technical Report 44, Max Planck Institute for Biological Cybernetics, Tbingen, Germany, 1996.

B. Schölkopf, A. Smola, and K.-R. Müller. Nonlinear component analysis as a kernel eigenvalue problem. *Neural Computation*, 10:1299–1319, 1998.

J. Shawe-Taylor, N. Cristianini, and J. Kandola. On the concentration of spectral properties. In T.G. Dietterich, S. Becker, and Z. Ghahramani, editors, *Advances in Neural Information Processing Systems 14*. MIT Press, 2002.

J. Shawe-Taylor and C.K.I. Williams. The stability of kernel principal components analysis and its relation to the process eigenspectrum. In S. Becker, S. Thrun, and K. Obermayer, editors, *Advances in Neural Information Processing Systems 15*. MIT Press, 2003.

J. Shi and J. Malik. Normalized cuts and image segmentation. In *Proc. IEEE Conf. Computer Vision and Pattern Recognition*, pages 731–737, 1997.

D. Spielman and S. Teng. Spectral partitioning works: planar graphs and finite element meshes. In *Proceedings of the 37th Annual Symposium on Foundations of Computer Science*, 1996.

G. Strang. *Linear Algebra and Its Applications*. Academic Press, New York, 1980.

Y. Whye Teh and S. Roweis. Automatic alignment of local representations. In S. Becker, S. Thrun, and K. Obermayer, editors, *Advances in Neural Information Processing Systems 15*. MIT Press, 2003.

J. Tenenbaum, V. de Silva, and J.C.L. Langford. A global geometric framework for nonlinear dimensionality reduction. *Science*, 290(5500):2319–2323, Dec. 2000.

W. Torgerson. Multidimensional scaling, 1: Theory and method. *Psychometrika*, 17:401–419, 1952.

A.W. van der Vaart and J. Wellner. *Weak Convergence and Empirical Processes with applications to Statistics*. Springer, New York, 1996.

Jakob J. Verbeek, Sam T. Roweis, and Nikos Vlassis. Non-linear cca and pca by alignment of local models. In S. Thrun, L. Saul, and B. Schölkopf, editors, *Advances in Neural Information Processing Systems 16*, Cambridge, MA, 2004. MIT Press.

P. Vincent and Y. Bengio. Manifold parzen windows. In S. Becker, S. Thrun, and K. Obermayer, editors, *Advances in Neural Information Processing Systems 15*, Cambridge, MA, 2003. MIT Press.

Michail Vlachos, Carlotta Domeniconi, Dimitrios Gunopulos, George Kollios, and Nick Koudas. Non-linear dimensionality reduction techniques for classification and visualization. In *Proc. of 8th SIGKDD*, Edmonton, Canada, 2002. URL citeseer.ist.psu.edu/573153.html.

Yair Weiss. Segmentation using eigenvectors: a unifying view. In *Proceedings IEEE International Conference on Computer Vision*, pages 975–982, 1999.

Christopher K. I. Williams and Matthias Seeger. Using the Nyström method to speed up kernel machines. In T.K. Leen, T.G. Dietterich, and V. Tresp, editors, *Advances in Neural Information Processing Systems 13*, pages 682–688, Cambridge, MA, 2001. MIT Press.

C.K.I. Williams. On a connection between kernel pca and metric multidimensional scaling. In T.K. Leen, T.G. Dietterich, and V. Tresp, editors, *Advances in Neural Information Processing Systems 13*, pages 675–681. MIT Press, 2001.

C.K.I. Williams and M. Seeger. The effect of the input density distribution on kernel-based classifiers. In *Proceedings of the Seventeenth International Conference on Machine Learning*. Morgan Kaufmann, 2000.

Laurent Zwald, Olivier Bousquet, and Gilles Blanchard. Statistical properties of kernel principal component analysis. Technical report, submitted, 2004.

Chapter 28

Constructing Orthogonal Latent Features for Arbitrary Loss

Michinari Momma[1] and Kristin P. Bennett[2]

[1] Fair Isaac Corporation, San Diego, CA 92130 USA mommam@alum.rpi.edu
[2] Department of Mathematical Sciences , Rensselaer Polytechnic Institute, Troy, NY 12180 USA bennek@rpi.edu

Summary. A boosting framework for constructing orthogonal features targeted to a given loss function is developed. Combined with techniques from spectral methods such as PCA and PLS, an orthogonal boosting algorithm for linear hypothesis is used to efficiently construct orthogonal latent features selected to optimize the given loss function. The method is generalized to construct orthogonal nonlinear features using the kernel trick. The resulting method, Boosted Latent Features (BLF) is demonstrated to both construct valuable orthogonal features and to be a competitive inference method for a variety of loss functions. For the least squares loss, BLF reduces to the PLS algorithm and preserves all the attractive properties of that algorithm. As in PCA and PLS, the resulting nonlinear features are valuable for visualization, dimensionality reduction, improving generalization by regularization, and use in other learning algorithms, but now these features can be targeted to a specific inference task/loss function. The data matrix is factorized by the extracted features. The low-rank approximation of the data matrix provides efficiency and stability in computation, an attractive characteristic of PLS-type methods. Computational results demonstrate the effectiveness of the approach on a wide range of classification and regression problems.

28.1 Introduction

We consider the problem of feature construction targeted towards a given inference task. The class of features considered are linear combinations of the input attributes. The quintessential unsupervised feature extraction method for such linear features is principal component analysis (PCA). PCA constructs orthogonal features consisting of linear combinations of the input vectors called principal components. In PCA, the principal components are chosen to maximize the explained variance and to be orthogonal to each other. The resulting principal components are useful for visualization, understanding importance of variables, and as a form of dimensionality reduction or regularization when used in inference functions, e.g principal component regression. PCA takes into account only the input variables and does not use the response

M. Momma and K.P. Bennett: *Constructing Orthogonal Latent Features for Arbitrary Loss*, StudFuzz **207**, 551–583 (2006)
www.springerlink.com

variables. While PCA is widely used, it is not appropriate for every task. The approach is entirely based on the least squares loss, which may not be appropriate when the underlying noise is non-Gaussian. PCA may not produce good features or may require more features for supervised learning tasks such as classification, regression, and ranking.

Our goal is to construct orthogonal features that are targeted toward a given inference task. The features are assumed to be linear combinations of the input data that are orthogonal to each other. The features constructed from factorizations of the data and hypothesis space are targeted toward a specific inference task as defined by the loss function. We seek a small set of features that span the portion of the hypothesis space of interest for the specific inference task. The features act as a form of regularization: the loss function is minimized with respect to the smaller set of features instead of the original typically much higher dimensional input space. This relatively small set of features can then be used for many purposes such as predictive models, visualization, and outlier detection.

The primary motivation for this approach comes from boosting, which can be viewed as a method for constructing features as well as predictive models. Ensemble methods such as AdaBoost (Freund and Shapire, 1997) and Gradient Boost (Friedman, 2001) construct a linear combination of hypotheses in order to optimize a specified loss function. But the resulting hypotheses do not meet our goals. Ensemble methods typically use many weak hypotheses. But our goal is to create a few orthogonal features that span the space of interest. The key missing property in ensemble methods, for the purpose of reducing dimensionality of the feature space, is orthogonality. Forcing orthogonality can dramatically increase the convergence speed of the boosting algorithm. Thus much fewer features need to be constructed to obtain the same decrease in loss function. By simply adding orthogonality to the ensemble method, the desirable properties of PCA are once again regained – a small set of orthogonal features are identified that explains properties of the data of interest; but now the definition of interest can go beyond explaining variance.

Thus, we propose a boosting method with orthogonal components as its weak hypotheses. Boosting has been shown to be equivalent to gradient descent projected into a hypothesis space. At each iteration, the hypothesis or feature is constructed to maximize the inner product with the gradient. In orthogonal boosting, we construct the feature that is orthogonal to all previous features that maximizes this inner product. For linear hypotheses, this constrained optimization problem can be very efficiently and exactly solved by projecting the data into the orthogonal subspace.

Orthogonal boosting of linear hypotheses is very closely related to spectral or data factorization methods such as PCA, canonical correlation analysis, factor analysis, and partial least squares regression (PLS) analysis (Johnson and Wichern, 1992, Wold, 1966). For least squares regression, orthogonal boosting with linear regression of linear hypotheses reduces exactly to PLS. The proposed approach generalizes PLS to an arbitrary loss function. In PLS, the linear hypotheses are called latent variables or features. Thus we call boosting of such linear hypotheses, Boosted Latent Features (BLF).

Many papers have been written on PLS and its properties from many different perspectives. The boosting perspective given here is novel but our goal is not to give another perspective on PLS. Our goal is to create an efficient orthogonal feature construction method that maintains the benefits of PLS. BLF and PLS both construct a set of orthogonal linear features that form a factorization of the input and response variables. The BLF algorithm shares with PLS efficiency and elegance for computing inference functions. PLS has been shown to be a conjugate gradient method applied to the least squares regression. BLF is also a subspace method, closely related to but not identical to nonlinear conjugate gradient algorithms. Previously CGBOOST used a nonlinear conjugate gradient algorithm in function space but it did not produce orthogonal hypotheses and computational results were mixed due to overfitting (Li et al., 2003). Like PLS, BLF is a form of regularization of the loss function that reduces the norm of the solution as compared to the unconstrained optimal solution, alleviating overfitting. An extensive discussion of regularization in PLS and alternative gradient based regularization strategies can be found in (Friedman and Popescu, 2004).

A major benefit of BLF is that it can be elegantly and efficiently extended to create nonlinear latent hypotheses and functions through the use of kernels. Using an approach very similar to that of kernel partial least squares (Rosipal and Trejo, 2001), data can be mapped into a feature space and the orthogonal features are constructed in the feature space. Kernel methods exist for construction of orthogonal nonlinear feature such as kernel principal component analysis, kernel partial least squares, and kernel canonical correlation analysis, but they are all based on the least squares loss (Shawe-Taylor and Cristianini, 2004). See(Borga et al., 1997) for discussions of their common ancestry. Neural network approaches have been used to generalize PLS to nonlinear functions (e.g. (Malthouse et al., 1997)). But these methods face the problems of neural networks: inefficiency, local minima, limited loss functions, and lack of generality that are largely eliminated in kernel methods. Iterative reweighted partial least squares generalized linear PLS to exponential loss functions (Marx, 1996) using a generalized linear model approach, but this approach is more computationally costly, less general, and theoretically somewhat more obtuse than BLF.

BLF is efficient and stable computationally for kernel logistic regression (KLR). KLR can be solved by a boosting-type method (Zhu and Hastie, 2002) with the regularization term explicitly included in its objective function but without orthogonalization. An open question is if there are additional advantages to orthogonalizing weak hypotheses in terms of generalization. Orthogonal least squares regression and its variants (Chen, 2003, Nair et al., 2002) utilize a similar approach. They orthogonalize weak hypotheses each consisting of a *single* support vector, not a linear combination of support vectors. The BLF can also be applied to extend their methodologies to other loss functions.

This chapter is organized as follows. We review AnyBoost (Mason et al., 1999) to provide a general framework for ensemble methods for differentiable loss functions, and use this to introduce the orthogonalized version of Any-Boost, OrthoAnyBoost. Section 3 shows how OrthoAnyBoost can be efficiently implemented in linear hypothesis spaces forming the Boosted Latent Feature Algorithm. Section 4 examines the convergence properties of BLF. In Section 5, PLS is shown to be the special case within the BLF framework of the squared loss functions with linear weak hypotheses. Section 28.6 provides variants of BLF for three more loss functions. The kernel version of BLF is developed in Section 28.7. Computational results found in Section 28.8 illustrate the potential of kernel BLF for feature construction. In addition, standard benchmarks demonstrate the approach is quite competitive with existing classification and regression algorithms.

The following explains the notation used throughout the paper. Assume we are given a training data set of size m with a single response, $((\mathbf{x}_1, y_1), \ldots, (\mathbf{x}_m, y_m))$, with the column vectors $\mathbf{x}_i \in R^n$ and $y_i \in R$. Although PLS and KPLS can be used for multivariate response, we focus on cases with univariate response variables. $\boldsymbol{A}^T$ denotes a vector/matrix transpose. The data matrix is $\boldsymbol{X} = [\mathbf{x}_1, \ldots, \mathbf{x}_m]^T$ and the response vector is $\mathbf{y} = [y_1, \ldots, y_m]^T$. After centering each of the predictive variables, the centered data matrix is denoted by either $\widetilde{\boldsymbol{X}}$ or $\boldsymbol{X}^1$. $||\mathbf{y}||$ denotes the 2-norm of $\mathbf{y}$. The dot product of two column vectors $\mathbf{u}$ and $\mathbf{v}$ is denoted by $\boldsymbol{u}^T \boldsymbol{v}$. The outer product of $\mathbf{u}$ and $\mathbf{v}$ is denoted by $\boldsymbol{u}\boldsymbol{v}^T$. The reader should be careful to distinguish the use of dot products from that of outer products. Iteration indices are expressed as superscripts. Subscripts indicate components of a matrix or vector. In a slight abuse of notation, we use $\nabla l(\mathbf{y}, \mathbf{f})$ to denote the gradient or the subgradient of the function l with respect to $\mathbf{f}$ if the gradient does not exist.

28.2 General Framework in BLF

This section investigates a boosting algorithm that combines properties of the orthogonal components of PCA and PLS with the flexibility of general ensemble and boosting methods with respect to a loss function. The goal is to create a set of orthogonal features or functions that explain the response according to some loss function. We adapt notation of the general ensemble method, AnyBoost (Mason et al., 1999).

28.2.1 AnyBoost

Boosting algorithms construct a set of features called hypotheses that try to span the response e.g. $\mathbf{y} \approx \boldsymbol{T}\mathbf{c}$ for regression or $\mathbf{y}^T \boldsymbol{T}\mathbf{c} > 0$ for classification.

AnyBoost (Algorithm 10) (Mason et al., 1999) constructs an optimal linear combination of hypotheses to fit the response by performing gradient descent in function or hypothesis space. Let $\mathcal{T}$ be a set of real-valued functions that are the hypotheses. The span of $\mathcal{T}$ forms a linear function space. The inner product in this function space is defined as $\mathbf{t}^T \mathbf{f} \equiv \sum_{i=1}^{m} t(\mathbf{x}_i) f(\mathbf{x}_i)$. Thus we can also think of $\mathbf{t}$ as an m-vector with $t_i = t(\mathbf{x}_i)$. A linear combination of the hypotheses, $\sum_{i=1}^{N} c^i \mathbf{t}^i$ for $\mathbf{t}^1, \mathbf{t}^2, \ldots, \mathbf{t}^N$, can be written as $\boldsymbol{T}\mathbf{c}$ where the columns of $\boldsymbol{T}$ are the hypotheses and the vector $\mathbf{c}$ contains their coefficients. We want to find the element, $\mathbf{t} \in \operatorname{span}(\mathcal{T})$ that approximately minimizes some loss function $l(\mathbf{y}, \mathbf{f})$. AnyBoost accomplishes this by doing gradient descent in the hypothesis space. Note $\nabla l(\mathbf{y}, \boldsymbol{T}\mathbf{c}) = \nabla_f l(\mathbf{y}, \boldsymbol{T}\mathbf{c})$ will be used to denote the gradient of the loss function in the function space. Ideally $\mathcal{T}$ spans the same space as $\mathbf{y}$ but this is generally not the case. Thus the linear functional is minimized with respect to the loss function to fit the response.

Given the current function $\mathbf{f} = \boldsymbol{T}\mathbf{c}$, a descent direction in the hypothesis space, $\mathbf{t}^{i+1}$ is constructed. Any function with a positive inner product with the negative gradient $(-\nabla l(\mathbf{y}, \boldsymbol{T}\mathbf{c})^T \mathbf{t}^{i+1})$ must be a descent direction. In AnyBoost, a weak learning algorithm is used to construct a hypothesis that approximately maximizes the inner product of the hypothesis with the negative gradient at each iteration. The hypothesis is added with an appropriate stepsize c^{i+1} to the current function to form $\boldsymbol{T}\mathbf{c} + c^{i+1}\mathbf{t}^{i+1}$. The algorithm terminates if the weak learner fails to produce a weak hypothesis that is a descent direction indicating convergence, or if it reaches the maximum number of iterations.

Algorithm 10: AnyBoost

Given: class of weak hypotheses $\mathcal{T}$, $l(\mathbf{y}, \mathbf{t})$ with gradient $\nabla l(\mathbf{y}, \mathbf{t})$, weak learner that finds $\mathbf{t} \in \mathcal{T}$ maximizing $\mathbf{u}^T \mathbf{t}$:

(1.) Let $\mathbf{f} =$ constant or null hypothesis
(2.) Compute $\mathbf{u}^1 = -\nabla l(\mathbf{y}, \mathbf{f})$
(3.) For $i = 1$ to N
(4.)   Let $\mathbf{t}^i \in \arg\max_{\mathbf{t} \in \mathcal{T}} \mathbf{u}^{i^T} \mathbf{t}$
(5.)   if $-\mathbf{u}^{i^T} \mathbf{t}^i < 0$ then return $\mathbf{f}$
(6.)   Choose c^i to reduce $l(\mathbf{y}, \mathbf{f} + c^i \mathbf{t}^i)$
(7.)   Let $\mathbf{f} = \sum_{j=1}^{i} \mathbf{t}^j c^j$
(8.)   Compute $\mathbf{u}^{i+1} = -\nabla l(\mathbf{y}, \mathbf{f})$
(9.) end for
(10.) return $\mathbf{f}$

Variations can be developed by specifying the weak learner in Step 4, the loss function, and the algorithm to optimize the step size c^i (Step 6). For example, in (Mason et al., 1999), it is shown that AdaBoost (Freund and

Shapire, 1997) minimizes the exponential loss with exact line search, Confidence Boost (Schapire and Singer, 1999) minimizes the exponential loss with an inexact line search, and LogitBoost (Friedman et al., 2000) minimizes the negative binomial log-likelihood with a single Newton step used to compute c. Cost sensitive decision tree algorithms are commonly used as the weak learner.

28.2.2 AnyBoost with Orthogonal Weak Hypotheses

Our goal is to create a set of orthogonal features that explains the response, i.e. for regression $\mathbf{y} \approx \mathbf{Tc} = \mathbf{XVc}$, $\mathbf{T}^T\mathbf{T} = \mathbf{I}$, where $\mathbf{V}$ is a projection matrix with $\mathbf{v}^i$, $i = 1, \ldots, N$ as its column vectors, in such a way that a given loss function is minimized. By changing AnyBoost to produce orthogonal hypotheses, we can force AnyBoost to produce features that both factorize and explain the input space like PCA and span the response or output space as in original AnyBoost. The resulting algorithm becomes:

Algorithm 11: OrthoAnyBoost

Same as Algorithm 10 except for the following steps:

4. Let $\mathbf{t}^i \in$ $\begin{array}{l} \arg\max_{\mathbf{t}\in\mathcal{T}} \ {\mathbf{u}^i}^T \mathbf{t} \\ subject\ to \quad \mathbf{t}^T \mathbf{t}^j = 0 \quad j = 1, \ldots, i-1. \end{array}$

6. Optimize $\mathbf{c}$ to reduce $l(\mathbf{y}, \sum_{j=1}^{i} \mathbf{t}^j c^j)$

This simple change has some immediate ramifications. We may now use more powerful hypotheses since the hypothesis subproblem is now regularized. Second, the approach is no longer a gradient descent algorithm. We have transformed it into a subspace algorithm (Nocedal and Wright, 1999) similar but not equivalent to a nonlinear conjugate gradient algorithm. At each iteration, the algorithm computes the optimal solution over the current subspace. This is a form of regularization of the original problem. Subspace methods such as conjugate gradient can be much faster than gradient methods particularly for ill-posed problems such as the ones that we are considering. Empirically BLF converges much faster than boosting without orthogonalization. In fact for linear hypotheses, we can prove finite termination of BLF at an optimal solution of the full problem. For the least squares loss with linear hypotheses, BLF reduces to the conjugate gradient method. We will not show this directly but instead show that the method reduces to partial least squares, which has been previously shown to be a conjugate gradient algorithm (Phatak and de Hoog, 2002). Note for finite termination, the regression coefficients $\mathbf{c}$ must be re-optimized (refitted) at every iteration, an approach not typical of most boosting or ensemble algorithms.

28.3 BLF with Linear Functions

The OrthoAnyBoost framework can be applied to linear hypotheses. Linear hypotheses are of the form $t_k = \mathbf{x}_k^T \mathbf{v}$ in input space or $t_k = \phi(\mathbf{x}_k)^T \mathbf{v}$ in feature space for use in a kernel framework. We reserve discussion of the natural extension to kernels to Section 7. To apply OrthoAnyBoost to linear hypotheses requires that the weak learner (Step 4) and the hypotheses weighting computation (Step 6) be modified accordingly. The big advantage of linear hypotheses is that the optimal orthogonal linear function found in Step 4 can be efficiently computed in closed form by recasting the problem into the null space of the constraints using a widely-used procedure commonly known as "deflation." For clarity we first provide a detailed description of deflation since it has proven a source of confusion and an understanding of deflation is necessary for all other parts of the algorithm.

28.3.1 Enforcing Orthogonality in Linear Spaces

The requirements that the new hypotheses be orthogonal to all previous constraints form linear equality constraints in hypothesis space. For linear hypotheses, $\mathbf{t} = \mathbf{X}\mathbf{v}$, Step 4 in OrthoAnyboost reduces to

$$\begin{aligned} \max_{\mathbf{v}} \quad & \mathbf{u}^{i^T} \mathbf{X} \mathbf{v} \\ subject\ to\ & \mathbf{t}^{j^T} \mathbf{X} \mathbf{v} = 0 \quad j = 1, \ldots, i-1 \\ & \mathbf{X}\mathbf{v} \in \mathcal{T} \end{aligned} \tag{28.1}$$

The linear equalities can be eliminated by mapping the problem into the null space of the linear constraints, as in power methods for computing principal components. This also provides a way to naturally bound the hypothesis space $\mathcal{T}$ so that Problem (28.1) has a solution. Define $\mathbf{T}^i = [\mathbf{t}^1, \ldots, \mathbf{t}^i]$. Let $\mathbf{Z}^i$ be any null space matrix for the matrix $\mathbf{T}^{i^T} \mathbf{X}$. Then for any $\mathbf{w}$, $\mathbf{v}^i = \mathbf{Z}^{i-1}\mathbf{w}^i$ implies $\mathbf{T}^{i^T} \mathbf{X} \mathbf{v}^i = 0$. Note when $i = 1$, $\mathbf{Z}^0$ is defined to be $\mathbf{I}$ as a special case and $\mathbf{X}^1 = \mathbf{X}$. Problem (28.1) can be reparameterized as

$$\begin{aligned} \max_{\mathbf{w}} \quad & \mathbf{u}^{i^T} \mathbf{X} \mathbf{Z}^{i-1} \mathbf{w} \\ subject\ to\ & \mathbf{w}^T \mathbf{w} = 1 \end{aligned} \tag{28.2}$$

Note that the constraint $\mathbf{w}^T \mathbf{w} = 1$ must be added to insure that the problem is bounded and that an optimal solution always exists. Problem (28.2) has a closed form solution $\mathbf{w}^i = \frac{\mathbf{Z}^{i-1^T} \mathbf{X}^T \mathbf{u}^i}{\|\mathbf{Z}^{i-1^T} \mathbf{X}^T \mathbf{u}^i\|}$.

An equivalent way to solve Problem (28.2) is to map the data into the null space $\mathbf{X}^i = \mathbf{X}\mathbf{Z}^{i-1}$ by "deflating" or orthogonalizing the data with respect to the constraints. At the ith iteration, the linear hypothesis $\mathbf{t}^i = \frac{\mathbf{X}^i \mathbf{w}^i}{\|\mathbf{X}^i \mathbf{w}^i\|}$ is chosen. The deflated data matrix is $\mathbf{X}^{i+1} = (\mathbf{I} - \mathbf{t}^i \mathbf{t}^{i^T})\mathbf{X}^i$.

While not obvious, deflating $\boldsymbol{X}$ in this way is equivalent to multiplying $\boldsymbol{X}$ by the appropriate null space matrix. At the ith iteration, Problem (28.2) is equal to

$$\max_{\mathbf{w}} \quad \mathbf{u}^{i^T} \boldsymbol{X}^i \mathbf{w}$$
$$subject\ to\ \mathbf{w}^T \mathbf{w} = 1. \tag{28.3}$$

A major advantage of this form of the problem is that the optimal solution has a closed form $\mathbf{w}^i \propto \boldsymbol{X}^{i^T} \mathbf{u}^i$. The optimal solution in the transformed space is $\mathbf{w}^i$. The optimal solution is the original space is $\mathbf{v}^i = \boldsymbol{Z}^{i-1} \mathbf{w}^i$.

The fact that orthogonality is enforced, i.e. $\boldsymbol{T}^{i^T} \boldsymbol{X}^{i+i} = 0$, can be proved by induction. Consider the second iteration, where $\mathbf{t}^1 = \frac{\boldsymbol{X}\mathbf{w}^1}{\|\boldsymbol{X}\mathbf{w}^1\|}$

$$\boldsymbol{X}^2 = (\boldsymbol{I} - \mathbf{t}^1 \mathbf{t}^{1^T})\boldsymbol{X} = \boldsymbol{X} - \frac{\boldsymbol{X}\mathbf{w}^1}{\|\boldsymbol{X}\mathbf{w}^1\|}\mathbf{t}^{1^T}\boldsymbol{X} = \boldsymbol{X}(\boldsymbol{I} - \frac{\mathbf{w}^1}{\|\boldsymbol{X}\mathbf{w}^1\|}\mathbf{t}^{1^T}\boldsymbol{X})$$

Define $\boldsymbol{Z}^1 = (\boldsymbol{I} - \frac{\mathbf{w}^1}{\|\boldsymbol{X}\mathbf{w}^i\|}\mathbf{t}^{1^T}\boldsymbol{X})$, then $\mathbf{t}^{1^T}\boldsymbol{X}^2 = \mathbf{t}^{1^T}\boldsymbol{X}\boldsymbol{Z}^1 = \mathbf{t}^{1^T}\boldsymbol{X} - \mathbf{t}^{1^T}\frac{\boldsymbol{X}\mathbf{w}^1}{\|\boldsymbol{X}\mathbf{w}^1\|}\mathbf{t}^{1^T}\boldsymbol{X} = 0$. Thus the optimal solution of Problem (28.3) for the second iteration (the first has no linear constraints) will satisfy the orthogonality constraints. For the ith iteration assume $\boldsymbol{X}^i = \boldsymbol{X}\boldsymbol{Z}^{i-1}$ and $\mathbf{t}^{j^T}\boldsymbol{X}\boldsymbol{Z}^{i-1} = 0 \quad j = 1, \ldots, i-1$. At iteration $i+1$,

$$\boldsymbol{X}^{i+1} = (\boldsymbol{I} - \mathbf{t}^i \mathbf{t}^{i^T})\boldsymbol{X}^i = \boldsymbol{X}^i - \frac{\boldsymbol{X}^i\mathbf{w}^i}{\|\boldsymbol{X}^i\mathbf{w}^i\|}\mathbf{t}^{i^T}\boldsymbol{X}^i = \boldsymbol{X}\boldsymbol{Z}^{i-1}(\boldsymbol{I} - \frac{\mathbf{w}^i}{\|\boldsymbol{X}^i\mathbf{w}^i\|}\mathbf{t}^{i^T}\boldsymbol{X}^i)$$

By the assumption, $\mathbf{t}^{j^T}\boldsymbol{X}^{i+1} = \mathbf{t}^{j^T}\boldsymbol{X}\boldsymbol{Z}^{i-1}(\boldsymbol{I} - \frac{\mathbf{w}^i}{\|\boldsymbol{X}^i\mathbf{w}^i\|}\mathbf{t}^{i^T}\boldsymbol{X}^i) = 0, \quad j = 1, \ldots, i-1$, thus we need only worry about $\mathbf{t}^i = \frac{\boldsymbol{X}^i\mathbf{w}^i}{\|\boldsymbol{X}^i\mathbf{w}^i\|}$. Since

$$\mathbf{t}^{i^T}\boldsymbol{X}^{i+1} = \mathbf{t}^{i^T}\boldsymbol{X}^i(\boldsymbol{I} - \frac{\mathbf{w}^i}{\|\boldsymbol{X}^i\mathbf{w}^i\|}\mathbf{t}^{i^T}\boldsymbol{X}^i) = \mathbf{t}^{i^T}\boldsymbol{X}^i - \mathbf{t}^{i^T}\frac{\boldsymbol{X}^i\mathbf{w}^i}{\|\boldsymbol{X}^i\mathbf{w}^i\|}\mathbf{t}^{i^T}\boldsymbol{X}^i = 0.$$

Thus $\mathbf{v}^i = \boldsymbol{Z}^{i-1}\mathbf{w}^i$ satisfies the orthogonality constraints of Problems (28.1) and (28.2). Note that the optimal $\mathbf{w}^i$ at each iteration is in the deflated space. Steps at the end of BLF map the solution back to the original data space. As shown in the next section, this can be done efficiently using the stored vectors $\mathbf{w}^i$ and $\mathbf{p}^i = \boldsymbol{X}^{i^T}\mathbf{t}^i$ without explicitly deflating the data.

28.3.2 Boosting Latent Features

Algorithm 12 provides the full algorithm for boosting orthogonal linear hypotheses or latent features. To start the algorithm, the initial or "zeroth" weak hypothesis is taken to be a constant hypothesis $\mathbf{t} \propto \mathbf{e}$, a vector of ones. The step size on the constant column is obtained by solving $\min_c l(\mathbf{y}, c\mathbf{e})$, For example for the least squares loss function, $c = \mu_y = mean(\mathbf{y})$. The first hypothesis is then $f = c\mathbf{e}$. The data is deflated by the scaled $\mathbf{t}^0 = \frac{\mathbf{e}}{\|\mathbf{e}\|}$ so $\boldsymbol{X}^1 = \boldsymbol{X} - \mathbf{t}^0\mathbf{t}^{0^T}\boldsymbol{X} = \boldsymbol{X} - \mathbf{e}\mu_X^T$ where μ_X is the mean of $\boldsymbol{X}$. Thus deflation

---

Algorithm 12: Boosted Latent Features (BLF)

Input: data X; response y; number of latent variables N.

(1.) Compute $\mu_y = \arg\min_{\mu_y} l(y, \mu_y e)$. Set $\mu_X = \frac{1}{m} X^T e$.
Deflate $X^1 = X - e\mu_X^T$. $u^1 = -\nabla l(y, \mu_y e)$, $T = [\ ]$.

(2.) For $i = 1$ to N

(3.)   Compute optimal solution to Problem (28.3): $w^i = X^{iT} u^i$

(4.)   Compute linear hypothesis: $t^i = X^i w^i$, $t^i \leftarrow t^i / \|t^i\|$, $T = [T \ t]$

(5.)   Deflate: $p^i = X^{iT} t^i$, $X^{i+1} = X^i - t^i p^{iT}$

(6.)   Compute function: $(\mu_y, c) = \arg\min_{(\mu_y, c)} l(y, \mu_y e + Tc)$

(7.)   Compute negative gradient: $u^{i+1} = -\nabla l(y, \mu_y e + Tc)$

(8.) end

(9.) Final features: $T(x) = (x - \mu_X)^T W (P^T W)^{-1}$ where W and P have
w^i and p^i as their columns, respectively.

(10.) Compute coefficients in original space: $g = W (P^T W)^{-1} c$

(11.) Final function: $f(x) = (x - \mu_X)^T g + \mu_y$.

---

centers each attribute. Frequently centering is done as preprocessing, thus we treat it as a distinct first step.

Using arguments much like those in Section 28.3.1, it can be shown that deflation forces orthogonality of many of the vectors constructed in the algorithm.

Lemma 1. *Orthogonality Properties of BLF. The following properties hold at termination of BLF:*

(1.) The vectors w^i are mutually orthogonal: $w^{iT} w^j = 0$, for $i \neq j$.

(2.) The vectors t^i are mutually orthogonal: $t^{iT} t^j = 0$, for $i \neq j$.

(3.) The vectors w^i are orthogonal to the vectors p^j for $i < j$. Thus $P^T W$ is an upper triangular matrix.

Proof. These properties also hold for PLS and the proofs for PLS in (Höskuldsson, 1988) apply to BLF without change. □

The end products of the algorithm are the final predictive function and the latent linear features. Recall that the weights for the linear hypotheses were computed in the null space using the deflated data, thus Steps 9 and 10 are added to create the linear functions and final predictive function to the original space. As shown in the following theorem, it is not necessary to explicitly compute the null matrices or deflate the data to compute the final coefficients of the latent features.

Theorem 1 (Mapping to original space). *The linear hypotheses $T = [t^1, \ldots, t^N]$ in the original centered space are*

$$T = \widetilde{X} W (P^T W)^{-1} \tag{28.4}$$

Proof. The algorithm can run until the centered data matrix $\widetilde{X}$ is fully factorized: $\widetilde{X} = \sum_{i=1}^{\text{rank}(\tilde{X})} \mathbf{t}^i \mathbf{p}^{i^T}$ and it can be divided into two parts.

$$\widetilde{X} = \sum_{i=1}^{N} \mathbf{t}^i \mathbf{p}^{i^T} + \sum_{i=N+1}^{\text{rank}(\tilde{X})} \mathbf{t}^i \mathbf{p}^{i^T}. \tag{28.5}$$

Multiplying by W on both sides yields:

$$\widetilde{X}W = \sum_{i=1}^{N} \mathbf{t}^i \mathbf{p}^{i^T} W + \sum_{i=N+1}^{\text{rank}(\tilde{X})} \mathbf{t}^i \mathbf{p}^{i^T} W = \sum_{i=1}^{N} \mathbf{t}^i \mathbf{p}^{i^T} W. \tag{28.6}$$

By Lemma 1, $\mathbf{p}^{i^T} \mathbf{w}^j = 0$ for $i = N+1, \ldots, \text{rank}(\widetilde{X})$ with $j = 1, \ldots, N$. So $\left(\widetilde{X} - \sum_{i=1}^{N} \mathbf{t}^i \mathbf{p}^{i^T}\right) W = \left(\widetilde{X} - TP^T\right) W = 0$ holds. Exploiting the fact that $P^T W$ is invertible (it is in fact a triangular matrix), solving for T yields the solution. $\square$

The final steps in Algorithm 12 give the final latent feature function and predictive functions in the original space. These steps are always the same regardless of the particular loss function used. Step 9 computes the latent features. The features may be scaled by the importance represented by the coefficients $\mathbf{c}$,

$$\hat{T}(\mathbf{x}) = (\mathbf{x} - \mu_X) W \left(P^T W\right)^{-1} diag(\mathbf{c}). \tag{28.7}$$

As written, BLF does full refitting, i.e. the problem in Step 6 is solved to optimality. This allows us to prove some additional properties of the algorithm.

Lemma 2. Optimal Refit *Let l be a convex continuously (sub)differentiable function. At each iteration after Step 6 is successfully completed*

$$T^T \nabla l(\mathbf{y}, \mu_y \mathbf{e} + T\mathbf{c}) = 0 \tag{28.8}$$

Proof. We prove this for the subgradient case since the differentiable functions are just a special case and let ∇ denote a subgradient. For convex unconstrained minimization with subgradients, a solution is optimal if and only if there exists a zero subgradient at that point (Bazaraa et al., 1993). The optimality condition for $\mathbf{c}$ is: $0 = \nabla_{\mathbf{c}} Loss(\mathbf{y}, \mu_y \mathbf{e} + T\mathbf{c}) = T^T \nabla Loss(\mathbf{y}, \mu_y \mathbf{e} + T\mathbf{c})$ by the chain rule. $\square$

Corollary 1. Orthogonality of U and T *At iteration i let $\mathbf{u}^{i+1} = -\nabla l(\mathbf{y}, \mu_y \mathbf{e} + T\mathbf{c})$, where $\nabla l(\mathbf{y}, \mu_y \mathbf{e} + T\mathbf{c})$ is the optimal subgradient found by refitting satisfying (28.8), then*

$$\mathbf{u}^{i^T} \mathbf{t}^j = 0, \text{ for } i > j. \tag{28.9}$$

Proof. By Lemma 1, at the ith step, $\mathbf{u}^{i+1^T} T = 0$. Since $T = [\mathbf{t}^1 \cdots \mathbf{t}^i]$, $\mathbf{u}^{i+1^T} \mathbf{t}^j = 0$, for $i \geq j$ holds in general. $\square$

This property permits W to be expressed in terms of the original variables.

Corollary 2. Alternate form of W *At iteration i let $\mathbf{u}^i = -\nabla l(\mathbf{y}, \mu_y \mathbf{e} + T\mathbf{c})$ where $\nabla l(\mathbf{y}, \mu_y \mathbf{e} + T\mathbf{c})$ is the optimal subgradient found by refitting satisfying (28.8), then $\mathbf{w}^i = \widetilde{X}^T \mathbf{u}^i$ for $i = 1, \ldots N$, where $\widetilde{X}$ is the centered data.*

For AdaBoost with discrete functions, the step sizes c^i are not reoptimized at later iterations. Step 6 in Algorithm 12 optimizes just one c^i. Therefore, the inequality $i > j$ in Equation (28.9) only holds with $i = j + 1$. The relation $\mathbf{u}^{j+1} \mathbf{t}^j = 0$ is known in AdaBoost as "the weak hypothesis will be exactly uncorrelated with the labels" (Schapire and Singer, 1999), where "labels" corresponds to the pseudo residual $\mathbf{u}$.

28.4 Convergence Properties of BLF

If BLF is not stopped early by restricting the number of latent variables, BLF constructs a solution of the full model

$$\min_{\mathbf{g}, \mu_y} l(\mathbf{y}, X\mathbf{g} + \mu_y \mathbf{e}) \qquad (28.10)$$

in a finite number of iterations. Without loss of generality, assume that the threshold is eliminated by augmenting X with a constant column and that data centering is optionally done as a preliminary step to form $\widetilde{X}$. Equivalently we can say BLF constructs an optimal solution of

$$\min_{\mathbf{g}} l(\mathbf{y}, \widetilde{X}\mathbf{g}) \qquad (28.11)$$

Theorem 2. Finite Convergence of BLF to Full Model *Let l be a convex (sub)differentiable function and $\mathbf{u}^i$ be the negative (sub)gradient in Step 7 that is optimal for refitting. Assume BLF is modified to check for termination conditions at each iterations, e.g. at each iteration BLF calculates $\mathbf{g} = W \left(P^T W \right)^{-1} \mathbf{c}$ and terminates if $\mathbf{g}$ is optimal for (28.11) or if $\widetilde{X} = TP^T$. Then BLF terminates in a finite number of iterations and $\mathbf{g}$ is optimal for (28.11).*

Proof. The algorithm will terminate finitely since the number of iterations is limited by the rank of $\widetilde{X}$. If $\mathbf{g}$ is optimal then the result holds. If the algorithm terminates because $\widetilde{X} = TP^T$ then due to refitting $0 = T^T \mathbf{u}^{i+1} = T^T \nabla l(\mathbf{y}, T\mathbf{c}) = T^T \nabla l(\mathbf{y}, \widetilde{X}\mathbf{g})$. Multiplying by P, $PT^T \nabla l(\mathbf{y}, \widetilde{X}\mathbf{g}) = \widetilde{X}^T \nabla l(\mathbf{y}, \widetilde{X}\mathbf{g}) = 0$, thus the (sub)gradient of the original problem is zero and thus optimal since the first order optimality conditions are satisfied and l is convex (Bazaraa et al., 1993). □

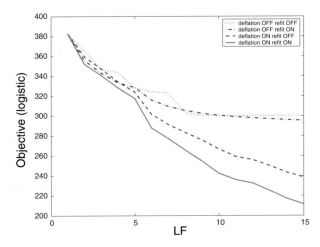

Fig. 28.1. Objective values of logistic loss for Diabetes data. Four possible options are examined: with/without deflation and with/without refit. Note the kernel trick is used to enrich the input space.

By introducing orthogonality, BLF goes from being a gradient descent algorithm, which may have slow convergence for ill-conditioned loss functions to a conjugate-gradient-like subspace method with finite termination. As we show in the next section, BLF with least squares loss reduces to PLS, which has been previously shown to be a conjugate gradient algorithm. Thus we can view BLF as a novel nonlinear conjugate gradient algorithm. Empirically the difference in convergence rate of the gradient/boosting method versus the orthogonal BLF method can be quite marked for some loss functions. Figure 28.1 illustrates the different convergence rates for the negative binomial or logistic loss evaluated on the training data with the kernel trick explained in detail in Section 28.7. Without deflation and refit the method is just boosting. Without deflation and with refit the method is a gradient method with a more extensive line search. Details about the logistic loss implementation are provided in Section 6.2. The figure shows that without deflation the algorithm stalls and fails to make progress, a classic problem with gradient descent methods. With deflation, convergence improves markedly, with the fastest algorithm being BLF with deflation and refitting.

In practice BLF would be halted after a few iterations and the difference between BLF with and without refitting is not large in the first five iterations. This suggestions that full refitting may not be necessary to achieve good testing set results, but we leave the computational study of the best BLF variant to future work.

28.5 PLS and BLF

For the least squares loss function, orthogonal boosting with linear hypotheses reduces to the well studied partial least squares regression (PLS) algorithm. We derive BLF with least squares loss and show it reduces to PLS (Algorithm 13).

Algorithm 13: Linear PLS

Input: X, $\mathbf{y}$, N.

(1.) Center data and compute mean response:
$$\boldsymbol{\mu}_X = \frac{1}{m}X^T\mathbf{e}, \; X^1 = X - \mathbf{e}\boldsymbol{\mu}_X^T, \; \mu_y = \frac{1}{m}\mathbf{e}^T\mathbf{y}.$$
$\mathbf{u}^1 = \mathbf{y} - \mu_y\mathbf{e} \; T = [\,]$.

(2.) For $i = 1$ to N

(3.)   Compute optimal solution to problem (28.3): $\mathbf{w}^i = {X^i}^T\mathbf{u}^i$

(4.)   Compute linear hypothesis: $\mathbf{t}^i = X^i\mathbf{w}^i$, $\mathbf{t}^i \leftarrow \mathbf{t}^i/\|\mathbf{t}^i\|$, $T = [T \; \mathbf{t}]$

(5.)   Deflate: $\mathbf{p}^i = {X^i}^T\mathbf{t}^i$, $X^{i+1} = X^i - \mathbf{t}^i{\mathbf{p}^i}^T$

(6.)   Compute function: $c^i = {\mathbf{u}^i}^T\mathbf{t}^i$

(7.)   Compute negative gradient: $\mathbf{u}^{i+1} = \mathbf{u}^i - \mathbf{t}^i c^i$

(8.) end

(9.) Final features: $T(\mathbf{x}) = (\mathbf{x} - \boldsymbol{\mu}_X)^T W \left(P^T W\right)^{-1}$ where W and P have $\mathbf{w}^i$ and $\mathbf{p}^i$ as their columns, respectively.

(10.) Compute coefficients in original space: $\mathbf{g} = W \left(P^T W\right)^{-1}\mathbf{c}$

(11.) Final function: $f(\mathbf{x}) = (\mathbf{x} - \boldsymbol{\mu}_X)^T\mathbf{g} + \mu_y$.

The PLS loss function is the squared loss: $l(\mathbf{y},\mathbf{f}) = \|\mathbf{y} - \mathbf{f}\|_2^2$. The negative gradient in Step 7 in Algorithm 12 is simply the residual: $\mathbf{u}^1 = -\nabla l(\mathbf{y},\mathbf{f}) = (\mathbf{y} - \mathbf{f})$. As discussed above, the first step of BLF uses the constant hypothesis so deflation is equivalent to centering the data. Also at the constant hypothesis, the negative gradient of the squared loss is $-\nabla l(\mathbf{y},\mu_y\mathbf{e}) = \mathbf{y} - \mu_y\mathbf{e}$, which is precisely $\mathbf{u}^1$ in PLS (Step 1), the step that mean centers the response. In general, at the ith iteration, the negative gradient is: $\mathbf{u}^i = -\nabla l(\mathbf{y},\mu_y\mathbf{e} + \sum_{j=1}^{i-1}\mathbf{t}^j c^j) = \mathbf{y} - \mu_y\mathbf{e} - \sum_{j=1}^{i-1}\mathbf{t}^j c^j = \mathbf{y} - \mathbf{f}$. The negative gradient is exactly the residual for the squared loss. At the ith iteration, the algorithm finds the descent direction by maximizing the inner product between $\mathbf{u}^i$ and the data projection $X^i\mathbf{w}^i$. Since, all the variables are centered, maximizing the inner product becomes equivalent to maximizing the covariance of the residual response with the linear hypothesis, which corresponds to the statistical interpretation of the PLS optimization problem.

Step 6 computes the regression coefficients. PLS optimizes one coefficient at a time in closed form, but because of the orthogonality of latent features and the least squares loss function, the coefficient is globally optimal for the loss function and there is no need for refit. PLS will be equivalent to ordinary least squares or the full model when the full rank of the original matrix is used in the extracted features. As noted above, for a generic loss function, refit is needed to insure convergence to the optimal solution of for the full-model and can increase the convergence rate.

BLF can be regarded as a generalization of PLS to an arbitrary loss function. The algorithmic and regularization properties of PLS have been extensively studied. PLS is in fact a classic conjugate gradient or Lanczos method (Phatak and de Hoog, 2002). Thus for least squares loss, BLF reduces to a conjugate gradient method that exploits second order properties of the loss function. BLF is *not* a classical nonlinear conjugate gradient algorithm. Conjugate gradient boosting in hypothesis space has been tried previously with mixed results (Li et al., 2003). BLF requires the hypotheses to be orthogonal in the hypothesis space. A nonlinear conjugate gradient algorithm would not enforce this property. BLF is a novel subspace optimization algorithm. It would be interesting to investigate the convergence rate properties of BLF and to compare BLF with a nonlinear conjugate algorithm but we leave these for future work.

PLS is known to regularize the solution and the extent of the regularization depends on the number of features constructed. See (Friedman and Popescu, 2004) for a nice analysis of these properties. When the number of feature is equal to the rank of the data, the PLS solution will coincide with the ordinary least squares solution. The norm of the regression weights increases as more latent features are added. This property is also maintained but we have not formally proven it. BLF is closely related to the gradient based regularization also proposed in (Friedman and Popescu, 2004).

28.6 BLF for Arbitrary Loss

As discussed, Algorithm 12 gives the general framework for BLF. Among many choices of loss functions, the following loss functions are examined:

(1.) Squared loss (PLS)

$$l(\mathbf{y}, \mathbf{f}) = \sum_{i=1}^{m} \gamma_i^2 \left(y_i - f(\mathbf{x}_i) \right)^2, \tag{28.12}$$

(2.) Least absolute deviation (1-norm loss)

$$l(\mathbf{y}, \mathbf{f}) = \sum_{i=1}^{m} \gamma_i |y_i - f(\mathbf{x}_i)|, \tag{28.13}$$

(3.) Exponential loss

$$l(\mathbf{y}, \mathbf{f}) = \sum_{i=1}^{m} \gamma_i \exp\left(-y_i f(\mathbf{x}_i)\right), \tag{28.14}$$

(4.) Negative binomial log-likelihood (logistic loss)

$$l(\mathbf{y}, \mathbf{f}) = \sum_{i=1}^{m} \gamma_i \ln\left(1 + e^{-2y_i f(\mathbf{x}_i)}\right).$$

Note we introduced local weighting γ_i (> 0) for each data point $\mathbf{x}_i$ to be applicable for more general settings such as cost sensitive learning. Additionally, there are many possible strategies for optimizing the loss function that effect how to optimize $\mathbf{w}$ for creating the weak hypotheses and the step size $\mathbf{c}$. Following OrthoAnyBoost, $\mathbf{w}$ is determined by maximizing the inner product between the latent feature determined by $\mathbf{w}$ and the negative gradient. The step size or hypothesis weight $\mathbf{c}$ is optimized by an exact line search or one iteration of a Newton method, ensuring the full model can eventually be obtained.

28.6.1 Least Absolute Deviation

The least absolute deviation loss is more robust to outlying points that the squared loss function. Here a Least Absolute Deviation (LAD) loss function is used for regression problems. Support Vector Machines (SVM) for regression is one example of a method that uses a more robust loss function than ordinary least squares (Drucker et al., 1997). The LAD loss can be seen as a special case of the ϵ-insensitive loss (Drucker et al., 1997) in which ϵ is set to zero. In principle, it is possible to use the ϵ-insensitive loss for BLF and it may make the model more tolerant to noise. We leave this to future work.

The LAD loss function is defined as

$$l(\mathbf{y}, \mathbf{f}) = \sum_{i=1}^{m} \gamma_i |y_i - f(\mathbf{x}_i)|, \tag{28.15}$$

where $f(\mathbf{x}_i)$ is the sum of weak hypotheses: $f(\mathbf{x}_i) = \sum_{j=1}^{N} f^j(\mathbf{x}_i)$, and $\mathbf{f}$ is a vector representation of the function f. The $\gamma_i > 0$ are fixed error weights for each point. The LAD loss function is not differentiable at every point, but an appropriate subgradient does exist. For BLF, the LAD subgradient is defined as:[3]

$$\nabla l(\mathbf{y}, \mathbf{f}) = -[\gamma_1 \mathrm{sign}(y_1 - f_1), \ldots, \gamma_m \mathrm{sign}(y_m - f_m)]^T, \tag{28.16}$$

where f_i is the ith element of the vector $\mathbf{f}$, $sign(\eta) = 1$ if $\eta > 0$, $sign(\eta) = -1$ if $\eta < 0$, and $sign(\eta) = 0$ if $\eta = 0$. Note all convergence and orthogonality properties are maintained and the algorithm performs quite well.

The first step in BLF is to optimize a constant hypothesis. The solution of the hypothesis weight problem $\mathrm{argmin}_{\mu_y} \sum_{i=1}^{m} \gamma_i |y_i - \mu_y|$ is the weighted median of $\{y_i\}_{i=1}^{m}$ with weights $\{\gamma_i\}_{i=1}^{m}$, see for example (Hoaglin et al., 1982). Once μ_y is obtained, the negative gradient $\mathbf{u}$ is computed. Thus, $\mathbf{u}^1$ at the first iteration is written by

$$\mathbf{u}^1 = [\gamma_1 \mathrm{sign}(y_1 - \mu_y), \ldots, \gamma_m \mathrm{sign}(y_m - \mu_y)]^T \tag{28.17}$$

---

[3]In a slight abuse of notation, we use ∇ to denote the subgradient whenever the gradient does not exist.

There are two options for optimizing the function coefficients **c**. We can optimize one c^i associated with $\mathbf{t}^i$ then fix the value of c^i for later iterations. Or we can refit, e.g. re-optimize all the regression coefficients $[\mu_y \ \mathbf{c}] = [\mu_y, c^1, \ldots, c^i]$ associated with the weak hypotheses selected so far:

$$(\mu_y, \mathbf{c}) = \arg\min \ \|\mathrm{diag}(\boldsymbol{\gamma})(\mathbf{y} - \sum_{j=1}^{i} \mathbf{t}^j c^j - \mu_y)\|_1, \qquad (28.18)$$

where $\boldsymbol{\gamma}$ is a vector representation of $\{\gamma_i\}_{i=1}^m$. In general, this problem can be solved using linear programming. Since $(\mu_y, \mathbf{c})$ are incrementally optimized, column generation methods can be used to efficiently update the solution at each iteration. Note that if reoptimization/refitting is not chosen, a closed-form solution can be utilized. See (Friedman, 2001) for more detail. Once the regression coefficients are solved, the negative gradient, or pseudo response, is updated:

$$\mathbf{u}^{i+1} = \left[\gamma_1 \mathrm{sign}(y_1 - \mu_y - \sum_{j=1}^{i} t_1^j c^j), \ldots, \gamma_m \mathrm{sign}(y_m - \mu_y - \sum_{j=1}^{i} t_m^j c^j)\right]^T,$$
$$(28.19)$$

where t_j^i is the jth element of the ith weak hypothesis $\mathbf{t}^i$. In summary, the steps in Algorithm 12 for LAD loss are specified as follows:

- Step 1: $\mu_y = \mathrm{median}_{\boldsymbol{\gamma}}(\mathbf{y})$ and $\mathbf{u}^1 = \left[\gamma_1 \mathrm{sign}(y_1 - \mu_y), \ldots, \gamma_m \mathrm{sign}(y_m - \mu_y)\right]^T$
- Step 6: LAD loss is minimized by linear program (28.18)
- Step 7: $\mathbf{u}^{i+1} = \left[\gamma_1 \mathrm{sign}(y_1 - \mu_y - \sum_{j=1}^{i} t_1^j c^j), \ldots, \gamma_m \mathrm{sign}(y_m - \mu_y - \sum_{j=1}^{i} t_m^j c^j)\right]^T$

28.6.2 Exponential Loss

The exponential loss function was used in AdaBoost (Freund and Shapire, 1997) for binary classification problems. AdaBoost changes the weights on the data points at each iteration; more difficult instances are given larger weights. The algorithm can be understood as minimizing the exponential loss functional (Friedman et al., 2000, Lafferty, 1999):

$$l(\mathbf{y}, \mathbf{f}) = \sum_{i=1}^{m} \gamma_i \exp\left(-y_i f(\mathbf{x}_i)\right), \qquad (28.20)$$

where the responses are given in binary coding: $y_i \in \{+1, -1\}$, $i = 1, \ldots, m$.

Computing Descent Directions in BLF

Let's discuss how to formulate the exponential loss in BLF. At first a constant weak hypothesis is added to the additive model. Then the regression coefficient μ_y on the constant hypothesis is determined by optimizing

$$\min_{\mu_y} l(\mathbf{y}, \mu_y \mathbf{e}) = \min_{\mu_y} \sum_{i=1}^{m} \gamma_i \exp\left(-y_i \mu_y\right)$$

$$= \min_{\mu_y} e^{-\mu_y} \sum_{i \in \mathcal{C}^+} \gamma_i + e^{\mu_y} \sum_{i \in \mathcal{C}^-} \gamma_i, \qquad (28.21)$$

where $\mathcal{C}^+$ is a set of positive data points and $\mathcal{C}^-$ is a set of negative data points. With this simplification, it is now easy to solve the optimality condition $\frac{\partial l(\mathbf{y}, \mu_y)}{\partial \mu_y} = 0$, to compute the solution:

$$\mu_y = \frac{1}{2} \ln \frac{\sum_{i \in \mathcal{C}^+} \gamma_i}{\sum_{i \in \mathcal{C}^-} \gamma_i}. \qquad (28.22)$$

The next step computes the negative gradient of the loss function:

$$u_k^{i+1} = \left(-\nabla l(\mathbf{y}, \mu_y + \sum_{j=1}^{i} \mathbf{t}^j c^j)\right)_k = \gamma_k e^{-y_k \left(\mu_y + \sum_{j=1}^{i} t_k^j c^j\right)} y_k. \qquad (28.23)$$

As seen in Equation (28.23), the weak hypotheses in previous iterations are absorbed in the "weight" defined as $\gamma_k e^{-y_k \left(\mu_y + \sum_{j=1}^{i} t_k^j c^j\right)}$. This weighted response becomes a new response variable to fit in creating a new weak hypothesis, which corresponds to the weight updates on each of the data points in the AdaBoost algorithm.

Solving for Function Coefficients

At every iteration i, BLF computes the function coefficients $\mathbf{c}$ when a new weak hypothesis or latent feature is added to the model. Refitting for exponential loss minimizes the following loss function with respect to μ_y and $\mathbf{c}$:

$$l(\mathbf{y}, \sum_{j=1}^{i} \mathbf{t}^j c^j) = \sum_{k=1}^{m} \gamma_k \exp\left(-y_k \left(\mu_y + \sum_{j=1}^{i} t_k^j c^j\right)\right). \qquad (28.24)$$

It is convenient to define a matrix notation. Let

$$\mathbf{d} = \left[\gamma_1 \exp\left(-y_1 (\mu_y + \sum_{j=1}^{i} t_1^j c^j)\right), \dots, \gamma_m \exp\left(-y_m (\mu_y + \sum_{j=1}^{i} t_m^j c^j)\right)\right]^T$$

$$(28.25)$$

Then the loss function is simply expressed by

$$l(\mathbf{y}, \mu_y + \sum_{j=1}^{i} \mathbf{t}^j c^j) = \mathbf{e}^T \mathbf{d}. \tag{28.26}$$

The gradient with respect to $(\mu_y, \mathbf{c})$ is given by

$$\nabla_{(\mu_y, \mathbf{c})} l(\mathbf{y}, \mu_y + \sum_{j=1}^{i} \mathbf{t}^j c^j) = \begin{bmatrix} -\sum_{k=1}^{m} \gamma_k e^{-y_k(\mu_y + \sum_{j=1}^{i} t_k^j c^j)} y_k \\ -\sum_{k=1}^{m} \gamma_k e^{-y_k(\mu_y + \sum_{j=1}^{i} t_k^j c^j)} y_k t_k^1 \\ \vdots \\ -\sum_{k=1}^{m} \gamma_k e^{-y_k(\mu_y + \sum_{j=1}^{i} t_k^j c^j)} y_k t_k^i \end{bmatrix} \tag{28.27}$$

$$= -[\mathbf{e}\ \boldsymbol{T}]^T \mathrm{diag}(\mathbf{d})\mathbf{y}.$$

The Hessian is given as follows:

$$\nabla^2_{(\mu_y, \mathbf{c})} l(\mathbf{y}, \mu_y + \sum_{j=1}^{i} \mathbf{t}^j c^j) = [\mathbf{e}\ \boldsymbol{T}]^T \mathrm{diag}(\mathbf{y})\mathrm{diag}(\mathbf{d})\mathrm{diag}(\mathbf{y})[\mathbf{e}\ \boldsymbol{T}] \tag{28.28}$$

Since $y_i \in \{+1, -1\}$ for binary classification, the Hessian can be written as

$$\nabla^2_{(\mu_y, \mathbf{c})} l = [\mathbf{e}\ \boldsymbol{T}]^T \mathrm{diag}(\mathbf{d})[\mathbf{e}\ \boldsymbol{T}]. \tag{28.29}$$

Thus the Newton step is given by:

$$\begin{bmatrix} \mu_y \\ \mathbf{c} \end{bmatrix} = ([\mathbf{e}\ \boldsymbol{T}]^T \mathrm{diag}(\mathbf{d})[\mathbf{e}\ \boldsymbol{T}])^{-1} [\mathbf{e}\ \boldsymbol{T}]^T \mathrm{diag}(\mathbf{d})\mathbf{y}, \tag{28.30}$$

which is just a weighted least squares problem with the vector $\mathbf{d}$ determining the weights on data points. Since μ_y and $\mathbf{c}$ are incrementally optimized, the Newton step can be started from the previously optimized value to reduce the number of iterations to converge. In practice, a few iterations are sufficient to get a good fit to the response variable. In summary, the steps in Algorithm 12 for the exponential loss are specified as follows:

- Step 1: $\mu_y = \frac{1}{2} \ln \frac{\sum_{i \in c+} \gamma_i}{\sum_{i \in c-} \gamma_i}$
- Step 1 and 7:

$$u_k^{i+1} = \gamma_k e^{-y_k\left(\mu_y + \sum_{j=1}^{i} t_k^j c^j\right)} y_k$$

- Steps 6: the step size $(\mu_y, \mathbf{c})$ optimized by the Newton step

28.6.3 The Negative Binomial Log-likelihood

The binomial likelihood is parameterized by

$$p(y = 1|\mathbf{x}) = \frac{e^{f(\mathbf{x})}}{e^{f(\mathbf{x})} + e^{-f(\mathbf{x})}} = \frac{e^{f(\mathbf{x})}}{2\cosh(f(\mathbf{x}))}$$

$$p(y = -1|\mathbf{x}) = \frac{e^{-f(\mathbf{x})}}{e^{f(\mathbf{x})} + e^{-f(\mathbf{x})}} = \frac{e^{-f(\mathbf{x})}}{2\cosh(f(\mathbf{x}))}. \tag{28.31}$$

Equivalently, $f(\mathbf{x})$ is expressed by $p(y = 1|\mathbf{x})$ and $p(y = -1|\mathbf{x})$:

$$f(\mathbf{x}) = \frac{1}{2}\ln\frac{p(y = 1|\mathbf{x})}{p(y = -1|\mathbf{x})}. \tag{28.32}$$

The negative log-likelihood loss, including local weighting γ_i, is given by

$$
\begin{aligned}
l(\mathbf{y}, \mathbf{f}) &= -\sum_{i \in \mathcal{C}^+}\gamma_i \ln\left(p(y = 1|\mathbf{x}_i)\right) - \sum_{i \in \mathcal{C}^-}\gamma_i \ln\left(p(y = -1|\mathbf{x}_i)\right) \\
&= -\sum_{i=1}^{m}\gamma_i\left[y_i f(\mathbf{x}_i) - \ln\left(2\cosh f(\mathbf{x}_i)\right)\right] \\
&= \sum_{i=1}^{m}\gamma_i \ln\left(1 + e^{-2y_i f(\mathbf{x}_i)}\right).
\end{aligned}
\tag{28.33}
$$

Note that this loss function is used for additive logistic regression (Friedman et al., 2000, Lafferty, 1999) and is just a factor of 2 different from the loss in logistic regression. It is also known that negative binomial loss is equivalent to the exponential loss up to the second order (Friedman et al., 2000, Lafferty, 1999).

Computing the Descent Directions in the Negative Binomial Log-likelihood

In Step 1, a constant hypothesis is added to the model:

$$
\begin{aligned}
\min l(\mathbf{y}, \mu_y \mathbf{e}) &= \sum_{i=1}^{m}\gamma_i \ln\left(1 + e^{-2y_i\mu_y}\right) \\
&= \ln\left(1 + e^{-2\mu_y}\right)\sum_{i \in \mathcal{C}^+}\gamma_i + \ln\left(1 + e^{2\mu_y}\right)\sum_{i \in \mathcal{C}^-}\gamma_i. \quad (28.34)
\end{aligned}
$$

Solving $\frac{\partial l(\mathbf{y},\mu_y\mathbf{e})}{\partial\mu_y} = 0$ yields

$$\mu_y = \frac{1}{2}\ln\frac{\sum_{i \in \mathcal{C}^+}\gamma_i}{\sum_{i \in \mathcal{C}^-}\gamma_i}. \tag{28.35}$$

The next step computes the negative gradient of the loss function.

$$u_k^{i+1} = \left(-\nabla l(\mathbf{y}, \mu_y\mathbf{e} + \sum_{j=1}^{i}\mathbf{t}^j c^j)\right)_k = \gamma_k\left(y_k - \tanh\left(\mu_y + \sum_{j=1}^{i}t_k^j c^j\right)\right). \tag{28.36}$$

This equation means that the new pseudo response is given by the residual between $\mathbf{y}$ and the hyperbolic tangent of the linear combination of the weak hypotheses, just like in neural networks.

Solving for Function Coefficients

Refitting re-optimizes the function coefficients $\mathbf{c}$ at each iteration. The loss function is written by

$$l(\mathbf{y}, \mu_y \mathbf{e} + \sum_{j=1}^{i} \mathbf{t}^j c^j) =$$

$$-\sum_{k=1}^{m} \gamma_k \left[y_k (\mu_y + \sum_{j=1}^{i} t_k^j c^j) - \ln \left(2 \cosh \left(\mu_y + \sum_{j=1}^{i} t_k^j c^j \right) \right) \right] .(28.37)$$

The gradient with respect to $(\mu_y, \mathbf{c})$ is

$$\nabla_{(\mu_y, \mathbf{c})} l(\mathbf{y}, \mu_y \mathbf{e} + \sum_{j=1}^{i} \mathbf{t}^j c^j)$$

$$= \begin{bmatrix} -\sum_{k=1}^{m} \gamma_k \left(y_k - \tanh(\mu_y + \sum_{j=1}^{i} t_k^j c^j) \right) \\ -\sum_{k=1}^{m} \gamma_k t_k^1 \left(y_k - \tanh(\mu_y + \sum_{j=1}^{i} t_k^j c^j) \right) \\ \vdots \\ -\sum_{k=1}^{m} \gamma_k t_k^i \left(y_k - \tanh(\mu_y + \sum_{j=1}^{i} t_k^j c^j) \right) \end{bmatrix} \tag{28.38}$$

$$= -[\mathbf{e}\ T]^T \mathbf{r},$$

where $r_k = \gamma_k \left(y_k - \tanh(\mu_y + \sum_{j=1}^{i} t_k^j c^j) \right)$. Furthermore, the Hessian is given by

$$\nabla^2_{(\mu_y, \mathbf{c})} l = [\mathbf{e}\ T]^T \operatorname{diag}(\mathbf{d}) [\mathbf{e}\ T], \tag{28.39}$$

where $\mathbf{d}$ is defined as follows:

$$\mathbf{d} = \left(\left[\gamma_1 \cosh^{-2}(\mu_y + \sum_{j=1}^{i} t_1^j c^j), \dots, \gamma_m \cosh^{-2}(\mu_y + \sum_{j=1}^{i} t_m^j c^j) \right]^T \right) \tag{28.40}$$

Thus the Newton step is given by

$$\begin{bmatrix} \mu_y \\ \mathbf{c} \end{bmatrix} = ([\mathbf{e}\ T]^T \operatorname{diag}(\mathbf{d})[\mathbf{e}\ T])^{-1} [\mathbf{e}\ T]^T \mathbf{r}. \tag{28.41}$$

This optimization is done in the same fashion as for exponential loss.

The Newton step may not always lead to a decrease in the objective function, leading to a condition that can be readily detected while training. A modified Newton step, which adds a multiple of the identity matrix to the Hessian before computing the Newton step, can be used. The following heuristic for a Modified Newton step was found to work well in practice based on the parameter $0 \leq \lambda \leq 1$, Hessian $\boldsymbol{H} = \nabla^2_{(\mu_y, \mathbf{c})} l$, and iteration number i:

$$\widehat{\boldsymbol{H}} = (1 - \lambda)\boldsymbol{H} + \frac{\lambda \, trace(\boldsymbol{H})}{i + 1} \boldsymbol{I} \tag{28.42}$$

Then $\widehat{\boldsymbol{H}}$ is used to compute a modified Newton step in Step 6 instead of the original Newton step.

In summary, the steps in Algorithm 12 for the negative binomial log-likelihood loss are specified as follows:

- Step 1: $\mu_y = \frac{1}{2} \ln \frac{\sum_{j \in c+} \gamma_j}{\sum_{j \in c-} \gamma_j}$
- Steps 1 and 7:

$$u_k^{i+1} = \gamma_k \left(y_k - \tanh \left(\mu_y + \sum_{j=1}^{i} t_k^j c^j \right) \right)$$

- Step 6: the step size $(\mu_y, \mathbf{c})$ optimized by the modified Newton step

28.7 Kernel BLF

The kernel extension of BLF is similar to that of kernel PLS (Rosipal and Trejo, 2001). BLF is expressed in terms of inner products of data points, or equivalently in terms of the Gram matrix $\boldsymbol{X}\boldsymbol{X}^T$. We then transform the input space into the kernel-defined feature space: $\boldsymbol{X}\boldsymbol{X}^T \mapsto \boldsymbol{K}$. With this basic rule in mind, we can modify the BLF algorithm (Algorithm 12). In Step 1, the deflation of the data matrix should be replaced by centering the kernel matrix. That is, $\boldsymbol{K}^1 = \widetilde{\boldsymbol{K}} = \left(\boldsymbol{I} - \frac{1}{m} \mathbf{e}\mathbf{e}^T \right) \boldsymbol{K} \left(\boldsymbol{I} - \frac{1}{m} \mathbf{e}\mathbf{e}^T \right)$. Steps 3 and 4 are combined to avoid expressing $\mathbf{w}$. Namely, the latent feature is computed by $t^i \in \{\mathbf{t} | \mathbf{u}^{iT} \boldsymbol{K}^i \mathbf{t} > 0\}$. Step 5 is replaced by the kernel deflation: $\boldsymbol{K}^{i+1} = \left(\boldsymbol{I} - \mathbf{t}^i \mathbf{t}^{iT} \right) \boldsymbol{K}^i \left(\boldsymbol{I} - \mathbf{t}^i \mathbf{t}^{iT} \right)$.

The step computing the function coefficients associated with the original data matrix $\boldsymbol{X}$ is now computed using the dual function coefficients, $\tilde{\boldsymbol{\beta}}$. The formula in Step 9 in Algorithm 12 is expressed in terms of dual variables $\mathbf{t}$, $\mathbf{u}$, and the kernel $\boldsymbol{K}$. As shown in Section 3, the properties $\boldsymbol{P} = \widetilde{\boldsymbol{X}}^T \boldsymbol{T}$ and $\boldsymbol{W} = \widetilde{\boldsymbol{X}}^T \boldsymbol{U}$ hold. Thus the formula $\mathbf{g} = \boldsymbol{W} \left(\boldsymbol{P}^T \boldsymbol{W} \right)^{-1} \mathbf{c}$ can be rewritten by $\mathbf{g} = \widetilde{\boldsymbol{X}}^T \boldsymbol{U} \left(\boldsymbol{T}^T \widetilde{\boldsymbol{X}} \widetilde{\boldsymbol{X}}^T \boldsymbol{U}^T \right)^{-1} \mathbf{c}$, which gives the formula for the kernel PLS regression coefficients: $\tilde{\boldsymbol{\beta}} = \boldsymbol{U} \left(\boldsymbol{T}^T \widetilde{\boldsymbol{K}} \boldsymbol{U}^T \right)^{-1} \mathbf{c}$. The first property $\boldsymbol{P} = \widetilde{\boldsymbol{X}}^T \boldsymbol{T}$

results from deflation. However, as shown in Corollary 2, $\boldsymbol{W} = \widetilde{\boldsymbol{X}}^T \boldsymbol{U}$ only holds if the step sizes are refitted. Thus, early termination of the Newton iteration to determine $\mathbf{c}^i$ or fixing the step size for earlier iterations will violate Equation (28.9). Therefore, we need to find a matrix $\boldsymbol{A}$ such that $\boldsymbol{W} = \widetilde{\boldsymbol{X}}^T \boldsymbol{A}$ in order to express the inference functions in terms of the original data. Using the deflation formula, we can readily find a formula for an ith column of the matrix $\boldsymbol{A}$:

$$\mathbf{a}^i = \mathbf{u}^i - \sum_{j=1}^{i-1} (\mathbf{t}^{j^T} \mathbf{u}^i) \mathbf{t}^j. \tag{28.43}$$

Using the matrix $\boldsymbol{A}$, we can write the formula for the function coefficients as follows:

$$\tilde{\boldsymbol{\beta}} = \boldsymbol{A} \left(\boldsymbol{T}^T \widetilde{\boldsymbol{K}} \boldsymbol{A} \right)^{-1} \mathbf{c}. \tag{28.44}$$

Using the dual regression coefficient $\tilde{\boldsymbol{\beta}}$, the final prediction is written by

$$f(\mathbf{x}) = \sum_{i=1}^{m} \tilde{K}(\mathbf{x}, \mathbf{x}_i) \tilde{\beta}_i + \mu_y, \tag{28.45}$$

where the centered test kernel is denoted by $\tilde{\mathbf{k}} \equiv [\tilde{K}(\mathbf{x}, \mathbf{x}_1), \dots, \tilde{K}(\mathbf{x}, \mathbf{x}_m)]^T$. The formula for the un-centered test kernel is easily derived first by considering a linear kernel, then extending to general kernels:

$$
\begin{aligned}
\tilde{\mathbf{k}}^T &= \left(\mathbf{x} - \frac{1}{m} \boldsymbol{X}^T \mathbf{e} \right)^T \boldsymbol{X}^T \left(\boldsymbol{I} - \frac{1}{m} \mathbf{e}\mathbf{e}^T \right) \\
&= \left(\mathbf{x}^T \boldsymbol{X}^T - \frac{1}{m} \mathbf{e}^T \boldsymbol{X} \boldsymbol{X}^T \right) \left(\boldsymbol{I} - \frac{1}{m} \mathbf{e}\mathbf{e}^T \right) \\
&\mapsto (\mathbf{k}^T - \boldsymbol{\mu}_K^T) \left(\boldsymbol{I} - \frac{1}{m} \mathbf{e}\mathbf{e}^T \right),
\end{aligned}
\tag{28.46}
$$

where $\boldsymbol{\mu}_K$ is a mean vector of the kernel matrix: $\boldsymbol{\mu}_K = \frac{1}{m} \boldsymbol{K} \mathbf{e}$. Thus, by putting $\boldsymbol{\beta} = \left(\boldsymbol{I} - \frac{1}{m} \mathbf{e}\mathbf{e}^T \right) \tilde{\boldsymbol{\beta}}$ and $\mu_K = \boldsymbol{\mu}_K^T \boldsymbol{\beta}$, the final regression function is simply expressed by

$$f(\mathbf{x}) = \sum_{i=1}^{m} K(\mathbf{x}, \mathbf{x}_i) \beta_i - \mu_K + \mu_y. \tag{28.47}$$

With Equation (28.47), prediction can be done using the uncentered kernel, modified dual regression coefficient $\boldsymbol{\beta}$ and a constant bias term $(-\mu_K + \mu_y)$.

28.8 Computational Results

This section presents computational results for linear and kernel BLF. Recall BLF can be used both to construct orthogonal features targeted to a

---

Algorithm 14: Kernel Boosted Latent Features

Input: K, y, N.

(1.) Compute $\mu_y = \arg\min_{\mu_y} l(y, \mu_y e)$.
Deflate $K^1 = \left(I - \frac{1}{m}ee^T\right) K \left(I - \frac{1}{m}ee^T\right)$. $u^1 = -\nabla l(y, \mu_y e)$, $T = [\ ]$.

(2.) For $i = 1$ to N

(3.)   Compute latent features:
$t^i = K^i u^i$, $t^i \leftarrow t^i / \|t^i\|$, $T = [T\ t]$

(4.)   Deflate: $K^{i+1} = \left(I - t^i t^{i^T}\right) K^i \left(I - t^i t^{i^T}\right)$

(5.)   Compute the function:
$(\mu_y, c) = \arg\min_{(\mu_y, c)} l(y, \mu_y e + Tc)$

(6.)   Compute: $u^{i+1} = -\nabla l(y, Tc)$

(7.) end

(8.) For $i = 1$ to N

(9.) $a^i = u^i - \sum_{j=1}^{i-1}(t^{j^T} u^i) t^j$

(10.) end

(11.) Final features: $T(x) = K^1(x, X) A \left(T^T K^1 A\right)^{-1}$, where A has a^i as its columns.

(12.) Compute the coefficients and bias:
$\beta = \left(I - \frac{1}{m}ee^T\right) A \left(T^T K^1 A\right)^{-1} c$, $\mu_K = \mu_K^T \beta$.

(13.) Final prediction function is:
$f(x) = \sum_{i=1}^{m} K(x, x_i)\beta - \mu_K + \mu_y$.

---

particular loss function and to create predictive models for a loss function. Figure 28.2 illustrates the value of orthogonal features targeted to a specific loss. The Cancer data set from the UCI machine learning repository (Blake and Merz, 1998) is used. The first two features of the nine dimensional Cancer data constructed by PCA, BLF with least squares (BLF-LS) loss, which is equivalent to PLS, and BLF with logistic loss (BLF-LOG) are shown. The PCA plot does not use information from the response and does not capture the class structure as well. The PCA plot looks rather noisy since the outliers have more influence; outliers account for the high variance. BLF-LS and BLF-LOG discriminate the two classes more clearly and could be used as input another modeling method or the BLF function could be used for classification. BLF-LOG constructs a nice Gaussian blob for each class.

The next two subsections are devoted to quantitative studies of BLF. The first set of experiments illustrates BLF's competitive performance as a predictive modeling tool on standard machine learning test beds using common cost functions. In the second case study, we investigate the use of kernel BLF on a very high dimensional unbalanced classification problem, *Thrombin*, from the KDD cup 2001 (also known as the Dorothea dataset in the NIPS 2003 Feature Selection Challenge).

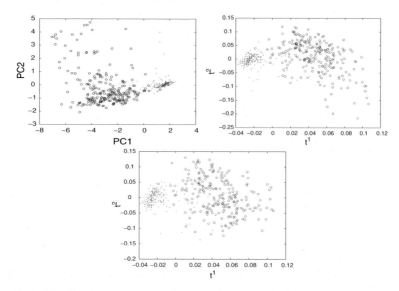

Fig. 28.2. The first two components of PCA (*upper − left*), BLF-LS (PLS) (*upper − right*) and BLF-LOG (*lower*). The positive examples are dots and the negative examples are circles.

28.8.1 The Benchmark Data Sets

In this subsection, real world data sets, Boston Housing, Cancer, Diabetes, Liver, and Wisconsin Breast Cancer (WBC) from the UCI machine learning repository (Blake and Merz, 1998), and Albumin (a drug discovery dataset) (Bennett et al., 2000), are examined. Boston Housing and Albumin are regression data sets. All others are binary classification problems. We report both the mean and the standard deviation of the mean squared error in case of regression or accuracy in percentages in case of classification. Except for Boston Housing, all the experiments are repeated 100 times. For Boston Housing, 10-fold cross validation (CV) is adopted. Uniform weighting of data is always used.

For each of the trials, 10% of the data are held out as a test set and the rest are used for training. Every experiment uses the same splits of training and test sets. The number of latent features N is determined by 10-fold CV inside the training set. N is selected to be the number of latent features that minimizes a moving average error curve, thus N is different among the different train-test splits. Since the error curve is based on random splits of the 10-fold training data CV, the error for a given latent feature i is averaged with the next smaller $i-1$ and next larger $i+1$ results for all the experiments.

For kernel BLF, the radial basis function (RBF) kernel,

$$k\left(\mathbf{x}_i, \mathbf{x}_j\right) = \exp(-\frac{\|\mathbf{x}_i - \mathbf{x}_j\|^2}{\sigma^2}),$$

is used. The kernel parameter σ is determined so that the value of σ creates reasonably good models for *all the loss functions* by 10-fold cross-validation runs over all the data points. Five to ten different σs are tried depending on the data set.

Refitting the step size is used with in all of the loss functions except least squares. But least squares, not refitting is mathematically equivalent to refitting. LAD refit was solved exactly by linear programming. For exponential and logistic loss, the Newton step is iterated only once for computational efficiency. The modified Newton step in Equation (28.42) with $\lambda = 0.1$ is used. For the RBF kernel, the kernel matrix has full rank m, and without the regularization provided by the single modified Newton Step the model overfits the training set quickly. We leave a more formal treatment of regularization within BLF to future work.

As a reference, we also perform the same experiments with SVMs (Boser et al., 1992). The experimental settings are the same as other BLF models except for a selection of trade-off parameter C. Depending on the data set, a wide range of values of C (about 10) is examined by 10-fold CV inside the training data and the one that minimizes the CV error is chosen. SVM^{light} (Joachims, 1999) is used for all the SVM experiments.

The paired student t-test is used to test significance. Since the baseline method is BLF-LS (PLS), the t-test is performed against the results of BLF-LS. Bold numbers indicate the method was significantly better than BLF-LS with quadratic loss using a significance level of 0.05. Note that SVM models are not compared in the t-test.

As shown in Table 28.1, BLF works well for all of the loss functions. For regression, BLF with LAD loss (BLF-LAD) and BLF-LS achieve results for Boston Housing that are not significantly different. Note that only 10 trials were performed for Boston Housing, which might be why the difference is

Table 28.1. Regression results for linear models for the quadratic and least absolute deviation (LAD) loss functions. SVM results provided for reference. Data sizes ($m \times n$) are also shown.

		Quadratic		LAD		SVM	
		Train	Test	Train	Test	Train	Test
Boston Housing	Mean	21.8173	23.4910	24.6214	25.6909	22.5446	24.1008
(506 × 13)	STD	1.1102	10.1094	1.4260	14.0496	1.1150	11.6537
	LF	8 (0)		6.8 (1.40)		—	
Albumin	Mean	0.1586	0.4044	0.1245	**0.3201**	0.0937	0.4331
(94 × 524)	STD	0.0999	0.2165	0.0243	0.2037	0.0443	0.2012
	LF	3.39 (1.39)		4.95 (0.77)		—	

Table 28.2. Classification results for linear models for the quadratic, exponential and logistic loss functions. SVM results provided for reference. Data sizes ($m \times n$) are also shown.

		Quadratic		Exponential		Logistic		SVM	
		Train	Test	Train	Test	Train	Test	Train	Test
Cancer	Mean	95.97	96.00	96.57	96.22	96.91	**96.74**	97.03	96.72
(699 × 9)	STD	0.30	0.24	0.30	2.25	0.22	2.13	0.24	1.95
	LF	4.08 (1.03)		5.84 (0.85)		5.41 (1.08)		—	
Diabetes	Mean	78.08	76.01	77.62	75.80	78.26	**76.33**	77.93	76.04
(768 × 8)	STD	0.51	4.03	0.58	3.90	0.52	4.02	0.62	4.08
	LF	5.20 (0.70)		5.15 (0.78)		5.39 (0.71)		—	
Ionosphere	Mean	90.18	85.86	91.64	85.97	93.47	**86.83**	93.12	86.06
(351 × 34)	STD	1.04	5.86	1.30	5.90	1.31	5.86	1.74	5.54
	LF	5.69 (1.62)		7.40 (1.26)		7.11 (1.27)		—	
Liver	Mean	69.93	**69.24**	69.40	68.53	96.98	69.00	70.93	69.15
(345 × 6)	STD	1.33	8.15	1.45	8.76	1.37	8.92	1.27	7.62
	LF	4 (0)		4 (0)		4 (0)		—	
WBC	Mean	96.64	95.91	98.10	**97.14**	98.65	**97.80**	98.50	97.52
(569 × 30)	STD	0.31	2.33	0.33	1.93	0.22	1.85	0.34	1.88
	LF	6.16 (0.75)		5.34 (0.54)		5.38 (0.49)		—	

not significant by the t-test. For Albumin, the results for the LAD loss show significantly better performance. For classification (Table 28.2), BLF-LOG performs better than BLF-LS on all of the data sets except for Liver. The exponential loss (BLF-EXP) is not as good as BLF-LOG for all of the data sets. Figure 28.3 shows accuracy versus number of latent features for BLF-LOG and BLF-LS on the cancer data. BLF-LS's quadratic loss function penalizes points that are classified "too well", which serves as a form of capacity control for BLF-LS. BLF-LS does not always fit training and test sets well but the behavior is stable. BLF-LOG, however, does not have such penalization via the loss function, so BLF-LOG can fit the training data better than BLF-LS. Eventually BLF-LOG ovefits. Selecting the number of latent features in BLF serves as regularization, and the experiments show that the tuning number of latent features by CV succeeds in avoiding over-fitting – an advantage over regular logistic regression.

The kernel version of BLF is also examined for the same data sets. The RBF kernel is used with parameters, σ, as follows: 50 for Albumin, 4.24 for

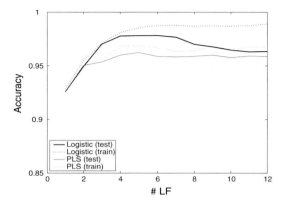

Fig. 28.3. Testing accuracy for Wisconsin Breast Cancer for logistic and least squares loss.

Boston Housing, 5 for Cancer, 5 for Diabetes, 3 for Ionosphere, 6 for Liver and 9 for WBC. In kernel models, especially with the RBF kernel, regularization plays a key role for capacity control, and capacity control is much more important for nonlinear models. As mentioned in linear models, BLF-LS has a natural penalization for overfitting, but the classification loss functions do not have such penalization except for early stopping. As seen in Table 28.4, the advantage observed in linear models seems to be reduced and the differences between the models are less significant. This observation suggests that the kernel BLF with explicit regularization in the objective function may be better to exploit the logistic and other loss function without overfitting.

Table 28.3. Regression results for kernel models for quadratic and least absolute deviation (LAD) loss functions. SVM results are also shown as a reference. Data sizes $(m \times n)$ are also shown.

		Quadratic		LAD		SVM	
		Train	Test	Train	Test	Train	Test
Boston Housing	Mean	3.5723	9.8334	4.8788	10.4530	3.8194	10.1425
(506 × 13)	STD	0.4464	4.3849	0.9215	7.6215	0.5040	7.3367
	LF	19.2 (1.81)		29.6 (4.97)		—	
Albumin	Mean	0.0854	0.3764	0.0992	**0.3266**	0.1017	0.4190
(94 × 524)	STD	0.0098	0.2054	0.0228	0.1983	0.0460	0.1894
	LF	4.14 (1.15)		4.95 (0.70)		—	

Table 28.4. Classification results for kernel models for quadratic, exponential, and logistic loss functions. Data sizes ($m \times n$) are also shown.

		Quadratic		Exponential		Logistic		SVM	
		Train	Test	Train	Test	Train	Test	Train	Test
Cancer	Mean	96.95	96.46	96.87	96.51	96.70	**96.71**	96.16	96.87
(699 × 9)	STD	0.47	1.99	0.35	2.15	0.28	1.98	0.27	1.82
	LF	2.94 (3.72)		3.98 (0.32)		1.17 (0.45)		—	
Diabetes	Mean	79.60	75.76	78.67	75.01	79.58	**76.36**	79.06	76.03
(768 × 8)	STD	0.66	4.63	1.55	4.68	0.62	4.64	0.82	4.01
	LF	4.17 (0.57)		3.28 (1.34)		4.14 (0.38)		—	
Ionosphere	Mean	99.18	94.11	99.45	**94.80**	98.99	**94.66**	97.94	94.14
(351 × 34)	STD	0.45	3.49	0.51	3.60	0.53	3.51	0.88	3.50
	LF	8.73 (2.56)		8.02 (3.71)		5.85 (1.42)		—	
Liver	Mean	76.45	72.62	76.03	72.26	76.54	72.94	75.90	73.26
(345 × 6)	STD	1.22	7.09	1.10	6.06	0.90	6.74	1.03	6.92
	LF	5.56 (1.48)		5.76 (1.13)		5.62 (0.65)		—	
WBC	Mean	98.80	97.88	99.45	97.37	98.87	97.70	98.67	97.95
(569 × 30)	STD	0.20	1.81	0.53	2.07	0.30	1.90	0.21	1.78
	LF	10.53 (2.05)		8.89 (4.35)		5.87 (0.92)		—	

Comparison with PCA

To evaluate the effectiveness of BLF at feature construction, we repeated the kernel classification experiments using principal components instead of latent features. The results show that much fewer latent features than PCA latent features are required to achieve comparable accuracies. The experiments for kernel models in Table 28.4 are repeated using the principal component from PCA instead of the latent features from BLF. For the least squares loss, this becomes the standard principal component regression algorithm. In terms of generalization error, the methods with PCA and BLF features were very similar: the paired t-test resulted in 2 wins (significantly better results) for BLF, 2 wins for PCA, and 8 ties for Cancer, Diabetes, Ionosphere, and Liver. For Wisconsin Breast Cancer, BLF wins for all the loss functions. Significantly fewer boosted latent features than principal components were required to achieve similar/slightly better performance. Figure 28.4 plots the number of latent features used for BLF and PCA. PCA always required more features.

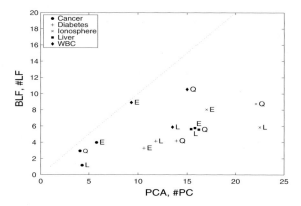

Fig. 28.4. Number of the principal components and latent features for kernel classification models. Quadratic (Q), exponential (E), and logistic (L) loss functions are used for both BLF and PCA. PCA requires more dimensions to obtain similar performance to BLF.

28.8.2 Case Study on High Dimensional Data

We use a very high dimensional data set, *Thrombin* from KDD Cup 2001, to illustrate BLF as a feature construction method. *Thrombin* is a classification problem to predict a compound's binding activity (active/inactive) to Thrombin. The original training/test split for the 2001 KDD Cup is very challenging: 42 active and 1867 inactive compounds (actives to inactives ratio is about 2.25%) with 139,351 binary features. However the test set is more balanced: 150 actives and 484 inactives (31%). In order to slightly reduce the difficulty in the training set, the original training and test sets are merged and randomly divided so that the new training/test split is 100 actives and 330 inactives. This gives the active/inactive ratio 4.55% in the new training split and 30% in the new test split. We repeat the experiment 100 times to compare models. Note that all 100 experiments use the same training/test splits. Because of the unbalanced class distribution, we adopt a local weighting γ_i for each data point $\mathbf{x}_i$: $\gamma_i = 1/|\mathcal{C}^+|, \forall i \in \mathcal{C}^+$. $\gamma_i = 1/|\mathcal{C}^-|, \forall i \in \mathcal{C}^-$. This weighting has been previously used in (Bennett and Mangasarian, 1992). For numerical stability in the weighted logistic loss function, the modified Newton step in equation (28.42) with $\lambda = 0.1$ was used to optimize the function coefficients in Step 6 of Algorithm 12.

In order to compare the quality of the orthogonal latent features created by different loss functions in BLF, SVM models are created using the BLF features. The quality of the features is evaluated by the performance of SVM on the test set. BLF with logistic loss and squared loss (BLF-LS) are used for constructing the features. Since our goal is to show BLF is an effective feature construction method, the full 139,351 binary features, without any preprocessing, are used for the input space of the BLF models. Figure 28.5

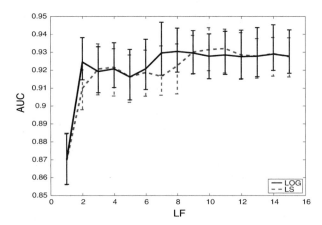

Fig. 28.5. Median of area under ROC curve with respect to the number of latent features for BLF with logistic loss and squared loss. The error bars correspond to 1/4-th and 3/4-th order statistics. The statistics are based on 100 randomized experiments.

shows the area under ROC curve (AUC) for BLF-LOG and BLF-LS, as a function of the number of latent features. As seen in the previous section, logistic loss fits the data faster than squared loss: the "peak" appears at a smaller number of latent features in logistic loss than for squared loss. We pick 5 and 15 as the number of orthogonal features that are used in the next stage of SVM experiments. Obviously, five features is before the peak and the dimensionality may have been reduced too much to have good predictive ability. But for the case with 15 latent features, the curves for both squared loss and logistic loss seem to be stabilized and we can expect better predictability. We use the nonlinear orthogonal features, $\boldsymbol{T}\mathrm{diag}(\mathbf{c})$ as in Equation (28.7), as the input data to a classic linear SVM (Cortes and Vapnik, 1995). Since features in $\boldsymbol{T}$ are normalized to have length one, we need to use feature weighting $\mathrm{diag}(\mathbf{c})$ for better performance especially for the relatively small dimensional space used here. Figure 28.6 illustrates the AUC for SVM models with a wide spectrum of the cost parameter C in the SVM. As baseline cases, results of SVM models trained with the full data set are also shown in the figure. SVM is given a better metric space trained by BLF so it can overfit as the value C gets large. However, the SVM models that use the original features stay at almost the same value of AUC within the value of C shown in Figure 28.6. As seen in Figure 28.5, at LF=5 and 15, BLF-LOG and BLF-LS have very similar performance. However, after training by SVM, the features from BLF-LOG are slightly better than those from BLF-LS. Overall, with a reasonably good choice of parameter C, SVM using reduced features by BLF can improve models with the original features. Further, features created by logistic

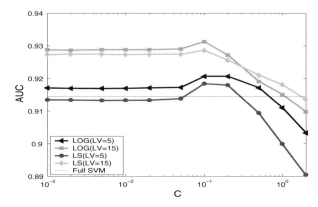

Fig. 28.6. Area under ROC curve for SVM models using features created by BLF with squared loss and logistic loss for LF=5 and 15. The dotted line shows SVM models using the original full features.

loss perform slightly better than those constructed using the squared loss over a wide range of C.

28.9 Conclusion

In this chapter, a framework for constructing orthogonal features by boosting, OrthoAnyBoost, was proposed. Using techniques from spectral methods such as PCA and PLS, OrthoAnyboost can be very efficiently implemented in linear hypothesis spaces. The resulting method, BLF, was demonstrated to both construct valuable orthogonal features and to be a competitive predictive method by itself for a variety of loss functions. BLF performs feature construction based on a given (sub)differentiable loss function. For the least squares loss, BLF reduces to PLS and preserves all the attractive properties of that algorithm. As in PCA and PLS, the resulting nonlinear features are valuable for visualization, dimensionality reduction, improving generalization, and use in other learning algorithms, but now these features can be targeted to a specific inference task. The data matrix is factorized by the extracted features. The low-rank approximation of the data matrix provides efficiency and stability in computation. The orthogonality properties of BLF guarantee that it converges to the optimal solution of the full model in a finite number of iterations. Empirically, orthogonality makes BLF converge much faster than gradient boosting. The predictive model is constructed in a reduced dimensionality space thus providing capacity control leading to good generalization. The method is generalized to nonlinear hypotheses using kernels.

Computational results demonstrate how BLF can be applied to a wide range of commonly used loss functions. The results illustrate differences in loss functions. As always, the best loss function depends on the data and inference task. The least absolute deviation is more robust than the squared loss and the version of BLF using the LAD loss showed some improvements for drug discovery data where attributes are inter-correlated and noisy. Classification loss functions such as exponential loss and logistic loss are more natural for classification problems and the loss functions can be weighted to handle problems with unbalanced data or unequal misclassification costs. Future work is need to investigate the theoretical properties of BLF from both optimization and learning points of view and to apply the approach to other learning tasks/loss functions such as ranking.

References

M. Bazaraa, H. Sherali, and C. Shetty. *Nonlinear Programming: Theory and Algorithms.* Wiley, 1993.

K. P. Bennett, C. M. Breneman, and M. J. Embrechts. DDASSL project, 2000. http://www.drugmining.com.

K. P. Bennett and O. L. Mangasarian. Robust linear programming discrimination of two linearly inseparable sets. *Optimization Methods and Software*, 1:23–34, 1992.

C. L. Blake and C. J. Merz. UCI Repository of machine learning databases, 1998. http://www.ics.uci.edu/~mlearn/MLRepository.html.

M. Borga, T. Landelius, and H. Knutsson. A unified approach to PCA, PLS, MLR and CCA. Report LiTH-ISY-R-1992, ISY, SE-581 83 Linköping, Sweden, November 1997.

Bernhard E. Boser, Isabelle Guyon, and Vladimir Vapnik. A training algorithm for optimal margin classifiers. In *Fifth Annual Workshop on Computational Learning Theory*, pages 144–152. ACM, 1992.

S. Chen. Local regularization assisted orthogonal least squares regression, 2003. http://www.ecs.soton.ac.uk/~sqc/.

Corinna Cortes and Vladimir Vapnik. Support-vector networks. *Machine Learning*, 20(3):273–297, 1995.

H. Drucker, C. Burges, L. Kaufman, A. Smola, , and V. Vapnik. Support vector regression machines. In *Advances in Neural Information Processing Systems 9*, pages 155–161. MIT Press, 1997.

Y. Freund and R. E. Shapire. A decision-theoretic generalization of on-line learning and an application to boosting. *Journal of Computer and System Sciences*, 55 (1):119–139, August 1997.

J. Friedman, T. Hastie, and R. Tibshirani. Additive logistic regression: a statistical view of boosting. *Annals of Statistics*, (28):337–307, 2000.

J. H. Friedman. Greedy function approximation: a gradient boosting machine. *Annals of Statistics*, 29(5):1189–1232, 2001.

Jerome H. Friedman and Bogdan E. Popescu. Gradient directed regularization of linear regression and classification. Technical report, Stanford University, 2004.

D. C. Hoaglin, F. Mosteller, and J. W. Tukey. *Understanding Robust and Exploratory Data Analysis.* John Wiley & Sons, 1982.

A. Höskuldsson. PLS regression methods. *Journal of Chemometrics*, 2:211–228, 1988.

T. Joachims. Making large-scale svm learning practical. In B. Schölkopf, C. Burges, and A. Smola, editors, *Advances in Kernel Methods - Support Vector Learning*, chapter 11. 1999.

Richard A. Johnson and Dean W. Wichern. *Applied Multivariate Statistical Analysis*. Prentice Hall, 1992.

John Lafferty. Additive models, boosting, and inference for generalized divergences. In *Proceedings of the twelfth annual conference on Computational learning theory*, pages 125–133, New York, NY, USA, 1999. ACM Press.

L. Li, Y. S. Abu-Mostafa, and A. Pratap. Cgboost:conjugate gradient in function space. Technical Report Computer Science Technical Report CaltechC-STR:2003.007, CalTech, 2003.

Edward Malthouse, Ajit Tamhane, and Richard Mah. Nonlinear partial least squares. *Computers in Chemical Engineering*, 12(8):875–890, 1997.

Brian Marx. Iteratively reweighted partial least squares for generalized linear models. *Technometrics*, 38(4):374–381, 1996.

L. Mason, J. Baxter, P. Bartlett, and M. Frean. Boosting algorithms as gradient descent in function space. Technical report, RSISE, Australian National University, 1999. URL `citeseer.nj.nec.com/mason99boosting.html`.

P. B. Nair, A. Choudhury, and A. J. Keane. Some greedy learning algorithms for sparse regression and classification with mercer kernels. *Journal of Machine Learning Research*, 3:781–801, 2002.

Jorge Nocedal and Stephen J. Wright. *Numerical Optimization*. Springer Verlag, 1999.

A. Phatak and F. de Hoog. Exploiting the connection between PLS, lanczos, and conjugate gradients: Alternative proofs of some properties of PLS. *Journal of Chemometrics*, 16:361–367, 2002.

R. Rosipal and L. T. Trejo. Kernel partial least squares regression in reproducing kernel hilbert space. *Journal of Machine Learning Research*, 2:97–123, 2001.

R. E. Schapire and Y. Singer. Improved boosting algorithms using confidence-rated predictions. *Machine Learning*, 37:297–336, 1999.

J. Shawe-Taylor and N. Cristianini. *Kernel Methods for Pattern Analysis*. Cambridge University Press, 2004.

H. Wold. Estimation of principal components and related models by iterative least squares. In *Multivariate Analysis*, pages 391–420, New York, 1966. Academic Press.

J. Zhu and T. Hastie. Kernel logistic regression and the import vector machine. In T. G. Dietterich, S. Becker, and Z. Ghahramani, editors, *Advances in Neural Information Processing Systems 14*, pages 1081–1088, Cambridge, MA, 2002. MIT Press.

Chapter 29

Large Margin Principles for Feature Selection

Ran Gilad-Bachrach[1], Amir Navot[2], and Naftali Tishby[1,2]

[1] School of Computer Science and Engineering The Hebrew University, Jerusalem, Israel.
[2] Interdisciplinary Center for Neural Computation, The Hebrew University, Jerusalem, Israel.
`ranb@cs.huji.ac.il`, `anavot@cs.huji.ac.il`, `tishby@cs.huji.ac.il`

In this paper we introduce a margin based feature selection criterion and apply it to measure the quality of sets of features. Using margins we devise novel selection algorithms for multi-class categorization problems and provide theoretical generalization bound. We also study the well known *Relief* algorithm and show that it resembles a gradient ascent over our margin criterion. We report promising results on various datasets.

29.1 Introduction

In many supervised learning tasks the input data are represented by a very large number of features, but only a few of them are relevant for predicting the label. Even state-of-the-art classification algorithms (e.g. Support Vector Machine (SVM) (Boser et al., 1992)) cannot overcome the presence of a large numbers of weakly relevant and redundant features. This is usually attributed to "the curse of dimensionality" (Bellman, 1961), or to the fact that irrelevant features decrease the signal-to-noise ratio. In addition, many algorithms become computationally intractable when the dimension is high. On the other hand, once a good small set of features has been chosen, even the most basic classifiers (e.g. 1-Nearest Neighbor (Fix and Hodges, 1951)) can achieve high performance levels. Therefore, *feature selection*, i.e. the task of choosing a small subset of features, which is sufficient to predict the target labels, is crucial for efficient learning.

In this paper we introduce the idea of measuring the quality of a set of features by the margin it induces. A margin (Boser et al., 1992, Schapire et al., 1998) is a geometric measure for evaluating the confidence of a classifier with respect to its decision. Margins already play a crucial role in current machine learning research. For instance, SVM (Boser et al., 1992) is a prominent large margin algorithm. The novelty of this paper is the use of large margin principles for feature selection.

R. Gilad-Bachrach et al.: *Large Margin Principles for Feature Selection*, StudFuzz **207**, 585–606 (2006)
`www.springerlink.com`

Throughout this paper we will use the 1-NN as the "study-case" predictor, but most of the results are relevant to other distance based classifiers (e.g. LVQ (Kohonen, 1995), SVM-RBF (Boser et al., 1992)) as well. To validate this, we compare our algorithms to the R2W2 algorithm (Weston et al., 2000), which was specifically designed as a feature selection scheme for SVM. We show that even in this setting our algorithms compare favorably to all the contesters we have used.

The margin for the Nearest-Neighbor was previously defined in (Crammer et al., 2002). The use of margins allows us to devise new feature selection algorithms as well as prove a PAC (Probably Approximately Correct) style generalization bound. The bound is on the generalization accuracy of 1-NN on a selected set of features, and guarantees good performance for any feature selection scheme which selects a small set of features while keeping the margin large. On the algorithmic side, we use a margin based criterion to measure the quality of sets of features. We present two new feature selection algorithms, *G-flip* and *Simba*, based on this criterion. The merits of these algorithms are demonstrated on various datasets.

Finally, we study the *Relief* feature selection algorithm ((Kira and Rendell, 1992), see [chapter 3, in part I]) in the large margin context. While *Relief* does not explicitly maximize any evaluation function, we show here that implicitly it maximizes the margin based evaluation function.

29.2 Margins

Margins play a crucial role in modern machine learning research. They measure the classifier confidence when making its decision. Margins are used both for theoretic generalization bounds and as guidelines for algorithm design.

29.2.1 Two Types of Margins

As described in (Crammer et al., 2002) there are two natural ways of defining the margin of an instance with respect to a classification rule. The more common type, *sample-margin*, measures the distance between the instance and the decision boundary induced by the classifier. *Support Vector Machines* (Boser et al., 1992), for example, finds the separating hyper-plane with the largest sample-margin. Bartlett (1998), also discusses the distance between instances and the decision boundary. He uses the sample-margin to derive generalization bounds.

An alternative definition, the *hypothesis-margin*, requires the existence of a distance measure on the hypothesis class. The margin of a hypothesis with respect to an instance is the distance between the hypothesis and the closest hypothesis that assigns an alternative label to the given instance. For example *AdaBoost* (Freund and Schapire, 1997) uses this type of margin with the L_1-norm as the distance measure between hypotheses.

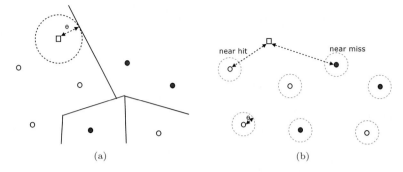

near hit

near miss

(a)   (b)

Fig. 29.1. The two types of margin for the Nearest Neighbor rule. We consider a sample of points (the circles) and measure the margin with respect to a new instance (the square). The sample margin 1(a) is the distance between the new instance and the decision boundary (the Voronoi tessellation). The hypothesis margin 1(b) is the largest distance the sample points can travel without altering the label of the new instance. In this case it is half the difference between the distance to the **nearmiss** and the distance to the **nearhit**.

Throughout this paper we will be interested in margins for 1-NN. For 1-NN, the classifier is defined by a set of training points (prototypes) and the decision boundary is the Voronoi tessellation. The sample margin in this case is the distance between the instance and the Voronoi tessellation, and therefore it measures the sensitivity to small changes of the instance position. The hypothesis margin for this case is the maximal distance θ such that the following condition holds: if we draw a ball with radius θ around each prototype, any change of the location of prototypes inside their θ ball will not change the assigned label of the instance. Therefore, the hypothesis margin measures the stability to small changes in the prototypes locations. See Figure 29.1 for illustration. The sample margin for 1-NN can be unstable, as shown in (Crammer et al., 2002) and thus the hypothesis margin is preferable in this case. In the same paper, the following results were proved:

(1.) The hypothesis-margin lower bounds the sample-margin.
(2.) It is easy to compute the hypothesis-margin of an instance x with respect to a set of points P by the following formula:

$$\theta_P(\mathbf{x}) = \frac{1}{2}\Big(\|\mathbf{x} - \mathbf{nearmiss}(\mathbf{x})\| - \|\mathbf{x} - \mathbf{nearhit}(\mathbf{x})\| \Big)$$

where $\mathbf{nearhit}(\mathbf{x})$ and $\mathbf{nearmiss}(\mathbf{x})$ denote the nearest point to $\mathbf{x}$ in P with the same and different label, respectively. Note that a chosen set of features affects the margin through the distance measure.

Therefore in the case of Nearest Neighbor a large hypothesis-margin ensures a large sample-margin, and a hypothesis-margin is easy to compute.

29.2.2 Margin Based Evaluation Function

A good generalization can be guaranteed if many sample points have a large margin (see section 29.4). We introduce an evaluation function, which assigns a score to sets of features according to the margin they induce. First we formulate the margin as a function of the selected set of features.

Definition 1. *Let P be a set of points and x be an instance. Let w be a weight vector over the feature set, then the margin of x is*

$$\theta_P^{\mathbf{w}}(\mathbf{x}) = \frac{1}{2}\left(\|\mathbf{x} - \boldsymbol{nearmiss}(\mathbf{x})\|_{\mathbf{w}} - \|\mathbf{x} - \boldsymbol{nearhit}(\mathbf{x})\|_{\mathbf{w}}\right) \qquad (29.1)$$

where $\|\mathbf{z}\|_{\mathbf{w}} = \sqrt{\sum_i w_i^2 z_i^2}$.

Definition 1 extends beyond feature selection and allows weight over the features. When selecting a set of features F we can use the same definition by identifying F with its indicating vector. Therefore, we denote by $\theta_P^F(\mathbf{x}) := \theta_P^{I_F}(\mathbf{x})$ where I_F is one for any feature in F and zero otherwise.

Since $\theta^{\lambda w}(\mathbf{x}) = |\lambda|\theta^{\mathbf{w}}(\mathbf{x})$ for any scalar λ, it is natural to introduce some normalization factor. The natural normalization is to require $\max w_i^2 = 1$, since it guarantees that $\|\mathbf{z}\|_{\mathbf{w}} \leq \|\mathbf{z}\|$ where the right hand side is the Euclidean norm of z.

Now we turn to defining the evaluation function. The building blocks of this function are the margins of all the sample points. The margin of each instance $\mathbf{x}$ is calculated with respect to the sample excluding $\mathbf{x}$ ("leave-one-out margin").

Definition 2. *Let $u(\cdot)$ be a utility function. Given a training set S and a weight vector w, the evaluation function is:*

$$e(\mathbf{w}) = \sum_{\mathbf{x} \in S} u\left(\theta_{S\backslash\mathbf{x}}^{\mathbf{w}}(\mathbf{x})\right) \qquad (29.2)$$

The utility function controls the contribution of each margin term to the overall score. It is natural to require the utility function to be non-decreasing; thus larger margin introduce larger utility. We consider three utility functions: linear, zero-one and sigmoid. The linear utility function is defined as $u(\theta) = \theta$. When the linear utility function is used, the evaluation function is simply the sum of the margins. The zero-one utility is equals 1 when the margin is positive and 0 otherwise. When this utility function is used the utility function is proportional to the leave-one-out error. The sigmoid utility is $u(\theta) = 1/(1 + exp(-\beta\theta))$. The sigmoid utility function is less sensitive to outliers than the linear utility, but does not ignore the magnitude of the margin completely as the zero-one utility does. Note also that for $\beta \to 0$ or $\beta \to \infty$ the sigmoid utility function becomes the linear utility function or the zero-one utility function respectively. In the *Simba* algorithm we assume that the utility function is differentiable, and therefore the zero-one utility cannot be used.

It is natural to look at the evaluation function solely for weight vectors **w** such that $\max w_i^2 = 1$. However, formally, the evaluation function is well defined for any **w**, a fact which we make use of in the *Simba* algorithm. We also use the notation $e(F)$, where F is a set of features to denote $e(I_F)$.

29.3 Algorithms

In this section we present two algorithms, which attempt to maximize the margin based evaluation function. Both algorithms can cope with multi-class problems. Our algorithms can be considered as filter methods for general classifiers. They are also close to wrapper for 1-NN. A Matlab implementation of these algorithms is available at http://www.cs.huji.ac.il/labs/learning/code/feature_selection/.

29.3.1 Greedy Feature Flip Algorithm (G-flip)

The *G-flip* (algorithm 15) is a greedy search algorithm for maximizing $e(F)$, where F is a set of features. The algorithm repeatedly iterates over the feature set and updates the set of chosen features. In each iteration it decides to remove or add the current feature to the selected set by evaluating the margin term (29.2) with and without this feature. This algorithm is similar to the zero-temperature Monte-Carlo (Metropolis) method. It converges to a local maximum of the evaluation function, as each step increases its value and the number of possible feature sets is finite. The computational complexity of one pass over all features of a naïve implementation of *G-flip* is $\Theta\left(N^2 m^2\right)$ where N is the number of features and m is the number of instances. However the complexity can be reduced to $\Theta\left(N m^2\right)$ since updating the distance matrix can be done efficiently after each addition/deletion of a feature from the current active set. Empirically *G-flip* converges in a few iterations. In all our experiments it converged after less than 20 epochs, in most of the cases in less than 10 epochs. A nice property of this algorithm is that once the utility function is chosen, it is *parameter free*. There is no need to tune the number of features or any type of threshold.

29.3.2 Iterative Search Margin Based Algorithm (Simba)

The *G-flip* algorithm presented in section 29.3.1 tries to find the feature set that maximizes the margin directly. Here we take another approach. We first find the weight vector **w** that maximizes $e(\mathbf{w})$ as defined in (29.2) and then use a threshold in order to get a feature set. Of course, it is also possible to use the weights directly by using the induced distance measure instead. Since

---

Algorithm 15: Greedy Feature Flip (G-flip)

(1.) Initialize the set of chosen features to the empty set: $F = \phi$
(2.) for $t = 1, 2, \ldots$
      a) pick a random permutation s of $\{1 \ldots N\}$
      b) for $i = 1$ to N,
          i. evaluate $e_1 = e\left(F \cup \{s(i)\}\right)$ and $e_2 = e\left(F \setminus \{s(i)\}\right)$
          ii. if $e_1 > e_2$, $F = F \cup \{s(i)\}$
              else-if $e_2 > e_1$, $F = F \setminus \{s(i)\}$
      c) if no change made in step (b) then break

$e(\mathbf{w})$ is smooth almost everywhere, whenever the utility function is smooth, we use gradient ascent in order to maximize it. The gradient of $e(\mathbf{w})$ when evaluated on a sample S is:

$$(\triangledown e(\mathbf{w}))_i = \frac{\partial e(\mathbf{w})}{\partial w_i} = \sum_{\mathbf{x} \in S} \frac{\partial u(\theta(\mathbf{x}))}{\partial \theta(\mathbf{x})} \frac{\partial \theta(\mathbf{x})}{\partial w_i} \tag{29.3}$$

$$= \frac{1}{2} \sum_{\mathbf{x} \in S} \frac{\partial u(\theta(\mathbf{x}))}{\partial \theta(\mathbf{x})} \left(\frac{(x_i - \mathbf{nearmiss}(\mathbf{x})_i)^2}{\|\mathbf{x} - \mathbf{nearmiss}(\mathbf{x})\|_{\mathbf{w}}} - \frac{(x_i - \mathbf{nearhit}(\mathbf{x})_i)^2}{\|\mathbf{x} - \mathbf{nearhit}(\mathbf{x})\|_{\mathbf{w}}} \right) w_i$$

In *Simba* (algorithm 16) we use a stochastic gradient ascent over $e(\mathbf{w})$ while ignoring the constraint $\|\mathbf{w}^2\|_\infty = 1$. In each step we evaluate only one term in the sum in (29.3) and add it to the weight vector $\mathbf{w}$. The projection on the constraint is done only at the end (step 3).

The computational complexity of *Simba* is $\Theta(TNm)$ where T is the number of iterations, N is the number of features and m is the size of the sample S. Note that when iterating over all training instances, i.e. when $T = m$, the complexity is $\Theta\left(Nm^2\right)$.

29.3.3 Comparison to Relief

Relief (Kira and Rendell, 1992) is a feature selection algorithm (see algorithm 17), which was shown to be very efficient for estimating feature quality. The algorithm holds a weight vector over all features and updates this vector according to the sample points presented. Kira & Rendell proved that under some assumptions, the expected weight is large for relevant features and small for irrelevant ones. They also explain how to choose the relevance threshold τ in a way that ensures the probability that a given irrelevant feature chosen is small. *Relief* was extended to deal with multi-class problems, noise and missing data by Kononenko (1994). For multi-class problems (Kononenko, 1994) also presents a version called *Relief-F* that instead of using the distance to the nearest point with an alternative label, looks at the distances to the nearest instance of any alternative class and takes the average. In the experiments we made *Relief-F* was inferior to the standard *Relief*.

Algorithm 16: Simba

(1.) initialize $\mathbf{w} = (1, 1, \ldots, 1)$
(2.) for $t = 1 \ldots T$
    a) pick randomly an instance $\mathbf{x}$ from S
    b) calculate $\mathbf{nearmiss}(\mathbf{x})$ and $\mathbf{nearhit}(\mathbf{x})$ with respect to $S \setminus \{\mathbf{x}\}$ and
       the weight vector $\mathbf{w}$.
    c) for $i = 1, \ldots, N$
       calculate

$$\triangle_i = \frac{1}{2} \frac{\partial u(\theta(\mathbf{x}))}{\partial \theta(\mathbf{x})} \left(\frac{(x_i - \mathbf{nearmiss}(\mathbf{x})_i)^2}{\|\mathbf{x} - \mathbf{nearmiss}(\mathbf{x})\|_\mathbf{w}} - \frac{(x_i - \mathbf{nearhit}(\mathbf{x})_i)^2}{\|\mathbf{x} - \mathbf{nearhit}(\mathbf{x})\|_\mathbf{w}} \right) w_i$$

    d) $\mathbf{w} = \mathbf{w} + \triangle$
(3.) $\mathbf{w} \leftarrow \mathbf{w}^2 / \|\mathbf{w}^2\|_\infty$ where $(\mathbf{w}^2)_i := (w_i)^2$.

Note that the update rule in a single step of *Relief* is similar to the one performed by *Simba* when the utility function is linear, i.e. $u(\theta) = \theta$ and thus $\partial u(\theta)/\partial \theta = 1$. Indeed, empirical evidence shows that *Relief* does increase the margin (see section 29.5). However, there is a major difference between *Relief* and *Simba*: *Relief* does not re-evaluate the distances according to the weight vector $\mathbf{w}$ and thus it is inferior to *Simba*. In particular, *Relief* has no mechanism for eliminating redundant features. *Simba* may also choose correlated features, but only if this contributes to the overall performance. In terms of computational complexity, *Relief* and *Simba* are equivalent.

Algorithm 17: RELIEF (Kira and Rendell, 1992)

(1.) initiate the weight vector to zero: $\mathbf{w} = 0$
(2.) for $t = 1 \ldots T$,
   • pick randomly an instance $\mathbf{x}$ from S
   • for $i = 1 \ldots N$,
     – $w_i = w_i + (x_i - \mathbf{nearmiss}(\mathbf{x})_i)^2 - (x_i - \mathbf{nearhit}(\mathbf{x})_i)^2$
(3.) the chosen feature set is $\{i | w_i > \tau\}$ where τ is a threshold

---

29.3.4 Comparison to R2W2

R2W2 (Weston et al., 2000) is the state-of-the-art feature selection algorithm for the *Support Vector Machines* (SVM) classifier. This algorithm is a sophisticated wrapper for SVM and therefore uses the maximal margin principle for feature selection indirectly. The goal of the algorithm is to find a weights vector over the features, which will minimize the objective function of the SVM optimization problem. This objective function can be written as $R^2 W^2$

where R is the radius of a ball containing all the training data and W is the norm of the linear separator. The optimization is done using gradient descent. After each gradient step a new SVM optimization problem is constructed and solved. Thus it becomes cumbersome for large scale data.

The derivation of R2W2 algorithm assumes that the data are linearly separable. Since this cannot be guaranteed in the general case we use the "ridge" trick of adding a constant value to the diagonal of the kernel matrix. Note also that R2W2 is designed for binary classification tasks only. There are several ways in which it can be extended to multi class problems. However, these extensions will make the algorithm even more demanding than its original version.

As in SVM, R2W2 can be used together with a kernel function. We chose to use the Radial Basis Function (RBF) kernel. The RBF kernel is defined to be

$$K(\mathbf{x}_1, \mathbf{x}_2) = e^{-\frac{\|\mathbf{x}_1 - \mathbf{x}_2\|}{2\sigma^2}}$$

where σ is a predefined parameter. The choice of the RBF kernel is due to the similarity between SVM with RBF kernel and the nearest-neighbor rule. Our implementation is based on the one in the *Spider package* (Weston et al., 2004).

29.4 Theoretical Analysis

In this section we use feature selection and large margin principles to prove finite sample generalization bound for 1-*Nearest Neighbor*. (Cover and Hart, 1967), showed that asymptotically the generalization error of 1-NN can exceed by at most a factor of 2 the generalization error of the Bayes optimal classification rule. However, on finite samples nearest neighbor can over-fit and exhibit poor performance. Indeed 1-NN will give zero training error, on almost any sample.

The training error is thus too rough to provide information on the generalization performance of 1-NN. We therefore need a more detailed measure in order to provide meaningful generalization bounds and this is where margins become useful. It turns out that in a sense, 1-NN is a maximum margin algorithm. Indeed once our proper definition of margin is used, i.e. sample-margin, it is easy to verify that 1-NN generates the classification rule with the largest possible margin.

The combination of a large margin and a small number of features provides enough evidence to obtain a useful bound on the generalization error. The bound we provide here is data-dependent (Shawe-Taylor et al., 1998, Bartlett, 1998). Therefore, the quality of the bound depends on our specific sample. It holds simultaneously for any possible method to select a set of features. If an algorithm selects a small set of features with a large margin, the bound guarantees it will generalize well. This is the motivation for *Simba* and *G-flip*.

We use the following notation in our theoretical results:

Definition 3. *Let $\mathcal{D}$ be a distribution over $\mathcal{X} \times \{\pm 1\}$ and $h : \mathcal{X} \longrightarrow \{\pm 1\}$ a classification function. We denote by $er_{\mathcal{D}}(h)$ the generalization error of h with respect to $\mathcal{D}$:*

$$er_{\mathcal{D}}(h) = \Pr_{\mathbf{x},y \sim \mathcal{D}}[h(\mathbf{x}) \neq y]$$

For a sample $S = \{(\mathbf{x}_k, y_k)\}_{k=1}^{m} \in (\mathcal{X} \times \{\pm 1\})^m$ and a constant $\gamma > 0$ we define the γ-sensitive training error to be

$$\hat{er}_S^{\gamma}(h) = \frac{1}{m}\left|\left\{(k : h(\mathbf{x}_k) \neq y_k) \quad or \quad (\mathbf{x}_k \text{ has sample-margin} < \gamma)\right\}\right|$$

Our main result is the following theorem[1]:

Theorem 1. *Let $\mathcal{D}$ be a distribution over $\mathbf{R}^N \times \{\pm 1\}$ which is supported on a ball of radius R in $\mathbf{R}^N$. Let $\delta > 0$ and let S be a sample of size m such that $S \sim \mathcal{D}^m$. With probability $1 - \delta$ over the random choice of S, for any set of features F and any $\gamma \in (0,1]$*

$$er_{\mathcal{D}}(h) \leq \hat{er}_S^{\gamma}(h) + \sqrt{\frac{2}{m}\left(d\ln\left(\frac{34em}{d}\right)\log_2(578m) + \ln\left(\frac{8}{\gamma\delta}\right) + (|F|+1)\ln N\right)}$$

Where h is the nearest neighbor classification rule when distance is measured only on the features in F and $d = (64R/\gamma)^{|F|}$.

A few notes about this bound; First the size of the feature space, N, appears only logarithmically in the bound. Hence, it has a minor effect on the generalization error of 1-NN. On the other hand, the number of selected features, F, appears in the exponent. This is another realization of the "curse of dimensionality" (Bellman, 1961). See appendix A for the proof of theorem 1.

A large margin for many sample points will make the first term of the bound small, while using a small set of features will make the second term of the bound small. This gives us the motivation to look for small sets of features that induce large margin, and that is what *G-flip* and *Simba* do. As this bound is a worst case bound, like all the PAC style bounds, it is very loose in most of the cases, and the empirical results are expected to be much better.

29.5 Empirical Assessment

We first demonstrate the behavior of *Simba* on a small synthetic problem. Then we compare the different algorithms on image and text classification tasks. The first task is pixel (feature) selection for discriminating between

---

[1]Note that the theorem holds when sample-margin is replaced by hypothesis-margin since the later lower bounds the former.

male and female face images. The second task is a word (feature) selection for multi-class document categorization. In order to demonstrate the ability of our algorithms to work with other classifiers (beside of Nearest Neighbor) we also report results with SVM with RBF kernel (see section 29.5.4). We also report the results obtained on some of the datasets of the *NIPS-2003 feature selection challenge* (Guyon and Gunn, 2003). For these comparisons we have used the following feature selection algorithms: *Simba* with both linear and sigmoid utility functions (referred as *Simba(lin)* and *Simba(sig)* respectively), *G-flip* with linear, zero-one and sigmoid utility functions (referred as *G-flip(lin)*, *G-flip(zero-one)* and *G-flip(sig)* respectively), *Relief*, *R2W2* and *Infogain*[2].

29.5.1 The Xor Problem

To demonstrate the quality of the margin based evaluation function and the ability of the *Simba* algorithm[3] to deal with dependent features we use a synthetic problem. The problem consisted of 1000 sample points with 10 real valued features. The target concept is a *xor* function over the first 3 features. Hence, the first 3 features are relevant while the other features are irrelevant. Notice that this task is a special case of parity function learning and is considered hard for many feature selection algorithms (Guyon and Elisseeff, 2003). Thus for example, any algorithm which does not consider functional dependencies between features fails on this task. The simplicity (some might say over-simplicity) of this problem, allows us to demonstrate some of the interesting properties of the algorithms studied.

Figures 29.2 present the results we obtained on this problem. A few phenomena are apparent in these results. The value of the margin evaluation function is highly correlated with the angle between the weight vector and the correct feature vector (see figures 29.2 and 29.3). This correlation demonstrates that the margins characterize correctly the quality of the weight vector. This is quite remarkable since our margin evaluation function can be measured empirically on the training data whereas the angle to the correct feature vector is unknown during learning.

As suggested in section 29.3.3 *Relief* does increase the margin as well. However, *Simba*, which maximizes the margin directly, outperforms *Relief* quite significantly. as shown in Figure 29.2.

29.5.2 Face Images

We applied the algorithms to the AR face database (Martinez and Benavente, 1998), which is a collection of digital images of males and females with various

---

[2] *Infogain* ranks features according to the mutual information between each feature and the labels (see part I, chapter 3, section 3). Recall that the mutual information between two random variables X, Y is $I(X,Y) = \sum_{x,y} p(x,y) \log \frac{p(x,y)}{p(x)p(y)}$

[3] The linear utility function was used in this experiment.

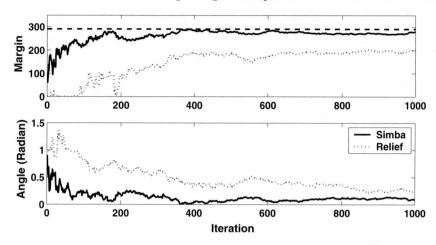

Fig. 29.2. The results of applying *Simba* (solid) and *Relief* (dotted) on the *xor* synthetic problem. **Top:** The margin value, $e(\mathbf{w})$, at each iteration. The dashed line is the margin of the correct weight vector. **Bottom:** the angle between the weight vector and the correct feature vector at each iteration (in Radians).

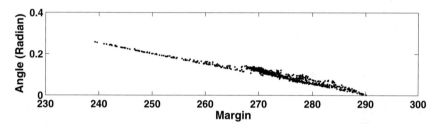

Fig. 29.3. The scatter plot shows the angle to the correct feature vector as function of the value of the margin evaluation function. The values were calculated for the *xor* problem using *Simba* during iterations 150 to 1000. Note the linear relation between the two quantities.

facial expressions, illumination conditions, and occlusions. We selected 1456 images and converted them to gray-scale images of 85×60 pixels, which are taken as our initial 5100 features. Examples of the images are shown in Figure 29.4. The task we tested is classifying the male vs. the female faces.

In order to improve the statistical significance of the results, the dataset was partitioned independently 20 times into training data of 1000 images and test data of 456 images. For each such partitioning (split) *Simba*[4] , *G-flip*, *Relief*, *R2W2* and *Infogain* were applied to select optimal features and the 1-NN algorithm was used to classify the test data points. We used 10 random starting points for *Simba* (i.e. random permutations of the train data)

---

[4] *Simba* was applied with both linear and sigmoid utility functions. We used $\beta = 0.01$ for the sigmoid utility.

Fig. 29.4. Excerpts from the face images dataset.

and selected the result of the single run which reached the highest value of the evaluation function. The average accuracy versus the number of features chosen, is presented in Figure 29.5. *G-flip* gives only one point on this plot, as it chooses the number of features automatically, although this number changes between different splits, it does not change significantly.

The features *G-flip (zero-one)* selected enabled 1-NN to achieve accuracy of 92.2% using about 60 features only, which is better than the accuracy obtained with the whole feature set (91.5%). *G-flip (zero-one)* outperformed any other alternative when only few dozens of features were used. *Simba(lin)* significantly outperformed *Relief*, *R2W2* and *Infogain*, especially in the small number of features regime. If we define a difference as significant if the one algorithm is better than the other in more than 90% of the partition, we can see that *Simba* is significantly better than *Relief*, *Infogain* and *R2W2* when fewer than couple of hundreds of features are being used (see Figure 29.6).

Moreover, the 1000 features that *Simba* selected enabled 1-NN to achieve an accuracy of 92.8%, which is better than the accuracy obtained with the whole feature set (91.5%). A closer look on the features selected by *Simba* and *Relief* (Figure 29.7) reveals the clear difference between the two algorithms. *Relief* focused on the hair-line, especially around the neck, and on other contour areas in a left-right symmetric fashion. This choice is suboptimal as those features are highly correlated to each other and therefore a smaller subset is sufficient. *Simba* on the other hand selected features in other informative facial locations but mostly on one side (left) of the face, as the other side is clearly highly correlated and does not contribute new information to this task. Moreover, this dataset is biased in the sense that more faces are illuminated from the right. Many of them are saturated and thus *Simba* preferred the left side over the less informative right side.

29.5.3 Reuters

We applied the different algorithms on a multi-class text categorization task. For these purpose we used a subset of the Reuters-21578 dataset[5]. We have used the documents, which are classified to exactly one of the following 4 topics: *interest*, *trade*, *crude* and *grain*. The obtained dataset contains 2066

---

[5]The dataset can be found at http://www.daviddlewis.com/resources/

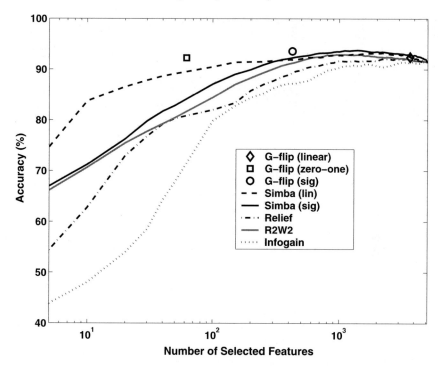

Fig. 29.5. Results for AR faces dataset. The accuracy achieved on the AR faces dataset when using the features chosen by the different algorithms. The results were averaged over the 20 splits of the dataset. For the sake of visual clarity, error bars are presented separately in Figure 29.6.

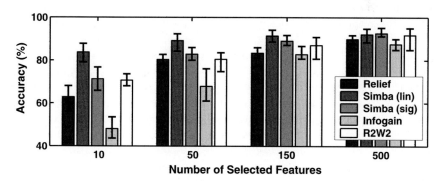

Fig. 29.6. Error intervals for AR faces dataset. The accuracy achieved on the AR faces dataset when using the features chosen by the different algorithms. The error intervals show the area were 90% of the results (of the 20 repeats) fell, i.e., the range of the results after eliminating the best and the worse iterations out of the 20 repeats.

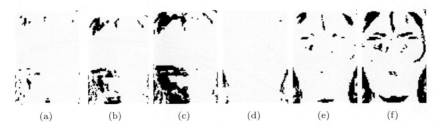

(a)   (b)   (c)   (d)   (e)   (f)

Fig. 29.7. The features selected (in black) by *Simba(lin)* and *Relief* for the face recognition task. 7(a), 7(b) and 7(c) shows 100, 500 and 1000 features selected by *Simba*. 7(d), 7(e) and 7(f) shows 100, 500 and 1000 features selected by *Relief*.

documents, which are approximately equally distributed between the four classes. Each document was represented as the vector of counts of the different words. Stop-words were omitted and numbers were converted to a predefined special character as a preprocessing.

To improve the statistical significance of our results, the corpus was partitioned 20 times into training set of 1000 documents and test set of 1066 documents. For each such partitioning, the words that appears less than 3 times in the training set were eliminated, which left ∼4000 words (features). For each partitioning *G-flip*, *Simba*, *Relief*, *Relief-F* and *Infogain* were applied to select an optimal set of features[6] and the 1-NN was used to classify the test documents. *Simba* and *G-flip* were applied with both linear and sigmoid utility functions[7]. *G-flip* was also applied with the zero-one utility function. We have used 10 random starting points for *Simba* and selected the results of the single run that achieved the highest value for the evaluation function. The average (over the 20 splits) accuracy versus the number of chosen features is presented in Figure 29.8. *G-flip* gives only one point on this plot, as it chooses the number of features automatically, although this number changes between different splits, it does not change significantly. Another look on the results is given in Figure 29.9. This figure shows error intervals around the average that allow to appreciate the statistical significance of the differences in the accuracy. The top ranked twenty features are presented in table 29.1.

The best overall accuracy (i.e. when ignoring the number of features used) was achieved by *G-flip(sig)*. *G-flip(sig)* got 94.09% generalization accuracy using ∼350 features. *Infogain* and *Simba(sig)* are just a little behind with 92.86% and 92.41% , that was achieved using ∼40 and ∼30 features only (respectively). *Relief* is far behind with 87.92% that was achieved using 250 features.

---

[6] *R2W2* was not applied for this problem as it is not defined for multi-class problems.

[7] The sigmoid utility function was used with $\beta = 1$ for both *G-flip* and *Simba*.

Table 29.1. The first 20 words (features) selected by the different algorithms for the Reuters dataset. Note that *Simba(sig)* is the only algorithm which selected the titles of all four classes (interest, trade, crude, grain) among the first twenty features

Simba(sig)		Simba(lin)		Relief		Infogain	
oil	rate	oil	corn	##	s	oil	trade
wheat	bpd	##	brazil	#	bank	tonnes	wheat
trade	days	tonnes	wish	###	rate	crude	bank
tonnes	crude	trade	rebate	oil	dlrs	rate	agriculture
corn	japanese	wheat	water	trade	barrels	grain	barrels
bank	deficit	bank	stocks	####	rates	petroleum	rates
rice	surplus	rice	certificates	billion	bpd	pct	corn
grain	indonesia	rates	recession	opec	pct	tariffs	japan
rates	interest	billion	got	tonnes	barrel	mt	deficit
ec	s	pct	ve	wheat	cts	energy	bpd

The advantage of *Simba(sig)* is very clear when looking on the accuracy versus the number of features used. Among the algorithms that were tested, *Simba(sig)* is the only algorithm that achieved (almost) best accuracy over the whole range. Indeed, when less than few dozens of features are used, *Infogain*, achieved similar accuracy, but when more features are used, it's accuracy drops dramatically. While *Simba(sig)* achieved accuracy above 90% for any number of features between ~20 and ~3000, the accuracy achieved by *Infogain* dropped below 90% when more than ~400 features are used, and below 85% when more than ~1500 are used.

The advantage of *Simba(sig)* over relief is also very clear. The accuracy achieved using the features chosen by *Simba(sig)* is notably and significantly better than the one achieved using the features chosen by *Relief*, for any number of selected features. Using only the top 10 features of *Simba(sig)* yields accuracy of 89.21%, which is about the same as the accuracy that can be achieved by any number of features chosen by *Relief*.

29.5.4 Face Images with Support Vector Machines

In this section we show that our algorithms works well also when using another distance based classifier, instead of the Nearest Neighbor classifier. We test the different algorithms together with SVM with RBF kernel classifier and show that our algorithms works as good as, and even better than the *R2W2* algorithm that was tailored specifically for this setting and is much more computationally demanding.

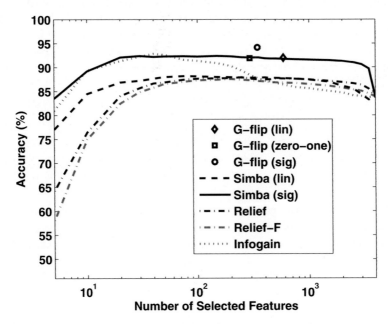

Fig. 29.8. Results for Reuters dataset. The accuracy achieved on the Reuters dataset when using the features chosen by the different algorithms. The results were averaged over the 20 splits of the dataset. For the sake of visual clarity, error bars are presented separately in Figure 29.9.

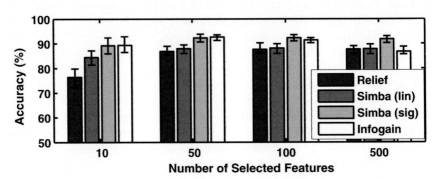

Fig. 29.9. Error intervals for Reuters dataset. The accuracy achieved on the Reuters dataset when using the features chosen by the different algorithms. The error intervals show the area were 90% of the results (of the 20 repeats) fell, i.e., the range of the results after eliminating the best and the worse iterations out of the 20 repeats.

We have used the AR face database (Martinez and Benavente, 1998). We have repeated the same experiment as described in section 29.5.2, the only difference being that once the features were chosen, we used SVM-RBF to classify the test data points. The sigma parameter used in the RBF kernel was selected to be 3500, the same parameter used in the R2W2 feature selection algorithm. The value for this parameter was tuned using cross-validation. See Figure 29.10 for a summary of the results.

Both *Simba* and *G-flip* perform well, especially in the small number of features regime. The results in the graph are the average over the 20 partitions of the data. Note that the only winnings which are 90% significant are those of *Simba (lin)* and *G-flip (zero-one)* when only few dozens of features are used.

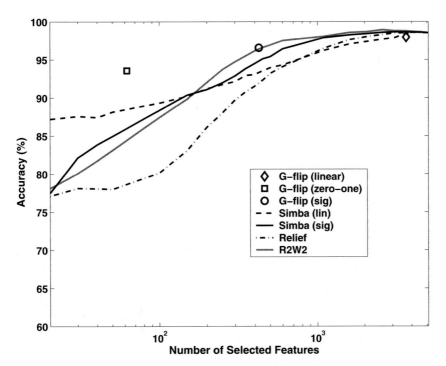

Fig. 29.10. Results for AR faces dataset with SVM-RBF classifier. The accuracy achieved on the AR faces dataset when using the features chosen by the different algorithms and SVM-RBF classifier. The results were averaged over the 20 splits of the dataset

29.5.5 The NIPS-03 Feature Selection Challenge

We applied *G-flip(lin)* as part of our experiments in the *NIPS-03 feature selection challenge* (Guyon and Gunn, 2003). The two datasets for which we have used *G-flip* are *ARCENE* and *MADELON*.

In *ARCENE* we first used *Principal Component Analysis* (PCA) as a preprocessing and then applied *G-flip* to select the principal components to be used for classification. *G-flip* selected only 76 features. These features were fed to a *Support Vector Machine* (SVM) with RBF kernel and ended up with 12.66% balanced error (the best result on this dataset was 10.76% balanced error).

In *MADELON* we did not apply any preprocessing. *G-flip* selected only 18 out of the 500 features in this dataset. Feeding these features to SVM-RBF resulted in 7.61% balanced error, while the best result on this dataset was 6.22% error.

Although *G-flip* is very simple and naïve algorithm, it ranked as one of the leading feature selection method on both *ARCENE* and *MADELON*. It is interesting to note that when 1-NN is used as the classification rule, instead of SVM-RBF, the error degrades only by ~1% on both datasets. However, on the other datasets of the feature selection challenge, we did not use *G-flip* either due to its computational requirements or due to poor performance. Note that we tried only the linear utility function for *G-flip*. We did not try *Simba* on any of the challenge datasets.

29.6 Discussion and Further Research Directions

A margin-based criterion for measuring the quality of a set of features has been presented. Using this criterion we derived algorithms that perform feature selection by searching for the set that maximizes it. We suggested two new methods for maximizing the margin based-measure, *G-flip*, which does a naïve local search, and *Simba*, which performs a gradient ascent. These are just representatives of the variety of optimization techniques (search methods) which can be used. We have also showed that the well known *Relief* algorithm (Kira and Rendell, 1992) approximates a gradient ascent algorithm that maximizes this measure. The nature of the different algorithms presented here was demonstrated on various feature selection tasks. It was shown that our new algorithm *Simba*, which is gradient ascent on our margin based measure, outperforms *Relief* on all these tasks. One of the main advantages of the margin based criterion is the high correlation that it exhibits with the features quality. This was demonstrated in figures 29.2 and 29.3.

The margin based criterion was developed using the 1-Nearest-Neighbor classifier but we expect it to work well for any distance based classifier. Additionally to the test we made with 1-NN, we also tested our algorithms with

SVM-RBF classifier and showed that they compete successfully with state-of-the-art algorithm that was designed specifically for SVM and is much more computationally demanding.

Our main theoretical result is a new rigorous bound on the finite sample generalization error of the 1-*Nearest Neighbor* algorithm. This bound depends on the margin obtained following the feature selection.

In the experiments we have conducted, the merits of the new algorithms were demonstrated. However, our algorithms use the Euclidean norm and assume that it is meaningful as a measure of similarity in the data. When this assumption fails, our algorithms might not work. Coping with other similarity measures will be an interesting extension to the work presented here.

The user of *G-flip* or *Simba* should chose a utility function to work with. In this paper we have demonstrated three such functions: linear, zero-one and sigmoid utility functions. The linear utility function gives equal weight to all points and thus might be sensitive to outliers. The sigmoid utility function suppress the influence of such outliers. We have also experimented that *G-flip* with zero-one utility uses less features than *G-flip* with linear utility where the sigmoid utility sits in between. It is still an open problem how to adapt the right utility for the data being studied. Never the less, much like the choice of kernel for SVM, using a validation set it is possible to find a reasonable candidate. A reasonable initial value for the parameter β of the sigmoid utility is something of the same order of magnitude as one over the average distance between training instances. As for the choosing between *G-flip* and *Simba*; *G-flip* is adequate when the goal is to chose the best feature subset, without a need to control its precise size. *Simba* is more adequate when ranking the features is required.

Several other research directions can be further investigated. One of them is to utilize a better optimization algorithm for maximizing our margin-based evaluation function. It is also possible to use the margin based criteria and the *Simba* algorithm to learn distance measures.

Another interesting direction is to link the feature selection algorithms to the *Learning Vector Quantization* (LVQ) (Kohonen, 1995) algorithm. As was shown in (Crammer et al., 2002), LVQ can be viewed as a maximization of the very same margin term. But unlike the feature selection algorithms presented here, LVQ does so by changing prototypes location and not the subset of the features. This way LVQ produces a simple but robust hypothesis. Thus, LVQ and our feature selection algorithms maximize the same margin criterion by controlling different (dual) parameters of the problem. In that sense the two algorithms are dual. One can combine the two by optimizing the set of features and prototypes location together. This may yield a winning combination.

References

P. Bartlett. The size of the weights is more important than the size of the network. *IEEE Transactions on Information Theory*, 44(2):525–536, 1998.

R. Bellman. *Adaptive Control Processes: A Guided Tour.* Princeton University Press, 1961.

B. Boser, I. Guyon, and V. Vapnik. Optimal margin classifiers. In *In Fifth Annual Workshop on Computational Learning Theory*, pages 144–152, 1992.

T.M. Cover and P.E. Hart. Nearest neighbor pattern classifier. *IEEE Transactions on Information Theory*, 13:21–27, 1967.

K. Crammer, R. Gilad-Bachrach, A. Navot, and N. Tishby. Margin analysis of the lvq algorithm. In *Proc. 17'th Conference on Neural Information Processing Systems (NIPS)*, 2002.

E. Fix and j. Hodges. Discriminatory analysis. nonparametric discrimination: Consistency properties. Technical Report 4, USAF school of Aviation Medicine, 1951.

Y. Freund and R. E. Schapire. A decision-theoretic generalization of on-line learning and an application to boosting. *Journal of Computer and System Sciences*, 55 (1):119–139, 1997.

I. Guyon and A. Elisseeff. An introduction to variable and feature selection. *Journal of Machine Learnig Research*, pages 1157–1182, Mar 2003.

I. Guyon and S. Gunn. Nips feature selection challenge. http://www.nipsfsc.ecs.soton.ac.uk/, 2003.

K. Kira and L. Rendell. A practical approach to feature selection. In *Proc. 9th International Workshop on Machine Learning*, pages 249–256, 1992.

T. Kohonen. *Self-Organizing Maps.* Springer-Verlag, 1995.

I. Kononenko. Estimating attributes: Analysis and extensions of RELIEF. In *Proc. European Conference on Machine Learning*, pages 171–182, 1994. URL citeseer.nj.nec.com/kononenko94estimating.html.

A.M. Martinez and R. Benavente. The ar face database. Technical report, CVC Tech. Rep. #24, 1998. http://rvl1.ecn.purdue.edu/~aleix/aleix_face_DB.html.

R. E. Schapire, Y. Freund, P. Bartlett, and W. S. Lee. Boosting the margin : A new explanation for the effectiveness of voting methods. *Annals of Statistics*, 1998.

J. Shawe-Taylor, P.L. Bartlett, R.C. Williamson, and M. Anthony. Structural risk minimization over data-dependent hierarchies. *IEEE transactions on Information Theory*, 44(5):1926–1940, 1998.

J. Weston, S. Mukherjee, O. Chapelle, M. Pontil, T. Poggio, and V. Vapnik. Feature selection for SVMs. In *Proc. 15th Conference on Neural Information Processing Systems (NIPS)*, pages 668–674, 2000. URL citeseer.nj.nec.com/article/weston01feature.html.

J. Weston, A. Elisseeff, G. BakIr, and F. Sinz. The spider, 2004. http://www.kyb.tuebingen.mpg.de/bs/people/spider/.

A Complementary Proofs

We begin by proving a simple lemma which shows that the class of nearest neighbor classifiers is a subset of the class of 1-Lipschitz functions. Let $\mathbf{nn}_F^S(\cdot)$

be a function such that the sign of $\mathbf{nn}_F^S(\mathbf{x})$ is the label that the nearest neighbor rule assigns to $\mathbf{x}$, while the magnitude is the sample-margin, i.e. the distance between $\mathbf{x}$ and the decision boundary.

Lemma 1. *Let F be a set of features and let S be a labeled sample. Then for any $\mathbf{x}_1, \mathbf{x}_2 \in \mathbf{R}^N$:*

$$\left| nn_F^S(\mathbf{x}_1) - nn_F^S(\mathbf{x}_2) \right| \leq \|F(\mathbf{x}_1) - F(\mathbf{x}_2)\|$$

where $F(\mathbf{x})$ is the projection of $\mathbf{x}$ on the features in F.

Proof. Let $\mathbf{x}_1, \mathbf{x}_2 \in \mathcal{X}$. We split our argument into two cases. First assume that $\mathbf{nn}_F^S(\mathbf{x}_1)$ and $\mathbf{nn}_F^S(\mathbf{x}_2)$ have the same sign. Let $\mathbf{z}_1, \mathbf{z}_2 \in \mathbf{R}^{|F|}$ be the points on the decision boundary of the 1-NN rule which are closest to $F(\mathbf{x}_1)$ and $F(\mathbf{x}_2)$ respectively. From the definition of $\mathbf{z}_{1,2}$ it follows that $\mathbf{nn}_F^S(\mathbf{x}_1) = \|F(\mathbf{x}_1) - \mathbf{z}_1\|$ and $\mathbf{nn}_F^S(\mathbf{x}_2) = \|F(\mathbf{x}_2) - \mathbf{z}_2\|$ and thus

$$\begin{aligned}
\mathbf{nn}_F^S(\mathbf{x}_2) &\leq \|F(\mathbf{x}_2) - \mathbf{z}_1\| \\
&\leq \|F(\mathbf{x}_2) - F(\mathbf{x}_1)\| + \|F(\mathbf{x}_1) - \mathbf{z}_1\| \\
&= \|F(\mathbf{x}_2) - F(\mathbf{x}_1)\| + \mathbf{nn}_F^S(\mathbf{x}_1)
\end{aligned} \tag{29.4}$$

By repeating the above argument while reversing the roles of $\mathbf{x}_1$ and $\mathbf{x}_2$ we get

$$\mathbf{nn}_F^S(\mathbf{x}_1) \leq \|F(\mathbf{x}_2) - F(\mathbf{x}_1)\| + \mathbf{nn}_F^S(\mathbf{x}_2) \tag{29.5}$$

Combining (29.4) and (29.5) we obtain

$$\left| \mathbf{nn}_F^S(\mathbf{x}_2) - \mathbf{nn}_F^S(\mathbf{x}_1) \right| \leq \|F(\mathbf{x}_2) - F(\mathbf{x}_1)\|$$

The second case is when $\mathbf{nn}_F^S(\mathbf{x}_1)$ and $\mathbf{nn}_F^S(\mathbf{x}_2)$ have alternating signs. Since $\mathbf{nn}_F^S(\cdot)$ is continuous, there is a point $\mathbf{z}$ on the line connecting $F(\mathbf{x}1)$ and $F(\mathbf{x}_2)$ such that $\mathbf{z}$ is on the decision boundary. Hence,

$$\begin{aligned}
\left| \mathbf{nn}_F^S(\mathbf{x}_1) \right| &\leq \|F(\mathbf{x}_1) - \mathbf{z}\| \\
\left| \mathbf{nn}_F^S(\mathbf{x}_2) \right| &\leq \|F(\mathbf{x}_2) - \mathbf{z}\|
\end{aligned}$$

and so we obtain

$$\begin{aligned}
\left| \mathbf{nn}_F^S(\mathbf{x}_2) - \mathbf{nn}_F^S(\mathbf{x}_1) \right| &= \left| \mathbf{nn}_F^S(\mathbf{x}_2) \right| + \left| \mathbf{nn}_F^S(\mathbf{x}_1) \right| \\
&\leq \|F(\mathbf{x}_2) - \mathbf{z}\| + \|F(\mathbf{x}_1) - \mathbf{z}\| \\
&= \|F(\mathbf{x}_2) - F(\mathbf{x}_1)\|
\end{aligned}$$

The main tool for proving theorem 1 is the following:

Theorem 2. *(Bartlett, 1998) Let $\mathcal{H}$ be a class of real valued functions. Let S be a sample of size m generated i.i.d. from a distribution $\mathcal{D}$ over $\mathcal{X} \times \{\pm 1\}$ then with probability $1 - \delta$ over the choices of S, every $h \in \mathcal{H}$ and every $\gamma \in (0, 1]$ let $d = fat_{\mathcal{H}}(\gamma/32)$:*

$$er_{\mathcal{D}}(h) \leq \hat{er}_S^\gamma(h) + \sqrt{\frac{2}{m}\left(d \ln\left(\frac{34em}{d}\right) \log_2(578m) + \ln\left(\frac{8}{\gamma\delta}\right)\right)}$$

We now turn to prove theorem 1:

Proof of theorem 1: Let F be a set of features such that $|F| = n$ and let $\gamma > 0$. In order to use theorem 2 we need to compute the fat-shattering dimension of the class of nearest neighbor classification rules which use the set of features F. As we saw in lemma 1 this class is a subset of the class of 1-Lipschitz functions on these features. Hence we can bound the fat-shattering dimension of the class of NN rules by the dimension of Lipschitz functions.

Since $\mathcal{D}$ is supported in a ball of radius R and $\|\mathbf{x}\| \geq \|F(\mathbf{x})\|$, we need to calculate the fat-shattering dimension of Lipschitz functions acting on points in $\mathbf{R}^n$ with norm bounded by R. The fat_γ-dimension of the 1-NN functions on the features F is thus bounded by the largest γ packing of a ball in $\mathbf{R}^n$ with radius R, which in turn is bounded by $(2R/\gamma)^{|F|}$.

Therefore, for a fixed set of features F we can apply to theorem 2 and use the bound on the fat-shattering dimension just calculated. Let $\delta_F > 0$ and we have according to theorem 2 with probability $1 - \delta_F$ over sample S of size m that for any $\gamma \in (0, 1]$

$$\mathrm{er}_\mathcal{D}\,(\text{nearest-neighbor}) \leq \hat{\mathrm{er}}_S^\gamma\,(\text{nearest-neighbor}) + \tag{29.6}$$

$$\sqrt{\frac{2}{m}\left(d\ln\left(\frac{34em}{d}\right)\log_2\left(578m\right) + \ln\left(\frac{8}{\gamma\delta_F}\right)\right)}$$

for $d = (64R/\gamma)^{|F|}$. By choosing $\delta_F = \delta/\left(N\binom{N}{|F|}\right)$ we have that $\sum_{F\subseteq[1...N]} \delta_F = \delta$ and so we can apply the union bound to (29.6) and obtain the stated result.

Chapter 30

Feature Extraction for Classification of Proteomic Mass Spectra: A Comparative Study

Ilya Levner, Vadim Bulitko, and Guohui Lin

University of Alberta
Department of Computing Science
Edmonton, Alberta, T6G 2E8, CANADA
ilya@cs.ualberta.ca, bulitko@cs.ualberta.ca, ghlin@cs.ualberta.ca

Summary. To satisfy the ever growing need for effective screening and diagnostic tests, medical practitioners have turned their attention to high resolution, high throughput methods. One approach is to use mass spectrometry based methods for disease diagnosis. Effective diagnosis is achieved by classifying the mass spectra as belonging to healthy or diseased individuals. Unfortunately, the high resolution mass spectrometry data contains a large degree of noisy, redundant and irrelevant information, making accurate classification difficult. To overcome these obstacles, feature extraction methods are used to select or create small sets of relevant features. This paper compares existing feature selection methods to a novel wrapper-based feature selection and centroid-based classification method. A key contribution is the exposition of different feature extraction techniques, which encompass dimensionality reduction and feature selection methods. The experiments, on two cancer data sets, indicate that feature selection algorithms tend to both reduce data dimensionality and increase classification accuracy, while the dimensionality reduction techniques sacrifice performance as a result of lowering the number of features. In order to evaluate the dimensionality reduction and feature selection techniques, we use a simple classifier, thereby making the approach tractable. In relation to previous research, the proposed algorithm is very competitive in terms of (i) classification accuracy, (ii) size of feature sets, (iii) usage of computational resources during both training and classification phases.
Keywords: feature extraction, classification, mining bio-medical data, mass spectrometry, dimensionality reduction.

30.1 Introduction

Early detection of diseases, such as cancer, is critical for improving patient survival rates and medical care. To satisfy the ever growing need for effective screening and diagnostic tests, medical practitioners have turned their attention to mass spectrometry based methods. While other proteomic methods

I. Levner et al.: *Feature Extraction for Classification of Proteomic Mass Spectra: A Comparative Study*, StudFuzz **207**, 607–624 (2006)
www.springerlink.com

exist, such as PAGE*, mass spectrometry (**MS**) based approaches provide high throughput, are widely applicable, and have the potential to be highly accurate. This paper examines supervised classification in **proteomic** applications. The term proteomics will be restricted to mean the study of protein spectra, acquired by mass spectrometry techniques, to classify disease and identify potentially useful protein biomarkers. A **biomarker** is an identified protein(s) whose abundance is correlated with the state of a particular disease or condition. Currently, single biomarkers, such as PSA[†] used to detect prostate cancer, are relied on for disease screening and diagnosis. The identification of each biomarker, tailored for a specific disease, is a time consuming, costly and tedious process. In addition, for many diseases it is suspected that no single biomarkers exit, which are capable of producing reliable diagnoses. The following quote further motivates the use of high resolution MS techniques:

> *"The ability to distinguish sera from an unaffected individual or an individual with [for example] ovarian cancer based upon a single serum proteomic m/z feature alone is **not possible** across the entire serum study set. Accurate histological distinction is only possible when the key m/z features and their intensities are considered en masse. A limitation of individual cancer biomarkers is the lack of sensitivity and specificity when applied to large heterogeneous populations."* (Conrads et al., 2003)

While high-resolution mass spectrometry techniques are thought to have potential for accurate diagnosis due to the vast amount of information captured, they are problematic for supervised training of classifiers. Specifically, the many thousands of raw attributes forming the spectra frequently contain a large amount of redundancy, information irrelevant to a particular disease, and measurement noise. Therefore, aggressive feature extraction techniques are crucial for learning high-accuracy classifiers and realizing the full potential of mass spectrometry based disease diagnosis.

The rest of the paper is organized as follows. We first motivate the task by presenting two important disease diagnosis problems and recent studies on them. A novel combination of feature selection and classification methods is subsequently proposed and empirically evaluated on ovarian and prostate cancer data sets. The paper is concluded with discussion and future research directions.

30.1.1 Ovarian Cancer Studies

In (Petricoin et al., 2002a), genetic algorithms together with self-organizing maps were used to distinguish between healthy women and those afflicted

---

*The acronym PAGE stands for polyacrylamide gel electrophoresis. It is also known as 2DE for two dimensional polyacrylamide gel electrophoresis (Patterson and Aebersold, 2003).

[†]PSA stands for prostate specific antigen.

with ovarian cancer. Although cross-validation studies were not conducted, the approach was able to correctly classify all cancer stricken patients and 95% of healthy women, on a single test set. Motivated by the need for greater recall and precision, in (Conrads et al., 2003), a low resolution mass spectrometry technique was compared with a high resolution technique using the same ovarian cancer data set. The goal was to determine whether sensitivity and PPV[‡] (i.e., recall and precision) scores would improve by using a higher resolution spectra provided by the SELDI TOF MS hardware[§]. Keeping all other parameters fixed (including the machine learning algorithm), classification based on high resolution data achieved 100% specificity and PPV scores on the ovarian cancer data set. In contrast, none of the models based on the low resolution mass spectra could achieve perfect precision and recall scores. The researchers, therefore, concluded that the 60-fold increase in resolution improved the performance of the pattern recognition method used. Due to the low prevalence of (ovarian) cancer (Kainz, 1996), a screen test would require a 99.6% specificity to achieve a clinically acceptable positive predictive value of 10%. As a result, high resolution mass spectrometry techniques have been adopted to increase classification accuracy.

Unfortunately, increasing data resolution proliferates "the curse of dimensionality", and thereby decreases the applicability of supervised classification techniques. As a result, **feature extraction** is needed to extract/select salient features in order to make classification feasible. In addition to making machine learning algorithms tractable, feature extraction can help identify the set(s) of proteins (i.e., features) that can be used as potential biomarkers. In turn, key protein identification can shed light on the nature of the disease and help develop clinical diagnostic tests and treatments.

Using the same data set, in (Lilien et al., 2003) the researchers used Principle Component Analysis (PCA) (Kirby, 2001) for dimensionality reduction and Linear Discriminant Analysis (LDA) for classification. For each of the various train/test data splits, 1000 cross-validation runs with re-sampling were conducted. When training sets were larger than 75% of the total sample size, perfect (100%) accuracy was achieved. Using only 50% of data for training, the performance dropped by 0.01%. We conclude that PCA appears to be an effective way to reduce data dimensionality.

In (Wu et al., 2003), the researchers compared two feature extraction algorithms together with several classification approaches. The T-statistic[¶] was used to rank features in terms of relevance. Then two feature subsets were greedily selected (respectively having 15 and 25 features each). Support vector machines (SVM), random forests, LDA, Quadratic Discriminant Analysis, k-nearest neighbors, and bagged/boosted decision trees were subsequently

---

[‡]PPV stands for Positive Predictive Value, see glossary for details.

[§]SELDI TOF MS stands for surface-enhanced laser desorption/ionization time-of-flight mass spectrometry.

[¶]The T-statistic is also known as the student-t test (Press et al., 2002).

used to classify the data. In addition, random forests were also used to select relevant features with previously mentioned algorithms used for classification. Again 15 and 25 feature sets were selected and classification algorithms applied. When the T-statistic was used as a feature extraction technique, SVM, LDA and random forests classifiers obtained the top three results (accuracy appears to be about 85%). On the other hand, classification accuracy improved to approximately 92% when random forests were used for both feature extraction and classification. Similar performance was also achieved using 1-nearest-neighbor.

The data from (Wu et al., 2003), was subsequently analyzed in (Tibshirani et al., 2004), using the nearest shrunken centroid algorithm. The ten-fold cross validated specificity was 74% with a corresponding sensitivity of 71%. Thus the balanced accuracy (BACC) of this algorithm was 72.5%. Although the accuracy of this algorithm is less than that of other methods presented, this approach used only seven features‖ out of 91360.

30.1.2 Prostate Cancer Studies

In (Adam et al., 2002), the researchers used a decision tree algorithm to differentiate between healthy individuals and those with prostate cancer. This study also used the SELDI TOF MS to acquire the mass spectra. Receiver Operating Characteristics (ROC) curves were used to identify informative peaks which were subsequently used by the decision tree classification algorithm. The researchers did not perform cross-validation, but on a single test set the classifier achieved an 81% sensitivity and a 97% specificity, yielding a balanced accuracy (BACC) of 89%.

In (Qu et al., 2002), the performance was improved from (Adam et al., 2002) by using ROC curves to identify relevant features. For classification, the researchers used decision trees together with AdaBoost and its variant, Boosted Decision Stump Feature Selection (BDSFS) method. AdaBoost achieved perfect accuracy on the single test set for the prostate cancer data set. However, a 10-fold cross validation performance yielded average sensitivity of 98.5% and a specificity of 97.9%, for an overall BACC of 98%. For the BDSFS, the results were worse, with a sensitivity of 91.1% and a specificity of 94.3%. The researchers informally report that other classifiers had similar accuracies but were more difficult to interpret.

In (Lilien et al., 2003), the researchers again used PCA for dimensionality reduction and LDA for classification. The data set was obtained from the authors of (Adam et al., 2002). In the same fashion as with the ovarian cancer set, the researchers conducted a detailed study using various train/test set sizes. For each train/test data split, 1000 cross-validation runs (with re-sampling)

‖It should be noted that peak extraction and clustering were used to preprocess the data and produced 192 peaks from which 7 were used by the shrunken centroid algorithm.

Comparison of three reports for prostate cancer diagnosis based on SELDI-TOF technology.

	Adam et al. (1)	Petricoin et al. (12)	Qu et al. (29)
Diagnostic sensitivity and specificity	83%; 97%	95%; 78–83%	97–100%; 97–100%
SELDI-TOF chip type	IMAC-Cu	Hydrophobic C-16	IMAC-Cu
Distinguishing peaks, m/z^a	4475, 5074, 5382, **7024**, 7820, 8141, 9149, 9507, **9656**	2092, 2367, 2582, 3080, 4819, 5439, 18220	Noncancer vs cancer: 3963, 4080, 6542, 6797, 6949, 6991, **7024**, 7885, 8067, 8356, **9656**, 9720
			Healthy individuals vs BPH:[b] 3486, 4071, 4580, 5298, 6099, 7054, 7820, 7844, 8943
Bioinformatic analysis	**Decision tree algorithm**	**Proprietary; based on genetic algorithms and cluster analysis**	**Boosted decision tree algorithm**

Fig. 30.1. Comparison of classification techniques for prostate cancer diagnosis (reproduced from (Diamandis, 2003).) Respectively, the accuracies for (Adam et al., 2002, Petricoin et al., 2002b, Qu et al., 2002) are 89%, 83%, 98%. This comparison demonstrates the wide classification variance due to different mass spectrometry and machine learning approaches.

were conducted. When training sets were larger than 75% of the total sample size, an average accuracy of 88% was achieved. Using only 50% of data for training, the performance dropped to 86%. In comparison to ovarian cancer sets the lower accuracy suggests that this data set is much more difficult to classify correctly.

In (Petricoin et al., 2002b, Wulfkuhle et al., 2003), researchers used Genetic Algorithms (GA's) for feature extraction and Self Organizing Maps (SOM's) for classification of prostate cancer. This approach achieved a 95% specificity and a 71% sensitivity, for a balanced accuracy of 83%. Although cross validation was carried out, the results were not presented.

In (Diamandis, 2003), the aforementioned studies on prostate cancer raised the following question: Why do the features and classification performance vary so drastically across studies? Indeed, results reproduced in Figure 30.1, indicate that different SELDI-TOF approaches combined with different machine learning techniques for pattern recognition produce highly variable results. This observation further motivates the need for comparative studies done on a regular basis using several mass spectrometry techniques in conjunction with a number of machine learning approaches. We attempt to carry out such a study in this paper.

30.2 Existing Feature Extraction and Classification Methods

Feature extraction is central to the fields of machine learning, pattern recognition and data mining. This section introduces algorithms used in this study. More details on the algorithms used within this study can be found in Part 1, Chapters 3 and 4.

30.2.1 Centroid Classification Method

A fast and simple algorithm for classification is the centroid method (Hastie et al., 2001, Park et al., 2003). This algorithm assumes that the target classes correspond to individual (single) clusters and uses the cluster means (or centroids) to determine the class of a new sample point. A prototype pattern for class C_j is defined as the arithmetic mean:

$$\boldsymbol{\mu}_{C_j} = \frac{1}{|C_j|} \sum_{\boldsymbol{x}_i \in C_j} \boldsymbol{x}_i$$

where $\boldsymbol{x}_i$'s are the training samples labeled as class C_j. Recall that the training sample is a MS spectra represented as a multi-dimensional vector (denoted in bold). In a similar fashion, we can obtain a prototypical vector for all the other classes. During classification, the class label of an unknown sample $\boldsymbol{x}$ is determined as:

$$C(\boldsymbol{x}) = \arg \min_{C_j}\ d(\boldsymbol{\mu}_{C_j}, \boldsymbol{x})$$

where $d(\boldsymbol{x}, \boldsymbol{y})$ is a distance function or:

$$C(\boldsymbol{x}) = \arg \max_{C_j}\ s(\boldsymbol{\mu}_{C_j}, \boldsymbol{x})$$

where $s(\boldsymbol{x}, \boldsymbol{y})$ is a similarity metric. This simple classifier will form the basis of our studies. It works with any number of features and its run-time complexity is proportional to the number of features and the complexity of the distance or similarity metric used. Preliminary experiments were conducted to establish which similarity/distance metric is most appropriate for the centroid classification algorithm**, and the L_1 distance metric was selected. Defined by:

$$L_1(\boldsymbol{x}, \boldsymbol{\mu}) = \|\boldsymbol{x} - \boldsymbol{\mu}\|_1 \qquad (30.1)$$

with $\|\boldsymbol{y}\|_1 = \sum_i^N |y(i)|$, and $y(i)$ being the value of the ith feature. The value $L_1(\boldsymbol{x}, \boldsymbol{\mu})$ has a linear cost in the number of features. In this study, data sets contain two classes and hence the number of calls to a metric is also two. Therefore, the centroid classifier, at run-time, is linear in the number of features. During training, two prototypes are computed and the cost of computing each prototype is $O(mN)$, where N is the number of features and m is the number of training samples which belong to a given class. Note that m only varies between data sets and not during training or feature selection processes. Thus, we can view m as a constant and conclude that the centroid classifier has $O(N)$ cost in the training phase.

**Due to space restrictions, the results are not shown. A companion technical report (Levner, 2004) provides experimental details and supplementary material.

30.2.2 Nearest Shrunken Centroid

A special purpose feature selection algorithm for the nearest centroid algorithm was developed by Tibshirani et al. and presented in (Hastie et al., 2001, Tibshirani et al., 2003, 2004). The algorithm, related to the lasso method described in Part 1, Chapter 1, Section 4, tries to shrink the class prototypes (μ_{C_j}) towards the overall mean:

$$\mu = \frac{1}{m} \sum_{i=1}^{m} x_i \qquad (30.2)$$

Briefly, the algorithm calculates:

$$d_j = \frac{\mu_{C_j} - \mu}{m_j(s)} \qquad (30.3)$$

where $m_j = \sqrt{\frac{1}{|C_j|} - \frac{1}{m}}$, s is a vector of pooled within class variances for each feature and division is done component wise. We can now view the class centroid as:

$$\mu_{C_j} = \mu + m_j(s \cdot d_j) \qquad (30.4)$$

where $\cdot$ denotes component wise multiplication. By decreasing d_j we can move the class centroid towards the overall centroid. When a component of the class centroid is equal to the corresponding component of the overall mean for all classes, the feature no longer plays a part in classification and is effectively removed. Hence as d_j shrinks progressively more features are removed.

30.2.3 Ordered and Sequential Feature Selection

Using the aforementioned centroid method as the base classifier, we can select features with SFS (Sequential Forward Selection) technique or via an ordered feature selection approach. Both of these wrapper-based techniques incrementally build a feature set by adding one feature at a time to the active (i.e., previously selected) set of features and invoking the nearest centroid classifier using the active feature set. Sequential Forward (respectively Backward) selection (SFS and SBS) methods start from an empty (respectively full) set of features and at each step add (respectively remove) a single feature that produces the greatest increase in performance. In contrast, the ordered feature selection approach first evaluates each of the N features independently of all others. The features are then ranked according to the performance of the base classifier (i.e., the nearest centroid classifier in our case). Once ranked and sorted, the ordered feature selection approach incrementally adds the topmost ranked feature to the active set. In total, N feature subsets are tried, where s_1 contains a single top ranked feature, s_2 contains the two top ranked features, and so on until s_N is tried. In contrast, to the SFS procedure, ordered feature selection is linear in the number of calls to the base classifier since at each stage the top ranked feature is added to the active set and the newly created active set is evaluated by the base classifier. Since there are only N features the total number of calls to the base classifier is $2N$, N initial calls to rank individual features, and N times to evaluate the ever larger subsets $s_1, ..., s_N$. Unlike the SFS algorithm, the greedy approach will not stop until all N sets have been tried. The final stage of the algorithm merely selects the feature set producing the best classification accuracy on the particular data set.

30.2.4 Univariate Statistical Tests

Instead of ranking features by invoking a classifier, one can use filter ranking based on statistical tests. In general, univariate statistical tests analyze each feature independently of others. The student-t (T-test) and the Kolmogorov-Smirnov (KS-test) (Press et al., 2002) algorithms are common examples. Both tests compare feature values from samples belonging to class i to feature values from samples belonging to class j. The goal is to determine if the feature values for class i come from a different distribution than those for class j. The key difference between the two tests are the assumptions they make. The T-test assumes that both distributions have identical variance, and makes no assumptions as to whether the two distributions are discrete or continuous. On the other hand, the KS-test assumes that the two distributions are continuous, but makes no other assumptions.

In the case of the T-test, the null hypothesis is $\mu_A = \mu_B$, representing that the mean of feature value for class A is the same as the mean of the feature values for class B. In the case of the KS-test, the null hypothesis is $cdf(A) = cdf(B)$, meaning that feature values from both classes have an identical cumulative distribution. Both tests determine if the observed differences are statistically significant and return a score representing the probability that the null hypothesis is true. Thus, features can be ranked using either of these statistics according to the significance score of each feature. In addition, the two tests can be combined together into a composite statistic. While many possible composition strategies exist, we limit our experiments to a simple multiplicative composition, whereby the T-test significance score is multiplied together with the KS-test significance score (referred to as the T*KS-test henceforth).

Both the benefits and drawbacks of these statistical tests stem from the assumption that features are independent. On one hand, the independence assumption makes these approaches very fast. On the other hand, the independence assumption may not hold for all data sets. Technical details on these and other statistical tests can be found in (Hastie et al., 2001, Press et al., 2002).

Recall that in (Wu et al., 2003), the T-test and random forests were used for feature extraction teamed with a number of classifiers. The researchers used the T-test to rank each feature but chose to test classification algorithms with 15 and 25 top-ranked features. Their line of research appears more focused on comparing classifiers rather than the two feature extractors (T-test and random forests). In contrast, we show that feature ranking coupled with ordered feature selection can automatically find a feature subset of arbitrary size that improves performance (with respect to using either a single best feature or using all features).

30.2.5 Dimensionality Reduction

Recall that feature selection algorithms attempt to select relevant features with respect to the performance task, or conversely remove redundant or irrelevant ones. In contrast, dimensionality reduction algorithms attempt to extract features capable of reconstructing the original high dimensional data. For example, PCA (Kirby, 2001) attempts to find a linear combination of principal components that preserves the data variance. In proteomic pattern recognition the most common technique is down sampling. This technique filters the spectra and sub-samples it to reduce the dimensionality. A common approach is to convolve the spectrum with a uniform filter at regular intervals (windows). This technique, essentially removes high frequency components. In order to test the conjecture made in (Conrads et al., 2003), that higher resolution data tends to improve classification performance, we will use this approach to test the merit of dimensionality reduction via down sampling.

30.3 Experimental Results

We conducted experiments on the ovarian and prostate data sets, previously used in (Petricoin et al., 2002a) and (Petricoin et al., 2002b). The ovarian cancer set includes sera from 91 controls and 162 ovarian cancers patients. Acquired from (Johann, 2003), each data sample contains 15,156 features. The prostate cancer data set is composed of 322 samples in total, and was also acquired from (Johann, 2003). There are 190 serum samples from patients with benign prostate whose PSA levels are greater than four, 63 samples with no evidence of disease and PSA level less than one, 26 samples with prostate cancer with PSA levels four through ten, and 43 samples with prostate cancer and PSA levels greater than ten. Again, each sample is a histogram with 15,156 bins, with each bin corresponding to a single feature.

For all experiments, each data set was split into three subsets of equal size. Each test fold used one of the three subsets with the remaining two subsets used for training. We ran two sets of experiments. The first optimized performance directly on the test set. For a given feature selection technique, this approach produces a single feature set and hence makes feature analysis possible. The drawback of this approach is that performance estimates are overly optimistic. To get a better performance estimate, a second set of experiments was carried out. It optimized performance on the training set.

Specifically, we used a leave-one-out cross-validation (LOOCV) internal loop based solely on the training set to select features using a subset of the most promising algorithms. The reported accuracy for all experiments is the average classification accuracy over the three test folds and the error bars represent one standard deviation. Accuracy is taken as the arithmetic mean of sensitivity and specificity. This measure is related to BER (Balanced Error Rate) and can be analogously thought of as balanced accuracy ($BACC$), where $BER = 1 - BACC$.

Dimensionality Reduction

We progressively down-sampled the spectra by averaging each sample spectra using a uniform filter. In other words, given a window of size w we averaged w adjacent features (i.e., m/z values) into a single new feature. The window was then shifted by w features and the process repeated. For each trial we increased size of the window w. This effectively produces data with progressively lower resolution and reduced dimensionality. For each down sampled data set we used the centroid classifier. The results, presented in Figure 30.2, show that classification performance decreases as the size of the filter increases. However, the decrease is clearly non-monotonic and, in essence, very noisy. This noise can be attributed to either the filtering or the sub-sampling stages of the down-sampling process. To determine which of the two components produced the oscillations in classification accuracy, another experiment was carried out.

In the second experiment we performed frequency based data filtering. The procedure first transformed each spectra into the frequency domain via the Fast Fourier Transform (FFT). Then a low pass filter was applied to the frequency coefficients in order to remove the high frequency components. The final stage transformed the filtered data back to the spatial domain. By varying the size of the low pass filter, the number of frequency coefficients used in reconstructing the MS spectra was varied and, in essence, considered feature selection in the frequency domain. Clearly the loss in accuracy, shown on the right side of Figure 30.2, is much more monotonic in comparison with the down-sampling method (the left-hand side). This suggests that the majority of oscillations result from the sub-sampling step rather than the frequency filtering step. This led us to the conjecture that down-sampling is in general detrimental to classification performance. To further investigate this hypothesis, we ran the centroid classifier on each individual feature for the down-sampled spectra and found the classification performance inferior to the performance of a single best feature from the non down-sampled spectra. This further supported the claim that down-sampling appears detrimental to classification accuracy. The conclusions drawn are in line with those in (Conrads et al., 2003) where changes in resolution created by different MS techniques produced similar results. Because the MS spectra are histograms describing the ion concentrations based on the mass-to-charge ratios, the low resolution techniques effectively aggregate distinct ion concentrations into a

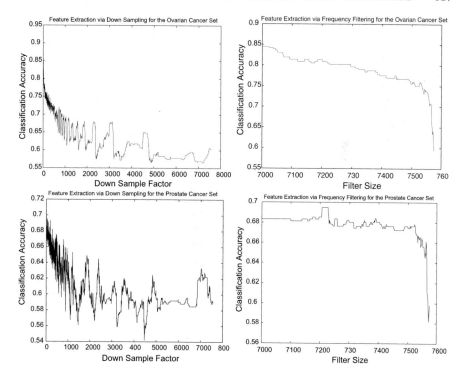

Fig. 30.2. Classification accuracy on progressively down-sampled data. **Top:** ovarian cancer data set. **Bottom:** prostate cancer data set. **Left:** Down-sampling Performance. The down-sample factor indicates the ratio of original number of features to the number of features after down-sampling. As the down-sample factor increases, the number of features decreases. **Right:** Frequency filtering. While all data sets exhibit oscillations, the performance nevertheless gradually declines as the dimensionality of the data is reduced as indicated by the increasing down-sampling factor on the x-axis.

single bin. Hence, down-sampling, whether due to low-resolution MS hardware or done deliberately in software to reduce data dimensionality, appears to lower diagnosis performance.

30.3.1 Ordered and Sequential Feature Selection

To compute the exact relevance of individual features, the centroid classifier was ran on individual features. Histogram plots for each data set are shown in Figure 30.3. Each plot represents the distribution of features with respect to classification accuracy and shows that a very large number of features are essentially irrelevant and/or redundant with respect to diagnosis. This provides further unfavorable evidence for the down-sampling approach, which in essence, aggregates individual features together. Such an approach would

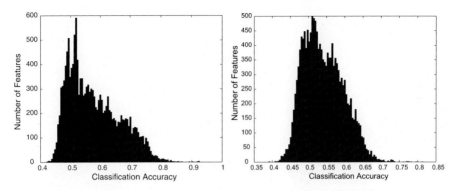

Fig. 30.3. Performance using individual features **Left:** ovarian cancer data set. **Right:** prostate cancer data set. The histograms show the number of features with a specific classification accuracy on a single test fold when individual features are used.

inevitably merge relevant and irrelevant (or redundant) features together and decrease the overall performance as evidenced by the experimental results of the previous section. Interestingly, there are a number of features within each data set that produce classification accuracies below 50%. These features can mislead and confuse the classifier.

Once each feature was ranked and the feature set sorted, ordered feature selection was used. In addition, the SFS procedure was also employed to select *relevant* feature sets. The results are presented in Figure 30.4 and are discussed in the next section.

30.3.2 Performance Comparison

Figure 30.4 presents the best performance for each feature extraction technique on each data set. Clearly, SFS coupled with the centroid algorithm produced superior results in comparison to the other algorithms tested in terms of feature set size and classification accuracy.

On the ovarian cancer data set, classification based on four features selected via SFS had the same accuracy of 98.0%, tieing with a set composed of 48 features created by the ordered feature selection. Previously, PCA coupled with LDA produced the only perfect cross-validated classification accuracy (Lilien et al., 2003). On the prostate cancer data set, the SFS classifier increased the base classification accuracy from 69.7% to 94% using only 11 of 15,154 features. In contrast, PCA coupled with LDA produced an accuracy of 88% (Lilien et al., 2003) . In (Qu et al., 2002), the boosted decision stumps produced an impressive 98% accuracy on the same data set. However, we were unable to get this set and used the data set from (Petricoin et al., 2002b), where the accuracy using GA's combined with SOM's was only 83%. Overall the SFS/centroid system appeared competitive with the previous approaches

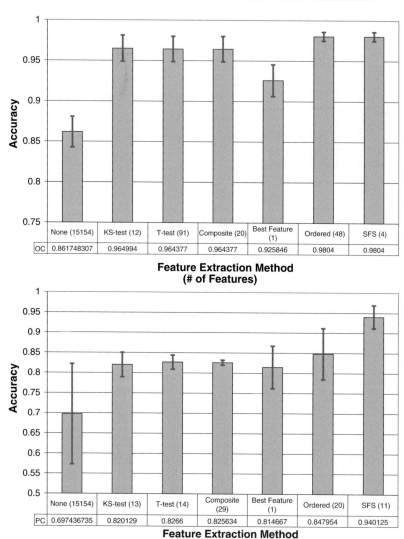

Fig. 30.4. Performance of Feature Extraction Algorithms optimized on the test sets. **Top:** ovarian cancer data set. **Bottom:** prostate cancer data set.

in terms of classification accuracy but produced considerably smaller feature sets. Note that the PCA/LDA approach always uses n features corresponding to n eigenvectors. Since the rank of the covariance matrix is bounded by the number of samples, n is necessarily upper bounded by the number of training samples, and was set to this upper bound in (Lilien et al., 2003). Furthermore, boosted decision stumps used to classify the prostate cancer data set needed

Table 30.1. Active feature set extracted by the SFS procedure. **Top:** Ovarian cancer data set. **Bottom:** Prostate cancer data set. Column 1 shows the order each feature was added to the active set. Column 2 contains the feature index. Column 3 shows classification accuracy using just the one feature. Column 4-6 present the rank of each feature using the T-test, KS-test, and T*KS-test with respect to topmost ranked feature. The SFS procedure does not appear to select the same features as any of the ordered FS methods.

Order Added	Feature Index	Individual Feature Accuracy	T-Test	KS-Test	T*KS -Test
1	1679	0.9258	5003	9289	14129
2	541	0.8303	8185	7502	9272
3	1046	0.62	9012	13276	5997
4	2236	0.9104	4855	5501	7953

Order Added	Feature Index	Individual Feature Accuracy	T-Test	KS-Test	T*KS -Test
1	2400	0.8147	2106	1880	1499
2	6842	0.6393	7823	14543	11650
3	2667	0.6246	1756	7601	13111
4	6371	0.5776	5600	609	4297
5	2005	0.5262	7128	11984	8482
6	1182	0.5147	12400	6180	890
7	7604	0.6328	7694	12788	5943
8	462	0.4531	11165	14343	11810
9	659	0.5868	13282	11766	11307
10	187	0.4994	14893	1807	5032
11	467	0.6036	12602	8744	2272

500 stumps to achieve the aforementioned accuracy. In contrast, the SFS/-centroid method selected only 5 and 11 features for the ovarian and prostate cancer data sets respectively, while producing comparable classification accuracy.

Active Feature Sets

The relationship between the features selected by the SFS procedure and the corresponding rankings based on statistical tests is illustrated in Table 30.1. Each table examines the features selected by the SFS procedure for the ovarian and prostate cancer data sets. In both cases, the features added to the active set are ranked far from first by the statistical tests. In addition, individual feature performance does not appear to be an effective indicator of classification performance within a set of features. In fact, the eighth and tenth features have individual classification accuracies of less than 50% on the prostate data. Furthermore, not a single feature selected by any of the ordered feature selection approaches appears in the active set produced by

Table 30.2. Active feature set extracted from the prostate cancer data set by the SFS procedure. Column 1 shows the order each feature was added into the active set. Column 2 contains the feature index. Column 3 provides the actual mass-to-charge ratio of each feature. The last three columns present nearby (±500 Da) features found previously in (Adam et al., 2002, Qu et al., 2002, Petricoin et al., 2002b). Clearly the SFS procedure found a set of features very different than the other algorithms.

Order Added	Feature Index	M/Z	Adam et al.	Qu et al.	Petricoin et al.
1	2400	500.8			
2	6842	4074.8	4475	3963; 4080; 4071	
3	2667	618.6			
4	6371	3533.0		3486	3080
5	2005	349.4			
6	1182	121.3			
7	7604	5033.3	5074	5289	4819; 5439
8	462	18.4			
9	659	37.6			
10	187	3.0			
11	467	18.8			

the SFS procedure. However, the ordered approaches do improve performance in comparison to the classification accuracy based on the full feature set. This indicates that there are a number of relevant features related to the presence/absence of cancer.

To further examine the features extracted by the SFS, we compared the active sets extracted by this procedure for the prostate cancer set to the features selected using other approaches surveyed in the previous research literature (refer to Figure 30.1). The results are summarized in Table 30.2. Clearly, very few common features are observed. As hypothesized in (Diamandis, 2003), it appears that different algorithms extract different relevant features based on their internal machinery and bias. A crucial goal for future research is therefore, to determine which, if any, features can serve as potential biomarkers, and shed light on the nature of cancer, and possibly even its cure.

30.3.3 LOOCV Performance

The previous section presented results with classification performance optimized directly on the test set. While this approach produces feature sets that can be analyzed easily, algorithm performance may be grossly optimistic. To produce a more realistic performance estimate we re-ran our experiments with feature selection done using leave-one-out cross-validation (LOOCV) within the training set. This procedure was also repeated 3 times for each external test set. Due to the increased cost of the LOOCV procedure, we selected SFS, KS-test, T-test and also the nearest shrunken centroid algorithm for comparison. Results are presented in Figure 30.5. The LOOCV performance estimates are similar to performance optimized on test sets for the ovarian cancer. However, for the prostate cancer LOOCV performance is substantially lower.

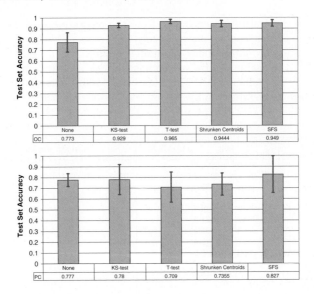

Fig. 30.5. Performance of Feature Extraction Algorithms optimized using LOOCV within the training set. The balanced accuracy is averaged over 3 test folds unseen during training. **Top:** ovarian cancer data set. **Bottom:** prostate cancer data set.

Table 30.3. Computational times, in CPU seconds, taken by each algorithm for the LOOCV feature selection.

CPU Time (sec)	None	KS-test	T-test	SFS	Shrunken Centroids
Ovarian	0.87	24.55	623.37	2175.75	33115.00
Prostate	1.31	25.4	639.14	3269.37	33115.00

The running times of each algorithm are presented in Table 30.3. Although the nearest centroid takes the greatest amount of time, the running time is dictated by the number of shrunken centroids examined during the LOOCV stage. Recall that decreasing d_j shrinks the class centroid (for each class). Hence the number of times we decrease d_j directly impacts performance. In our case we used 200 progressively shrunken centroid sets and picked the best one using LOOCV.

30.4 Conclusion

Mass spectrometry disease diagnosis is an emerging field poised to improve the quality of medical diagnosis. However, the large dimensionality of the data requires the use of feature extraction techniques prior to data mining and classification. This paper analyzed statistical and wrapper-based approaches to feature selection as well as dimensionality reduction via down-sampling. Experimental results indicate that down-sampling appears detrimental to classification performance, while feature selection techniques, in particular sequential forward selection coupled with a fast but simple nearest centroid classifier, can greatly reduce the dimensionality of the data and improve classification accuracy. Future research will investigate how the selected features impact classification accuracy when used in conjunction with more sophisticated classifiers, such as Artificial Neural Networks and Support Vector Machines. From a biological perspective, it is of interest to investigate the nature of the selected features. As potential biomarkers, these features may shed light on the cause or even the cure to cancer and other disease.

30.5 Acknowledgements

We would like to thank Lihong Li and Greg Lee for their various contributions. Ovarian and prostate cancer data sets provided by the National Cancer Institute, Clinical Proteomics Program Databank (Johann, 2003). Funding for this research was provided by University of Alberta, National Science and Engineering Research Council, and Alberta Ingenuity Center for Machine Learning.

Glossary

In this section we define the various measures used. Respectively, TP, TN, FP, FN, stand for the number of true positive, true negative, false positive, false negative samples at classification time.

Sensitivity $\frac{TP}{TP+FN}$ is also known as Recall.

Specificity $\frac{TN}{TN+FP}$

PPV (Positive Predictive Value) $\frac{TP}{TP+FP}$. is also known as Precision.

NPV (Negative Predictive Value) $\frac{TN}{TP+FP}$

Accuracy defined as $\frac{1}{2}(\frac{TP}{TP+FN} + \frac{TN}{TN+FP})$ in this paper.

References

B. Adam, Y. Qu, J. W. Davis, M. D. Ward, M. A. Clements, L. H. Cazares, O. J. Semmes, P. F. Schellhammer, Y. Yasui, Z. Feng, and Jr. G. L. Wright. Serum protein fingerprinting coupled with a pattern-matching algorithm distinguishes prostate cancer from benign prostate hyperplasia and healthy men. *Cancer Research*, 62(13):3609–3614, 2002.

T. P. Conrads, M. Zhou, E. F. Petricoin III, L. Liotta, and T. D. Veenstra. Cancer diagnosis using proteomic patterns. *Expert Reviews in Molecular Diagnostics*, 3 (4):411–420, 2003.

E. Diamandis. Proteomic patterns in biological fluinds: Do they represent the future of cancer diagnostics. *Clinical Chemistry (Point/CounterPoint)*, 48(8):1272–1278, 2003.

T. Hastie, R. Tibshirani, and J. Friedman. *The Elements of Statistical Learning*. Springer Series in Statistics. Springer Verlag, New York, 2001.

D. Johann. Clinical proteomics program databank. Technical report, National Cancer Institute, Center for Cancer Research, NCI-FDA Clinical Proteomics Program, 2003. http://ncifdaproteomics.com/ppatterns.php.

C. Kainz. Early detection and preoperative diagnosis of ovarian carcinoma (article in german). *Wien Med Wochenschr*, 146(1–2):2–7, 1996.

Michael Kirby. *Geometric Data Analysis: An Empirical Approach to Dimensionality Reduction and the Study of Patterns*. John Wiley & Sons, New York, 2001.

I. Levner. Proteomic pattern recognition. Technical report, University of Alberta, April 2004. No: TR04-10.

R.H. Lilien, H. Farid, and B. R. Donald. Probabilistic disease classification of expression-dependent proteomic data from mass spectrometry of human serum. *Computational Biology*, 10(6), 2003.

H. Park, M. Jeon, and J. B. Rosen. Lower dimensional representation of text data based on centroids and least squares. *BIT*, 43(2):1–22, 2003.

S. D. Patterson and R. H. Aebersold. Proteomics: The first decade and beyond. *Nature, Genetics Supplement*, 33:311–323, 2003.

E. F. Petricoin, A. M. Ardekani, B. A. Hitt, P. J. Levine, V. A. Fusaro, S. M. Steinberg, G. B. Mills, C. Simone, D. A. Fishman, E. C. Kohn, and L. A. Liotta. Use of proteomic patterns in serum to identify ovarian cancer. *The Lancet*, 359 (9306):572–577, 2002a.

E. F. Petricoin, D.K. Ornstein, C. P. Paweletz, A. Ardekani, P.S. Hackett, B. A. Hitt, A. Velassco, C.Trucco, L. Wiegand, K. Wood, C. Simone, P. J. Levine, W. M. Linehan, M. R. Emmert-Buck, S. M. Steinberg, E. C. Kohn, and L. A. Liotta. Serum preteomic patterns for detection of prostate cancer. *Journal of the National Cancer Institute*, 94(20):1576–1578, 2002b.

W.H. Press, S.A. Teukolsky, W.T. Vetterling, and B.P. Flannery. *Numerical Recipes in C: The Art of Scientifi Computing, Second Edition*. Cambridge University Press, 2002.

Y. Qu, B. Adam, Y. Yasui, M. D. Ward, L. H. Cazares, P. F. Schellhammer, Z. Feng, O. J. Semmes, and Jr. G. L. Wright. Boosted decision tree analysis of surface-enhanced laser desorption/ionization mass spectral serum profiles discriminates prostate cancer from noncancer patients. *Clinical Chemistry*, 48(10):1835–1843, 2002.

R. Tibshirani, T. Hastie, B. Narasimhan, and G. Chu. Class prediction by nearest shrunken centroids, with applications to dna microarrays. *Statistical Science*, 18 (1):104–117, 2003.

R. Tibshirani, T. Hastiey, B. Narasimhanz, S. Soltys, G. Shi, A. Koong, and Q. Le. Sample classifcation from protein mass spectrometry by 'peak probability contrasts'. *BioInformatics*, 2004.

B. Wu, T. Abbott, D. Fishman, W. McMurray, G. Mor, K. Stone, D. Ward, K. Williams, and H. Zhao. Comparison of statistical methods for classifcation of ovarian cancer using mass spectrometry data. *BioInformatics*, 19(13), 2003.

J. D. Wulfkuhle, L. A. Liotta, and E. F. Petricoin. Proteomic applications for the early detection of cancer. *Nature Reviews*, 3:267–275, 2003.

Chapter 31

Sequence Motifs: Highly Predictive Features of Protein Function

Asa Ben-Hur[1] and Douglas Brutlag[2]

[1] Department of Computer Science, Colorado State University
`asa@cs.colostate.edu`
[2] Department of Biochemistry, Stanford University `brutlag@stanford.edu`

Summary. Protein function prediction, i.e. classification of proteins according to their biological function, is an important task in bioinformatics. In this chapter, we illustrate that the presence of sequence motifs – elements that are conserved across different proteins – are highly discriminative features for predicting the function of a protein. This is in agreement with the biological thinking that considers motifs to be the building blocks of protein sequences. We focus on proteins annotated as enzymes, and show that despite the fact that motif composition is a very high dimensional representation of a sequence, that most classes of enzymes can be classified using a handful of motifs, yielding accurate and interpretable classifiers. The enzyme data falls into a large number of classes; we find that the one-against-the-rest multi-class method works better than the one-against-one method on this data.

31.1 Introduction

Advances in DNA sequencing are yielding a wealth of sequenced genomes. And yet, understanding the function of the proteins coded by a specific genome is still lagging. The determination of the function of genes and gene products is performed mainly on the basis of sequence similarity (homology) (Domingues and Lengauer, 2003). This leaves the function of a large percentage of genes undetermined: close to 40% of the known human genes do not have a functional classification by sequence similarity (Lander et al., 2001, Venter et al., 2001).

The most commonly used methods for measuring sequence similarity are the Smith-Waterman algorithm (Smith and Waterman, 1981), and BLAST (Altschul et al., 1997). These assign a similarity by aligning a pair of sequences. Other commonly used methods measure similarity to a family of proteins: PSI-BLAST, profiles, or HMM methods (Altschul et al., 1997, Gribskov et al., 1987, Sonnhammer et al., 1998). Motif methods on the other hand, represent short, highly conserved *regions* of proteins (Falquet et al., 2002, Nevill-Manning et al., 1998, Huang and Brutlag, 2001). Sequence motifs often correspond to functional regions of a protein – catalytic sites, binding sites,

A. Ben-Hur and D. Brutlag: *Sequence Motifs: Highly Predictive Features of Protein Function,*
StudFuzz **207**, 625–645 (2006)
`www.springerlink.com`

structural motifs etc .(Falquet et al., 2002). The presence of such protein motifs often reveals important clues to a protein's role even if it is not globally
similar to any known protein. The motifs for most catalytic sites and binding
sites are conserved over much larger taxonomic distances and evolutionary
time than the rest of the sequence. However, a single motif is often not sufficient to determine the function of a protein. The catalytic site or binding site
of a protein might be composed of several regions that are not contiguous in
sequence, but are close in the folded protein structure (for example, the catalytic site in serine proteases requires three conserved regions). In addition, a
motif representing a binding site might be common to several protein families
that bind the same substrate. Therefore, a pattern of motifs is required in
general to classify a protein into a certain family of proteins. Manually constructed fingerprints are provided by the PRINTS database (Attwood et al.,
2002). We suggest an automatic method for the construction of such fingerprints by representing a protein sequence in a feature space of motif counts,
and performing feature selection in this feature space. Our experiments show
that motifs are highly predictive of enzyme function; using feature selection
we find small sets of motifs that characterize each class of enzymes; classifiers trained on those feature sets have reduced error rates compared to SVM
classifiers trained on all the features. Representing protein sequences using a
"bag of motifs" representation is analogous to the bag of words representation used in text categorization. This type of approach was suggested in the
context of remote homology detection (Ben-Hur and Brutlag, 2003) (see also
the unpublished manuscript (Logan et al., 2001)).

Our work should be compared with several other approaches for protein
classification. Leslie and co-authors have focused on various flavors of kernels
that represent sequences in the space of k-mers, allowing gaps and mismatches;
these include the spectrum and mismatch kernels (Leslie et al., 2002a,b). The
k-mers used by these methods are analogous to the discrete motifs used here.
k-mers are less flexible than motifs, but can provide a result in cases when a
sequence does not contain known motifs. When it comes to remote homology
detection, discriminative approaches based on these kernels and kernels based
on HMM models of sequence families (Fisher kernels) (Jaakkola and Haussler,
1999, Jaakkola et al., 1999) yield state of the art performance.

An alternative approach is to represent a sequence by a set of high-level
descriptors such as amino acid counts (1-mers), predicted secondary structure content, molecular weight, average hydrophobicity, as well as annotations of the sequence that document its cellular location, tissue specificity etc.
(Syed and Yona, 2003, des Jardins et al., 1997). These approaches are complementary to sequence-similarity based approaches such as our motif-based
approach.

SVMs are typically used for multi-class problems with either the one-
against-the-rest or one-against-one methods (Schölkopf and Smola, 2002). The
large number of classes in the data considered in this chapter makes the one-
against-one method infeasible. Moreover, we find that the accuracy of the

one-against-the rest method is better, which we attribute to the large number of classes. Other studies of multi-class classification using SVMs (see (Rifkin and Klautau, 2004) and references therein) have not addressed datasets with such a large number of classes.

In this chapter we consider the problem of classifying proteins according to their enzymatic activity using a motif-based representation. In Section 31.2 we introduce the Enzyme Commission (EC) numbering system used to classify enzymes. In Section 31.3 we describe sequence motifs and the classification and feature selection methods used in this chapter. Finally, we show results of these methods, illustrating that SVM-based feature selection methods yield accurate low dimensional predictors of enzyme function.

31.2 Enzyme Classification

Enzymes represent about a third of the proteins in the Swiss-Prot database (O'Donovan et al., 2002), and have a well established system of annotation. The function of an enzyme is specified by a name given to it by the Enzyme Commission (EC) (of the International Union of Biochemistry and (NC-IUBMB), 1992). The name corresponds to an *EC number*, which is of the form: n1.n2.n3.n4, e.g. 1.1.3.13 for alcohol oxidase. The first number is between 1 and 6, and indicates the general type of chemical reaction catalyzed by the enzyme; the main categories are oxidoreductases, transferases, hydrolases, lyases, isomerases and ligases. The remaining numbers have meanings that are particular to each category. Consider for example, the oxidoreductases (EC number starting with 1), which involve reactions in which hydrogen or oxygen atoms or electrons are transferred between molecules. In these enzymes, n2 specifies the chemical group of the (electron) donor molecule, n3 specifies the (electron) acceptor, and n4 specifies the substrate. The EC classification system specifies over 750 enzyme names; a particular protein can have several enzymatic activities. Therefore, at first glance, this is not a standard multi-class problem, since each pattern can have more than one class label; this type of problem is sometimes called a *multi-label* problem (Elisseeff and Weston, 2001). In order to reduce this multi-label problem into a multi-class problem consider the biological scenarios in which an enzyme has multiple functions:

(1.) The enzyme can catalyze different reactions using the same catalytic site.
(2.) The enzyme is a multi-enzyme, an enzyme with multiple catalytic functions that are contributed by distinct subunits/domains of the protein (McNaught and Wilkinson, 1997).

In both cases it is reasonable to consider an enzyme that catalyzes more than one reaction as distinct from enzymes that catalyze only one reaction. This is clear for multi-enzymes; in the other case, a catalytic site that can catalyze more than one reaction might have different sequence characteristics than a catalytic site that only catalyzes one of the reactions. We found that this multi-label problem can be reduced to a regular multi-class problem by considering a group of enzyme that have several activities as a class by itself; for example, there are 22 enzymes that have EC numbers `1.1.1.1` and `1.2.1.1`, and these can be perfectly distinguished from enzymes with the single EC number `1.1.1.1` using a classifier that uses the motif composition of the proteins. When looking at annotations in the Swiss-Prot database we then found that these two groups are indeed recognized as distinct.

31.3 Methods

We propose to use the motif composition of a protein to define a similarity measure or *kernel* function that can be used with various kernel based classification methods such as Support Vector Machines (SVMs).

31.3.1 The Discrete Motif Composition Kernel

In this chapter we use discrete sequence motifs extracted using the eMOTIF method (Nevill-Manning et al., 1998, Huang and Brutlag, 2001), which is described here briefly. A motif is a simple regular expression specifying the allowed amino acids in each position of the motif. Consider for example the motif [as].dkf[filmv]..[filmv]...l[ast]. A sequence matches (or contains) this motif if it has either an a or an s in some position, followed by any amino acid, then d, k, f and so on, matching until the end of the motif. A group of amino acids in brackets is called a *substitution group*. A formal definition is as follows:

Definition 1. *Denote by $\mathcal{A}$ the alphabet of amino acids. A substitution group $S = \{s_1, \ldots, s_k\}$ is a subset of $\mathcal{A}$, written as $[s_1 \ldots s_k]$. Let $\bar{S}$ be a set of substitution groups, and let '.' denote the wildcard character.*
A motif m is a sequence over $\mathcal{A} \cup \bar{S} \cup \{.\}$.
A sequence $s = s_1 s_2 \ldots s_{|s|} \in \mathcal{A}^*$ is said to contain *a motif m at position i if for $j = 1, \ldots, |m|$, if $m_j \in \mathcal{A}$ then $s_{i+j-1} = m_j$; if m_j is a substitution group S then $s_{i+j-1} \in S$; if m_j is the wildcard character, then s_{i+j-1} can be any character. A sequence s contains a motif m, if s contains m at some position.*

Protein sequence motifs are typically extracted from ungapped regions (blocks) of a multiple sequence alignment (see Figure 31.1 for an illustration of the process). Each position in the motif represents the variability in a column of the block. A substitution group such as [filmv] denotes the appearance of several amino acids in a particular column in a block. Motifs generated by the eMOTIF method contain only a limited number of substitution groups that reflect chemical and physical properties of amino acids and their tendency to co-occur in multiple sequence alignments. If the pattern of amino acids that appear in a column of a block does not match any substitution group, then the motif contains the wildcard symbol, '.'.

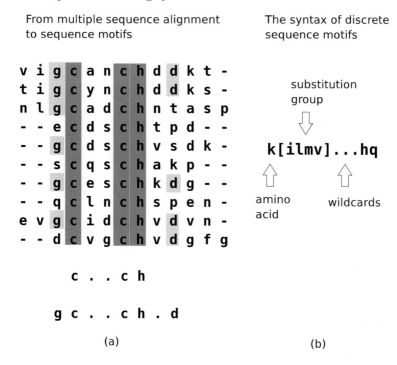

Fig. 31.1. (a) The pipeline from multiple sequence alignment to the construction of sequence motifs: we use discrete motifs that are simple regular expressions that represent the variability in conserved columns in ungapped regions of a multiple sequence alignment. (b) The syntax of a discrete motif: each position in the motif is either an amino acid, a substitution group (a set of amino acids) or the wildcard symbol. Our motif databases use only a limited set of substitution groups; substitution groups are sets of amino acids that tend to substitute for each other in column of a multiple sequence alignment. Considering a limited set of substitution groups helps avoid overfitting.

Motifs can often be associated with specific functional sites of a protein: catalytic sites, DNA binding sites, protein-protein interactions sites, small molecule binding sites etc. We give a few examples that illustrate this in the context of the enzyme data.

Example 1. The motif k[kr][iv]a[iv][iv]g.g.sgl..[ilv][kr] appears in 19 out of 19 enzymes belonging to the enzyme class 1.14.13.11. It characterizes a binding site for an FAD molecule.

Example 2. In many cases a binding site motif is not specific to one class of enzymes, but characterizes a similar functional site in several enzyme classes that constitute a broader family of proteins: The motif dp.f...h.....[ilmv]... [fwy] has 57 hits in the Swiss-Prot database, and is specific to the EC classes 1.14.18.1, 1.10.3.1 and 5.3.3.12. The histidine residue, represented by the h in the pattern, binds one of two copper atoms that act as co-factors for these enzymes.

We are currently undertaking the task of automatically characterizing the function of motifs in our database based on annotations available in the Swiss-Prot database.

A sequence s can be represented in a vector space indexed by a set of motifs $\mathcal{M}$:

$$\Phi(s) = (\phi_m(s))_{m \in \mathcal{M}}, \tag{31.1}$$

where $\phi_m(s)$ is the number of occurrences of the motif m in s. Now define the *motif kernel* as:

$$K(s, s') = \Phi(s) \cdot \Phi(s'). \tag{31.2}$$

Since in most cases a motif appears only once in a sequence, this kernel essentially counts the number of motifs that are common to both sequences. The computation of the kernel can be performed efficiently by representing the motif database in a TRIE structure: Let m be a motif over the alphabet $\mathcal{A} \cup \bar{\mathcal{S}} \cup \{.\}$. Every prefix of m has a node; let m_1 and m_2 be prefixes of m; there is an edge from m_1 to m_2 if $|m_2| = |m_1| + 1$. The motifs are stored in the leaf nodes of the TRIE. To find all motifs that are contained in a sequence x at a certain position, traverse the TRIE using DFS and record all the leaf nodes encountered during the traversal (see Figure 31.2 for an illustration).

To find all motifs that are contained in a sequence s at any position, this search is started at each position of s. Thus the computation time of the motif content of a sequence is linear in its length. Unlike a standard TRIE searching the motif TRIE has a complexity that depends on the size of the database; this is the result of the presence of wildcards and substitution groups. A trivial upper bound is linear in the size of the database; numerical experiments indicate that the complexity is sub-linear in practice.

The motif kernel is analogous to the "bag of words" representation that is commonly used in information retrieval, where a document is represented as a vector of (weighted) counts of the number of occurrences of each word in the document (Joachims, 2002, 1998). In a recent study we found that this "bag of motifs" representation of a protein sequence provides state of the art performance in detecting remote homologs (Ben-Hur and Brutlag, 2003). Like the bag of words representation, our motif composition vector is both high dimensional and sparse: the eBlocks database of motifs (Su et al., 2004) used in this work contains close to 500,000 motifs, while a sequence typically contains only a handful of conserved regions. Motifs are often very specific as features: We found that using feature selection we could reduce the number of motifs to a few tens at the most, while maintaining classification accuracy (see Section 31.4 for details).

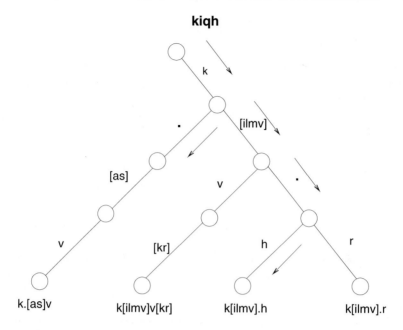

Fig. 31.2. Motifs are stored in the leaves of a TRIE. The figure shows a TRIE storing the motifs k.[as]v, k[ilmv]v[kr], k[ilmv].h and k[ilmv].r. To find the motif content, the tree is traversed, matching at each position a letter from the sequence with the same letter, a substitution group containing the letter, or the wildcard symbol. Traversing the tree shows that the sequence kiqh contains the motif k[ilmv].h.

31.3.2 The PSSM Kernel

An alternative way of representing the pattern of conservation in the columns of an ungapped block from a multiple sequence alignment is by Position Specific Scoring Matrices (PSSMs). A PSSM is a matrix with a column for each column of the block, and a row for each amino acid. A column in the PSSM represents the frequency of each amino acid in a column in the block. The raw score of a sequence with respect to a PSSM is the product of the entries that correspond to the sequence. The raw score can then be converted to a p-value or an E-value that reflects how likely that score to arise by chance, or to be observed in a database of a given size. Given a database of PSSMs we can represent a protein sequence by a vector where each component is the score with respect to a given scoring matrix. The score we used is the negative of the logarithm of the p-value, keeping only entries whose p-value is better than 10^{-6}. This keeps the representation sparse, as for the motif kernel.

While PSSMs capture more information than a given motif about the sequence variability in a block, it is much more time consuming to compute the PSSM composition vector, since each scoring matrix needs to be considered, whereas in the case of the discrete motifs, the computation time is sub-linear in the number of motifs because of the way they are represented in a TRIE. In our experiments we use PSSMs constructed using the eMATRIX method, that also provide an efficient method for scoring a PSSM, by avoiding scanning the complete PSSM if its score is unlikely to exceed the significance threshold.

31.3.3 Classification Methods

In what follows, we assume that our data are vectors $\mathbf{x}_i$ representing the motif content of the input sequences. In this chapter, we report results using two classification methods: SVMs and k-Nearest-Neighbors (kNN). A linear SVM is a two-class classifiers with a decision function of the form

$$f(\mathbf{x}) = \mathbf{w} \cdot \mathbf{x} + b, \qquad (31.3)$$

where $\mathbf{w}$ is a weight vector, and b is a constant, and a pattern $\mathbf{x}$ is classified according to the sign of $f(\mathbf{x})$. The vector $\mathbf{w}$ and the bias, b, are chosen to maximize the margin between the decision surface (hyperplane) and the positive examples on one side, and negative examples on the other side, in the case of linearly separable data; in the case of non-separable data some slack is introduced (Boser et al., 1992, Schölkopf and Smola, 2002, Cristianini and Shawe-Taylor, 2000). As a consequence of the optimization process, the weight vector can be expressed as a weighted sum of the Support Vectors (SV):

$$\mathbf{w} = \sum_{i \in SV} \beta_i \mathbf{x}_i . \qquad (31.4)$$

The decision function is now written as:

$$f(\mathbf{x}) = \sum_{i \in SV} \beta_i \mathbf{x}_i \cdot \mathbf{x} + b . \qquad (31.5)$$

To extend the usefulness of SVMs to include nonlinear decision functions, and non-vector data one proceeds by mapping the data into a feature space, typically high dimensional, using a map Φ, and then considering a linear SVM in the high dimensional feature space (Schölkopf and Smola, 2002, Cristianini and Shawe-Taylor, 2000). Since the SVM optimization problem can be expressed in terms of dot products, this approach is practical if the so called *kernel function*, $K(\mathbf{x}, \mathbf{x}') = \Phi(\mathbf{x}) \cdot \Phi(\mathbf{x}')$, can be computed efficiently. In terms of the kernel function, the decision function is expressed as:

$$f(\mathbf{x}) = \sum_{i \in SV} \beta_i K(\mathbf{x}_i, \mathbf{x}) + b . \qquad (31.6)$$

A kNN classifier classifies a pattern according to the class label of the training set patterns that are most similar to it. We use a kNN classifier with a continuous valued decision function that assigns a score for class j defined by:

$$f_j(\mathbf{x}) = \sum_{i \in kNN_j(\mathbf{x})} K(\mathbf{x}_i, \mathbf{x}), \qquad (31.7)$$

where $kNN_j(\mathbf{x})$ is the set of k nearest neighbors of $\mathbf{x}$ in class j; a pattern $\mathbf{x}$ is classified to the highest scoring class.

31.3.4 Feature Scoring and Selection

In order to show that motifs are highly predictive of the class of an enzyme we compute for each motif feature the following statistics. The Positive-Predictive Value (PPV) of a feature is the fraction of the predictions made on the basis of a motif m that are correct, namely

$$\mathrm{ppv}(m) = \frac{\mathrm{count}(m|C)}{\mathrm{count}(m)}, \qquad (31.8)$$

where $\mathrm{count}(m)$ ($\mathrm{count}(m|C)$) is the number of occurrences of the motif m (in class C). Note that this is referred to as *precision* in information retrieval. A motif has PPV which is equal to 1 in a class C if it occurs only in proteins from class C. On the other end of the spectrum we consider the sensitivity (or *recall* in information retrieval terms) of a motif m in picking members of a class C:

$$\mathrm{sens}(m) = \frac{\mathrm{count}(m|C)}{|C|}, \qquad (31.9)$$

where $|C|$ is the size (number of members) of class C.

The motifs in the database we use are often highly redundant in their pattern of occurrence in a group of proteins. Feature selection methods that are based on ranking individual features do not handle redundancy, and are therefore not suitable for producing a small subset of features without an additional filter for redundancy.

In this chapter we focus on SVM-based feature selection methods and show their effectiveness. Recall that the weight vector of an SVM, $\mathbf{w}$, is a weighted sum of a subset of the motif composition vectors (the support vectors). In most cases the number of support-vectors was rather small when training a classifier to distinguish one enzyme class from all the others, so the weight vector is typically very sparse; discarding features that are not represented in the weight vector already yields a significant reduction in the number of features, without modifying the decision function. The idea of using the magnitude of the weight vector to perform feature selection is implemented in the Recursive Feature Elimination (RFE) method (Guyon et al., 2002), which alternates between training an SVM and discarding a subset of the features with small components of the weight vector. A backward selection method such as RFE requires a halting condition. A halting condition based on cross-validation is expensive to compute; furthermore, using a validation set is not always practical for our data, since many classes have few positive examples (not more than 10). Therefore we use a halting condition that is based on a simple bound on classifier error, namely the number of support vectors. We observe the following behavior of the number of support vectors in successive iterations of RFE: initially, most of the features removed are noise; when these are eliminated the data is simpler to describe, requiring less support vectors. At a later stage essential features are removed, making the features insufficient to describe the data, so many data points will be misclassified, making them bounded support vectors. We choose the smallest number of features for which the number of support vectors is minimal.

A related method to RFE is the zero-norm method of Weston *et al.* (Weston et al., 2003). They formulate the feature selection problem as a search for the smallest set of features such that a dataset is still linearly separable (with slack variables added for the case of data that is not linearly separable). In other words, minimizing the zero norm of the weight vector of a linear decision boundary, subject to the constraints that the decision boundary separates the two classes (the zero norm of a vector is its number of nonzero coefficients). They show that this difficult combinatorial problem can be relaxed into a problem that is solved by an algorithm similar to RFE: alternate between training an SVM and multiplying the data by the absolute value of the weight vector (feature i of each pattern is multiplied by $|w_i|$). This is iterated until convergence.

In general, minimizing the zero norm might not be an optimal strategy: the method minimizes the number of variables that separate the two classes, without considering the margin of the separation. However, the data underlying the motif kernel is discrete; therefore, if a set of motifs separates the data, it does so with large margin. This can explain the good performance obtained with this method on the motif data: the accuracy of classifiers trained on features selected using the zero-norm method is higher than that of classifiers trained on all the features. For comparison we tested the zero-norm method on the problem of predicting protein function on the basis of gene expression data (we used the data analyzed in (Brown et al., 2000) and considered the five functional classes shown to be predictable using SVMs). In this case of continuous data in 79 dimensions, SVMs trained on all features outperformed SVMs trained using the zero-norm method (data not shown).

31.3.5 Multi-Class Classification

Protein function prediction is a classification problem with a large number of classes (hundreds of classes in the EC classification scheme alone). This poses a computational challenge when using a two-class classifier such as SVM. The two standard approaches for using a two-class classifier for multi-class data are the one-against-one method and one-against-the-rest method (Schölkopf and Smola, 2002). The one-against-the-rest method trains c classifiers, where

c is the number of classes in the data, and classifier i is trained on class i against the rest of the data. A test example is then classified to the class that receives the highest value of the decision function. The one-against-one method requires training $c(c-1)/2$ classifiers on all pairs of classes; an unseen pattern is tested by all these classifiers and is classified to the class that receives the highest number of votes. In our case this amounts to training 651 * 650/2 = 2,111,575 classifiers, which makes this too computationally intensive. Moreover, the data in Table 31.3 shows that the one-against-the-rest method works better on the enzyme function prediction task. It can be argued that the large number of "irrelevant" tests performed by the one-against-one method may be the cause of this.

A recent paper by Rifkin and Klautau (Rifkin and Klautau, 2004) argues that the one-against-the-rest method should work as well as other multi-class methods, and presents experimental results to support their arguments, including a critical analysis of previous studies. The datasets they considered are UCI datasets that have a small number of classes compared to the enzyme data. Yeang *et al.* (Yeang et al., 2001) studied a gene expression dataset with 14 classes corresponding to patients with various types of cancer; they also obtained higher accuracy with one-against-the-rest than with one-against-one. Our results further support their findings in the case of a multi-class problem with a much larger number of classes.

31.3.6 Assessing Classifier Performance

In some of our analyses we will consider two-class problems that are highly unbalanced, i.e. one class is much larger than the other; in such cases the standard error rate is not a good measure of classifier performance. Therefore we consider two alternative metrics for assessing the performance of a classifier: the area under the Receiver Operator Characteristic (ROC) curve (Egan, 1975), and the balanced success rate. The balanced success rate is:

$$1 - \sum_i P(err|C_i) , \qquad (31.10)$$

where $P(err|C)$ is a shorthand for a classifier's error on patterns that belong to class C. The ROC curve describes the trade-off between sensitivity and specificity; it is a plot of the true positive rate as a function of the false positive rate for varying classification thresholds (Egan, 1975). The area under the ROC curve (AUC) is commonly used to summarize the ROC curve. The AUC is a measure of how well the classifier works at *ranking* patterns: it quantifies the extent to which positive examples are ranked above the negative examples. The AUC is a useful metric for assessing a classifier used in the context of protein classification: a user will typically be interested in the most promising patterns, i.e. patterns that are most likely to belong to a class of interest. The AUC score however, can be problematic for highly unbalanced data: one can obtain a very high AUC even when the ranking produced by the classifier is almost useless from the point of view of the user if for example 500 out of 30,000 patterns from the negative class are ranked above the real members of the class. Therefore we consider the ROC50 curve, which counts true positives only up to the first 50 false positives (Leslie et al., 2002b). A classifier that correctly classifies all the data has an ROC50 score (AUC50) equal to 1, while if the top 50 values of the decision function are false positives, the AUC50 is 0.

31.4 Results

We extracted protein sequences annotated with EC numbers from the Swiss-Prot database Release 40.0 (O'Donovan et al., 2002). EC numbers were taken from the description lines; we removed sequence fragments, and sequences where the assigned EC number was designated as "putative" or assigned by homology. Sequences with an incompletely specified EC number were discarded as well. Enzyme classes with a small number of representatives (less than 10) were not considered in our analysis. The resulting dataset has 31117 enzymes in 651 classes. In some cases we focus on oxidoreductases – enzymes that have an EC number starting with 1. The statistics of the two datasets are summarized in Table 1.

Table 31.1. The enzyme sequence data; oxidoreductases are enzymes with EC number that starts with 1.

	number of sequences	number of classes	number of motifs
Oxidoreductases	5911	129	59783
All enzymes	31117	651	178450

In order to illustrate that the eBLOCKS database (Su et al., 2004) contains many motifs that are predictive of enzyme function we consider their positive predictive value (PPV) and sensitivity in picking members of each enzyme class. Out of the 651 enzyme classes, 600 classes had a motif that was perfectly specific to that class, i.e. had a PPV equal to 1. To see the sensitivity of such perfectly specific motifs, for each class we find the set of motifs with maximum PPV and find the one with maximum sensitivity. The distribution of the sensitivity of these motifs is shown in Figure 31.3. We observe that 89 enzyme classes have a motif that covers all its proteins and has no hits outside the class. In general we do not expect to find motifs that cover all members of an enzyme class, since it might be heterogeneous, composed of several clusters in sequence space. We considered aggregating motifs with perfect PPV to form predictors of EC classes, but only a limited number of classes had sets of

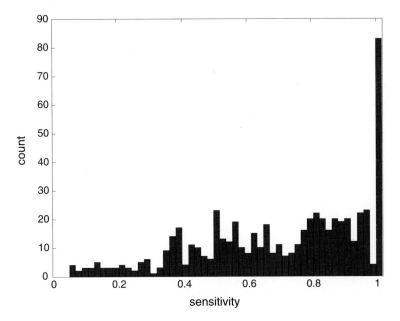

Fig. 31.3. The motif database contains motifs whose occurrences are highly correlated with the EC classes. For each class we computed the highest PPV, and the
highest sensitivity of a motif with that value of PPV. 600 out of 651 classes had
a motif with PPV equal to 1. The distribution of the sensitivity of these motifs is
shown. There are 89 EC classes that have a "perfect motif": a motif that cover all
enzymes of the class, and appears only in that class, i.e. the class can be predicted
on the basis of a single motif.

motifs that cover the entire class, so a less strict form of feature selection is
required.

Next, we report experiments using discrete motifs and PSSMs as features
for predicting the EC number of an enzyme. In these experiments we used
the PyML package (see section 31.4.3). We used a linear kernel in motif space
in view of its high dimensionality. SVM performance was not affected by
normalizing the patterns to unit vectors, but was critical for the kNN classifier.
On other datasets we observed that classification accuracy did not vary much
when changing the SVM soft margin constant, so we kept it at its default
value. All the reported results are obtained using 5-fold cross-validation, with
the same split used in all cases. In order to reduce the computational effort
in comparing multiple approaches we focus on enzymes whose EC number
starts with 1 (oxidoreductases); this yields a dataset with 129 classes and
5911 enzymes. We compare classifiers trained on all 129 one-against-the-rest
problems. Results are found in Figure 31.4 that shows the number of classes
which have a given level of performance for two metrics: the AUC50 (area
under ROC50 curve) and the balanced success rate.

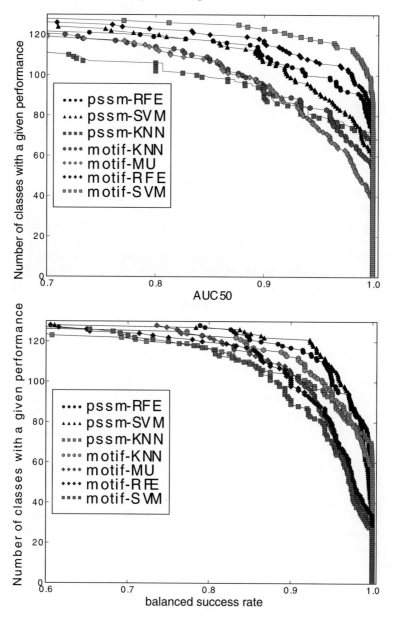

Fig. 31.4. Performance in 5 fold cross-validation for the 129 enzyme classes in the oxidoreductase data. The number of classes with a given level of performance is plotted for the various methods: Top: Area under ROC50 curve (AUC50). Bottom: balanced success rate.

The results in Figure 31.4 show that in most cases feature selection with RFE or the zero-norm method lead to performance that is at least as good as that of an SVM that uses all features for both the motif and PSSM kernels. For the motif kernel we see a significant improvement with respect to the balanced success rate under feature selection, but a degradation in performance performance with respect to the AUC50 metric. This may be explained by the fact that the feature selection process is governed by the objective of the underlying SVM, which is good performance under the balanced success rate (since we account for the class imbalance by introducing a misclassification cost that is inversely proportional to the class size); optimizing the balanced success rate can then come at the expense of the AUC50. The merit of the RFE feature selection halting condition based on the number of support vectors is illustrated in Figure 31.5, that shows the typical behavior of performance as features are eliminated. We note that the RFE method performed better than the zero-norm method. We attribute this to the difference in the halting conditions used, and the resulting number of features — the zero-norm method yielded 10 features on average, whereas RFE yielded 77 features on average over the 129 classes. Using a polynomial kernel after the feature selection stage using RFE yielded worse results than using a linear kernel (data not shown).

The kNN classifier works well, outperforming the motif-SVM with respect to the balanced success rate. The fact that kNN works so well despite the high dimensionality of the data is the result of the presence of highly informative features, coupled with the sparsity of the data and the discreteness of the representation. When using the PSSM kernel kNN no longer performs as well as the SVM method; in this case the data is still highly sparse, but the magnitude of the feature plays an important part. The SVM's better performance with respect to the AUC50 metric can be explained by the fact that SVM training explicitly optimizes the decision function while for the kNN classifier a continuous valued decision function is an after-thought, rather than an integral part of the design.

We also ran an experiment with a filter method that ranks a feature m according to abs $\left(\frac{\text{count}(m|C)}{|C|} - \frac{\text{count}(m|\bar{C})}{|\bar{C}|} \right)$, where $\bar{C}$ is the set of patterns outside of class C. Features whose score was less two standard deviations above the average score obtained on randomly labeled datasets were discarded. Due to the high redundancy of the motif features, the method is not successful in reducing the dimensionality in a significant way – over 5000 features on average over the 129 enzyme classes were chosen, and the performance was almost identical to an SVM trained on all the features.

31.4.1 Enzymes with Multiple Functions

In our analysis we considered a set of enzymes with multiple functionalities as a unique class; for example enzymes with EC numbers 1.1.1.1 and 1.2.1.1 were considered a distinct class. The data contains 27 classes with multiple functionalities. In order to quantify the degree of success in predicting a

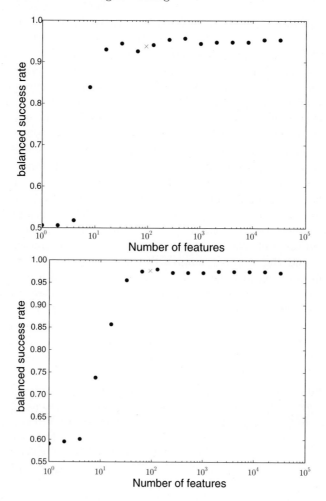

Fig. 31.5. Performance as a function of the number of features with RFE feature selection. We illustrate the dependence of the balanced success rate on the number of features in two typical runs of the method on the classes 1.9.3.1 (top) and 1.6.5.3 (bottom). Red X's denote the number of features chosen according to the number of support vectors. In each case 70% of the members of the class were used for training and 30% were used for testing.

class with multiple functionalities, for each class with multiple functionalities we assessed the accuracy of a classifier trained to distinguish between the multiple-functionality class and the classes with which it shares a function. The average balanced success rate in this experiment was 0.95, and the average AUC50 was 0.95. These results support our reduction of the multi-label problem to a multi-class problem.

31.4.2 Multi-Class Classification

The results of multi-class experiments appear in Table 31.2. The BLAST method assigns the class label according to the class of the enzyme with which an input sequence has the best BLAST E-value (a nearest neighbor BLAST). The one-against-the-rest motif-SVM method worked slightly better than the BLAST-based method and better than the nearest-neighbor motif method. When using the nearest neighbor motif method normalization of the kernel was critical: normalizing the kernel by dividing each pattern by its L_1-norm improved the results significantly. Most of the misclassifications in the case of the non-normalized kernel occurred by classifying a pattern into a class of large proteins that contain many motifs; such classes "attract" members of other classes. It is interesting to note that the PSSM based kernel performed poorly when used for the multi-class classification, despite being better for individual two-class problems. We attribute that to the fact the PSSM kernel is sensitive to the value of a variable as opposed to the motif kernel that is discrete. This results in decision functions that are incommensurate across classes; this is supported by the observation that normalizing the output of each classifier into a probability using Platt's method (Platt, 1999) provided big improvement, although it is still worse than the motif kernel.

Table 31.2. Success rate in Multi-class classification of the enzyme data, estimated by 5-fold CV. The standard deviation in 10 repeats of the experiment was 0.002 and below. The kernel used with the kNN classifier was normalized by dividing each motif composition vector by its L_1 norm. The pssm-Platt-SVM method converts the SVM decision function values into probabilities

Method	success rate	balanced success rate
motif-kNN-L1	0.94	0.92
motif-SVM	0.96	0.94
pssm-SVM	0.84	0.70
pssm-Platt-SVM	0.89	0.82
BLAST	0.96	0.93

A comparison of the performance of multi-class methods that use the motif kernel is provided in Table 31.3 for an increasing number of classes. It shows an advantage for the one-against-the-rest method that increases as the number of classes increases. The better performance of the one-against-the-rest method may be explained by the large number of "irrelevant" comparisons made by the one-against-one method.

Table 31.3. Success rate in multi-class classification measured using 5-fold cross-validation for the motif-SVM method when varying the number of classes.

Number of classes	one-against-rest	one-against-one
10	0.99	0.97
20	0.98	0.96
40	0.98	0.96
60	0.98	0.94
80	0.98	0.95

31.4.3 Data and Software

A license for the motif database used in this work is freely available for academic users; see `http://motif.stanford.edu`. The machine learning experiments were performed using PyML, which is an object oriented environment for performing machine learning experiments, available at `http://pyml.sourceforge.net`.

31.5 Discussion

Several databases of conserved regions reviewed in the introduction are constructed by experts who use known annotations to group protein sequences that are then modeled by motifs, profiles, or HMMs, namely PROSITE, BLOCKS+ and Pfam (Falquet et al., 2002, Henikoff et al., 1999, Sonnhammer et al., 1998). The use of such patterns as features to train classifiers that predict protein function can lead to biased results since knowledge about function is often incorporated by the experts in the course of developing these databases. The only way to avoid bias in evaluating classifier performance is to use as testing examples proteins that were not used in the development of the database, which can be difficult to achieve. The eBLOCKs database used in this study, on the other hand, is constructed in an unsupervised way by aligning clusters of similar sequences in Swiss-Prot (Su et al., 2004), so our results are free from such bias.

Although results of classification using motifs did not offer a significant advantage over BLAST in terms of accuracy, our examples suggest that motifs can offer greater interpretability. Since manually curating the function of motifs is infeasible, we are working on automating the process to produce annotations for as many motifs as possible using sequence annotations available in the Swiss-Prot database.

SVM-based multi-class methods for protein function prediction are expensive to train in view of the large number of classes. The data is very sparse in sequence space: most classes are well separated. One can take advantage of this property in many ways. We performed experiments using a method that uses one-class SVMs to filter a small number of "candidate" classes, and then deciding among that smaller number of classes using a one-against-one approach. Since the number of classes that are not well separated is small, this resulted in the need to train only a small number of classifiers, with accuracy similar to the one-against-the-rest method.

In this work we represented conserved regions by either discrete motifs or PSSMs. There was no clear winner in terms of accuracy: PSSMs gave better performance with respect to the balanced success rate whereas motifs gave better AUC50 scores. So although PSSMs offer greater flexibility in describing a pattern of conservation, the fact that the eMOTIF method generates multiple motifs out of a conserved sequence block, compensates for the loss in expressive power of a single motif. The advantage of using discrete motifs over PSSMs is the efficient search methods for computing motif hits. Using a hybrid approach – choosing a small number of features, be it PSSMs or motifs, one can avoid the computational burden.

31.6 Conclusion

In this chapter we have illustrated that the motif composition of a sequence is a very "clean" representation of a protein; since a motif compactly captures the features from a sequence that are essential for its function, we could obtain accurate classifiers for predicting enzyme function using a small number of motifs. We plan to develop the motif-based classifiers as a useful resource that can help in understanding protein function.

References

S.F. Altschul, T.L. Madden, A.A. Schaffer, J. Zhang, Z. Zhang, W. Miller, and D. J. Lipman. Gapped BLAST and PSI-BLAST: A new generation of protein database search programs. *Nucleic Acids Research*, 25:3389–3402, 1997.

T.K. Attwood, M. Blythe, D.R. Flower, A. Gaulton, J.E. Mabey, N. Maudling, L. McGregor, A. Mitchell, G. Moulton, K. Paine, and P. Scordis. PRINTS and PRINTS-S shed light on protein ancestry. *Nucleic Acids Research*, 30(1):239–241, 2002.

A. Ben-Hur and D. Brutlag. Remote homology detection: A motif based approach. In *Proceedings, eleventh international conference on intelligent systems for molecular biology*, volume 19 suppl 1 of *Bioinformatics*, pages i26–i33, 2003.

B. E. Boser, I. M. Guyon, and V. N. Vapnik. A training algorithm for optimal margin classifiers. In D. Haussler, editor, *5th Annual ACM Workshop on COLT*, pages 144–152, Pittsburgh, PA, 1992. ACM Press. URL http://www.clopinet.com/isabelle/Papers/colt92.ps.

Michael P. S. Brown, William Noble Grundy, David Lin, Nello Cristianini, Charles Walsh Sugnet, Terence S. Furey, Jr. Manuel Ares, and David Haussler. Knowledge-based analysis of microarray gene expression data by using support vector machines. *Proc. Natl. Acad. Sci. USA*, 97:262–267, 2000.

N. Cristianini and J. Shawe-Taylor. *An Introduction to Support Vector Machines.* Cambridge UP, 2000.

M. des Jardins, P.D. Karp, M. Krummenacker, T.J. Lee, and C.A. Ouzounis. Prediction of enzyme classification from protein sequence without the use of sequence similarity. In *Intelligent Systems for Molecular Biology*, pages 92–99, 1997.

F.S. Domingues and T. Lengauer. Protein function from sequence and structure. *Applied Bioinformatics*, 2(1):3–12, 2003.

J.P. Egan. *Signal detection theory and ROC analysis.* Series in Cognition and Perception. Academic Press, New York, 1975.

A. Elisseeff and J. Weston. A kernel method for multi-labelled classification. In *Advances in Neural Information Processing Systems*, 2001.

L. Falquet, M. Pagni, P. Bucher, N. Hulo, C.J. Sigrist, K. Hofmann, and A. Bairoch. The PROSITE database, its status in 2002. *Nucliec Acids Research*, 30:235–238, 2002.

M. Gribskov, A.D. McLachlan, and D. Eisenberg. Profile analysis: Dectection of distantly related proteins. *Proc. Natl. Acad. Sci. USA*, 84:4355–4358, 1987.

I. Guyon, J. Weston, S. Barnhill, and V. Vapnik. Gene selection for cancer classification using support vector machines. *Machine Learning*, 46:389–422, 2002.

S. Henikoff, J.G. Henikoff, and S. Pietrokovski. Blocks+: A non-redundant database of protein alignment blocks derived from multiple compilations. *Bioinformatics*, 15(6):471–479, 1999.

J.Y. Huang and D.L. Brutlag. The eMOTIF database. *Nucleic Acids Research*, 29 (1):202–204, 2001.

T. Jaakkola, M. Diekhans, and D. Haussler. Using the Fisher kernel method to detect remote protein homologies. In *Proceedings of the Seventh International Conference on Intelligent Systems for Molecular Biology*, pages 149–158, 1999.

T.S. Jaakkola and D. Haussler. Exploiting generative models in discriminative classifiers. In *Advances in Neural Information Processing Systems 11*, 1999. URL http://www.cse.ucsc.edu/research/ml/papers/Jaakola.ps.

T. Joachims. Text categorization with support vector machines: Learning with many relevant features. In Claire Nédellec and Céline Rouveirol, editors, *Proceedings of the European Conference on Machine Learning*, pages 137–142, Berlin, 1998. Springer. URL http://www-ai.cs.uni-dortmund.de/DOKUMENTE/joachims_98a.ps.gz.

T. Joachims. *Learning to Classify Text using Support Vector Machines.* Kluwer Academic Publishers, 2002.

E.S. Lander, L.M. Linton, and B. Birren. Initial sequencing and analysis of the human genome. *Nature*, 409(6822):860–921, 2001.

C. Leslie, E. Eskin, and W.S. Noble. The spectrum kernel: A string kernel for SVM protein classification. In *Proceedings of the Pacific Symposium on Biocomputing*, pages 564–575. World Scientific, 2002a.

C. Leslie, E. Eskin, J. Weston, and W. Stafford Noble. Mismatch string kernels for svm protein classification. In *Advances in Neural Information Processing Systems*, 2002b.

B. Logan, P. Moreno, B. Suzek, Z. Weng, and S. Kasif. A study of remote homology detection. Technical report, Cambridge Research Laboratory, June 2001.

A.D. McNaught and A. Wilkinson. *IUPAC Compendium of Chemical Terminology.* Royal Society of Chemistry, Cambridge, UK, 1997.

C.G. Nevill-Manning, T.D. Wu, and D.L. Brutlag. Highly specific protein sequence motifs for genome analysis. *Proc. Natl. Acad. Sci. USA*, 95(11):5865–5871, 1998.

C. O'Donovan, M.J. Martin, A. Gattiker, E. Gasteiger, A. Bairoch A., and R. Apweiler. High-quality protein knowledge resource: SWISS-PROT and TrEMBL. *Brief. Bioinform.*, 3:275–284, 2002.

Nomenclature Committee of the International Union of Biochemistry and Molecular Biology (NC-IUBMB). *Enzyme Nomenclature. Recommendations 1992*. Academic Press, 1992.

J. C. Platt. Probabilities for support vector machines. In A. Smola, P. Bartlett, B. Schölkopf, and D. Schuurmans, editors, *Advances in Large Margin Classifiers*, pages 61–74. MIT Press, 1999.

R. Rifkin and A. Klautau. In defense of one-vs-all classification. *Journal of Machine Learning Research*, 5:101–141, 2004.

B. Schölkopf and A.J. Smola. *Learning with Kernels: Support Vector Machines, Regularization, Optimization and Beyond*. MIT Press, Cambridge, MA, 2002.

T. Smith and M. Waterman. Identification of common molecular subsequences. *Journal of Molecular Biology*, 147:195–197, 1981.

E.L. Sonnhammer, S.R. Eddy, and E. Birney. Pfam: multiple sequence alignments and hmm-profiles of protein domains. *Nucleic Acids Research*, 26(1):320–322, 1998.

Q. Su, S. Saxonov, L. Liu, and D.L. Brutlag. eBLOCKS: Automated database of protein conserved regions maximizing sensitivity and specificity. *Nucleic Acids Research*, 33:In Press, 2004.

U. Syed and G. Yona. Using a mixture of probabilistic decision trees for direct prediction of protein function. In *RECOMB*, 2003.

J.C. Venter, M.D. Adams, E.W. Myers, and P.W. Li. The sequence of the human genome. *Science*, 2901(16):1304–1351, 2001.

J. Weston, A. Elisseeff, M. Tipping, and B. Schölkopf. Use of the zero-norm with linear models and kernel methods. *Journal of Machine Learning Research*, 3 (7-8):1439–1461, 2003. URL http://www.ai.mit.edu/projects/jmlr/papers/volume3/weston03a/abstract.html.

C.H. Yeang, S. Ramaswamy, P. Tamayo, S. Mukherjee, R. Rifkin, M. Angelo, M. Reich, E. Lander, J. Mesirov, and T. Golub. Molecular classification of multiple tumor types. In *Proceedings, eleventh international conference on intelligent systems for molecular biology*, volume 17 suppl 1 of *Bioinformatics*, pages S316–S322, 2001.

Elementary Statistics

Elementary Statistics

Gérard Dreyfus[1]

École Supérieure de Physique et de Chimie Industrielles (ESPCI-Paristech),
Laboratoire d'Électronique (CNRS UMR 7084), 10 rue Vauquelin, 75005 Paris -
FRANCE Gerard.Dreyfus@espci.fr

Statistics is the art of dealing quantitatively with uncertainty and fluctuations. *Random variables* are the central concept of statistics: a quantity whose value cannot be predicted with absolute certainty can be considered as the *realization* of a random variable. That does not mean that the phenomenon that produces the value of interest is random, i.e. is not subject to the usual determinism of macroscopic physics; it means that the observer is not aware of all the factors that control the result, hence cannot control them. Typically, the outcome of throwing a die can be considered as a (discrete) random variable; however, a good mechanical engineer could design a die-throwing machine that would produce any desired outcome, repeatedly and on request. That is a feasible task, provided the initial conditions (position, angular and linear velocity of the die) and the subsequent factors (friction on the table, elasticity of the latter, etc.) are carefully controlled. Similarly, statistical studies can predict the outcome of elections within some accuracy by considering votes as random variables, although each individual citizen does *not* cast a random vote.

There is a wealth of textbooks on statistics, among which the novice reader can find his own favorite. Some of the notations used in the present appendix are borrowed from (Mood et al., 1974).

1 Basic Principles

We first start with some definitions and the basic concepts of random variables and distributions.

1.1 What is a Random Variable?

Definition 1. *A random variable is fully defined by its probability density or probability distribution function (pdf). A random variable Y has a probability* *density $p_Y(y)$ if the probability that a realization y of the random variable Y lies in an interval $[y, y + dy]$ is equal to $p_Y(y)dy$.*

Alternatively, a random variable can be defined by its cumulative distribution function $F_Y(y)$, which is the probability that a realization z of Y be smaller than or equal to y: $F_Y(y) = Pr(z \le y)$. Therefore, the probability density is the derivative of the cumulative distribution function: $P_Y(y) = \frac{dF_y(y)}{dy}$. As a consequence, if y is a real variable, one has $F_Y(+\infty) = 1$, $F_Y(-\infty) = 0$ and $\int_{-\infty}^{+\infty} p_Y(y)\, dy = 1$.

1.2 Examples of Probability Distributions

(1.) *Uniform distribution*: a random variable Y has a uniform distribution if its density probability is $p_Y(y) = 1/(b-a)$ on a given interval $[a, b]$, and is zero elsewhere.

(2.) *Gaussian distribution*: $p_Y(y) = \frac{1}{\sqrt{2\pi\sigma^2}} \exp\left(-\frac{(y-\mu)^2}{2\sigma^2}\right)$ where μ is the mean of the distribution and σ its standard deviation. As a particular case, a standard normal variable has a Gaussian distribution with $\mu = 0$ and $\sigma = 1$.

(3.) The probability distribution of a certain variable of value y_0 is a Dirac delta distribution $\delta(y - y0)$.

Additional useful probability distributions will be defined below.

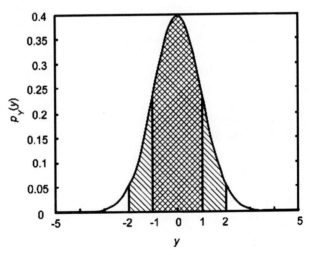

Fig. 1. Probability distribution of a standard normal random variable (Gaussian distribution with = 0 and = 1). Approximately 68% of the distribution lies between -1 and +1, and approximately 96% of the distribution lies between -2 and +2.

1.3 Joint Distribution – Independent Variables

Denoting by $p_{X,Y}(x,y)$ the joint density of two random variables, the probability of a realization of X lying between x and $x + dx$ and of a realization of Y lying between y and $y + dy$ is $p_{X,Y}(x,y)dxdy$.

   Two random variables X and Y are independent if the probability of a realization of one variable taking some value is independent of the value of the realization of the other variable. Then one has: $p_{X,Y}(x,y) = p_X(x)p_Y(y)$.

1.4 Expectation Value of a Random Variable

Definition 2. *The expectation value of a random variable Y is defined as:* $\int_{-\infty}^{+\infty} p_Y(y)\, dy = 1$. *Therefore, it is the first moment of the pdf.*

1.5 Properties

(1.) The expectation value of the sum of random variables is the sum of the expectation values of the random variables.
(2.) The expectation value of a uniformly distributed variable Y in interval $[a, b]$ is $(a + b)/2$.
(3.) The expectation value of a Gaussian distribution of mean μ is μ.
(4.) The expectation value of a certain variable of value y_0 is y_0.

2 Estimating and Learning

2.1 The Simplest Machine Learning Problem

In machine learning, it is desired to learn from data. Consider the following simple problem: data is available from measurements of a quantity that we know to be constant (i.e. an object has been weighed on different scales, ranging from a 19th century Roberval scale to a fancy digital one). We want to learn the "true" weight from the data.

   We first postulate a model whereby the result of each measurement is the sum of the "true" weight and of an unpredictable noise. Therefore, our statistical model is the following: the results of the measurements are realizations of a random variable Y, which is the sum

- of a *certain* variable (with probability distribution $P_Y(y) = \delta(y - \theta)$, which we want to learn from data,
- and of a random variable N with zero expectation value (e.g. Gaussian distributed with zero mean and standard deviation σ).

Then the expectation value of Y is given by: $E_Y = E_\theta + E_N = \theta$ since the expectation value of N is equal to zero.

Therefore, our machine learning problem is: how can we estimate the single parameter θ of our model

$$g(\theta) = \theta \tag{1}$$

from the available data? In other words, what learning algorithm should we use? To this end, the concept of *estimator* is useful.

2.2 Estimators

As mentioned in the previous section, when measurements of a quantity of interest are modeled as realizations of a random variable Y, one is primarily concerned with estimating the expectation value of the quantity of interest from the experimental results. Therefore, we would like to find a quantity that can be *computed* from the measurements, and that is "as close as possible" to the (unknown) true value; that quantity itself can be regarded as a random variable, function of the observable random variable Y.

Definition 3. *An estimator is a random variable, which is a function one or more* observable *random variables.*

2.3 Unbiased Estimator of a Certain Variable

An estimator of a certain variable is *unbiased* if its expectation value is equal to the value of the certain variable. Therefore, learning algorithms that estimate the parameters of a model from data actually try to find unbiased estimators of those parameters, the latter being considered as *certain* variables. That view of parameter estimation is called the *frequentist* approach. By contrast, the *Bayesian* approach considers the parameters themselves as random variables that are not certain. The frequentist approach is the simplest and most popular, although Bayesian approaches may provide excellent results, as exemplified in the results of the feature selection competition (Neal and Zhang, 2005).

2.4 The Mean is an Unbiased Estimator of the Expectation Value

If m examples that are to be used for training are results of measurements that were performed independently of each other, under conditions that are assumed to be identical, each measured value y_i can be viewed as a realization of a random variable Y_i. Since the experiments were performed independently under identical conditions, those random variables are independent and have the same probability distribution, hence the same expectation value E_Y (they are said to be "i.i.d.", which stands for independent identically distributed).

The random variable "mean" is defined as $M = \frac{1}{m}\sum_{i=1}^{m} Y_i$. Since the expectation value of the sum is the sum of the expectation values, one has $E_M = \frac{1}{m}\sum_{i=1}^{m} E_Y = E_Y$: the expectation value of the mean is equal to the certain variable E_Y, hence the mean is an unbiased estimator of the expectation value of Y. Therefore, the realization μ of the estimator M, which can be computed as $\mu = \frac{1}{m}\sum_{i=1}^{m} y_i$, is an estimate of the "true" value of the quantity of interest.

Therefore, the training algorithm for finding the single parameter of the model described in the previous section is: compute the mean value μ of the data, and set the estimated value of θ equal to μ.

However, the fact that an unbiased estimate of the "true" value of the quantity of interest can be computed does not say anything about the accuracy of that estimate, which depends both on the "amount of noise" present in the data, and on the number of observations. In that context, the *variance* is another quantity of interest.

2.5 Variance of a Random Variable

The variance of a random variable Y is the quantity $var_Y = \sigma^2 = \int_{-\infty}^{+\infty}(y - E_Y)^2 p(y)\, dy$. Therefore, it is the centered, second moment of the probability density p_Y.

2.6 Properties

(1.) $var_Y = E_{Y^2} - E_Y^2$
(2.) $var_{aY} = a^2 var_Y$
(3.) The variance of a uniformly distributed random variable in $[a, b]$ is $(b - a)^2/12$.
(4.) The variance of a Gaussian distributed random variable of standard deviation σ is equal to σ^2.

2.7 Unbiased Estimator of the Variance

If the m measurements of the quantity of interest are performed independently and under identical conditions, each measurement can be considered as a realization of a random variable Y_i, all variables Y_i having the same distribution, hence the same variance. Then, it can be shown that the random variable $S^2 = \frac{1}{m-1}\sum_{i=1}^{m}(Y_i - M)^2$, where M is the estimator of the expectation value of Y, is an unbiased estimator of the variance of Y.

Therefore, given m experimental observations y_i, an estimate of the variance of the quantity of interest can be obtained by

- computing the mean $\mu = \frac{1}{m}\sum_{i=1}^{m} y_i$
- and computing the estimate of the variance as $s^2 = \frac{1}{m-1}\sum_{i=1}^{m}(y_i - \mu)^2$

Thus, the computed quantity s^2 provides a quantitative assessment of how the measured values are "scattered" around the mean value. That is interesting information, because it gives an idea of the amount of "noise" present in the measurements. Still, it does not tell us how close the mean is to the "true" value.

In order to get a better insight into that question, one should keep in mind that the mean is just a single realization of the random variable M, which has a variance. That variance can be estimated by performing m' sets of m experiments: that would provide m' realizations μ_j of the mean. Then the mean μ' of the means μ_j can be computed, and an estimate of the variance of the mean can be computed as $s_\mu^2 = \frac{1}{m'-1} \sum_{j=1}^{m'} (\mu_j - \mu')^2$. That computation requires a relatively heavy procedure. A more elegant way of assessing the quality of an estimate will be provided in Section 4.

3 Some Additional Useful Probability Distributions

Before investigating further the problem of assessing the acceptability of using the mean as an estimate of the expectation value, we define three additional probability distributions of interest in the book.

3.1 The χ^2 (Pearson) Distribution

If a random variable X is the sum of the squares of m random independent Gaussian variables, then it has a χ^2 *(or Pearson) distribution with m degrees of freedom*. It can be shown that $E(X) = m$ and that $var(X) = 2m$.

3.2 Student Distribution

If Y_1 is a normal variable, and if Y_2 is a random variable, which is independent of Y_1 and which has a (Pearson) distribution with m degrees of freedom, then the random variable $Z = \frac{Y_1}{\sqrt{Y_2/m}}$ has a *Student distribution with m degrees of freedom*.

3.3 Fisher Distribution

If Y_1 is a Pearson variable with m_1 degrees of freedom, and if Y_2 is a Pearson variable with m_2 degrees of freedom, then the random variable $Z = \frac{Y_1/m_1}{Y_2/m_2}$ has a *Fisher distribution* with m_1 and m_2 degrees of freedom.

4 Confidence Intervals

In the previous sections, we have shown that the accuracy of the estimation of the "true" value by the mean value computed from a set of experiments depends both on the number of experiments and on the noise present in the data. Confidence intervals are an elegant way of combining size of the experiment and variability, in order to assess whether the discrepancy between the mean and the expectation value is "acceptable".

Definition 4. *A confidence interval, with confidence threshold $1 - \alpha$ around the mean of a random variable Y, contains the value of the expectation of Y with probability $1 - \alpha$.*

For instance, assume that a confidence interval with confidence threshold 0.95 ($\alpha = 0.05$) is computed; if 100 different sets of experiments are performed, 100 means, 100 variance estimates, and 100 confidence intervals with confidence threshold 95 % can be computed; for 95% of those data sets, the confidence interval will contain the mean. Of course, for a particular data set, there is no way to guarantee that the true value is within the specific confidence interval computed from that specific data set: the risk of getting wrong by chance cannot be altogether eliminated.

4.1 How to Design a Confidence Interval

In order to design a confidence interval for a random variable Y, one seeks an appropriate random variable Z, function of Y, whose distribution $p_Z(z)$ is known and independent of Y. Since the distribution $p_Z(z)$ is known, the equation $Pr(z_1 < z < z_2) = \int_{z_1}^{z_2} p_Z(z)\,dz = 1 - \alpha$ can be solved easily: one just has to invert the cumulative distribution function of Z, i.e. compute the value of z_1 such that $Pr(z < z_1) = \alpha/2$, and the value of z_2 such that $Pr(z > z_2) = \alpha/2$. When z_1 and z_2 are found, function $Z(Y)$ is inverted in order to find the values of a and b such that $Pr(a < y < b) = 1 - \alpha$.

4.2 An Example: A Confidence Interval for the Parameter of the Model Described by Equation 1

Confidence intervals can be applied to the modeling problem that has been described above, i.e. the validity of the estimation of the expectation value by the mean value. As before, we assume that m experiments have been performed, that the result y_i of each experiment is a realization of a random variable Y_i, and that the experiments are designed in such a way that the m random variables are independent and identically distributed. In addition, we assume that distribution to be Gaussian with mean μ and standard deviation σ.

The sum of independent Gaussian variables is Gaussian distributed, with mean equal to the sum of the means and variance equal to the sum of the variances. In the present case, the m variables are identically distributed: their sum is Gaussian distributed with mean $m\mu$ and variance $m\sigma^2$. Therefore, their mean M is Gaussian distributed with mean μ_0 and variance σ^2/m, hence the random variable $\frac{M-\mu_0}{\sigma/\sqrt{m}}$ has zero mean and unit variance. The probability distribution of that variable is known and independent of M as required, but it contains the two unknowns μ_0 and σ, so that it cannot be used for designing a confidence interval.

Instead, consider the random variable $\sum_{i=1}^{m} \frac{(Y_i-M)^2}{\sigma^2}$; being a sum of Gaussian variables, it is a Pearson variable with $m-1$ degrees of freedom (there are only $m-1$ *independent* variables in the sum since M depends on the variables Y_i).

Therefore, from the definition provided in Section 3, the variable $Z = \frac{M-\mu_0}{\sigma/\sqrt{m}} \big/ \sum_{i=1}^{m} \frac{(Y_i-M)^2}{\sigma^2(m-1)} = \frac{M-\mu_0}{\sqrt{S^2/m}}$, where S^2 is the estimator of the variance, has a Student distribution with $m-1$ degrees of freedom. By inverting the cumulative distribution function of the Student distribution, two numbers z_1 and z_2 can be computed, such that a realization z of Z lies between those two values with probability $1-\alpha$. Since the Student distribution is symmetrical, one can choose z_1 and z_2 such that $|z_1| = |z_2| = z_0$, e.g. $z_1 = -z_0$ and $z_2 = z_0$. The set of experiments provides a realization of variable Z: $z = \frac{\mu-\mu_0}{\sqrt{s^2/m}}$ with $\mu = \frac{1}{m}\sum_{i=1}^{m} y_i$ and $s = \sqrt{\frac{1}{m-1}\sum_{i=1}^{m}(y_i-\mu)^2}$. The two linear inequalities $z_1 < z < z_2$ can easily be solved to provide the two required boundary values a and b for μ_0: $a = \mu - \sqrt{\frac{s^2}{m}}z_0$ and $b = \mu + \sqrt{\frac{s^2}{m}}z_0$, where μ, s, and m depend on the experiments, and z_0 depends on the confidence threshold $1-\alpha$ only.

As expected, the width $2\sqrt{\frac{s^2}{m}}z_0$ of the confidence interval depends both on the number of experiments m and on the noise (through the estimated variance s). The larger the number of experiments, the smaller the confidence interval, hence the more reliable the estimation of the expectation value μ_0 by the mean μ. Conversely, the larger the variability of the results, as expressed by s, the larger the confidence interval, hence the less reliable the estimation of the expectation value μ_0 by the mean μ. If a maximum value of the confidence interval is specified in the requirements of the problem, then a large variability must be compensated for by a large number of experiments.

Therefore, a confidence interval for the single parameter of model Equation 1 has been derived here. Similarly, confidence intervals can be derived both for the parameters of models trained from data, and for the predictions of such models. If the model is assumed to be true, e.g. a model based on physical knowledge, containing parameters that have a physical significance, it is very important to have confidence intervals on those parameters. On the other hand, when black-box models are designed (e.g. SVM's, neural nets, etc.), the parameters have no specific significance, so that confidence intervals on their values are not very useful; by contrast, confidence intervals on the prediction of the models are absolutely necessary, and may also be very useful for experimental planning. Section 6 relates PAC learning to the confidence interval on the error rate for classification.

5 Hypothesis Testing

On the basis of the observed data, one can make an assumption and test it, i.e. assess whether that assumption - called the null hypothesis - is true, or whether the fact that the data seems to support that assumption is just a matter of chance, due to the small size of the data set and/or to the variability of the data. Statisticians have derived a very wide variety of statistical tests, for use in many different situations, see for instance (Lehmann, 1993).

5.1 Definitions: Test Statistic, Risk, p-Value

In order to test a hypothesis (termed traditionally the **"null" hypothesis**), one must find a random variable Z (termed **test "statistic"**), whose distribution (termed the **"null" distribution**) is known if that hypothesis is true, and a realization z of which can be computed from the available data. If the probability of that realization lying in a given interval is "too low" (in a sense to be defined below), the probability of the null hypothesis being true is too low, hence that hypothesis should be rejected.

In addition to defining the null hypothesis, a threshold probability must be chosen, which depends on the *risk* that is deemed admissible of rejecting the null hypothesis although it is actually true (that is called a Type-I error). That threshold probability is usually denoted as α.

The results of hypothesis tests may also be analyzed in terms of the **p-value** of the realization z of Z: the p-value of z is the probability of a realization of Z being further from the mean of the pdf of Z than z, if the null hypothesis is true. Therefore, a small p-value sheds doubt on the validity of the null hypothesis and α can be viewed as an upper bound of the p-value of z.

In a **two-tailed test**, the risk is split on either side of the *pdf* while in a **one-tailed test**, the risk is blocked on one side (see example below.)

The probability of making a Type-II error (i.e. accepting the null hypothesis while it is actually wrong) is unknown; making such an error based on a given test means that the available data has led us to miss some small difference, either because data are too sparse, or because the variance of the examples is too large. That issue is related to the **power of the test**.

In most hypothesis tests, the acceptance or rejection criterion was defined with respect to a threshold α that is an upper bound of the probability of a Type-I error. If a quantity β is defined as the probability of a Type-II error, then the power of the test is $1 - \beta$. Therefore, among all possible values of α, one should choose the value that provides the best tradeoff between the probability of the two types of errors, i.e. the smallest α and the highest power.

5.2 Example: The Student T-Test

As a first illustration, assume that a model predicts that a quantity y is equal to μ_0. A set of m examples is available for testing the validity of the model, and we wish to assess whether the examples support the prediction of the model. We make the assumption that the examples have been gathered in such a way that they can be modeled as realizations of m random independent variables with the same probability distribution. We further assume that distribution to be Gaussian with expectation value E_Y and standard deviation σ. Then the null hypothesis (denoted traditionally as H_0) is: H_0: $E_Y = \mu_0$, which is to be tested against the alternative hypothesis H_1: $E_Y \neq \mu_0$.

If the null hypothesis is true, then the random variable $Z = \frac{M-\mu_0}{\sqrt{S^2/m}}$, where M is the estimator of the expectation value, S^2 is the estimator of the variance, and m is the number of examples, has a Student distribution with $m-1$ degrees of freedom. A realization z of that variable can be computed from the examples as $z = \frac{\mu-\mu_0}{\sqrt{s^2/m}}$ where $\mu = \frac{1}{m}\sum_{i=1}^{N} y_i$ (the mean) and $s = \sqrt{\frac{1}{m-1}\sum_{i=1}^{N}(y_i - \mu)^2}$ (the estimator of the variance). Since the Student distribution is known, the probability of a realization of that variable being smaller than or equal to z if the null hypothesis is true can be computed. Furthermore, by inverting the cumulative distribution function of the Student variable, one can find z_1 such that the probability of z being smaller than z_1 is $\alpha/2$, and z_2 such that the probability of z being larger than z_2 is $\alpha/2$. Therefore, if the value of z, computed from the experiments, lies between z_1 and z_2, then one can conclude that, with probability $1 - \alpha$, the null hypothesis is supported by the experiments. Conversely, if the realization of Z lies outside that interval, the null hypothesis will be rejected (although it may be true), because we consider that the experiments do not provide conclusive evidence that the null hypothesis is true. Therefore, α is an upper bound on the probability of a Type-I error, subject to the Gaussian assumption for the examples (Figure 2). It is traditionally taken equal to 0.05.

As an illustration, assume that one has 100 examples with mean value $\mu = 2$ while the prediction of the model is $\mu_0 = 1$. Then $\mu - \mu_0 = 1$. If the examples have a large variance, i.e. are very much scattered around the mean, e.g. $s = 10$, then $z = 1$. From the Student distribution, one can compute that the probability of the magnitude of z being larger than or equal to 1 is equal to 0.32. If the value $\alpha = 0.05$ has been chosen, one can safely conclude that the null hypothesis is true: the difference found between the mean and its theoretical value supports the hypothesis that the theoretical mean is indeed equal to 1, given the scatter of the data, and the amount of data available. Conversely, if the same difference between the mean and its predicted value is found, but with data that are much less scattered, e.g. $s = 3$, then $z = 3.3$, and the probability of the magnitude of z being larger than or equal to 3.3 is 8×10^{-3}, so that the null hypothesis is rejected by the test: given the scatter

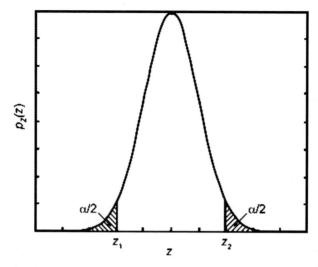

Fig. 2. The hatched area is the probability of z lying outside the interval $[z_1, z_2]$ if the null hypothesis is true; it is equal to the risk α rejecting the null hypothesis although it is true.

of the data and the number of examples, it cannot be concluded safely that the predicted value of 1 is correct.

We can also analyze the test in terms of p-value. In the present case, the p-value is called a two-tail p-value since its definition involves two inequalities ($z < z_1$ or $z > z_2$). If the realization of z, computed from the examples, is further from the mean than z_1 or z_2, i.e. if the p-value of z is smaller than α, the null hypothesis is rejected. In the above example, the two-tail p-value of $z = 1$ is 0.32, whereas the p-value of $z = 3.3$ is 8×10^{-3}. There are many hypothesis tests where the validity of the null hypothesis is questioned if a realization of the random variable obeys a single inequality (e.g. it is larger than a given value z_0); then α is the upper bound of the *one-tail* p-value of z (i.e. the probability of z being larger than z_0 if the null hypothesis is true, where the p-value of z_0 is α) (Figure 3).

5.3 Parametric vs. Nonparametric Tests

In the above example, the underlying assumption was that the examples were Gaussian distributed. That assumption is not necessarily true, especially when data are sparse. *Non-parametric tests* do not rely on any assumption about the distribution of the examples, but the price to pay is a smaller power than parametric tests. The Wilcoxon rank sum test presented in the text is an example of non-parametric test that may be substituted to the T-test to test the equality of the mean of two populations when the Gaussian assumption cannot be relied upon.

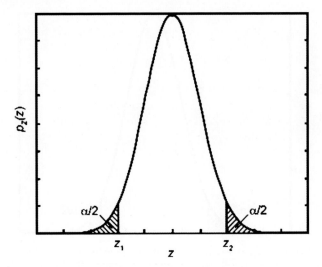

Fig. 3. One-tail hypothesis test.

6 Probably Approximately Correct (PAC) Learning and Guaranteed Estimators

This section briefly introduces the new branch of statistics known as learning theory, stemming from the work of Vapnik, Valiant, and others.

6.1 Confidence Intervals on the Error Rate

Probably Approximately Correct (PAC) learning is a general framework for the problem of learning classification tasks from examples, i.e. tasks in which a mapping from a set of real numbers to the set of binary variables $\{0, 1\}$ is sought (Valiant, 1984). More specifically, given a training set of m examples, and a test set drawn from the same probability distribution, a set of classifiers $\{g\}$ is said to be PAC learnable if there is a learning algorithm such that the following relation holds for all classifiers of that set:

$$Pr\left[Pr(classification\ error) > \varepsilon\right] < \alpha \qquad (2)$$

where ε and α are predefined goals. Clearly, the smaller ε and α, the better the classifiers found by the training algorithm. $\{g\}$ is said to be efficiently PAC learnable if there exists a learning algorithm, polynomial in ε, α and $\ln(m)$, such that all classifiers of $\{g\}$ comply with Equation 2. The complexity of that learning algorithm is defined as the smallest training set size that makes $\{g\}$ learnable.

Now assume that, for a given family a classifiers $\{g\}$, we would like to know how far the error rate e_V computed on the validation set (an estimate of the "empirical risk") is from the unknown generalization error ("true risk") e. A confidence interval ε with confidence level $1 - \alpha$ is such that

$$Pr\left[(e_V - e) > \varepsilon\right] < \alpha \tag{3}$$

Formally, there is a clear similarity between Equation 2 and Equation 3, but they do not provide the same information. An analytic expression of ε involves the Vapnik-Chervonenkis dimension of the class of classifiers $\{g\}$, the number m of examples, and α (Vapnik, 1982). Note, however, that most present feature selection procedures do not make use of the bounds derived by either approach.

6.2 Guaranteed Estimators

The issue of the number of observations required in order to get a statistically significant result is essential in experimental planning. Assume that we have an estimator Z of a quantity of interest w, e.g. the mean as an estimator of the expectation value of the unknown distribution of an observed variable X. How many observations should we gather in order to be sure that, with probability $(1 - \alpha)$, the difference between the realization z of the estimator and the "true" value w of the quantity of interest does not exceed a given ε? In other words, we want to find the number of examples m such that the following inequality holds:

$$Pr\left[|z - w| \geq \varepsilon\right] \leq \alpha \tag{4}$$

In order to solve that problem, one must resort to distribution-independent inequalities if the distribution of the observed variable is unknown, or to tighter distribution-dependent equalities if the distribution is known.

The simplest distribution-independent inequality is the Chebyshev inequality (see for instance Mood et al. (1974)): for a random variable Y

$$Pr\left[|y - E_Y| \geq r \; var_Y\right] \leq \frac{1}{r^2} \tag{5}$$

As an example, assume that it is desired to design an experiment such that the probability of the difference between the mean of a variable X and the expectation of the quantity of interest being larger than ε is smaller than a given value α. Then the estimator Z is the mean M. Since the mean is an unbiased estimator of the expectation value, one has $E_M = E_X$. Since $M = \frac{1}{m} \sum_{i=1}^{m} X_i$, the variance of the mean is $var_M = (1/m^2) \sum_{i=1}^{m} var_x = (1/m)var_x$. Then the Chebyshev inequality Equation 5 can be written as $Pr\left[|\mu - E_X| \geq \varepsilon\right] \leq \left(\frac{var_M}{\varepsilon}\right)^2 = \left(\frac{var_X}{m\varepsilon}\right)^2$ where μ is a realization of M, i.e. the value of the mean computed from the observations.

Therefore, one has from Equation 4: $\alpha = \left(\frac{var_X}{m\varepsilon}\right)^2$, hence $m = \frac{var_X}{\varepsilon\sqrt{\alpha}}$. In order to make use of that relation, an estimate of the variance must be available, e.g. from preliminary experiments.

The above bound is distribution-independent. If the distribution is known, tighter bounds can be derived. If the variable X is known to be Gaussian distributed with expectation value E_X and standard deviation σ, M is Gaussian distributed with mean E_X and standard deviation $\frac{\sigma}{\sqrt{m}}$. Therefore, the variable $Y = \frac{M-\mu}{\sigma/\sqrt{m}}$ obeys a standard normal law (Gaussian with $\mu = 0$ and $\sigma = 1$). Therefore, Equation 4 can be written as:

$$Pr\left[|y| \geq \varepsilon \frac{\sqrt{m}}{\sigma}\right] \leq \alpha,$$

from which a threshold value y_α is derived by inverting the *cdf* of the normal distribution, as shown on Figure 2. Therefore, the required number of examples is given by $m = \left(\frac{y_\alpha \sigma}{\varepsilon}\right)^2$.

As an illustration, the number of examples required in a test set in order to have a statistically significant assessment of the classification error was derived in (Guyon et al., 1998).

References

I. Guyon, J. Makhoul, R. Schwartz, and V. Vapnik. What size test set gives good error rate estimates? *IEEE Transactions on Pattern Analysis and Machine Intelligence*, 20:52–64, 1998.

E.L. Lehmann. *Testing statistical hypotheses*. Chapman and Hall, New York, 1993.

A.M. Mood, F.A. Graybill, and D.C. Boes. *Introduction to the theory of statistics*. McGraw-Hill, Singapore, 1974.

R. Neal and J. Zhang. *Feature Extraction, Foundations and Applications*, chapter High dimensional classification with Bayesian neural networks and Dirichlet diffusion trees. Springer, 2005.

L.G. Valiant. A theory of the learnable. *Communications of the ACM*, 27(11): 1134–1142, 1984.

V.N. Vapnik. *Estimation of dependencies based on empirical data*. Springer, New-York, 1982.

Appendix B

Feature Selection Challenge Datasets

Experimental Design

Isabelle Guyon[1]

ClopiNet, 955 Creston Rd., Berkeley, CA 94708, USA. `isabelle@clopinet.com`

Background

Results published in the field of feature or variable selection (see e.g. the special issue of JMLR on variable and feature selection: `http://www.jmlr.org/papers/special/feature.html`) are for the most part on different data sets or used different data splits, which make them hard to compare. We formatted a number of datasets for the purpose of benchmarking variable selection algorithms in a controlled manner*. The data sets were chosen to span a variety of domains (cancer prediction from mass-spectrometry data, handwritten digit recognition, text classification, and prediction of molecular activity). One dataset is artificial. We chose data sets that had sufficiently many examples to create a large enough test set to obtain statistically significant results. The input variables are continuous or binary, sparse or dense. All problems are two-class classification problems. The similarity of the tasks allows participants to enter results on all data sets. Other problems will be added in the future.

Method

Preparing the data included the following steps:

- Preprocessing data to obtain features in the same numerical range (0 to 999 for continuous data and 0/1 for binary data).
- Adding "random" features distributed similarly to the real features. In what follows we refer to such features as probes to distinguish them from the real features. This will allow us to rank algorithms according to their ability to filter out irrelevant features.

---

*In this document, we do not make a distinction between features and variables. The benchmark addresses the problem of selecting input variables. Those may actually be features derived from the original variables using a preprocessing.

- Randomizing the order of the patterns and the features to homogenize the data.
- Training and testing on various data splits using simple feature selection and classification methods to obtain baseline performances.
- Determining the approximate number of test examples needed for the test set to obtain statistically significant benchmark results using the rule-of-thumb $n_{test} = 100/p$, where p is the test set error rate (see What size test set gives good error rate estimates? I. Guyon, J. Makhoul, R. Schwartz, and V. Vapnik. PAMI, 20 (1), pages 52–64, IEEE. 1998, http://www.clopinet.com/isabelle/Papers/test-size.ps.Z). Since the test error rate of the classifiers of the benchmark is unknown, we used the results of the baseline method and added a few more examples.
- Splitting the data into training, validation and test set. The size of the validation set is usually smaller than that of the test set to keep as much training data as possible.

Both validation and test set truth-values (labels) are withheld during the benchmark. The validation set serves as development test set. During the time allotted to the participants to try methods on the data, participants are allowed to send the validation set results (in the form of classifier outputs) and obtain result scores. Such score are made available to all participants to stimulate research. At the end of the benchmark, the participants send their *test set results*. The scores on the test set results are disclosed simultaneously to all participants after the benchmark is over.

Data Formats

All the data sets are in the same format and include 8 files in ASCII format:

dataname.param Parameters and statistics about the data.

dataname.feat Identities of the features (in the order the features are found in the data).

dataname_train.data Training set (a sparse or a regular matrix, patterns in lines, features in columns).

dataname_valid.data Validation set.

dataname_test.data Test set.

dataname_train.labels Labels (truth values of the classes) for training examples.

dataname_valid.labels Validation set labels (withheld during the benchmark).

dataname_test.labels Test set labels (withheld during the benchmark).

The matrix data formats used are:

Regular Matrices a space delimited file with a new-line character at the end of each line.

Sparse Binary Matrices for each line of the matrix, a space delimited list of indices of the non-zero values. A new-line character at the end of each line.

Sparse Non-Binary Matrices for each line of the matrix, a space delimited list of indices of the non-zero values followed by the value itself, separated from it index by a colon. A new-line character at the end of each line.

The results on each dataset should be formatted in 7 ASCII files:

dataname_train.resu ± 1 classifier outputs for training examples (mandatory for final submissions).

dataname_valid.resu ± 1 classifier outputs for validation examples (mandatory for development and final submissions).

dataname_test.resu ± 1 classifier outputs for test examples (mandatory for final submissions).

dataname_train.conf confidence values for training examples (optional).

dataname_valid.conf confidence values for validation examples (optional).

dataname_test.conf confidence values for test examples (optional).

dataname.feat list of features selected (one integer feature number per line, starting from one, ordered from the most important to the least important if such order exists). If no list of features is provided, it will be assumed that all the features were used.

Format for classifier outputs:

.resu files should have one ± 1 integer value per line indicating the prediction for the various patterns.

.conf files should have one decimal positive numeric value per line indicating classification confidence. The confidence values can be the absolute discriminant values. They do not need to be normalized to look like probabilities. They will be used to compute ROC curves and Area Under such Curve (AUC).

Result Rating

The classification results are rated with the balanced error rate (the average of the error rate on training examples and on test examples). The area under the ROC curve is also be computed, if the participants provide classification confidence scores in addition to class label predictions. But the relative strength of classifiers is judged only on the balanced error rate. The participants are invited to provide the list of features used. For methods having performance differences that are not statistically significant, the method using the smallest number of features wins. If no feature set is provided, it is assumed that all the features were used. The organizers may then provide the participants with one or several test sets containing only the features selected to verify the accuracy of the classifier when it uses those features only. The proportion of random probes in the feature set is also be computed. It is used to assess the relative strength of method with non-statistically significantly different error rates and a relative difference in number of features that is less than 5%. In that case, the method with smallest number of random probes in the feature set wins.

Datasets

The following sections describe the five datatsets provided for the challenge in detail. These datasets were prepared for the NIPS 2003 variable and feature selection benchmark in August 2003.

ARCENE

The task of ARCENE is to distinguish cancer versus normal patterns from mass-spectrometric data. This is a two-class classification problem with continuous input variables.

Original Owners

The data were obtained from two sources: The National Cancer Institute (NCI) and the Eastern Virginia Medical School (EVMS). All the data consist of mass-spectra obtained with the SELDI technique. The samples include patients with cancer (ovarian or prostate cancer), and healthy or control patients.

NCI ovarian data: The data were originally obtained from `http://clinical proteomics.steem.com/download-ovar.php`. We use the 8/7/02 data set:
`http://clinical proteomics.steem.com/Ovarian_Dataset_8-7-02.zip`.
The data includes 253 spectra, including 91 controls and 162 cancer spectra. Number of features: 15154. **NCI prostate cancer data**: The data were originally obtained from `http://clinical proteomics.steem.com/JNCI_Data_7-3-02.zip` on the web page `http://clinical proteomics.steem.com/download-prost.php`. There are a total of 322 samples: 63 samples with no evidence of disease and PSA level less than 1; 190 samples with benign prostate with PSA levels greater than 4; 26 samples with prostate cancer with PSA levels 4 through 10; 43 samples with prostate cancer with PSA levels greater than 10. Therefore, there are 253 normal samples and 69 disease samples. The original training set is composed of 56 samples:

- 25 samples with no evidence of disease and PSA level less than 1 ng/ml.
- 31 biopsy-proven prostate cancer with PSA level larger than 4 ng/ml.

But the exact split is not given in the paper or on the web site. The original test set contains the remaining 266 samples (38 cancer and 228 normal). Number of features: 15154.

EVMS prostate cancer data: The data are downloadable from: `http://www.evms.edu/vpc/seldi/`. The training data data includes 652 spectra from 326 patients (spectra are in duplicate) and includes 318 controls and 334 cancer spectra. Study population: 167 prostate cancer (84 state 1 and 2; 83 stage 3 and 4), 77 benign prostate hyperplasia, and 82 age-matched normals. The test data includes 60 additional patients. The labels for the test set are not provided with the data, so the test spectra are not used for the benchmark. Number of features: 48538.

Past Usage

NCI ovarian cancer original paper: "Use of proteomic patterns in serum to identify ovarian cancer Emanuel F Petricoin III, Ali M Ardekani, Ben A Hitt, Peter J Levine, Vincent A Fusaro, Seth M Steinberg, Gordon B Mills, Charles Simone, David A Fishman, Elise C Kohn, Lance A Liotta. THE LANCET, Vol. 359, February 16, 2002, `www.thelancet.com`" are so far not reproducible. Note: The data used is a newer set of spectra obtained after the publication of the paper and of better quality. 100% accuracy is easily achieved on the test set using various data splits on this version of the data.

NCI prostate cancer original paper: Serum proteomic patterns for detection of prostate cancer. Petricoin et al. Journal of the NCI, Vol. 94, No. 20, Oct. 16, 2002. The test results of the paper are shown in Table 1.

EVMS prostate cancer original paper: Serum Protein Fingerprinting

Table 1. Results of Petricoin et al. on the NCI prostate cancer data. Fp=false positive, FN=false negative, TP=true positive, TN=true negative. Error=(FP+FN)/(FP+FN+TP+TN), Specificity=TN/(TN+FP), Sensitivity=TP/(TP+FN).

FP	FN	TP	TN	Error	1-error	Specificity	Sensitivity
51	2	36	177	20.30%	79.70%	77.63%	94.74%

Coupled with a Pattern-matching Algorithm Distinguishes Prostate Cancer from Benign Prostate Hyperplasia and Healthy Men, Bao-Ling Adam, et al., CANCER RESEARCH 62, 3609-3614, July 1, 2002. In the following excerpt from the original paper some baseline results are reported:

Surface enhanced laser desorption/ionization mass spectrometry protein profiles of serum from 167 PCA patients, 77 patients with benign prostate hyperplasia, and 82 age-matched unaffected healthy men were used to train and develop a decision tree classification algorithm that used a nine-protein mass pattern that correctly classified 96% of the samples. A blinded test set, separated from the training set by a stratified random sampling before the analysis, was used to determine the sensitivity and specificity of the classification system. A sensitivity of 83%, a specificity of 97%, and a positive predictive value of 96% for the study population and 91% for the general population were obtained when comparing the PCA versus noncancer (benign prostate hyperplasia/healthy men) groups.

Experimental Design

We merged the datasets from the three different sources ($253 + 322 + 326 = 901$ samples). We obtained $91 + 253 + 159 = 503$ control samples (negative class) and $162 + 69 + 167 = 398$ cancer samples (positive class). The motivations for merging datasets include:

- Obtaining enough data to be able to cut a sufficient size test set.
- Creating a problem where possibly non-linear classifiers and non-linear feature selection methods might outperform linear methods. The reason is that there will be in each class different clusters corresponding differences in disease, gender, and sample preparation.
- Finding out whether there are features that are generic of the separation cancer vs. normal across various cancers.

We designed a preprocessing that is suitable for mass-spec data and applied it to all the data sets to reduce the disparity between data sources. The preprocessing consists of the following steps:

Limiting the mass range We eliminated small masses under m/z=200 that include usually chemical noise specific to the MALDI/SELDI process (influence of the "matrix"). We also eliminated large masses over m/z=10000 because few features are usually relevant in that domain and we needed to compress the data.

Averaging the technical repeats In the EVMS data, two technical repeats were available. We averaged them because we wanted to have examples in the test set that are independent so that we can apply simple statistical tests.

Removing the baseline We subtracted in a window the median of the 20% smallest values. An example of baseline detection is shown in Figure 1.

Smoothing The spectra were slightly smoothed with an exponential kernel in a window of size 9.

Re-scaling The spectra were divided by the median of the 5% top values.

Taking the square root The square root of the all values was taken.

Aligning the spectra We slightly shifted the spectra collections of the three datasets so that the peaks of the average spectrum would be better aligned (Figure 2). As a result, the mass-over-charge (m/z) values that identify the features in the aligned data are imprecise. We took the NCI prostate cancer m/z as reference.

Limiting more the mass range To eliminate border effects, the spectra border were cut.

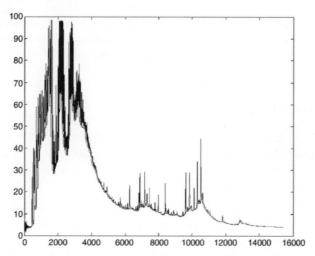

Fig. 1. Example of baseline detection (EVMS data).

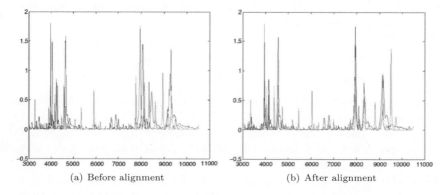

(a) Before alignment   (b) After alignment

Fig. 2. Central part of the spectra. (We show in red the average NCI ovarian spectra, in blue the average NCI prostate spectra, and in green the average EVMS prostate spectra.)

Soft thresholding the values After examining the distribution of values in the data matrix, we subtracted a threshold and equaled to zero all the resulting values that were negative. In this way, we kept only about 50% of non-zero value, which represents significant data compression (see Figure 3).

Quantizing We quantized the values to 1000 levels.

The resulting data set including all training and test data merged from the three sources has 901 patterns from 2 classes and 9500 features. We remove one pattern to obtain the round number 900. At every step, we checked that the change in performance of a linear SVM classifier trained and tested on

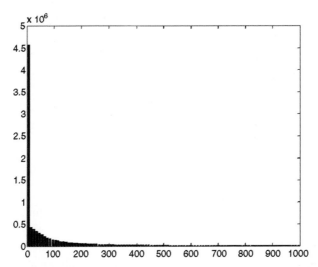

Fig. 3. Distributions of the values in the ARCENE data after preprocessing.

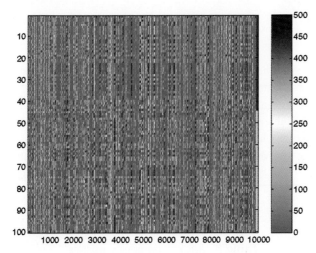

Fig. 4. Heat map of the training set of the ARCENE data. We represent the data matrix (patients in line and features in columns). The values are clipped at 500 to increase the contrast. The values are then mapped to colors according to the color-map on the right. The stripe beyond the 10000 feature index indicated the class labels: +1=red, -1=green.

a random split of the data was not significant. On that basis, we have some confidence that our preprocessing did not alter significantly the information content of the data. We further manipulated the data to add random "probes":

- We identified the region of the spectra with least information content using an interval search for the region that gave worst prediction performance of a linear SVM (indices 2250-4750). We replaced the features in that region by "random probes" obtained by randomly permuting the values in the columns of the data matrix.
- We identified another region of low information content: 6500-7000. We added 500 random probes that are permutations of those features.

After such manipulations, the data had 10000 features, including 7000 real features and 3000 random probes. The reason for not adding more probes is purely practical: non-sparse data cannot be compressed sufficiently to be stored and transferred easily in the context of a benchmark.

Data Statistics

Statistics about the data are presented in Tables 2 and 3.

Table 2. Arcene Datasets: Number of examples and class distribution.

	Positive ex.	Negative ex.	Total	Check sum
Training set	44	56	100	70726744
Validation set	44	56	100	71410108
Test set	310	390	700	493023349
All	398	502	900	635160201

Table 3. Arcene variable statistics.

Real variables	Random probes	Total
7000	3000	10000

All variables are integer quantized on 1000 levels. There are no missing values. The data are not very sparse, but for data compression reasons, we thresholded the values. Approximately 50% of the entries are non zero. The data was saved as a non-sparse matrix.

Results of the Run of the Lambda Method and Linear SVM

Before the benchmark, we ran some simple methods to determine what an appropriate number of examples should be. The "lambda" method (provided with the sample code) had approximately a 30% test error rate ans a linear SVM trained on all features a 15% error rate. The rule of thumb number_of_test_examples $= 100/\text{test\_errate} = 100/.15 = 667$ led us to keep 700 examples for testing. The best benchmark error rates are of the order 15%, which confirms that our estimate was correct.

GISETTE

The task of GISETTE is to discriminate between to confusable handwritten digits: the four and the nine. This is a two-class classification problem with sparse continuous input variables.

Original Owners

The data set was constructed from the MNIST data that is made available by Yann LeCun of the NEC Research Institute at `http://yann.lecun.com/exdb/mnist/`. The digits have been size-normalized and centered in a fixed-size image of dimension 28×28. We show examples of digits in Figure 1.

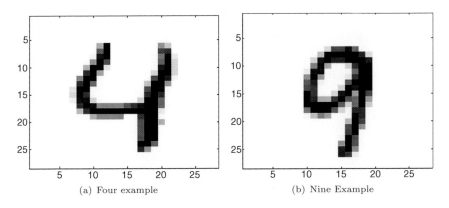

(a) Four example   (b) Nine Example

Fig. 1. Two examples of digits from the MNIST database. We used only examples of fours and nines to prepare our dataset.

Past Usage

Many methods have been tried on the MNIST database. An abbreviated list from http://yann.lecun.com/exdb/mnist/ is shown in Table 1.

Table 1. Previous results on MNIST database.

Method	Test Error Rate (%)
linear classifier (1-layer NN)	12.0
linear classifier (1-layer NN) [deskewing]	8.4
pairwise linear classifier	7.6
K-nearest-neighbors, Euclidean	5.0
K-nearest-neighbors, Euclidean, deskewed	2.4
40 PCA + quadratic classifier	3.3
1000 RBF + linear classifier	3.6
K-NN, Tangent Distance, 16×16	1.1
SVM deg 4 polynomial	1.1
Reduced Set SVM deg 5 polynomial	1.0
Virtual SVM deg 9 poly [distortions]	0.8
2-layer NN, 300 hidden units	4.7
2-layer NN, 300 HU, [distortions]	3.6
2-layer NN, 300 HU, [deskewing]	1.6
2-layer NN, 1000 hidden units	4.5
2-layer NN, 1000 HU, [distortions]	3.8
3-layer NN, 300+100 hidden units	3.05
3-layer NN, 300+100 HU [distortions]	2.5
3-layer NN, 500+150 hidden units	2.95
3-layer NN, 500+150 HU [distortions]	2.45
LeNet-1 [with 16×16 input]	1.7
LeNet-4	1.1
LeNet-4 with K-NN instead of last layer	1.1
LeNet-4 with local learning instead of ll	1.1
LeNet-5, [no distortions]	0.95
LeNet-5, [huge distortions]	0.85
LeNet-5, [distortions]	0.8
Boosted LeNet-4, [distortions]	0.7
K-NN, shape context matching	0.67

Reference: Y. LeCun, L. Bottou, Y. Bengio, and P. Haffner. "Gradient-based learning applied to document recognition." Proceedings of the IEEE, 86(11):2278-2324, November 1998. http://yann.lecun.com/exdb/publis/index.html#lecun-98

Experimental Design

To construct the dataset, we performed the following steps:

- We selected a random subset of the "four" and "nine" patterns from the training and test sets of the MNIST.
- We normalized the database so that the pixel values would be in the range [0, 1]. We thresholded values below 0.5 to increase data sparsity.
- We constructed a feature set, which consists of the original variables (normalized pixels) plus a randomly selected subset of products of pairs of variables. The pairs were sampled such that each pair member is normally distributed in a region of the image slightly biased upwards. The rationale beyond this choice is that pixels that are discriminative of the "four/nine" separation are more likely to fall in that region (See Figure 2).

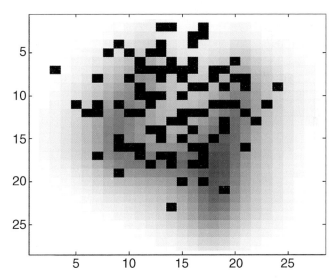

Fig. 2. Example of a randomly selected subset of pixels in the region of interest. Pairs of pixels used as features in dataset B use pixels drawn randomly according to such a distribution.

- We eliminated all features that had only zero values.
- Of the remaining features, we selected all the original pixels and complemented them with pairs to attain the number of 2500 features.

- Another 2500 pairs were used to construct "probes": the values of the features were individually permuted across patterns (column randomization). In this way we obtained probes that are similarly distributed to the other features.
- We randomized the order of the features.
- We quantized the data to 1000 levels.
- The data set was split into training, validation, and test set, by putting an equal amount of patterns of each class in every set.

In spite of the fact that the data are rather sparse (about 13% of the values are non-zero), we saved the data as a non-sparse matrix because we found that it can be compressed better in this way.

Data Statistics

Statistics about the data are presented in Tables 2 and 3.

Table 2. GISETTE Datasets: Number of examples and class distribution.

	Positive ex.	Negative ex.	Total	Check sum
Training set	3000	3000	6000	3197297133
Validation set	500	500	1000	529310977
Test set	3250	3250	6500	3404549076
All	6750	6750	13500	7131157186

Table 3. GISETTE variable statistics.

Real variables	Random probes	Total
2500	2500	5000

All variables are integer quantized on 1000 levels. There are no missing values. The data are rather sparse. Approximately 13% of the entries are non zero. The data was saved as a non-sparse matrix, because it compresses better in that format.

Results of the Run of the Lambda Method and Linear SVM

Before the benchmark, we ran some simple methods to determine what an appropriate number of examples should be. The "lambda" method (provided with the sample code) had approximately a 30% test error rate and a linear SVM trained on all features a 3.5% error rate. The rule of thumb number_of_test_examples $= 100/$test_errate $= 100/0.035{=}2857$. However, other explorations we made with on-linear SVMs and the examination of previous performances obtained on the entire MNIST dataset indicate that the best error rates could be below 2%. A test set of 6500 example should allow error rates as low as 1.5%. This motivated our test set size choice. The best benchmark error rates confirmed that our estimate was just right.

DEXTER

The task of DEXTER is to filter texts about "corporate acquisitions". This is a two-class classification problem with sparse continuous input variables.

Original Owners

The original data set we used is a subset of the well-known Reuters text categorization benchmark. The data was originally collected and labeled by Carnegie Group, Inc. and Reuters, Ltd. in the course of developing the CONSTRUE text categorization system. It is hosted by the UCI KDD repository: `http://kdd.ics.uci.edu/databases/reuters21578/reuters21578.html`. David D. Lewis is hosting valuable resources about this data (see `http://www.daviddlewis.com/resources/testcollections/reuters21578/`). We used the "corporate acquisition" text classification class pre-processed by Thorsten Joachims ¡thorsten@joachims.org¿. The data are one of the examples of the software package SVM-Light., see `http://svmlight.joachims.org/`. The example can be downloaded from `ftp://ftp-ai.cs.uni-dortmund.de/pub/Users/thorsten/svm_light/examples/example1.tar.gz`.

Past Usage

Hundreds of articles have appeared on this data. For a list see: `http://kdd.ics.uci.edu/databases/reuters21578/README.txt` Also, 446 citations including "Reuters" were found on CiteSeer: `http://citeseer.nj.nec.com`.

Experimental Design

The original data formatted by Thorsten Joachims is in the *ag-of-words* representation. There are 9947 features (of which 2562 are always zeros for all the examples) that represent frequencies of occurrence of word stems in text. Some normalizations have been applied that are not detailed by Thorsten

Joachims in his documentation. The task is to learn which Reuters articles are about *corporate acquisitions*.

The frequency of appearance of words in text is known to follow approximately Zipf's law (for details, see e.g. http://linkage.rockefeller.edu/ wli/zipf/). According to that law, the frequency of occurrence of words, as a function of the rank k when the rank is determined by the frequency of occurrence, is a power-law function $P_k \sim 1/k^a$ with the exponent a close to unity. We estimated that $a = 0.9$ gives us a reasonable approximation of the distribution of the data (see Figure 1). The following steps were taken to

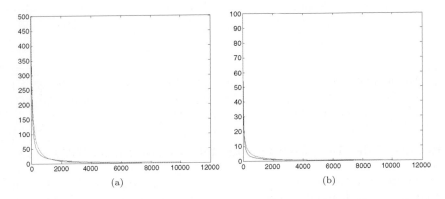

(a)   (b)

Fig. 1. Comparison of the real data and the random probe data distributions. We plot as a function of the rank of the feature: (a) the **number** of non-zero values of a given feature and (b) the **sum** of non-zero values of a given feature. The rank is given by the number of non-zero features. Red: real data. Blue: simulated data.

prepare our version of the dataset:

- We concatenated the original training set (2000 examples, class balanced) and test set (600 examples, class balanced).
- We added to the original 9947 features, 10053 features drawn at random according to Zipf law, to obtain a total of 20000 features. Fraction of non-zero values in the real data: 0.46%. Fraction of non-zero values in the simulated data: 0.5%.
- The feature values were quantized to 1000 levels.
- The order of the features and the order of the patterns were randomized.
- The data was split into training, validation, and test sets, with balanced numbers of examples of each class in each set.

Data Statistics

Statistics about the data are presented in Tables 1 and 2.

Table 1. Dexter Datasets: Number of examples and class distribution.

	Positive ex.	Negative ex.	Total	Check sum
Training set	150	150	300	2885106
Validation set	150	150	300	2887313
Test set	1000	1000	2000	18992356
All	1300	1300	2600	24764775

Table 2. Dexter variable statistics.

Real variables	Random probes	Total
9947	10053	20000

All variables are integer quantized on 1000 levels. There are no missing values. The data are very sparse. Approximately 0.5% of the entries are non zero. The data was saved as a sparse-integer matrix.

Results of the Run of the Lambda Method and Linear SVM

Before the benchmark, we ran some simple methods to determine what an appropriate number of examples should be. The "lambda" method (provided with the sample code) had approximately a 20% test error rate and a linear SVM trained on all features a 5.8% error rate. The rule of thumb number_of_test_examples = 100/test_errate = 100/0.058= 1724 made it likely that 2000 test examples will be sufficient to obtains statistically significant results. The benchmark test results confirmed that this estimate was correct.

Dorothea

The task of DOROTHEA is to predict which compounds bind to Thrombin. This is a two-class classification problem with sparse binary input variables.

Original Owners

The dataset with which DOROTHEA was created is one of the KDD (Knowledge Discovery in Data Mining) Cup 2001. The original dataset and papers of the winners of the competition are available at: http://www.cs.wisc.edu/~dpage/kddcup2001/. DuPont Pharmaceuticals graciously provided this data set for the KDD Cup 2001 competition. All publications referring to analysis of this data set should acknowledge DuPont Pharmaceuticals Research Laboratories and KDD Cup 2001.

Past Usage

There were 114 participants to the competition that turned in results. The winner of the competition was Jie Cheng (Canadian Imperial Bank of Commerce). His presentation is available at: http://www.cs.wisc.edu/~dpage/kddcup2001/Hayashi.pdf. The data was also studied by Weston and collaborators: J. Weston, F. Perez-Cruz, O. Bousquet, O. Chapelle, A. Elisseeff and B. Schoelkopf. "Feature Selection and Transduction for Prediction of Molecular Bioactivity for Drug Design". Bioinformatics. A lot of information is available from Jason Weston's web page, including valuable statistics about the data: http://www.kyb.tuebingen.mpg.de/bs/people/weston/kdd/kdd.html.

One binary attribute (active A or inactive I) must be predicted. Drugs are typically small organic molecules that achieve their desired activity by binding to a target site on a receptor. The first step in the discovery of a new drug is usually to identify and isolate the receptor to which it should bind, followed by testing many small molecules for their ability to bind to the target site. This leaves researchers with the task of determining what separates the active (binding) compounds from the inactive (non-binding) ones. Such a determination can then be used in the design of new compounds that not only bind, but also have all the other properties required for a drug (solubility, oral absorption, lack of side effects, appropriate duration of action, toxicity, etc.). The original training data set consisted of 1909 compounds tested for their ability to bind to a target site on thrombin, a key receptor in blood clotting. The chemical structures of these compounds are not necessary for our analysis and were not included. Of the training compounds, 42 are active (bind well) and the others are inactive. To simulate the real-world drug design environment, the test set contained 634 additional compounds that were in fact generated based on the assay results recorded for the training set. Of the test compounds, 150 bind well and the others are inactive. The compounds in the test set were made after chemists saw the activity results for the training set, so the test set had a higher fraction of actives than did the training set in the original data split. Each compound is described by a single feature vector comprised of a class value (A for active, I for inactive) and 139,351 binary features, which describe three-dimensional properties of the molecule. The definitions of the individual bits are not included we only know that they were generated in an internally consistent manner for all 1909 compounds. Biological activity in general, and receptor binding affinity in particular, correlate with various structural and physical properties of small organic molecules. The task is to determine which of these properties are critical in this case and to learn to accurately predict the class value. In evaluating the accuracy, a differential cost model was used, so that the sum of the costs of the actives will be equal to the sum of the costs of the inactives.

To outperform these results, the paper of Weston et al., 2002, utilizes the combination of an efficient feature selection method and a classification strategy that capitalizes on the differences in the distribution of the training and the test set. First they select a small number of relevant features (less than 40) using an unbalanced correlation score:

$$ f_j = \sum_{y_i=1} X_{i,j} - \lambda \sum_{y_i=-1} X_{i,j} $$

where the score for feature j is f_j, the training data are a matrix X where the columns are the features and the examples are the rows, and a larger score is assigned to a higher rank. The coefficient λ is a positive constant. The authors suggest to take $\lambda > 3$ to select features that have non-zero entries only for positive examples. This score encodes the prior information that the data are unbalanced and that only positive correlations are likely to be useful. The score has an information theoretic motivation, see the paper for details.

Experimental Design

The original data set was modified for the purpose of the feature and variable selection benchmark:

- The original training and test sets were merged.
- The features were sorted according to the f_j criterion with $\lambda = 3$, computed using the original test set (which is richer is positive examples).
- Only the top ranking 100000 original features were kept.
- The all zero patterns were removed, except one that was given label -1.
- For the second half lowest ranked features, the order of the patterns was individually randomly permuted (in order to create "random probes").
- The order of the patterns and the order of the features were globally randomly permuted to mix the original training and the test patterns and remove the feature order.
- The data was split into training, validation, and test set while respecting the same proportion of examples of the positive and negative class in each set.

We are aware that out design biases the data in favor of the selection criterion f_j. It remains to be seen however whether other criteria can perform better, even with that bias.

Data Statistics

Statistics about the data are presented in Tables 1 and 2.

Table 1. DOROTHEA Datasets: Number of examples and class distribution.

	Positive ex.	Negative ex.	Total	Check sum
Training set	78	722	800	713978
Validation set	34	316	350	330556
Test set	78	722	800	731829
All	190	1760	1950	1776363

We mapped Active compounds to the target value +1 (positive examples) and Inactive compounds to the target value -1 (negative examples). We provide in the last column the total number of non-zero values in the data sets.

Table 2. DOROTHEA variable statistics.

Real variables	Random probes	Total
50000	50000	100000

All variables are binary. There are no missing values. The data are very sparse. Less than 1% of the entries are non zero (1776363/(1950*100000)). The data was saved as a sparse-binary matrix.

The following table summarizes the number of non-zero features in various categories of examples in the entire data set.

Table 3. DOROTHEA Non-zero features.

Type	Min	Max	Median
Positive examples	687	11475	846
Negative examples	653	3185	783
All	653	11475	787

Results of the Run of the Lambda Method and Linear SVM

Before the benchmark, we ran some simple methods to determine what an appropriate number of examples should be. The "lambda" method (provided with the sample code) had a 21% test error rate. We chose this method because it outperformed methods used in the KDD benchmark on this dataset, according to the paper of Weston et al and we could not outperform it with the linear SVM. The rule of thumb number_of_test_examples = 100/test_errate = 100/0.21 = 476 made it likely that 800 test examples will be sufficient to obtains statistically significant results. This was slightly underestimated: the best benchmark results are around 11% error, thus 900-1000 test examples would have been better.

Madelon

Summary. The task of Madelon is to classify random data. This is a two-class classification problem with sparse binary input variables.

Original Owners

The data are synthetic. It was generated by the Matlab program hypercube_data.m, which is appended.

Past Usage

None, although the idea of the program is inspired by: Grafting: Fast, Incremental Feature Selection by Gradient Descent in Function Space Simon Perkins, Kevin Lacker, James Theiler; JMLR, 3(Mar):1333-1356, 2003. http://www.jmlr.org/papers/volume3/perkins03a/perkins03a.pdf

Experimental Design

To draw random data, the program takes the following steps:

- Each class is composed of a number of Gaussian clusters. N(0,1) is used to draw for each cluster num_useful_feat examples of independent features.
- Some covariance is added by multiplying by a random matrix A, with uniformly distributed random numbers between -1 and 1.
- The clusters are then placed at random on the vertices of a hypercube in a num_useful_feat dimensional space. The hypercube vertices are placed at values class_sep.
- Redundant features are added. They are obtained by multiplying the useful features by a random matrix B, with uniformly distributed random numbers between -1 and 1.

- Some of the previously drawn features are repeated by drawing randomly from useful and redundant features.
- Useless features (random probes) are added using N(0,1).
- All the features are then shifted and rescaled randomly to span 3 orders of magnitude.
- Random noise is then added to the features according to N(0,0.1).
- A fraction flip_y of labels are randomly exchanged.

To illustrate how the program works, we show a small example generating a XOR-type problem. There are only 2 relevant features, 2 redundant features, and 2 repeated features. Another 14 random probes were added. A total of 100 examples were drawn (25 per cluster). Ten percent of the labels were flipped.

In Figure 1, we show all the scatter plots of pairs of features, for the useful and redundant features. For the two first features, we recognize a XOR-type pattern. For the last feature, we see that after rotation, we get a feature that alone separates the data pretty well. In Figure 2, we show the heat map of the data matrix. In Figure 3, we show the same matrix after random permutations of the rows and columns and grouping of the examples per class. We notice that the data looks pretty much like white noise to the eye.

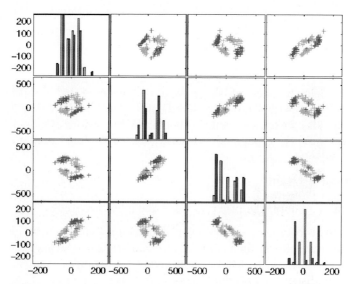

Fig. 1. Scatter plots of the XOR-type example data for pairs of useful and redundant features. Histograms of the examples for the corresponding features are shown on the diagonal.

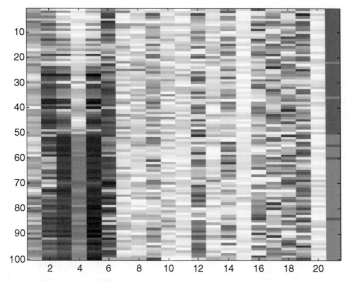

Fig. 2. Heat map of the XOR-type example data. We show all the coefficients of the data matrix. The intensity indicates the magnitude of the coefficients. The color indicates the sign. In lines, we show the 100 examples drawn (25 per cluster). I columns, we show the 20 features. Only the first 6 ones are relevant: 2 useful, 2 redundant, 2 repeated. The data have been shifted and scaled by column to look "more natural". The last column shows the target values, with some "flipped" labels.

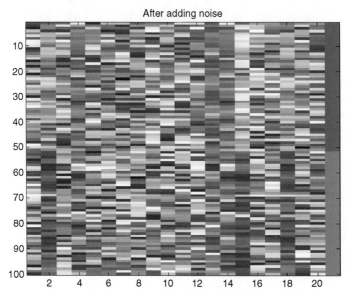

Fig. 3. Heat map of the XOR-type example data. This is the same matrix as the one shown in Figure 2. However, the examples have been randomly permuted and grouped per class. The features have also been randomly permuted. Consequently, after normalization, the data looks very uninformative to the eye.

We then drew the data used for the benchmark with the following choice of parameters:

```
num_class=2;                    % Number of classes.
num_pat_per_cluster=250;        % Number of patterns per cluster.
num_useful_feat=5;              % Number of useful features.
num_clust_per_class=16;         % Number of cluster per class.
num_redundant_feat=5;           % Number of redundant features.
num_repeat_feat=10;             % Number of repeated features.
num_useless_feat=480;           % Number of useless features.
class_sep=2;                    % Cluster separation scaling factor.
flip_y = 0.01;                  % Fraction of flipped labels.
```

Figure 4 and Figure 5 show the appearance of the data.

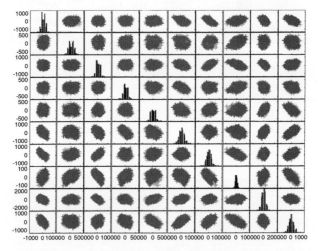

Fig. 4. Scatter plots of the benchmark data for pairs of useful and redundant features. We can see that the two classes overlap completely in all pairs of features. This is normal because 5 dimensions are needed to separate the data.

Data Statistics

Statistics about the data are presented in Tables 1 and 2.

Two additional test sets of the same size were drawn similarly and reserved to be able to test the features selected by the benchmark participants, in case it becomes important to make sure they trained only on those features.

All variables are integer. There are no missing values. The data are not sparse. The data was saved as a non-sparse matrix.

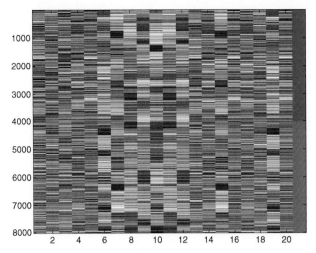

Fig. 5. Heat map of the benchmark data for the relevant features (useful, redundant, and repeated). We see the clustered structure of the data.

Table 1. Madelon Datasets: Number of examples and class distribution.

	Positive ex.	Negative ex.	Total	Check sum
Training set	1000	1000	2000	488026911
Validation set	300	300	600	146425645
Test set	900	900	1800	439236341
All	2200	2200	4400	1073688897

Table 2. Madelon variable statistics.

Real variables	Random probes	Total
20	480	500

Results of the Run of the Lambda Method and Linear SVM

Before the benchmark, we ran some simple methods to determine what an appropriate number of examples should be. The "lambda" method (provided with the sample code) performs rather poorly on this highly non-linear problem (41% error). We used the K-nearest neighbor method, with K=3, with only the 5 useful features. With the 2000 training examples and 2000 test examples, we obtained 10% error. The rule of thumb number_of_test_examples = 100/test_errate = 100/0.1 = 1000 makes it likely that 1800 test examples will be sufficient to obtains statistically significant results. The benchmark results confirmed that this was a good (conservative) estimate.

MATLAB Code of the Lambda Method

```
function idx=lambda_feat_select(X, Y, num)
%idx=lambda_feat_select(X, Y, num)
% Feature selection method that ranks according to the dot
% product with the target vector. Note that this criterion
% may not deliver good results if the features are not
% centered and normalized with respect to the example distribution.

% Isabelle Guyon -- August 2003 -- isabelle@clopinet.com

fval=Y'*X; [sval, si]=sort(-fval); idx=si(1:num);

function [W,b]=lambda_classifier(X, Y)
%[W,b]=lambda_classifier(X, Y)
% This simple but efficient two-class linear classifier
% of the type Y_hat=X*W'+b
% was invented by Golub et al.
% Inputs:
% X -- Data matrix of dim (num examples, num features)
% Y -- Output matrix of dim (num examples, 1)
% Returns:
% W -- Weight vector of dim (1, num features)
% b -- Bias value.

% Isabelle Guyon -- August 2003 -- isabelle@clopinet.com

Posidx=find(Y>0); Negidx=find(Y<0); Mu1=mean(X(Posidx,:));
Mu2=mean(X(Negidx,:)); Sigma1=std(X(Posidx,:),1);
Sigma2=std(X(Negidx,:),1); W=(Mu1-Mu2)./(Sigma1+Sigma2);
B=(Mu1+Mu2)/2; b=-W*B';
```

MATLAB Code Used to Generate Madelon

```
function [XP,YP,ixrp,iyrp, xrp,yrp,all_C,A,B,rf,shift,scale ] =
hypercube_data(num_class, num_useful_feat, num_clust_per_class,
num_pat_per_cluster, num_redundant_feat, num_repeat_feat,
num_useless_feat, class_sep, flip_y, num_repeat_val, rnd, debug,
xrp, yrp,all_C,A,B,rf,shift,scale)
%[XP,YP,ixrp,iyrp, xrp,yrp,all_C,A,B,rf,shift,scale ] =
hypercube_data(num_class, num_useful_feat, num_clust_per_class,
num_pat_per_cluster, num_redundant_feat, num_repeat_feat,
num_useless_feat, class_sep, flip_y, num_repeat_val, rnd, debug,
xrp, yrp,all_C,A,B,rf,shift,scale)
% Draws a pattern recognition problem at random, for a
% num_class-class problem.
% Useful features:
%   Each class is composed of a number of Gaussian clusters that
%   are on the vertices of a hypercube in a subspace of dimension
%   num_useful_feat. N(0,1) is used to draw the examples of
%   independent features for each cluster.
%   Some covariance is added by multiplying by a random matrix A,
%   with uniformly distributed random numbers between -1 and 1.
%   The clusters are then placed on the hypercube vertices.
%   The hypercube vertices are placed at values +-class_sep.
% Redundant features:
%   Useful features are multiplied by a random matrix B,
%   with uniformly distributed random numbers between -1 and 1.
% Repeated features:
%   Drawn randomly from useful and redundant features.
% Useless features:
%   Additional features drawn at random not related to the concept.
% Features are then shifted and rescaled randomly to span 3 orders
% of magnitude. Random noise is then added to the features according
% to N(0,.1) to create several replicates. If flip_y is provided, a
% random fraction flip_y of labels are randomly exchanged.
```

```
% Aknowledgement: The idea is inspired by the work of Simon Perkins.
% Inputs:
%  num_class              -- Number of classes
%  num_useful_feat        -- Number of features initially drawn to
%                            explain the concept
%  num_clust_per_class    -- Number of cluster per class
%  num_pat_per_cluster    -- Number of patterns per cluster (all
%                            balanced for now, can be generalized to
%                            imbalanced classes by taking a subset of
%                            samples of each class)
%  num_redundant_feat     -- Number of features linearly dependent
%                            upon the useful features
%  num_repeat_feat        -- Number of features repeating the
%                            previous ones (drawn at random)
%  num_useless_feat       -- Number of features dran at random
%                            regardless of class label information
%  class_sep              -- Factor multiplying the hypercube
%                            dimension
%  flip_y                 -- Fraction of y labels to be randomly
%                            exchanged
%  num_repeat_val         -- number of times each entry is repeated
%                            (modulo some noise)
%  rnd                    -- Flag to enable or disable random
%                            permutations
%  debug                  -- 0/1 flag.
% Returns:
%  XP                     -- Matrix (num_pat, num_feat,
%                            num_repeat_val) of randomly permuted
%                            features
%  YP                     -- Vector of 0,1...num_class target class
%                            labels (in random order, to be used
%                            eventually for clustering)
%  ixrp                   -- permutation matrix to be used to
%                            restore the original feature order
%  iyrp                   -- permutation matrix to be used to restore
%                            the original pattern order (class labels
%                            of the same class are consecutive and
%                            there are the same number of example
%                            per class, before label corruption)
%                            Y=YP(iyrp); X=XP(iyrp,ixrp);
%  all_C                  -- A matrix (2^num_useful_feat,
%                            num_useful_feat) of hypercube vertices
%                            where to place the cluter centers.
%  A                      -- Matrix used to correlate the useful
%                            features.
%  B                      -- Matrix used to create dependent
%                            (redundant) features.
%  rf                     -- Indices of repeated features.
%  shift                  -- Shift applied.
```

```
% scale                    -- Scale applied.

% Isabelle Guyon -- July 2003 -- isabelle@clopinet.com

if nargin<8, class_sep=1; end
if nargin<9, flip_y=0; end
if nargin<10, num_repeat_val=1; end
if nargin<11, rnd=0; end % disable random permutation
if nargin<12, debug=0; end
if nargin<13, xrp=[]; end
if nargin<14, yrp=[]; end
if nargin<15, all_C=[]; end
if nargin<16, A={}; end
if nargin<17, B=[]; end
if nargin<18, rf=[]; end
if nargin<19, shift=[]; end
if nargin<20, scale=[]; end

% Count features and patterns
num_feat=num_useful_feat + num_repeat_feat + ...
         num_redundant_feat + num_useless_feat;
num_pat_per_class=num_pat_per_cluster*num_clust_per_class;
num_pat=num_pat_per_class*num_class;
X=zeros(num_pat, num_feat);

% Attribute class labels
y=0:num_class-1;
Y=repmat(y, num_pat_per_class, 1);
Y=Y(:);

% Hypercube design
is_XOR=0;
if num_useful_feat==2 & num_class==2 & num_clust_per_class==2,
    is_XOR=1;
    all_C=[-1 -1; 1 1; 1 -1; -1 1]; % XOR
else
    if isempty(all_C)
        fprintf('New C\n');
        all_C=2*ff2n(num_useful_feat)-1;
        rndidx=randperm(size(all_C,1));
        all_C=all_C(rndidx,:);
    end
end

% Draw A
if isempty(A)
    fprintf('New A\n');
    for k=1:num_class*num_clust_per_class
        A{k} = 2*rand(num_useful_feat, num_useful_feat)-1;
```

```
        end
end
% Loop over all clusters
for k=1:num_class*num_clust_per_class
    % define the range of patterns of that cluster
    kmin=(k-1)*num_pat_per_cluster+1;
    kmax=kmin+num_pat_per_cluster-1;
    kidx=kmin:kmax;
    % Draw n features independently at random
    X(kidx,1:num_useful_feat)= ...
    random('norm', 0, 1, num_pat_per_cluster, num_useful_feat);
    % Multiply by a random matrix to add feature co-variance
    X(kidx,1:num_useful_feat)=X(kidx,1:num_useful_feat)*A{k};
    % Shift the center off zero to separate the clusters
    C=all_C(k,:)*class_sep;
    X(kidx,1:num_useful_feat) = ...
    X(kidx,1:num_useful_feat) + repmat(C, num_pat_per_cluster, 1);
end

% Create redundant features by multiplying by a random matrix
if isempty(B),
    fprintf('New B\n');
    B = 2*rand(num_useful_feat, num_redundant_feat)-1;
end
X(:,num_useful_feat+1:num_useful_feat+num_redundant_feat)= ...
X(:,1:num_useful_feat)*B;

% Repeat num_repeat_feat features, chosen at random among
% useful and redundant feat
nf=num_useful_feat+num_redundant_feat;
if isempty(rf)
    fprintf('New rf\n');
    rf=round(1+rand(num_repeat_feat,1)*(nf-1));
end
X(:,nf+1:nf+num_repeat_feat)=X(:,rf);

% Add useless features : these are uncorrelated with one another,
% but could be correlated :=)
X(:,num_feat-num_useless_feat+1:num_feat)= ...
random('norm', 0, 1,num_pat, num_useless_feat);

% Add random y label errors
num_err_pat = round(num_pat*flip_y);
rp=randperm(num_pat);
fi=rp(1:num_err_pat);
Y(fi)=mod(Y(fi)+round(rand(num_err_pat,1)*(num_class-1)),num_class);

% Randomly shift and scale
if isempty(shift)
```

```
    fprintf('New shift\n');
    shift=rand(num_feat,1);
end
if isempty(scale)
    fprintf('New scale\n');
    scale=1+100*rand(num_feat,1);
end
X=X+repmat(shift',num_pat,1); X=X.*repmat(scale',num_pat,1);

% Randomly permute the features and patterns
if isempty(xrp)
    fprintf('New xrp, yrp\n');
    if rnd
        xrp=randperm(num_feat);
        yrp=randperm(num_pat);
    else
        xrp=1:num_feat;
        yrp=1:num_pat;
    end
end
XP0=X(yrp,xrp); YP=Y(yrp);

% Create inverse random indices
ixrp(xrp)=1:num_feat;
iyrp(yrp)=1:num_pat;

% Create several replicates by adding a little bit of random noise
XP=zeros(num_pat, num_feat, num_repeat_val);
for k=1:num_repeat_val
    N=random('norm', 0, .1*sqrt(num_repeat_val), num_pat, num_feat);
    XP(:,:,k)=XP0.*(1+N);
end
```

Feature Selection Challenge Fact Sheets

Fact Sheet for Chapter 10

High Dimensional Classification with Bayesian Neural Networks and Dirichlet Diffusion Trees

Radford M. Neal[1] and Jianguo Zhang[2]

[1] Dept. of Statistics and Dept. of Computer Science, University of Toronto
   radford@stat.utoronto.ca, http://www.cs.utoronto.ca/~radford/
[2] Dept. of Statistics, University of Toronto
   jianguo@stat.utoronto.ca

Methods

For `BayesNN-DFT-combo`, dimensionality was reduced to no more than about a thousand by Principal Component Analysis or by feature selection based on univariate significance tests, sometimes after transforming the original features to increase correlation with the class. Classification was then done by either a Bayesian neural network or by a method based on Dirichlet diffusion trees, using Automatic Relevance Determination priors to adjust the influence of the features or principal components (and sometimes to further prune the set of features to be used). Results on the validation set were used to make some model choices (eg, whether to use a neural network or Dirichlet diffusion tree model, and whether to use a feature subset or principal components). There was also a considerable element of human judgement involved. For three of the datasets, the method used was the same as for `BayesNN-large`, and for one dataset, it was also the same as `BayesNN-small`.

For `BayesNN-small`, a subset of no more than about a thousand features was chosen based on univariate significance tests, sometimes after transforming the original features to increase correlation with the class. Classification was then done by a Bayesian neural network, using Automatic Relevance Determination priors to adjust the influence of the features (and sometimes to further prune the set of features to be used). Results on the validation set were used to make some model choices (eg, among different size feature sets). There was also a considerable element of human judgement involved.

For `BayesNN-large` dimensionality was reduced to no more than about a thousand by Principal Component Analysis or by feature selection based on univariate significance tests, sometimes after transforming the original features to increase correlation with the class. Classification was then done by a Bayesian neural network, using Automatic Relevance Determination priors to adjust the influence of the features or principal components (and sometimes to further prune the set of features to be used). Results on the validation set were used to make some model choices (eg, whether to use a feature subset or principal components). There was also a considerable element of human judgement involved. For two of the datasets, the method used was the same as for `BayesNN-small`.

Results

`BayesNN-DFT-combo` was the winning entry, among both the December 1st and December 8th entries. Its performance was the best or close to the best for all datasets.

`BayesNN-small` was the second-place entry, among both the December 1st and December 8th entries. Its overall balanced error rate was very close to that of the third-place entry, `BayesNN-large`, but it used many fewer features (only 4.74%). None of the other original entries using such a small number of features had performance comparable to `BayesNN-small`.

`BayesNN-large` was the third-place entry, among both the December 1st and December 8th entries. It's overall balanced error rate was very close to that of the second-place entry, `BayesNN-small`.

Code

The Bayesian neural network and Dirichlet diffusion tree programs are part of the Software for Flexible Bayesian Modeling available from `http://www.cs.utoronto.ca/~radford/`. Preprocessing scripts are not available, but those for the neural network methods were similar (albeit more complex and ad hoc) to those for the "New-Bayes" methods, which are available from the same web site.

Keywords

Significance tests, Principal Components Analysis, Bayesian learning, Neural networks, Dirichlet diffusion trees, Automatic Relevance Determination.

Fact Sheet for Chapter 11

Ensembles of Regularized Least Squares Classifiers for High-Dimensional Problems

Kari Torkkola[1] and Eugene Tuv[2]

[1] Motorola, Intelligent Systems Lab, Tempe, AZ, USA,
Kari.Torkkola@motorola.com
[2] Intel, Analysis and Control Technology, Chandler, AZ, USA,
eugene.tuv@intel.com

Method

Centering was used as preprocessing for all data, and standardization was used in those cases where it improved the cross-validation (CV) results. For one data set, we used mutual information to weigh the variables. For feature selection, we used feature ranking using importances from a Random Forest. The number of features was selected with 10-fold CV. We used ensembles of RLSCs (Regularized Least Squares Classifier) with Gaussian kernels for classification. All hyperparameters are adjusted after feature selection, using again 10-fold CV with the same training data.

Results

In the challenge, for the December 1st submissions, our best entry is the 5th, using the criterion of the organizers. We also conducted comparison experiments with single RLSC as classifier and found that ensemble techniques reduce the amount of hyperparameters.

Code

Random Forest experiments were run using Intel's IDEAL (not available). RLSC was written using MATLAB (one-liner), and all cross-validation experimentation was done using MATLAB (code not available).

Keywords

standardization, embedded feature selection, Random Forest, feature ranking, L2 norm regularization, RLSC, bagging, 10-fold cross-validation, ensemble method.

Fact Sheet for Chapter 12

Combining SVMs with Various Feature Selection Strategies

Yi-Wei Chen and Chih-Jen Lin

Department of Computer Science, National Taiwan University, Taipei 106, Taiwan

Method

We linearly scale each feature to $[0, 1]$ as preprocessing. No other preprocessing is conducted. For feature selection, we use 1) no selection, 2) F-score, and 3) random forest. After ranking the features using the whole training data set, we select the number of features by either human eye or 5-fold cross-validation (CV). For classification, we use support vector machines (SVM). All hyperparameters are adjusted after feature selection, using again 5-fold CV with the same training data. In addition to these filter-type approaches, we also consider radius-margin-bound based SVM, which is a selection method optimizing directly the prediction performance.

Results

For the December 1st submissions, we rank 3rd as a group as our best entry is the 6st, using the criterion of the organizers. In addition, we rank 1st on GISETTE. For the December 8st submissions, we rank 2nd as a group and our best entry is the 4th. Besides good performance, our strategies are simple and easy to implement.

Code

For SVM we use the package LIBSVM, which is available at http://www.csie.ntu.edu.tw/~cjlin/libsvm/. The random forest implementation is available through the R software: http://www.r-project.org/. The radius-margin bound SVM is available at http://www.csie.ntu.edu.tw/~cjlin/libsvmtools/. Our scripts reproducing the competition results are available at http://www.csie.ntu.edu.tw/~b88052/NIPS03/.

Keywords

linear scaling, filter, F-score, Random Forest, radius-margin bound, feature ranking, SVM, grid-search, K-fold cross-validation.

Fact Sheet for Chapter 13

Feature Selection with Transductive Support Vector Machines

Zhili Wu[1] and Chunhung Li[2]

[1] Department of Computer Science, Hong Kong Baptist University
vincent@comp.hkbu.edu.hk
[2] Department of Computer Science, Hong Kong Baptist University
chli@comp.hkbu.edu.hk

Method

We mainly normalize features to be with zero mean and unit standard deviation, and PCA is used to generate features for the ARCENE dataset. For the DOROTHEA and DEXTER datasets we divide each data matrix entry by the square root of the product of row sum and column sum. For feature selection, our first approach is to use filtering scores like Fisher scores or odds-ratio, and through iterative backward search to select a number of features which are ranked high by the scores. The quality of the selected feature subset is indicated by the (five / ten / leave-one-out) cross-validation (CV) accuracy of training data. The wrapper we used is a transductive SVM trained upon all available data including the validation and testing datasets as unlabeled data. We also implement the recursive-feature-elimination (RFE) in conjugation with TSVMs. At each iteration, we use feature weights approximated from the TSVM model at last iteration, to pruning some features with small scores. The number of features is also specified by CV accuracy of TSVM models. Another method we use is the multiplicative updates (MU) with TSVMs. The feature weights for multiplicative updates are obtained by using the technique of RFE. And at the beginning the filtering scores can also be used as feature weights for MU. All hyperparameters are adjusted after feature pruning/deactivation at each iteration, using again CV with the whole data/ only training dataset, guided by grid search on the hyperprameter space. And the final classifiers are mainly TSVMs and SVMs.

Results

For the challenge due on the December 1st, we rank 4th as a group and our best entry is the 7th using the criterion of the organizers. We further submit post-challenge submissions of the Dexter data and demonstrate the TSVM-RFE for Dexter ranks 2nd.

Other Datasets

In addition to the challenge datasets, we used a toy dataset (30 training examples in each of 50 trials, 500 fixed testing examples. 100 features with 94 random probes, continuous, 2 classes).

Code

Our implementation was done in Matlab, together with function calls to SVMs (SVMLight and LibSVM). The python tool for model selection provided by Libsvm is also used. Our scripts are available at `http://www.comp.hkbu.edu.hk/vincent/nipsTransFS.htm`.

Keywords

centering, scaling, standardization, PCA, filter, wrapper, embedded feature selection, correlation coefficient, SVM, feature ranking, ordered feature selection, backward elimination, multiplicative updates, leave-one-out, K-fold cross-validation, transductive SVM, grid-search, cross-validation, K-fold.

Fact Sheet for Chapter 14

Variable Selection using Correlation and SVC Methods: Applications

Amir Reza Saffari Azar Alamdari

Electrical Engineering Department, Sahand University of Technology, Mellat Blvd., Tabriz, Iran `amir@ymer.org`

Method

- First of all, constant variables, which their values do not change over the training set, are detected and removed from the dataset.
- The variables are normalized to have zero mean values and also to fit in the $[-1, 1]$ range, except the Dorothea (in Dorothea only zero values are converted to -1).
- For each dataset, using a k-fold cross-validation (k depends on the dataset), a MLP neural network with one hidden layer is trained to estimate the number of neurons in the hidden layer.
- The correlation and SVC values are calculated and sorted for each variable in the dataset.
- The first estimation for the number of good variables in each dataset is computed using a simple cross-validation method for the MLP predictor in step 2. Since an online validation test was provided through the challenge website, these numbers were optimized in next steps to be consistent with the actual preprocessing and also predictors.
- 25 MLP networks with different randomly chosen initial weights are trained on the selected subset using SCG algorithm. The transfer function of each neuron is selected to be tangent sigmoid for all predictors. The number of neurons in the hidden layer is selected on the basis of the experimental results of the variable selection step, but is tuned manually according to the online validation tests.
- After the training, those networks with acceptable training error performances are selected as committee members (because in some cases the networks are stuck to the local minima during the training sessions). This selection procedure is carried out by filtering out low performance networks using a threshold on the training error.
- For validation/test class prediction, the output values of the committee networks are averaged to give the overall confidence about the class labels. The sign of this confidence value gives the final predicted class label.

- The necessity of a linear PCA preprocessing method usage is also determined for each dataset by applying the PCA to the selected subset of variables and then comparing the validation classification results to the non-preprocessing system.
- These procedures are applied for both correlation and SVC ranking methods in each dataset, and then one with higher validation performance (lower classification error) and also lower number of variables is selected as the basic algorithm for the variable selection in that dataset.
- Using online validation utility, the number of variables and also the number of neurons in the hidden layer of MLPs are tuned manually to give the best result.

Results

- In the challenge, for the December 1st submissions, the Collection2 entry ranked 5th as a group and the 14rd as the best entry, using the criterion of the organizers.
- The correlation and SVC variable ranking methods are very simple, easy to implement, and computational time efficient algorithms which have relatively good performance compared to other complex methods. These methods are very useful when the variable space dimension is large and other methods using exhaustive search in subset of possible variables need much more computations.
- Another point is the benefits of using a simple ensemble averaging method over single predictors, especially in situations where generalization is not satisfactory, due to the complexity of the problem, or low number of training examples.

Code

- Our implementation was done in MATLAB 6.5 software `http://www.mathworks.com`
- Some useful programs are available at `http://www.ymer.org/research/variable.htm`

Keywords

Normalization, PCA, Filter Methods, Correlation, Single Variable Classifier (SVC), Ranking, K-fold Cross-Validation, MLP Neural Networks, Ensemble Averaging, Voting System.

Fact Sheet for Chapter 15

Tree-Based Ensembles with Dynamic Soft Feature Selection

Alexander Borisov, Victor Eruhimov and Eugene Tuv

Intel, {alexander.borisov,victor.eruhimov,eugene.tuv}@intel.com

Method

MART-WS: stage-wise stochastic boosting of shallow random trees built on a small intelligently sampled subset of variables. The variable sampling distribution is modified at every iteration to up-weight more relevant features based on dynamically learned importances. The same method was used for all datasets. No preprocessing/normalization was done. All parameters (number of experts, regularization parameter, number of variables sampled, fixed tree depth, etc) were selected using test portion of training data.

Results

7th group, 16th entry. We also compared MART-WS with Freidman's MART (gradient tree boosting) on 3 out of 5 challenge datasets (where it was feasible to run standard MART), UCI datasets, artificial data. In all cases MART-WS showed a substantial reduction in computational complexity without loss of accuracy (it often outperformed MART). For the challenge data MART-WS was up to 100 times faster than MART, and more accurate.

Other Datasets

In addition to challenge data, we used UCI data, and simulated artificial data.

Code

All experiments were run using Intel's IDEAL (not available).

Keywords

tree based ensemble, gradient tree boosting, random forest, variable importance, embedded feature selection, dynamic feature selection.

Fact Sheet for Chapter 16

Sparse, Flexible and Efficient Modeling using L_1 Regularization

Saharon Rosset[1] and Ji Zhu[2]

[1] IBM T.J. Watson Research Center, P.O. Box 218, Yorktown Heights, NY 10598
`srosset@us.ibm.com`
[2] Department of Statistics, University of Michigan, Ann Arbor, MI 48109-1092
`jizhu@umich.edu`

Method

As preprocessing, for each input variable, we computed the fraction τ of non-zero entries and p-value of the univariate t-statistic, then we selected the variable only if the corresponding τ is big enough or p is small enough. For the ARCENE dataset and DOROTHEA dataset, we also computed the principal components (PCA) as features after the pre-selection.

We then applied the regularized optimization scheme, i.e. equation (1) in the chapter by Rosset and Zhu, via different (loss L, penalty J) pairs for different datasets:

- ARCENE: L is the Huberized hinge loss, and J is the L_2-norm penalty, i.e the linear support vector machine.
- DEXTER: L is the Huberized hinge loss, and J is the L_1-norm penalty.
- DOROTHEA: L is the Huberized hinge loss, and J is the L_1-norm penalty.
- GISETTE: L is the exponential loss, J is the L_1-norm penalty, i.e. the AdaBoost.
- MADELON: L is the hinge loss, and J is the L_2-norm penalty, i.e. the support vector machine (using the radial basis kernel).

All hyper-parameters (including the number of features) were selected using 5-fold cross-validation.

Results

In the challenge, for the December 1st submissions, we rank 6th as a group and our best entry is the 16th, using the criterion of the organizers. For three of the five datasets, we used the L_1-norm regularization, which does a kind of automatic continuous variable/feature selection. We also exploited various characteristics of the (loss L, penalty J) pairs so that we could compute the whole solution path efficiently. As we mentioned in the chapter, for most of the datasets, we are simply using the original input variables as the basis functions. Hence the classification boundaries are simply linear hyper-planes, which are obviously not flexible enough. This may help explain why our results are not so satisfactory. One of our future work will be on how to generalize the L_1 regularization to non-parametric settings.

Code

Our implementation was done in R. Our scripts reproducing the results are available at `http://www.stat.lsa.umich.edu/~jizhu/Feature`.

Keywords

PCA, cross-validation, automatic feature selection, L_1-norm regularization, L_2-norm regularization, SVM.

Fact Sheet for Chapter 17

Margin Based Feature Selection and Infogain with Standard Classifiers

Ran Gilad-Bachrach and Amir Navot

The Hebrew University, Jerusalem, Israel
ranb@cs.huji.ac.il,anavot@cs.huji.ac.il

Method

We used a mixture of known methods and one new margin based selection method. For each dataset we chose the combination of methods that we found to be best for this dataset. We did not perform any exhaustive tuning for parameters. The parameters were roughly tuned manually.

No preprocessing was done on some of the datasets; for others we used simple normalization. For one dataset (ARCENE) the data were converted using PCA. For feature selection we used two variants of Infogain and a novel margin-based selection algorithm. For classification we used SVM, transductive SVM, Naïve Bayes (with Good-Turing zero correction) and Aggressive Perceptron.

Results

In the challenge, using the average balance error rate criterion (BER) , which was the criterion during the challenge, we ranked 2^{ed} as a team both for December 1st and 8th, and our best entry was 6th for December 1st and 4th for December 8th. Using the criterion of the organizers, for the December 1st submissions, we ranked 8th as a team and our best entry was 20th. For December 8th, we ranked 7th as a team and our best entry was 15th.

Other Datasets

Aside from the challenge data, we used publicly available AR face dataset (http://rvl1.ecn.purdue.edu/~aleix/aleix_face_DB.html) and Reuters-21578 (can be found at http://www.daviddlewis.com/resources/) to test our new margin-based feature selection algorithms

Code

Our implementation was done in Matlab. A Matlab code for a mature version of our novel margin-based feature selection method is available at `http://www.cs.huji.ac.il/labs/learning/code/feature_selection/` For SVM, We used the SVM tool-box by Gavin Cawley see `http://theoval.sys.uea.ac.uk/~gcc/svm/toolbox/` for code and details. We made some minor changes in this code to make it faster for sparse data. Other Matlab scripts that we have used can be found at `www.cs.huji.ac.il/labs/learning/code/fsc`.

Keywords

PCA, filter, mutual information, margin, SVM, perceptron, optimal Bayes, Good-Turing zero correction, transduction

Fact Sheet for Chapter 18

Bayesian Support Vector Machines
for Feature Ranking and Selection

Wei Chu[1], S. Sathiya Keerthi[2], Chong Jin Ong[3], and Zoubin Ghahramani[1]

[1] Gatsby Computational Neuroscience Unit, University College London, London,
   WC1N 3AR, UK. chuwei@gatsby.ucl.ac.uk, zoubin@gatsby.ucl.ac.uk
[2] Yahoo! Research Lab., Pasadena, CA 91105, USA.
   sathiya.keerthi@overture.com
[3] Department of Mechanical Engineering, National University of Singapore,
   Singapore, 119260. mpeongcj@nus.edu.sg

Method

The feature vectors were normalized to have zero mean and unit variance coordinate-wise. An ARD (automatic relevance determination) Gaussian kernel was employed in Bayesian support vector machines. The ARD parameters were optimized by maximizing the model evidence in the Bayesian framework, and then the features were ranked in descending order using the optimal ARD values. For feature selection, support vector machines with $L-1$ loss function and Gaussian kernel was used as the learning algorithm. A forward selection was carried out to determine the minimal subset of relevant features. The top-ranked features were added into the subset one by one, and the validation error of support vector machines was calculated by 5-fold cross validation. This procedure was repeated as long as adding the next top-ranked feature into the subset did not increase the validation error significantly. This feature subset was then used along with all the training data for modelling.

Results

In the challenge, for the December 1st submissions, we rank 8th as a group and our best entry is the 22rd, using the criterion of the organizers. On the Madelon dataset, our entry was assigned the highest score by the organizers, as we selected a very compact subset of relevant features and yielded very competitive predictive results. The AUC performance of our entries is also very close to the winning entries. We carried out the forward feature selection using Fisher score in feature ranking for comparison purpose. Our approach using ARD variables for feature ranking achieved a more compact feature set than Fisher score ranking, along with better generalization performance.

Code

Our implementation of Bayesian support vector machines was done in ANSI C, which can be downloaded at `http://guppy.mpe.nus.edu.sg/~chuwei/btsvc.htm`.

Keywords

Bayesian support vector machines, Gaussian processes, automatic relevance determination, relevance variable, feature ranking, forward feature selection, K-fold cross-validation.

Fact Sheet for Chapter 19

Nonlinear Feature Selection with the Potential Support Vector Machine

Sepp Hochreiter and Klaus Obermayer

Technische Universität Berlin
Fakultät für Elektrotechnik und Informatik
Franklinstraße 28/29, 10587 Berlin, Germany
{hochreit,oby}@cs.tu-berlin.de

Method

We applied the following protocol.

1. *Feature pre-selection ("present call")*. We removed features which contained more than 70 % constant values across the training set.

2. *Kernel choice.* We checked whether an RBF-kernel or a linear kernel should be used. For the RBF-kernel we used $k_\sigma(\boldsymbol{x}_i, \boldsymbol{x}_j) = \exp\left(\frac{1}{2\,\sigma^2}\,\|\boldsymbol{x}_i - \boldsymbol{x}_j\|^2\right)$. For the kernel check we use a ν-SVM with fixed $\nu = 0.3$ and 3 different σ values for the RBF-kernel: $\sigma_1 =$ "average distance of data points", $\sigma_2 = 0.1\ \sigma_1$, $\sigma_3 = 10.0\ \sigma_1$ on simple centered and rescaled data (standardization). If one of these σ values for the RBF-kernel outperformed the linear kernel on cross-validation runs (see next step), we switched to the RBF-kernel. Only for MADELON we used an RBF-kernel.

3. *P-SVM feature selection combined with s-fold cross-validation hyperparameter optimization.* As classifier we used the ν-SVM with zero offset value and a linear kernel, except for MADELON, where we used an RBF-kernel. In this step we determine (1) a feature ranking, (2) the number of top ranked features to use for classification, and (3) the ν.

   We performed s-fold cross validation on the training set, where for each cross-validation run we determined a feature ranking with the P-SVM. Using this fold-specific feature ranking, we performed a ν-SVM classification, where we used the highest ranked 10, 100, 500, and 1000 features and ν-values from $\{0.1, 0.2, 0.3, 0.5\}$. The combination of feature number and ν which performed best in the s-fold cross validation procedure is considered as optimal.

   In a second phase the number of features was refined with previously chosen optimal ν by testing 4 feature numbers which are equidistant between the two feature numbers with lowest cross-validation error of the first phase. For s we choose

- $s = 10$ for ARCENE and DEXTER,
- $s = 8$ for DOROTHEA, and
- $s = 3$ for GISETTE and MADELON.

Note, that the cross validation is a simplified version of the protocol described in (Hochreiter and Obermayer, 2004).

4. *Feature ranking and model selection.* The features are finally ranked according to the s-fold cross-validation runs. To rank the features we first sorted them according to how often they were selected in a fold. Equally scoring features are sorted according to their average ranking position in the folds. Still equally scoring features are sorted according to their maximal relevance value. Note, that this feature ranking reintroduces redundancies among features because different P-SVM runs may select different features. The optimal number of features and the optimal value of ν was determined in Step 3. With these parameters a final model is selected and applied to the validation and the test set.

Results

For the December 1st submissions, we rank 10th as a group and our best entry is the 26th, using the criterion of the organizers. We further analyzed the results of the challenge with other objectives and demonstrated that our method is one of the best methods for compact feature subsets ($<10\%$ of the features and non-negative score). We conducted other experiments on gene expression data showing that our method yields features sets considerably more compact that others reported in the literature, for similar to better performances (Hochreiter and Obermayer, 2004).

Other Datasets

In addition to challenge data, we used the Weston benchmark dataset, another data set from `http://www.kyb.tuebingen.mpg.de/bs/people/weston/10`, and publicly available gene expression data sets (see Hochreiter and Obermayer, 2004).

Code

We used for the ν-SVM and the C-SVM the "libsvm" software from `http://www.csie.ntu.edu.tw/~cjlin/libsvm/` and the SPIDER software from `http://www.kyb.tuebingen.mpg.de/bs/people/spider/` for the different feature selection techniques (R2W2, Fisher, etc.). Our implementation of the P-SVM was done in C.

Keywords

P-SVM, Potential Support Vector machine, redundancy reduction, compact feature sets, K-fold cross-validation, neural networks.

References

S. Hochreiter and K. Obermayer. Gene selection for microarray data. In B. Schölkopf, K. Tsuda, and J.-P. Vert, editors, *Kernel Methods in Computational Biology*, pages 319–355. MIT Press, 2004.

Keywords

References

Fact Sheet for Chapter 20

Combining a Filter Method with SVMs

Thomas Navin Lal, Olivier Chapelle, and Bernhard Schölkopf

Max-Planck-Institute for Biological Cybernetics, Tübingen, Germany.
`navin@tuebingen.mpg.de`, `olivier.chapelle@tuebingen.mpg.de`,
`bs@tuebingen.mpg.de`

Method

In a first step all data was normalized for numerical reasons. Classification was done using support vector machines. For each data set the kernel function as well as SVM parameters were chosen using heuristics as well as cross-validation prior to feature selection. The obtained parameters were not changed afterwards. Features were ranked with correlation coefficients. The number of features used for classification was obtained in a cross-validation scheme.

Results

For the December 1st data sets, our method was ranked 28th and our group was 14th. We were ranked 11th for the December 8 submission; as a group we were ranked fifths. On the DEXTER data set of December 8 we were ranked first.

Code

All calculations were done with the machine learning toolbox *Spider*, which is developed at the Max-Planck-Institute for Biological Cybernetics in Tübingen Germany. It is written in Matlab® and C++ and is available at `http://www.kyb.tuebingen.mpg.de/bs/people/spider/`.

Keywords

filter, correlation coefficient, support vector machines, feature ranking, K-fold cross-validation.

Fact Sheet for Chapter 21

Feature Selection via Sensitivity Analysis with Direct Kernel PLS

Mark J. Embrechts[1], Robert A. Bress[1], and Robert H. Kewley[2]

[1] Department of Decision Sciences and Engineering Systems Rensselaer Polytechnic Institute, Troy, NY, embrem@rpi.edu
[2] Center for Army Analysis, Washington D.C.

Method

- ARCENE: The preprocessing of the data was based on centering and retaining the 2000 most correlated features. The modeling method is kernel partial least squares (K-PLS) with 5 latent variables and s = 1600, based on cross-validation. Feature selection was based on a repetitive sensitivity analysis, were during each step 10% of the features were dropped (based on 100 bootstrap models with leave 10% out) using K-PLS with s = 1600 for all steps. 514 or 5.14% of the original features were retained based on cross-validation.
- DEXTER: The DEXTER data used the original inputs as features, and the inputs that had more than 1% nonzero entries were retained. In a next phase the 1000 most correlated inputs were retained. Feature selection was based on a repetitive sensitivity analysis, were during each step 10% of the features were dropped (based on 100 bootstrap models with leave 10% out) using K-PLS with 5 latent variables and s = 900. 205 or 1.57% of the original features were retained based on multiple bootstrap cross-validation.
- DOROTHEA: The DOROTHEA data used the original inputs as features, and the inputs that had more than 3% nonzero entries were retained. The data were centered and in a next phase feature selection was based on a repetitive sensitivity analysis, were during each step 10% of the features were dropped (based on 100 bootstrap models with leave 10% out) using K-PLS with 5 latent variables and s = 15. 540 or 0.54% of the original features were retained based on multiple bootstrap cross-validation.
- GISETTE: The data were centered and the 2000 most correlated inputs were retained as features. Feature selection was based on a repetitive sensitivity analysis, were during each step 10% of the features were dropped (based on 10 bootstrap models with leave 10% out) using K-PLS with 5 latent variables and s = 20. 1300 or 26% of the original features were retained based on multiple bootstrap cross-validation.

- MADELON: The data were centered and the 2000 most correlated inputs were retained as features. Feature selection was based on a repetitive sensitivity analysis, were during each step 10% of the features were dropped (based on 100 bootstrap models with leave 10% out) using K-PLS with 7 latent variables and s = 100. 13 or 2.6% of the original features were retained based on bootstrap cross-validation. A classification cut-off was determined based on cross-validation.

Results

We entered results only for the December 1st submission.

- ARCENE: We ranked 4th as a group, with an overall ranking of 12th.
- DEXTER: We ranked 10th as a group, with an overall ranking of 27th.
- DOROTHEA: We ranked 13rd as a group, with an overall raking of 40th.
- GISETTE:We ranked 9th as a group, with an overall raking of 25th .
- MADELON: We ranked 13th as a group, with an overall raking of 33th.

Code

The K-PLS methodology for feature selection was standard procedure with the Analyze/StripMiner code (www.drugmining.com).

Keywords

Centering, multiple-bootstrap cross-validation, K-PLS, sensitivity analysis, feature ranking.

Fact Sheet for Chapter 22

Information Gain, Correlation and Support Vector Machines

Danny Roobaert, Grigoris Karakoulas, and Nitesh V. Chawla

Customer Behavior Analytics
Retail Risk Management
Canadian Imperial Bank of Commerce (CIBC)
Toronto, Canada
{danny.roobaert,grigoris.karakoulas,nitesh.chawla}@cibc.ca

Method

As part of preprocessing we rescaled all feature values between 0 and 1. Also, a feature that had a constant value over the dataset was removed. In all datasets, except MADELON, we used information gain to quantify the relevance of each feature relative to the target. Only features with a non-zero information gain value were selected to be relevant. For the MADELON dataset, the correlation matrix was calculated of all the features and the target. We selected those features that had an average absolute column correlation exceeding about 10 times the global absolute correlation level. For classification, a Support Vector Machine (SVM) was trained using an enhanced SMO algorithm. For all datasets, except MADELON, a linear kernel was used. For the dataset MADELON, an RBF kernel was used. For the imbalanced dataset Dorothea, we used asymmetrical regularization by over-weighting the minority class with a factor equal to the number of data-elements in the majority class over the number of data-elements in the minority class. For SVM hyperparameter optimization we used pattern search as proposed by Momma and Bennett. For performance evaluation, BER was estimated using random 10 fold cross validation. It should be pointed out that this is the method that yielded the Dec. 8th results. This method was still under development on Dec 1st. The method of the submission on Dec. 1st was abandoned after that submission.

Results

For the December 8th submissions, we rank 6th as a group and our best entry on ARCENE and MADELON was not statistically different from the winner. In terms of number of features used, we rank 2nd (within the top 20 submissions), using only 12.78% features on average. We rank 1st overall in identifying probes across all datasets. We also conducted experiments with learning SVM with all the original feature sets. Our results show that feature selection improves SVM performance on all the challenge datasets.

Code

Our implementation was done in Matlab and we used the LIBSVM software package for training SVMs http://www.csie.ntu.edu.tw/~cjlin/libsvm.

Keywords

scaling, filter, information gain, correlation coefficient, feature relevance, K-fold cross-validation, Support Vector Machines, pattern search.

Fact Sheet for Chapter 23

Mining for Complex Models Comprising Feature Selection and Classification

Krzysztof Grabczewski and Norbert Jankowski

Department of Informatics, Nicolaus Copernicus University, Toruń, Poland
kgrabcze@phys.uni.torun.pl, norbert@phys.uni.torun.pl

Method

Different classification tasks require different learning schemes to be satisfactorily solved. Most real-world datasets can be modeled only by complex structures resulting from deep data exploration with a number of different classification and data transformation methods. The search through the space of complex structures must be augmented with reliable validation strategies. All these techniques were necessary to build accurate models for the five high-dimensional datasets of the NIPS 2003 Feature Selection Challenge. Several feature selection algorithms (e.g. based on variance, correlation coefficient, decision trees) and several classification schemes (e.g. nearest neighbors, Normalized RBF, Support Vector Machines) were used to build complex models which transform the data and then classify. Committees of feature selection models and ensemble classifiers were also very helpful to construct models of high generalization abilities.

Results

The models were submitted to the second part of the contest (December 8th). We assessed the second stage as more important. Thus, for real, we did not take part in the stage of December 1st. In the final contest we reached the group rank of 3 and best entry rank of 7. All our computations have been run on personal computers including notebooks - thanks to the algorithms no supercomputers or clusters are necessary to obtain interesting results in data mining.

Code

Our implementation was done in Borland C++ Builder 6. We hope that the system with all described methods will be available in one year, please browse the `http://www.phys.uni.torun.pl/kis/DataMining.html`.

Keywords

centering, scaling, standardization, PCA, filter, wrapper, embedded feature selection, feature selection wrapper, correlation coefficient, mutual information, miscellaneous classifiers, including neural network, decision trees, machine learning, SVM, kernel-method, SSV, kNN, NRBF, LVQ, feature ranking, forward selection, training error, balanced error rate, K-fold cross-validation, regularization, ensemble method, committee, grid-search, quadratic programming, complex models, meta-learning.

Fact Sheet for Chapter 24

Combining Information-Based Supervised and Unsupervised Feature Selection

Sang-Kyun Lee, Seung-Joon Yi, and Byoung-Tak Zhang

Biointelligence Laboratory
School of Computer Science and Engineering
Seoul National University
Seoul 151-742, Korea
sklee@bi.snu.ac.kr,sjlee@bi.snu.ac.kr,btzhang@bi.snu.ac.kr

Method

We preprocess the data by converting continuous values to binary ones (discretization) and filtering out uninformative features w.r.t. the threshold defined as the average mutual information value of randomly permutated feature vectors (for each dataset).

For feature selection, we adopted a filter method using information theoretic ranking criteria but divided it into two steps: supervised and unsupervised. In the supervised step we approximately find the Markov blanket of the class variable by selecting relevant and independent features with a heuristic multi-objective optimization process. To compensate the possible biases due to misleading casual relationships estimated in the supervised stage, we generate another feature set in unsupervised way using a hierarchical agglomerative clustering. Finally we combined the two subsets by the union-set operation.

We use naïve Bayes classifier and support vector machine (SVM) with linear kernel for classification with 10-fold cross validation. To investigate the pure performance gain acquired by the feature selection, we intentionally omit hyper-parameter tuning and use the default parameters of WEKA v3.4 for the SVM (C=10).

Results

In the challenge, for the December 1st submissions, we rank 15th as a group and our best entry is the 40th, using the criterion of the organizers. For compact feature subsets (¡5% of the total number of features) we rank 6th. Note that we did not perform parameter tuning which can affect classification performance greatly. The performance using the combined feature subset is significantly better than using only the supervised feature subset, in case of naïve

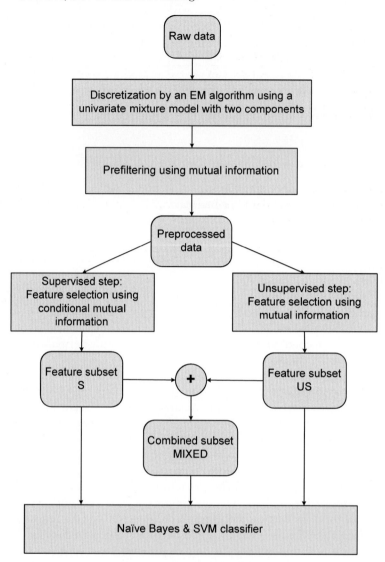

Fig. 24.1. Method Description

Bayes classifier (p-value = 0.0214). But in case of SVM there is little evidence of performance gain (p-value = 0.2398), although we can observe improvements in a part of datasets. Considering the computational efficiency (feature selection for all dataset took less than two hours in the 2.4GHz Pentium 4 PC), simplicity (our method consist of linear procedures) and no hyper-parameter tuning, our method showed noteworthy result.

Code

We implemented preprocessing and feature selection in C++. For classification we used the open source data mining software WEKA ver. 3.4 (available at `http://www.cs.waikato.ac.nz/ml/weka/html`)

Keywords

discretization, expectation-maximization, filter, conditional mutual information, forward selection, training error, naïve Bayes, SVM.

Fact Sheet for Chapter 25

An Enhanced Selective Naïve Bayes Method with Optimal Discretization

Marc Boullé

France Telecom R&D, 2 avenue Pierre Marzin, 22307 Lannion Cedex, France
marc.boulle@francetelecom.com

Method

Our method is the Enhanced Selected Naïve Bayes (ESNB) method, an extension of the wrapper approach applied to the Naïve Bayes predictor. It is also use as a filter for multi-layer perceptron (ESNB+NN).

Preprocessing

- new Bayes optimal discretization method called MODL

Feature selection

- sequential forward selection using the Naïve Bayes classifier enclosed in a wrapper approach
- the feature subset selection is the area under the lift curve (a variant of the ROC curve)
- the selection process stops when this area does not increase anymore

Classification

- ESNB: Naïve Bayes classifier, with a correction of the predicted output labels probabilities
- ESNB+NN: use of the feature set selected by the ESNB method; non-linear multi-layer perceptron with a hidden layer containing 1, 5, 10, 25 or 50 neurons; the hyper-parameter is chosen based on the validation set

Results

In the challenge, for the December 1st submissions, we rank 17th as a group and our best entry is 52 th, using the criterion of the organizers. For the December 8th, we got an average 18% BER (balanced error rate) with our ESNB submission and 12% with the ESNB+NN submission (the best challenge result is 6%). The ESNB method is very fast, simple and does not require any parameter. It produces very compact feature sets (on average 1% of the input features): no other challenge entry had better results with fewer selected features. The MODL discretization method is very efficient at detecting probes (only 1 probe among the total 321 selected features in our December 8th submission). The ESNB methods compute the posterior probability of the output labels. The MODL discretization method is a very efficient univariate selection method, with strong theoretical foundations: it produces Bayes optimal discretizations. (http://www.nipsfsc.ecs.soton.ac.uk/datasets/).

Code

The implementation is not available. The MODL discretization method is patented (contact the author for conditions of use).

Keywords

discretization, naïve bayes, wrapper, forward feature selection, lift curve area selection criterion, multi-layer perceptron.

Fact Sheet for Chapter 26

An Input Variable Importance Definition based on Empirical Data Probability Distribution

V. Lemaire and F. Clérot

Statistical Information Processing group
France Telecom Research and Development
FTR&D/SUSI/TSI
2 Avenue Pierre Marzin
22307 Lannion cedex FRANCE
vincent.lemaire@rd.francetelecom.com,fabrice.clerot@rd.francetelecom.com

Method

The method can be qualified as a variable ranking method 'in the context of other variables'. The method consists in replacing a variable value by another value obtained by randomly choosing among other values of that variable in the training set. The impact of this change on the output is measured and averaged over all training examples and changes of that variable for a given training example. As a search strategy, backward elimination is used.

The preprocessing used is only a zero-mean, unit-variance standardization. The variables importance is measured using only the training set.

As we wish to investigated the performance of our variable importance measurement, we chose to use a single learning machine for all datasets (no bootstrap method): a MLP neural network with 1 hidden layer, tangent hyperbolic activation function and stochastic back-propagation of the squared error as training algorithm. We added a regularization term active only on directions in weight space, which are orthogonal to the training update.

As usual, for each dataset we split the training set in two sets: a training (70%) and a validation set (30%). The training set is used to train and the validation set is used to stop training (early stopping method).

Results:
What we expect from a variable selection technique is to adapt itself in such situation by removing as many features as possible. Therefore, what we can expect from the combination of our simple model and our selection technique is to keep a BER reasonably close to the average while using significantly less features on all datasets.

The results obtained show that restricting ourselves to a simple model with no bootstrap techniques cannot allow us to reach very good BER, particularly on databases as ARCENE where the number of examples is quite small. Several methods obtain a better error rate, up to twice better than our method, but at the expense of a significantly higher number of selected variables. Although admittedly not the most adapted for accuracy on some datasets, this simple model indeed reaches a 'reasonable' BER. The proposed method, combined with backward elimination using only one neural network, selects very few variables compared with the other methods. The proposed variable selection technique exhibits the expected behavior by both keeping the BER to a reasonable level and dramatically reducing the number of features on all datasets. The number of selected probes decreases when the number of the training examples increases.

Other challengers conducted comparison experiments with single neural networks as classifiers and found that ensemble techniques significantly improve the results

Code

Our implementation was done in C the code is only a development code not available.

Keywords

Standardization, wrapper, feature ranking, backward elimination, neural networks.

Feature Selection Challenge Results Tables

Result Tables of the NIPS2003 Challenge

Isabelle Guyon[1] and Steve Gunn[2]

[1] ClopiNet, 955 Creston Rd., Berkeley, CA 94708, USA. isabelle@clopinet.com
[2] School of Electronics and Computer Science, University of Southampton,
Southampton, United Kingdom. s.r.gunn@ecs.soton.ac.uk

In each of the following tables, we show the BER in %, the AUC times 100, the percentage of features used (Ffeat), the percentage of probes in the features selected (Fprob), and whether the BER differs significantly from the best BER (Sig). The significance of the difference is obtained by the McNemar paired test, with risk 5%.

ARCENE

Table 1. December 1st 2003 challenge results for ARCENE.

Method	People	Score	BER	AUC	Ffeat	Fprob	Sig
BayesNN-DF	Radford Neal an	98.18	13.30 (1)	93.48 (1)	100.00	30.00	0
BayesNN-DF	Radford Neal +	98.18	13.30 (2)	93.48 (2)	100.00	30.00	0
inf5	Amir Reza Saffa	85.45	17.30 (17)	82.70 (38)	5.00	0.00	1
RF+RLSC	Kari Torkkola +	81.82	15.14 (3)	84.86 (32)	100.00	30.00	0
KPLS	Mark J. Embrech	81.82	16.71 (12)	83.67 (34)	5.14	8.56	1
BayesNN-sm	Radford Neal	78.18	16.59 (10)	91.15 (8)	10.70	1.03	1
Bayesian S	Chu Wei	78.18	15.17 (4)	91.52 (6)	100.00	30.00	0
final 2	Yi-Wei Chen	74.55	15.27 (5)	84.73 (33)	100.00	30.00	0
Bayesian +	Chu Wei	70.91	15.55 (6)	91.25 (7)	100.00	30.00	0
multi23	Sang-Kyun Lee	67.27	18.41 (20)	81.59 (40)	1.85	0.00	1
svm linear	Ran Bachrach	60.00	15.97 (7)	91.57 (4)	100.00	30.00	1
svm linear	Ran Bachrach	60.00	15.97 (8)	91.57 (5)	100.00	30.00	1
inf2	Amir Reza Saffa	60.00	18.99 (21)	81.01 (42)	2.00	0.00	1
svm linear	Ran Bachrach	60.00	15.97 (9)	91.58 (3)	100.00	30.00	1
cbamethod3	CBAGroup	41.82	16.62 (11)	90.87 (10)	100.00	30.00	1
Modified R	Vivian Ng; Leo	34.55	19.64 (32)	86.72 (25)	3.60	8.06	1
GhostMiner	GhostMiner Team	27.27	16.94 (13)	83.06 (35)	100.00	30.00	1
Collection	Amir Reza Saffa	25.45	19.24 (26)	80.76 (43)	20.18	2.92	1
Collection	Amir Reza Saffa	25.45	19.24 (27)	80.76 (44)	20.18	2.92	1
GhostMiner	GhostMiner Team	23.64	17.07 (14)	82.93 (36)	100.00	30.00	1
mixed	Sang-Kyun Lee	21.82	20.65 (41)	79.35 (46)	4.17	0.72	1
multi33	Sang-Kyun Lee	21.82	20.65 (42)	79.35 (47)	4.17	0.72	1
The Best o	Amir Navot + Ra	18.18	17.20 (15)	90.13 (12)	100.00	30.00	1
greatest h	Amir Navot + Ra	18.18	17.20 (16)	90.13 (13)	100.00	30.00	1
P-SVM (pre	Sepp Hochreiter	16.36	20.55 (39)	87.75 (21)	7.00	61.00	1
BayesNN-E	Radford Neal	16.36	18.11 (18)	90.21 (11)	96.56	29.70	1
final 1	Yi-Wei Chen	16.36	20.59 (40)	79.41 (45)	6.61	0.61	1
FS + SVM	Thomas Navin La	12.73	18.20 (19)	81.80 (39)	47.00	13.55	1
Depends I	Saharon Rosset	0.00	19.62 (28)	88.91 (15)	30.00	5.70	1
Depends II	Saharon Rosset	0.00	19.62 (29)	88.91 (16)	30.00	5.70	1
Depends IV	Saharon Rosset	0.00	19.62 (30)	88.91 (17)	30.00	5.70	1
Depends V	Saharon Rosset	0.00	19.62 (31)	88.91 (18)	30.00	5.70	1
BayesNN-la	Radford Neal	-5.45	19.19 (25)	89.88 (14)	96.56	29.70	1
Depends II	Saharon Rosset	-9.09	19.78 (33)	88.79 (19)	30.00	5.70	1
IDEAL	BorisovEruhimov	-27.27	19.12 (22)	87.33 (22)	100.00	30.00	1
IDEAL	Borisov Eruhimo	-27.27	19.12 (23)	87.33 (23)	100.00	30.00	1
IDEAL	BorisovEruhimov	-27.27	19.12 (24)	87.33 (24)	100.00	30.00	1
transSVMba	wu zhili	-41.82	20.01 (34)	85.97 (27)	100.00	30.00	1
SVMbased3	wu zhili + C.H.	-41.82	20.01 (35)	85.97 (28)	100.00	30.00	1
myBestVali	wu zhili	-41.82	20.01 (36)	85.97 (29)	100.00	30.00	1
TransSVMba	wu zhili	-41.82	20.01 (37)	85.97 (30)	100.00	30.00	1
svmBased4	wuzhili+C.H.Li	-41.82	20.01 (38)	85.97 (31)	100.00	30.00	1
IDEAL	BorisovEruhimov	-52.73	20.82 (43)	88.45 (20)	100.00	30.00	1
CBAMethod1	cba	-56.36	23.95 (45)	76.61 (49)	0.99	1.01	1
DIMACS	Alexander Genki	-63.64	27.46 (49)	78.16 (48)	0.86	29.07	1
P-SVM (pre	Sepp Hochreiter	-63.64	23.45 (44)	86.44 (26)	3.00	60.33	1
P-SVM (pre	Sepp Hochreiter	-67.27	25.77 (46)	82.73 (37)	1.35	62.22	1
SVCR-n-m	Amir Reza Saffa	-70.91	26.24 (48)	73.76 (52)	2.00	2.00	1
P-SVM / nu	Sepp Hochreiter	-70.91	26.05 (47)	81.37 (41)	1.70	64.12	1
Enhanced S	M. B.	-78.18	31.25 (52)	75.93 (50)	0.05	40.00	1
ROBELON	Vincent Lemaire	-85.45	29.65 (51)	70.35 (54)	1.50	60.00	1
multi32	Sang-Kyun Lee	-85.45	27.50 (50)	72.50 (53)	4.17	0.72	1
P-SVM (few	Sepp Hochreiter	-85.45	31.27 (53)	75.85 (51)	0.20	60.00	1
multi31	Sang-Kyun Lee	-92.73	32.05 (54)	67.95 (55)	4.17	0.72	1
CBAMethod1	cba	-96.36	33.50 (55)	90.91 (9)	100.00	30.00	1
Final	Fan Li	-100.00	34.45 (56)	65.55 (56)	100.00	30.00	1

Table 2. December 8th 2003 challenge results for **ARCENE.**

Method	People	Score	BER	AUC	Ffeat	Fprob	Sig
BayesNN-sm	Radford Neal	94.29	11.86 (7)	95.47 (1)	10.70	1.03	0
RF with fe	Vivian Ng + Leo	88.57	12.63 (10)	93.79 (6)	3.80	0.79	1
CBAMethod3	CBAGroup	85.71	11.12 (4)	94.89 (2)	28.25	0.28	0
CBAMethod3	CBAGroup	85.71	11.12 (5)	94.89 (3)	28.25	0.28	0
RF+RLSC	Kari Torkkola +	71.43	11.12 (3)	88.88 (18)	99.20	29.96	0
final2-2	Yi-Wei Chen	68.57	10.73 (1)	90.63 (12)	100.00	30.00	0
final 2-3	Yi-Wei Chen	68.57	10.73 (2)	90.63 (13)	100.00	30.00	0
FS+SVM	Thomas Navin La	65.71	12.76 (12)	87.24 (22)	47.00	5.89	1
RF+RLSC	Kari Torkkola +	65.71	11.60 (6)	88.40 (19)	99.20	29.96	0
BayesNN-DF	Radford Neal +	48.57	12.25 (8)	93.01 (9)	100.00	30.00	0
Bayesian +	Chu Wei	42.86	12.47 (9)	93.65 (7)	100.00	30.00	1
Nameless -	Amir Navot + Ra	37.14	12.66 (11)	93.37 (8)	100.00	30.00	1
IDEAL	BorisovEruhimov	28.57	13.04 (13)	94.81 (4)	100.00	30.00	1
IDEAL	BorisovEruhimov	28.57	13.04 (14)	94.81 (5)	100.00	30.00	1
GhostMiner	GhostMiner Team	14.29	13.53 (15)	86.47 (23)	100.00	30.00	1
GhostMiner	GhostMiner Team	8.57	13.76 (16)	86.24 (24)	100.00	30.00	1
GhostMiner	GhostMiner Team	2.86	14.05 (17)	85.95 (26)	100.00	30.00	1
A shot in	Amir + Ran	-2.86	15.82 (19)	84.18 (29)	44.00	12.48	1
BayesNN-la	Radford Neal	-8.57	15.88 (20)	92.34 (10)	96.56	29.70	1
test	Yi-Wei Chen	-14.29	15.27 (18)	84.73 (28)	100.00	30.00	1
P-SVM / nu	Sepp Hochreiter	-17.14	19.69 (27)	88.32 (20)	3.00	67.33	1
P-SVM / nu	Sepp Hochreiter	-17.14	19.69 (28)	88.32 (21)	3.00	67.33	1
P-SVM / nu	Sepp Hochreiter	-25.71	18.24 (24)	89.14 (14)	7.00	64.43	1
MyFinal	Amir Reza Saffa	-25.71	17.29 (21)	82.71 (31)	20.18	2.92	1
P-SVM / nu	Sepp Hochreiter	-25.71	18.24 (25)	89.14 (15)	7.00	64.43	1
P-SVM / nu	Sepp Hochreiter	-25.71	18.24 (26)	89.14 (16)	7.00	64.43	1
BayesNN-la	Radford Neal	-48.57	17.94 (22)	91.47 (11)	96.56	29.70	1
METHOD2	CBA	-54.29	17.94 (23)	88.97 (17)	100.00	30.00	1
ESNB+NN	Marc Boulle + V	-60.00	22.92 (32)	83.78 (30)	0.14	7.14	1
originalFi	wu zhili	-65.71	20.01 (29)	85.97 (25)	100.00	30.00	1
ESNB	Marc Boulle	-71.43	23.14 (33)	84.83 (27)	0.14	7.14	1
METHOD1	CBA	-80.00	21.48 (30)	78.52 (32)	100.00	30.00	1
METHOD3	CBA	-80.00	21.48 (31)	78.52 (33)	100.00	30.00	1
Sparse Bay	DIMACS	-88.57	26.99 (34)	72.78 (34)	1.17	68.38	1
ntmce	D. C. Martins J	-94.29	30.41 (36)	69.59 (36)	0.07	28.57	1
ROBELON	Vincent Lemaire	-100.00	29.65 (35)	70.35 (35)	1.50	60.00	1

Dexter

Table 1. December 1st 2003 challenge results for DEXTER.

Method	People	Score	BER	AUC	Ffeat	Fprob	Sig
BayesNN-DF	Radford Neal an	96.36	3.90 (1)	99.01 (2)	1.52	12.87	0
BayesNN-la	Radford Neal	96.36	3.90 (2)	99.01 (3)	1.52	12.87	0
BayesNN-DF	Radford Neal +	96.36	3.90 (3)	99.01 (4)	1.52	12.87	0
BayesNN-sm	Radford Neal	89.09	4.00 (4)	99.03 (1)	1.52	12.87	0
FS + SVM	Thomas Navin La	85.45	4.20 (5)	95.80 (34)	18.57	49.78	0
transSVMba	wu zhili	70.91	4.40 (6)	97.92 (15)	29.47	59.71	0
SVMbased3	wu zhili + C.H.	70.91	4.40 (7)	97.92 (16)	29.47	59.71	0
myBestVali	wu zhili	70.91	4.40 (8)	97.92 (17)	29.47	59.71	0
TransSVMba	wu zhili	70.91	4.40 (9)	97.92 (18)	29.47	59.71	0
svmBased4	wuzhili+C.H.Li	70.91	4.40 (10)	97.92 (19)	29.47	59.71	0
Collection	Amir Reza Saffa	63.64	4.95 (11)	95.05 (35)	5.01	36.86	1
IDEAL	BorisovEruhimov	58.18	5.50 (18)	98.35 (10)	1.00	25.50	1
IDEAL	BorisovEruhimov	58.18	5.50 (19)	98.35 (11)	1.00	25.50	1
IDEAL	Borisov Eruhimo	58.18	5.50 (20)	98.35 (12)	1.00	25.50	1
IDEAL	BorisovEruhimov	58.18	5.50 (21)	98.35 (13)	1.00	25.50	1
RF+RLSC	Kari Torkkola +	49.09	5.40 (16)	94.60 (38)	2.50	28.40	1
The Best o	Amir Navot + Ra	40.00	5.25 (12)	98.80 (5)	7.00	43.71	1
greatest h	Amir Navot + Ra	40.00	5.25 (13)	98.80 (6)	7.00	43.71	1
Collection	Amir Reza Saffa	34.55	5.40 (15)	94.60 (37)	38.76	61.30	1
BayesNN-E	Radford Neal	30.91	5.45 (17)	98.54 (9)	87.18	57.65	1
Final	Fan Li	27.27	5.40 (14)	94.60 (36)	100.00	50.27	1
P-SVM (few	Sepp Hochreiter	23.64	7.40 (36)	97.26 (24)	0.10	0.00	1
Depends II	Saharon Rosset	12.73	6.90 (28)	96.28 (29)	0.56	44.64	1
Depends I	Saharon Rosset	12.73	6.90 (29)	96.28 (30)	0.56	44.64	1
Depends II	Saharon Rosset	12.73	6.90 (30)	96.28 (31)	0.56	44.64	1
Depends IV	Saharon Rosset	12.73	6.90 (31)	96.28 (32)	0.56	44.64	1
Depends V	Saharon Rosset	12.73	6.90 (32)	96.28 (33)	0.56	44.64	1
final 2	Yi-Wei Chen	0.00	6.50 (24)	93.50 (39)	1.04	10.53	1
final 1	Yi-Wei Chen	0.00	6.50 (26)	93.50 (41)	1.04	10.53	1
KPLS	Mark J. Embrech	-5.45	6.80 (27)	93.47 (42)	1.57	28.12	1
SVCR-n-m	Amir Reza Saffa	-12.73	6.50 (25)	93.50 (40)	2.00	21.50	1
Modified R	Vivian Ng; Leo	-20.00	7.35 (35)	97.97 (14)	1.70	20.29	1
P-SVM (pre	Sepp Hochreiter	-20.00	7.10 (33)	97.82 (20)	4.00	51.63	1
Bayesian S	Chu Wei	-21.82	6.35 (22)	98.57 (7)	36.04	60.15	1
Bayesian +	Chu Wei	-21.82	6.35 (23)	98.57 (8)	36.04	60.15	1
inf5	Amir Reza Saffa	-23.64	7.10 (34)	92.90 (43)	5.00	42.10	1
DIMACS	Alexander Genki	-30.91	8.00 (37)	91.53 (44)	0.64	9.45	1
cbamethod3	CBAGroup	-34.55	8.70 (38)	91.04 (46)	0.38	0.00	1
CBAMethod1	cba	-34.55	8.70 (39)	91.04 (47)	0.38	0.00	1
CBAMethod1	cba	-34.55	8.70 (43)	91.04 (48)	0.38	0.00	1
Enhanced S	M. B.	-45.45	9.80 (48)	96.42 (25)	0.17	0.00	1
ROBELON	Vincent Lemaire	-49.09	9.70 (47)	90.30 (49)	0.61	29.51	1
inf2	Amir Reza Saffa	-52.73	8.85 (44)	91.15 (45)	2.00	39.50	1
P-SVM (pre	Sepp Hochreiter	-60.00	8.70 (40)	96.39 (26)	2.50	46.60	1
P-SVM (pre	Sepp Hochreiter	-60.00	8.70 (41)	96.39 (27)	2.50	46.60	1
P-SVM / nu	Sepp Hochreiter	-60.00	8.70 (42)	96.39 (28)	2.50	46.60	1
svm linear	Ran Bachrach	-67.27	9.55 (45)	97.37 (23)	100.00	50.27	1
svm linear	Ran Bachrach	-70.91	9.65 (46)	97.39 (22)	100.00	50.27	1
GhostMiner	GhostMiner Team	-74.55	10.95 (49)	89.05 (50)	100.00	50.27	1
GhostMiner	GhostMiner Team	-78.18	11.20 (50)	88.80 (51)	100.00	50.27	1
multi32	Sang-Kyun Lee	-81.82	14.60 (52)	85.40 (52)	5.09	37.46	1
mixed	Sang-Kyun Lee	-87.27	14.85 (53)	85.15 (53)	5.09	37.46	1
multi31	Sang-Kyun Lee	-87.27	14.85 (54)	85.15 (54)	5.09	37.46	1
svm linear	Ran Bachrach	-92.73	13.65 (51)	97.49 (21)	100.00	50.27	1
multi33	Sang-Kyun Lee	-96.36	15.40 (55)	84.60 (55)	5.09	37.46	1
multi23	Sang-Kyun Lee	-100.00	17.60 (56)	82.40 (56)	4.14	32.29	1

Table 2. December 8th 2003 challenge results for DEXTER.

Method	People	Score	BER	AUC	Ffeat	Fprob	Sig
FS+SVM	Thomas Navin La	100.00	3.30 (1)	96.70 (23)	18.57	42.14	0
BayesNN-DF	Radford Neal +	85.71	4.05 (5)	99.09 (1)	1.52	12.87	1
BayesNN-la	Radford Neal	85.71	4.05 (6)	99.09 (2)	1.52	12.87	1
BayesNN-sm	Radford Neal	85.71	4.05 (7)	99.09 (3)	1.52	12.87	1
BayesNN-la	Radford Neal	85.71	4.05 (8)	99.09 (4)	1.52	12.87	1
GhostMiner	GhostMiner Team	71.43	3.50 (2)	96.50 (24)	100.00	50.27	0
GhostMiner	GhostMiner Team	65.71	3.60 (3)	96.40 (25)	100.00	50.27	1
GhostMiner	GhostMiner Team	54.29	3.80 (4)	96.20 (26)	100.00	50.27	1
Sparse Bay	DIMACS	54.29	5.05 (14)	94.37 (29)	0.93	6.49	1
RF+RLSC	Kari Torkkola +	48.57	4.65 (10)	95.35 (27)	2.50	28.40	1
RF+RLSC	Kari Torkkola +	31.43	4.85 (11)	95.15 (28)	2.50	28.40	1
originalFi	wu zhili	31.43	4.40 (9)	97.92 (15)	29.47	59.71	1
final2-2	Yi-Wei Chen	22.86	5.35 (15)	96.86 (21)	1.21	2.90	1
final 2-3	Yi-Wei Chen	22.86	5.35 (16)	96.86 (22)	1.21	2.90	1
A shot in	Amir + Ran	11.43	5.00 (12)	98.65 (7)	7.00	37.14	1
Nameless -	Amir Navot + Ra	11.43	5.00 (13)	98.65 (8)	7.00	37.14	1
CBAMethod3	CBAGroup	0.00	6.00 (20)	98.47 (11)	0.60	0.00	1
CBAMethod3	CBAGroup	0.00	6.00 (21)	98.47 (12)	0.60	0.00	1
IDEAL	BorisovEruhimov	-5.71	6.25 (23)	98.70 (5)	0.50	6.00	1
IDEAL	BorisovEruhimov	-5.71	6.25 (24)	98.70 (6)	0.50	6.00	1
ESNB	Marc Boulle	-8.57	6.50 (28)	97.90 (19)	0.33	0.00	1
P-SVM / nu	Sepp Hochreiter	-11.43	5.40 (17)	98.47 (13)	4.50	53.56	1
P-SVM / nu	Sepp Hochreiter	-11.43	5.40 (18)	98.47 (14)	4.50	53.56	1
RF with fe	Vivian Ng + Leo	-25.71	6.25 (22)	98.49 (10)	1.00	5.00	1
ESNB+NN	Marc Boulle + V	-25.71	7.20 (30)	97.49 (20)	0.33	0.00	1
Bayesian +	Chu Wei	-48.57	5.45 (19)	98.54 (9)	36.04	60.15	1
test	Yi-Wei Chen	-48.57	6.50 (29)	93.50 (30)	1.04	10.53	1
P-SVM / nu	Sepp Hochreiter	-54.29	6.40 (25)	97.91 (16)	1.50	36.33	1
P-SVM / nu	Sepp Hochreiter	-54.29	6.40 (26)	97.91 (17)	1.50	36.33	1
P-SVM / nu	Sepp Hochreiter	-54.29	6.40 (27)	97.91 (18)	1.50	36.33	1
METHOD2	CBA	-77.14	7.75 (31)	92.25 (31)	100.00	50.27	1
METHOD1	CBA	-77.14	7.75 (32)	92.25 (32)	100.00	50.27	1
METHOD3	CBA	-77.14	7.75 (33)	92.25 (33)	100.00	50.27	1
ROBELON	Vincent Lemaire	-88.57	9.70 (34)	90.30 (34)	0.61	29.51	1
MyFinal	Amir Reza Saffa	-94.29	12.60 (35)	87.40 (35)	5.33	38.65	1
ntmce	D. C. Martins J	-100.00	15.80 (36)	84.20 (36)	0.04	0.00	1

DOROTHEA

Table 1. December 1st 2003 challenge results for DOROTHEA.

Method	People	Score	BER	AUC	Ffeat	Fprob	Sig
BayesNN-DF	Radford Neal an	98.18	8.54 (1)	95.92 (2)	100.00	50.00	0
BayesNN-la	Radford Neal	98.18	8.54 (2)	95.92 (3)	100.00	50.00	0
BayesNN-E	Radford Neal	92.73	8.61 (3)	95.98 (1)	100.00	50.00	0
BayesNN-DF	Radford Neal +	89.09	8.68 (4)	95.86 (4)	100.00	50.00	0
greatest h	Amir Navot + Ra	85.45	10.86 (6)	92.19 (13)	0.30	0.00	1
BayesNN-sm	Radford Neal	81.82	10.63 (5)	93.50 (5)	0.50	0.40	1
SVMbased3	wu zhili + C.H.	78.18	11.52 (11)	88.48 (20)	0.50	18.88	1
svmBased4	wuzhili+C.H.Li	74.55	12.45 (12)	87.55 (24)	0.50	18.88	1
IDEAL	BorisovEruhimov	65.45	10.98 (7)	93.26 (9)	100.00	50.00	1
IDEAL	BorisovEruhimov	65.45	10.98 (8)	93.26 (10)	100.00	50.00	1
IDEAL	Borisov Eruhimo	65.45	10.98 (9)	93.26 (11)	100.00	50.00	1
IDEAL	BorisovEruhimov	65.45	10.98 (10)	93.26 (12)	100.00	50.00	1
Modified R	Vivian Ng; Leo	56.36	13.72 (16)	91.67 (16)	0.40	20.75	1
transSVMba	wu zhili	49.09	13.46 (13)	86.54 (25)	0.50	18.88	1
myBestVali	wu zhili	49.09	13.46 (14)	86.54 (26)	0.50	18.88	1
TransSVMba	wu zhili	49.09	13.46 (15)	86.54 (27)	0.50	18.88	1
P-SVM (pre	Sepp Hochreiter	34.55	17.06 (36)	90.56 (17)	0.14	60.00	1
Collection	Amir Reza Saffa	32.73	13.93 (17)	86.07 (28)	1.25	13.22	1
Collection	Amir Reza Saffa	32.73	13.93 (18)	86.07 (29)	1.25	13.22	1
P-SVM (pre	Sepp Hochreiter	29.09	16.21 (27)	88.00 (22)	0.24	29.58	1
P-SVM (pre	Sepp Hochreiter	29.09	16.21 (28)	88.00 (23)	0.24	29.58	1
RF+RLSC	Kari Torkkola +	23.64	16.23 (29)	83.77 (37)	0.28	29.23	1
cbamethod3	CBAGroup	16.36	16.61 (31)	75.57 (47)	0.30	0.00	1
CBAMethod1	cba	16.36	16.61 (32)	75.57 (48)	0.30	0.00	1
CBAMethod1	cba	16.36	16.61 (33)	75.57 (49)	0.30	0.00	1
mixed	Sang-Kyun Lee	10.91	15.26 (19)	84.74 (30)	0.77	27.68	1
multi31	Sang-Kyun Lee	10.91	15.26 (20)	84.74 (31)	0.77	27.68	1
The Best o	Amir Navot + Ra	9.09	16.33 (30)	88.20 (21)	0.30	8.67	1
final 2	Yi-Wei Chen	-3.64	16.82 (34)	83.18 (38)	0.45	2.70	1
final 1	Yi-Wei Chen	-3.64	16.82 (35)	83.18 (39)	0.45	2.70	1
Depends II	Saharon Rosset	-14.55	15.69 (23)	84.51 (32)	5.21	76.99	1
Depends I	Saharon Rosset	-14.55	15.69 (24)	84.51 (33)	5.21	76.99	1
Depends IV	Saharon Rosset	-14.55	15.69 (25)	84.51 (34)	5.21	76.99	1
Depends V	Saharon Rosset	-14.55	15.69 (26)	84.51 (35)	5.21	76.99	1
Bayesian S	Chu Wei	-25.45	15.47 (21)	92.06 (14)	100.00	50.00	1
Bayesian +	Chu Wei	-25.45	15.47 (22)	92.06 (15)	100.00	50.00	1
KPLS	Mark J. Embrech	-34.55	19.18 (40)	82.10 (40)	0.54	92.04	1
P-SVM / nu	Sepp Hochreiter	-38.18	18.10 (37)	90.38 (18)	1.50	91.07	1
FS + SVM	Thomas Navin La	-41.82	19.68 (41)	80.32 (42)	1.00	8.90	1
Depends II	Saharon Rosset	-41.82	18.31 (38)	84.16 (36)	5.21	76.99	1
Final	Fan Li	-45.45	18.47 (39)	81.53 (41)	100.00	50.00	1
Enhanced S	M. B.	-45.45	21.03 (42)	89.43 (19)	0.05	1.89	1
P-SVM (few	Sepp Hochreiter	-49.09	22.61 (45)	78.29 (44)	0.01	0.00	1
ROBELON	Vincent Lemaire	-56.36	22.24 (44)	77.76 (45)	0.07	12.31	1
multi32	Sang-Kyun Lee	-56.36	21.50 (43)	78.50 (43)	0.77	27.68	1
multi23	Sang-Kyun Lee	-63.64	24.69 (47)	75.31 (50)	0.24	0.00	1
inf2	Amir Reza Saffa	-67.27	25.03 (48)	74.97 (51)	2.00	3.55	1
GhostMiner	GhostMiner Team	-70.91	24.08 (46)	75.92 (46)	100.00	50.00	1
SVCR-n-m	Amir Reza Saffa	-74.55	29.11 (49)	70.89 (52)	2.00	1.05	1
inf5	Amir Reza Saffa	-78.18	30.53 (51)	69.47 (54)	5.00	10.44	1
GhostMiner	GhostMiner Team	-81.82	30.32 (50)	69.68 (53)	100.00	50.00	1
multi33	Sang-Kyun Lee	-85.45	34.10 (52)	65.90 (55)	0.77	27.68	1
svm linear	Ran Bachrach	-92.73	39.38 (53)	93.42 (8)	100.00	50.00	1
svm linear	Ran Bachrach	-92.73	39.38 (54)	93.43 (6)	100.00	50.00	1
svm linear	Ran Bachrach	-92.73	39.38 (55)	93.42 (7)	100.00	50.00	1
DIMACS	Alexander Genki	-100.00	45.46 (56)	57.38 (56)	6.06	49.45	1

Table 2. December 8th 2003 challenge results for DOROTHEA.

Method	People	Score	BER	AUC	Ffeat	Fprob	Sig
BayesNN-DF	Radford Neal +	97.14	8.61 (1)	95.92 (2)	100.00	50.00	0
BayesNN-la	Radford Neal	97.14	8.61 (2)	95.92 (3)	100.00	50.00	0
IDEAL	BorisovEruhimov	85.71	8.92 (3)	94.80 (4)	100.00	50.00	0
IDEAL	BorisovEruhimov	85.71	8.92 (4)	94.80 (5)	100.00	50.00	0
BayesNN-la	Radford Neal	77.14	9.11 (5)	95.98 (1)	100.00	50.00	0
A shot in	Amir + Ran	68.57	11.40 (7)	93.10 (7)	0.40	0.00	1
Nameless -	Amir Navot + Ra	68.57	11.40 (8)	93.10 (8)	0.40	0.00	1
BayesNN-sm	Radford Neal	60.00	11.07 (6)	93.42 (6)	0.50	0.40	1
ESNB+NN	Marc Boulle + V	54.29	14.59 (17)	91.50 (13)	0.07	0.00	1
RF with fe	Vivian Ng + Leo	42.86	14.24 (16)	91.40 (14)	0.32	4.38	1
originalFi	wu zhili	42.86	13.46 (14)	86.54 (22)	0.50	18.88	1
P-SVM / nu	Sepp Hochreiter	31.43	12.54 (9)	90.84 (15)	0.80	62.50	1
P-SVM / nu	Sepp Hochreiter	31.43	12.54 (10)	90.84 (16)	0.80	62.50	1
P-SVM / nu	Sepp Hochreiter	31.43	12.54 (11)	90.84 (17)	0.80	62.50	1
GhostMiner	GhostMiner Team	17.14	13.11 (12)	86.89 (20)	100.00	50.00	1
GhostMiner	GhostMiner Team	17.14	13.11 (13)	86.89 (21)	100.00	50.00	1
final 2-3	Yi-Wei Chen	8.57	15.61 (20)	77.56 (34)	0.20	0.00	1
P-SVM / nu	Sepp Hochreiter	-5.71	16.45 (25)	89.21 (18)	0.27	74.07	1
P-SVM / nu	Sepp Hochreiter	-5.71	16.45 (26)	89.21 (19)	0.27	74.07	1
GhostMiner	GhostMiner Team	-8.57	14.08 (15)	85.92 (23)	100.00	50.00	1
RF+RLSC	Kari Torkkola +	-14.29	15.66 (21)	84.34 (24)	0.28	29.23	1
RF+RLSC	Kari Torkkola +	-20.00	16.28 (23)	83.72 (26)	0.28	29.23	1
test	Yi-Wei Chen	-25.71	16.82 (28)	83.18 (28)	0.45	2.70	1
CBAMethod3	CBAGroup	-28.57	15.26 (18)	92.34 (10)	0.57	0.00	1
CBAMethod3	CBAGroup	-28.57	15.26 (19)	92.34 (11)	0.57	0.00	1
MyFinal	Amir Reza Saffa	-42.86	15.75 (22)	84.25 (25)	1.25	13.22	1
FS+SVM	Thomas Navin La	-42.86	16.34 (24)	83.66 (27)	1.00	3.20	1
Bayesian +	Chu Wei	-54.29	16.68 (27)	93.03 (9)	100.00	50.00	1
final2-2	Yi-Wei Chen	-60.00	19.47 (29)	70.10 (35)	0.39	0.51	1
ntmce	D. C. Martins J	-65.71	22.17 (33)	77.83 (32)	0.01	0.00	1
ROBELON	Vincent Lemaire	-71.43	22.24 (34)	77.76 (33)	0.07	12.31	1
ESNB	Marc Boulle	-77.14	23.75 (35)	92.32 (12)	0.07	0.00	1
METHOD2	CBA	-88.57	21.97 (30)	78.03 (29)	100.00	50.00	1
METHOD1	CBA	-88.57	21.97 (31)	78.03 (30)	100.00	50.00	1
METHOD3	CBA	-88.57	21.97 (32)	78.03 (31)	100.00	50.00	1
Sparse Bay	DIMACS	-100.00	27.03 (36)	56.86 (36)	0.16	16.67	1

GISETTE

Table 1. December 1st 2003 challenge results for GISETTE.

Method	People	Score	BER	AUC	Ffeat	Fprob	Sig
final 2	Yi-Wei Chen	98.18	1.37 (8)	98.63 (31)	18.26	0.00	0
final 1	Yi-Wei Chen	98.18	1.37 (9)	98.63 (32)	18.26	0.00	0
Depends II	Saharon Rosset	87.27	1.34 (4)	98.26 (34)	30.00	0.00	0
Depends I	Saharon Rosset	87.27	1.34 (5)	98.26 (35)	30.00	0.00	0
Depends II	Saharon Rosset	87.27	1.34 (6)	98.26 (36)	30.00	0.00	0
Depends V	Saharon Rosset	87.27	1.34 (7)	98.26 (37)	30.00	0.00	0
BayesNN-DF	Radford Neal an	70.91	1.29 (1)	99.90 (1)	100.00	50.00	0
BayesNN-la	Radford Neal	70.91	1.29 (2)	99.90 (2)	100.00	50.00	0
BayesNN-DF	Radford Neal +	70.91	1.29 (3)	99.90 (3)	100.00	50.00	0
transSVMba	wu zhili	56.36	1.58 (11)	99.84 (9)	15.00	0.00	1
SVMbased3	wu zhili + C.H.	56.36	1.58 (12)	99.84 (10)	15.00	0.00	1
myBestVali	wu zhili	56.36	1.58 (13)	99.84 (11)	15.00	0.00	1
Depends IV	Saharon Rosset	56.36	1.48 (10)	98.26 (38)	30.00	0.00	0
TransSVMba	wu zhili	56.36	1.58 (14)	99.84 (12)	15.00	0.00	1
svmBased4	wuzhili+C.H.Li	56.36	1.58 (15)	99.84 (13)	15.00	0.00	1
FS + SVM	Thomas Navin La	49.09	1.69 (16)	98.31 (33)	14.00	0.00	1
RF+RLSC	Kari Torkkola +	45.45	1.89 (19)	98.11 (39)	6.14	0.00	1
BayesNN-sm	Radford Neal	38.18	2.03 (26)	99.79 (14)	7.58	0.26	1
IDEAL	BorisovEruhimov	32.73	1.89 (17)	99.85 (5)	12.00	0.00	1
IDEAL	BorisovEruhimov	32.73	1.89 (18)	99.85 (6)	12.00	0.00	1
IDEAL	Borisov Eruhimo	32.73	1.89 (20)	99.85 (7)	12.00	0.00	1
IDEAL	BorisovEruhimov	32.73	1.89 (21)	99.85 (8)	12.00	0.00	1
P-SVM (pre	Sepp Hochreiter	18.18	2.06 (27)	99.76 (16)	12.00	36.50	1
P-SVM (pre	Sepp Hochreiter	18.18	2.06 (28)	99.76 (17)	12.00	36.50	1
P-SVM (few	Sepp Hochreiter	10.91	1.98 (23)	99.74 (18)	16.00	41.88	1
P-SVM / nu	Sepp Hochreiter	10.91	1.98 (24)	99.74 (19)	16.00	41.88	1
KPLS	Mark J. Embrech	5.45	2.02 (25)	97.92 (40)	26.00	0.00	1
P-SVM (pre	Sepp Hochreiter	1.82	2.12 (29)	99.71 (21)	30.00	54.40	1
BayesNN-E	Radford Neal	-1.82	1.91 (22)	99.86 (4)	100.00	50.00	1
The Best o	Amir Navot + Ra	-12.73	2.31 (34)	99.76 (15)	50.00	15.20	1
GhostMiner	GhostMiner Team	-14.55	2.12 (30)	97.88 (41)	100.00	50.00	1
GhostMiner	GhostMiner Team	-14.55	2.12 (31)	97.88 (42)	100.00	50.00	1
DIMACS	Alexander Genki	-16.36	2.42 (36)	97.32 (46)	11.48	52.44	1
Final	Fan Li	-23.64	2.29 (32)	97.71 (43)	100.00	50.00	1
Collection	Amir Reza Saffa	-23.64	2.58 (37)	97.42 (45)	10.10	0.00	1
cbamethod3	CBAGroup	-27.27	2.38 (35)	99.73 (20)	100.00	50.00	1
Collection	Amir Reza Saffa	-30.91	2.31 (33)	97.69 (44)	99.10	50.07	1
mixed	Sang-Kyun Lee	-40.00	2.74 (43)	97.26 (47)	9.30	0.00	1
multi33	Sang-Kyun Lee	-40.00	2.74 (44)	97.26 (48)	9.30	0.00	1
Bayesian S	Chu Wei	-47.27	2.62 (38)	99.67 (23)	100.00	50.00	1
Bayesian +	Chu Wei	-47.27	2.62 (39)	99.67 (24)	100.00	50.00	1
CBAMethod1	cba	-49.09	3.00 (47)	96.63 (51)	5.98	0.00	1
Modified R	Vivian Ng; Leo	-52.73	2.89 (45)	99.59 (29)	9.60	0.00	1
Enhanced S	M. B.	-56.36	3.12 (49)	99.49 (30)	3.02	0.00	1
svm linear	Ran Bachrach	-60.00	2.63 (40)	99.67 (26)	100.00	50.00	1
svm linear	Ran Bachrach	-60.00	2.63 (41)	99.67 (27)	100.00	50.00	1
inf5	Amir Reza Saffa	-60.00	3.12 (50)	96.88 (50)	5.00	0.00	1
svm linear	Ran Bachrach	-60.00	2.63 (42)	99.67 (25)	100.00	50.00	1
ROBELON	Vincent Lemaire	-70.91	3.48 (52)	96.52 (52)	1.80	5.56	1
multi23	Sang-Kyun Lee	-74.55	3.11 (48)	96.89 (49)	7.24	0.00	1
greatest h	Amir Navot + Ra	-78.18	3.00 (46)	99.63 (28)	100.00	50.00	1
SVCR-n-m	Amir Reza Saffa	-85.45	3.68 (53)	96.32 (53)	2.00	0.00	1
CBAMethod1	cba	-89.09	3.29 (51)	99.69 (22)	100.00	50.00	1
inf2	Amir Reza Saffa	-92.73	4.82 (54)	95.18 (54)	2.00	0.00	1
multi32	Sang-Kyun Lee	-96.36	6.29 (55)	93.71 (55)	9.30	0.00	1
multi31	Sang-Kyun Lee	-100.00	8.51 (56)	91.49 (56)	9.30	0.00	1

Table 2. December 8th 2003 challenge results for GISETTE.

Method	People	Score	BER	AUC	Ffeat	Fprob	Sig
final2-2	Yi-Wei Chen	97.14	1.35 (7)	98.71 (22)	18.32	0.00	0
final 2-3	Yi-Wei Chen	97.14	1.35 (8)	98.71 (23)	18.32	0.00	0
test	Yi-Wei Chen	88.57	1.37 (9)	98.63 (27)	18.26	0.00	0
FS+SVM	Thomas Navin La	82.86	1.31 (6)	98.69 (26)	34.00	0.18	0
BayesNN-DF	Radford Neal +	71.43	1.26 (1)	99.92 (1)	100.00	50.00	0
BayesNN-la	Radford Neal	71.43	1.26 (2)	99.92 (2)	100.00	50.00	0
BayesNN-la	Radford Neal	71.43	1.26 (3)	99.92 (3)	100.00	50.00	0
GhostMiner	GhostMiner Team	57.14	1.31 (4)	98.69 (24)	100.00	50.00	0
GhostMiner	GhostMiner Team	57.14	1.31 (5)	98.69 (25)	100.00	50.00	0
P-SVM / nu	Sepp Hochreiter	37.14	1.82 (19)	99.79 (10)	4.00	0.50	1
P-SVM / nu	Sepp Hochreiter	37.14	1.82 (20)	99.79 (11)	4.00	0.50	1
P-SVM / nu	Sepp Hochreiter	37.14	1.82 (21)	99.79 (12)	4.00	0.50	1
GhostMiner	GhostMiner Team	25.71	1.42 (10)	98.58 (28)	100.00	50.00	0
RF+RLSC	Kari Torkkola +	22.86	1.77 (17)	98.23 (29)	6.14	0.00	1
RF+RLSC	Kari Torkkola +	22.86	1.77 (18)	98.23 (30)	6.14	0.00	1
P-SVM / nu	Sepp Hochreiter	11.43	1.75 (15)	99.79 (13)	9.90	19.19	1
P-SVM / nu	Sepp Hochreiter	11.43	1.75 (16)	99.79 (14)	9.90	19.19	1
originalFi	wu zhili	8.57	1.58 (11)	99.84 (9)	15.00	0.00	1
CBAMethod3	CBAGroup	-2.86	1.60 (12)	99.85 (4)	30.46	0.00	1
METHOD2	CBA	-2.86	1.60 (13)	99.85 (6)	30.46	0.00	1
CBAMethod3	CBAGroup	-2.86	1.60 (14)	99.85 (5)	30.46	0.00	1
BayesNN-sm	Radford Neal	-20.00	2.09 (25)	99.78 (17)	7.58	0.26	1
MyFinal	Amir Reza Saffa	-25.71	2.08 (24)	97.92 (31)	10.10	0.00	1
IDEAL	BorisovEruhimov	-34.29	1.89 (22)	99.85 (7)	12.00	0.00	1
IDEAL	BorisovEruhimov	-34.29	1.89 (23)	99.85 (8)	12.00	0.00	1
ESNB+NN	Marc Boulle + V	-42.86	2.46 (29)	99.64 (19)	3.26	0.00	1
RF with fe	Vivian Ng + Leo	-54.29	2.51 (30)	99.62 (20)	9.60	0.00	1
ESNB	Marc Boulle	-60.00	2.68 (32)	99.52 (21)	3.26	0.00	1
Sparse Bay	DIMACS	-60.00	2.46 (28)	97.64 (32)	13.48	48.37	1
A shot in	Amir + Ran	-62.86	2.23 (26)	99.79 (15)	50.00	13.96	1
Nameless -	Amir Navot + Ra	-62.86	2.23 (27)	99.79 (16)	50.00	13.96	1
Bayesian +	Chu Wei	-77.14	2.58 (31)	99.67 (18)	100.00	50.00	1
ROBELON	Vincent Lemaire	-82.86	3.48 (33)	96.52 (33)	1.80	5.56	1
ntmce	D. C. Martins J	-88.57	6.43 (36)	93.57 (36)	0.20	0.00	1
METHOD1	CBA	-97.14	6.15 (34)	93.85 (34)	100.00	50.00	1
METHOD3	CBA	-97.14	6.15 (35)	93.85 (35)	100.00	50.00	1

MADELON

Table 1. December 1st 2003 challenge results for MADELON.

Method	People	Score	BER	AUC	Ffeat	Fprob	Sig
Bayesian +	Chu Wei	100.00	7.17 (5)	96.95 (7)	1.60	0.00	0
RF+RLSC	Kari Torkkola +	96.36	6.67 (3)	93.33 (33)	3.80	0.00	0
final 2	Yi-Wei Chen	90.91	6.61 (1)	93.39 (31)	4.80	16.67	0
final 1	Yi-Wei Chen	90.91	6.61 (2)	93.39 (32)	4.80	16.67	0
BayesNN-DF	Radford Neal an	76.36	7.17 (4)	97.82 (1)	100.00	96.00	0
P-SVM (pre	Sepp Hochreiter	76.36	8.67 (20)	96.46 (12)	1.40	0.00	1
BayesNN-DF	Radford Neal +	76.36	7.17 (6)	97.82 (2)	100.00	96.00	0
P-SVM (pre	Sepp Hochreiter	76.36	8.67 (21)	96.46 (13)	1.40	0.00	1
Bayesian S	Chu Wei	74.55	7.89 (8)	96.93 (8)	1.60	0.00	1
P-SVM (pre	Sepp Hochreiter	67.27	8.89 (24)	96.39 (14)	1.40	0.00	1
BayesNN-sm	Radford Neal	56.36	7.72 (7)	97.11 (4)	3.40	0.00	1
transSVMba	wu zhili	45.45	8.56 (13)	95.78 (20)	2.60	0.00	1
SVMbased3	wu zhili + C.H.	45.45	8.56 (14)	95.78 (21)	2.60	0.00	1
myBestVali	wu zhili	45.45	8.56 (15)	95.78 (22)	2.60	0.00	1
TransSVMba	wu zhili	45.45	8.56 (16)	95.78 (23)	2.60	0.00	1
svmBased4	wuzhili+C.H.Li	45.45	8.56 (17)	95.78 (24)	2.60	0.00	1
Collection	Amir Reza Saffa	41.82	9.44 (30)	90.56 (36)	2.00	0.00	1
SVCR-n-m	Amir Reza Saffa	41.82	9.44 (31)	90.56 (37)	2.00	0.00	1
Collection	Amir Reza Saffa	41.82	9.44 (32)	90.56 (38)	2.00	0.00	1
BayesNN-la	Radford Neal	38.18	8.11 (10)	97.12 (3)	3.40	0.00	1
cbamethod3	CBAGroup	30.91	8.83 (23)	96.53 (11)	2.60	0.00	1
The Best o	Amir Navot + Ra	21.82	8.61 (18)	96.97 (5)	3.60	0.00	1
greatest h	Amir Navot + Ra	21.82	8.61 (19)	96.97 (6)	3.60	0.00	1
Modified R	Vivian Ng; Leo	16.36	8.72 (22)	96.93 (9)	4.00	0.00	1
BayesNN-E	Radford Neal	9.09	8.06 (9)	96.90 (10)	100.00	96.00	1
Depends II	Saharon Rosset	3.64	9.06 (25)	96.05 (16)	4.20	9.52	1
Depends I	Saharon Rosset	3.64	9.06 (26)	96.05 (17)	4.20	9.52	1
Depends II	Saharon Rosset	3.64	9.06 (27)	96.05 (18)	4.20	9.52	1
Depends IV	Saharon Rosset	3.64	9.06 (28)	96.05 (19)	4.20	9.52	1
KPLS	Mark J. Embrech	1.82	9.83 (33)	90.16 (39)	2.60	0.00	1
GhostMiner	GhostMiner Team	0.00	8.28 (11)	91.72 (34)	100.00	96.00	1
GhostMiner	GhostMiner Team	0.00	8.28 (12)	91.72 (35)	100.00	96.00	1
Depends V	Saharon Rosset	-9.09	9.28 (29)	96.07 (15)	4.20	9.52	1
P-SVM (few	Sepp Hochreiter	-21.82	11.00 (34)	95.12 (25)	2.40	0.00	1
P-SVM / nu	Sepp Hochreiter	-21.82	11.00 (35)	95.12 (26)	2.40	0.00	1
IDEAL	BorisovEruhimov	-32.73	12.56 (36)	94.73 (27)	100.00	96.00	1
IDEAL	BorisovEruhimov	-32.73	12.56 (37)	94.73 (28)	100.00	96.00	1
IDEAL	Borisov Eruhimo	-32.73	12.56 (38)	94.73 (29)	100.00	96.00	1
IDEAL	BorisovEruhimov	-32.73	12.56 (39)	94.73 (30)	100.00	96.00	1
FS + SVM	Thomas Navin La	-41.82	14.06 (40)	85.94 (40)	4.00	35.00	1
inf5	Amir Reza Saffa	-45.45	14.83 (41)	85.17 (41)	5.00	44.00	1
ROBELON	Vincent Lemaire	-49.09	16.78 (42)	83.22 (42)	1.60	0.00	1
CBAMethod1	cba	-54.55	18.00 (43)	82.58 (43)	2.60	0.00	1
CBAMethod1	cba	-54.55	18.00 (44)	82.58 (44)	2.60	0.00	1
inf2	Amir Reza Saffa	-60.00	21.06 (45)	78.94 (45)	2.00	0.00	1
Enhanced S	M. B.	-63.64	34.06 (46)	68.51 (46)	1.80	11.11	1
mixed	Sang-Kyun Lee	-67.27	38.50 (50)	61.50 (51)	2.40	91.67	1
multi33	Sang-Kyun Lee	-70.91	38.39 (48)	61.61 (50)	7.60	78.95	1
svm linear	Ran Bachrach	-78.18	38.33 (47)	62.90 (48)	100.00	96.00	1
multi32	Sang-Kyun Lee	-80.00	40.56 (54)	59.44 (55)	7.60	78.95	1
multi23	Sang-Kyun Lee	-81.82	39.28 (52)	60.72 (52)	7.80	82.05	1
svm linear	Ran Bachrach	-85.45	38.50 (49)	62.92 (47)	100.00	96.00	1
multi31	Sang-Kyun Lee	-87.27	40.56 (55)	59.44 (56)	7.60	78.95	1
svm linear	Ran Bachrach	-89.09	38.83 (51)	62.64 (49)	100.00	96.00	1
Final	Fan Li	-96.36	39.89 (53)	60.11 (53)	100.00	96.00	1
DIMACS	Alexander Genki	-100.00	41.06 (56)	59.62 (54)	100.00	96.00	1

Table 2. December 8th 2003 challenge results for MADELON.

Method	People	Score	BER	AUC	Ffeat	Fprob	Sig
Bayesian +	Chu Wei	94.29	7.11 (13)	96.95 (10)	1.60	0.00	1
BayesNN-la	Radford Neal	85.71	6.56 (3)	97.62 (2)	3.40	0.00	0
BayesNN-sm	Radford Neal	85.71	6.56 (4)	97.62 (3)	3.40	0.00	0
final2-2	Yi-Wei Chen	71.43	7.11 (12)	92.89 (25)	3.20	0.00	1
RF+RLSC	Kari Torkkola +	71.43	6.67 (6)	93.33 (22)	3.80	0.00	0
GhostMiner	GhostMiner Team	65.71	7.44 (14)	92.56 (26)	3.00	0.00	1
BayesNN-la	Radford Neal	60.00	6.78 (9)	97.46 (6)	3.40	0.00	1
BayesNN-DF	Radford Neal +	54.29	6.22 (1)	98.07 (1)	100.00	96.00	0
CBAMethod3	CBAGroup	51.43	6.72 (7)	97.57 (4)	4.00	0.00	0
CBAMethod3	CBAGroup	51.43	6.72 (8)	97.57 (5)	4.00	0.00	0
final 2-3	Yi-Wei Chen	48.57	6.50 (2)	93.50 (20)	4.80	16.67	0
RF+RLSC	Kari Torkkola +	48.57	7.00 (11)	93.00 (24)	3.80	0.00	1
METHOD2	CBA	42.86	6.83 (10)	97.23 (9)	4.00	0.00	0
GhostMiner	GhostMiner Team	37.14	7.67 (17)	92.33 (27)	3.00	0.00	1
test	Yi-Wei Chen	31.43	6.61 (5)	93.39 (21)	4.80	16.67	0
GhostMiner	GhostMiner Team	8.57	8.28 (18)	91.72 (28)	3.00	0.00	1
A shot in	Amir + Ran	5.71	7.61 (15)	97.25 (7)	4.00	0.00	1
Nameless -	Amir Navot + Ra	5.71	7.61 (16)	97.25 (8)	4.00	0.00	1
originalFi	wu zhili	-2.86	8.56 (19)	95.78 (17)	2.60	0.00	1
P-SVM / nu	Sepp Hochreiter	-20.00	9.67 (21)	96.31 (12)	1.40	0.00	1
P-SVM / nu	Sepp Hochreiter	-20.00	9.67 (22)	96.31 (13)	1.40	0.00	1
P-SVM / nu	Sepp Hochreiter	-20.00	9.67 (24)	96.31 (14)	1.40	0.00	1
P-SVM / nu	Sepp Hochreiter	-20.00	9.67 (25)	96.31 (15)	1.40	0.00	1
P-SVM / nu	Sepp Hochreiter	-20.00	9.67 (26)	96.31 (16)	1.40	0.00	1
MyFinal	Amir Reza Saffa	-31.43	9.44 (20)	90.56 (29)	2.00	0.00	1
RF with fe	Vivian Ng + Leo	-42.86	9.67 (23)	96.76 (11)	4.00	0.00	1
FS+SVM	Thomas Navin La	-48.57	11.22 (27)	88.78 (30)	4.00	35.00	1
IDEAL	BorisovEruhimov	-57.14	12.56 (28)	94.73 (18)	100.00	96.00	1
IDEAL	BorisovEruhimov	-57.14	12.56 (29)	94.73 (19)	100.00	96.00	1
ESNB+NN	Marc Boulle + V	-65.71	14.94 (30)	93.22 (23)	1.40	0.00	1
ROBELON	Vincent Lemaire	-71.43	16.78 (33)	83.22 (33)	1.60	0.00	1
METHOD1	CBA	-80.00	16.06 (31)	83.94 (31)	100.00	96.00	1
METHOD3	CBA	-80.00	16.06 (32)	83.94 (32)	100.00	96.00	1
ntmce	D. C. Martins J	-88.57	24.44 (34)	75.56 (34)	2.20	45.45	1
ESNB	Marc Boulle	-94.29	35.17 (35)	68.56 (35)	1.40	0.00	1
Sparse Bay	DIMACS	-100.00	38.00 (36)	62.75 (36)	0.60	0.00	1

Overall Results

Table 1. December 1st 2003 overall challenge results.

Method	People	Score	BER	AUC	Ffeat	Fprob	Sig
BayesNN-DF	Radford Neal an	88.00	6.84 (1)	97.22 (1)	80.30	47.77	0.00
BayesNN-DF	Radford Neal +	86.18	6.87 (2)	97.21 (2)	80.30	47.77	0.00
BayesNN-sm	Radford Neal	68.73	8.20 (3)	96.12 (5)	4.74	2.91	0.80
BayesNN-la	Radford Neal	59.64	8.21 (4)	96.36 (3)	60.30	28.51	0.40
RF+RLSC	Kari Torkkola +	59.27	9.07 (7)	90.93 (29)	22.54	17.53	0.60
final 2	Yi-Wei Chen	52.00	9.31 (9)	90.69 (31)	24.91	11.98	0.40
SVMbased3	wu zhili + C.H.	41.82	9.21 (8)	93.60 (16)	29.51	21.72	0.80
svmBased4	wuzhili+C.H.Li	41.09	9.40 (10)	93.41 (18)	29.51	21.72	0.80
final 1	Yi-Wei Chen	40.36	10.38 (23)	89.62 (34)	6.23	6.10	0.60
transSVMba	wu zhili	36.00	9.60 (13)	93.21 (20)	29.51	21.72	0.80
myBestVali	wu zhili	36.00	9.60 (14)	93.21 (21)	29.51	21.72	0.80
TransSVMba	wu zhili	36.00	9.60 (15)	93.21 (22)	29.51	21.72	0.80
BayesNN-E	Radford Neal	29.45	8.43 (5)	96.30 (4)	96.75	56.67	0.80
Collection	Amir Reza Saffa	28.00	10.03 (20)	89.97 (32)	7.71	10.60	1.00
Collection	Amir Reza Saffa	20.73	10.06 (21)	89.94 (33)	32.26	25.50	1.00
IDEAL	BorisovEruhimov	19.27	10.01 (17)	94.71 (11)	62.60	40.30	1.00
IDEAL	Borisov Eruhimo	19.27	10.01 (18)	94.71 (12)	62.60	40.30	1.00
IDEAL	BorisovEruhimov	19.27	10.01 (19)	94.71 (13)	62.60	40.30	1.00
Depends I	Saharon Rosset	17.82	10.52 (25)	92.80 (24)	13.99	27.37	0.80
greatest h	Amir Navot + Ra	17.45	8.99 (6)	95.54 (8)	42.18	24.74	1.00
Depends II	Saharon Rosset	16.00	10.55 (27)	92.78 (26)	13.99	27.37	0.80
Bayesian +	Chu Wei	15.27	9.43 (11)	95.70 (7)	67.53	38.03	0.60
The Best o	Amir Navot + Ra	15.27	9.94 (16)	94.77 (10)	32.18	19.52	1.00
Depends V	Saharon Rosset	15.27	10.57 (28)	92.81 (23)	13.99	27.37	0.80
IDEAL	BorisovEruhimov	14.18	10.35 (22)	94.93 (9)	62.60	40.30	1.00
P-SVM (pre	Sepp Hochreiter	14.18	11.28 (32)	93.66 (15)	4.63	34.74	1.00
FS + SVM	Thomas Navin La	12.73	11.56 (33)	88.44 (40)	16.91	21.45	0.80
Depends II	Saharon Rosset	12.36	11.04 (31)	92.73 (27)	13.99	27.37	0.80
Bayesian S	Chu Wei	11.64	9.50 (12)	95.75 (6)	67.53	38.03	0.80
Depends IV	Saharon Rosset	11.64	10.55 (26)	92.80 (25)	13.99	27.37	0.80
KPLS	Mark J. Embrech	9.82	10.91 (30)	89.46 (35)	7.17	25.74	1.00
Modified R	Vivian Ng; Leo	6.91	10.46 (24)	94.58 (14)	3.86	9.82	1.00
cbamethod3	CBAGroup	5.45	10.63 (29)	90.75 (30)	40.65	16.00	1.00
P-SVM (pre	Sepp Hochreiter	5.09	12.14 (35)	93.46 (17)	7.38	45.65	1.00
P-SVM (pre	Sepp Hochreiter	0.00	11.82 (34)	93.41 (19)	3.83	34.60	1.00
P-SVM (few	Sepp Hochreiter	-24.36	14.85 (41)	89.25 (36)	3.74	20.38	1.00
inf5	Amir Reza Saffa	-24.36	14.58 (40)	85.42 (45)	5.00	19.31	1.00
GhostMiner	GhostMiner Team	-26.55	12.47 (36)	87.53 (42)	100.00	55.25	1.00
GhostMiner	GhostMiner Team	-30.18	13.80 (38)	86.20 (43)	100.00	55.25	1.00
mixed	Sang-Kyun Lee	-32.36	18.40 (46)	81.60 (50)	4.34	31.51	1.00
CBAMethod1	cba	-35.64	14.05 (39)	84.49 (47)	2.05	0.20	1.00
P-SVM / nu	Sepp Hochreiter	-36.00	13.17 (37)	92.60 (28)	4.82	48.73	1.00
SVCR-n-m	Amir Reza Saffa	-40.36	14.99 (42)	85.01 (46)	2.00	4.91	1.00
inf2	Amir Reza Saffa	-42.55	15.75 (43)	84.25 (48)	2.00	8.61	1.00
Final	Fan Li	-47.64	20.10 (48)	79.90 (51)	100.00	55.25	1.00
svm linear	Ran Bachrach	-48.36	21.19 (50)	88.99 (37)	100.00	55.25	1.00
svm linear	Ran Bachrach	-49.09	21.21 (51)	88.99 (38)	100.00	55.25	1.00
multi23	Sang-Kyun Lee	-50.55	20.62 (49)	79.38 (52)	4.25	22.87	1.00
CBAMethod1	cba	-51.64	16.02 (44)	87.96 (41)	40.65	16.00	1.00
multi33	Sang-Kyun Lee	-54.18	22.26 (55)	77.74 (55)	5.38	28.96	1.00
svm linear	Ran Bachrach	-54.91	22.09 (53)	88.96 (39)	100.00	55.25	1.00
Enhanced S	M. B.	-57.82	19.85 (47)	85.96 (44)	1.02	10.60	1.00
ROBELON	Vincent Lemaire	-62.18	16.37 (45)	83.63 (49)	1.12	21.47	1.00
DIMACS	Alexander Genki	-62.18	24.88 (56)	76.80 (56)	23.81	47.28	1.00
multi31	Sang-Kyun Lee	-71.27	22.25 (54)	77.75 (54)	5.38	28.96	1.00
multi32	Sang-Kyun Lee	-80.00	22.09 (52)	77.91 (53)	5.38	28.96	1.00

Table 2. December 8st 2003 overall challenge results.

Method	People	Score	BER	AUC	Ffeat	Fprob	Sig
BayesNN-DF	Radford Neal +	71.43	6.48 (1)	97.20 (1)	80.30	47.77	0.20
BayesNN-la	Radford Neal	66.29	7.27 (3)	96.98 (3)	60.30	28.51	0.40
BayesNN-sm	Radford Neal	61.14	7.13 (2)	97.08 (2)	4.74	2.91	0.60
final 2-3	Yi-Wei Chen	49.14	7.91 (8)	91.45 (25)	24.91	9.91	0.40
BayesNN-la	Radford Neal	49.14	7.83 (5)	96.78 (4)	60.30	28.51	0.60
final2-2	Yi-Wei Chen	40.00	8.80 (17)	89.84 (29)	24.62	6.68	0.60
GhostMiner	GhostMiner Team	37.14	7.89 (7)	92.11 (21)	80.60	36.05	0.80
RF+RLSC	Kari Torkkola +	35.43	8.04 (9)	91.96 (22)	22.38	17.52	0.80
GhostMiner	GhostMiner Team	35.43	7.86 (6)	92.14 (20)	80.60	36.05	0.80
RF+RLSC	Kari Torkkola +	34.29	8.23 (12)	91.77 (23)	22.38	17.52	0.60
FS+SVM	Thomas Navin La	31.43	8.99 (19)	91.01 (27)	20.91	17.28	0.60
GhostMiner	GhostMiner Team	26.29	8.24 (13)	91.76 (24)	80.60	36.05	0.60
CBAMethod3	CBAGroup	21.14	8.14 (10)	96.62 (5)	12.78	0.06	0.60
CBAMethod3	CBAGroup	21.14	8.14 (11)	96.62 (6)	12.78	0.06	0.60
Nameless -	Amir Navot + Ra	12.00	7.78 (4)	96.43 (9)	32.28	16.22	1.00
test	Yi-Wei Chen	6.29	9.31 (21)	90.69 (28)	24.91	11.98	0.60
A shot in	Amir + Ran	4.00	8.41 (14)	94.59 (15)	21.08	12.72	1.00
IDEAL	BorisovEruhimov	3.43	8.53 (15)	96.58 (7)	62.50	36.40	0.80
IDEAL	BorisovEruhimov	3.43	8.53 (16)	96.58 (8)	62.50	36.40	0.80
originalFi	wu zhili	2.86	9.60 (23)	93.21 (18)	29.51	21.72	1.00
RF with fe	Vivian Ng + Leo	1.71	9.06 (20)	96.01 (11)	3.74	2.03	1.00
P-SVM / nu	Sepp Hochreiter	-2.86	9.52 (22)	94.91 (12)	4.72	39.94	1.00
P-SVM / nu	Sepp Hochreiter	-4.57	10.02 (25)	94.63 (14)	2.14	33.33	1.00
P-SVM / nu	Sepp Hochreiter	-6.29	9.73 (24)	94.80 (13)	2.94	32.75	1.00
Bayesian +	Chu Wei	-8.57	8.86 (18)	96.37 (10)	67.53	38.03	1.00
P-SVM / nu	Sepp Hochreiter	-10.29	10.30 (26)	94.58 (16)	4.61	42.25	1.00
P-SVM / nu	Sepp Hochreiter	-12.00	10.80 (27)	94.31 (17)	2.03	35.65	1.00
ESNB+NN	Marc Boulle + V	-28.00	12.42 (30)	93.12 (19)	1.04	1.43	1.00
METHOD2	CBA	-36.00	11.22 (28)	91.27 (26)	66.89	26.05	0.80
MyFinal	Amir Reza Saffa	-44.00	11.43 (29)	88.57 (31)	7.77	10.96	1.00
Sparse Bay	DIMACS	-58.86	19.90 (36)	76.88 (36)	3.27	27.98	1.00
ESNB	Marc Boulle	-62.29	18.25 (34)	88.63 (30)	1.04	1.43	1.00
ROBELON	Vincent Lemaire	-82.86	16.37 (33)	83.63 (34)	1.12	21.47	1.00
METHOD1	CBA	-84.57	14.68 (31)	85.32 (32)	100.00	55.25	1.00
METHOD3	CBA	-84.57	14.68 (32)	85.32 (33)	100.00	55.25	1.00
ntmce	D. C. Martins J	-87.43	19.85 (35)	80.15 (35)	0.50	14.81	1.00

Index